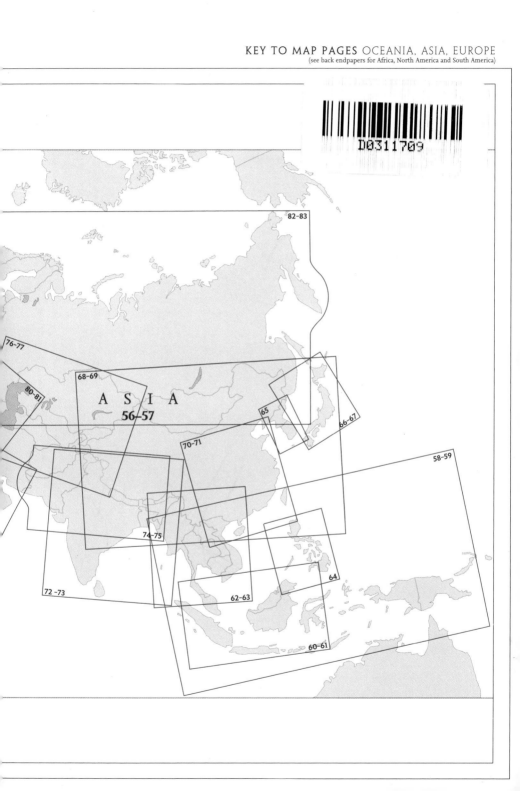

82-83

76-77

68-69

80-81

A S I A
56-57

65

66-67

70-71

58-59

74-75

72 -73

62-63

64

60-61

COMPACT

ATLAS

OF THE

WORLD

THE TIMES COMPACT ATLAS OF THE WORLD

Times Books, 77-85 Fulham Palace Road, London W6 8JB

The Times is a registered trademark of Times Newspapers Ltd

First published 1994
Second Edition 2000
Third Edition 2004
Fourth Edition 2007
Fifth Edition 2010

Sixth Edition 2013

Copyright © Times Books Group Ltd 2012

Maps © Collins Bartholomew Ltd 2012

Printed in Hong Kong

British Library Cataloguing in Publication Data.
A catalogue record for this book is
available from the British Library.

ISBN 978-0-00-748105-7

Imp 001

All mapping in this atlas is generated from Collins Bartholomew digital databases. Collins Bartholomew, the UK's leading independent geographical information supplier, can provide a digital, custom, and premium mapping service to a variety of markets.

For further information:
Tel: +44 (0) 208 307 4515
e-mail: collinsbartholomew@harpercollins.co.uk
or visit our website at: www.collinsbartholomew.com

If you would like to comment on any aspect of this atlas, please write to
Times Atlases, HarperCollins Publishers, Westerhill Road, Bishopbriggs, Glasgow, G64 2QT
email: timesatlases@harpercollins.co.uk
or visit our website at: www.timesatlases.com
or follow us on Twitter @TimesAtlas

THE TIMES

COMPACT

ATLAS

OF THE

WORLD

TIMES BOOKS
LONDON

Pages	Title	Scale

THE WORLD TODAY

6–29	Countries of the World	
30–33	Continents and Oceans	
34–35	Climate	
36–37	Environment	
38–39	Population and Cities	
40–41	Telecommunications	

ATLAS OF THE WORLD

42–43	Introduction to Atlas of the World	
44–45	World Physical Features	
46–47	World Countries	

48–49	OCEANIA	1:50 000 000
50–51	Australia	1:20 000 000
52–53	Australia Southeast	1:7 500 000
54	New Zealand	1:7 500 000
55	Antarctica	1:45 000 000

56–57	ASIA	1:55 000 000
58–59	Southeast Asia	1:25 000 000
60–61	Malaysia and Indonesia West	1:12 000 000
62–63	Continental Southeast Asia	1:12 000 000
64	Philippines	1:12 000 000
65	North Korea and South Korea	1:6 500 000
66–67	Japan	1:7 500 000
68–69	China and Mongolia	1:25 000 000
70–71	China Central	1:12 000 000
72–73	South Asia	1:15 000 000
74–75	Pakistan, India North and Bangladesh	1:12 000 000
76–77	Central Asia	1:15 000 000
78–79	Arabian Peninsula	1:12 000 000
80–81	East Mediterranean	1:12 000 000
82–83	Russian Federation	1:30 000 000

84–85	EUROPE	1:25 000 000
86–87	European Russian Federation	1:15 000 000
88–89	Northeast Europe	1:6 000 000
90–91	Ukraine and Moldova	1:6 000 000
92–93	Scandinavia and Iceland	1:7 500 000
94–95	British Isles	1:6 000 000

CONTENTS

Pages	Title	Scale
96	Scotland	1:3 000 000
97	Ireland	1:3 000 000
98–99	England and Wales	1:3 000 000
100–101	Northwest Europe	1:3 000 000
102–103	Central Europe	1:6 000 000
104–105	France and Switzerland	1:6 000 000
106–107	Spain and Portugal	1:6 000 000
108–109	Italy and The Balkans	1:6 000 000
110–111	Southeast Europe	1:6 000 000
112–113	AFRICA	1:45 000 000
114–115	Northwest Africa	1:20 000 000
116–117	Northeast Africa	1:20 000 000
118–119	Central Africa	1:15 000 000
120–121	Southern Africa	1:15 000 000
122–123	Republic of South Africa	1:7 500 000
124–125	NORTH AMERICA	1:40 000 000
126–127	Canada	1:25 000 000
128–129	Canada West	1:12 000 000
130–131	Canada East	1:12 000 000
132–133	United States of America	1:20 000 000
134–135	USA West	1:8 000 000
136–137	USA North Central	1:8 000 000
138–139	USA Northeast	1:8 000 000
140–141	USA Southeast	1:8 000 000
142–143	USA South Central	1:8 000 000
144–145	Mexico	1:12 000 000
146–147	Central America and the Caribbean	1:15 000 000
148–149	SOUTH AMERICA	1:35 000 000
150–151	South America North	1:20 000 000
152–153	South America South	1:20 000 000
154–155	Brazil Southeast	1:7 500 000
	OCEANS	
156–157	Pacific Ocean	1:90 000 000
158	Atlantic Ocean	1:90 000 000
159	Indian Ocean	1:90 000 000
160	Arctic Ocean	1:45 000 000
161–240	INDEX AND ACKNOWLEDGEMENTS	

All independent countries and populated dependent and disputed territories are included in this list of the states and territories of the world; the list is arranged in alphabetical order by the conventional name form. For independent states, the full name is given below the conventional name, if this is different; for territories, the status is given. The capital city name is the same form as shown on the reference maps.

The statistics used for the area and population are the latest available and include estimates. The information on languages and religions is based on the latest information on 'de facto' speakers of the language or 'de facto' adherents to the religion. The information available on languages and religions varies greatly from country to country. Some countries include questions in censuses, others do not, in which case best estimates are used. The order of the languages and religions reflect their relative importance within the country; generally, languages or religions are included when more than one per cent of the population are estimated to be speakers or adherents.

Membership of selected international organizations is shown for each independent country. Territories are not shown as having separate memberships of these organizations.

ABBREVIATIONS

CURRENCIES

| CFA | Communauté Financière Africaine |
| CFP | Comptoirs Français du Pacifique |

ORGANIZATIONS

APEC	Asia-Pacific Economic Cooperation
ASEAN	Association of Southeast Asian Nations
CARICOM	Caribbean Community
CIS	Commonwealth of Independent States
Comm.	The Commonwealth
EU	European Union
OECD	Organization of Economic Co-operation and Development
OPEC	Organization of Petroleum Exporting Countries
SADC	Southern African Development Community
UN	United Nations

Abkhazia
Disputed Territory (Georgia)

Area Sq Km	8 700
Area Sq Miles	3 360
Population	180 000
Capital	Sokhumi (Aq"a)

AFGHANISTAN
Islamic State of Afghanistan

Area Sq Km	652 225	Religions	Sunni Muslim,
Area Sq Miles	251 825		Shi'a Muslim
Population	32 358 000	Currency	Afghani
Capital	Kābul	Organizations	UN
Languages	Dari, Pushtu, Uzbek,Turkmen	Map page	76–77

ALBANIA
Republic of Albania

Area Sq Km	28 748	Religions	Sunni Muslim,
Area Sq Miles	11 100		Albanian Orthodox
Population	3 216 000		Roman Catholic
Capital	Tirana (Tiranë)	Currency	Lek
Languages	Albanian, Greek	Organizations	UN
		Map page	109

ALGERIA
People's Democratic Republic of Algeria

Area Sq Km	2 381 741	Religions	Sunni Muslim
Area Sq Miles	919 595	Currency	Algerian dinar
Population	35 980 000	Organizations	OPEC, UN
Capital	Algiers (Alger)	Map page	114–115
Languages	Arabic, French, Berber		

American Samoa
United States Unincorporated Territory

Area Sq Km	197	Religions	Protestant, Roman
Area Sq Miles	76		Catholic
Population	70 000	Currency	United States do
Capital	Fagatogo	Map page	49
Languages	Samoan, English		

ANDORRA
Principality of Andorra

Area Sq Km	465	Religions	Roman Catholic
Area Sq Miles	180	Currency	Euro
Population	86 000	Organizations	UN
Capital	Andorra la Vella	Map page	104
Languages	Spanish, Catalan, French		

ANGOLA
Republic of Angola

Area Sq Km	1 246 700	Religions	Roman Catholic,
Area Sq Miles	481 354		Protestant,
Population	19 618 000		traditional beliefs
Capital	Luanda	Currency	Kwanza
Languages	Portuguese, Bantu, local languages	Organizations	OPEC, SADC, UN
		Map page	120

Anguilla
United Kingdom Overseas Territory

Area Sq Km	155	Religions	Protestant, Roman
Area Sq Miles	60		Catholic
Population	16 000	Currency	East Caribbean
Capital	The Valley		dollar
Languages	English	Map page	147

ANTIGUA AND BARBUDA

Area Sq Km	442	Religions	Protestant,
Area Sq Miles	171		Roman Catholic
Population	90 000	Currency	East Caribbean
Capital	St John's		dollar
Languages	English, creole	Organizations	CARICOM,
			Comm., UN
		Map page	147

ARGENTINA
Argentine Republic

Area Sq Km	2 766 889	Religions	Roman Catholic,
Area Sq Miles	1 068 302		Protestant
Population	40 765 000	Currency	Argentinian peso
Capital	Buenos Aires	Organizations	UN
Languages	Spanish, Italian,	Map page	152–153
	Amerindian		
	languages		

ARMENIA
Republic of Armenia

Area Sq Km	29 800	Religions	Armenian Orthodox
Area Sq Miles	11 506	Currency	Dram
Population	3 100 000	Organizations	CIS, UN
Capital	Yerevan (Erevan)	Map page	81
Languages	Armenian, Azeri		

Aruba
Self-governing Netherlands Territory

Area Sq Km	193	Religions	Roman Catholic,
Area Sq Miles	75		Protestant
Population	108 000	Currency	Aruban florin
Capital	Oranjestad	Map page	147
Languages	Papiamento, Dutch,		
	English		

Ascension
part of St Helena, Ascension and Tristan da Cunha

Area Sq Km	88	Religions	Protestant,
Area Sq Miles	34		Roman Catholic
Population	884	Currency	St Helena pound
Capital	Georgetown	Map page	113
Languages	English		

AUSTRALIA
Commonwealth of Australia

Area Sq Km	7 692 024	Religions	Protestant, Roman
Area Sq Miles	2 969 907		Catholic, Orthodox
Population	22 606 000	Currency	Australian dollar
Capital	Canberra	Organizations	APEC, Comm.,
Languages	English, Italian,		OECD, UN
	Greek	Map page	50–51

Australian Capital Territory (Federal Territory)

Area Sq Km	2 358	Population	359 700
Area Sq Miles	910	Capital	Canberra

Jervis Bay Territory (Territory)

Area Sq Km	73	Population	611
Area Sq Miles	28		

New South Wales (State)

Area Sq Km	800 642	Population	7 253 400
Area Sq Miles	309 130	Capital	Sydney

Northern Territory (Territory)

Area Sq Km	1 349 129	Population	230 200
Area Sq Miles	520 902	Capital	Darwin

Queensland (State)

Area Sq Km	1 730 648	Population	4 532 300
Area Sq Miles	668 207	Capital	Brisbane

South Australia (State)

Area Sq Km	983 482	Population	1 647 800
Area Sq Miles	379 725	Capital	Adelaide

Tasmania (State)

Area Sq Km	68 401	Population	508 500
Area Sq Miles	26 410	Capital	Hobart

Victoria (State)

Area Sq Km	227 416	Population	5 567 100
Area Sq Miles	87 806	Capital	Melbourne

Western Australia (State)

Area Sq Km	2 529 875	Population	2 306 200
Area Sq Miles	976 790	Capital	Perth

AUSTRIA
Republic of Austria

Area Sq Km	83 855	Religions	Roman Catholic,
Area Sq Miles	32 377		Protestant
Population	8 413 000	Currency	Euro
Capital	Vienna (Wien)	Organizations	EU, OECD, UN
Languages	German, Croatian,	Map page	102–103
	Turkish		

AZERBAIJAN
Republic of Azerbaijan

Area Sq Km	86 600	Religions	Shi'a Muslim, Sunni
Area Sq Miles	33 436		Muslim, Russian
Population	9 306 000		and Armenian
Capital	Baku (Bakı)		Orthodox
Languages	Azeri, Armenian,	Currency	Azerbaijani manat
	Russian, Lezgian	Organizations	CIS, UN
		Map page	81

Azores (Arquipélago dos Açores)
Autonomous Region of Portugal

Area Sq Km	2 300	Religions	Roman Catholic,
Area Sq Miles	888		Protestant
Population	245 374	Currency	Euro
Capital	Ponta Delgada	Map page	112
Languages	Portuguese		

THE BAHAMAS
Commonwealth of The Bahamas

Area Sq Km	13 939	Religions	Protestant,
Area Sq Miles	5 382		Roman Catholic
Population	347 000	Currency	Bahamian dollar
Capital	Nassau	Organizations	CARICOM,
Languages	English, creole		Comm., UN
		Map page	146–147

BAHRAIN
Kingdom of Bahrain

Area Sq Km	691	Religions	Shi'a Muslim,
Area Sq Miles	267		Sunni Muslim,
Population	1 324 000		Christian
Capital	Manama	Currency	Bahraini dinar
	(Al Manāmah)	Organizations	UN
Languages	Arabic, English	Map page	79

BANGLADESH
People's Republic of Bangladesh

Area Sq Km	143 998	Religions	Sunni Muslim,
Area Sq Miles	55 598		Hindu
Population	150 494 000	Currency	Taka
Capital	Dhaka (Dacca)	Organizations	Comm., UN
Languages	Bengali, English	Map page	75

BARBADOS

Area Sq Km	430	Religions	Protestant,
Area Sq Miles	166		Roman Catholic
Population	274 000	Currency	Barbados dollar
Capital	Bridgetown	Organizations	CARICOM,
Languages	English, creole		Comm., UN
		Map page	147

BELARUS
Republic of Belarus

Area Sq Km	207 600	Religions	Belorussian
Area Sq Miles	80 155		Orthodox,
Population	9 559 000		Roman Catholic
Capital	Minsk	Currency	Belarus rouble
Languages	Belorussian,	Organizations	CIS, UN
	Russian	Map page	88–89

BELGIUM
Kingdom of Belgium

Area Sq Km	30 520		
Area Sq Miles	11 784	Religions	Roman Catholic,
Population	10 754 000		Protestant
Capital	Brussels (Bruxelles)	Currency	Euro
Languages	Dutch (Flemish),	Organizations	EU, OECD, UN
	French (Walloon),	Map page	100
	German		

BELIZE

Area Sq Km	22 965	Religions	Roman Catholic,
Area Sq Miles	8 867		Protestant
Population	318 000	Currency	Belize dollar
Capital	Belmopan	Organizations	CARICOM,
Languages	English, Spanish,		Comm., UN
	Mayan, creole	Map page	147

BENIN
Republic of Benin

Area Sq Km	112 620	Religions	Traditional beliefs
Area Sq Miles	43 483		Roman Catholic,
Population	9 100 000		Sunni Muslim
Capital	Porto-Novo	Currency	CFA franc
Languages	French, Fon,	Organization	UN
	Yoruba, Adja,	Map page	114
	local languages		

Bermuda
United Kingdom Overseas Territory

Area Sq Km	54	Religions	Protestant,
Area Sq Miles	21		Roman Catholic
Population	65 000	Currency	Bermuda dollar
Capital	Hamilton	Map page	125
Languages	English		

BHUTAN
Kingdom of Bhutan

Area Sq Km	46 620	Religions	Buddhist, Hindu
Area Sq Miles	18 000	Currency	Ngultrum,
Population	738 000		Indian rupee
Capital	Thimphu	Organizations	UN
Languages	Dzongkha,	Map page	75
	Nepali, Assamese		

BOLIVIA
Plurinational State of Bolivia

Area Sq Km	1 098 581	Religions	Roman Catholic,
Area Sq Miles	424 164		Protestant, Baha'i
Population	10 088 000	Currency	Boliviano
Capital	La Paz/Sucre	Organizations	UN
Languages	Spanish, Quechua,	Map page	152
	Aymara		

Bonaire
Netherlands Special Municipality

Area Sq Km	288	Religions	Roman Catholic,
Area Sq Miles	111		Protestant
Population	13 389	Currency	US dollar
Capital	Kralendijk	Map page	147
Languages	Dutch, Papiamento		

Bonin Islands (Ogasawara-shotō)
part of Japan

Area Sq Km	104	Religions	Shintoist, Buddhist,
Area Sq Miles	40		Christian
Population	2 772	Currency	Yen
Capital	Ōmura	Map page	69
Languages	Japanese		

BOSNIA-HERZEGOVINA
Republic of Bosnia and Herzegovina

Area Sq Km	51 130	Religions	Sunni Muslim,
Area Sq Miles	19 741		Serbian Orthodox,
Population	3 752 000		Roman Catholic,
Capital	Sarajevo		Protestant
Languages	Bosnian, Serbian,	Currency	Marka
	Croatian	Organizations	UN
		Map page	109

BOTSWANA
Republic of Botswana

Area Sq Km	581 370	Religions	Traditional beliefs,
Area Sq Miles	224 468		Protestant,
Population	2 031 000		Roman Catholic
Capital	Gaborone	Currency	Pula
Languages	English, Setswana,	Organizations	Comm., SADC, UN
	Shona, local	Map page	120
	languages		

BRAZIL
Federative Republic of Brazil

Area Sq Km	8 514 879	Religions	Roman Catholic,
Area Sq Miles	3 287 613		Protestant
Population	196 655 000	Currency	Real
Capital	Brasília	Organizations	UN
Languages	Portuguese	Map page	150–151

BRUNEI
State of Brunei Darussalam

Area Sq Km	5 765	Religions	Sunni Muslim,
Area Sq Miles	2 226		Buddhist, Christian
Population	406 000	Currency	Brunei dollar
Capital	Bandar Seri Begawan	Organizations	APEC, ASEAN,
Languages	Malay, English,		Comm., UN
	Chinese	Map page	61

BULGARIA
Republic of Bulgaria

Area Sq Km	110 994	Religions	Bulgarian
Area Sq Miles	42 855		Orthodox,
Population	7 446 000		Sunni Muslim
Capital	Sofia (Sofiya)	Currency	Lev
Languages	Bulgarian, Turkish,	Organizations	EU, UN
	Romany, Macedonian	Map page	110

BURKINA FASO
Democratic Republic of Burkina Faso

Area Sq Km	274 200	Religions	Sunni Muslim,
Area Sq Miles	105 869		traditional beliefs,
Population	16 968 000		Roman Catholic
Capital	Ouagadougou	Currency	CFA franc
Languages	French, Moore	Organizations	UN
	(Mossi), Fulani, local	Map page	114
	languages		

BURUNDI
Republic of Burundi

Area Sq Km	27 835	Religions	Roman Catholic,
Area Sq Miles	10 747		traditional beliefs,
Population	8 575 000		Protestant
Capital	Bujumbura	Currency	Burundian franc
Languages	Kirundi (Hutu,	Organizations	UN
	Tutsi), French	Map page	119

CAMBODIA
Kingdom of Cambodia

Area Sq Km	181 000	Religions	Buddhist, Roman
Area Sq Miles	69 884		Catholic, Sunni
Population	14 305 000		Muslim
Capital	Phnom Penh	Currency	Riel
Languages	Khmer, Vietnamese	Organizations	ASEAN, UN
		Map page	63

CAMEROON
Republic of Cameroon

Area Sq Km	475 442	Religions	Roman Catholic,
Area Sq Miles	183 569		traditional beliefs,
Population	20 030 000		Sunni Muslim,
Capital	Yaoundé		Protestant
Languages	French, English,	Currency	CFA franc
	Fang, Bamileke,	Organizations	Comm., UN
	local languages	Map page	118

CANADA

Area Sq Km	9 984 670	Religions	Roman Catholic,
Area Sq Miles	3 855 103		Protestant, Eastern
Population	34 350 000		Orthodox, Jewish
Capital	Ottawa	Currency	Canadian dollar
Languages	English, French,	Organizations	APEC, Comm.,
	local languages		OECD, UN
		Map page	126–127

Alberta (Province)

Area Sq Km	661 848	Population	3 742 753
Area Sq Miles	255 541	Capital	Edmonton

British Columbia (Province)

Area Sq Km	944 735	Population	4 554 085
Area Sq Miles	364 764	Capital	Victoria

Manitoba (Province)

Area Sq Km	647 797	Population	1 243 653
Area Sq Miles	250 116	Capital	Winnipeg

New Brunswick (Province)

Area Sq Km	72 908	Population	753 232
Area Sq Miles	28 150	Capital	Fredericton

Newfoundland and Labrador (Province)

Area Sq Km	405 212	Population	509 148
Area Sq Miles	156 453	Capital	St John's

Northwest Territories (Territory)

Area Sq Km	1 346 106	Population	43 554
Area Sq Miles	519 734	Capital	Yellowknife

Nova Scotia (Province)

Area Sq Km	55 284	Population	943 414
Area Sq Miles	21 345	Capital	Halifax

Nunavut (Territory)

Area Sq Km	2 093 190	Population	33 303
Area Sq Miles	808 185	Capital	Iqaluit
			(Frobisher Bay)

CANADA

Ontario (Province)

Area Sq Km	1 076 395	Population	13 282 444
Area Sq Miles	415 598	Capital	Toronto

Prince Edward Island (Province)

Area Sq Km	5 660	Population	143 481
Area Sq Miles	2 185	Capital	Charlottetown

Québec (Province)

Area Sq Km	1 542 056	Population	7 942 983
Area Sq Miles	595 391	Capital	Québec

Saskatchewan (Province)

Area Sq Km	651 036	Population	1 052 050
Area Sq Miles	251 366	Capital	Regina

Yukon (Territory)

Area Sq Km	482 443	Population	34 306
Area Sq Miles	186 272	Capital	Whitehorse

Canary Islands (Islas Canarias)
Autonomous Community of Spain

Area Sq Km	7 447	Religions	Roman Catholic
Area Sq Miles	2 875	Currency	Euro
Population	2 103 992	Map page	114
Capital	Santa Cruz de Tenerife/Las Palmas		
Languages	Spanish		

CAPE VERDE
Republic of Cape Verde

Area Sq Km	4 033	Religions	Roman Catholic,
Area Sq Miles	1 557		Protestant
Population	501 000	Currency	Cape Verde escudo
Capital	Praia	Organizations	UN
Languages	Portuguese, creole	Map page	46

Cayman Islands
United Kingdom Overseas Territory

Area Sq Km	259	Religions	Protestant, Roman
Area Sq Miles	100		Catholic
Population	57 000	Currency	Cayman Islands
Capital	George Town		dollar
Languages	English	Map page	146

CENTRAL AFRICAN REPUBLIC

Area Sq Km	622 436	Religions	Protestant,
Area Sq Miles	240 324		Roman Catholic,
Population	4 487 000		traditional beliefs,
Capital	Bangui		Sunni Muslim
Languages	French, Sango,	Currency	CFA franc
	Banda, Baya, local	Organizations	UN
	languages	Map page	118

Ceuta
Autonomous Community of Spain

Area Sq Km	19	Religions	Roman Catholic,
Area Sq Miles	7		Muslim
Population	78 674	Currency	Euro
Capital	Ceuta	Map page	106
Languages	Spanish, Arabic		

CHAD
Republic of Chad

Area Sq Km	1 284 000	Religions	Sunni Muslim,
Area Sq Miles	495 755		Roman Catholic,
Population	11 525 000		Protestant,
Capital	Ndjamena		traditional beliefs
Languages	Arabic, French,Sara,	Currency	CFA franc
	local languages	Organizations	UN
		Map page	115

Chatham Islands
part of New Zealand

Area Sq Km	963	Religions	Protestant
Area Sq Miles	372	Currency	New Zealand dol
Population	640	Map page	49
Capital	Waitangi		
Languages	English		

CHILE
Republic of Chile

Area Sq Km	756 945	Religions	Roman Catholic,
Area Sq Miles	292 258		Protestant
Population	17 270 000	Currency	Chilean peso
Capital	Santiago	Organizations	APEC, OECD, UN
Languages	Spanish, Amerindian	Map page	152–153
	languages		

CHINA
People's Republic of China

Area Sq Km	9 584 492	Religions	Confucian, Taois
Area Sq Miles	3 700 593		Buddhist, Christi
Population	1 332 079 000		Sunni Muslim
Capital	Beijing (Peking)	Currency	Yuan, Hong Kon,
Languages	Mandarin (Putonghua),		dollar, Macao pat
	Wu, Cantonese,	Organizations	APEC, UN
	Hsiang,	Map page	68–69
	regional languages		

Anhui (Province)

Area Sq Km	139 000	Population	61 350 000
Area Sq Miles	53 668	Capital	Hefei

Bejing (Municipality)

Area Sq Km	16 800	Population	16 950 000
Area Sq Miles	6 487	Capital	Beijing (Peking)

Chongqing (Municipality)

Area Sq Km	23 000	Population	28 390 000
Area Sq Miles	8 880	Capital	Chongqing

Taiwan: the People's Republic of China claims Taiwan as its 23rd Province

ujian (Province)

Area Sq Km	121 400	Population	36 040 000
rea Sq Miles	46 873	Capital	Fuzhou

ansu (Province)

Area Sq Km	453 700	Population	26 280 000
rea Sq Miles	175 175	Capital	Lanzhou

uangdong (Province)

Area Sq Km	178 000	Population	95 440 000
rea Sq Miles	68 726	Capital	Guangzhou
			(Canton)

uangxi Zhuangzu Zizhiqu (Autonomous Region)

Area Sq Km	236 000	Population	48 160 000
rea Sq Miles	91 120	Capital	Nanning

uizhou (Province)

Area Sq Km	176 000	Population	37 930 000
rea Sq Miles	67 954	Capital	Guiyang

ainan (Province)

Area Sq Km	34 000	Population	8 540 000
rea Sq Miles	13 127	Capital	Haikou

ebei (Province)

Area Sq Km	187 700	Population	69 890 000
rea Sq Miles	72 471	Capital	Shijiazhuang

eilongjiang (Province)

Area Sq Km	454 600	Population	38 250 000
rea Sq Miles	175 522	Capital	Harbin

enan (Province)

Area Sq Km	167 000	Population	94 290 000
rea Sq Miles	64 479	Capital	Zhengzhou

ong Kong (Special Administrative Region)

Area Sq Km	1 075	Population	6 978 000
rea Sq Miles	415	Capital	Hong Kong

ubei (Province)

Area Sq Km	185 900	Population	57 110 000
rea Sq Miles	71 776	Capital	Wuhan

unan (Province)

Area Sq Km	210 000	Population	63 800 000
rea Sq Miles	81 081	Capital	Changsha

angsu (Province)

Area Sq Km	102 600	Population	76 770 000
rea Sq Miles	39 614	Capital	Nanjing

angxi (Province)

Area Sq Km	166 900	Population	44 000 000
rea Sq Miles	64 440	Capital	Nanchang

ilin (Province)

Area Sq Km	187 000	Population	27 340 000
rea Sq Miles	72 201	Capital	Changchun

Liaoning (Province)

Area Sq Km	147 400	Population	43 150 000
Area Sq Miles	56 911	Capital	Shenyang

Macao (Special Administrative Region)

Area Sq Km	17	Population	552 000
Area Sq Mile	7		

Nei Mongol Zizhiqu (Inner Mongolia) (Autonomous Region)

Area Sq Km	1 183 000	Population	24 140 000
Area Sq Miles	456 759	Capital	Hohhot

Ningxia Huizu Zizhiqu (Autonomous Region)

Area Sq Km	66 400	Population	6 180 000
Area Sq Miles	25 637	Capital	Yinchuan

Qinghai (Province)

Area Sq Km	721 000	Population	5 540 000
Area Sq Miles	278 380	Capital	Xining

Shaanxi (Province)

Area Sq Km	205 600	Population	37 620 000
Area Sq Miles	79 383	Capital	Xi'an

Shandong (Province)

Area Sq Km	153 300	Population	94 170 000
Area Sq Miles	59 189	Capital	Jinan

Shanghai (Municipality)

Area Sq Km	6 300	Population	18 880 000
Area Sq Miles	2 432	Capital	Shanghai

Shanxi (Province)

Area Sq Km	156 300	Population	34 110 000
Area Sq Miles	60 348	Capital	Taiyuan

Sichuan (Province)

Area Sq Km	569 000	Population	81 380 000
Area Sq Miles	219 692	Capital	Chengdu

Tianjin (Municipality)

Area Sq Km	11 300	Population	11 760 000
Area Sq Miles	4 363	Capital	Tianjin

Xinjiang Uygur Zizhiqu (Sinkiang) (Autonomous Region)

Area Sq Km	1 600 000	Population	21 310 000
Area Sq Miles	617 763	Capital	Ürümqi

Xizang Zizhiqu (Tibet) (Autonomous Region)

Area Sq Km	1 228 400	Population	2 870 000
Area Sq Miles	474 288	Capital	Lhasa

Yunnan (Province)

Area Sq Km	394 000	Population	45 430 000
Area Sq Miles	152 124	Capital	Kunming

Zhejiang (Province)

Area Sq Km	101 800	Population	51 200 000
Area Sq Miles	39 305	Capital	Hangzhou

11

Christmas Island
Australian External Territory

Area Sq Km	135	Religions	Buddhist, Sunni
Area Sq Miles	52		Muslim, Protestant,
Population	1 403		Roman Catholic
Capital	The Settlement	Currency	Australian dollar
Languages	English	Map page	58

Cocos Islands (Keeling Islands)
Australian External Territory

Area Sq Km	14	Religions	Sunni Muslim,
Area Sq Miles	5		Christian
Population	621	Currency	Australian dollar
Capital	West Island	Map page	58
Languages	English		

COLOMBIA
Republic of Colombia

Area Sq Km	1 141 748	Religions	Roman Catholic,
Area Sq Miles	440 831		Protestant
Population	46 927 000	Currency	Colombian peso
Capital	Bogotá	Organizations	UN
Languages	Spanish, Amerindian	Map page	150
	languages		

COMOROS
United Republic of the Comoros

Area Sq Km	1 862	Religions	Sunni Muslim,
Area Sq Miles	719		Roman Catholic
Population	754 000	Currency	Comoros franc
Capital	Moroni	Organizations	UN
Languages	Shikomor (Comorian),	Map page	121
	French, Arabic		

CONGO
Republic of the Congo

Area Sq Km	342 000	Religions	Roman Catholic,
Area Sq Miles	132 047		Protestant,
Population	4 140 000		traditional beliefs,
Capital	Brazzaville		Sunni Muslim
Languages	French, Kongo,	Currency	CFA franc
	Monokutuba, local	Organizations	UN
	languages	Map page	118

CONGO, DEMOCRATIC REPUBLIC OF THE

Area Sq Km	2 345 410	Religions	Christian, Sunni
Area Sq Miles	905 568		Muslim
Population	67 758 000	Currency	Congolese franc
Capital	Kinshasa	Organizations	SADC, UN
Languages	French, Lingala,	Map page	118–119
	Swahili, Kongo,		
	local languages		

Cook Islands
Self-governing New Zealand Overseas Territory

Area Sq Km	293	Religions	Protestant, Roman
Area Sq Miles	113		Catholic
Population	20 000	Currency	New Zealand doll
Capital	Avarua	Map page	49
Languages	English, Maori		

COSTA RICA
Republic of Costa Rica

Area Sq Km	51 100	Religions	Roman Catholic,
Area Sq Miles	19 730		Protestant
Population	4 727 000	Currency	Costa Rican colón
Capital	San José	Organizations	UN
Languages	Spanish	Map page	146

CÔTE D'IVOIRE (IVORY COAST)
Republic of Côte d'Ivoire

Area Sq Km	322 463	Religions	Sunni Muslim,
Area Sq Miles	124 504		Roman Catholic,
Population	20 153 000		traditonal beliefs,
Capital	Yamoussoukro		Protestant
Languages	French, creole, Akan,	Currency	CFA franc
	local languages	Organizations	UN
		Map page	114

CROATIA
Republic of Croatia

Area Sq Km	56 538	Religions	Roman Catholic,
Area Sq Miles	21 829		Serbian Orthodox
Population	4 396 000		Sunni Muslim
Capital	Zagreb	Currency	Kuna
Languages	Croatian, Serbian	Organizations	UN
		Map page	109

CUBA
Republic of Cuba

Area Sq Km	110 860	Religions	Roman Catholic,
Area Sq Miles	42 803		Protestant
Population	11 254 000	Currency	Cuban peso
Capital	Havana (La Habana)	Organizations	UN
Languages	Spanish	Map page	146

Curaçao
Self-governing Netherlands Territory

Area Sq Km	444	Religions	Roman Catholic,
Area Sq Miles	171		Protestant
Population	142 180	Currency	Caribbean guilder
Capital	Willemstad	Map page	147
Languages	Dutch, Papiamento		

CYPRUS
Republic of Cyprus

Area Sq Km	9 251	Religions	Greek Orthodox,
Area Sq Miles	3 572		Sunni Muslim
Population	1 117 000	Currency	Euro
Capital	Nicosia (Lefkosia)	Organizations	Comm., EU, UN
Languages	Greek, Turkish,	Map page	80
	English		

CZECH REPUBLIC

Area Sq Km	78 864	Religions	Roman Catholic,
Area Sq Miles	30 450		Protestant
Population	10 534 000	Currency	Koruna
Capital	Prague (Praha)	Organizations	EU, OECD, UN
Languages	Czech, Moravian,	Map page	102–103
	Slovakian		

DENMARK
Kingdom of Denmark

Area Sq Km	43 075	Religions	Protestant
Area Sq Miles	16 631	Currency	Danish krone
Population	5 573 000	Organizations	EU, OECD, UN
Capital	Copenhagen	Map page	93
	(København)		
Languages	Danish		

DJIBOUTI
Republic of Djibouti

Area Sq Km	23 200	Religions	Sunni Muslim,
Area Sq Miles	8 958		Christian
Population	906 000	Currency	Djibouti franc
Capital	Djibouti	Organizations	UN
Languages	Somali, Afar, French,	Map page	117
	Arabic		

DOMINICA
Commonwealth of Dominica

Area Sq Km	750	Religions	Roman Catholic,
Area Sq Miles	290		Protestant
Population	68 000	Currency	East Caribbean
Capital	Roseau		dollar
Languages	English, creole	Organizations	CARICOM, Comm.,
			UN
		Map page	147

DOMINICAN REPUBLIC

Area Sq Km	48 442	Religions	Roman Catholic,
Area Sq Miles	18 704		Protestant
Population	10 056 000	Currency	Dominican peso
Capital	Santo Domingo	Organizations	UN
Languages	Spanish, creole	Map page	147

Easter Island (Isla de Pascua)
part of Chile

Area Sq Km	171	Religions	Roman Catholic
Area Sq Miles	66	Currency	Chilean peso
Population	4 888	Map page	157
Capital	Hanga Roa		
Languages	Spanish		

EAST TIMOR (TIMOR-LESTE)
Democratic Republic of Timor-Leste

Area Sq Km	14 874	Religions	Roman Catholic
Area Sq Miles	5 743	Currency	United States dollar
Population	1 154 000	Organisations	UN
Capital	Dili	Map page	59
Languages	Portuguese, Tetun,		
	English		

ECUADOR
Republic of Ecuador

Area Sq Km	272 045	Religions	Roman Catholic
Area Sq Miles	105 037	Currency	United States dollar
Population	14 666 000	Organizations	OPEC, UN
Capital	Quito	Map page	150
Languages	Spanish, Quechua,		
	Amerindian		
	languages		

EGYPT
Arab Republic of Egypt

Area Sq Km	1 000 250	Religions	Sunni Muslim,
Area Sq Miles	386 199		Coptic Christian
Population	82 537 000	Currency	Egyptian pound
Capital	Cairo (Al Qāhirah)	Organizations	UN
Languages	Arabic	Map page	116

EL SALVADOR
Republic of El Salvador

Area Sq Km	21 041	Religions	Roman Catholic,
Area Sq Miles	8 124		Protestant
Population	6 227 000	Currency	El Salvador colón,
Capital	San Salvador		United States dollar
Languages	Spanish	Organizations	UN
		Map page	146

EQUATORIAL GUINEA
Republic of Equatorial Guinea

Area Sq Km	28 051	Religions	Roman Catholic,
Area Sq Miles	10 831		traditional beliefs
Population	720 000	Currency	CFA franc
Capital	Malabo	Organizations	UN
Languages	Spanish, French,	Map page	118
	Fang		

ERITREA
State of Eritrea

Area Sq Km	117 400	Religions	Sunni Muslim,
Area Sq Miles	45 328		Coptic Christian
Population	5 415 000	Currency	Nakfa
Capital	Asmara	Organizations	UN
Languages	Tigrinya, Tigre	Map page	116

ESTONIA
Republic of Estonia

Area Sq Km	45 200	Religions	Protestant, Estonian
Area Sq Miles	17 452		and Russian
Population	1 341 000		Orthodox
Capital	Tallinn	Currency	Euro
Languages	Estonian, Russian	Organizations	EU, OECD, UN
		Map page	88

ETHIOPIA
Federal Democratic Republic of Ethiopia

Area Sq Km	1 133 880	Religions	Ethiopian
Area Sq Miles	437 794		Orthodox,
Population	84 734 000		Sunni Muslim,
Capital	Addis Ababa		traditional beliefs
	(Ādīs Ābeba)	Currency	Birr
Languages	Oromo, Amharic,	Organizations	UN
	Tigrinya, local	Map page	117
	languages		

13

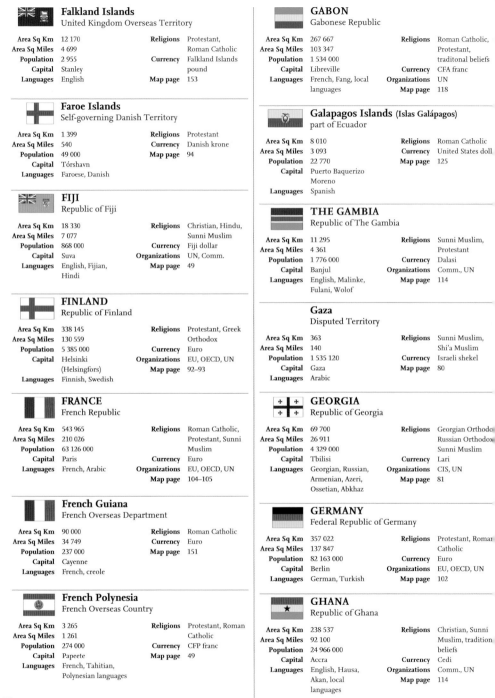

Falkland Islands
United Kingdom Overseas Territory

Area Sq Km	12 170	**Religions**	Protestant,
Area Sq Miles	4 699		Roman Catholic
Population	2 955	**Currency**	Falkland Islands
Capital	Stanley		pound
Languages	English	**Map page**	153

Faroe Islands
Self-governing Danish Territory

Area Sq Km	1 399	**Religions**	Protestant
Area Sq Miles	540	**Currency**	Danish krone
Population	49 000	**Map page**	94
Capital	Tórshavn		
Languages	Faroese, Danish		

FIJI
Republic of Fiji

Area Sq Km	18 330	**Religions**	Christian, Hindu,
Area Sq Miles	7 077		Sunni Muslim
Population	868 000	**Currency**	Fiji dollar
Capital	Suva	**Organizations**	UN, Comm.
Languages	English, Fijian,	**Map page**	49
	Hindi		

FINLAND
Republic of Finland

Area Sq Km	338 145	**Religions**	Protestant, Greek
Area Sq Miles	130 559		Orthodox
Population	5 385 000	**Currency**	Euro
Capital	Helsinki	**Organizations**	EU, OECD, UN
	(Helsingfors)	**Map page**	92–93
Languages	Finnish, Swedish		

FRANCE
French Republic

Area Sq Km	543 965	**Religions**	Roman Catholic,
Area Sq Miles	210 026		Protestant, Sunni
Population	63 126 000		Muslim
Capital	Paris	**Currency**	Euro
Languages	French, Arabic	**Organizations**	EU, OECD, UN
		Map page	104–105

French Guiana
French Overseas Department

Area Sq Km	90 000	**Religions**	Roman Catholic
Area Sq Miles	34 749	**Currency**	Euro
Population	237 000	**Map page**	151
Capital	Cayenne		
Languages	French, creole		

French Polynesia
French Overseas Country

Area Sq Km	3 265	**Religions**	Protestant, Roman
Area Sq Miles	1 261		Catholic
Population	274 000	**Currency**	CFP franc
Capital	Papeete	**Map page**	49
Languages	French, Tahitian,		
	Polynesian languages		

GABON
Gabonese Republic

Area Sq Km	267 667	**Religions**	Roman Catholic,
Area Sq Miles	103 347		Protestant,
Population	1 534 000		traditonal beliefs
Capital	Libreville	**Currency**	CFA franc
Languages	French, Fang, local	**Organizations**	UN
	languages	**Map page**	118

Galapagos Islands (Islas Galápagos)
part of Ecuador

Area Sq Km	8 010	**Religions**	Roman Catholic
Area Sq Miles	3 093	**Currency**	United States doll
Population	22 770	**Map page**	125
Capital	Puerto Baquerizo		
	Moreno		
Languages	Spanish		

THE GAMBIA
Republic of The Gambia

Area Sq Km	11 295	**Religions**	Sunni Muslim,
Area Sq Miles	4 361		Protestant
Population	1 776 000	**Currency**	Dalasi
Capital	Banjul	**Organizations**	Comm., UN
Languages	English, Malinke,	**Map page**	114
	Fulani, Wolof		

Gaza
Disputed Territory

Area Sq Km	363	**Religions**	Sunni Muslim,
Area Sq Miles	140		Shi'a Muslim
Population	1 535 120	**Currency**	Israeli shekel
Capital	Gaza	**Map page**	80
Languages	Arabic		

GEORGIA
Republic of Georgia

Area Sq Km	69 700	**Religions**	Georgian Orthodo
Area Sq Miles	26 911		Russian Orthodox
Population	4 329 000		Sunni Muslim
Capital	Tbilisi	**Currency**	Lari
Languages	Georgian, Russian,	**Organizations**	CIS, UN
	Armenian, Azeri,	**Map page**	81
	Ossetian, Abkhaz		

GERMANY
Federal Republic of Germany

Area Sq Km	357 022	**Religions**	Protestant, Roman
Area Sq Miles	137 847		Catholic
Population	82 163 000	**Currency**	Euro
Capital	Berlin	**Organizations**	EU, OECD, UN
Languages	German, Turkish	**Map page**	102

GHANA
Republic of Ghana

Area Sq Km	238 537	**Religions**	Christian, Sunni
Area Sq Miles	92 100		Muslim, tradition
Population	24 966 000		beliefs
Capital	Accra	**Currency**	Cedi
Languages	English, Hausa,	**Organizations**	Comm., UN
	Akan, local	**Map page**	114
	languages		

COUNTRIES OF THE WORLD

Gibraltar
United Kingdom Overseas Territory

Area Sq Km	7	Religions	Roman Catholic,
Area Sq Miles	3		Protestant, Sunni
Population	29 000		Muslim
Capital	Gibraltar	Currency	Gibraltar pound
Languages	English, Spanish	Map page	106

GREECE
Hellenic Republic

Area Sq Km	131 957	Religions	Greek Orthodox,
Area Sq Miles	50 949		Sunni Muslim
Population	11 390 000	Currency	Euro
Capital	Athens (Athina)	Organizations	EU, OECD, UN
Languages	Greek	Map page	111

Greenland
Self-governing Danish Territory

Area Sq Km	2 175 600	Religions	Protestant
Area Sq Miles	840 004	Currency	Danish krone
Population	57 000	Map page	127
Capital	Nuuk (Godthåb)		
Languages	Greenlandic, Danish		

GRENADA

Area Sq Km	378	Religions	Roman Catholic,
Area Sq Miles	146		Protestant
Population	105 000	Currency	East Caribbean
Capital	St George's		dollar
Languages	English, creole	Organizations	CARICOM, Comm.,
			UN
		Map page	147

Guadeloupe
French Overseas Department

Area Sq Km	1 780	Religions	Roman Catholic
Area Sq Miles	687	Currency	Euro
Population	463 000	Map page	147
Capital	Basse-Terre		
Languages	French, creole		

Guam
United States Unincorporated Territory

Area Sq Km	541	Religions	Roman Catholic
Area Sq Miles	209	Currency	United States dollar
Population	182 000	Map page	59
Capital	Hagåtña		
Languages	Chamorro, English,		
	Tagalog		

GUATEMALA
Republic of Guatemala

Area Sq Km	108 890	Religions	Roman Catholic,
Area Sq Miles	42 043		Protestant
Population	14 757 000	Currency	Quetzal, United
Capital	Guatemala City		States dollar
Languages	Spanish, Mayan	Organizations	UN
	languages	Map page	146

Guernsey
United Kingdom Crown Dependency

Area Sq Km	78	Religions	Protestant, Roman
Area Sq Miles	30		Catholic
Population	65 264	Currency	Pound sterling
Capital	St Peter Port	Map page	95
Languages	English, French		

GUINEA
Republic of Guinea

Area Sq Km	245 857	Religions	Sunni Muslim,
Area Sq Miles	94 926		traditional beliefs,
Population	10 222 000		Christian
Capital	Conakry	Currency	Guinea franc
Languages	French, Fulani,	Organizations	UN
	Malinke, local	Map page	114
	languages		

GUINEA-BISSAU
Republic of Guinea-Bissau

Area Sq Km	36 125	Religions	Traditional beliefs,
Area Sq Miles	13 948		Sunni Muslim,
Population	1 547 000		Christian
Capital	Bissau	Currency	CFA franc
Languages	Portuguese, crioulo,	Organizations	UN
	local languages	Map page	114

GUYANA
Co-operative Republic of Guyana

Area Sq Km	214 969	Religions	Protestant, Hindu,
Area Sq Miles	83 000		Roman Catholic,
Population	756 000		Sunni Muslim
Capital	Georgetown	Currency	Guyana dollar
Languages	English, creole,	Organizations	CARICOM, Comm.,
	Amerindian		UN
	languages	Map page	150

HAITI
Republic of Haiti

Area Sq Km	27 750	Religions	Roman Catholic,
Area Sq Miles	10 714		Protestant, Voodoo
Population	10 124 000	Currency	Gourde
Capital	Port-au-Prince	Organizations	CARICOM, UN
Languages	French, creole	Map page	147

HONDURAS
Republic of Honduras

Area Sq Km	112 088	Religions	Roman Catholic,
Area Sq Miles	43 277		Protestant
Population	7 755 000	Currency	Lempira
Capital	Tegucigalpa	Organizations	UN
Languages	Spanish, Amerindian	Map page	147
	languages		

HUNGARY

Area Sq Km	93 030	Religions	Roman Catholic,
Area Sq Miles	35 919		Protestant
Population	9 966 000	Currency	Forint
Capital	Budapest	Organizations	EU, OECD, UN
Languages	Hungarian	Map page	103

15

ICELAND
Republic of Iceland

Area Sq Km	102 820	Religions	Protestant
Area Sq Miles	39 699	Currency	Icelandic króna
Population	324 000	Organizations	OECD, UN
Capital	Reykjavík	Map page	92
Languages	Icelandic		

INDIA
Republic of India

Area Sq Km	3 064 898	Religions	Hindu,
Area Sq Miles	1 183 364		Sunni Muslim,
Population	1 241 492 000		Shi'a Muslim,
Capital	New Delhi		Sikh, Christian
Languages	Hindi, English, many	Currency	Indian rupee
	regional languages	Organizations	Comm., UN
		Map page	72–73

INDONESIA
Republic of Indonesia

Area Sq Km	1 919 445	Religions	Sunni Muslim,
Area Sq Miles	741 102		Protestant, Roman
Population	242 326 000		Catholic, Hindu,
Capital	Jakarta		Buddhist
Languages	Indonesian, local	Currency	Rupiah
	languages	Organizations	APEC, ASEAN,
			OPEC, UN
		Map page	58–59

IRAN
Islamic Republic of Iran

Area Sq Km	1 648 000	Religions	Shi'a Muslim,
Area Sq Miles	636 296		Sunni Muslim
Population	74 799 000	Currency	Iranian rial
Capital	Tehrān	Organizations	OPEC, UN
Languages	Farsi, Azeri, Kurdish,	Map page	81
	regional languages		

IRAQ
Republic of Iraq

Area Sq Km	438 317	Religions	Shi'a Muslim,
Area Sq Miles	169 235		Sunni Muslim,
Population	32 665 000		Christian
Capital	Baghdād	Currency	Iraqi dinar
Languages	Arabic, Kurdish,	Organizations	OPEC, UN
	Turkmen	Map page	81

IRELAND
Republic of Ireland

Area Sq Km	70 282	Religions	Roman Catholic,
Area Sq Miles	27 136		Protestant
Population	4 526 000	Currency	Euro
Capital	Dublin	Organizations	EU, OECD, UN
	(Baile Átha Cliath)	Map page	97
Languages	English, Irish		

Isle of Man
United Kingdom Crown Dependency

Area Sq Km	572	Religions	Protestant, Roman
Area Sq Miles	221		Catholic
Population	83 000	Currency	Pound sterling
Capital	Douglas	Map page	98
Languages	English		

ISRAEL
State of Israel

Area Sq Km	20 770	Religions	Jewish, Sunni
Area Sq Miles	8 019		Muslim, Christia
Population	7 562 000		Druze
Capital	Jerusalem*	Currency	Shekel
	(Yerushalayim)	Organizations	OECD, UN
	(El Quds)	Map page	80
Languages	Hebrew, Arabic		

*De facto capital. Disputed.

ITALY
Italian Republic

Area Sq Km	301 245	Religions	Roman Catholic
Area Sq Miles	116 311	Currency	Euro
Population	60 789 000	Organizations	EU, OECD, UN
Capital	Rome (Roma)	Map page	108–109
Languages	Italian		

JAMAICA

Area Sq Km	10 991	Religions	Protestant, Roma
Area Sq Miles	4 244		Catholic
Population	2 751 000	Currency	Jamaican dollar
Capital	Kingston	Organizations	CARICOM, Com
Languages	English, creole		UN
		Map page	146

JAPAN

Area Sq Km	377 727	Religions	Shintoist, Buddh
Area Sq Miles	145 841		Christian
Population	126 497 000	Currency	Yen
Capital	Tōkyō	Organizations	APEC, OECD, UN
Languages	Japanese	Map page	66–67

Jersey
United Kingdom Crown Dependency

Area Sq Km	116	Religions	Protestant, Roma
Area Sq Miles	45		Catholic
Population	92 500	Currency	Pound sterling
Capital	St Helier	Map page	95
Languages	English, French		

JORDAN
Hashemite Kingdom of Jordan

Area Sq Km	89 206	Religions	Sunni Muslim,
Area Sq Miles	34 443		Christian
Population	6 330 000	Currency	Jordanian dinar
Capital	'Ammān	Organizations	UN
Languages	Arabic	Map page	80

Juan Fernández Islands
part of Chile

Area Sq Km	179	Religions	Roman Catholic,
Area Sq Miles	69		Protestant
Population	832	Currency	Chilean peso
Capital	San Juan Bautista	Map page	157
Languages	Spanish, Amerindian		
	languages		

KAZAKHSTAN
Republic of Kazakhstan

Area Sq Km	2 717 300	Religions	Sunni Muslim,
Area Sq Miles	1 049 155		Russian Orthodox,
Population	16 207 000		Protestant
Capital	Astana (Akmola)	Currency	Tenge
Languages	Kazakh, Russian,	Organizations	CIS, UN
	Ukrainian, German,	Map page	76–77
	Uzbek, Tatar		

KENYA
Republic of Kenya

Area Sq Km	582 646	Religions	Christian,
Area Sq Miles	224 961		traditional beliefs
Population	41 610 000	Currency	Kenyan shilling
Capital	Nairobi	Organizations	Comm., UN
Languages	Swahili, English,	Map page	119
	local languages		

KIRIBATI
Republic of Kiribati

Area Sq Km	717	Religions	Roman Catholic,
Area Sq Miles	277		Protestant
Population	101 000	Currency	Australian dollar
Capital	Bairiki	Organizations	Comm., UN
Languages	Gilbertese, English	Map page	49

KOSOVO
Republic of Kosovo

Area Sq Km	10 908	Religions	Sunni Muslim,
Area Sq Miles	4 212		Serbian Orthodox
Population	2 180 686	Currency	Euro
Capital	Prishtinë (Priština)	Map page	109
Languages	Albanian, Serbian		

KUWAIT
State of Kuwait

Area Sq Km	17 818	Religions	Sunni Muslim,
Area Sq Miles	6 880		Shi'a Muslim,
Population	2 818 000		Christian, Hindu
Capital	Kuwait (Al Kuwayt)	Currency	Kuwaiti dinar
Languages	Arabic	Organizations	OPEC, UN
		Map page	78

KYRGYZSTAN
Kyrgyz Republic

Area Sq Km	198 500	Religions	Sunni Muslim,
Area Sq Miles	76 641		Russian Orthodox
Population	5 393 000	Currency	Kyrgyz som
Capital	Bishkek (Frunze)	Organizations	CIS, UN
Languages	Kyrgyz, Russian, Uzbek	Map page	77

LAOS
Lao People's Democratic Republic

Area Sq Km	236 800	Religions	Buddhist,
Area Sq Miles	91 429		traditional beliefs
Population	6 288 000	Currency	Kip
Capital	Vientiane (Viangchan)	Organizations	ASEAN, UN
Languages	Lao, local languages	Map page	62–63

LATVIA
Republic of Latvia

Area Sq Km	64 589	Religions	Protestant,
Area Sq Miles	24 938		Roman Catholic,
Population	2 243 000		Russian Orthodox
Capital	Rīga	Currency	Lats
Languages	Latvian, Russian	Organizations	EU, UN
		Map page	88

LEBANON
Republic of Lebanon

Area Sq Km	10 452	Religions	Shi'a Muslim, Sunni
Area Sq Miles	4 036		Muslim, Christian
Population	4 259 000	Currency	Lebanese pound
Capital	Beirut (Beyrouth)	Organizations	UN
Languages	Arabic, Armenian,	Map page	80
	French		

LESOTHO
Kingdom of Lesotho

Area Sq Km	30 355	Religions	Christian,
Area Sq Miles	11 720		traditional beliefs
Population	2 194 000	Currency	Loti, South African
Capital	Maseru		rand
Languages	Sesotho, English,	Organizations	Comm., SADC, UN
	Zulu	Map page	123

LIBERIA
Republic of Liberia

Area Sq Km	111 369	Religions	Traditional beliefs,
Area Sq Miles	43 000		Christian,
Population	4 129 000		Sunni Muslim
Capital	Monrovia	Currency	Liberian dollar
Languages	English, creole,	Organizations	UN
	local languages	Map page	114

LIBYA

Area Sq Km	1 759 540	Religions	Sunni Muslim
Area Sq Miles	679 362	Currency	Libyan dinar
Population	6 423 000	Organizations	OPEC, UN
Capital	Tripoli (Ṭarābulus)	Map page	115
Languages	Arabic, Berber		

LIECHTENSTEIN
Principality of Liechtenstein

Area Sq Km	160	Religions	Roman Catholic,
Area Sq Miles	62		Protestant
Population	36 000	Currency	Swiss franc
Capital	Vaduz	Organizations	UN
Languages	German	Map page	105

LITHUANIA
Republic of Lithuania

Area Sq Km	65 200	Religions	Roman Catholic,
Area Sq Miles	25 174		Protestant,
Population	3 307 000		Russian Orthodox
Capital	Vilnius	Currency	Litas
Languages	Lithuanian, Russian,	Organizations	EU, UN
	Polish	Map page	88

17

Lord Howe Island
part of Australia

Area Sq Km	17
Area Sq Miles	6
Population	364
Languages	English

Religions	Protestant, Roman Catholic
Currency	Australian dollar
Map page	51

LUXEMBOURG
Grand Duchy of Luxembourg

Area Sq Km	2 586
Area Sq Miles	998
Population	516 000
Capital	Luxembourg
Languages	Letzeburgish, German, French

Religions	Roman Catholic
Currency	Euro
Organizations	EU, OECD, UN
Map page	100

MACEDONIA (F.Y.R.O.M.)
Republic of Macedonia

Area Sq Km	25 713
Area Sq Miles	9 928
Population	2 064 000
Capital	Skopje
Languages	Macedonian, Albanian, Turkish

Religions	Macedonian Orthodox, Sunni Muslim
Currency	Macedonian denar
Organizations	UN
Map page	111

MADAGASCAR
Republic of Madagascar

Area Sq Km	587 041
Area Sq Miles	226 658
Population	21 315 000
Capital	Antananarivo
Languages	Malagasy, French

Religions	Traditional beliefs, Christian, Sunni Muslim
Currency	Malagasy ariary, Malagasy franc
Organizations	SADC, UN
Map page	121

Madeira
Autonomous Region of Portugal

Area Sq Km	779
Area Sq Miles	301
Population	247 399
Capital	Funchal
Languages	Portuguese

Religions	Roman Catholic, Protestant
Currency	Euro
Map page	114

MALAWI
Republic of Malawi

Area Sq Km	118 484
Area Sq Miles	45 747
Population	15 381 000
Capital	Chichewa, English, local languages

Religions	Christian, traditional beliefs, Sunni Muslim
Currency	Malawian kwacha
Organizations	Comm., SADC, UN
Map page	121

MALAYSIA
Federation of Malaysia

Area Sq Km	332 965
Area Sq Miles	128 559
Population	28 859 000
Capital	Kuala Lumpur/ Putrajaya
Languages	Malay, English, Chinese, Tamil, local languages

Religions	Sunni Muslim, Buddhist, Hindu, Christian, traditional beliefs
Currency	Ringgit
Organizations	APEC, ASEAN, Comm., UN
Map page	60–61

MALDIVES
Republic of the Maldives

Area Sq Km	298
Area Sq Miles	115
Population	320 000
Capital	Male
Languages	Divehi (Maldivian)

Religions	Sunni Muslim
Currency	Rufiyaa
Organizations	Comm., UN
Map page	56

MALI
Republic of Mali

Area Sq Km	1 240 140
Area Sq Miles	478 821
Population	15 840 000
Capital	Bamako
Languages	French, Bambara, local languages

Religions	Sunni Muslim, traditional beliefs, Christian
Currency	CFA franc
Organizations	UN
Map page	114

MALTA
Republic of Malta

Area Sq Km	316
Area Sq Miles	122
Population	418 000
Capital	Valletta
Languages	Maltese, English

Religions	Roman Catholic
Currency	Euro
Organizations	Comm., EU, UN
Map page	84

MARSHALL ISLANDS
Republic of the Marshall Islands

Area Sq Km	181
Area Sq Miles	70
Population	55 000
Capital	Delap-Uliga-Djarrit
Languages	English, Marshallese

Religions	Protestant, Roman Catholic
Currency	United States dollar
Organizations	UN
Map page	48

Martinique
French Overseas Department

Area Sq Km	1 079
Area Sq Miles	417
Population	407 000
Capital	Fort-de-France
Languages	French, creole

Religions	Roman Catholic, traditional beliefs
Currency	Euro
Map page	147

MAURITANIA
Islamic Arab and African Republic of Mauritania

Area Sq Km	1 030 700
Area Sq Miles	397 955
Population	3 542 000
Capital	Nouakchott
Languages	Arabic, French, local languages

Religions	Sunni Muslim
Currency	Ouguiya
Organizations	UN
Map page	114

MAURITIUS
Republic of Mauritius

Area Sq Km	2 040
Area Sq Miles	788
Population	1 307 000
Capital	Port Louis
Languages	English, creole, Hindi, Bhojpuri, French

Religions	Hindu, Roman Catholic, Sunni Muslim
Currency	Mauritius rupee
Organizations	Comm., SADC, U
Map page	113

Mayotte
French Overseas Department

Area Sq Km	373	Religions	Sunni Muslim,
Area Sq Miles	144		Christian
Population	211 000	Currency	Euro
Capital	Dzaoudzi	Map page	121
Languages	French, Mahorian		

Melilla
Autonomous Community of Spain

Area Sq Km	13	Religions	Roman Catholic,
Area Sq Miles	5		Muslim
Population	76 034	Currency	Euro
Capital	Melilla	Map page	114
Languages	Spanish, Arabic		

MEXICO
United Mexican States

Area Sq Km	1 972 545	Religions	Roman Catholic,
Area Sq Miles	761 604		Protestant
Population	114 793 000	Currency	Mexican peso
Capital	Mexico City	Organizations	APEC, OECD, UN
Languages	Spanish, Amerindian languages	Map page	144–145

MICRONESIA, FEDERATED STATES OF

Area Sq Km	701	Religions	Roman Catholic,
Area Sq Miles	271		Protestant
Population	112 000	Currency	United States dollar
Capital	Palikir	Organizations	UN
Languages	English, Chuukese, Pohnpeian, local languages	Map page	48

MOLDOVA
Republic of Moldova

Area Sq Km	33 700	Religions	Romanian
Area Sq Miles	13 012		Orthodox,
Population	3 545 000		Russian Orthodox
Capital	Chişinău (Kishinev)	Currency	Moldovan leu
Languages	Romanian, Ukrainian, Gagauz, Russian	Organizations	CIS, UN
		Map page	90

MONACO
Principality of Monaco

Area Sq Km	2	Religions	Roman Catholic
Area Sq Miles	1	Currency	Euro
Population	35 000	Organizations	UN
Capital	Monaco-Ville	Map page	105
Languages	French, Monégasque, Italian		

MONGOLIA

Area Sq Km	1 565 000	Religions	Buddhist,
Area Sq Miles	604 250		Sunni Muslim
Population	2 800 000	Currency	Tugrik (tögrög)
Capital	Ulan Bator (Ulaanbaatar)	Organizations	UN
		Map page	68–69
Languages	Khalka (Mongolian), Kazakh, local languages		

MONTENEGRO

Area Sq Km	13 812	Religions	Montenegrin,
Area Sq Miles	5 333		Orthodox,
Population	632 000		Sunni Muslim
Capital	Podgorica	Currency	Euro
Languages	Serbian, (Montenegrin), Albanian	Organizations	UN
		Map page	109

Montserrat
United Kingdom Overseas Territory

Area Sq Km	100	Religions	Protestant, Roman
Area Sq Miles	39		Catholic
Population	4 655	Currency	East Caribbean
Capital	Brades*		dollar
Languages	English	Organizations	CARICOM
	*Temporary capital	Map page	147

MOROCCO
Kingdom of Morocco

Area Sq Km	446 550	Religions	Sunni Muslim
Area Sq Miles	172 414	Currency	Moroccan dirham
Population	32 273 000	Organizations	UN
Capital	Rabat	Map page	114
Languages	Arabic, Berber, French		

MOZAMBIQUE
Republic of Mozambique

Area Sq Km	799 380	Religions	Traditional beliefs,
Area Sq Miles	308 642		Roman Catholic,
Population	23 930 000		Sunni Muslim
Capital	Maputo	Currency	Metical
Languages	Portuguese, Makua, Tsonga, local languages	Organizations	Comm., SADC, UN
		Map page	121

MYANMAR (BURMA)
Republic of the Union of Myanmar

Area Sq Km	676 577	Religions	Buddhist,
Area Sq Miles	261 228		Christian,
Population	48 337 000		Sunni Muslim
Capital	Nay Pyi Taw	Currency	Kyat
Languages	Burmese, Shan, Karen, local languages	Organizations	ASEAN, UN
		Map page	62–63

Nagorno-Karabakh (Dağlıq Qarabağ)
Disputed Territory (Azerbaijan)

Area Sq Km	6 000
Area Sq Miles	2 317
Population	140 000
Capital	Xankändi (Stepanakert)

NAMIBIA
Republic of Namibia

Area Sq Km	824 292	Religions	Protestant,
Area Sq Miles	318 261		Roman Catholic
Population	2 324 000	Currency	Namibian dollar
Capital	Windhoek	Organizations	Comm., SADC, UN
Languages	English, Afrikaans, German, Ovambo, local languages	Map page	121

19

NAURU
Republic of Nauru

Area Sq Km	21	Religions	Protestant, Roman
Area Sq Miles	8		Catholic
Population	10 000	Currency	Australian dollar
Capital	Yaren	Organizations	Comm., UN
Languages	Nauruan, English	Map page	48

NEPAL
Federal Democratic Republic of Nepal

Area Sq Km	147 181	Religions	Hindu, Buddhist,
Area Sq Miles	56 827		Sunni Muslim
Population	30 486 000	Currency	Nepalese rupee
Capital	Kathmandu	Organizations	UN
Languages	Nepali, Maithili,	Map page	75
	Bhojpuri, English,		
	local languages		

NETHERLANDS
Kingdom of the Netherlands

Area Sq Km	41 526	Religions	Roman Catholic,
Area Sq Miles	16 033		Protestant, Sunni
Population	16 665 000		Muslim
Capital	Amsterdam/	Currency	Euro
	The Hague	Organizations	EU, OECD, UN
	('s-Gravenhage)	Map page	100
Languages	Dutch, Frisian		

New Caledonia
French Overseas Collectivity

Area Sq Km	19 058	Religions	Roman Catholic,
Area Sq Miles	7 358		Protestant, Sunni
Population	255 000		Muslim
Capital	Nouméa	Currency	CFP franc
Languages	French, local	Map page	48
	languages		

NEW ZEALAND

Area Sq Km	270 534	Religions	Protestant, Roman
Area Sq Miles	104 454		Catholic
Population	4 415 000	Currency	New Zealand dollar
Capital	Wellington	Organizations	APEC, Comm.,
Languages	English, Maori		OECD, UN
		Map page	54

NICARAGUA
Republic of Nicaragua

Area Sq Km	130 000	Religions	Roman Catholic,
Area Sq Miles	50 193		Protestant
Population	5 870 000	Currency	Córdoba
Capital	Managua	Organizations	UN
Languages	Spanish, Amerindian	Map page	146
	languages		

NIGER
Republic of Niger

Area Sq Km	1 267 000	Religions	Sunni Muslim,
Area Sq Miles	489 191		traditional beliefs
Population	16 069 000	Currency	CFA franc
Capital	Niamey	Organizations	UN
Languages	French, Hausa,	Map page	115
	Fulani, local		
	languages		

NIGERIA
Federal Republic of Nigeria

Area Sq Km	923 768	Religions	Sunni Muslim,
Area Sq Miles	356 669		Christian,
Population	162 471 000		traditional belief
Capital	Abuja	Currency	Naira
Languages	English, Hausa,	Organizations	Comm., OPEC, U
	Yoruba, Ibo, Fulani,	Map page	115
	local languages		

Niue
Self-governing New Zealand Overseas Territory

Area Sq Km	258	Religions	Christian
Area Sq Miles	100	Currency	New Zealand dol
Population	1 496	Map page	48
Capital	Alofi		
Languages	English, Nivean		

Norfolk Island
Australian External Territory

Area Sq Km	35	Religions	Protestant, Roma
Area Sq Miles	14		Catholic
Population	2 523	Currency	Australian dollar
Capital	Kingston	Map page	48
Languages	English		

Northern Mariana Islands
United States Commonwealth

Area Sq Km	477	Religions	Roman Catholic
Area Sq Miles	184	Currency	United States do
Population	61 000	Map page	59
Capital	Capitol Hill		
Languages	English, Chamorro,		
	local languages		

NORTH KOREA
Democratic People's Republic of Korea

Area Sq Km	120 538	Religions	Traditional belie
Area Sq Miles	46 540		Chondoist,
Population	24 451 000		Buddhist
Capital	P'yŏngyang	Currency	North Korean wo
Languages	Korean	Organizations	UN
		Map page	65

NORWAY
Kingdom of Norway

Area Sq Km	323 878	Religions	Protestant, Roma
Area Sq Miles	125 050		Catholic
Population	4 925 000	Currency	Norwegian krone
Capital	Oslo	Organizations	OECD, UN
Languages	Norwegian	Map page	92–93

OMAN
Sultanate of Oman

Area Sq Km	309 500	Religions	Ibadhi Muslim,
Area Sq Miles	119 499		Sunni Muslim
Population	2 846 000	Currency	Omani riyal
Capital	Muscat (Masqaṭ)	Organizations	UN
Languages	Arabic, Baluchi,	Map page	79
	Indian languages		

PAKISTAN
Islamic Republic of Pakistan

Area Sq Km	803 940	Religions	Sunni Muslim,
Area Sq Miles	310 403		Shi'a Muslim,
Population	176 745 000		Christian, Hindu
Capital	Islamabad	Currency	Pakistani rupee
Languages	Urdu, Punjabi,	Organizations	Comm., UN
	Sindhi, Pushtu,	Map page	74
	English, Balochi		

PALAU
Republic of Palau

Area Sq Km	497	Religions	Roman Catholic,
Area Sq Miles	192		Protestant,
Population	21 000		traditional beliefs
Capital	Melekeok (Ngerulmud)	Currency	United States dollar
Languages	Palauan, English	Organizations	UN
		Map page	59

PANAMA
Republic of Panama

Area Sq Km	77 082	Religions	Roman Catholic,
Area Sq Miles	29 762		Protestant, Sunni
Population	3 571 000		Muslim
Capital	Panama City	Currency	Balboa
Languages	Spanish, English,	Organizations	UN
	Amerindian	Map page	146
	languages		

PAPUA NEW GUINEA
Independent State of Papua New Guinea

Area Sq Km	462 840	Religions	Protestant,
Area Sq Miles	178 704		Roman Catholic,
Population	7 014 000		traditional beliefs
Capital	Port Moresby	Currency	Kina
Languages	English, Tok Pisin	Organizations	APEC, Comm., UN
	(creole), local	Map page	59
	languages		

PARAGUAY
Republic of Paraguay

Area Sq Km	406 752	Religions	Roman Catholic,
Area Sq Miles	157 048		Protestant
Population	6 568 000	Currency	Guaraní
Capital	Asunción	Organizations	UN
Languages	Spanish, Guaraní	Map page	152

PERU
Republic of Peru

Area Sq Km	1 285 216	Religions	Roman Catholic,
Area Sq Miles	496 225		Protestant
Population	29 400 000	Currency	Nuevo sol
Capital	Lima	Organizations	APEC, UN
Languages	Spanish, Quechua,	Map page	150
	Aymara		

PHILIPPINES
Republic of the Philippines

Area Sq Km	300 000	Religions	Roman Catholic,
Area Sq Miles	115 831		Protestant, Sunni
Population	94 852 000		Muslim, Aglipayan
Capital	Manila	Currency	Philippine peso
Languages	English, Filipino,	Organizations	APEC, ASEAN, UN
	Tagalog, Cebuano,	Map page	64
	local languages		

Pitcairn Islands
United Kingdom Overseas Territory

Area Sq Km	45	Religions	Protestant
Area Sq Miles	17	Currency	New Zealand
Population	48		dollar
Capital	Adamstown	Map page	49
Languages	English		

POLAND
Polish Republic

Area Sq Km	312 683	Religions	Roman Catholic,
Area Sq Miles	120 728		Polish Orthodox
Population	38 299 000	Currency	Złoty
Capital	Warsaw (Warszawa)	Organizations	EU, OECD, UN
Languages	Polish, German	Map page	103

PORTUGAL
Portuguese Republic

Area Sq Km	88 940	Religions	Roman Catholic,
Area Sq Miles	34 340		Protestant
Population	10 690 000	Currency	Euro
Capital	Lisbon (Lisboa)	Organizations	EU, OECD, UN
Languages	Portuguese	Map page	106

Puerto Rico
United States Commonwealth

Area Sq Km	9 104	Religions	Roman Catholic,
Area Sq Miles	3 515		Protestant
Population	3 746 000	Currency	United States
Capital	San Juan		dollar
Languages	Spanish, English	Map page	147

QATAR
State of Qatar

Area Sq Km	11 437	Religions	Sunni Muslim
Area Sq Miles	4 416	Currency	Qatari riyal
Population	1 870 000	Organizations	OPEC, UN
Capital	Doha (Ad Dawḩah)	Map page	79
Languages	Arabic		

Réunion
French Overseas Department

Area Sq Km	2 551	Religions	Roman Catholic
Area Sq Miles	985	Currency	Euro
Population	856 000	Map page	113
Capital	St-Denis		
Languages	French, creole		

Rodrigues Island
part of Mauritius

Area Sq Km	104	Religions	Christian
Area Sq Miles	40	Currency	Rupee
Population	37 837	Map page	159
Capital	Port Mathurin		
Languages	English, creole		

ROMANIA

Area Sq Km	237 500	Religions	Romanian
Area Sq Miles	91 699		Orthodox,
Population	21 436 000		Protestant,
Capital	Bucharest (Bucureşti)		Roman Catholic
Languages	Romanian,	Currency	Romanian leu
	Hungarian	Organizations	EU, UN
		Map page	110

RUSSIAN FEDERATION

Area Sq Km	17 075 400	Religions	Russian Orthodox,
Area Sq Miles	6 592 849		Sunni Muslim,
Population	142 836 000		Protestant
Capital	Moscow (Moskva)	Currency	Russian rouble
Languages	Russian, Tatar,	Organizations	APEC, CIS, UN
	Ukrainian, local	Map page	82–83
	languages		

RWANDA
Republic of Rwanda

Area Sq Km	26 338	Religions	Roman Catholic,
Area Sq Miles	10 169		traditional beliefs,
Population	10 943 000		Protestant
Capital	Kigali	Currency	Rwandan franc
Languages	Kinyarwanda,	Organizations	UN
	French, English	Map page	119

Saba
Netherlands Special Municipality

Area Sq Km	13	Religions	Roman Catholic,
Area Sq Miles	5		Protestant
Population	1 737	Currency	US dollar
Capital	The Bottom	Map page	147
Languages	Dutch, English		

St-Barthélémy
French Overseas Collectivity

Area Sq Km	21	Religions	Roman Catholic
Area Sq Miles	8	Currency	Euro
Population	8 823	Map page	147
Capital	Gustavia		
Languages	French		

St Helena
part of St Helena, Ascension and Tristan da Cunha

Area Sq Km	122	Religions	Protestant, Roman
Area Sq Miles	47		Catholic
Population	4 257	Currency	St Helena pound
Capital	Jamestown	Map page	113
Languages	English		

St Helena, Ascension and Tristan da Cunha
United Kingdom Overseas Territory

Area Sq Km	410	Religions	Protestant, Roman
Area Sq Miles	158		Catholic,
Population	5 404	Currency	St Helena pound,
Capital	Jamestown		Pound sterling
Languages	English	Map page	113

ST KITTS AND NEVIS
Federation of Saint Kitts and Nevis

Area Sq Km	261	Religions	Protestant, Roman
Area Sq Miles	101		Catholic
Population	53 000	Currency	East Caribbean
Capital	Basseterre		dollar
Languages	English, creole	Organizations	CARICOM,
			Comm., UN
		Map page	147

ST LUCIA
Saint Lucia

Area Sq Km	616	Religions	Roman Catholic,
Area Sq Miles	238		Protestant
Population	176 000	Currency	East Caribbean
Capital	Castries		dollar
Languages	English, creole	Organizations	CARICOM, Comm
			UN
		Map page	147

St-Martin
French Overseas Collectivity

Area Sq Km	54	Religions	Roman Catholic
Area Sq Miles	21	Currency	Euro
Population	37 163	Map page	147
Capital	Marigot		
Languages	French		

St Pierre and Miquelon
French Territorial Collectivity

Area Sq Km	242	Religions	Roman Catholic
Area Sq Miles	93	Currency	Euro
Population	6 290	Map page	131
Capital	St-Pierre		
Languages	French		

ST VINCENT AND THE GRENADINES

Area Sq Km	389	Religions	Protestant, Roman
Area Sq Miles	150		Catholic
Population	109 000	Currency	East Caribbean
Capital	Kingstown		dollar
Languages	English, creole	Organizations	CARICOM, Comm
			UN
		Map page	147

SAMOA
Independent State of Samoa

Area Sq Km	2 831	Religions	Protestant, Roman
Area Sq Miles	1 093		Catholic
Population	184 000	Currency	Tala
Capital	Apia	Organizations	Comm., UN
Languages	Samoan, English	Map page	49

SAN MARINO
Republic of San Marino

Area Sq Km	61	Religions	Roman Catholic
Area Sq Miles	24	Currency	Euro
Population	32 000	Organizations	UN
Capital	San Marino	Map page	108
Languages	Italian		

SÃO TOMÉ AND PRÍNCIPE
Democratic Republic of São Tomé and Príncipe

Area Sq Km	964	Religions	Roman Catholic,
ea Sq Miles	372		Protestant
Population	169 000	Currency	Dobra
Capital	São Tomé	Organizations	UN
Languages	Portuguese, creole	Map page	113

SAUDI ARABIA
Kingdom of Saudi Arabia

Area Sq Km	2 200 000	Religions	Sunni Muslim,
ea Sq Miles	849 425		Shi'a Muslim
Population	28 083 000	Currency	Saudi Arabian riyal
Capital	Riyadh (Ar Riyāḍ)	Organizations	OPEC, UN
Languages	Arabic	Map page	78–79

SENEGAL
Republic of Senegal

Area Sq Km	196 720	Religions	Sunni Muslim,
ea Sq Miles	75 954		Roman Catholic,
Population	12 768 000		traditional beliefs
Capital	Dakar	Currency	CFA franc
Languages	French, Wolof,	Organizations	UN
	Fulani, local	Map page	114
	languages		

SERBIA
Republic of Serbia

Area Sq Km	77 453	Religions	Roman Catholic,
ea Sq Miles	29 904		Serbian Orthodox,
Population	7 306 677		Sunni Muslim
Capital	Belgrade (Beograd)	Currency	Serbian dinar
Languages	Serbian, Hungarian	Organizations	UN
		Map page	109

SEYCHELLES
Republic of the Seychelles

Area Sq Km	455	Religions	Roman Catholic,
ea Sq Miles	176		Protestant
Population	87 000	Currency	Seychelles rupee
Capital	Victoria	Organizations	Comm., SADC, UN
Languages	English, French,	Map page	113
	creole		

SIERRA LEONE
Republic of Sierra Leone

Area Sq Km	71 740	Religions	Sunni Muslim,
ea Sq Miles	27 699		traditional beliefs
Population	5 997 000	Currency	Leone
Capital	Freetown	Organizations	Comm., UN
Languages	English, creole,	Map page	114
	Mende, Temne,		
	local languages		

SINGAPORE
Republic of Singapore

Area Sq Km	639	Religions	Buddhist, Taoist,
rea Sq Miles	247		Sunni Muslim,
Population	5 188 000		Christian, Hindu
Capital	Singapore	Currency	Singapore dollar
Languages	Chinese, English,	Organizations	APEC, ASEAN,
	Malay, Tamil		Comm., UN
		Map page	60

Sint Eustatius
Netherlands Special Municipality

Area Sq Km	21	Religions	Protestant, Roman
Area Sq Miles	8		Catholic
Population	2 886	Currency	US dollar
Capital	Oranjestad	Map page	147
Languages	Dutch, English		

Sint Maarten
Self-governing Netherlands Territory

Area Sq Km	34	Religions	Protestant, Roman
Area Sq Miles	13		Catholic
Population	37 429	Currency	Caribbean guilder
Capital	Philipsburg	Map page	147
Languages	Dutch, English		

SLOVAKIA
Slovak Republic

Area Sq Km	49 035	Religions	Roman Catholic,
Area Sq Miles	18 933		Protestant,
Population	5 472 000		Orthodox
Capital	Bratislava	Currency	Euro
Languages	Slovak,	Organizations	EU, OECD, UN
	Hungarian, Czech	Map page	103

SLOVENIA
Republic of Slovenia

Area Sq Km	20 251	Religions	Roman Catholic,
Area Sq Miles	7 819		Protestant
Population	2 035 000	Currency	Euro
Capital	Ljubljana	Organizations	EU, OECD, UN
Languages	Slovene, Croatian,	Map page	108–109
	Serbian		

SOLOMON ISLANDS

Area Sq Km	28 370	Religions	Protestant, Roman
Area Sq Miles	10 954		Catholic
Population	552 000	Currency	Solomon Islands
Capital	Honiara		dollar
Languages	English, creole,	Organizations	Comm., UN
	local languages	Map page	48

SOMALIA
Somali Republic

Area Sq Km	637 657	Religions	Sunni Muslim
Area Sq Miles	246 201	Currency	Somali shilling
Population	9 557 000	Organizations	UN
Capital	Mogadishu	Map page	117
	(Muqdisho)		
Languages	Somali, Arabic		

Somaliland
Disputed Territory (Somalia)

Area Sq Km	140 000	Map page	117
Area Sq Miles	54 054		
Population	3 500 000		
Capital	Hargeysa		

23

SOUTH AFRICA, REPUBLIC OF

Area Sq Km	1 219 080	Religions	Protestant, Roman
Area Sq Miles	470 689		Catholic, Sunni
Population	50 460 000		Muslim, Hindu
Capital	Pretoria (Tshwane)/	Currency	Rand
	Cape Town	Organizations	Comm., SADC, UN
Languages	Afrikaans, English,	Map page	122–123
	nine official local		
	languages		

SOUTH KOREA
Republic of Korea

Area Sq Km	99 274	Religions	Buddhist,
Area Sq Miles	38 330		Protestant,
Population	48 391 000		Roman Catholic
Capital	Seoul (Sŏul)	Currency	South Korean won
Languages	Korean	Organizations	APEC, OECD, UN
		Map page	65

South Ossetia
Disputed Territory (Georgia)

Area Sq Km	4 000
Area Sq Miles	1 544
Population	70 000
Capital	Tskhinvali

SOUTH SUDAN
Republic of South Sudan

Area Sq Km	644 329	Religions	Traditional beliefs,
Area Sq Miles	248 775		Christian
Population	8 260 490	Currency	South Sudan pound
Capital	Juba	Map page	117
Languages	English, Arabic,		
	Dinka, Nuer, local		
	languages		

SPAIN
Kingdom of Spain

Area Sq Km	504 782	Religions	Roman Catholic
Area Sq Miles	194 897	Currency	Euro
Population	46 455 000	Organizations	EU, OECD, UN
Capital	Madrid	Map page	106–107
Languages	Spanish, Castilian,		
	Catalan, Galician,		
	Basque		

SRI LANKA
Democratic Socialist Republic of Sri Lanka

Area Sq Km	65 610	Religions	Buddhist, Hindu,
Area Sq Miles	25 332		Sunni Muslim,
Population	21 045 000		Roman Catholic
Capital	Sri Jayewardenepura	Currency	Sri Lankan rupee
	Kotte	Organizations	Comm., UN
Languages	Sinhalese, Tamil,	Map page	73
	English		

SUDAN
Republic of the Sudan

Area Sq Km	1 861 484	Religions	Sunni Muslim,
Area Sq Miles	718 725		traditional beliefs,
Population	36 371 510		Christian
Capital	Khartoum	Currency	Sudanese pound
Languages	Arabic, Dinka,		(Sudani)
	Nubian, Beja, Nuer,	Organizations	UN
	local languages	Map page	116–117

SURINAME
Republic of Suriname

Area Sq Km	163 820	Religions	Hindu, Roman
Area Sq Miles	63 251		Catholic, Protestan
Population	529 000		Sunni Muslim
Capital	Paramaribo	Currency	Suriname guilder
Languages	Dutch,	Organizations	CARICOM, UN
	Surinamese,	Map page	151
	English, Hindi		

Svalbard
part of Norway

Area Sq Km	61 229	Religions	Protestant
Area Sq Miles	23 641	Currency	Norwegian krone
Population	2 400	Map page	82
Capital	Longyearbyen		
Languages	Norwegian		

SWAZILAND
Kingdom of Swaziland

Area Sq Km	17 364	Religions	Christian,
Area Sq Miles	6 704		traditional beliefs
Population	1 203 000	Currency	Emalangeni,
Capital	Mbabane		South African rand
Languages	Swazi, English	Organizations	Comm., SADC, UN
		Map page	123

SWEDEN
Kingdom of Sweden

Area Sq Km	449 964	Religions	Protestant,
Area Sq Miles	173 732		Roman Catholic
Population	9 441 000	Currency	Swedish krona
Capital	Stockholm	Organizations	EU, OECD, UN
Languages	Swedish	Map page	92–93

SWITZERLAND
Swiss Confederation

Area Sq Km	41 293	Religions	Roman Catholic,
Area Sq Miles	15 943		Protestant,
Population	7 702 000	Currency	Swiss franc
Capital	Bern	Organizations	OECD, UN
Languages	German, French,	Map page	105
	Italian, Romansch		

SYRIA
Syrian Arab Republic

Area Sq Km	185 180	Religions	Sunni Muslim, Shi
Area Sq Miles	71 498		Muslim, Christian
Population	20 766 000	Currency	Syrian pound
Capital	Damascus (Dimashq)	Organizations	UN
Languages	Arabic, Kurdish,	Map page	80
	Armenian		

TAIWAN
Republic of China

Area Sq Km	36 179	Religions	Buddhist, Taoist,
Area Sq Miles	13 969		Confucian,
Population	23 146 000		Christian
Capital	T'aipei	Currency	Taiwan dollar
Languages	Mandarin, Min,	Organizations	APEC
	Hakka, local	Map page	71
	languages		The People's Republic of China claims
			Taiwan as its 23rd province

TAJIKISTAN
Republic of Tajikistan

Area Sq Km	143 100	Religions	Sunni Muslim
Area Sq Miles	55 251	Currency	Somoni
Population	6 977 000	Organizations	CIS, UN
Capital	Dushanbe	Map page	77
Languages	Tajik, Uzbek, Russian		

TANZANIA
United Republic of Tanzania

Area Sq Km	945 087	Religions	Shi'a Muslim, Sunni
Area Sq Miles	364 900		Muslim, traditional
Population	46 218 000		beliefs, Christian
Capital	Dodoma	Currency	Tanzanian shilling
Languages	Swahili, English,	Organizations	Comm., SADC, UN
	Nyamwezi, local	Map page	119
	languages		

THAILAND
Kingdom of Thailand

Area Sq Km	513 115	Religions	Buddhist, Sunni
Area Sq Miles	198 115		Muslim
Population	69 519 000	Currency	Baht
Capital	Bangkok	Organizations	APEC, ASEAN, UN
	(Krung Thep)	Map page	62–63
Languages	Thai, Lao, Chinese,		
	Malay, Mon-Khmer		
	languages		

TOGO
Republic of Togo

Area Sq Km	56 785	Religions	Traditional beliefs,
Area Sq Miles	21 925		Christian, Sunni
Population	6 155 000		Muslim
Capital	Lomé	Currency	CFA franc
Languages	French, Ewe, Kabre,	Organizations	UN
	local languages	Map page	114

Tokelau
New Zealand Overseas Territory

Area Sq Km	10	Religions	Christian
Area Sq Miles	4	Currency	New Zealand dollar
Population	1 466	Map page	49
Capital	none		
Languages	English, Tokelauan		

TONGA
Kingdom of Tonga

Area Sq Km	748	Religions	Protestant, Roman
Area Sq Miles	289		Catholic
Population	105 000	Currency	Pa'anga
Capital	Nuku'alofa	Organizations	Comm., UN
Languages	Tongan, English	Map page	49

Transnistria
Disputed Territory (Moldova)

Area Sq Km	4 200	Map page	90
Area Sq Miles	1 622		
Population	520 000		
Capital	Tiraspol		

TRINIDAD AND TOBAGO
Republic of Trinidad and Tobago

Area Sq Km	5 130	Religions	Roman Catholic,
Area Sq Miles	1 981		Hindu, Protestant,
Population	1 346 000		Sunni Muslim
Capital	Port of Spain	Currency	Trinidad and
Languages	English, creole,		Tobago dollar
	Hindi	Organizations	CARICOM,
			Comm., UN
		Map page	147

Tristan da Cunha
part of St Helena, Ascension and Tristan da Cunha

Area Sq Km	200	Religions	Protestant, Roman
Area Sq Miles	77		Catholic
Population	263	Currency	Pound sterling
Capital	Settlement of	Map page	113
	Edinburgh		
Languages	English		

TUNISIA
Tunisian Republic

Area Sq Km	164 150	Religions	Sunni Muslim
Area Sq Miles	63 379	Currency	Tunisian dinar
Population	10 594 000	Organizations	UN
Capital	Tunis	Map page	115
Languages	Arabic, French		

TURKEY
Republic of Turkey

Area Sq Km	779 452	Religions	Sunni Muslim,
Area Sq Miles	300 948		Shi'a Muslim
Population	73 640 000	Currency	Lira
Capital	Ankara	Organizations	OECD, UN
Languages	Turkish, Kurdish	Map page	80

TURKMENISTAN
Republic of Turkmenistan

Area Sq Km	488 100	Religions	Sunni Muslim,
Area Sq Miles	188 456		Russian Orthodox
Population	5 105 000	Currency	Turkmen manat
Capital	Aşgabat (Ashkhabad)	Organizations	UN
Languages	Turkmen, Uzbek,	Map page	76
	Russian		

Turks and Caicos Islands
United Kingdom Overseas Territory

Area Sq Km	430	Religions	Protestant
Area Sq Miles	166	Currency	United States
Population	39 000		dollar
Capital	Grand Turk	Map page	147
	(Cockburn Town)		
Languages	English		

TUVALU

Area Sq Km	25	Religions	Protestant
Area Sq Miles	10	Currency	Australian dollar
Population	10 000	Organizations	Comm., UN
Capital	Vaiaku	Map page	49
Languages	Tuvaluan, English		

25

UGANDA
Republic of Uganda

Area Sq Km	241 038	Religions	Roman Catholic,
Area Sq Miles	93 065		Protestant, Sunni
Population	34 509 000		Muslim, traditional
Capital	Kampala		beliefs
Languages	English, Swahili,	Currency	Ugandan shilling
	Luganda, local	Organizations	Comm., UN
	languages	Map page	119

UKRAINE
Republic of Ukraine

Area Sq Km	603 700	Religions	Ukrainian
Area Sq Miles	233 090		Orthodox,
Population	45 190 000		Ukrainian Catholic,
Capital	Kiev (Kyiv)		Roman Catholic
Languages	Ukrainian, Russian	Currency	Hryvnia
		Organizations	CIS, UN
		Map page	90–91

UNITED ARAB EMIRATES
Federation of Emirates

Area Sq Km	77 700	Religions	Sunni Muslim,
Area Sq Miles	30 000		Shi'a Muslim
Population	7 891 000	Currency	United Arab
Capital	Abu Dhabi		Emirates dirham
	(Abū Ẓaby)	Organizations	OPEC, UN
Languages	Arabic, English	Map page	79

Abu Dhabi (Abū Ẓaby) (Emirate)

Area Sq Km	67 340	Population	1 628 000
Area Sq Miles	26 000	Capital	Abu Dhabi
			(Abū Ẓaby)

Ajman (Emirate)

Area Sq Km	259	Population	250 000
Area Sq Miles	100	Capital	Ajman

Dubai (Emirate)

Area Sq Km	3 885	Population	1 722 000
Area Sq Miles	1 500	Capital	Dubai

Fujairah (Emirate)

Area Sq Km	1 165	Population	152 000
Area Sq Miles	450	Capital	Fujairah

Ra's al Khaymah (Emirate)

Area Sq Km	1 684	Population	241 000
Area Sq Miles	650	Capital	Ra's al Khaymah

Sharjah (Emirate)

Area Sq Km	2 590	Population	1 017 000
Area Sq Miles	1 000	Capital	Sharjah

Umm al Qaywayn (Emirate)

Area Sq Km	777	Population	56 000
Area Sq Miles	300	Capital	Umm al Qaywayn

UNITED KINGDOM
of Great Britain and Northern Ireland

Area Sq Km	243 609	Religions	Protestant, Roma
Area Sq Miles	94 058		Catholic, Muslim
Population	62 417 000	Currency	Pound sterling
Capital	London	Organizations	Comm., EU, OEC
Languages	English, Welsh,		UN
	Gaelic	Map page	94–95

England (Constituent country)

Area Sq Km	130 433	Population	51 809 700
Area Sq Miles	50 360	Capital	London

Northern Ireland (Province)

Area Sq Km	13 576	Population	1 788 900
Area Sq Miles	5 242	Capital	Belfast

Scotland (Constituent country)

Area Sq Km	78 822	Population	5 194 000
Area Sq Miles	30 433	Capital	Edinburgh

Wales (Principality)

Area Sq Km	20 778	Population	2 999 300
Area Sq Miles	8 022	Capital	Cardiff

UNITED STATES OF AMERICA
Federal Republic

Area Sq Km	9 826 635	Religions	Protestant, Roma
Area Sq Miles	3 794 085		Catholic, Sunni
Population	313 085 000		Muslim, Jewish
Capital	Washington D.C.	Currency	United States do
Languages	English, Spanish	Organizations	APEC, OECD, UI
		Map page	132–133

Alabama (State)

Area Sq Km	135 765	Population	4 708 708
Area Sq Miles	52 419	Capital	Montgomery

Alaska (State)

Area Sq Km	1 717 854	Population	698 473
Area Sq Miles	663 267	Capital	Juneau

Arizona (State)

Area Sq Km	295 253	Population	6 595 778
Area Sq Miles	113 998	Capital	Phoenix

Arkansas (State)

Area Sq Km	137 733	Population	2 889 450
Area Sq Miles	53 179	Capital	Little Rock

California (State)

Area Sq Km	423 971	Population	36 961 664
Area Sq Miles	163 696	Capital	Sacramento

olorado (State)			
Area Sq Km	269 602	Population	5 024 748
ea Sq Miles	104 094	Capital	Denver

Louisiana (State)			
Area Sq Km	134 265	Population	4 492 076
Area Sq Miles	51 840	Capital	Baton Rouge

onnecticut (State)			
Area Sq Km	14 356	Population	3 518 288
ea Sq Miles	5 543	Capital	Hartford

Maine (State)			
Area Sq Km	91 647	Population	1 318 301
Area Sq Miles	35 385	Capital	Augusta

elaware (State)			
Area Sq Km	6 446	Population	885 122
ea Sq Miles	2 489	Capital	Dover

Maryland (State)			
Area Sq Km	32 134	Population	5 699 478
Area Sq Miles	12 407	Capital	Annapolis

istrict of Columbia (District)			
Area Sq Km	176	Population	599 657
ea Sq Miles	68	Capital	Washington

Massachusetts (State)			
Area Sq Km	27 337	Population	6 593 587
Area Sq Miles	10 555	Capital	Boston

orida (State)			
Area Sq Km	170 305	Population	18 537 969
ea Sq Miles	65 755	Capital	Tallahassee

Michigan (State)			
Area Sq Km	250 493	Population	9 969 727
Area Sq Miles	96 716	Capital	Lansing

eorgia (State)			
Area Sq Km	153 910	Population	9 829 211
rea Sq Miles	59 425	Capital	Atlanta

Minnesota (State)			
Area Sq Km	225 171	Population	5 266 214
Area Sq Miles	86 939	Capital	St Paul

awaii (State)			
Area Sq Km	28 311	Population	1 295 178
rea Sq Miles	10 931	Capital	Honolulu

Mississippi (State)			
Area Sq Km	125 433	Population	2 951 996
Area Sq Miles	48 430	Capital	Jackson

laho (State)			
Area Sq Km	216 445	Population	1 545 801
rea Sq Miles	83 570	Capital	Boise

Missouri (State)			
Area Sq Km	180 533	Population	5 987 580
Area Sq Miles	69 704	Capital	Jefferson City

linois (State)			
Area Sq Km	149 997	Population	12 910 409
rea Sq Miles	57 914	Capital	Springfield

Montana (State)			
Area Sq Km	380 837	Population	974 989
Area Sq Miles	147 042	Capital	Helena

ndiana (State)			
Area Sq Km	94 322	Population	6 423 113
rea Sq Miles	36 418	Capital	Indianapolis

Nebraska (State)			
Area Sq Km	200 346	Population	1 796 619
Area Sq Miles	77 354	Capital	Lincoln

owa (State)			
Area Sq Km	145 744	Population	3 007 856
rea Sq Miles	56 272	Capital	Des Moines

Nevada (State)			
Area Sq Km	286 352	Population	2 643 085
Area Sq Miles	110 561	Capital	Carson City

ansas (State)			
Area Sq Km	213 096	Population	2 818 747
rea Sq Miles	82 277	Capital	Topeka

New Hampshire (State)			
Area Sq Km	24 216	Population	1 324 575
Area Sq Miles	9 350	Capital	Concord

entucky (State)			
Area Sq Km	104 659	Population	4 314 113
rea Sq Miles	40 409	Capital	Frankfort

New Jersey (State)			
Area Sq Km	22 587	Population	8 707 739
Area Sq Miles	8 721	Capital	Trenton

UNITED STATES OF AMERICA
Federal Republic

New Mexico (State)
Area Sq Km	314 914	Population	2 009 671
Area Sq Miles	121 589	Capital	Santa Fe

New York (State)
Area Sq Km	141 299	Population	19 541 453
Area Sq Miles	54 556	Capital	Albany

North Carolina (State)
Area Sq Km	139 391	Population	9 380 884
Area Sq Miles	53 819	Capital	Raleigh

North Dakota (State)
Area Sq Km	183 112	Population	646 844
Area Sq Miles	70 700	Capital	Bismarck

Ohio (State)
Area Sq Km	116 096	Population	11 542 645
Area Sq Miles	44 825	Capital	Columbus

Oklahoma (State)
Area Sq Km	181 035	Population	3 687 050
Area Sq Miles	69 898	Capital	Oklahoma City

Oregon (State)
Area Sq Km	254 806	Population	3 825 657
Area Sq Miles	98 381	Capital	Salem

Pennsylvania (State)
Area Sq Km	119 282	Population	12 604 767
Area Sq Miles	46 055	Capital	Harrisburg

Rhode Island (State)
Area Sq Km	4 002	Population	1 053 209
Area Sq Miles	1 545	Capital	Providence

South Carolina (State)
Area Sq Km	82 931	Population	4 561 242
Area Sq Miles	32 020	Capital	Columbia

South Dakota (State)
Area Sq Km	199 730	Population	812 383
Area Sq Miles	77 116	Capital	Pierre

Tennessee (State)
Area Sq Km	109 150	Population	6 296 254
Area Sq Miles	42 143	Capital	Nashville

Texas (State)
Area Sq Km	695 622	Population	24 782 302
Area Sq Miles	268 581	Capital	Austin

Utah (State)
Area Sq Km	219 887	Population	2 784 572
Area Sq Miles	84 899	Capital	Salt Lake City

Vermont (State)
Area Sq Km	24 900	Population	621 760
Area Sq Miles	9 614	Capital	Montpelier

Virginia (State)
Area Sq Km	110 784	Population	7 882 590
Area Sq Miles	42 774	Capital	Richmond

Washington (State)
Area Sq Km	184 666	Population	6 664 195
Area Sq Miles	71 300	Capital	Olympia

West Virginia (State)
Area Sq Km	62 755	Population	1 819 777
Area Sq Miles	24 230	Capital	Charleston

Wisconsin (State)
Area Sq Km	169 639	Population	5 654 774
Area Sq Miles	65 498	Capital	Madison

Wyoming (State)
Area Sq Km	253 337	Population	544 270
Area Sq Miles	97 814	Capital	Cheyenne

URUGUAY
Oriental Republic of Uruguay

Area Sq Km	176 215	Religions	Roman Catholic,
Area Sq Miles	68 037		Protestant, Jewish
Population	3 380 000	Currency	Uruguayan peso
Capital	Montevideo	Organizations	UN
Languages	Spanish	Map page	153

UZBEKISTAN
Republic of Uzbekistan

Area Sq Km	447 400	Religions	Sunni Muslim,
Area Sq Miles	172 742		Russian Orthodox
Population	27 760 000	Currency	Uzbek som
Capital	Tashkent	Organizations	CIS, UN
Languages	Uzbek, Russian,	Map page	76–77
	Tajik, Kazakh		

VANUATU
Republic of Vanuatu

Area Sq Km	12 190	Religions	Protestant,
Area Sq Miles	4 707		Roman Catholic,
Population	246 000		traditional beliefs
Capital	Port Vila	Currency	Vatu
Languages	English, Bislama	Organizations	Comm., UN
	(creole), French	Map page	48

VATICAN CITY
Vatican City State or Holy See

Area Sq Km	0.5	Religions	Roman Catholic
Area Sq Miles	0.2	Currency	Euro
Population	800	Map page	108
Capital	Vatican City		
Languages	Italian		

VENEZUELA
Bolivarian Republic of Venezuela

Area Sq Km	912 050	Religions	Roman Catholic,
Area Sq Miles	352 144		Protestant
Population	29 437 000	Currency	Bolívar fuerte
Capital	Caracas	Organizations	OPEC, UN
Languages	Spanish, Amerindian	Map page	150
	languages		

VIETNAM
Socialist Republic of Vietnam

Area Sq Km	329 565	Religions	Buddhist, Taoist,
Area Sq Miles	127 246		Roman Catholic,
Population	88 792 000		Cao Dai, Hoa Hoa
Capital	Ha Nôi (Hanoi)	Currency	Dong
Languages	Vietnamese, Thai,	Organizations	APEC, ASEAN, UN
	Khmer, Chinese,	Map page	62–63
	local languages		

Virgin Islands (U.K.)
United Kingdom Overseas Territory

Area Sq Km	153	Religions	Protestant, Roman
Area Sq Miles	59		Catholic
Population	23 000	Currency	United States dollar
Capital	Road Town	Map page	147
Languages	English		

Virgin Islands (U.S.)
United States Unincorporated Territory

Area Sq Km	352	Religions	Protestant,
Area Sq Miles	136		Roman Catholic
Population	109 000	Currency	United States dollar
Capital	Charlotte Amalie	Map page	147
Languages	English, Spanish		

Wallis and Futuna Islands
French Overseas Collectivity

Area Sq Km	274	Religions	Roman Catholic
Area Sq Miles	106	Currency	CFP franc
Population	13 000	Map page	49
Capital	Matā'utu		
Languages	French, Wallisian,		
	Futunan		

West Bank
Disputed Territory

Area Sq Km	5 860	Religions	Sunni Muslim,
Area Sq Miles	2 263		Jewish,
Population	2 513 283		Shi'a Muslim,
Capital	none		Christian
Languages	Arabic, Hebrew	Currency	Jordanian dinar,
			Israeli shekel
		Map page	80

Western Sahara
Disputed Territory (Morocco)

Area Sq Km	266 000	Religions	Sunni Muslim
Area Sq Miles	102 703	Currency	Moroccan dirham
Population	548 000	Map page	114
Capital	Laâyoune		
Languages	Arabic		

YEMEN
Republic of Yemen

Area Sq Km	527 968	Religions	Sunni Muslim,
Area Sq Miles	203 850		Shi'a Muslim
Population	24 800 000	Currency	Yemeni riyal
Capital	Şan'ā'	Organizations	UN
Languages	Arabic	Map page	78–79

ZAMBIA
Republic of Zambia

Area Sq Km	752 614	Religions	Christian,
Area Sq Miles	290 586		traditional beliefs
Population	13 475 000	Currency	Zambian kwacha
Capital	Lusaka	Organizations	Comm., SADC, UN
Languages	English, Bemba,	Map page	120–121
	Nyanja, Tonga,		
	local languages		

ZIMBABWE
Republic of Zimbabwe

Area Sq Km	390 759	Religions	Christian,
Area Sq Miles	150 873		traditional beliefs
Population	12 754 000	Currency	Zimbabwean dollar
Capital	Harare		(suspended)
Languages	English, Shona,	Organizations	SADC, UN
	Ndebele	Map page	121

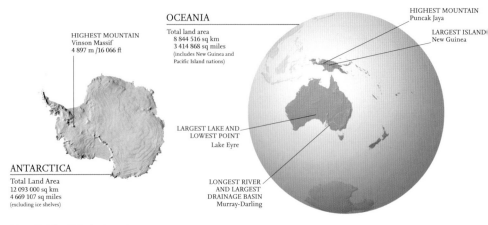

HIGHEST MOUNTAIN
Vinson Massif
4 897 m /16 066 ft

OCEANIA

Total land area
8 844 516 sq km
3 414 868 sq miles
(includes New Guinea and
Pacific Island nations)

HIGHEST MOUNTAIN
Puncak Jaya

LARGEST ISLAND
New Guinea

LARGEST LAKE AND
LOWEST POINT
Lake Eyre

ANTARCTICA

Total Land Area
12 093 000 sq km
4 669 107 sq miles
(excluding ice shelves)

LONGEST RIVER
AND LARGEST
DRAINAGE BASIN
Murray-Darling

HIGHEST MOUNTAINS	metres	feet	LARGEST ISLANDS	sq km	sq miles	LARGEST LAKES	sq km	sq miles	LONGEST RIVERS	km	miles
Puncak Jaya	5 030	16 502	New Guinea	808 510	312 166	Lake Eyre	0–8 900	0–3 436	Murray-Darling	3 672	2 282
Puncak Trikora	4 730	15 518	South Island	151 215	58 384	Lake Torrens	0–5 780	0–2 232	Darling	2 844	1 767
Puncak Mandala	4 700	15 420	North Island	115 777	44 701				Murray	2 375	1 476
Puncak Yamin	4 595	15 075	Tasmania	67 800	26 178				Murrumbidgee	1 485	923
Mt Wilhelm	4 509	14 793							Lachlan	1 339	832
Mt Kubor	4 359	14 301							Cooper Creek	1 113	692

ASIA

Total Land Area
45 036 492 sq km
17 388 589 sq miles

LARGEST DRAINAGE BASIN
Ob'-Irtysh

LARGEST LAKE
Caspian Sea

LONGEST RIVER
Yangtze

LOWEST POINT
Dead Sea

HIGHEST MOUNTAIN
Mount Everest

LARGEST ISLAND
Borneo

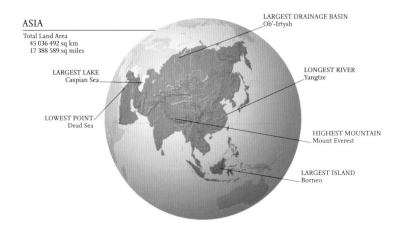

HIGHEST MOUNTAINS	metres	feet	LARGEST ISLANDS	sq km	sq miles	LARGEST LAKES	sq km	sq miles	LONGEST RIVERS	km	miles
Mt Everest	8 848	29 028	Borneo	745 561	287 861	Caspian Sea	371 000	143 243	Yangtze	6 380	3 965
K2	8 611	28 251	Sumatra	473 606	182 859	Lake Baikal	30 500	11 776	Ob'-Irtysh	5 568	3 460
Kangchenjunga	8 586	28 169	Honshū	227 414	87 805	Lake Balkhash	17 400	6 718	Yenisey-Angara-Selenga	5 550	3 449
Lhotse	8 516	27 939	Celebes	189 216	73 056	Aral Sea	17 158	6 625	Yellow	5 464	3 395
Makalu	8 463	27 765	Java	132 188	51 038	Ysyk-Köl	6 200	2 394	Irtysh	4 440	2 759
Cho Oyu	8 201	26 906	Luzon	104 690	40 421						

EUROPE

Total Land Area
9 908 599 sq km
3 825 710 sq miles

LARGEST ISLAND
Great Britain

LONGEST RIVER AND
LARGEST DRAINAGE BASIN
Volga

LARGEST LAKE AND
LOWEST POINT
Caspian Sea

HIGHEST MOUNTAIN
El'brus

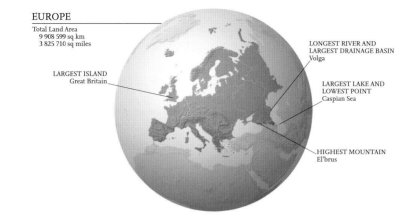

HIGHEST MOUNTAINS	metres	feet	LARGEST ISLANDS	sq km	sq miles	LARGEST LAKES	sq km	sq miles	LONGEST RIVERS	km	miles
El'brus	5 642	18 510	Great Britain	218 476	84 354	Caspian Sea	371 000	143 243	Volga	3 688	2 292
Gora Dykh-Tau	5 204	17 073	Iceland	102 820	39 699	Lake Ladoga	18 390	7 100	Danube	2 850	1 771
Shkhara	5 201	17 063	Ireland	83 045	32 064	Lake Onega	9 600	3 707	Dnieper	2 285	1 420
Kazbek	5 047	16 558	Ostrov Severnyy (part of Novaya Zemlya)	47 079	18 177	Vänern	5 585	2 156	Kama	2 028	1 260
Mont Blanc	4 810	15 781				Rybinskoye Vodokhranilishche	5 180	2 000	Don	1 931	1 200
Dufourspitze	4 634	15 203	Spitsbergen	37 814	14 600				Pechora	1 802	1 120

AFRICA

Total Land Area
30 343 578 sq km
11 715 655 sq miles

LONGEST RIVER
Nile

LOWEST POINT
Lake Assal

LARGEST LAKE
Lake Victoria

HIGHEST MOUNTAIN
Kilimanjaro

LARGEST DRAINAGE BASIN
Congo

LARGEST ISLAND
Madagascar

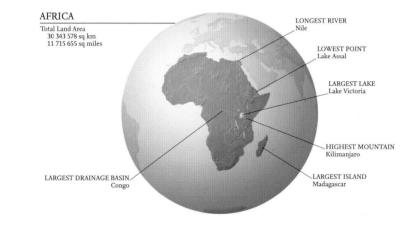

HIGHEST MOUNTAINS	metres	feet	LARGEST ISLANDS	sq km	sq miles	LARGEST LAKES	sq km	sq miles	LONGEST RIVERS	km	miles
Kilimanjaro	5 892	19 330	Madagascar	587 040	226 656	Lake Victoria	68 870	26 591	Nile	6 695	4 160
Mt Kenya	5 199	17 057				Lake Tanganyika	32 600	12 587	Congo	4 667	2 900
Margherita Peak	5 110	16 765				Lake Nyasa	29 500	11 390	Niger	4 184	2 600
Meru	4 565	14 977				Lake Volta	8 482	3 275	Zambezi	2 736	1 700
Ras Dejen	4 533	14 872				Lake Turkana	6 500	2 510	Webi Shabeelle	2 490	1 547
Mt Karisimbi	4 510	14 796				Lake Albert	5 600	2 162	Ubangi	2 250	1 398

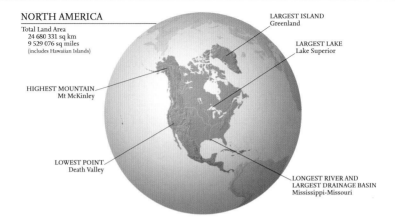

NORTH AMERICA

Total Land Area
24 680 331 sq km
9 529 076 sq miles
(includes Hawaiian Islands)

LARGEST ISLAND
Greenland

LARGEST LAKE
Lake Superior

HIGHEST MOUNTAIN
Mt McKinley

LOWEST POINT
Death Valley

LONGEST RIVER AND
LARGEST DRAINAGE BASIN
Mississippi-Missouri

HIGHEST MOUNTAINS	metres	feet	LARGEST ISLANDS	sq km	sq miles	LARGEST LAKES	sq km	sq miles	LONGEST RIVERS	km	miles
Mt McKinley	6 194	20 321	Greenland	2 175 600	839 999	Lake Superior	82 100	31 699	Mississippi-Missouri	5 969	3 709
Mt Logan	5 959	19 550	Baffin Island	507 451	195 927	Lake Huron	59 600	23 012	Mackenzie-Peace-Finlay	4 241	2 635
Pico de Orizaba	5 610	18 405	Victoria Island	217 291	83 896	Lake Michigan	57 800	22 317	Missouri	4 086	2 539
Mt St Elias	5 489	18 008	Ellesmere Island	196 236	75 767	Great Bear Lake	31 328	12 096	Mississippi	3 765	2 340
Volcán Popocatépetl	5 452	17 887	Cuba	110 860	42 803	Great Slave Lake	28 568	11 030	Yukon	3 185	1 979
			Newfoundland	108 860	42 031	Lake Erie	25 700	9 923			

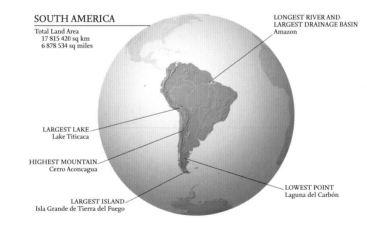

SOUTH AMERICA

Total Land Area
17 815 420 sq km
6 878 534 sq miles

LONGEST RIVER AND
LARGEST DRAINAGE BASIN
Amazon

LARGEST LAKE
Lake Titicaca

HIGHEST MOUNTAIN
Cerro Aconcagua

LOWEST POINT
Laguna del Carbón

LARGEST ISLAND
Isla Grande de Tierra del Fuego

HIGHEST MOUNTAINS	metres	feet	LARGEST ISLANDS	sq km	sq miles	LARGEST LAKES	sq km	sq miles	LONGEST RIVERS	km	miles
Cerro Aconcagua	6 959	22 831	Isla Grande de Tierra del Fuego	47 000	18 147	Lake Titicaca	8 340	3 220	Amazon	6 516	4 049
Nevado Ojos del Salado	6 908	22 664	Isla de Chiloé	8 394	3 241				Río de la Plata-Paraná	4 500	2 796
Cerro Bonete	6 872	22 546	East Falkland	6 760	2 610				Purus	3 218	2 000
Cerro Pissis	6 858	22 500	West Falkland	5 413	2 090				Madeira	3 200	1 988
Cerro Tupungato	6 800	22 309							São Francisco	2 900	1 802

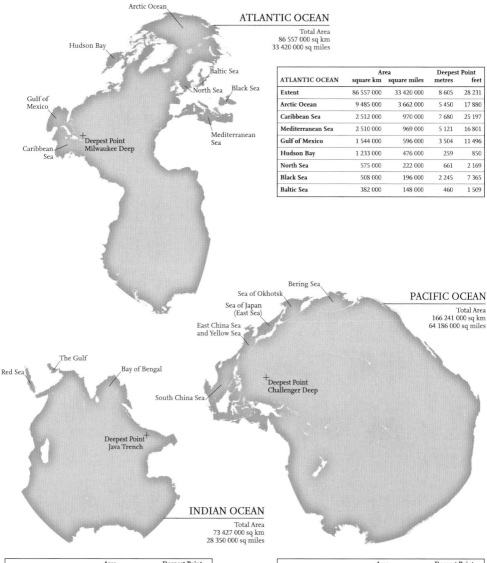

ATLANTIC OCEAN

Total Area
86 557 000 sq km
33 420 000 sq miles

Arctic Ocean

Hudson Bay

Baltic Sea

North Sea   Black Sea

Gulf of Mexico

Mediterranean Sea

Caribbean Sea

Deepest Point
Milwaukee Deep

ATLANTIC OCEAN	Area		Deepest Point	
	square km	square miles	metres	feet
Extent	86 557 000	33 420 000	8 605	28 231
Arctic Ocean	9 485 000	3 662 000	5 450	17 880
Caribbean Sea	2 512 000	970 000	7 680	25 197
Mediterranean Sea	2 510 000	969 000	5 121	16 801
Gulf of Mexico	1 544 000	596 000	3 504	11 496
Hudson Bay	1 233 000	476 000	259	850
North Sea	575 000	222 000	661	2 169
Black Sea	508 000	196 000	2 245	7 365
Baltic Sea	382 000	148 000	460	1 509

Bering Sea

Sea of Okhotsk

Sea of Japan
(East Sea)

East China Sea
and Yellow Sea

PACIFIC OCEAN

Total Area
166 241 000 sq km
64 186 000 sq miles

The Gulf

Red Sea

Bay of Bengal

Deepest Point
Challenger Deep

South China Sea

Deepest Point
Java Trench

INDIAN OCEAN

Total Area
73 427 000 sq km
28 350 000 sq miles

INDIAN OCEAN	Area		Deepest Point	
	square km	square miles	metres	feet
Extent	73 427 000	28 350 000	7 125	23 376
Bay of Bengal	2 172 000	839 000	4 500	14 764
Red Sea	453 000	175 000	3 040	9 974
The Gulf	238 000	92 000	73	239

PACIFIC OCEAN	Area		Deepest Point	
	square km	square miles	metres	feet
Extent	166 241 000	64 186 000	10 920	35 826
South China Sea	2 590 000	1 000 000	5 514	18 090
Bering Sea	2 261 000	873 000	4 150	13 615
Sea of Okhotsk	1 392 000	538 000	3 363	11 033
Sea of Japan (East Sea)	1 013 000	391 000	3 743	12 280
East China Sea and Yellow Sea	1 202 000	464 000	2 717	8 914

MAJOR CLIMATIC REGIONS AND SUB-TYPES

Winkel Tripel Projection
1:155 000 000

Köppen classification system

Polar

| EF | Ice cap |
| ET | Tundra |

Dry

| BS | Steppe |
| BW | Desert |

Cooler humid

Dc Dd	Subarctic
Db	Continental cool summer
Da	Continental warm summer

Tropical humid

| Aw As | Savanna |
| Af Am | Rain forest |

Warmer humid

Cb Cc	Temperate
Ca	Humid subtropical
Cs	Mediterranean

| o | Weather extreme location |

A Rainy climate with no winter: coolest month above 18°C (64.4°F).

B Dry climates; limits are defined by formulae based on rainfall effectiveness:
  BS Steppe or semi-arid climate.
  BW Desert or arid climate.

*C Rainy climates with mild winters: coolest month above 0°C (32°F), but below 18°C (64.4°F); warmest month above 10°C (50°F).

*D Rainy climates with severe winters: coldest month below 0°C (32°F); warmest month above 10°C (50°F).

E Polar climates with no warm season: warmest month below 10°C (50°F).
  ET Tundra climate: warmest month below 10°C (50°F) but above 0°C (32°F).
  EF Perpetual frost: all months below 0°C (32°F).

a Warmest month above 22°C (71.6°F).

b Warmest month below 22°C (71.6°F).

c Less than four months over 10°C (50°F).

d As 'c', but with severe cold: coldest month below -38°C (-36.4°F).

f Constantly moist rainfall throughout the year.

*h Warmer dry: all months above 0°C (32°F).

*k Cooler dry: at least one month below 0°C (32°F).

m Monsoon rain: short dry season, but is compensated by heavy rains during rest of the year.

n Frequent fog.

s Dry season in summer.

w Dry season in winter.

* Modification of Köppen definition

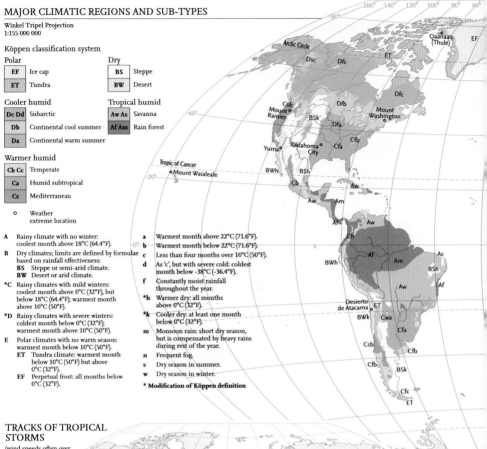

TRACKS OF TROPICAL STORMS

(wind speeds often over 160 km per hour)
1:300 000 000

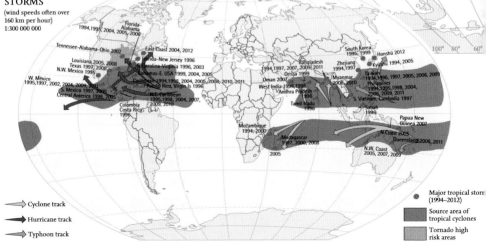

⇨ Cyclone track

➡ Hurricane track

⇨ Typhoon track

● Major tropical storm (1994–2012)

▮ Source area of tropical cyclones

▮ Tornado high risk areas

WORLD WEATHER EXTREMES

	Location			Location
Highest shade temperature	57.8°C/136°F Al ʿAzīzīyah, Libya (13th September 1922)	**Highest surface wind speed**		
Hottest place — Annual mean	34.4°C/93.9°F Dalol, Ethiopia	High altitude	372 km per hour/231 miles per hour Mount Washington, New Hampshire, USA (12th April 1934)
Driest place — Annual mean	0.1 mm/0.004 inches Atacama Desert, Chile		
Most sunshine — Annual mean	90% Yuma, Arizona, USA (over 4 000 hours)	Low altitude	333 km per hour/207 miles per hour Qaanaaq (Thule), Greenland (8th March 1972)
Least sunshine	Nil for 182 days each year, South Pole	Tornado	512 km per hour/318 miles per hour Oklahoma City, Oklahoma, USA (3rd May 1999)
Lowest screen temperature	-89.2°C/-128.6°F Vostok Station, Antarctica (21st July 1983)	**Greatest snowfall**	31 102 mm/1 224.5 inches Mount Rainier, Washington, USA (19th February 1971 — 18th February 1972)
Coldest place — Annual mean	-56.6°C/-69.9°F Plateau Station, Antarctica		
Wettest place — Annual mean	11 873 mm/467.4 inches Meghalaya, India	**Heaviest hailstones**	1 kg/2.21 lb Gopalganj, Bangladesh (14th April 1986)
Most rainy days	Up to 350 per year Mount Waialeale, Hawaii, USA	**Thunder-days average**	251 days per year Tororo, Uganda
Windiest place	322 km per hour/200 miles per hour in gales, Commonwealth Bay, Antarctica	**Highest barometric pressure**	1 083.8 mb Agata, Siberia, Rus. Fed. (31st December 1968)
		Lowest barometric pressure	870 mb 483 km/300 miles west of Guam, Pacific Ocean (12th October 1979)

WORLD LAND COVER

Winkel Tripel Projection
1:155 000 000

Irrigated croplands

Rain fed croplands

Mosaic croplands/vegetation

Mosaic vegetation/croplands

Closed to open broadleaved evergreen
or semi-deciduous forest

Closed broadleaved deciduous forest

Open broadleaved deciduous forest

Closed needle leaved evergreen forest

Open needle leaved deciduous or
evergreen forest

Closed to open mixed broadleaved and
needle leaved forest

Mosaic forest – shrubland/grassland

Mosaic grassland – forest/shrubland

Closed to open shrubland

Closed to open grassland

Sparse vegetation

Closed to open broadleaved forest,
regularly flooded (fresh-brackish water)

Closed broadleaved forest, permanently
flooded (saline-brackish water)

Closed to open vegetation, regularly flooded

Artificial areas

Bare areas

Water bodies

Permanent snow
and ice

No data

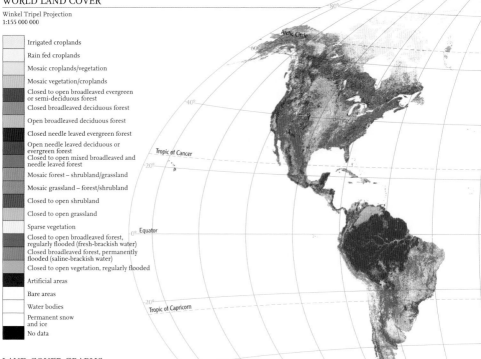

LAND COVER GRAPHS - CLASSIFICATION

Class description	Map classes
Forest/Woodland	Evergreen needleleaf forest
	Evergreen broadleaf forest
	Deciduous needleleaf forest
	Deciduous broadleaf forest
	Mixed forest
Shrubland	Closed shrublands
	Open shrublands
Grass/Savanna	Woody savannas
	Savannas
	Grasslands
Wetland	Permanent wetlands
Crops/Mosaic	Croplands
	Cropland/Natural vegetation mosaic
Urban	Urban and build-up
Snow/Ice	Snow and Ice
Barren	Barren or sparsely vegetated

GLOBAL LAND COVER COMPOSITION

Wetland 0.2%
Snow/Ice 11.6%
Urban 0.1%
Barren 12.5%
Forest/Woodland 22.1%
Crops/Mosaic 12.7%
Grass/Savanna 20.9%
Shrubland 19.9%

CONTINENTAL LAND COVER COMPOSITION

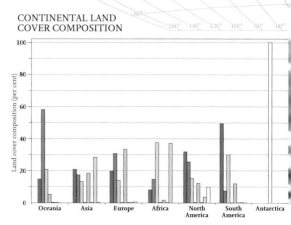

Land cover composition (per cent)

Oceania   Asia   Europe   Africa   North America   South America   Antarctica

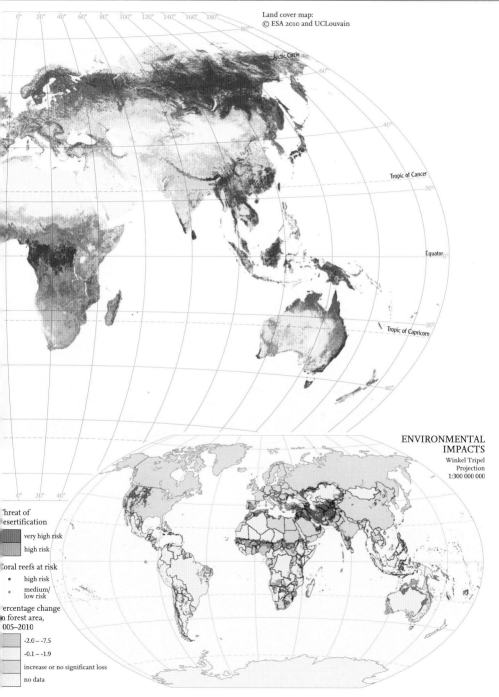

Land cover map:
© ESA 2010 and UCLouvain

Arctic Circle

Tropic of Cancer

Equator

Tropic of Capricorn

ENVIRONMENTAL
IMPACTS
Winkel Tripel
Projection
1:300 000 000

Threat of
desertification

very high risk

high risk

Coral reefs at risk

• high risk

• medium/
low risk

Percentage change
in forest area,
2005–2010

-2.0 – -7.5

-0.1 – -1.9

increase or no significant loss

no data

WORLD POPULATION DISTRIBUTION AND THE WORLD'S MAJOR CITIES

Winkel Tripel Projection
1:155 000 000

Major Urban Agglomerations

- over 20 million
- 10 million – 20 million
- 5 million – 10 million

Density of inhabitants

per sq km	per sq mile
1 000	2 500
500	1 250
250	625
100	250
50	125
25	62.5
5	12.5
1	2.5
0	0

Uninhabited

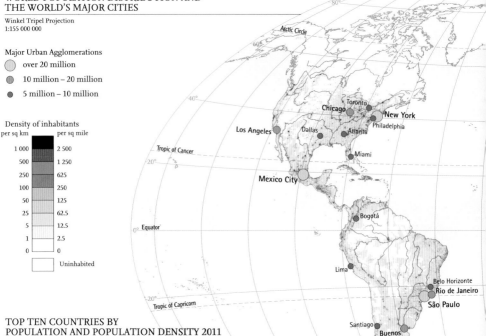

TOP TEN COUNTRIES BY POPULATION AND POPULATION DENSITY 2011

Total population	Country	Rank	Country*	Inhabitants per sq mile	Inhabitants per sq km
1 332 079 000	China	1	Bangladesh	2 707	1 045
1 241 492 000	India	2	Taiwan	1 658	640
313 085 000	USA	3	South Korea	1 262	487
242 326 000	Indonesia	4	Rwanda	1 076	415
196 655 000	Brazil	5	India	1 049	405
176 745 000	Pakistan	6	Netherlands	1 039	401
162 471 000	Nigeria	7	Haiti	945	365
150 494 000	Bangladesh	8	Belgium	913	352
142 836 000	Russian Federation	9	Japan	867	335
126 497 000	Japan	10	Sri Lanka	831	321

*Only countries with a population of over 10 million are considered.

KEY POPULATION STATISTICS FOR MAJOR REGIONS

	Population 2011 (millions)	Growth (per cent)	Infant mortality rate	Total fertility rate	Life expectancy (years)	% aged 60 and over 2010	% aged 60 and over 2050
World	6 974	1.1	42	2.45	69	11	22
More developed regions	1 240	0.3	6	1.7	78	22	32
Less developed regions	5 774	1.3	46	2.6	67	8	20
Africa	1 046	2.3	71	4.4	55	6	10
Asia	4 207	1.0	37	2.2	70	10	24
Europe	739	0.1	6	1.6	77	22	34
Latin America and the Caribbean	597	1.1	19	2.2	75	10	25
North America	348	0.9	6	2.0	79	19	27
Oceania	37	1.5	19	2.5	78	15	24

Except for population and % aged 60 and over figures the data are annual averages projected for the period 2010–2015.

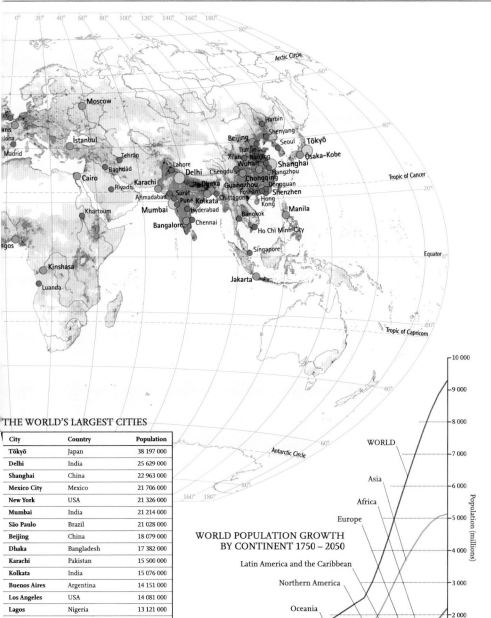

THE WORLD'S LARGEST CITIES

City	Country	Population
Tōkyō	Japan	38 197 000
Delhi	India	25 629 000
Shanghai	China	22 963 000
Mexico City	Mexico	21 706 000
New York	USA	21 326 000
Mumbai	India	21 214 000
São Paulo	Brazil	21 028 000
Beijing	China	18 079 000
Dhaka	Bangladesh	17 382 000
Karachi	Pakistan	15 500 000
Kolkata	India	15 076 000
Buenos Aires	Argentina	14 151 000
Los Angeles	USA	14 081 000
Lagos	Nigeria	13 121 000
Manila	Philippines	12 856 000
İstanbul	Turkey	12 459 000
Guangzhou	China	12 385 000
Rio de Janeiro	Brazil	12 380 000
Shenzhen	China	12 337 000
Moscow	Russian Federation	12 144 000

WORLD POPULATION GROWTH BY CONTINENT 1750 – 2050

WORLD

Asia

Africa

Europe

Latin America and the Caribbean

Northern America

Oceania

© Collins Bartholomew Ltd

MOBILE CELLULAR SUBSCRIBERS 2011

Winkel Tripel Projection
1:155 000 000

Cellular mobile subscribers
per 100 inhabitants 2011

- over 150
- 120–150
- 90–119.9
- 60–89.9
- 30–59.9
- 0–29.9
- no data

Total mobile cellular
subscribers 2011
Top ten countries

China
986 253 000

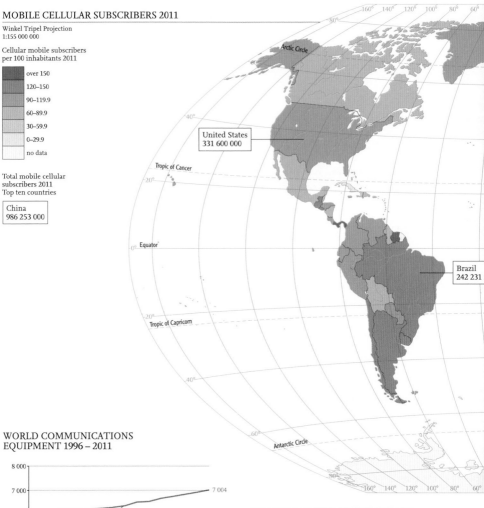

United States
331 600 000

Brazil
242 231

WORLD COMMUNICATIONS EQUIPMENT 1996 – 2011

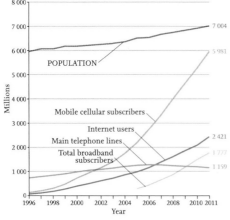

POPULATION

Mobile cellular subscribers

Internet users

Main telephone lines

Total broadband subscribers

7 004

5 981

2 421

1 777

1 159

Millions

1996 1998 2000 2002 2004 2006 2008 2010 2011
Year

TOP BROADBAND ECONOMIES 2011

Countries with the highest broadband penetration rate –
subscribers per 100 inhabitants

Top Economies	Fixed Broadband rate		Top Economies	Mobile Broadband rate
Netherlands	38.1	1	South Korea	91.0
Switzerland	37.9	2	Japan	87.8
Denmark	37.7	3	Sweden	84.0
South Korea	35.7	4	Australia	82.7
Norway	35.3	5	Finland	78.1
Iceland	34.1	6	Hong Kong, China	74.5
France	33.9	7	Portugal	72.5
Luxembourg	33.2	8	Luxembourg	72.1
Sweden	31.8	9	Singapore	69.7
Germany	31.7	10	Austria	67.4

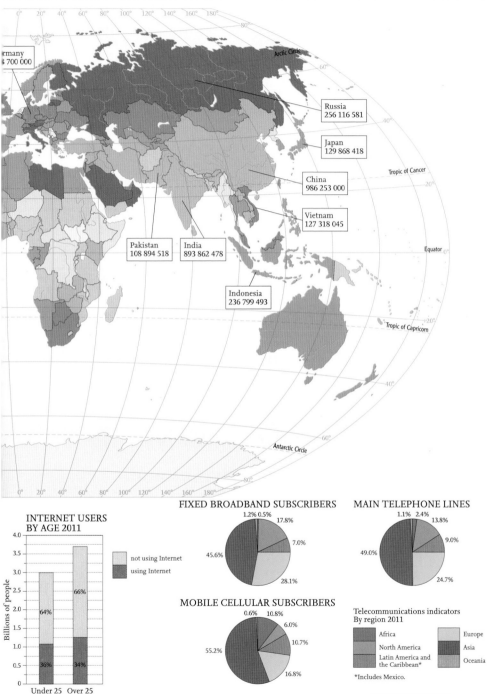

Germany
700 000

Russia
256 116 581

Japan
129 868 418

China
986 253 000

Vietnam
127 318 045

Pakistan
108 894 518

India
893 862 478

Indonesia
236 799 493

Arctic Circle

Tropic of Cancer

Equator

Tropic of Capricorn

Antarctic Circle

INTERNET USERS BY AGE 2011

Billions of people

not using Internet

using Internet

Under 25: 64% / 36%
Over 25: 66% / 34%

FIXED BROADBAND SUBSCRIBERS

1.2% 0.5%
17.8%
7.0%
45.6%
28.1%

MOBILE CELLULAR SUBSCRIBERS

0.6% 10.8%
6.0%
10.7%
55.2%
16.8%

MAIN TELEPHONE LINES

1.1% 2.4%
13.8%
9.0%
49.0%
24.7%

Telecommunications indicators
By region 2011

Africa
North America
Latin America and
the Caribbean*

Europe
Asia
Oceania

*Includes Mexico.

© Collins Bartholomew Ltd

41

MAP POLICIES

PLACE NAMES

The spelling of place names on maps has always been a matter of great complexity, because of the variety of the world's languages and the systems used to write them down. There is no standard way of spelling names or of converting them from one alphabet, or symbol set, to another. Instead, conventional ways of spelling have evolved in each of the world's major languages, and the results often differ significantly from the name as it is spelled in the original language. Familiar examples of English conventional names include Munich (München), Florence (Firenze) and Moscow (from the transliterated form, Moskva).

In this atlas, local name forms are used where these are in the Roman alphabet, though for major cities and main physical features, conventional English names are given first. The local forms are those which are officially recognized by the government of the country concerned, usually as represented by its official mapping agency. This is a basic principle laid down by the United Kingdom government's Permanent Committee on Geographical Names (PCGN) and the equivalent United States Board on Geographic Names (BGN). Prominent English-language and historic names are not neglected, however. These, and significant superseded names and alternate spellings, are included in brackets on the maps where space permits, and are cross-referenced in the index.

Country names are shown in conventional English form and include any recent changes promulgated by national governments and adopted by the United Nations. The names of continents, oceans, seas and under-water features in international waters also appear in English throughout the atlas, as do those

of other international features where such an English form exists and is in common use. International features are defined as features crossing one or more international boundary.

BOUNDARIES

The status of nations, their names and their boundaries, are shown in this atlas as they are at the time of going to press, as far as can be ascertained. Where an international boundary symbol appears in the sea or ocean it does not necessarily infer a legal maritime boundary, but shows which offshore islands belong to which country. The extent of island nations is shown by a short boundary symbol at the extreme limits of the area of sea or ocean within which all land is part of that nation.

Where international boundaries are the subject of dispute it may be that no portrayal of them will meet with the approval of any of the countries involved, but it is not seen as the function of this atlas to try to adjudicate between the rights and wrongs of political issues. Although reference mapping at atlas scales is not the ideal medium for indicating the claims of many separatist and irredentist movements, every reasonable attempt is made to show where an active territorial dispute exists, and where there is an important difference between 'de facto' (existing in fact, on the ground) and 'de jure' (according to law) boundaries. This is done by the use of a different symbol where international boundaries are disputed, or where the alignment is unconfirmed, to that used for settled international boundaries. Ceasefire lines are also shown by a separate symbol. For clarity, disputed boundaries and areas are annotated where this is considered necessary. The atlas aims to take a strictly neutral viewpoint of all such cases, based on advice from expert consultants.

MAP PROJECTIONS

Map projections have been selected specifically for the area and scale of each map, or suite of maps. As the only way to show the Earth with absolute accuracy is on a globe, all map projections are compromises. Some projections seek to maintain correct area relationships (equal area projections), true distances and bearings from a point (equidistant projections) or correct angles and shapes (conformal projections); others attempt to achieve a balance between these properties. The choice of projections used in this atlas has been made on an individual continental and regional basis. Projections used, and their individual parameters, have been defined to minimize distortion and to reduce scale errors as much as possible. The projection used is indicated at the bottom left of each map page.

SCALE

In order to directly compare like with like throughout the world it would be necessary to maintain a single scale throughout the atlas. However, the desirability of mapping the more densely populated areas of the world at larger scales, and other geographical considerations, such as the need to fit a homogeneous physical region within a uniform rectangular page format, mean that a range of scales have been used. Scales for continental maps range between 1:25 000 000 and 1:55 000 000, depending on the size of the continental land mass being covered. Scales for regional maps are typically in the range of 1:15 000 000 to 1:25 000 000. Mapping for most countries is at scales between 1:6 000 000 and 1:12 000 000, although for the more densely populated areas of Europe the scale increases to 1:3 000 000.

ABBREVIATIONS

Arch.	Archipelago				Mts	Mountains		
B.	Bay					Monts	French	hills, mountains
	Bahia, Baía	Portuguese	bay		N.	North, Northern		
	Bahía	Spanish	bay		O.	Ostrov	Russian	island
	Baie	French	bay		Pt	Point		
C.	Cape				Pta	Punta	Italian, Spanish	cape, point
	Cabo	Portuguese, Spanish	cape, headland		R.	River		
	Cap	French	cape, headland			Rio	Portuguese	river
Co	Cerro	Spanish	hill, peak, summit			Río	Spanish	river
E.	East, Eastern					Rivière	French	river
Est.	Estrecho	Spanish	strait		Ra.	Range		
Gt	Great				S.	South, Southern		
I.	Island, Isle					Salar, Salina, Salinas	Spanish	saltpan, saltpans
	Ilha	Portuguese	island		Sa	Serra	Portuguese	mountain range
	Islas	Spanish	island			Sierra	Spanish	mountain range
Is	Islands, Isles				Sd	Sound		
	Islas	Spanish	islands		S.E.	Southeast,		
Khr.	Khrebet	Russian	mountain range			Southeastern		
L.	Lake				St	Saint		
	Loch	(Scotland)	lake			Sankt	German	saint
	Lough	(Ireland)	lake			Sint	Dutch	saint
	Lac	French	lake		Sta	Santa	Italian, Portuguese,	
	Lago	Portuguese, Spanish	lake				Spanish	saint
M.	Mys	Russian	cape, point		Ste	Sainte	French	saint
Mt	Mount				Str.	Strait		
	Mont	French	hill, mountain		W.	West, Western		
Mt.	Mountain					Wadi, Wādī	Arabic	watercourse

MAP SYMBOLS

LAND AND WATER FEATURES

Lake

Impermanent lake

Salt lake or lagoon

Impermanent salt lake

Dry salt lake or salt pan

River

Impermanent river

Ice cap / Glacier

123 Pass
⊃⊂ Height in metres

∴ Site of special interest

᠃ Oasis

ᴨᴨᴨ Wall

TRANSPORT

Motorway

Main road

Track

Main railway

Canal

Main airport

BOUNDARIES

International boundary

Disputed international boundary
or alignment unconfirmed

Disputed territory boundary

Undefined international
boundary in the sea.
All land within this boundary is part
of state or territory named.

Administrative boundary
Shown for selected countries only.

Ceasefire line or other
boundary described on
the map

RELIEF

Contour intervals used in layer-colouring, for
land height and sea depth

METRES FEET		Ocean pages

METRES FEET	
5000	16404
3000	9843
2000	6562
1000	3281
500	1640
200	656
0	0

and below
sea level

200	656
4000	13124
6000	19686
M	FT

METRES FEET	
0	0
200	656
2000	6562
3000	9843
4000	13124
5000	16404
6000	19686
7000	22967
9000	29529
M	FT

123 Ocean deep
⁖ In metres.

1234 Summit
△ Height in metres

1234 Volcano
▲ Height in metres

STYLES OF LETTERING

Cities and towns are explained separately

Country	**FRANCE**
Overseas Territory/Dependency	**Guadeloupe**
Disputed Territory	WESTERN SAHARA
Administrative name Shown for selected countries only.	**SCOTLAND**
Area name	PATAGONIA

Physical features

Island	*Gran Canaria*
Lake	*Lake Erie*
Mountain	*Mt Blanc*
River	*Thames*
Region	*LAPPLAND*

CITIES AND TOWNS

Population	National Capital	Administrative Capital Shown for selected countries only	Other City or Town
over 10 million	**DHAKA** ▣	**Karachi** ⊙	**New York** ◉
5 million to 10 million	**MADRID** ▣	**Toronto** ⊙	**Philadelphia** ⊙
1 million to 5 million	**KĀBUL** □	**Sydney** ○	**Kaohsiung** ○
500 000 to 1 million	**BANGUI** □	**Winnipeg** ○	**Jeddah** ○
100 000 to 500 000	**WELLINGTON** □	Edinburgh ○	Apucarana ○
50 000 to 100 000	PORT OF SPAIN □	Bismarck ○	Invercargill ○
under 50 000	MALABO ▫	Charlottetown ○	Ceres ○

CONTINENTAL MAPS

BOUNDARIES	—— International boundary	------ Disputed international boundary	•••••••• Ceasefire line

CITIES AND TOWNS	National Capital	**Beijing** □	Other City or Town	**New York** ○

© Collins Bartholomew Ltd

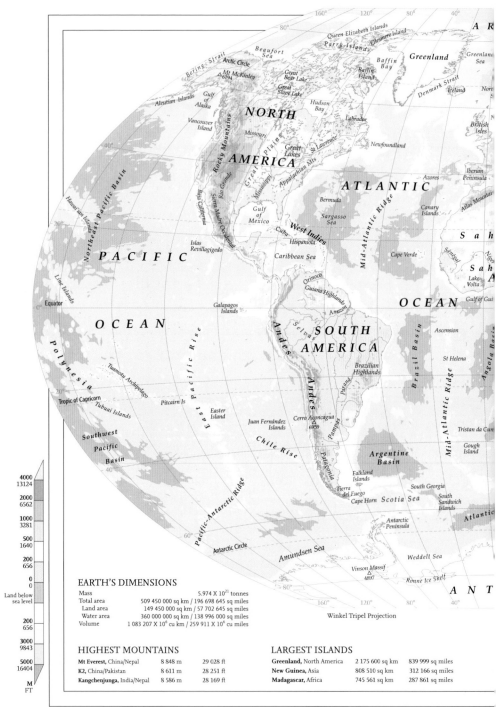

A R
Queen Elizabeth Islands
Parry Islands Ellesmere Island
Bering Strait
Beaufort Sea
Baffin Bay
Greenland
Greenland Sea
Arctic Circle
Mt McKinley △6194
Baffin Island
Great Bear Lake
Iceland
Nor
Denmark Strait
Aleutian Islands
Gulf of Alaska
Great Slave Lake
Hudson Bay
T
Vancouver Island
NORTH
Missouri
Labrador
Newfoundland
British Isles
N
Rocky Mountains
Great Lakes
St Lawrence
AMERICA
Appalachian Mts
Iberian Peninsula
Great Plains
Azores
ATLANTIC
Rio Grande
Mississippi
Baja California
Bermuda
Sierra Madre Occidental
Gulf of Mexico
Sargasso Sea
Canary Islands
Atlas Mountain
Mid-Atlantic Ridge
S a h
Islas Revillagigedo
West Indies
Cuba
Hispaniola
Northeast Pacific Basin
Hawaiian Islands
PACIFIC
Caribbean Sea
Cape Verde
Senegal
S a h
A
Niger
Line Islands
Orinoco
Guiana Highlands
Lake Volta
Equator
Galapagos Islands
Amazon
OCEAN
Gulf of Gui
OCEAN
Selvas
SOUTH
Andes
Ascension
Angola Basin
Polynesia
East Pacific Rise
Tuamotu Archipelago
AMERICA
Brazilian Highlands
St Helena
Brazil Basin
Mid-Atlantic Ridge
Tropic of Capricorn
Tubuai Islands
Pitcairn Is
Easter Island
Juan Fernández Islands
Cerro Aconcagua 6959
Andes
Paraná
Pampas
Patagonia
Tristan da Cun
Gough Island
Southwest Pacific Basin
Chile Rise
Argentine Basin
Falkland Islands
Pacific-Antarctic Ridge
Tierra del Fuego
Cape Horn
Scotia Sea
South Georgia
South Sandwich Islands
Atlantic
Antarctic Circle
Antarctic Peninsula
Amundsen Sea
Weddell Sea
Vinson Massif △4897
Ronne Ice Shelf
A N T

EARTH'S DIMENSIONS

Mass	5.974×10^{21} tonnes
Total area	509 450 000 sq km / 196 698 645 sq miles
Land area	149 450 000 sq km / 57 702 645 sq miles
Water area	360 000 000 sq km / 138 996 000 sq miles
Volume	$1\ 083\ 207 \times 10^{6}$ cu km / $259\ 911 \times 10^{6}$ cu miles

Winkel Tripel Projection

HIGHEST MOUNTAINS

Mt Everest, China/Nepal	8 848 m	29 028 ft
K2, China/Pakistan	8 611 m	28 251 ft
Kangchenjunga, India/Nepal	8 586 m	28 169 ft

LARGEST ISLANDS

Greenland, North America	2 175 600 sq km	839 999 sq miles
New Guinea, Asia	808 510 sq km	312 166 sq miles
Madagascar, Africa	745 561 sq km	287 861 sq miles

4000 13124
2000 6562
1000 3281
500 1640
200 656
0 0
Land below sea level
200 656
3000 9843
5000 16404
M FT

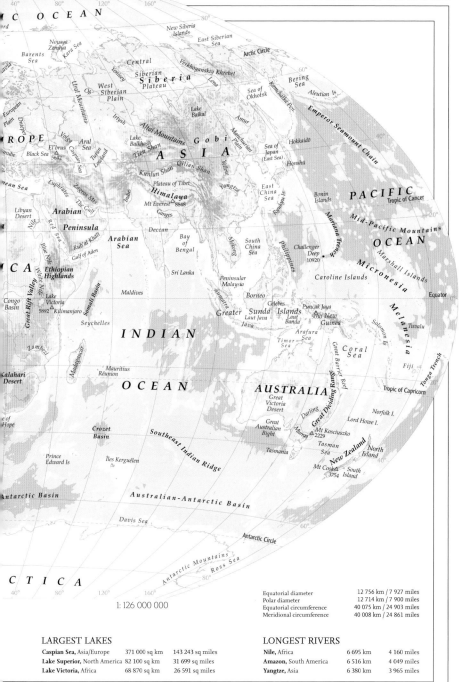

C O C E A N

New Siberia
Islands
East Siberian
Sea

Novaya
Zemlya
Kara Sea

Barents
Sea

Arctic Circle

Central
Siberian
Plateau

West
Siberian
Plain

Ob'
Yenisey
Lena
Verkhoyanskiy Khrebet

S i b e r i a

Bering
Sea

Sea of
Okhotsk

Komchatskaya Pen.

Kamchatka Pen.

Aleutian Is

Ural Mountains
Irtysh

Lake
Baikal

Altai Mountains

Amur

Manchurian
Plain

Emperor Seamount Chain

R O P E

Volga
El'brus
5642
Aral
Sea
Turan
Lowland

Lake
Balkhash

Tien Shan

G o b i

A S I A

Sea of
Japan
(East Sea)

Hokkaidō

Honshū

40°

Danube
Black Sea
Caspian Sea

Kunlun Shan
Qilian Shan

Yellow

East
China
Sea

Bonin
Islands

P A C I F I C

Tropic of Cancer

ean Sea

Euphrates
Zagros Mts
The Gulf

Plateau of Tibet

H i m a l a y a
Mt Everest 8848

Yangtze

Ryukyu Is

Mid-Pacific Mountains

O C E A N

Libyan
Desert

Nile
Red Sea

A r a b i a n
P e n i n s u l a

Ganges

Deccan

South
China
Sea

Challenger
Deep
10920

Mariana Trench

Marshall Islands

C A

Gulf of Aden

Blue Nile
White Nile

Rub' al Khali

A r a b i a n
S e a

Bay
of
Bengal

Mekong

Philippines

M i c r o n e s i a

Ethiopian
Highlands

Sri Lanka

Peninsular
Malaysia

Caroline Islands

Equator

Congo
Basin

Great Rift Valley

Lake
Victoria
5892 Kilimanjaro

Somali Basin

Maldives

Sumatra

Borneo

Celebes

Puncak Jaya
5030

New
Guinea

M e l a n e s i a

Seychelles

Greater Sunda Islands
Laut Java
Java

Laut
Banda

Arafura
Sea

Solomon Is

Tuvalu

Zambezi

Madagascar

I N D I A N

Timor
Sea

C o r a l
S e a

Fiji

Tonga Trench

Kalahari
Desert

Mauritius
Réunion

O C E A N

Great Dividing Range
Great Barrier Reef

Tropic of Capricorn

e of
Hope

Prince
Edward Is

Crozet
Basin

Southeast Indian Ridge

Îles Kerguélen

A U S T R A L I A

Great
Victoria
Desert

Great
Australian
Bight

Murray
Darling

Mt Kosciuszko
2229

Norfolk I.

Lord Howe I.

Tasman
Sea

New Zealand

North
Island

Tasmania

Mt Cook
3754

South
Island

Antarctic Basin

A u s t r a l i a n - A n t a r c t i c   B a s i n

Davis Sea

Antarctic Circle

C T I C A

Antarctic Mountains
Ross Sea

1: 126 000 000

Equatorial diameter	12 756 km / 7 927 miles	
Polar diameter	12 714 km / 7 900 miles	
Equatorial circumference	40 075 km / 24 903 miles	
Meridional circumference	40 008 km / 24 861 miles	

LARGEST LAKES

Caspian Sea, Asia/Europe	371 000 sq km	143 243 sq miles
Lake Superior, North America	82 100 sq km	31 699 sq miles
Lake Victoria, Africa	68 870 sq km	26 591 sq miles

LONGEST RIVERS

Nile, Africa	6 695 km	4 160 miles
Amazon, South America	6 516 km	4 049 miles
Yangtze, Asia	6 380 km	3 965 miles

160°   120°   80°   40°

A R

80°

Greenland
(Denmark)   Jan Maye
(Norwa)

Nuuk □   Reykjavik □ICELAND

C A N A D A

Vancouver   Edmonton

UNITED
KINGDOM
REP. OF
IRELAND
London
Paris
FRANC

Arctic Circle
U.S.A.
Anchorage

40°

Ottawa   Montreal
UNITED STATES   Toronto
OF Chicago   NewYork
San Francisco   Denver   Washington   Philadelphia
AMERICA   D.C.

Azores
(Portugal)   PORTUGAL   SPAIN
Algiers

Los Angeles

Rabat
MOROCCO

Bermuda
(U.K.)

Tropic of Cancer

Monterrey   Houston
THE
Miami   BAHAMAS
Nassau

A T L A N T I C Laâyoune   ALGE

20°
Hawai'ian
Islands
(U.S.A.)

MEXICO   Havana   CUBA
Mexico City □   HAITI   DOMINICAN
REP.

WESTERN
SAHARA

MAURITANIA
Nouakchott
MALI

P A C I F I C

BELIZE   JAMAICA   Puerto Rico
GUATEMALA   HONDURAS   (U.S.A.)
EL SALVADOR   NICARAGUA   TRINIDAD AND
COSTA RICA   San José   TOBAGO
PANAMA   Caracas   Port of Spain
Bogotá   VENEZUELA   Paramaribo
Georgetown   Cayenne
COLOMBIA   FR. G.

CAPE VERDE   SENEGAL
THE GAMBIA   Dakar   BUR
GUINEA-BISSAU   GUINEA   GH
Conakry   D'I   L
SIERRA LEONE   Monrovia   Accra   L
LIBERIA

Galapagos
Islands
(Ecuador)   Quito
ECUADOR

Equator

0°

O C E A N

KIRIBATI

B R A Z I L

O C E A N

Ascension
(U.K.)

American
Samoa

French
Polynesia
(France)

Lima   PERU
BOLIVIA   Brasília
La Paz   Sucre

St Helena
(U.K.)

Cook
Islands
(New Zealand)   Tahiti

PARAGUAY   Rio de Janeiro
São Paulo

St Helena, Ascensio
and Tristan da Cunha
(U.K.)

20°
Tropic of Capricorn

Pitcairn Islands
(U.K.)   Easter
Island
(Chile)

Asunción

Buenos
Aires   URUGUAY
Santiago   Montevideo

Tristan
da Cunha
(U.K.)

ARGENTINA

40°

Falkland
Islands
(U.K.)

South Georgia and
the South Sandwich
Islands
(U.K.)

Bouvetøy
(Norwa

60°

Antarctic Circle

A N T

160°   120°   80°   40°

Winkel Tripel Projection

ABBREVIATIONS

A.	ANDORRA	**BE.**	BENIN	**C.A.R.**	CENTRAL AFRICAN	**CZ.R.**	CZECH REPUBLIC
AL.	ALBANIA	**BEL.**	BELGIUM		REPUBLIC	**DEN.**	DENMARK
ARM.	ARMENIA	**B.H.**	BOSNIA-HERZEGOVINA	**C.D'I.**	CÔTE D'IVOIRE	**EQ.G.**	EQUATORIAL GUINEA
AUS.	AUSTRIA	**BN.**	BAHRAIN		(IVORY COAST)	**FR.G.**	FRENCH GUIANA
AZ.	AZERBAIJAN	**BUR.**	BURKINA FASO	**CR.**	CROATIA	**GEOR.**	GEORGIA
B.	BURUNDI	**CAM.**	CAMEROON	**CYP.**	CYPRUS	**GER.**	GERMANY

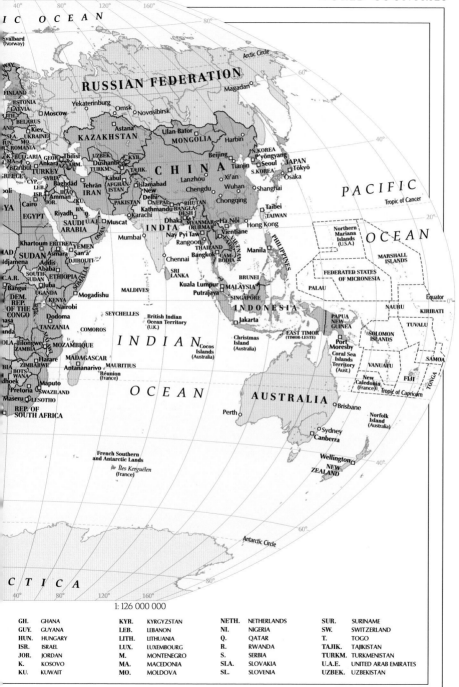

1: 126 000 000

GH.	GHANA	KYR.	KYRGYZSTAN	NETH.	NETHERLANDS	SUR.	SURINAME
GUY.	GUYANA	LEB.	LEBANON	NI.	NIGERIA	SW.	SWITZERLAND
HUN.	HUNGARY	LITH.	LITHUANIA	Q.	QATAR	T.	TOGO
ISR.	ISRAEL	LUX.	LUXEMBOURG	R.	RWANDA	TAJIK.	TAJIKISTAN
JOR.	JORDAN	M.	MONTENEGRO	S.	SERBIA	TURKM.	TURKMENISTAN
K.	KOSOVO	MA.	MACEDONIA	SLA.	SLOVAKIA	U.A.E.	UNITED ARAB EMIRATES
KU.	KUWAIT	MO.	MOLDOVA	SL.	SLOVENIA	UZBEK.	UZBEKISTAN

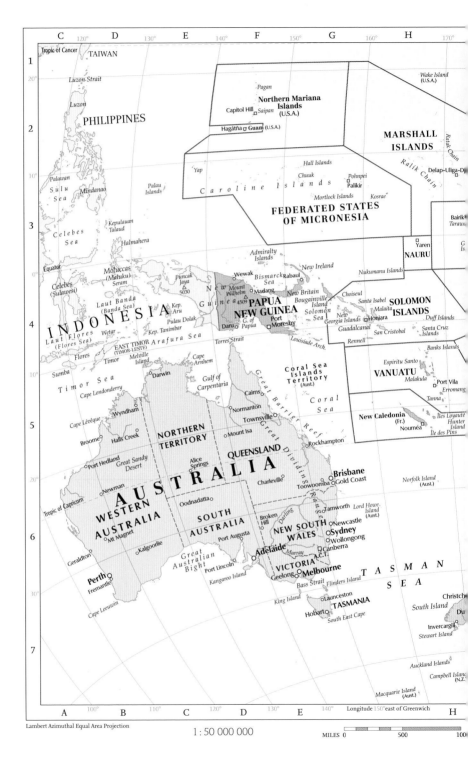

Lambert Azimuthal Equal Area Projection

1 : 50 000 000

MILES 0   500   100

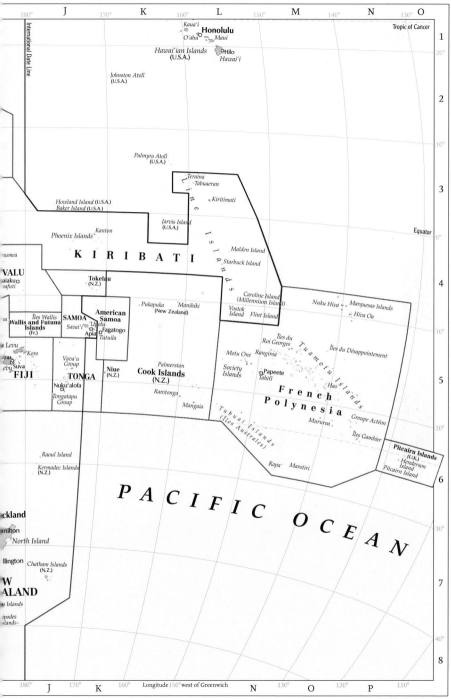

Tropic of Cancer

180° J 170° K 160° L 150° M 140° N 130° O

International Date Line

20°

Honolulu
Kaua'i
O'ahu○ ○Maui
Hawai'ian Islands
(U.S.A.) ○Hilo
Hawai'i

Johnston Atoll
(U.S.A.)

10°

Palmyra Atoll
(U.S.A.)

Teraina
Tabuaeran

Kiritimati

Howland Island (U.S.A.)
Baker Island (U.S.A.)

Jarvis Island
(U.S.A.)

Equator 0°

Kanton
Phoenix Islands

K I R I B A T I

Malden Island
Starbuck Island

uumea

VALU
aiaku□
nafuti

Tokelau
(N.Z.)

Caroline Island
(Millennium Island)

Nuku Hiva *Marquesas Islands*

Pukapuka *Manihiki*
(New Zealand)

Vostok
Island *Flint Island*

Hiva Oa

Îles Wallis
Wallis and Futuna
Islands
(Fr.)

SAMOA
Savai'i○□
Apia

American
Samoa
Upolu
□Fagatogo
Tutuila

Îles du
Roi Georges

Îles du Désappointement

a Levu
inu
ebu□ Koro
Suva
FIJI

Motu One
Rangiroa

TONGA
Nuku'alofa
□
Tongatapu
Group

Vava'u
Group

Niue
(N.Z.)

Palmerston

Cook Islands
(N.Z.)

Rarotonga

Society
Islands
□Papeete
Tahiti

Hao

F r e n c h
P o l y n e s i a

Mangaia

Tubuai Islands
(*Îles Australes*)

Groupe Actéon

Mururoa

Îles Gambier

Raoul Island

Kermadec Islands
(N.Z.)

Rapa *Marotiri*

Pitcairn Islands
(U.K.)
Henderson
Island
Pitcairn Island

ckland

milton

North Island

P A C I F I C

llington *Chatham Islands*
(N.Z.)

O C E A N

W
ALAND

u Islands

ipodes
slands

180° J 170° K 160° Longitude 150° west of Greenwich N 130° O 120° P 110°

20°

10°

20°

30°

40°

1

2

3

4

5

6

7

8

0 500 1000 1500 KM

© Collins Bartholomew Ltd

49

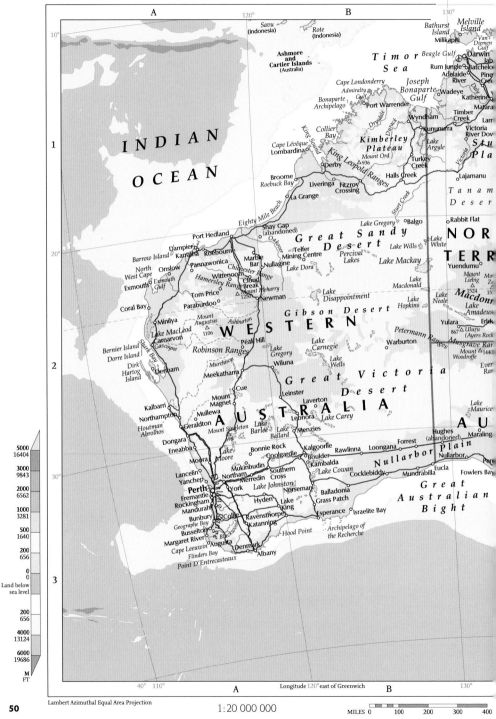

INDIAN

OCEAN

Savu
(Indonesia)

Rote
(Indonesia)

Ashmore
and
Cartier Islands
(Australia)

Timor
Sea

Beagle Gulf

Bathurst
Island
Milikapiti

Melville
Island

Van
Diemen
Gulf

Darwin
Jab
Batchelo
Ping
Cree

Adelaide
River

Katherine

Matara

Larr

Rum Jungle

Cape Londonderry

Admiralty
Gulf

Bonaparte
Archipelago

Joseph
Bonaparte
Gulf

Port Warrender

Wadeye

Timber
Creek

Victoria
River Dov

S t u
P l a

Collier
Bay

Kimberley
Plateau

Wyndham

Kununurra

Lake
Argyle

Cape Lévêque
Lombardina

Mount Ord
936

King Leopold Ranges

Turkey
Creek

Lajamanu

Derby

Halls Creek

Broome
Roebuck Bay

Liveringa

Fitzroy
Crossing

T a n a m
D e s e r

La Grange

Sturt Creek

Eighty Mile Beach

Shay Gap
(abandoned)

Lake Gregory

Balgo

Rabbit Flat

Great Sandy
Desert

Lake
White

N O R

Port Hedland

Oakover

Telfer
Mining Centre

Lake Wills

T E R R

Barrow Island

Dampier

Karratha

Roebourne

Percival
Lakes

Lake Mackay

Yuendumu

North
West Cape

Onslow

Pannawonica

Marble
Bar

Nullagine

Chichester Range

Lake Dora

Mount
Liebig
1524

Mo
Ze
15

Exmouth
Gulf

Wittenoom

Cloud
Break

Macdon

Exmouth

Hamersley Range

Mount Meharry
1250

Newman

Lake
Disappointment

Lake
Macdonald

Lake
Neale

Lake
Amadeus

Coral Bay

Tom Price

Paraburdoo

Gibson Desert

Lake
Hopkins

Yulara

Uluru
(Ayers Rock)

Erld

Minilya

Mount
Augustus
1106

Ashburton

W E S T E R N

Petermann Ranges

Warburton

Musgrave Ran

Mount
Woodroffe
1440

867

Bernier Island
Dorre Island

Carnarvon

Gascoyne

Peak Hill

Lake
Gregory

Lake
Carnegie

Ever
Ran

Dirk
Hartog
Island

Lake MacLeod

Robinson Ranges

Murchison

Wiluna

Lake
Wells

Great Victoria

Denham

Meekatharra

A U S T R A L I A

D e s e r t

Shark Bay

Cue

Mount
Magnet

Leinster

Laverton

Lake
Maurice

Kalbarri

Mullewa

Leonora

Lake Carey

A U

Northampton

Geraldton

Mount Singleton

995

Lake
Barlee

Menzies

Houtman
Abrolhos

Dongara

Lake
Moore

Bonnie Rock

Kalgoorlie

Rawlinna

Loongana

Forrest

Hughes
(abandoned)

Maraling

Eneabba

Moora

Coolgardie

Boulder

Kambalda

Nullarbor
Plain

Nullarbor

Pe

Lancelin

Yanchep

Mukinbudin

Northam

Southern
Cross

Merredin

Lake Cowan

Cocklebiddy

Mundrabilla

Eucla

Fowlers Bay

Perth

Fremantle

York

Lake Johnston

Norseman

Great
Australian
Bight

Rockingham

Mandurah

Hyden

Lake
King

Balladonia

Grass Patch

Bunbury

Collie

Ravensthorpe

Esperance

Israelite Bay

Geographe Bay

Busselton

Katanning

Margaret River

Augusta

Hood Point

Archipelago of
the Recherche

Cape Leeuwin

Flinders Bay

Denmark

Albany

Point D'Entrecasteaux

5000	16404
3000	9843
2000	6562
1000	3281
500	1640
200	656
0	
Land below sea level	
200	656
4000	13124
6000	19686
M	FT

Lambert Azimuthal Equal Area Projection

1 : 20 000 000

MILES 0　100　200　300　400

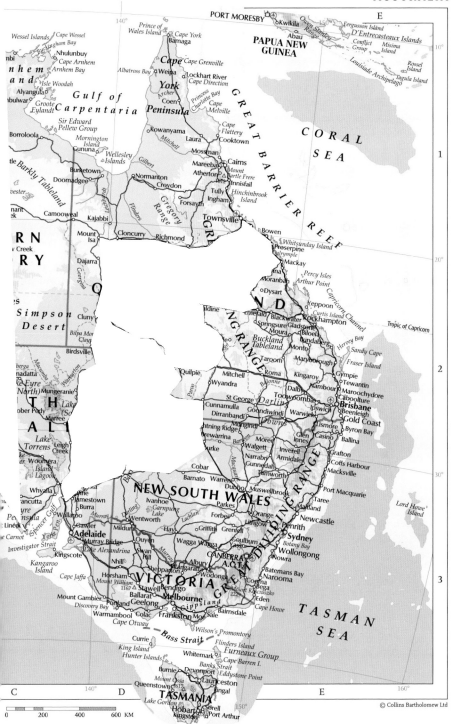

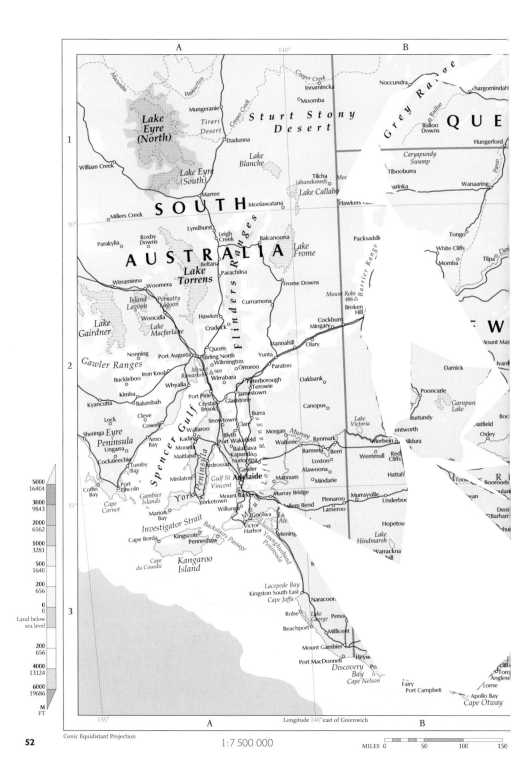

Macumba
Warburton
Copper Creek
Innamincka
Noccundra
thargomindah
Mungeranie
Tirari
Desert
Moomba
Grey Range
Bulloo
Lake
Eyre
(North)
Sturt Stony
Desert
Bulloo
Downs
QUE
Etadunna
Hungerford
1
Lake
Blanche
Caryapundy
Swamp
William Creek
Tibooburra
Lake Eyre
(South)
Tilcha
(abandoned)
Lake Callabo
arinka
Wanaaring
Marree
Moolawatana
Hawkers
SOUTH
Millers Creek
Lyndhurst
Leigh
Creek
Balcanoona
Packsaddl
Tongo
30°
Parakylia
Roxby
Downs
White Cliffs
Da
AUSTRALIA
Beltana
Lake
Frome
Momba
Tilpa
Wirraminna
Woomera
Lake
Torrens
Parachilna
Barrier Range
Island
Lagoon
Pernatty
Lagoon
Frome Downs
Mount Robe
486 △
Broken
Hill
W
Woocalla
Curnamona
Lake
Gairdner
Lake
Macfarlane
Hawker
Cradock
Cockburn
Mingary
Mount Mai
Quorn
Mannahill
Olary
Nonning
Port Augusta
Stirling North
Wilmington
Yunta
2   Gawler Ranges
Mount
Remarkable
△ 969
Orroroo
Paratoo
Darnick
Ivan
Buckleboo
Iron Knob
Wirrabara
Peterborough
Terowie
Oakbank
Pooncarie
Whyalla
Jamestown
Kimba
Balumbah
Port Pirie
Gladstone
Crystal
Brook
Canopus
Garmpun
Lake
Bo
Kyancutta
Cleve
Cowell
Snowtown
Burra
Lake
Victoria
Burtundy
atfield
Lock
Wallaroo
Clare
Morgan
Murray
Renmark
entworth
Oxley
Sheringa
Eyre
Peninsula
Arno
Bay
Kadina
Blyth
Port Wakefield
Waikerie
Merbein
ildura
Ungarra
Moonta
Balaklava
Barmera
Berri
Werrimull
Red
Cliffs
Cockaleechie
Maitland
Kapunda
Nuriootpa
Loxton
Tumby
Bay
Ardrossan
Gawler
Alawoona
Hattah
Too
Booroorb
Minlaton
Gulf St
Vincent
Adelaide
Mannum
Mindarie
ulan
wan
Coffin
Bay
Port
Lincoln
Gambier
Islands
Murray Bridge
Pinnaroo
Murrayville
Underboo
Deni
Cape
Carnot
Yorketown
Mount Barker
Tailem Bend
Lameroo
Barham
Marion
Bay
Willung
Goolwa
La
Ale
Hopetou
Investigator Strait
Backstairs Passage
Victor
Harbor
Menin
Lake
Hindmarsh
hu
Cape Borda
Kingscote
Penneshaw
Warrackna
hill
Cape
du Couedic
Kangaroo
Island
B
Lacepede Bay
Kingston South East
Cape Jaffa
Naracoor
3
Robe
Lake
George
Peno
Beachport
Millicent
Mount Gambier
Heyw
Port MacDonnell
Po
cliff
Discovery
Bay
Cape Nelson
Toro
Anglese
Fairy
Port Campbell
Lorne
Apollo Bay
Cape Otway

Spencer Gulf
York Peninsula
Yorke Peninsula
Flinders Ranges
Lofty Range
Mount Lofty Ranges
Younghusband Peninsula

5000
16404

3000
9843

2000
6562

1000
3281

500
1640

200
656

0
0
Land below
sea level

200
656

4000
13124

6000
19686

M
FT

Conic Equidistant Projection

1:7 500 000

MILES 0   50   100   150

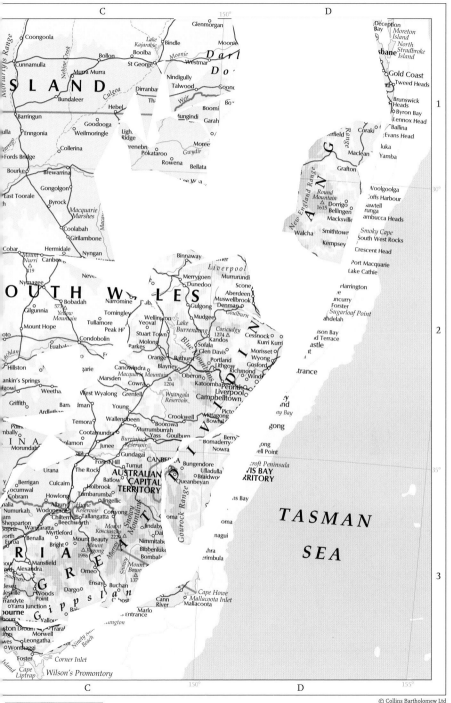

C

D

Deception Bay
Moreton Island
North Stradbroke Island
sbane
Glenmorgan
Coongoola
Lake Kajarabie
Bindle
Moonie
Bollon
Boolba
Cunnamulla
St George
Westmar
Murra Murra
Moonie
D a r l
D o
Gold Coast
Tweed Heads
Nindigully
Talwood
Goon
Brunswick Heads
Byron Bay
Lennox Head
Ballina
Evans Head
Dirranba
S L A N D
Bundaleer
Hebel
Boomi
Th
Weir
Culgoa
Goodooga
Weilmoringle
Ligh.
Ridge
Jungindi
Garah
Bo
erfield
Range
Coraki
Iuka
Yamba
Barringun
ulla
Enngonia
Collerina
renebri
Pokataroo
Gwydir
Moree
O/
Maclean
Fords Bridge
Rowena
Bellata
Grafton
Bourke
Brewarrina
ee W
Gongolgon
Round Mountain
Woolgoolga
East Toorale
h
Byrock
Dorrigo
1615
Bellingen
Macksville
Coffs Harbour
sawtell
runga
ambucca Heads
Macquarie Marshes
Coolabah
Girilambone
New England Range
Walcha
Smithtown
Kempsey
Smoky Cape
South West Rocks
Crescent Head
Cobar
Hermidale
Nyngan
Binnaway
mer
Port Macquarie
Lake Cathie
Mount Nurri
Canbe
119
Neve
Merrygoen
Dunedoo
Liverpool
Murrurundi
Scone
Harrington
.e
ncurry
Forster
Sugarloaf Point
ahdelah
Nymagee
Bobadah
Narromine
573
Gilgunnia
Yellow Mountain
Tullamore
Peak Hi
Tomingley
Wellington
Yeoval
Aberdeen
Muswellbrook
Gulgong
Denman
Mudgee
Goulburn
Mount Hope
oto
Condobolin
Euaba
Stuart Town
Molong
Parkes
Lake Burrendong
Coricudgy
1274
Kandos
Cessnock
Kurri Kurri
ison Bay
d Terrace
astle
Hillston
zarie
Marsden
Canowindra
Blayney
Macquarie Mountains
Orange
Bathurst
Oberon
Blue Mountains
Sofala
Glen Davis
Portland
Lithgow
Richmond
Morisset
Wyong
Gosford
Wind
trance
ankin's Springs
gowi
Weetha
West Wyalong
Grenfell
Cowra
1204
Wyangala Reservoir
Katoomba
Penrith
Liverpool
Campbelltown
y
nd
ny Bay
Griffith
Barn
Ardletha
Iman
Young
Crookwell
Picte
Bowral
gong
Temora
Wallendbeen
Boorowa
Mittagong
Pom
mbally
Cootamundra
Murrumburrah
Yass
Berry
Bomaderry
ong
ell Point
I N A
Morundah
lamon
Junee
Burrinjuck Reservoir
Goulburn
Nowra
a
Gundagai
Forest Hill
Tumut
CANBE A
Bungendore
croft Peninsula
Urana
The Rock
Batlow
Ulladulla
Braidwo
VIS BAY
RRITORY
y
Berrigan
ocumwal
Cobram
Culcairn
Howlong
Holbrook
Tumbarumba
AUSTRALIAN CAPITAL TERRITORY
Queanbeyan
ns Bay
nalia
Numurkah
am
Shepparton
ocha
Wangaratta
orth
Euoa
Albury
Wodonga
Chiltern
Myrtleford
Benalla
Jingellic
Hume Reservoir
Tallangatta
Beechworth
Corryong
Mount Kosciuszko
2228
Mount Beauty
Jindaby
Dal
oma
oma
nagui
TASMAN
SEA
Bright
Mansfield
Alexandra
R I A
Omeo
Mount Bogong
1986
Mount Bowe
137
Bibbenluke
Bombala
hra
rimbula
leseg
ilesville
randyte
G
Woods Point
Dargo
Ensay
Buchan
G i p p s l a n
Yarra Junction
bourn
G
Yallo
ston Droml
ngs
Wonthaggi
Traral
Morwell
Leongatha
Cann River
Marlo
Cape Howe
Mallacoota Inlet
Mallacoota
Foster
Corner Inlet
Island
Cape Liptrap
Wilson's Promontory

150°

150°

155°

1

2

3

30°

35°

C

D

53

0   100   200 KM

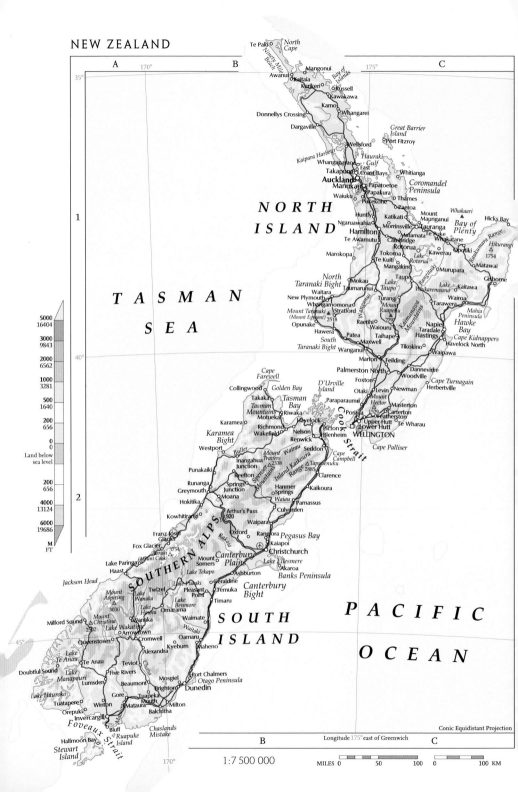

NEW ZEALAND

1:7 500 000

Longitude 175° east of Greenwich

MILES 0 50 100

0 100 KM

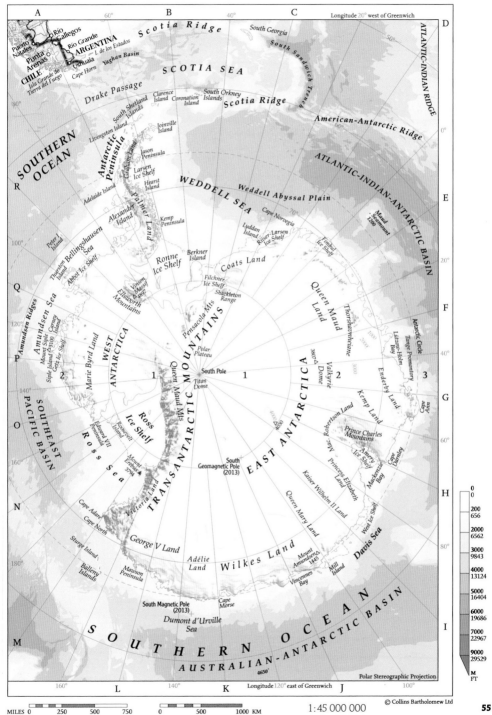

A 60° B 40° C

South Georgia

Scotia Ridge

ATLANTIC-INDIAN RIDGE

Puerto Natales
Río Gallegos
Río Grande
ARGENTINA
Punta Arenas
I. de los Estados
CHILE
Ushuaia
Yaghan Basin
Cape Horn
Isla Grande de
Tierra del Fuego

Scotia Ridge

SCOTIA SEA

South Sandwich Trench

Drake Passage

Clarence
Island
South Orkney
Islands
Coronation
Island
Scotia Ridge

American-Antarctic Ridge

SOUTHERN
OCEAN

South Shetland
Islands
Joinville
Island

R

Livingston Island
Antarctic
Peninsula
Graham Land
Jason
Peninsula
Larsen
Ice Shelf
Hearst
Island

WEDDELL SEA

Weddell Abyssal Plain

Cape Norvegia

ATLANTIC-INDIAN-ANTARCTIC BASIN

0°

Adelaide Island
Alexander
Island
Palmer Land
Kemp
Peninsula
Lyddan
Island
Kaiser
Wilhelm II
Larsen
Ice Shelf

Maud
Seamount
1200

E

20°

Peter I
Island
Bellingshausen
Sea
Thurston
Island
Abbot Ice Shelf
Ronne
Ice Shelf
Berkner
Island
Coats Land

Fimbul
Ice Shelf

Q

Amundsen Ridges

Amundsen Sea
Carney
Island
Ellsworth
Mountains
Vinson
Massif
4897
Filchner
Ice Shelf
Pensacola Mts
Shackleton
Range

Queen Maud
Land

Thorshavnheane

Antarctic
Circle

Lützow-Holm
Bay

F

P

Mount Siple
3100
Getz Ice Shelf
Siple Island
Marie Byrd Land
WEST
ANTARCTICA
Queen Maud Mts
Polar
Plateau
South Pole
1
1
3807
Valkyrie
Dome

Enderby Land

Targe Promontory

40°

3

SOUTHEAST
PACIFIC BASIN

2
Ross
Ice Shelf
Roosevelt
Island
Titan
Dome

EAST ANTARCTICA

2
Kemp Land

G

Cape
Ann

60°

O

Ross Sea
Edward VII
Peninsula
Mount
Erebus
3794
TRANSANTARCTIC MOUNTAINS
South
Geomagnetic Pole
(2013)

Mac.
Robertson Land
Prince Charles
Mountains
Amery
Ice Shelf

Mackenzie
Bay

Cape
Darnley

160°
Cape Adare
Cape North
Victoria Land
Kaiser Wilhelm II Land

Princess Elizabeth
Land

H

N

Sturge Island
George V Land

Queen Mary Land

West Ice Shelf
Davis Sea

80°

180°
Balleny
Islands
Manson
Peninsula
Adélie
Land
Wilkes Land
Mount
Amundsen
1445
Mill
Island

I

South Magnetic Pole
(2013)
Cape
Morse
Vincennes
Bay

M

Dumont d'Urville
Sea

SOUTHERN
OCEAN

AUSTRALIAN-ANTARCTIC BASIN

4650

Polar Stereographic Projection

160° L 140° K Longitude 120° east of Greenwich J 100°

0	0
200	656
2000	6562
3000	9843
4000	13124
5000	16404
6000	19686
7000	22967
9000	29529
M	FT

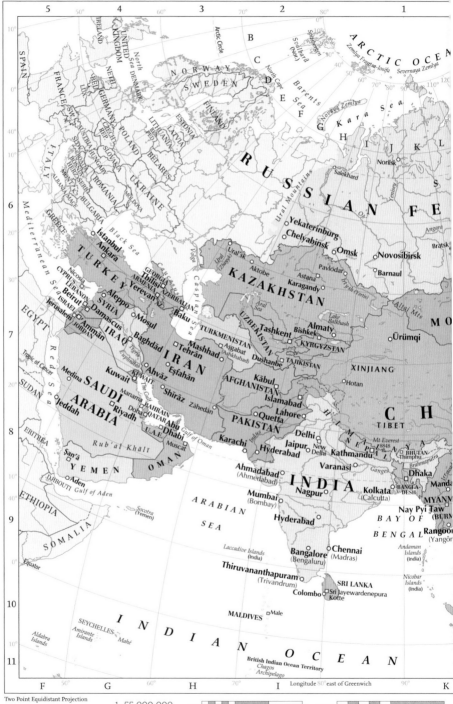

Two Point Equidistant Projection

1 : 55 000 000

KM 0   500   1000   1500   MILES 0   500   1000

F   50°   G   60°   H   70°   I   Longitude 80° east of Greenwich   90°   K

90°

Khulna

Chittagong

Cox's Bazar

Sittwe

Kyaukpyu

Thandwe

Hinthada

Bassein

Rangoon (Yangon)

MYANMAR (BURMA)

Mouths of the Irrawaddy

North Andaman

Middle Andaman

Andaman Islands (India)

South Andaman

Port Blair

Nachuge

Little Andaman

Car Nicobar

Nicobar Islands (India)

Great Nicobar

105°

CHINA

Yuxi

Kaiyuan

Gejiu

HANOI (Hanoi)

Louangphabang

Nam Dinh

Thanh Hoa

Vinh

Ha Tinh

Dong Hoi

Hue

Da Nang

Quang Ngai

THAILAND

BANGKOK (Krung Thep)

CAMBODIA

PHNOM PENH

VIETNAM

Ho Chi Minh City

B

Xiamen

Hong Kong

Gaoxiong

Hainan Dao

SOUTH CHINA SEA

MALAYSIA

KUALA LUMPUR

SINGAPORE

BORNEO

SUMATRA

JAKARTA (Java Sea)

INDONESIA

JAVA (JAWA)

INDIAN OCEAN

Cocos Islands (Australia)

Christmas Island (Australia)

5000
16404

3000
9843

2000
6562

1000
3281

500
1640

200
656

0
0

Land below sea level

200
656

4000
13124

6000
19686

M
FT

58

Albers Equal Area Conic Projection

1 : 25 000 000

Longitude 105° east of Greenwich

MILES 0   250

A   B

C
135°
D
150°
E

Tropic of Cancer

Ryukyu Islands
(Nansei-shotō)
(Japan)

IWAN

he People's Republic
f China claims Taiwan
s its 23rd Province.

Philippine
Sea

P A C I F I C

O C E A N

Northern
Mariana
Pagan
Islands
(U.S.A.)

15°

CAPITOL HILL Saipan
Tinian

Rota

HAGÅTÑA
Guam
(U.S.A.)

PHILIPPINES

Catanduanes
Legazpi
Sorsogon
Catarman
Samar
Roxas Catbalogan
Tacloban
Bacolod
Cebu Surigao
Bohol
Butuan
Cagayan de Oro
Iligan *Mindanao*
Cotabato **Davao**
mboanga Mati
General Santos

Moro
Gulf

FEDERATED STATES
OF MICRONESIA

Ulithi Fais

Yap *Sorol*
Colonia

Ngulu *Sorol*

Fauripik *C a r o l i n e*
Islands

PALAU Babeldaob
MELEKEOK

Kepulauan
Talaud
Sangir

Kepulauan
Sangire
Morotai

E a s t C a r o l i n e
Basin

Equator
0°

enanjung Manado Daruba
nahasa Tondano Tobelo
okwandang
Gorontalo Ternate *Halmahera*
epulauan Sao-Siu
ngian *Laut Maluku*
Luwuk *(Molucca Sea)* *Waigeo*
Labuna
ba Bacan Selat Dampir Kwoka Manokwari Biak
Banggai Misoöl *3000* Jazirah Numfoor *Selat Yapen*
pulauan Todeli Mangole Sorong Doberai Yapen
anggai Taliabu Dofa Sulawati Inanwatan *Teluk* Sarmi
Banggai *Kepulauan* Siula *Teluk Berau* *Cenderawasih*
Moluccas Fakfak Nabire
Namlea Piru *(Maluku)*
S *3019* Bula **A** Kaimana
Kendari Buru Ambon *Kepulauan* Adi Enarotali
Wowoni *Kepulauan* Amamapare
ka Saparua *Kepulauan* Tual Dobo Wokam
Buton *Banda* Kai Benjina
ibau *Kepulauan* Kai Keiil Besar Kobroör
Tukangbesi *L a u t B a n d a* *Kepulauan* Sia Trangan
(B a n d a S e a) Aru *Tanjung Deyong* Digul

Pelleluhu
Islands

Wuvulu
Island

Vanimo
Aitape
Jayapura
PAPUA *Puncak* Wewak
Trikora Sepik
4730

Admiralty
Islands Lorengau
Hermit Islands Manus Is
Schouten Islands

Maprik
NEW
GUINEA Mendi

Bogia

St Matthias
Group
Mussau Island

New Hanover
Umbukul Kavieng
Isabel Channel

Bismarck Archipelago *New*
Bismarck Sea Rabaul *Ireland*
2498
Manam Island Kimbe
Long Madang
Island Ulamona
Bogadjim **New Britain** Lau
Huon Kimbe
Peninsula Gasmata
Mount
PAPUA Goroka
Mt Mt Wilhelm Gazelle
Hagen *4509*
NEW GUINEA Lae Morobe

Pegunungan Van Rees
Tanjung d'Urville

Pegunungan Maoke
Puncak Jaya *Central Ra.*
5030 *4700*
Puncak Mandala

Trobriand
Islands

D'Entrecasteaux Is
Goschen Strait

Kiunga Kikori Kerema Wau
Balimo Bereina Bolubolu
Morehead Kikori **Gulf** Samarai
Merauke Daru **of Papua**
Thursday **PORT** Kwikila
Island **MORESBY** Abau Alotau

Laut *Kepulauan Barat Daya*
ores *(Sea)* Wuliaru
Sea) *Kepulauan* *Damar* Romang
Kalabahi Alor Babar Saumlakki
rantuka Huaki *Kepulauan* *Kepulauan Tanimbar*
Maliana Kawatu Selaru *Tanjung Vals*
Manatuto *Kepulauan* *Sermata*
Leti Selaru
DILI Kefamenanu
EAST TIMOR (TIMOR-LESTE)
OCUSSI *2960*
wu Manufui
Timor *Melville*
Kupang *Island*
Rote

A r a f u r a S e a

AUSTRALIA

Van Diemen Milikapiti
Bathurst Island Jabiru
Beagle Gulf Darwin
Batchelor *Arnhem*
Adelaide River Pine Creek *Land*
Croker Island Milingimbi *Gulf*
Cape Wessel Alyangula *of*
Wessel Islands *Carpentaria*
Nhulunbuy
Cape Arnhem

Prince of Wales Bamaga
Island Cape Grenville
Cape York Weipa Lockhart River

Cape Melville
Coen *Cape*
Cape York *Flattery*
Peninsula Cooktown
Laura

T i m o r S e a

1

2

15°

150°

15°

C
135°
D

0   250   500   750 KM

59

A   100°   B

Phangnga
Ban Khok Kloi   Thung   Nakhon Si Thammarat   Mui Ca Mau   Nam Căn   Đảo Côn Sơn
Thalang   Krabi Song   Khao Chum Thong   VIETNAM
Phuket   THAILAND
                      Phatthalung
         Trang   Thale Luang
                   Hat   Songkhla   S O U T H   C H I
                   Yai   Pattani
         Satun   Sadao
Andaman   Kangar Yala   Narathiwat
Sea   Langkawi   Rangae   Kota
Pulau   Kota
We   Sabang   Alor Star   Bharu
         Sungai Petani   Pasir
    Banda Aceh   Putih
Sigli   Pinang   Butterworth   Kuala Kerai
    Bireun   Lhokseumawe   George   Kuala   Kuala
Calang 2280   Takengon Peureula   Town   Taiping MALAYSIA Terengganu
                   Kuala   Tasik
Gunung Abongabong   Langsa   Kangsar Gunung Kenyir Dungun
   △2985   Pangkalansusu   Ipoh Tahan △2189
Blangkejeren   Belawan   PENINSULAR Kuala Lipis   Cukai
Gunung Leuser   Kampar Teluk Intan   MALAYSIA
   △3145   Binjai   Bagan
Tapaktuan   Datuk   KUALA   Kuantan
   Medan   Tebingtinggi   LUMPUR   Pekan
Simeulue   Pematangsiantar Nisaran Tanjungbalai Klang Temerluh
      Sidikalang   PUTRAJAYA Seremban   Padang Endau
Sinabang   Prapat Danau Bahau
         Singkil Toba Labuhanbilik Melaka Segamat   Mersing
Pulau-pulau   Balige
Baniak   Rantauprapat   Bagansiapiapi Muar   Keluang
   Sibolga   Batu Pahat
Nias   Gunungsitoli Gunungtua   Dumai Duri Bengkalis Johor Bahru
Sirombu   Padangsidimpuan   Daludalu   SINGAPORE
Telukdalam   Hutanopan   Minas   Bintan
   Natal   Airbangis   Talu Pekanbaru   Tanjungpinang
                Bangkinang
   Equator   Telo   Kampar
Tanahmasa Pulau-   Payakumbuh   Tembilahan Lingga
Tanahbala pulau Batu   Padangpanjang   Rengat Singkep
      Bukittinggi   Sijunjung   Kualatungal
Kagologolo   Padang Solok   Simpang
Siberut   Painan   Muarabungo   Batanghari Jambi
   Muarasiberut   Gunung Muaratembesi   Belinyu
Sipura   3805 Kerinci
   Kaliet   Sungaipenuh Bangko Sarolangun   Mentok   Sungailiat
Pagai   Mukomuko   Surulangun   Pangkalpinang Bangka
Utara   Pagai   Plaju   Rajik Koba
Burai Selatan   Sekayu Lubuklinggau Musi Palembang Tanjungpandan Toboali
5000   Curup Tebingtinggi Prabumulih   Kayuagung
16404   Bengkulu   Lahat
3000   Mega   Lempur   Martapura   Menggala
9843   3umo Muaradua   Kotabumi
2000   Gunung
6562   Bintuhan   Resag Metro
            2232 Kotaagung Bandar Lampung
1000
3281   2   Krui   Kotaagung
500   Enggano   Tanjung Cina
1640   Krakatau   JAKARTA
200   Tanjung Cina Selat Sunda Serang
656   Panaitan   Rangkasbitung Karawang
0   Deli   Sukabumi Bogor
0   Land below   Bandung Garut
   sea level
200
656   *I N D I A N*
4000
13124
6000
19686

M
FT

60   Albers Equal Area Conic Projection

A   Longitude 100° east of Greenwich   B

1:12 000 000   MILES 0   100   200   300

C · 120° · D

Palawan ○ Rio Tuba
○ *Bugsuk*
Balabac○
○ *Balabac*

S U L U
Balabac Strait
Banggi **S E A**
Kudat○ ○ *Cagayan de Tawi-Tawi*

Presidente Manuel A Roxas○ ○ Oroquieta○
Liloy○ ○ Ozamis○ ○Iligan
Siocon○ ○ Pagadian
Zamboanga Peninsula
Zamboanga○ ○ *Moro* Datu Piang
○Isabela *Gulf* Lebak○
○ *Basilan*

Kanibongan○ *Sulu Archipelago* ○Jolo
Kota Belud○ *Turtle Islands (Philippines)*
Gunung Kinabalu Sandakan○ Jolo○ **PHILIPPINES**
Kota ○ △ 4095
Kinabalu *Gunung Trus Madi* Tambisan○
Beaufort○ △ 2649 Lamag○ Siasi○
Labuan○ Lahad Panglima○
BANDAR SERI Tenom○ Kuamut Datu Sugala○ Tawi-Tawi
BEGAWAN Lawas○ ○ *Sibutu*
BRUNEI Tomani○ **SABAH**
Kuala Belait○ Pensiangan○
Lutong○ ○Seria *Lumbis* Semporna○
Miri○ △ Bukit Harden Tawau○
2136 Mensalong○ **C E L E B E S**

Bintulu○ Labang○ Long○ Kubuang○ Tarakan○ **S E A**
Akah
ng Igan○ Mukah○ Tanjungselor○
irik ○ ○ Belaga
Sibu○ **S A R A W A K** Tanjungredeb○
arikei Kapit○ Datadian○
○Saratok *Rajang*
○Debak Putusibau○ Sepinang○
rahan Sri Aman○ *Longwai* Tolitoli○ Kwandang○
Lubok Sangkulirang○ *Tanjung* **Semenanjung Minahasa**
Antu ○ △ 2988 *Mangkalihat* Gorontalo○
Semitau○ **B O R N E O** Bontang○ Moutong○ *Kepulauan*
au ○Sintang *Togian* *Tanjung*
Longgai○ △ 2000 Sidoan○ **Teluk** *Togian* *Pangkalsiang*
ahpinol Muaralaung○ Tenggarong○ Tomali○ **Tomini** 0°
Peguntungan Schwaner Tewah○ **Samarinda** Donggala○ *Batudaka*
atayap Rantaupanjang○ Muarateweh○ Palu○ Luwuk○
oangtiti Balikpapan○ Mapane○ Poso○ Uekuli○ Tataba○ Banggai○
Palangkaraya○ Tanahgrogot○ Babalo○ Tentena○ Kolonedale *Kepulauan*
Sampit○ Tanjung○ Mamuju○ Masamba○ Rantepang○ *Teluk Towori* *Banggai*
ngkalanbun Amuntai○ *Gandadiwata* △ 3074 Palopo○ Manui○
Kualapembuang○ Kandangan○ Somba○ Makale○ Malamala○ Wowoni○
Tanjung Banjarmasin○ Martapura○ Majene○ Polewali○ Anabanua○ Kendari○
Puting ○ Pagatan *Sebuku* Parepare○ Sengkang○ Kolaka○
Laut Watampone○ Raha○

O N E S I A Maros○ Sinjai○ *Kabaena* *Muna* Buton○ 2
A W A *Kepulauan Laut Kecil* **Makassar** *Gunung Lompobattang* Baubau○
E A) (Ujung Pandang) ○ Bulukumba○
-pulau ○ *Bawean* *Masalembu* Bontosunggu○
unjawa *Besar* *Pulau Selayar*
Tanjung Benteng○ *Batuata*
Bugel *Kepulauan Kangean* *Sabalana* *Tanahjampea* Kalao○ *Kalaotoa*
○*Pati* Tuban○ *Madura* *Kepulauan Tengah* *Kepulauan Bonerate*
Purwodadi○ Bangkalan○ Sumenep○ Arjasa○
arang Jombang○ **Surabaya** *Genteng* Raas○ **Laut Bali** *Kepulauan* **Laut Flores** *Kepulauan*
karta Pasuruan○ *Selat Madura* (*Bali Sea*) Bonerate (*Flores Sea*) *Solor*
Madiun○ **Malang** Situbondo○ **Sumbawa** Reo○ **Flores** Larantuka○
akarta *Gunung Semeru Gunung Raung* Banyuwangi○ *Gunung Tambora* △ 2821 Maumere○ Labala○
citan Ngunut○ Lumajang○ Jember○ △ 3142 Raba○ Labuhanbajo○ Bajawa○ Ende○
Barung Singaraja○ Dompu○ Ruteng○ Ende○
Denpasar Gianyar○ Mataram○ Sumbawabesar○ *Selat Sumba* **Laut Sawu**
d s **Bali** Praya○ Taliwang○ Plampang○ **Sumba** (*Savu Sea*)
Lombok *Selat Lombok* Waikabubak○ Memboro○ Waingapu○

C · 120° · D

© Collins Bartholomew Ltd

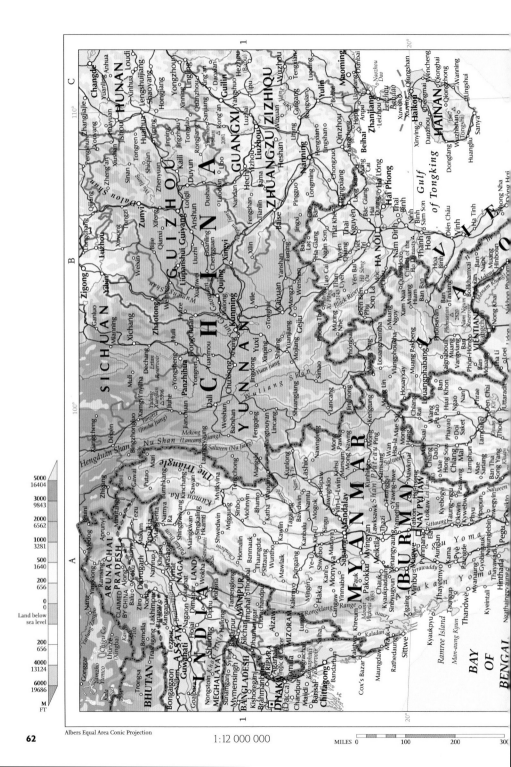

Albers Equal Area Conic Projection

1 : 12 000 000

MILES 0   100   200   300

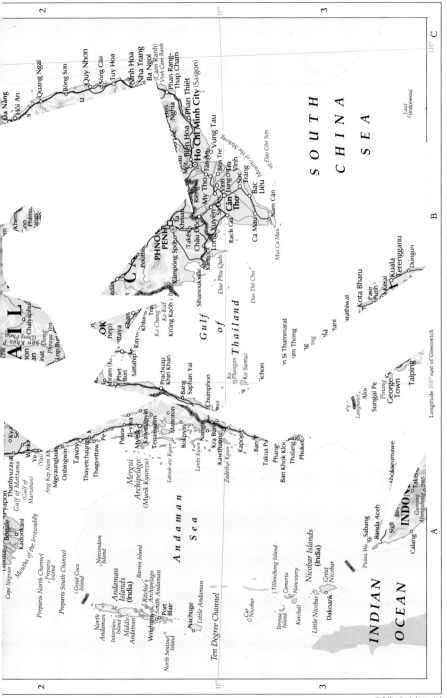

0   200   400 KM

PHILIPPINES

A 120° **B**

Luzon Strait

Dongsha
Qundao

20°

Itbayat
Batan Islands
Basco
Batan

Balintang Channel

Babuyan

Calayan
Babuyan Islands
Fuga
Camiguin

Babuyan Channel

Bangui
Laoag City
San Vicente
Aparri

PHILIPPINE SEA

Bangued
Vigan
Mount Chico
Sapocoy
Tuguegarao
Ilagan
Palanan

Tagudin
Bontoc
San Fernando
Mount Pulog
La Trinidad
Santiago
Bayombong
Dagupan
Baguio
LUZON

Lingayen
San Carlos
San Jose
Tarlac
Mount Pinatubo
Iba 1660
Cabanatuan
Gapan
Angeles
San Fernando
Olongapo
Valenzuela
Polillo Islands
Balanga
Quezon City
Pasig
MANILA
Tagaytay City
Santa Cruz
Labo
Daet
Pandan
Lubang Islands
San Pablo
Lucena
Libmanan
Catanduanes
Batangas
Lopez
Naga
Virac
Mount Halton
Calapan
Boac
Oas
Tabaco
Mayon
Mamburao
2585
Naujan
Legazpi
Sorsogon

Mindoro
Roxas
Burias
Irosin
Catarman
New
Sibuyan
Busuanga
San Jose
Romblon
Masbate
Calbayog
Calamian Group
Coron
Tablas
Sibuyan Sea
Masbate
Samar
Culion
Pandan
Roxas
Catbalogan
El Nido
Limapacan
Culasi
Visayan Sea
Tacloban
Taytay
Cuyo Islands
Panay
Pototan
Cadiz
Ormoc
Guiuan
Dalanganem Islands
San Jose de Buenavista
Iloilo
Bacolod
Cebu
Leyte
Dumaran
2450
Cebu
Dinagat
Palawan
Roxas
Negros
Talisay
Maasin
Siargao
Puerto Princesa
Cauayan
Bohol
Dapa
Apurahuan
Tanjay
Tagbilaran
Surigao
Aborlan
Bayawan
Siquijor
Mambajao
Tandag
Dumaguete
Camiguin
Mount Mantalingajan
2054
Brooke's Point
Presidente
Dipolog
Cagayan de Oro
Gingoog
Butuan
Rio Tuba
Manuel A Roxas
Oroquieta
Iligan
Malaybalay
Bislig
Bugsuk
SULU SEA
Liloy
Ozamis
MINDANAO
Balabac
Pagadian
Mount Ragang
Baganga
Balabac
Siocon
2815
Tagum
Balabac Strait
Zamboanga Peninsula
Cotabato
Mount Apo
Davao
Banggi
Zamboanga
Moro Gulf
Datu Piang
2954
Digos
Mati
Kudat
Cagayan de Tawi-Tawi
Kanibongan
Isabela
Lebak
Banga
Davao Gulf

5000
16404

3000
9843

2000
6562

1000
3281

500
1640

200
656

0
0
Land below sea level

200
656

4000
13124

6000
19686

M
FT

SOUTH CHINA SEA

Scarborough Shoal

Cordillera Central

Mindoro Strait

Palawan Passage

Sibuyan Ranges

Kota Belud
Kota Kinabalu
Gunung Kinabalu 4095
Ranau
Trus Madi 2649
Sandakan
Lamag
Tambisan
Basilan
Jolo
Jolo
Siasi
Kiamba
General Santos
Batulaki
MALAYSIA
Tenom
Kuamut
Lahad Datu
Panglima
Sugala
Tawi-Tawi
Sarangani Islands
Lawas
Tomani
SABAH
Pensiangan
Sibutu
Karakelong
Kepulauan Talaud
Lumbis
Semporna
CELEBES SEA
Pulutan
INDONESIA
Mensalong
Tawau
Kaburuang
Kubuang
Tarakan
Sangir
Tahuna

Sulu Archipelago

Turtle Islands (Philippines)

Cagayan de Tawi-Tawi

Bohol Sea

Sangir

Kepulauan Nanusa

INDONESIA

Longitude 120° east of Greenwich

A **B**

Albers Equal Area Conic Projection

1:12 000 000

KM 0
200

MILES 0
100
200

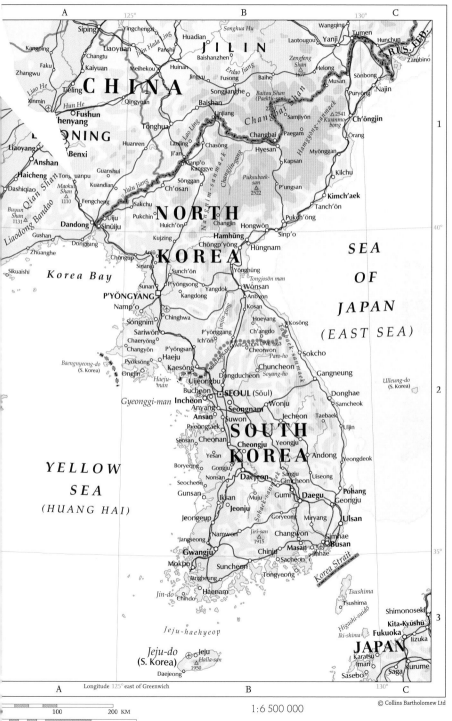

SEA

OF

JAPAN

(EAST SEA)

YELLOW

SEA

(HUANG HAI)

Korea Bay

NORTH

KOREA

CHINA

JILIN

SOUTH
KOREA

JAPAN

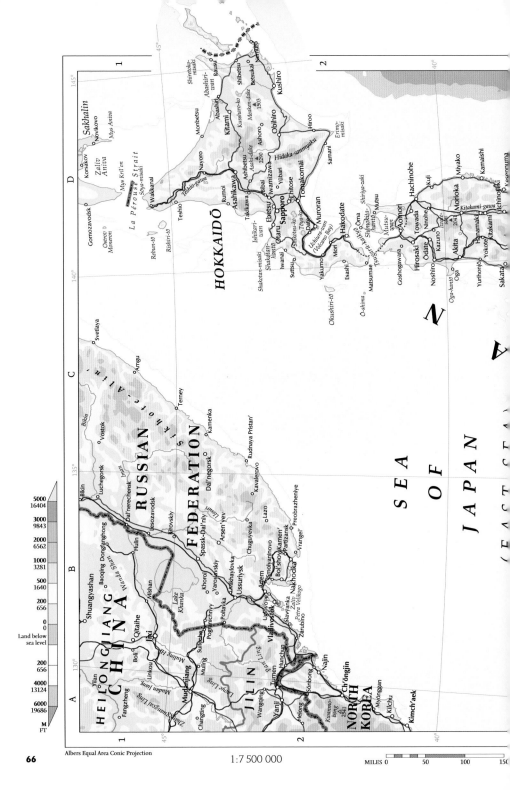

66

Albers Equal Area Conic Projection

1:7 500 000

MILES 0   50   100   150

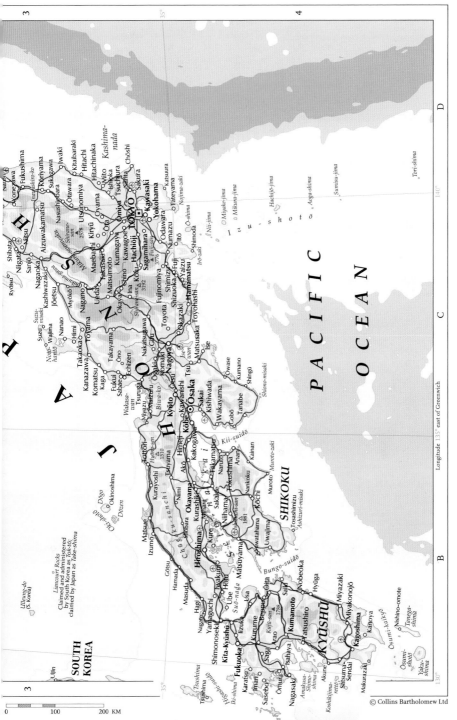

3

4

D

© Collins Bartholomew Ltd

PACIFIC

OCEAN

SHIKOKU

KYŪSHŪ

SOUTH
KOREA

Izu-shotō

Tori-shima

Sumisu-jima

Aoga-shima

Hachijō-jima

Mikura-jima

Miyake-jima

Nii-jima

O-shima

Nojima-zaki

Tateyama

Katsura

Katsuura

Shimoda

Iro-zaki

Numazu

Odawara

Yokohama

Kawasaki

TOKYO

Chōshi

Sakura

Sawara

Ishioka

Mito

Hitachinaka

Hitachi

Kitaibaraki

Iwaki

Kōriyama

Fukushima

Konosu

Kumagaya

Maebashi

Kiryū

Utsunomiya

Nasushiobara

Inawashiro-ko

Ōtawara

Sukagawa

Aizuwakamatsu

Aizu-wakamatsu

Shirane-
san
2578

Nikkō

Kirishima

Oyama

Tsuchiura

Tatsura

Hachiōji

Sagamihara

Fuji
3776

Fujinomiya

Shimizu

Yaizu

Hamamatsu

Shizuoka

Kakegawa

Toyohashi

Ōgaki

Toyota

Nagoya

Komaki

Ichinomiya

Gifu

Ōta

Ina

Nagano

Matsumoto

Takasaki

Ueda

Suwa

Shirane-san
3192

Kōfu

Kiso-sanmyaku
3190

Takayama

Toyama

Kanazawa

Komatsu

Kaga

Fukui

Echizen

Tsuruga

Obama

Maizuru

Wakasa-
wan

Biwa-ko

Kyoto

Ōtsu

Osaka

Kōbe

Himeji

Akashi

Kakogawa

Takasago

Kishiwada

Wakayama

Tanabe

Kainan

Cōbō

Shiono-misaki

Muroto-zaki

Muroto

Nankoku

Kōchi

Tokushima

Komatsushima

Naruto

Naka

Anan

Takamatsu

Marugame

Sakaide

Niihama

Imabari

Matsuyama

Uwajima

Yawatahama

Ōzu

Ishizuchi-san
1982

Tosashimizu

Ashizuri-misaki

Sukumo

Bungo-suidō

Saiki

Beppu

Ōita

Usa

Nakatsu

Usuki

Saga-
seki

Nobeoka

Hyūga

Miyazaki

Miyakonojō

Kaimon-
dake

Kagoshima

Ibusuki

Satsuma-
Sendai

Makurazaki

Osumi-
shotō

Yaku-
shima

Tanega-
shima

Nishino-omote

Kaimoya

Osumi-kaikyō

Kanoya

Kushira

Kumamoto

Yatsushiro

Amakusa-
Shimo-
shima

Minamata

Izumi

Akune

Koshikijima-
rettō

Shimabara

Isahaya

Omuta

Nagasaki

Sasebo

Sagā

Imari

Karatsu

Fukuoka

Iizuka

Kurume

Kita-Kyūshū

Tagawa

Nōgata

Yukuhashi

Shimonoseki

Ube

Onoda

Yamaguchi

Hōfu

Iwakuni

Hagi

Nagato

Masuda

Hamada

Gōtsu

Izumo

Matsue

Hiroshima

Kure

Takehara

Onomichi

Fukuyama

Kurashiki

Okayama

Niimi

Kurayoshi

Tottori

Sakaiminato

Yonago

Hyōno-sen
1510

Daisen
1711

Tsuyama

Ashō

Sōja

Akō

Chūgoku-sanchi

Kibi-kōgen

Dōgo

Oki-shotō

Nishinoshima

Okinoshima

Dōzen

Ullŭng-do
(S. Korea)

Ullŭng-do

Ullŭng

Liancourt Rocks
Claimed and administered
by South Korea as Tokdo;
claimed by Japan as Take-shima

Tsushima

Tsushima-kaikyō

Iki-shima

Higashi-suidō

Ryōtsu

Sado

Niigata

Shibata

Sanjō

Nagaoka

Kashiwazaki

Jōetsu

Myōkō

Itoigawa

Suzu

Noto-
hantō

Suzu-
misaki

Wajima

Nanao

Himi

Notojima

Kanazawa

HONSHŪ

Longitude 135° east of Greenwich

140°

130°

35°

35°

0   100   200 KM

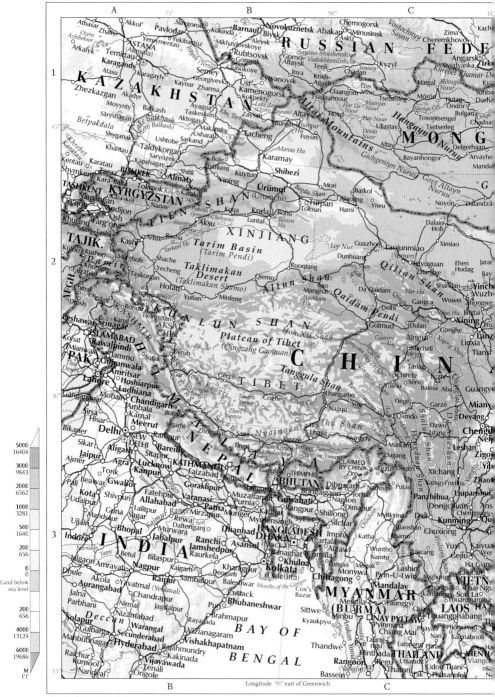

Albers Equal Area Conic Projection

1 : 25 000 000

Longitude 90° east of Greenwich

MILES 0   250   500

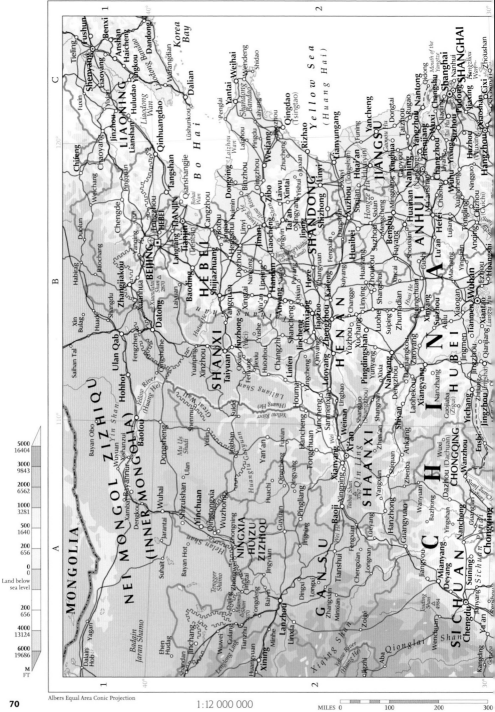

Albers Equal Area Conic Projection

1:12 000 000

MILES 0   100   200   300

5000
16404

3000
9843

2000
6562

1000
3281

500
1640

200
656

0
0
Land below
sea level

200
656

4000
13124

6000
19686

M
FT

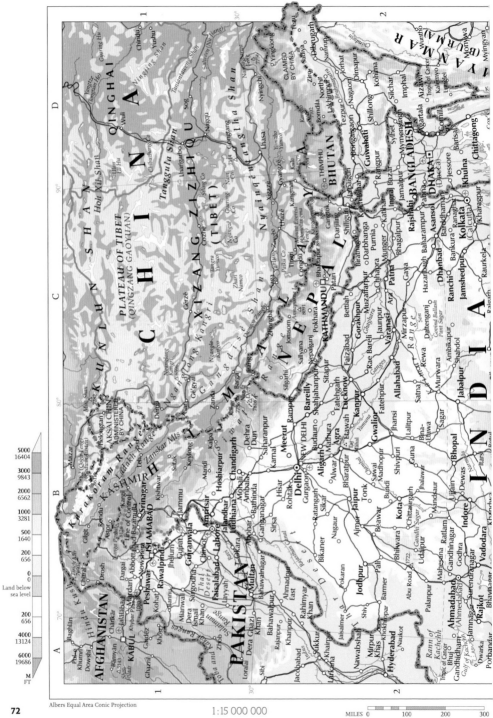

Albers Equal Area Conic Projection

1:15 000 000

MILES 0   100   200   300

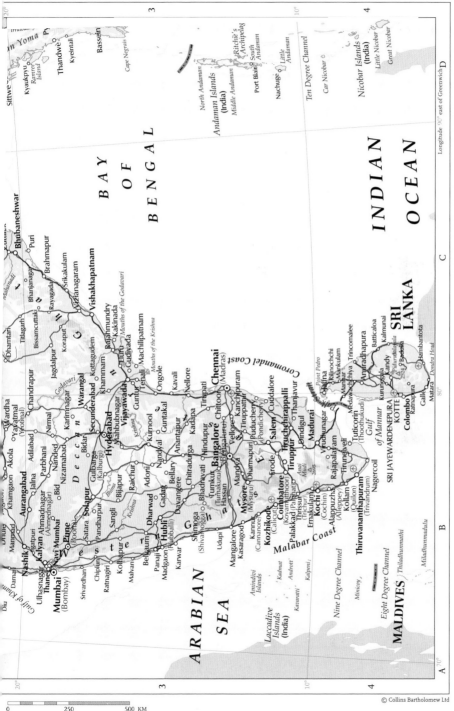

ARABIAN SEA

INDIAN OCEAN

BAY OF BENGAL

SRI LANKA

MALDIVES

Laccadive Islands (India)

Andaman Islands (India)

Nicobar Islands (India)

Ten Degree Channel

Nine Degree Channel

Eight Degree Channel

Coromandel Coast

Malabar Coast

Gulf of Mannar

0   250   500 KM

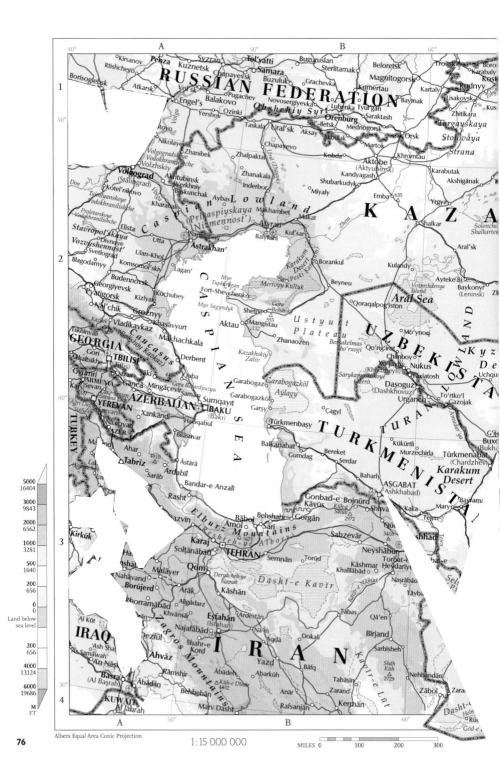

Albers Equal Area Conic Projection

1:15 000 000

MILES 0   100   200   300

Petropavlovskoye
Taiyrsha Kishkenekol'
Saumalkol' Kokshetau
Ruzayevka Makinsk
'sil' Atbasar Balkashino Akkol'
Zhaltyr Yereymentau
Derzhavinsk ASTANA (Akmola)
Ozero Osakarovka
Azhibeksor
nkel'dy Arkalyk Temirtau
Kypshak
Satpayev Karagandy
Satpaev Zhayrem Akadyr
ezkazgan Zhezkazgan
Moyynty Konyrat
Balkash
Saryshagan
Shyganak
Khantau
Shu
Moyynkum
Kentau Karatau
m Turkistan Taraz
Shymkent
TASHKENT
(Toshkent)
Jizzax
(Dzhizak)

RUSSIAN
FEDERATION

Karasuk
Ozero
Slavgorod Kulundinskoye
Kulunda Biysk Gorno-Altaysk
Aleysk
Pavlodar Mikhaylovskoye Rubtsovsk Altai Mountains
Yekibastuz Gornyak Ridder
Semey Ust'-Kamenogo Zyryanovsk
Karagayly Kaynar Zharma
Ayagoz Taskesu
Aktogay
Lepsi Usharal
Sarkand Karamay
Bole
Yining Borohoro Shan
Xinyuan

XINJIANG UYGUR ZIZHIQU
(SINKIANG)
Tarim Basin (Tarim Pendi)
Taklimakan Desert
(Taklimakan Shamo)

CHINA

KUNLUN SHAN

PLATEAU OF
TIBET
(QINGZANG GAOYUAN)

XIZANG ZIZHIQU
(TIBET)

HIMALAYA

INDIA

NEPAL

Longitude 70° east of Greenwich

0   250   500 KM

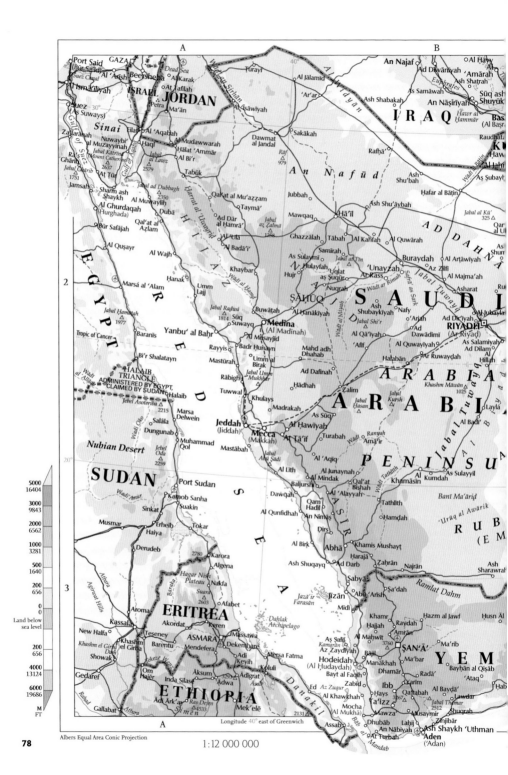

Albers Equal Area Conic Projection

1:12 000 000

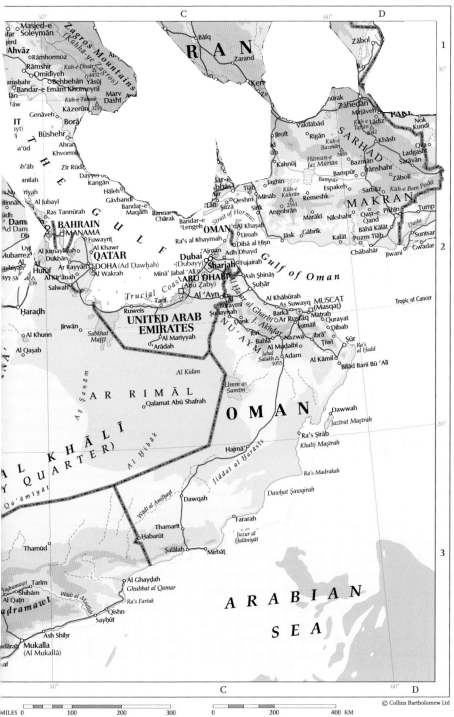

MILES 0 · 100 · 200 · 300

0 · 200 · 400 KM

© Collins Bartholomew Ltd

79

Albers Equal Area Conic Projection

1:12 000 000

Longitude 30° east of Greenwich

MILES 0   100   200   300

A   **B**

M FT	
5000	16404
3000	9843
2000	6562
1000	3281
500	1640
200	656
0	0
Land below sea level	
200	656
4000	13124
6000	19686

Conic Equidistant Projection

1:30 000 000

Longitude 75° east of Greenwich

MILES 0   200   400   600

Chamberlin Trimetric Projection

1 : 25 000 000

MILES 0   250   500

K.   KOSOVO
LIE.   LIECHTENSTEIN
MACE.   MACEDONIA
MONT.   MONTENEGRO

BARENTS SEA

Novaya Zemlya

Ostrov Kolguyev

kapp

o

Murmansk

White Sea

Archangel

2

Ural Mountains

Ob'

RUSSIAN FEDERATION

3

NLAND

Lake Onega

Syktyvkar

Lake Ladoga

Perm'

50°

elsinki

St Petersburg

Nizhniy Novgorod

Kazan'

allinn

ONIA

Yaroslavl'

Volga

ga

VIA

NIA

Moscow

Samara

Orenburg

us

Minsk

Ryazan'

Saratov

BELARUS

Voronezh

KAZAKHSTAN

4

Homyel'

Kiev

Kharkiv

Don

Volgograd

Aral Sea

UKRAINE

Dnipropetrovs'k

Donets'k

Volga

Astrakhan'

UZBEKISTAN

MOLDOVA

Chişinău

Rostov na-Donu

Dniep

Odessa

Sea of Azov

40°

Caspian Sea

MANIA

Krasnodar

Groznyy

Caucasus

TURKMENISTAN

Bucharest

El'brus 5642

GEORGIA

AZERBAIJAN

ofia

ARMENIA

LGARIA

Black Sea

ZER

İstanbul

5

essaloniki

egean Sea

TURKEY

IRAN

CE

Athens

Euphrates

Crete

CYPRUS

SYRIA

IRAQ

Tigris

30°

LEBANON

G

30°

H

40°

I

50°

J

© Collins Bartholomew Ltd

0   250   500   750 KM

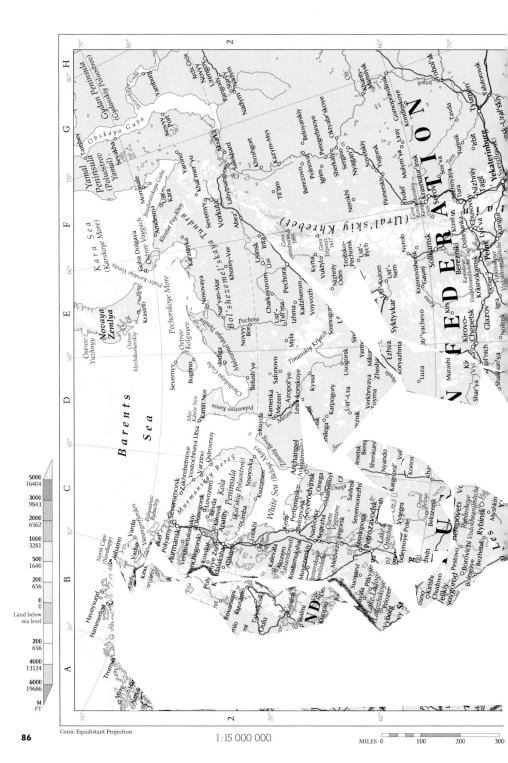

Conic Equidistant Projection

1:15 000 000

MILES 0 100 200 300

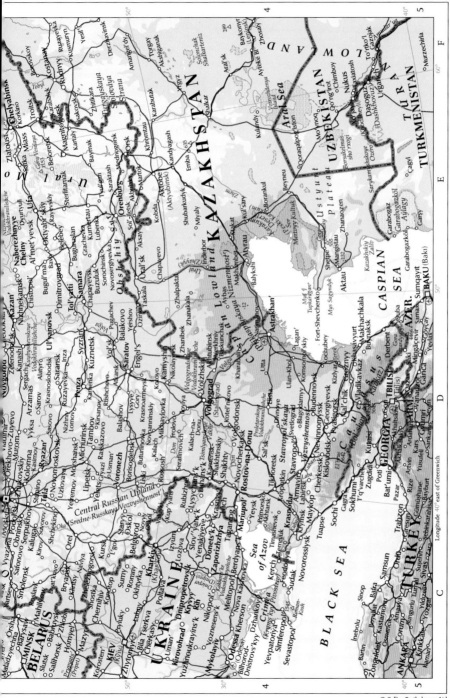

Conic Equidistant Projection

1 : 6 000 000

Longitude 25° east of Greenwich

MILES 0   50   100   150

5000
16404

3000
9843

2000
6562

1000
3281

500
1640

200
656

0
0

Land below
sea level

200
656

4000
13124

6000
19686

M
FT

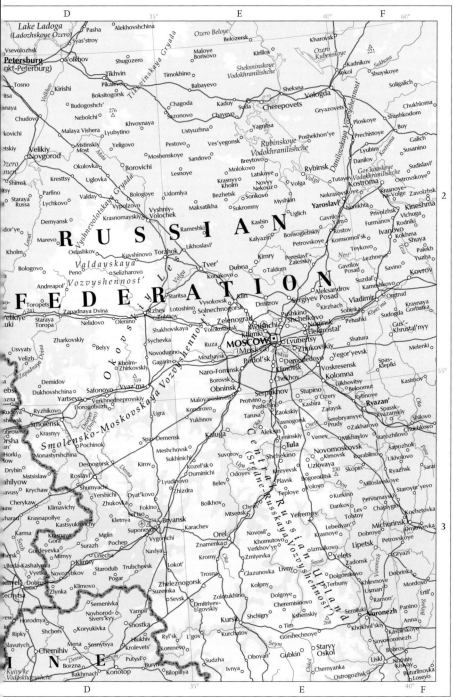

Lake Ladoga
(Ladozhskoye Ozero)
Pasha
Alekhovshchina
Ozero Beloye
Belozersk
Kharovsk
221

Vsevolozhsk
Syas'stroy
Volkhov
Shugozero
Maloye Borisovo
Kirillov
Ozero Kubenskoye
Kadnikov
Sukhona
Shuyskoye

Petersburg
(Sankt-Peterburg)
Tikhvin
Timokhino
Sheksninskoye Vodokhranilishche
Sokol
Soligalich

Tosno
Kirishi
Pikalevo
Babayevo
Sheksna
Vologda
Chukhloma

Budogoshch'
Boksitogorsk
Chagoda
Kaduy
Suda
Cherepovets
Gryazovets
Ploskoye
Shishkodom

naya
Chudovo
Nebolchi
276
Khvoynaya
Sazonovo
Chayevo
Yagnitsa
Danilov
Buy

kovichi
Malaya Vishera
Lyubytino
Ustyuzhna
Ves'yegonsk
Poshekhon'ye
Prechistoye
Galich

etskiy
Velikiy Novgorod
Mstinskiy Most
Yeligovo
Moshenskoye
Pestovo
Sandovo
Rybinskoye Vodokhranilishche
Breytovo
Lyubim
Kostroma
Susanino

Ozero men
Okulovka
Borovichi
Lesnoye
Molokovo
Latskoye
Rybinsk
Gor'kovskoye Vodokhranilishche
Sudislavl'

Shimsk
Kresttsy
Uglovka
Krasnyy Kholm
Novyy Nekouz
Volga
Tutayev
Kostroma
Ostrovskoye

sy
Staraya Russa
Parfino
Valday
Bologoye
Udomlya
Bezhetsk
Sonkovo
Myshkin
Nekrasovskoye
na-Volge
Zavolzhsk
8

Lychkovo
Vypolzovo
Vyshniy-Volochek
Maksatikha
Sukromny
Uglich
Yaroslavl'
Nerekhta
Privolzhsk
Kineshma
2

Demyansk
Krasnomayskiy
Rameshki
Kashin
Gavrilov-Yam
Furmanov
Vichuga
Rodniki

Marevo
Ostashkov
Kuvshinovo
Torzhok
Likhoslavl'
Kalyazin
Borisoglebskiy
Rostov
Komsomol'sk
Ivanovo
Kokhma
Shuya

Kholm
Bologovo
Peno
Selizharovo
Tver'
Dubna
Taldom
Pereslavl'-Zalesskiy
Nerl'
Teykovo
Savino
Palekh
Yuzha

Valdayskaya
Vozvyshennost'
Andreapol
Toropets
Zapadnaya Dvina
Staritsa
Vysokovsk
Klin
Dmitrov
Gavrilov Posad
Suzdal'
Kameshkovo
Kovrov

241
Staraya Toropa
Nelidovo
Olenino
Rzhev
Zubtsov
Lotoshino
Solnechnogorsk
Sergiyev Posad
Kirzhach
Vladimir
Orgtrud
Krasnaya Gorbatka

Zharkovskiy
Belyy
Shakhovskaya
Volokolamsk
Zelenograd
Myṭishchi
Shchelkovo
Noginsk
Petushki
Sudogda
Gus'-Khrustal'nyy

Usvyaty
Velizh
Novodugino
Sychevka
Ruza
Khimki
MOSCOW
(Moskva)
Lyubertsy
Elektrostal'
Shatura
Melenki

Kholm-Zhirkovskiy
Gagarin
Mozhaysk
Podol'sk
Zhukovskiy
Yegor'yevsk
Spas-Klepiki

Demidov
Safonovo
Vyaz'ma
319
Naro-Fominsk
Borovsk
Chekhov
Domodedovo
Klimovsk
Voskresensk
Kolomna
Kasimov

55°

ebsk
Dukhovshchina
Yartsevo
Verkhnedneprovskiy
Maloyaroslavets
Obninsk
Serpukhov
Stupino
Likhovitsy
Beloomut
Rybnoye

ozna
Ryzhikovo
Dorogobuzh
Ugra
Kondrovo
Protvino
Pushchino
Ozery
Kashira
Zarsysk
Prudy
Zakharovo
Ryazan'
Spass-Ryazanskiy
Ola

Rudnya
shevsk
Smolensk
Krasnyy
Pochinok
Spas-Demensk
Yukhnov
Tarusa
Yasnogorsk
Venev
Serebryanyye Prudy
Starozhilovo
Shilovo

browna
Orsha
Monastyrshchina
Desnogorsk
Meshchovsk
Kaluga
Aleksin
Leninskiy
Mikhaylov
Sapozhok

Horki
Drybin
Mstsislaw
Roslavl'
Kirov
Sukhinichi
Kozel'sk
Suvorov
Shchekino
Tula
Novomoskovsk
Uzlovaya
Koroblino
Ukholovo
Ryazhsk
Sarai

hilyow
Krychaw
Shumyachi
Dyat'kovo
Lyudinovo
Duminichi
Odoyev
Plavsk
Bogoroditsk
230
Skopin
Miloslavskoye
Staroyur'yevo

avusy
Cherykaw
Klimavichy
Yershichi
Zhukovka
Fokino
Belev
Teploye
Volovo
Don
Kurkino
Dankov
Pervomayskiy
Chaplygin
Kochetovka

harad'
Krasnapollye
Kletnya
Sel'tso
344
Bryansk
Bolkhov
Chern'
Mtsensk
Yefremov
Lev Tolstoy
Dobroye

Karma
Kastsyukovichy
Mglin
Suponevo
Karachev
Orel
Novosil'
Khomutovo
Verkhov'ye
Lebedyan'
Dobroye
Michurinsk
Dmitriyevka
3

Krasnaya Gora
Surazh
Pochep
Vygonichi
Znamenka
Kromy
Zmiyevka
Izmalkovo
Krasnoye
Lipetsk
Petrovskoye
Gryazi

Gordeyevka
Mirnyy
Unecha
Navlya
Lokot'
Livny
Yelets
Zadonsk
Dobrinka

ersk
Buda-Kashalyova
Klintsy
Starodub
Trubchevsk
Trosna
Glazunovka
Dolgorukovo
Terbuny
Khlevnoye
Usman'
Mordovo
Ertil'

myel'
Dobrush
Vyetka
Novozybkov
Pogar
Zheleznogorsk
Suzemka
Kolpny
Dolgoye
Cheremisinovo
Ramon'
Panino
Anna

echytsa
Zlynka
Klimovo
Semenivka
Sevsk
Dmitriyev-L'govskiy
Zolotukhino
Shchigry
Kshenskiy
Semiluki
Voronezh

ew
Ripky
Horodnya
Novhorod-Sivers'kyy
Yampil'
Shostka
Kursk
Tim
268
Khokhol'skiy
Kaspiiskoye
Novovoronezh

Slavutych
Chernihiv
Shchors
Koryukivka
Hlukhiv
Krolevets
Ryl'sk
L'gov
Kurchatov
Gorshechnoye
Staryy Oskol
Liski
Nizhniy Kislyay
Buturlinovka

INE
Mena
Sosnytsya
Korenevo
Oboyan'
Gubkin
Chernyanka
Bobrov

Kyïvs'ke Vodoskhovyshche
Borzna
Bakhmach
Putyvl'
Buryn
Sudzha
Ivnya
Ostrogozhsk
Loseyo

A 25° B 30°

Ciechanów   Zambrów   Vawkavysk   Zel'va   Baranavichy   o Nyasvizh   Asipovichy   Babruysk
Ostrów   Białystok   Svislach   Slonim   Lyakhavichy   Klyetsk   Kapyl'   Starraya   Rahachow   Chachersk
Mazowiecka   Hajnówka   Ivatsevichy   Hantsavichy   Salihorsk   Darrohi   Lyuban'   Hlusk   Zhlobin
Legionowo   Wyszków   Pruzhany   Byarroza   Tsyelyakhany   Mal'kavichy   Aktsyabrski   Svyetlahorsk
Vistula   WARSAW   Kamyanyets   Zhabinka   Kobryn   Drahichyn   Lyunynyets   Dzyatlavichy   Zhytkavichy   Kalinkavichy   Vasilyevichy   Homy
(Wisła)   (Warszawa)   Brest   Malaryta   Ivanava   Pina   Pinsk   Davyd-Haradok   Prypyats' (Pripet)   Mazyr   Rech
Minsk   Siedlce   Biała   Łuków   Podlaska   Parczew   Ratne   Lyubeshiv   Stolin   Lyel'chytsy   Yel'sk   Khoyniki   Lovew
Mazowiecki   POLAND   Radom   Pionki   Puławy   Lubartów   Lyubomil'   Kamin'-   Dubrovytsya   Moarshes   Narowlya   Brahin
Skarzysko-   Lublin   Chełm   Kovel'   Kashyrs'kyy   Sarny   Klesiv   Rokytne   Ovruch   Polis'ke   Chornob
Kamienna   Starachowice   Ostrowiec   Manevychi   Volodymyrets'   Narodychi   Uzh   Vodoskhovyshche
Lysica   Świętokrzyski   Krasnystaw   Turiys'ke   Berezne   Luhyny   Korosten'   Ivankiv
Sandomierz   Zamość   Volodymyr-   Kivertsi   Kostopil'   Yemil'chyne   Koze
Staszów   Tarnobrzeg   Stalowa   Volyns'kyy   Rivne   Korets'   Volodars'k-   Malyn   Borodyanka
Vistula   Wola   Biłgoraj   Novovolyns'k   Zdolbuniv   Novohrad-   Volyns'kyy   Chernyakhiv   Radomyshl'   Irpin'   Vyshho
(Wisła)   Mielec   Tomaszów   Sokal   Horokhiv   Mlyniv   Dubno   Ostroh   Volyns'kyy   Makariv   KIEV
Debica   Lubelski   Chervonohrad   Radyvyliv   Slavuta   Baranivka   Korostyshiv   (Kyiv)   Bor
Tarnów   Rzeszów   Przeworsk   Kum''yanka-   Brody   Krerenets'   Shepetivka   Polonne   Zhytomyr   Vasyl'kiv   Fastiv
Jasło   Jarosław   Zhovkva   Buz'ka   Pochayiv   Izyaslav   Chudniv   Berdychiv   Andrushivka   Bila   Tserkva   Kaharlyk
Gorlice   Krosno   Yavoriv   Zolochiv   Zbarazh   Starokostyantyniv   Kozyatyn   Skvyra   Myron
Przemyśl   Lviv   Peremyshlyany   Ternopil'   Krasyliv   Pivdennyy Buh   Tarashch
Sanok   Horodok   (L'vov)   Mykolayiv   Berezhany   Volochys'k   Khmel'nyts'kyy   Lityn   Vinnytsya   Tetiyiv   Zhashkiv
Ustrzyki   Sambir   Drohobych   Zhydachiv   Stryy   Kalush   Buchach   Horodok   Bar   Zhmerynka   Illintsi   Zvenyhorod
Dolne   Borysław   Turka   Dolyna   Ivano-   Terebovlya   Dunayivtsi   Nemyriv   Haysyn   Monastyryshche
Prešov   Humenné   Mizhhir''ya   Frankivs'k   Nadvirna   Horodenka   Borshchiv   Kam''yanets'-   Sharhorod   Khrystynivka   Tal'n
SLOVAKIA   Uzhhorod   Svalyava   Kolomyya   Zalishchyky   Podil's'kyy   Mohyliv-   Kryzhopil'   Teplyk   Uman'
Košice   Michalovce   Mukacheve   Verkhovyna   Khotyn   Sokyryany   Podil's'kyy   Yampil'   Bershad'   Ul'yanov
Trebišov   Berehove   Khust   Chernivtsi   Briceni   Ocnita   Kodyma   Pervomays'k
Sárospatak   Vynohradiv   Hora Hoverla   Storozhynets'   Edinet   Soroca   Balta   Lyubashivka
Szerencs   Kisvárda   Rakhiv   2061   Putyla   Dorohoi   Drochia   Floresti   Ananyiv
Nyíregyháza   Sighetu   Viseu   Rădăuți   Botoșani   Bălți   Rîbnița   Kotovs'k
HUNGARY   Marmatiei   de Sus   Suceava   Fălesti   Dubăsari   Shyryaye
Hajdúböszörmény   Satu   Borsa   Fălticeni   Hârlău   Ungheni   CHISINAU   Berez
Debrecen   Carei   Mare   Pietrosa   Vatra Dornei   Pascani   Targu   Mărulul   (Kishinev)   Tiraspol
Hajdúszoboszló   Acăs   Baia   2305   Frumos   Magura   Hincesti   ROzdil'na   Kominternivs'k
Săcueni   Șimleu   Mare   Pietrosu   Piatra   Iași   Dispoteni   Tighina   Căuseni   Bilyayivka   Ode
Oradea   Silvaniei   Zalău   Dej   2100   Bistrita   Neamt   Roman   Buhusi   Huși   Leova   Comrat   Illichivs'k
Salonta   Aleșd   Gherla   Reghin   Pașcani   Bacău   Vaslui   Bârlad   Comrat   Dnistrovs'kyy
Crișul Alb   Ștei   Cluj-   Turda   Bucin   Gheorgheni   Moinesti   Comănesti   Târgu   Ciadir-Lunga   Tarutyne   Bilhorod-
Sântana   Napoca   Ocna Mureș   Vârful   1273   Harghita-Mădăraș   Onesti   Ocna   Adjud   Tecuci   Cahul   Sarata   Dnistrovs
Lipova   Mureșul   Aiud   Târnăveni   1800   Miercurea-   Comănesti   Mărăsesti   Artsyz
ROMANIA   Alba   Sighisoara   Ciuc   Focsani   Ivesti   Bolhrad   Tatarbunary
Deva   Iulia   Mediaș   Agnita   Sfântu   Vulcanesti   Suvorove
Hunedoara   Sebeș   Sibiu   Gheorghe   Ramnicu   Galați   Reni   Kiliya
Lugoj   Orăstie   Cisnădie   Moldoveanu   Brașov   Vârful Ciucas   Sărat   Brăila   Vylkove
Caransebes   Hateg   2544   1954   Râmnicu   Buzău   Măcin   Izmail
Boca   Petroșani   Râsnov   Vârful Omu   Câmpulung   Vâlcea   Tulcea   Babadag
Resita   Vârful   2505   Câmpina   Urlati   Danube   Delta
Vârful   Parângul Mare   Câmpulung   Buzău   Ianca   Lacul Razim
Svinecu   2519   Târgu Jiu   Motru   Drăgăsani   Pitesti   Târgoviște   Ploiesti   Insurătei   Harsova   Baia
Mare   1224   Orșova   Strehaia   Costesti   Voluntari   Slobozia   Tândărei   Cernavoda
SERBIA   Drobeta-   Zagotin   Scornicesti   Gaesti   Titu   Buftea   Urziceni   Medgidia   Constanta
Turnu Severin   Craiova   Bals   Slatina   Bolintin-Vale   BUCHAREST   Fetesti   Năvodari
Bor   Vidin   Băilești   Caracal   Drăgănesti-Olt   Rosiori   Videle   (București)   Basarabi
Zaječar   Polina   Calafat   Danube (Dunărea)   Dăbuleni   de Vede   Oltenita   Silistra   Calarasi
Lom   Alexandria   Giurgiu   BULGARIA

A   Longitude 25° east of Greenwich   B   30°

90   Conic Equidistant Projection   1:6 000 000   MILES 0   50   100   150

5000 / 16404
3000 / 9843
2000 / 6562
1000 / 3281
500 / 1640
200 / 656
0 / 0
Land below sea level
200 / 656
4000 / 13124
6000 / 19686
M / FT

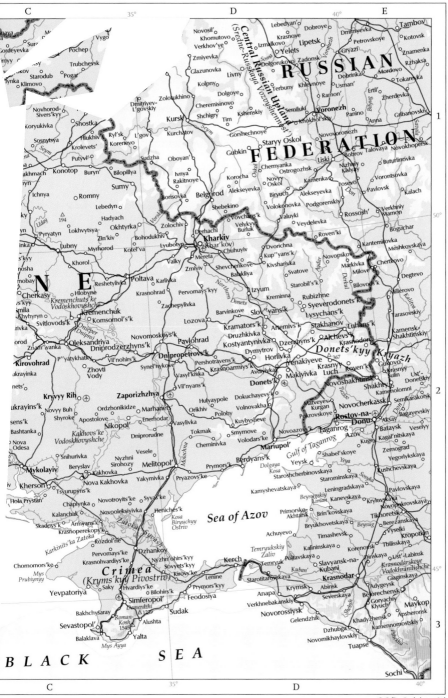

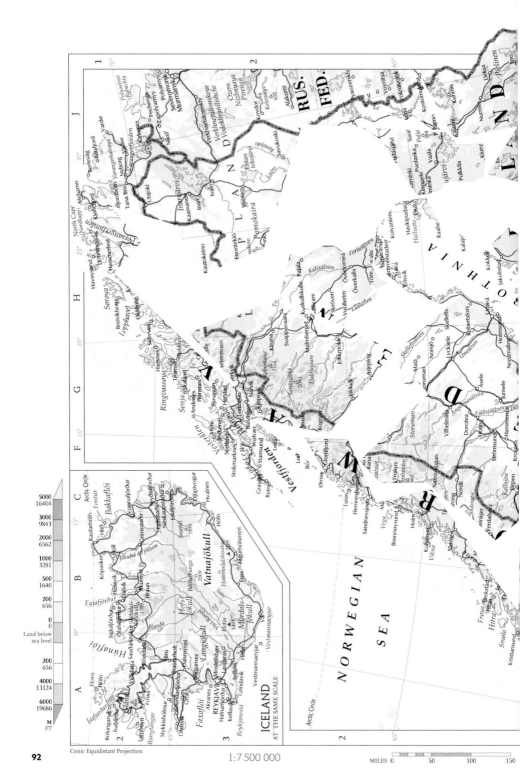

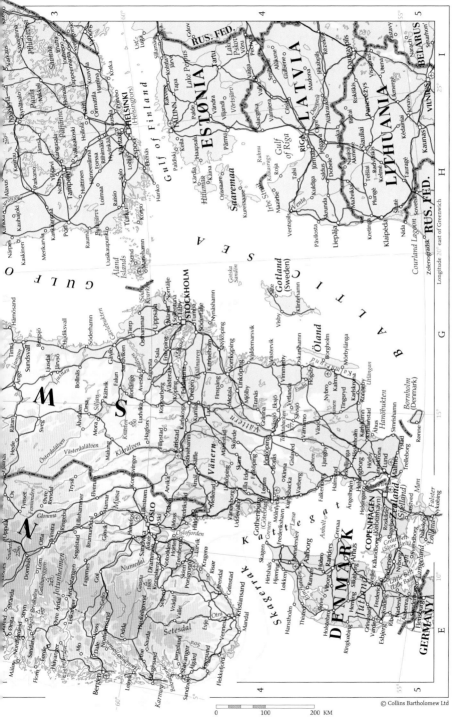

Conic Equidistant Projection

1:6 000 000

MILES 0   50   100   150

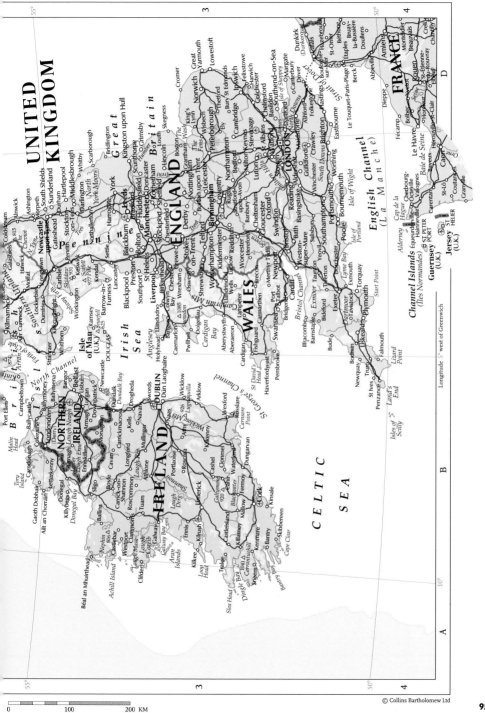

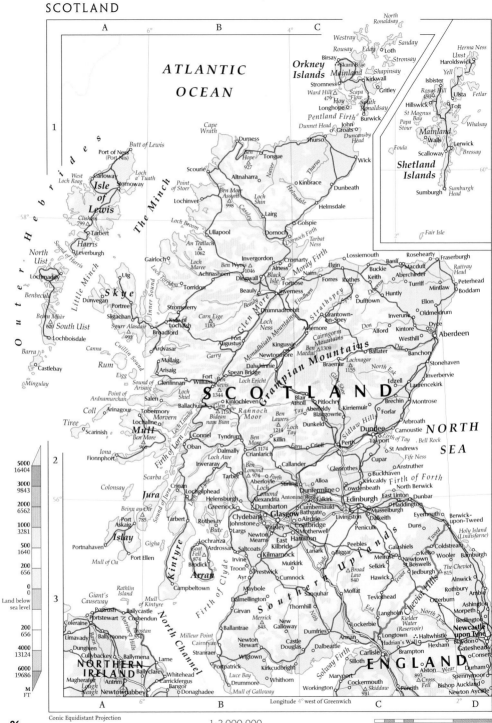

SCOTLAND

ATLANTIC

OCEAN

Orkney
Islands Mainland

North
Ronaldsay

Westray
Rousay Eday Loch Sanday
Birsay Stronsay
Skara Brae Shapinsay
Kirkwall
Stromness
Ward Hill Scapa
479 Flow South
Hoy Ronaldsay
Longhope Burwick
Pentland Firth
John
Dunnet Head o'Groats Duncansby
Head

Herma Ness
Unst
Haroldswick
Yell
Isbister
Rona Hill Ulsta Fetlar
450
Hillswick Toft
St Magnus
Bay Papa Whalsay
Stour
Foula Mainland
Walls
Lerwick
Shetland Scalloway Bressay

Shetland
Islands 60°

Sumburgh Sumburgh
Head

Fair Isle

Cape
Wrath
Durness Thurso Wick

Butt of Lewis
Port of Ness
(Port Nis)

Ben
Hope Tongue
927
Scourie
Altnaharra Kinbrace
Point of
Stoer Ben More Loch Dunbeath
Assynt Shin
Lochinver 998 Lairg Helmsdale

West
Loch Roag Carloway
Stornoway
Isle
of
Lewis

Loch
a' Tuath

Loch Broom

An Teallach
1062

Ullapool

Golspie

Dornoch Dornoch Firth Tarbat
Ness

Clisham
799 Tarbert

Harris
Leverburgh

North
Uist

Lochmaddy

Benbecula

Benn Mhòr
620 South Uist

Lochboisdale

Barra

Castlebay

Mingulay

Sound of Harris

Gairloch
Loch
Maree
Achnasheen

Torridon

Strome ferry

Invergordon
Ben Wyvis
1046
Dingwall Nairn
Black
Beauly Isle
Inverness Forres

Cromarty
Moray Firth
Alness
Fortrose Elgin

Lossiemouth Rosehearty
Banff Fraserburgh
Buckie Macduff
Rattray
Keith Head
Rothes Turriff Peterhead
Mintlaw
Dufftown Boddam
Huntly Ellon

Rosehearty

Uig

Dunvegan
Portree
Sligachan

Sgurr Alasdair
993

Ardvasar

Skye

Inner Sound

Carn Eige
1183
Kyle of
Lochalsh
Broadford

Drumnadrochit
Loch
Ness
Fort
Augustus

Aviemore
Grantown-
on-Spey
Cairngorm
Mountains
Ben 1309
Macdui

Inverurie Oldmeldrum
Alford Kintore Dyce
Westhill Aberdeen
Don

Clishain Mallaig
Garry
Arisaig

Rum

Eigg

Canna

Point of
Ardnamurchan

Coll

Tiree

Scarinish

Iona
Fionnphort

Mull

Ben More
966

Dunvegan

Cuillin Sound

Sound of
Arisaig

Fort
William
Ben
Nevis
1344
Glen Coe
Loch Linnhe

Glenfinnan
Loch
Shiel
Ballachulish

Spean Bridge
Dalwhinnie

Loch Ericht

Blair
Atholl

Kinlochleven
Bidean
nam Bian
Ben
Lawers
1214

Rannoch
Moor

Newtonmore
Kingussie

Monadhliath Mountains
Findhorn

Strathspey
Deveron

Grampian Mountains

Braemar
Ballater
1155
Lochnagar

Banchory
Stonehaven

Inverbervie
Laurencekirk

SCOTLAND

Pitlochry
Aberfeldy
Blairgowrie
Kirriemuir

Dee

North Esk

Brechin
Forfar

Montrose

Arbroath

Edzell

Sidlaw Hills
Carnoustie

NORTH

Tobermory
Morvern
Lochaline

Arinagour

Scarba

Colonsay

Jura

Beinn an Oir
785

Port
Askaig

Islay

Portnahaven

Mull of Oa

Morvern

Sound of Jura

Crinan
Lochgilphead

Tarbert

Rothesay
Bute

Gigha

Connel
Tyndrum
Dalmally
Loch Awe
Oban

Inveraray

Crianlarich

Ben
More
1174
Killin

Dunkeld

Loch
Tay

Ben
Lomond
974

Callander
Stirling

Loch
Lomond
Alexandria
Antonine Wall
Greenock Dumbarton

Helensburgh

Clydebank **Glasgow**
Johnstone Airdrie
Paisley Coatbridge
Largs Motherwell
Newton
Mearns East
Hamilton
Kilbride

Perth
Crieff

Cupar

Anstruther

Glenrothes

Alloa
Kirkcaldy
Cowdenbeath
Dunfermline
Falkirk
Cumbernauld
Bathgate
Livingston
Dalkeith

St Andrews
Fife Ness

Bell Rock

SEA

Firth of Tay
Taybort
Dundee

Firth of Forth
Buckhaven
North Berwick
Dunbar
Edinburgh East Linton
Haddington
Musselburgh Eyemouth Berwick-
upon-Tweed

Coldstream

Duns Holy Island
(Lindisfarne)

Penicuik

Lanark
Peebles

Galashiels
Melrose
Biggar St Boswells
Selkirk
Hawick

Kelso
Wooler The Cheviot
Jedburgh 815
The Cheviot

Alnwick

Amble

Rothbury

Ashington
Morpeth
Bedlington

Portnahaven

Kintyre

Goat
Fell
874

Campbeltown

Mull of
Kintyre

Rathlin
Island

Giant's
Causeway

Portrush
Coleraine
Portstewart Coshendun
Ballymoney Antrim Hills
Limavady
Dungiven
Cullybackey Ballymena
Magherafelt
Lough Antrim
Neagh Newtownabbey

Arran
Brodick

Ardrossan
Saltcoats
Kilmarnock
Irvine
Troon
Prestwick
Ayr

Maybole
Dalmellington

Girvan

Ballantrae

Ballycastle

Ballycastle

North Channel

Milleur Point

Cairnryan

Stranraer

Larne

Whitehead
Carrickfergus
Bangor
Donaghadee

Firth of Clyde

Kilmarnock
Muirkirk
Cumnock
Sanquhar

Moffat

Thornhill

New
Galloway

Newton
Stewart

Wigtown

Mull of Galloway

Merrick
843

Dalmellington

Teviothead

Southern Uplands

Broad
Law
840

Langholm

Dumfries
Lockerbie
Annan

Castle
Douglas
Kirkcudbright

Luce Bay Whithorn

Dalbeattie

Portpatrick

Solway Firth
Silloth
Maryport

Workington

Cockermouth
Skiddaw
931

Carlisle
Brampton

Tweed

Tyne

Jed
Coquet

The Cheviot
815

Cheviot Hills

North Tyne

Kielder
Water
(Reservoir)

Longtown
Haltwhistle
Hadrian's Wall Hexham

Esk

Nith

ENGLAND

Penrith

Alston

Wear

Cross
Fell 893

Bishop Auckland

Outerburn

Kielder
Water

Newcastle
upon Tyne

Blaydon Gateshead
Consett

Durham
Spennymoor
Newton Aycliffe

NORTHERN
IRELAND

Atlantic

0 6° 4° Longitude 4° west of Greenwich C 2° D

5000	
16404	
3000	
9843	
2000	
6562	
1000	
3281	
500	
1640	
200	
656	
0	
0	
Land below	
sea level	
200	
656	
4000	
13124	
6000	
19686	
M	
FT	

Conic Equidistant Projection

1:3 000 000

MILES 0 20 40 60

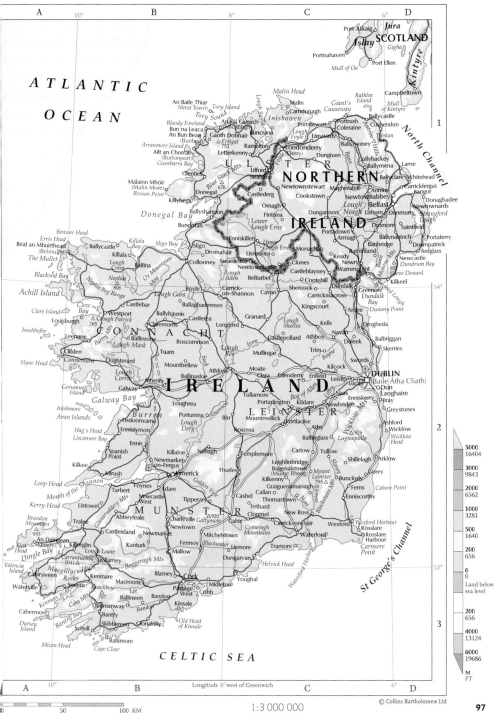

			M	FT
			5000	16404
			3000	9843
			2000	6562
			1000	3281
			500	1640
			200	656
			0	0
			Land below sea level	
			200	656
			4000	13124
			6000	19686

1 : 3 000 000

0   50   100 KM

© Collins Bartholomew Ltd

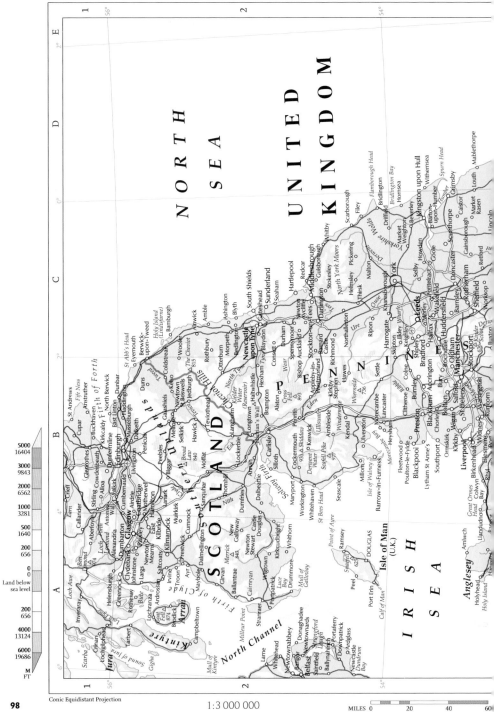

Conic Equidistant Projection

1:3 000 000

MILES 0   20   40   60

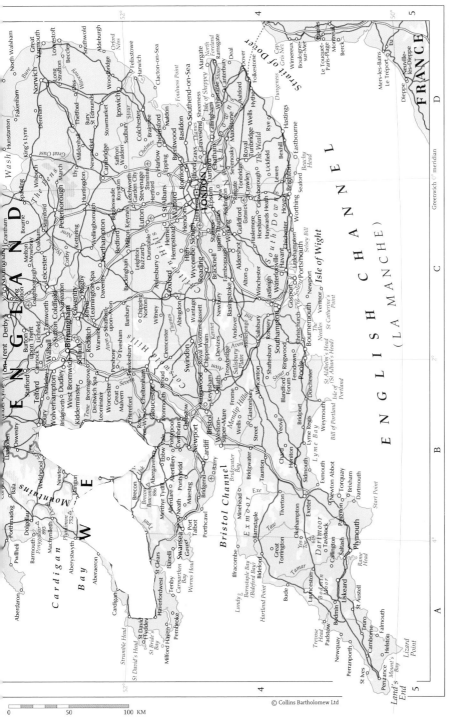

0   50   100 KM

NORTH

SEA

West Frisian Islands

East Frisian Islands Spiekeroog
Norderney Langeoog Langeoog Langer
Juist Norderney
Borkum Norden Wittr
Terschelling Schiermonnikoog Borkum Westerholt
West- Ameland OSTFRIESLA
Terschelling Hollum Uithuizen Hinte Aurich Wies
Oost- Ferwert Eenrum Delfzijl Emden (Os
Vlieland Dokkum Bedum Appingedam Leer Weste
Vlieland Burdaard Kollum Oenkerk Groningen Winschoten (Ostfriesland)
Harlingen Leeuwarden Eemskanaal Strucklingen
Texel Franeker Hoogezand- (Saterland)
Den Burg Witmarsum Reduzum Drachten Sappemeer Papenbu
Bolsward Veendam
Marchep Sneek Heerenveen Assen Frieso
Den Helder Sloten Walchum
Wieringerwerf Stadskanaal Walchum
IJsselmeer Wolvega Beilen Emmen Sustrum
Schagen Steenwijk Haren (Ems)
Nieuwe-Niedorp Creil Hoogeveen Löninger
Heerhugoward Enkhuizen Emmeloord Meppel
Bergen Urk Coevorden Meppen
Alkmaar Kraggenburg Groß-Hesepe
Castricum Berkhout Hoorn Hardenberg Lingen
Beverwijk Purmerend Markermeer Kampen Zwolle Ommen (Ems) Fürste
IJmuiden Lelystad Dronten Kloosterhaar
Zandvoort Zaandam Heerde Raalte Vriezenveen Nordhorn
Haarlem AMSTERDAM Vriezenveen Almelo
Hillegom Edmeden Harderwijk Nijverdal Oldenzaal Rheine
Noordwijk-Binnen Amstelveen Naarden Torenberg Deventer Borne Gronau
Katwijk aan Zee Leiden Nijkerk 107 Hengelo (Westfalen) Ibben
Alphen aan den Rijn Hilversum Amersfoort Apeldoorn Enschede Emsdetten
THE HAGUE Maarssen Utrecht Barneveld Zutphen Eibergen Steinfurt Greven
('s-Gravenhage) Waddinxveen Veenendaal Ede Doesburg Hoog- Ahaus
Hook of Holland Delft Gouda Nieuwegein Wageningen Keppel Winterswijk Havixbeck
(Hoek van Holland) Rotterdam Schoonhoven Neder Rijn Arnhem Coesfeld Mü
Vlaardingen Capelle aan Culemborg Andelst Zevenaar Doetinchem Borken
den IJssel Spijkenisse Gorinchem Nijmegen Velen
Helleveetsluis Oss Wichen Bocholt Asche
Scharendijke Dordrecht Maas Kleve MÜNSTERLAND Al
Burgh Middelharnis Hertogenbosch Wesel Dorsten Marl Hamm
Haamstede Zierikzee Oosterhout Waalwijk Uden Goch Keveler Gelsenkirchen Recklinghauser
Westkapelle Middelburg Roosendaal Erp St Anthonis Dinslaken Herne Lüne
Koudekerke Halsteren Boxtel Venray Duisburg Bottrop Dortn
Knokke- Bergen op Zoom Breda Tilburg Helmond Moers Essen Bochum
Zeebrugge Heist Hoogstraten Eindhoven Deurne Mülheim an der Ruhr Mayen
Blankenberge Sluis Veldhoven Asten Krefeld Hattingen Iserl
Ostend Breskens Zandvliet Brecht Turnhout Weert Venlo Viersen Wuppertal
(Oostende) Meetkerke St-Laureins Kapellen Schilde Westmalle Arendonk Kessel Mönchengladbach Düsseldorf Lüdenscheid
Nieuwpoort Maldegem Schilde Lille Lommel Roermond Wegberg Neuss Hilden Remscheid
Brugge Eeklo Antwerp Geel Bocholt Dormagen Solingen Attend
Veurne Zedelgem (Bruges) Wommelgem Lier Maaseik Hückelhoven Grevenbroich Leverkusen Gummersb
Diksmuide Torhout Wingene Antwerpen Hechtel Bergisch Gladbach
Roeselare Tielt (Anvers) Willebroek Beringen Sittard Heinsberg Cologne (Köln)
Ieper Deinze Ghent Dendermonde Mechelen Aarschot Diest Geleen Heerlen Kerkrade (Erft) Troisdorf
Kortrijk (Gent) Wichelen Vilvoorde Hasselt Heerlen Alsdorf Siegburg
Menen Zulte Aalst Scharbeek Leuven Tienen Maastricht Stolberg Bonn Bad
Mouscron Oudenaarde Anderlecht BRUSSELS Mechelen Aachen Düren Kreuzau Königswinter Altenki
Roubaix Ronse Uccle (Bruxelles) Tongeren Raeren Zülpich Bad Neuenahr- Meckenheim (Wester)
Lille Tournai Halle Waterloo Borgloon Eupen (Rheinland) Ahrweiler Rhein We
Villeneuve Pecq Ath Nivelles Fleurus Liège Verviers Mechernich Blankenheim Neuwied
d'Ascq Lens Soignies Eghezée Seraing Huy Kallo Dahlem Adenau Bad
Douai Hornu Mons Charleroi Namur Spa Malmédy Mayen Koblenz Monta
Valenciennes Framer Thuin Châtelet Ciney Durbuy Vielsalm St-Vith Hillesheim Lahr
Maubeuge Montignies- Assesse Marche- Dahlem Gerolstein Koblenz
Cambrai Aulnoye- le-Tilleul Hastière- en-Famenne St-Hubert Prüm Cochem
Aymeries Beaumont Lavaux Dinant Rochefort La Roche- Thommen Daun Emmelsh
Bohain-en- Avesnes- Philippeville en-Ardenne Houffalize Arzfeld Manderscheid Blankenrath Simmer
Vermandois sur-Helpe Couvin Beauraing St-Hubert Neuerburg Wittlich (Hu
Péronne La Capelle Hirson Momignies Chimay Bastogne Bitburg Bernkastel-Kues Bad
St-Quentin Guise Oise Rocroi Monthermé Libramont Wiltz Echternach Saltmal am Mül
Vervins Bogny-sur-Meuse Vresse Paliseul Bouillon Neufchâteau Ettelbruck Bad Kreuznac
Chauny Marle Serre Charleville- Sedan Carignan Mersch Trier Morbach
Tergnier Montcornet Rozoy- Mézières Bouillon Arlon Konz Idar-Oberstein
Laon sur-Serre Signy- 316 Virton LUXEMBOURG Reinsfeld Erbeskopf Idar-Oberstein
Noyon l'Abbaye Omont Mouzon Remich 818 Sohenheim Donner
Rethel Stenay Longuyon Saarburg St Wendel
Attichy Soissons Vouziers Aubange Esch-sur- Mettlach St Wendel Wolfferstin
Courmelles Aisne Bethony Dun-sur- Spincourt Alzette Merzig Neunkirchen Kaisersla
Villers- Meuse Haybes Orange Thionville Saarlouis Homburg
Cotterêts Fismes Tinqueux Reims Consenvoye Rombas

FRANCE

5000 16404
3000 9843
2000 6562
1000 3281
500 1640
200 656
0 0
Land below sea level
200 656
4000 13124
6000 19686
M FT

1

52°

2

50°

3

A 4° B 6° C

Longitude 6° east of Greenwich

Conic Equidistant Projection
1:3 000 000
MILES 0   20   40   60

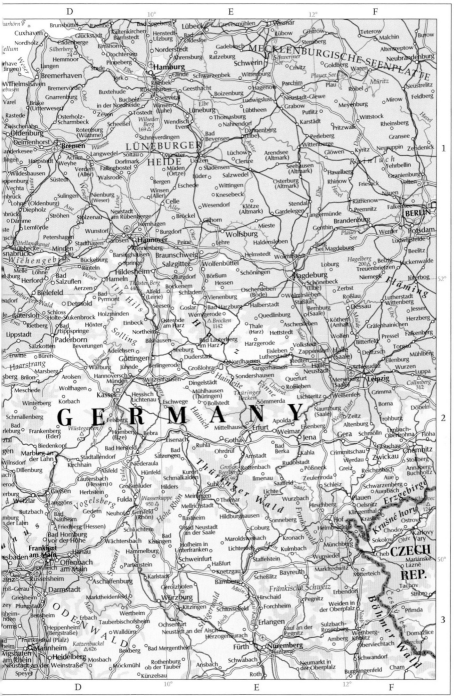

Conic Equidistant Projection

1:6 000 000

MILES 0   50   100   150

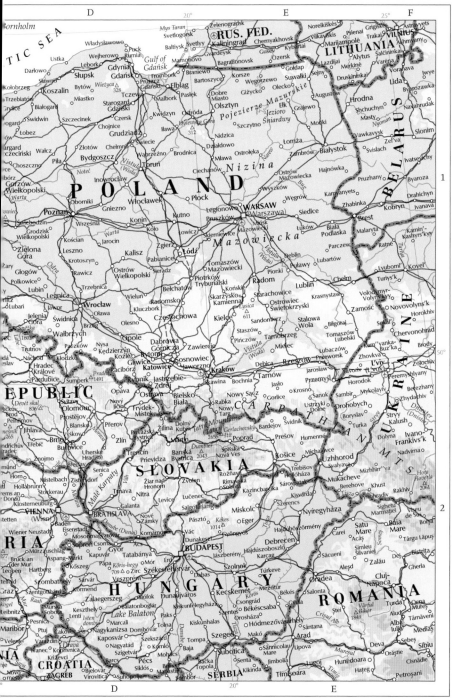

© Collins Bartholomew Ltd

0   100   200 KM

Conic Equidistant Projection

1 : 6 000 000

MILES 0   50   100   150

Greenwich 0° meridian

A		5°	B		0°

A   **B**

Bristol Channel
Ilfracombe
Barnstaple
Bideford
Bude
Tiverton
Exmoor
Taunton
Salisbury
Weston-super-Mare
Basingstoke
Reading
Bath
Winchester
Aldershot
Guildford
Maidstone
Gillingham
Dartford
Isle of Sheppey
Margat
Canterbury
Dover
LONDON
UNITED KINGDOM
Newquay
Tavistock
Liskeard
Dartmoor
Exeter
Dorchester
Yeovil
Southampton
Worthing
Brighton
Crawley
Folkestone
Ashford
St Ives
Truro
Bodmin
Plymouth
Torquay
Lyme Bay
Poole
Bournemouth
Portsmouth
Eastbourne
Hastings
Boulogne-sur-Mer
Penzance
Falmouth
Start Point
Isle of Wight
Le Touquet-Paris-Plage
Strait of Dover
Land's End
Lizard Point
Isles of Scilly
English Channel
(La Manche)
Berck
Dieppe
Abbe
Alderney
Cap de la Hague
Fécamp
Neufchâtel-en-Bray
PIC
Guernsey (U.K.)
ST PETER PORT
Equeurdreville-Hainneville
Tourlaville
Bourg-Octeville
Valogne
Le Havre
Yvetot
Bolb
Deauville
St-Étienne-du-Rouvray
Ch
Channel Islands
(Îles Normandes)
Jersey (U.K.)
ST HELIER
Baie de Seine
H
Ponto
Évreux
Jois-le
Station
Boulogne-Billancourt
Versailles
P
Roscoff
Lannion
Golfe de St-Malo
St-Lô
Coutan
Dreux
L'Aigle
Rambouillet
Île d'Ouessant
Lesneven
Guipavas
Cap Fréhel
Granville
Mantes-la-Jolie
Nogent-le-B
Morlaix
Dinard
St-Malo
Alençon
Chartres
Plouzane
Châteaulin
Douarnenez
Montas
Guingamp
Pointe du Raz
Quimper
Quimp
Concarneau
Lon
Ploemeur
Auray
Carnac
Île de Groix
Quiberon
Guérande
St-Na
Belle-Île
La Baule-Escoublac
Pornic
Nantes
Ver
Noirmoutier-en-l'Île
Île de Noirmoutier
Challan
St-Jean-de-Monts
Île d'Yeu
La Roche-sur-Yon
Les Sables-d'Olonne
Fou
Talmont-St-Hilaire
Île de Ré
B A Y
Pointe de Chassiron
St-Pierre-d'Oléron
Roch
O F
Pointe de la Coubre
Sain
B I S C A Y
Pointe de Grave
Soulac-sur-Mer
Roy
Gulf of Gascony
Montenr
Pauillac
Ambares
et-Lagrave
Mérignac
Pessac
Arcachon
La Teste-de-Buch
Gradignan
Bazas
Gironde
Montignac
Le Bugue
Bergerac
Gujan-
Mestras
Langon
Marmande
Villeneuve-sur-Lot
Souillac
la-Canéda
Figeac
Gourdon
Sarl
Mimizan
Lot
Nérac
Agen
Castsalou
Moiss
Limoges
Plateau
Limousin
AU
Aubus
Bourganeuf
Aha
Guéret
Argenton-sur-Creuse
Monthé
Vatan
Châteauroux
Argent
lou
Tulle
Egle
Pleau
Au
ahors
Re
franche-Aveyre
ouergue
Carmaux
Alb
Vale
Vitoria-Ga
Miranda de Ebro
Briviese
Osorno
Sahagún
P
Mar Cantábrico
Cabo de Peñas
Avilés
Gijón (Xixón)
Santander
Algorta
San Sebastián
Biarritz
Barak
Baroc
Torrecerredo
2648
Torelavega
Oviedo
Mieres
del Camin
Pena Ubiña
Cabanaquinta
Salas
AS
Cord
Peña
2417
Villablino
San Andrés
de Barado
Guardo
Aguilar
de Campo
Soustons
Aire-sur-l'Adour
Condom
Lectoure
Castelsarras
Grenade
M
auban
Toulouse
Colomiers
Auch
Union
Puylaurens
C
Gaillac
Grenade
Carmaux
Mimizan
Maubourguet
Muret
Tarbes
Pamiers
Limoux
Quillan
Carcas
Mazai
ORRA
RA
des Pr
Elea de los Caballeros
Alfaro
Tudela
Huesca
Graus
Tremp
La Seu d'Urgell
Berga
Ripoll
Medina
Roseco
encia
Lerma
Sierra de la Demanda
Arnedo
Caparroso

5000	16404
3000	9843
2000	6562
1000	3281
500	1640
200	656
0	0
Land below sea level	
200	656
4000	13124
6000	19686
M	FT

50°

45°

1

2

3

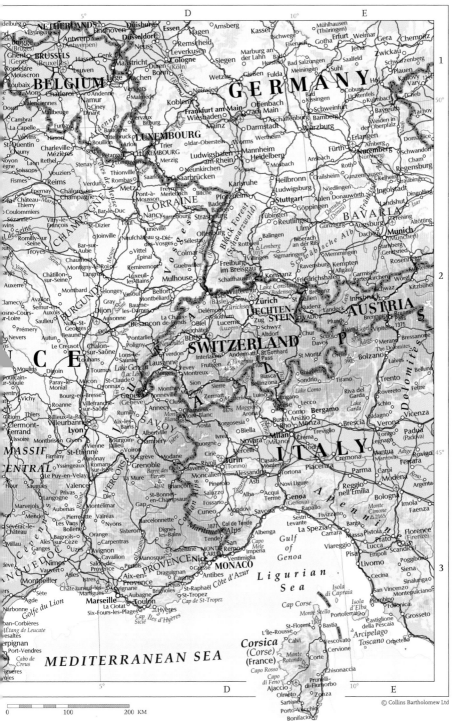

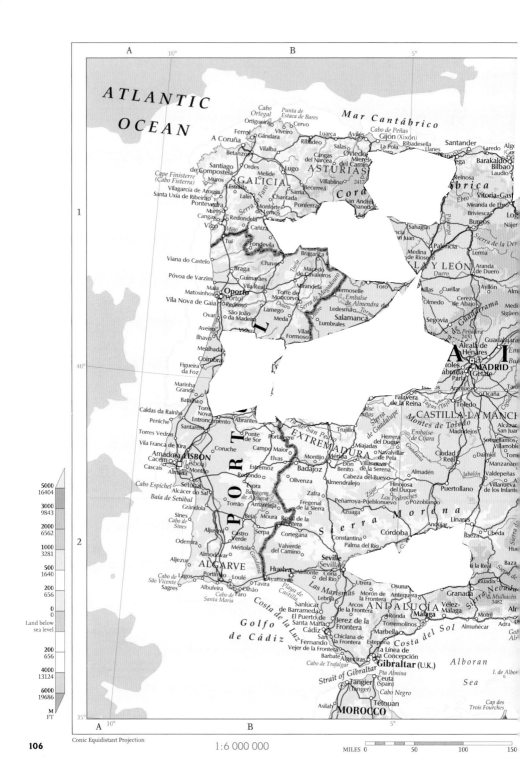

Conic Equidistant Projection

1:6 000 000

MILES 0   50   100   150

D
E

0°
5°

Arcachon
Gradignan
Garonne
Figeac
Marvejols
Mende
Pierrelatte
Valréas
Teste-de-Buch
Gujan-
Langon
Mestras
Marmande
Cahors
Villefranche-
de-Rouergue
Rodez
Espalion
Les Vans
Onyons
Digne-
les-Bains
Gulf
Bazas
Villeneuve-sur-Lot
Lot
Aveyron
Florac
Bagnols-
sur-Cèze
Bollène
Sisteron
of
Mimizan
Labouheyre
Castetjaloux
Agen
Moissac
Sévérac-
le-Château
Millau
Ales
Uzès
Orange
Carpentras
scony
Morcenx
Nérac
Roquefort
Castelsarrasin
Montauban
Carmaux
Tarn
Ganges
Nîmes
Avignon
Cavaillon
Manosque
Lectoure
Condom
Gaillac
Albi
Lodève
Salon-de-
Provence
Pertuis
Verdon
Draguignan
Mont-de-Marsan
FRANCE
Vauvert
Arles
Aix-en-
Provence
Fréjus
Soustons
Tartas
Adour
Colomiers
Toulouse
Castres
LANGUEDOC
Montpellier
Istres
Avignane
Brignoles
Dax
Aire-sur-
l'Adour
Auch
Cugnaux
Puylaurens
Sète
Châteauneuf-les-
Martigues
Aubagne
St-Tropez
Biarritz
Bayonne
Orthez
Maubourguet
Muret
Mazamet
Béziers
Marseille
La Ciotat
Toulon
Hyères
Irun
St-Jean-de-Luz
Gave de Pau
Pau
Tarbes
Carcassonne
Agde
Golfe du Lion
Six-Fours-les-Plages
Cap
Sicié
stián
Billère
Oloron-
Ste-Marie
Lourdes
Soulom
St-Gaudens
Pamiers
Narbonne
ostia)
Bagnères-
de-Luchon
Foix
Quillan
Limoux
Durban-Corbières
Étang de Leucate
Etxarri-Aranatz
PYRÉNÉES
Rivesaltes
ANDORRA
Perpignan
Pamplona
Aragón
Jaca
Monte
Perdido
Aneto
3404
ANDORRA
LA VELLA
Prades
Port-Vendres
afalla
3348
Les Escaldes
Céret
Cabo
de Creus
1
NAVARRA
Arguis
La Seu
d'Urgell
Figueres
Cap
de Begur
Costa Brava
borra
Sádaba
Ejea de los
Caballeros
Huesca
Graus
Tremp
Berga
Ripoll
Olot
Banyoles
Torroella de Montgrí
Alfaro
Tudela
Barbastro
Monzón
Torelló
Salt
Vic
Girona
Tarazona
Binéfar
CATALUÑA
Palamós
2516
Alagón
Zaragoza
Tàrrega
Igualada
Manresa
Blanes
ARAGÓN
Quinto
Fraga
Lleida
Martorell
Sabadell
Mataró
Costa
latayud
Cariñena
Escatrón
Caspe
Vall
Santa Coloma
de Gramenet
Daroca
Alcañiz
Gandesa
Reus
Barcelona
El Prat de Llobregat
dinaceta
Calamocha
Tortosa
Costa Dorada
Vilanova i la Geltrú
Molina
de Aragón
Monreal del Campo
Amposta
Golf de Sant Jordi
ordesilos
Perales
del Alfambra
Morella
Sant Carles
de la Ràpita
Teruel
Peñarroya
Vinaròs
Minorca
(Menorca)
2019
Torreblanca
Punta
Nati
Ciutadella
Es Mercadal
40°
Serranía
de Cuenca
Sarrión
L'Alcora
Castellón
de la
Plana
Majorca
(Mallorca)
Cap de
Formentor
Pollença
Alcúdia
Maó
Santa Cruz
de Moya
La Vall d'Uixó
Borriana
Sagunto
Sóller
Sa Pobla
Cap des Freu
inbalse
Alarcón
Utiel
Llíria
Sa Dragonera
Calvià
Sa Cabaneta
Manacor
tilla
Minglanilla
Manises
Burjassot
Palma de
Mallorca
Felanitx
a Roda
Riu Xúquer
Requena
Valencia
Golfo
Cap des Salines
Torrent
Catarroja
Sueca
de Valencia
Ibiza
(Eivissa)
Illa de
Cabrera
Albacete
Carcaixent
Cullera
Sant Joan de Labritja
BALEARIC ISLANDS
Almansa
Gandia
Oliva
Sant Antoni
de Portmany
Santa Eulalia del Río
(ISLAS BALEARES)
caraz
Yecla
Villena
Ibi
Alcoy-Alcoi
Dénia
Ontinyent
Cabo de
la Nao
Ibiza (Eivissa)
Sant Francesc
de Formentera
(Spain)
gura
Hellín
Jumilla
Novelda
Benidorm
Villajoyosa
Altea
Formentera
2
avaca
Cieza
Crevillent
Elda
Elche
Elx
MEDITERRANEAN SEA
ral
Alcantarilla
Segura
Orihuela
Torrevieja
Murcia
Lorca
Mazarrón
Cabo de Palos
Cartagena
Huércal-
Overa
Golfo de Mazarrón
Águilas
MEDITERRANEAN SEA
vera
Aïn
Dellys
Jijel
Cabo
de Gata
ALGIERS
(Alger)
Taya
Boumerdès
Bejaïa
Ténès
Djebel
Bissa
Tipasa
Kolea
Larba
Tizi
Ouzou
Bouira
Bougaa
Sétif
Sidi
Ali
Gouraya
Blida
Médéa
Berrouaghia
Bordj Bou
Arréridj
1157
Aïn Defla
Miliana
Sour el
Ghozlane
Aïn
Azel
Mostaganem
Oued Chlef
Chlef
Khemis
Miliana
Sidi
Aïssa
M'Sila
Cap Carbon
Aïn
Tédelès
Relizane
Bordj Bounaama
Ksar el
Boukhari
Oran
Oued
Tlélat
Zemmora
Barika
ALGERIA
Beni Saf
Aïn
Temouchent
Sig
Mohammadia
Mascara
Tissemsilt
Mahdia
Tiaret
Zenzach
Bou Saâda
M'Doukal

0   100   200 KM

© Collins Bartholomew Ltd

107

A

P

Merano
3736
Ortles Bolzano Dolom d'Ampezzo Cortina
3905 Adige Tarvisio
Tolmezzo Gemona del Friuli 2864 L
Chiavenna Tirano Maniago Cividale Tolmin SL
Bonneville Cluses Marigny Matterhorn Bellinzona Lake Como Sondrio Vittorio del Friuli
Rumilly Chamonix Verbania Lugano Riva del Veneto Udine LJUBLJAN
Annecy 4810 Mont-Blanc 4477 Locarno Garda Pordenone Gorizia Logatec
Ax-les-Bains Albertville Aosta Maggiore Varese Lake Portogruaro Monfalcone
Voiron Chambéry Borgosesia Arona Como Lecco Bergamo Garda Trieste
1 Novara Biella Busto Arsizio Rho Monza Brescia Venice Gulf of
Cuorgnè Ivrea Milan Treviglio Manerbio Venice Gulf of
Grenoble Modane Cirie Vercelli (Milano) Crema ua Laguna Veneta Venice Koper
45° Barre des Sorlex Rivoli Turin Po Vigevano Pavia Cremona dova Chioggia Poreč Pazin Rijeł
Écrins 4102 Glavero Moncalieri (Torino) Casale Monferrato Mantua Rovigo Rovinj Istria
La Mure Gap en-Champsaur Pinerolo Asti Tortona Piacenza Mantova Po Porto Tolle Pula
Barcelonnette Saluzzo Alba Alessandria Parma Adigo Portomaggiore Rt Kamenjak Cr
Fossano Acqui Terme Novi Ligure Genoa Parr Reno Argent Comacchio Veli Lo
Sisteron Digne-les-Bains Cuneo Mondovi Savona Genova Bologna Ravenna Loš
Manosque Castellane Tende Col de Tende 1871 Rapallo Sestri Levante Forli Cesenatico
Verdon Grasse MONTE-CARLO San Remo Albenga Gulf of Genoa La Spezia Rimini D
Draguignan Ventimiglia Capo Mele Imperia Pesaro R
Brignoles Fréjus Antibes Côte d'Azur Viareggio SAN Fano
Toulon St-Raphaël Nice MONACO Pisa MARINO Senigallia Ancona
Hyères St-Tropez Cannes Ligurian Livorno Sepolcro sepolcro Ancona
Cap Îles d'Hyères Cap de St-Tropez Sea Civitanova
Sicié Sea Isola di Capraia Piom bbio Perù Marche
2 Monte Isola gno San Bene del Tron
L'île-Rousse St-Florent 1307 Portoferraio d'Elba Folo Giuliar
Bastia Castiglione Teramo
Corsica Calvi Piargsa della Pescaia Penne Pes
(Corse) Vescovato Arcipelago Orbetello L'Aquila Ch
(France) Monte Cervione Toscano Monte Amaro
Rotondo Corte Isola di Montecristo Tarquinia Avezzano 2793 V
Capo Rosso 2622 Civitavecchia Tivoli
Capo di Feno Ghisonaccia VATICAN CIT ROME Sora Tive
Ajaccio Prunelli-di-Fiumorbo (Roma) Velletri
Olmeto Zonza Pomezia Aprilia Frosinone Campol
Sartène Punta d'Ovace Porto-Vecchio Anzio Sezze Cassino Vena
Capo Pertusato Bonifacio Sabaudia Latina Fondi Sessa Aurunc
Strait of Bonifacio Gaeta Cas
Punta Caprara Arzachena La Maddalena Golfo di Gaeta Naples
Isola Asinara Golfo dell' Capo Ferro Isole Ponziane Pozzuoli (Napoli)
Porto Torres Asinara Punta Olbia Isola d'Ischia Sorrento
Sassari Balestrieri Budoni Isola di Capri
Alghero Oschiri Capo Comino
Sardinia Ploaghe Budusò Siniscola
(Sardegna) Bonorva Nuoro Orosei
(Italy) Macomer Golfo di Orosei
Abbasanta Punta La Capo di Monte Santu
Oristano Marmora Tortoli TYRRHENIAN
40° Capo della Frasca Laconi 1834 SEA
Mandas Tertenia
Guspini San Gavino Monreale
Monte Linas Serramanna Villaputzu
Iglesias 1236 Assemini
Portoscuso Quartu Sant'Elena
Isola di San Pietro Punta Maxia Cagliari Capo Carbonara
Sant'Antioco 1017 Golfo di
Isola di Sant'Antioco Pula Cagliari Iso
Isola di Ustica Lip
3 Isola Fili
MEDITER San Bene del Tron
Sicily
(Sicilia)
Capo San Vito
Monte Sparagio Partinico Palermo
Trapani Rocca Cefalu
La Galite Alcamo Busambra Termini Imerese
Marsala 1613
Mazara del Vallo Partanna Leonforte
Collo Cap de Fer Chetaibi Menzel Bourguiba Bizerte Capo Granitola Castelvetrano Caltanissetta
Skikda Nefza Rass Jebel Sciacca Canicatti
Annaba El Kala Mateur Agrigento Caltagi
Azzaba El Hadjar El Tarf Tabarka Golfe de Licata Niscemi
ALGERIA TUNISIA Tunis Gela
M Jedeida Cap Golfo di
FT Bon
A Longitude 10° east of Greenwich B

Conic Equidistant Projection
1:6 000 000
MILES 0   50   100   150

5000 16404
3000 9843
2000 6562
1000 3281
500 1640
200 656
0 0
Land below sea level
200 656
4000 13124
6000 19686
M
FT

Conic Equidistant Projection

1:6 000 000

MILES 0   50   100   150

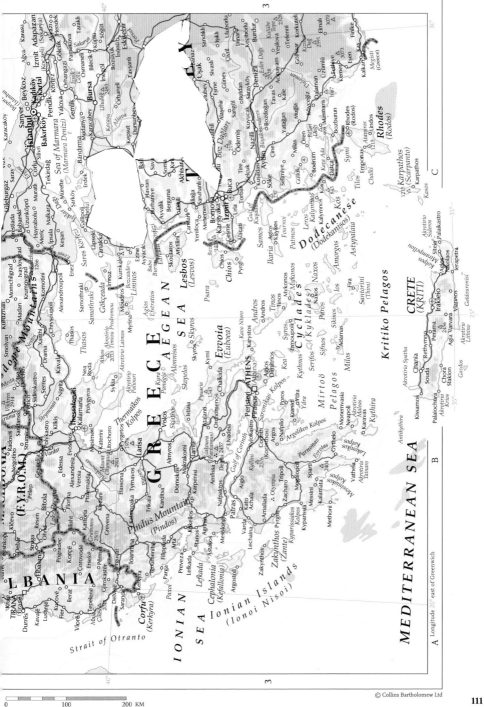

0   100   200 KM

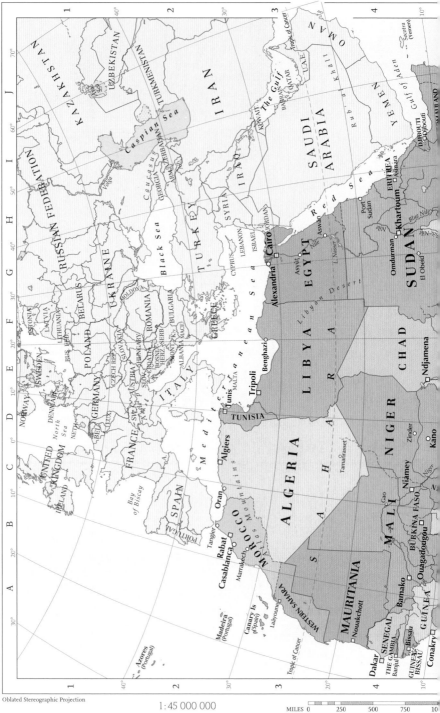

Oblated Stereographic Projection

1 : 45 000 000

MILES 0   250   500   750   10

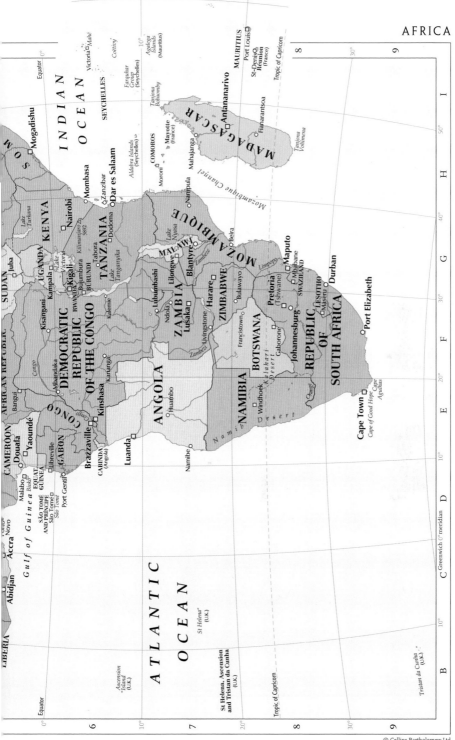

INDIAN OCEAN

ATLANTIC OCEAN

Equator

MAURITIUS
Port Louis
St-Denis
Réunion
(France)
Tropic of Capricorn

Aigalega
Islands
(Mauritius)

Victoria Mahé
Coëtivy
SEYCHELLES

Farquhar
Group
(Seychelles)

Tanjona
Bobaomby
Antananarivo
MADAGASCAR
Fianarantsoa
Tanjona
Vohimena

Mahajanga

Mayotte
(France)
COMOROS
Moroni

Aldabra Islands
(Seychelles)

Mogadishu

MOS...

Mombasa
Zanzibar
Dar es Salaam

Nairobi
KENYA
Lake
Turkana
Juba
SUDAN
UGANDA
Kampala
Kisangani
Lake
Victoria
Kigali
RWANDA
Bujumbura
BURUNDI
Kilimanjaro
5892
Dodoma
Tabora
TANZANIA
Lake
Tanganyika
Kalemie

Nampula
Lake
Nyasa
MALAWI
Lilongwe
Blantyre
MOZAMBIQUE
Beira
Mozambique Channel

CENTRAL AFRICAN REPUBLIC
Bangui
CAMEROON
Douala
Yaoundé
Malabo
Bioko
EQUAT.
GUINEA
SÃO TOMÉ
AND PRÍNCIPE
São Tomé
Libreville
GABON
Port Gentil
CONGO
Brazzaville
Mbandaka
Congo
DEMOCRATIC
REPUBLIC
OF THE CONGO
Kinshasa
CABINDA
(Angola)
Luanda

Lubumbashi
Kananga
Ndola
Lusaka
ZAMBIA
Livingstone
Zambezi

Harare
ZIMBABWE
Bulawayo
Limpopo

Maputo
Mbabane
SWAZILAND
Pretoria
(Tshwane)
Johannesburg
LESOTHO
Maseru
Durban

ANGOLA
Huambo

Namibe

NAMIBIA
Windhoek
Namib
Desert
Kalahari
Desert
Francistown
BOTSWANA
Gaborone
Orange

REPUBLIC
OF
SOUTH AFRICA
Cape Town
Cape of Good Hope
Cape
Agulhas
Port Elizabeth

LIBERIA
Abidjan
Accra
Gulf of Guinea
Novo

St Helena
(U.K.)

Ascension
Island
(U.K.)

St Helena, Ascension
and Tristan da Cunha
(U.K.)

Tropic of Capricorn

Tristan da Cunha
(U.K.)

Equator
Greenwich 0° meridian

0° 10° A B C D E F G H I
10° 20° 30°

6
7
8
9

0   500   1000   1500 KM

© Collins Bartholomew Ltd

113

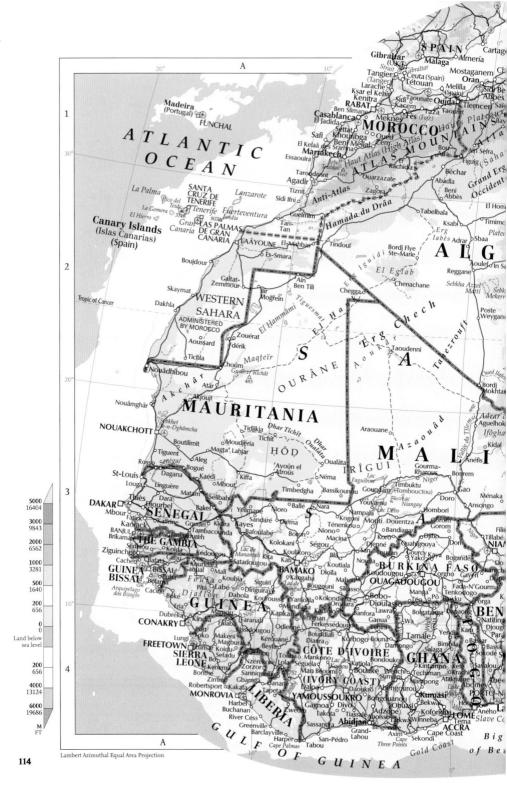

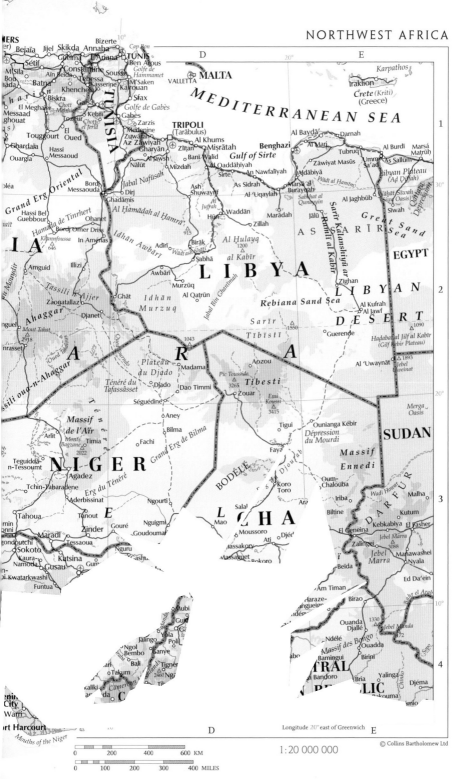

ERS
er) Bejaïa Jijel Skikda Annaba Bizerte
Sétif   Guelma L'Ariana **TUNIS** Cap Bon
M'Sila   Constantine Soussa Ben Arous
Bou   Aïn Beïda   Sousse Golfe de
hada Batna Khenchela M'Saken Hammamet **MALTA**
ir a   Biskra Kasserine Kairouan VALLETTA
El Meghaïer   Tebessa Sfax
Messaad   Tozeur Kebili Golfe de Gabès
ghouat El   Gafsa Chott Gabès
 s)   Oued Medenine Zarzis
Ghardaïa Hassi   Az Zāwiyah **TRIPOLI**
Ouargla   Messaoud   Gharyān Zuwārah (Tarābulus) Al Khums Darnah
oléa   Nālūt Al Jawsh Mizdah Banī Walīd Gulf of Sirte Zāwiyat Masūs Tubruq
Bordj   Jabal Nafūsah   Al Qaddāhiyah An Nawfalīyah Ajdābiyā
Hassi Bel   Messaouda Ghadāmis Ash   Sirte As Sidrah Marsá al
Guebbour Dirj   Shuwayrif Al 'Uqaylah Burayqah
Ohanet   Al Ḥamādah al Ḥamrā' Al   Waddān Marādah Jālū
Bordj Omer Driss   Jufrah   Zillah
In Amenas Adr 'ı Birāk Al Ḥulayq

1 : 20 000 000

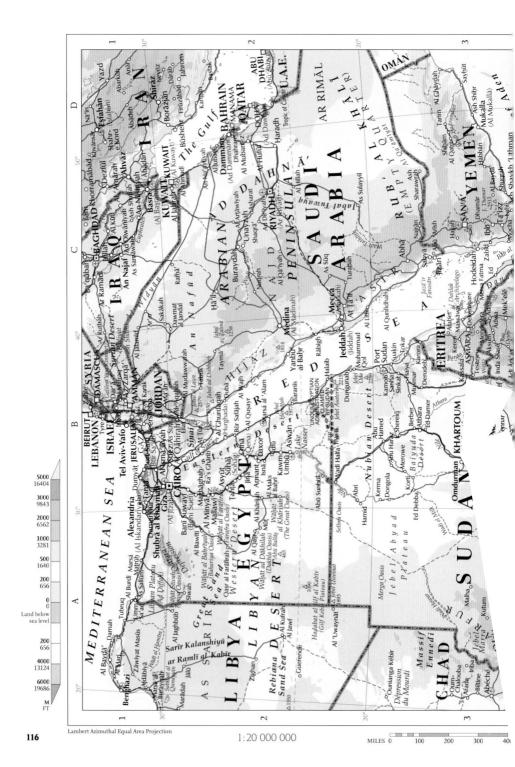

Lambert Azimuthal Equal Area Projection

1:20 000 000

MILES 0   100   200   300   400

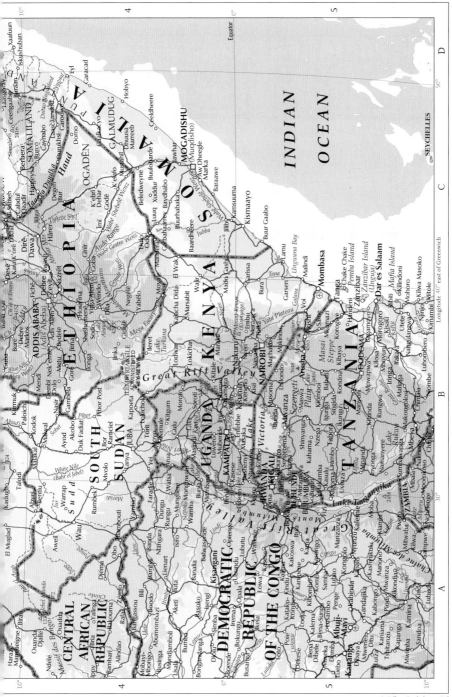

© Collins Bartholomew Ltd

0   200   400   600 KM

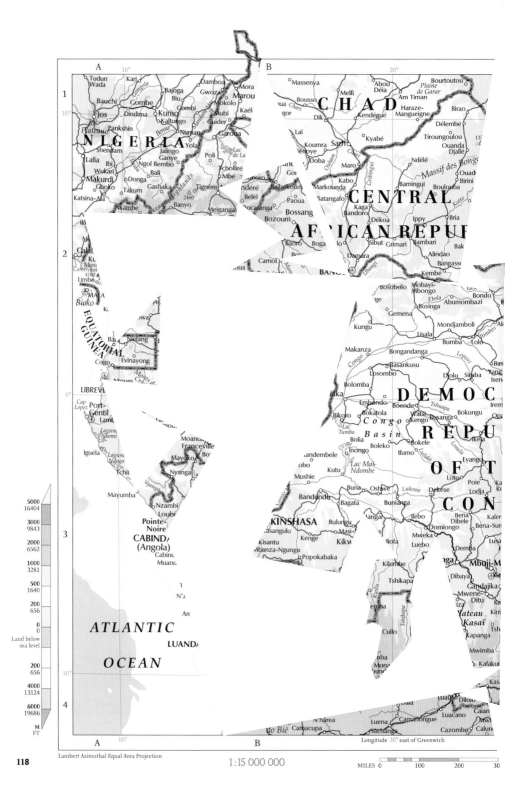

A 10°   B   20°

1 Tudun Kari   Damboa Mora   Massenya   Abou Bourtoutou
   Wada   Bajoga Gwoza   Marou   Déia   Am Timan   Plaine
        Bauchi Gombe   Biu   Mokolo   Bousso Melfi   de Garar
   Jos Dindima Kumo Combi Mubi Kaél   ua Chari Dik Kendégué Haraze- Birao
        Pankshin   Kaltungo Guider   Mangueigne
   Plateau Benue Numan Garoua   Laï   Kyabé   Tiroungoulou Délembé 133
   NIGERIA   Yola   Koumra Sarh   Ouanda
   Shendam Jalingo   Lac   èroye Doba Maro   Ndélé Djallé
   Lafia Ganye   de La   Massif des Bongo
   Ibi Ngol Bembo Poli   Goi   Kabo Bamingui Ouad Birini
   Wukari Bali Tchollíré   Batangafo Markounda Boulouba Kotto
   Makurdi Donga Mbé Tignère   ndéré Baibokoum Kaga CENTRAL
   Gboko Takum Gashaka 2460   Bélel Paoua Bandoro Dékoa Ippy Bria
   Katsina-Ala Nkambe Banyo Meiganga cabanga Bossang Sibut Grimari Bambari AFRICAN REPUI
2   Baoro Boga lo   Bak
   Calat   Carnot   Damara Alindao Bangass
   Mont   ua Mann   BANG   Ubangi Kembé
   Cameroun 4100   Bosobolo Mobayi- Bondo
   Limbe   Ebola Mbongo Abumombazi
   MALA   Gemena Businga
   Bioko Ki   Kungu Mondjamboli Ak
   EQUATORIAL   war   Makanza Lisala
   GUINEA Nielang   Bongandanga Bumba Lolo
   Evinayong Bau   Congo Basankusu Diolu Simba Isen
   Congo   laka Lbombo Djolu
   Monts d. Bolomba   DEMOC
   LIBREVI Ntoum Cra.   Embondo Boende Tshuapa Irem
   Cap   Bikoro Bokatola Watsi Bokungu Opa
   Lopez Port-   Congo Bolla Kengo Busanga Ikela REPU
   Gentil Lamf   Lac Boleko Bokele
   Lagune Moanda Tumba Basin   Eyangu OF T
   Nkomi Franceville Inongo Ifumo Lindaka Loto
   Iguéla Lagune Mayoko Bo andembele   Poie C O N
   Ndogo Nyanga   obo Kutu Lac Mai Dekese Lodja
   Tchil   Mushie Ndombe Lukenie Poie Ka
   Mayumba   Buna Oshwe   Bena Kaler
   Nzambi   Bandundu Bagata Bunianga Dibele Bena-Sun
   Loubi   angu Ilebo Domionio
3 Pointe- KINSHASA Bulurigu   Mweka Demba Lusa
   Noire asangulu Masi-   liofa Luebo
   CABINDA Kisantu Kenge Kikw   Kilembe Mbuji-M
   (Angola) Mbanza-Ngungu Popokabaka   Dibaya Gandajika
   Cabinc   Tshikapa Mwene-
   Muanc   Ditu
   N'z   Cuilo   lateau Kim
   An   Kasaï Tsh
   Mayumba   Kapanga
   Mona Kafakun
4 ATLANTIC LUANDA   Mwimba
   OCEAN   nba   Mona
                  harea Camacupa Luena Camanongue Luacano Caian
   to Bié   Cazombo Calun

A 10°   B   Longitude 20° east of Greenwich

Lambert Azimuthal Equal Area Projection   1:15 000 000   MILES 0   100   200   30

5000   16404
3000   9843
2000   6562
1000   3281
500   1640
200   656
0   0
Land below sea level
200   656
4000   13124
6000   19686
M FT

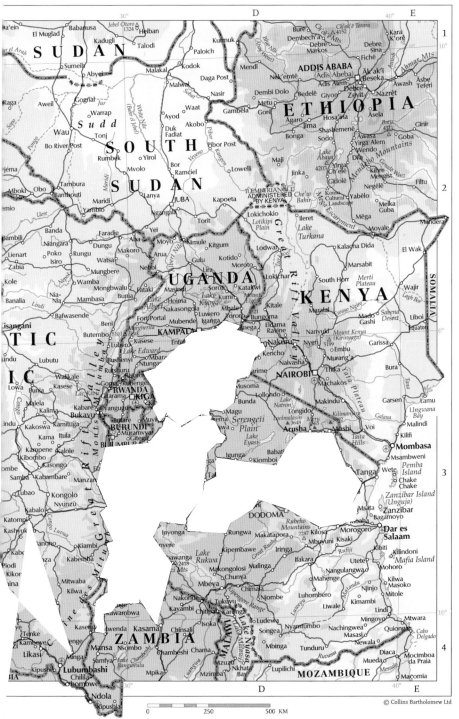

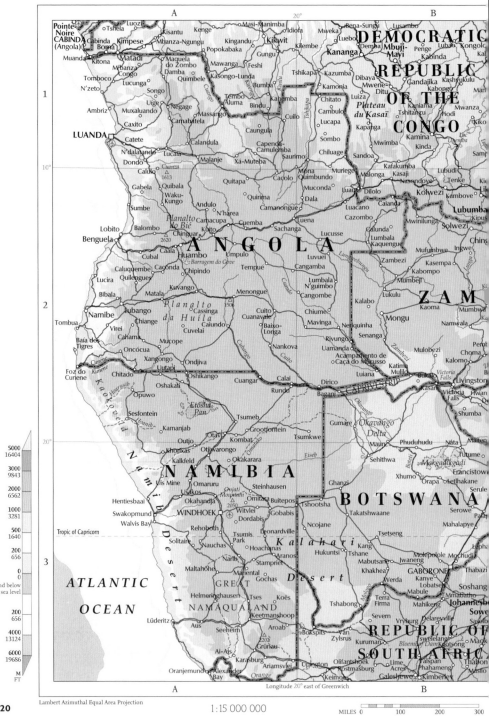

Lambert Azimuthal Equal Area Projection

1:15 000 000

MILES 0   100   200   300

Kondoa
Kibaya
Meia Meia · Handeni
mba Island
nake Chake
Zanzibar Island
(Unguja)
ODOMA
Rubeho
Mountains · Bagamoyo
Zanzibar
ungwa · Kilosa
Kibaha
Dar es Salaam
NZANIA
Kibiti
Mafia Island
Kipembawe
Isaki · Rufiji
Kilindoni
ngolosi · Utete
ohoro
Chunya · Nangulangwa
Mahenge
Kilwa Masoko
he
Luhombero · Njr
Kiman
itole

1

INDIAN
OCEAN

Liwale
litipa · Mi
Isoka · Ludewa · Nachingwea
Mtwara
Quionga
Rumphi · Songea · Masasi · Newala
Cabo Delgado
Mbinga · Tunduru · Diaca
Mocimboa
da Praia
COMOROS
Mzuzu · Lupilichi · Mueda
Messalo
MORONI
Ngazidja
(Grande Comore)
Mzimba · Macaloge · Salimo
Macomia
undazi · Maniamba · Marrupa · Montepuez
Pemba
DZAOUDZI
Nkhotakota · Lichinga · Litunde
Me
Mayotte
(France)
Kasungu · Namuno · Lúrio
erenje
Chipata · Dowa · Salima
Muite
Nacala
iri Mposhi
Petauke · Katete · Mchinji
Mangochi · Mutuali · Ribáue
Nami
çambique
abwe · Chifunde · Dedza · Cuamba · Alto
Tanjona
Bobaomby
Rufunsa · Bene · Machinga · Ligonha · Liupo
Antsirañana
hongwe
LUSAKA · Songo · Zomba · Murru · Chalaua
Tanjona
Anorontany
Ambilobe
hirundu · Magoé · Lake · Tete · Milange
Iharaña
Cabora Bassa
Angoche
Nosy Bé
Andoany
Guruve · Rushinga · Macatanja · ucubela
Ambanja · Massif du
Tsaratanana
Bindura · Pebane
MSambava
Quelimane
Bealanava · Andapa · Antalaha
Marromeu
Analalava · Antsohihy · Mahalevona
Chinde
Befandriana · Avaratra · Maroantsetra
Mahajanga · Mandritsata · Tanjona
Masoala
Soalala · Marovoay · Ankofa · Nosy
Dondo · Madirovalo · Ambato · Avaratra · Boraha
Beira · Boeny · Soanierana-
Buzi · (Ivongo)
Besalampy · Maevatanana · Andilamena
Machanga · Kandreho · Fenoarivo
Morafenobe · Vavatenina · Atsinanana
Ambatondrazaka
Maintirano · Ankazobe · Andilahatoby · Toamasina
Pambarra · Tsiroanomandidy · Ambohidratrimo · Ampasimanolotra
Antsalova · Miarinarivo · Moramanga
alburo · Miandrivazo · Betafo · ANTANANARIVO
Belo Tsiribihina · Ambatolampy

2

Antsirabe
Mahanoro
ambane · Morondava · Malaimbandy · Fandriana
Ambositra
avala · Mandabe · Ambato Finandrahana
njacaze · Manja · Berorona · Fianarantsoa · Ambohimahasoa
Xai-Xai · Mangoki · Ifanadiana · Mananjary
anhica · Morombe · Ankazoabo · Ambalavao · Ikongo
MAPUTO · Ihosy · Manakara
Bela Vista · Toliara · Vohipeno
Farafangana

3

INDIAN
OCEAN

Tropic of Capricorn

Betroka · Vangaindrano
Betioky · Isoanala
Bekily
Ejeda · Ampanihy
eni · Ampanihy
Mondlo · Ulundi · Androka · Amboasary · Tôlañaro
adysmith · Ampangeni · Beloha · Ambovombe
uthe · Ezakheni · Richards ay
Tanjona Vohimena

0 · 250 · 500 KM

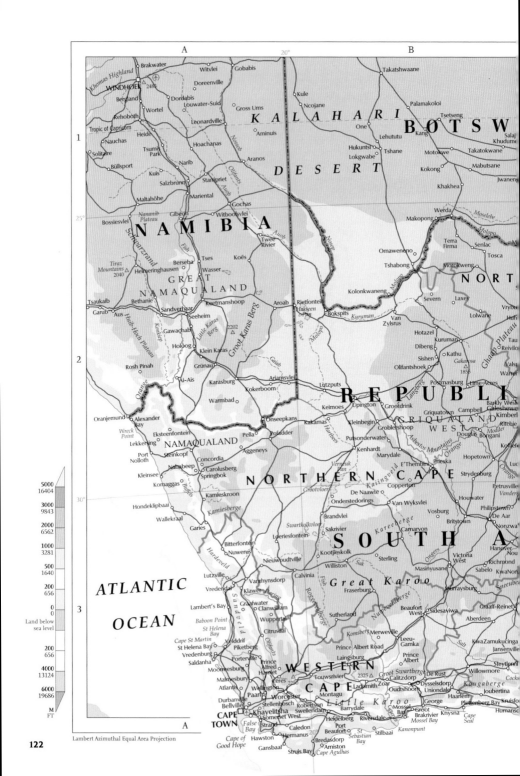

Lambert Azimuthal Equal Area Projection

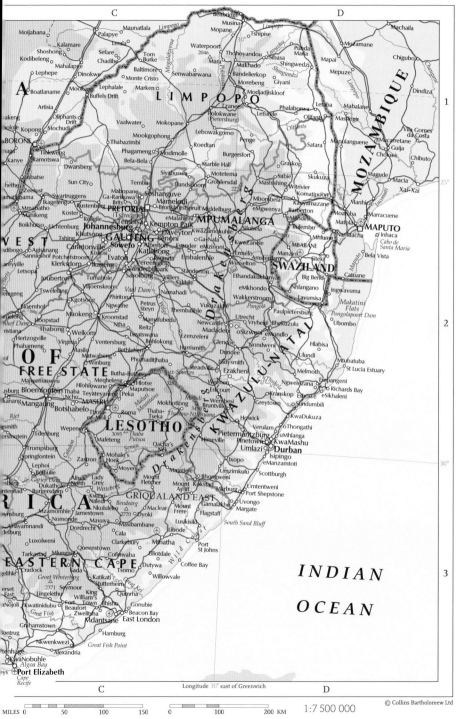

INDIAN

OCEAN

Longitude 30° east of Greenwich

1:7 500 000

MILES 0   50   100   150   0   100   200 KM

123

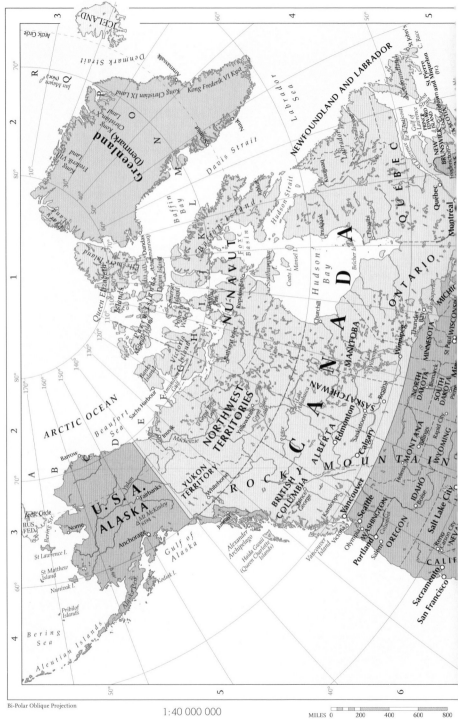

ICELAND

Arctic Circle

Denmark Strait

Ammassalik

Kong Frederik VI Kyst

R

Jan Mayen (Nor.)

Kong Christian IX Land

Q

Kong Christian X Land

P

O

Nuuk

Labrador Sea

NEWFOUNDLAND AND LABRADOR

Newfoundland

St John's
C. Race

S. Pierre
St Pierre and Miquelon (Fr.)

Cabot Strait

Gulf of St Lawrence

PRINCE EDWARD ISLAND

NOVA SCOTIA

NEW BRUNSWICK

Charlottetown

Fredericton

N

Greenland (Denmark)

Kong Frederik VIII Land

Davis Strait

Baffin Bay

M

Baffin Island

Labrador

Kuujjuaq

Hudson Strait

Inukjuak

Schefferville
Smallwood Res.

L

QUÉBEC

Québec

Montréal

K

Ellesmere Island

Dundas (Uummannaq)

Thule (Qaanaaq)

Devon Island

J

Foxe Basin

Bélcher Is.

Repulse Bay

Southampton I.

Coats I.

Mansel I.

Igloolik

ONTARIO

Lake Superior

MICHIGAN

WISCONSIN

Queen Elizabeth Islands

Axel Heiberg I.

Cornwallis I.

Somerset I.

Prince of Wales I.

I

Hudson Bay

Churchill

Sanikiluaq

Thunder Bay

MINNESOTA

St Paul

MICH

120°

130°

110°

Bathurst I.

Melville I.

H

NUNAVUT

Rankin Inlet

Winnipeg

NORTH DAKOTA

SOUTH DAKOTA

Bismarck

Pierre

Mi

140°

Banks Island

Victoria Island

G

Gjoa Haven

MANITOBA

Lake Winnipeg

Regina

SASKATCHEWAN

Lake Athabasca

CANADA

Rapid City

WYOMING

ARCTIC OCEAN

Beaufort Sea

Sachs Harbour

F

Amundsen Gulf

Coppermine

Yellowknife

Lake Winnipegosis

MONTANA

Billings

IDAHO

160°

170°

150°

Barrow

D

E

Inuvik

Mackenzie

Great Bear Lake

Great Slave Lake

NORTHWEST TERRITORIES

Fort Nelson

Lake Athabasca

Edmonton

ALBERTA

Calgary

Saskatoon

Medicine Hat

Boise

80°

70°

C

B

A

Union

Fairbanks

YUKON TERRITORY

Whitehorse

ROCKY

Prince George

BRITISH COLUMBIA

MOUNTAIN

Vancouver

Kamloops

Seattle

WASHINGTON

Spokane

SALT LAKE CITY

UTAH

Arctic Circle

RUS. FED.

Nome

U.S.A.

ALASKA

Mt McKinley 6194

Anchorage

Juneau

Gulf of Alaska

Alexander Archipelago

Haida Gwaii (Queen Charlotte Islands)

Vancouver Island

Victoria

Olympia

Columbia

OREGON

Salem

Portland

Reno

Carson City

NEVA

Bering Str.

St Lawrence I.

St Matthew Island

Kodiak I.

Pribilof Islands

Nunivak I.

Bering Sea

Aleutian Islands

Unimak I.

CALIF

Sacramento

San Francisco

60°

50°

40°

2

3

4

5

1

6

Bi-Polar Oblique Projection

1:40 000 000

MILES 0   200   400   600   800

Lambert Azimuthal Equal Area Projection

1:25 000 000

MILES 0   250   5

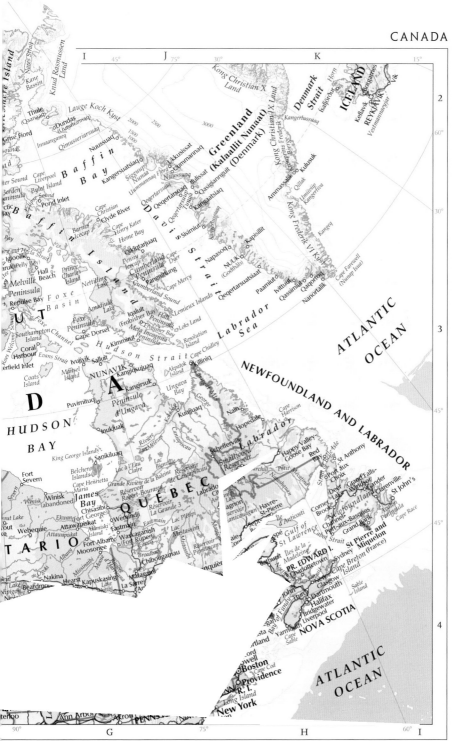

Lambert Azimuthal Equal Area Projection

1:12 000 000

MILES 0   100   200   300

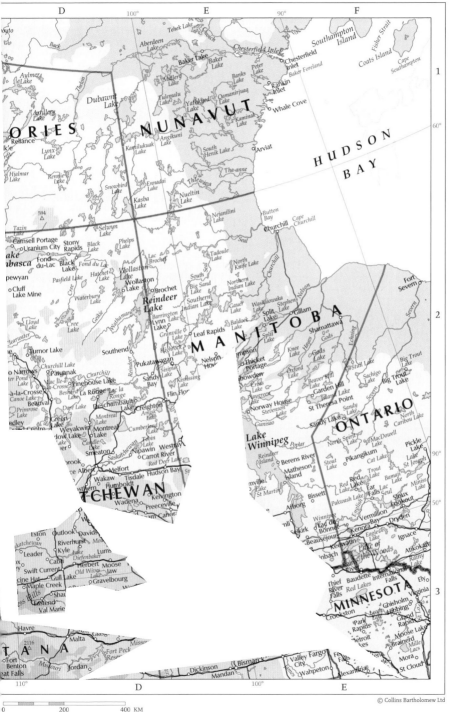

Lambert Azimuthal Equal Area Projection

1:12 000 000

Longitude 80° west of Greenwich

MILES 0   100   200   300

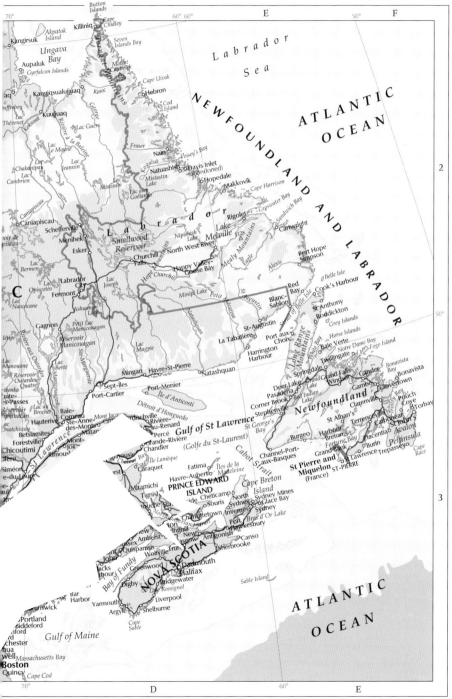

Button Islands
70°
Akpatok Island
Killiniq
Cape Chidley
60° 60°
E
50°
F

Kangirsuk
Aupaluk
Gyrfalcon Islands
Ungava Bay
Seven Islands Bay
Mount Caubvick

Labrador Sea

aq
uffinboy
Lac Thévenet
Kangiqsualujjuaq
Korok
Kuujjuaq
Koksoak
Lac Guer
Rivière à la Baleine
George
Cape Uivak
Hebron
Cod Island

ATLANTIC OCEAN

2

oir de
piscau
Lac Bermen
Caniapiscau
Scheffervile
Menihek
Esker
Fraser
Nain
Voisey's Bay
Natuashish
Davis Inlet (abandoned)
Hopedale
Makkovik
Cape Harrison
Kogaluk
Mistinibi
Lac aux Goélands

Rosewater Bay

Sandwich Bay

Lac Opiscotéo
Lac Naococane
Labrador City
Fermont
Ashuanipi Lake
Smallwood Reservoir
Churchill Falls
North West River
Nipishish Lake
Lake Melville
Rigolet
Medly Mountains
Eagle
Alexis
Cartwright

C
Lac Joseph
Hope-Churchill
Happy Valley-Goose Bay
Minipi Lake
Petit lac
Naskaupi
Martina
St-Augustin
Belle Isle
Blanc-Sablon
Red Bay
Cook's Harbour
St Anthony
Roddickton
Grey Islands
Belle Isle
Port Hope Simpson
3128

50°

léupi
Lac aux Outardes
Gagnon
Réservoir Manicouagan
Petit lac Manicouagan
Ste-Marguerite
Lac Magpie
La Tabatière
St-Augustin
Harrington Harbour
Port aux Choix
Baie-Verte
Long Range Mountains
Horse Islands
Notre Dame Bay
Fogo Island
King's Bay
Twillingate
Springdale

Lac Manouane
Réservoir Outardes Quatre
Lac Berté
Mingan
Havre-St-Pierre
Natashquan
Deer Lake
Grand Lake
Grand Falls-Windsor
Gander
Bonavista Bay
Bonavista

ribonka
iute-
es-Passes
éservoir
ipmuacan
Lac à Brochet
Baie-Comeau
Sept-Îles
Port-Cartier
Île d'Anticosti
Détroit d'Honguedo
Corner Brook
Stephenville
Pasadena
Grand Lake
St Alban's
Terrenceville
Clarenville
Trinity Bay
Pouch Cove
Torbay

Hauterive
ac
Onatchiway
Chicoutimi
tière
Betsiamites
Forestville
Ste-Anne-des-Monts
Mont-Joli
Mont Jacqu.
Cap
Rivière-au-Renard
Percé
Gulf of St Lawrence
(Golfe du St-Laurent)
St George's Bay
Burgeo
Harbour Breton
Fortune Bay
Placentia Bay
Avalon
St John's
Carbonear
Peninsula

St Lawrence
siméon
e-du-Loup
e-
aul
ton-
Grande-Rivière
Chandler
Île Lamèque
Fatima
Îles de la Madeleine
Havre-Auberto.
Caraquet
Chaleur Bay
Channel-Port-aux-Basques
Grand Bank
St Lawrence
Trepassey
Cape Race

St Pierre and Miquelon (France)
ST-PIERRE

Mitamichi
Tignish
touche
Cabot Strait
Prince Edward Island
Cheticamp
Souris
North Sydney
Sydney Mines
Inverness
Sydney
Glace Bay
Cape Breton Island
Bras d'Or Lake

3

John
Quispamsis
Sussex
Amherst
Bay of Fundy
Greenwood
Nova Scotia
Dartmouth
Halifax
Bridgewater
Lunenburg
Truro
Antigonish
Sherbrooke
Canso
Port Hawkesbury
Charlottetown
New Glasgow
Digby
Liverpool
Shelburne
Sable Island

ATLANTIC OCEAN

Brunswick
Portland
Biddeford
nford
rd
chester
hua
well
Boston
Quincy
Bar Harbor
Yarmouth
Argyle
Cape Sable
Lake Rossignol
Massachusetts Bay
Cape Cod
Gulf of Maine

70°
D
60°
E

© Collins Bartholomew Ltd

0   200   400 KM

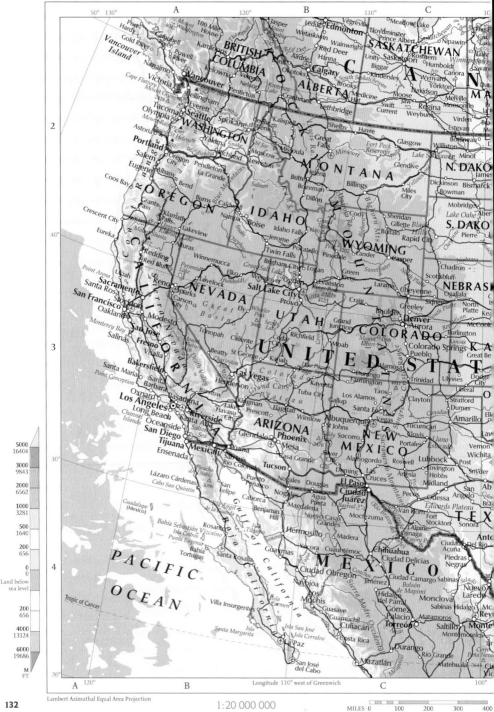

CANADA

BRITISH COLUMBIA

Vancouver Island

Gold River
Hardy
Port
Campbell River
Powell River
Nanaimo
Victoria
Cape Flattery
Mount Olympus
Olympia
Tacoma
Seattle
Everett
Bellingham
Vancouver
Penticton
Nelson
Cranbrook

ALBERTA

Jasper
Edson
Leduc
Edmonton
Vegreville
Lloydminster
Wetaskiwin
Wainwright
Red Deer
Airdrie
Calgary
Okotoks
Brooks
Hanna
Medicine Hat
Lethbridge
Swift Current

SASKATCHEWAN

Prince Albert
Saskatoon
Unity
Biggar
Kindersley
Rosetown
Humboldt
Wynyard
Yorkton
Davidson
Melville
Virden
Estevan
Moose Jaw
Regina
Weyburn
Bottineau
Minot

MONTANA

Great Falls
Helena
Butte
Bozeman
Billings
Miles City
Glendive
Dickinson
Bowman
Bismarck

N. DAKO
Mobridge
Lake Oahe
S. DAKO
Pierre

WASHINGTON

Olympia
Portland
Salem
Eugene
Albany
Oregon City
Pendleton
La Grande
Coos Bay
Bend
Burns

OREGON

Crescent City
Grants Pass
Klamath Falls
Lakeview
Alturas
Eureka
Redding
Red Bluff

IDAHO

Boise
Nampa
Caldwell
Idaho Falls
Pocatello
Twin Falls
Jerome

WYOMING

Cody
Lander
Casper
Sheridan
Gillette
Buffalo
Rapid City
Chadron
Scottsbluff

NEBRASK
Ogallala

Winnemucca
Reno
Sparks
Carson City
Lovelock
Elko
Wendover
Ogden
Logan
Brigham City
Salt Lake City
Provo
Evanston
Laramie
Cheyenne
Greeley
Boulder
Denver
Aurora

NEVADA
Tonopah
Ely

UTAH
Richfield
Grand Junction
Moab

COLORADO
Colorado Springs
Pueblo
Burlington

Sacramento
Stockton
San Francisco
Oakland
San Jose
Modesto
Fresno
Visalia
Bakersfield

CALIFORNIA

Santa Rosa
Santa Cruz
Salinas
Monterey Bay
Point Conception
Santa Barbara
Oxnard
Los Angeles
Long Beach
Pasadena
Santa Ana
Riverside
San Diego
Oceanside

Las Vegas
Henderson
Barstow
Kingman
Prescott
Flagstaff
Winslow

ARIZONA
Phoenix
Mesa
Tucson
Nogales
Douglas

NEW MEXICO
Albuquerque
Santa Fe
Gallup
Socorro
Las Cruces
El Paso
Roswell
Clovis
Portales
Hobbs
Artesia
Lubbock
Midland

UNITED STATES

Tijuana
Mexicali
Ensenada

MEXICO

Hermosillo
Guaymas
Ciudad Obregón
Navojoa
Los Mochis
Culiacán
Mazatlán
La Paz
San José del Cabo

Chihuahua
Ciudad Delicias
Torreón
Saltillo
Monclova
Matamoros

PACIFIC OCEAN

Tropic of Cancer

5000 / 16404
3000 / 9843
2000 / 6562
1000 / 3281
500 / 1640
200 / 656
0 / 0
Land below sea level
200 / 656
4000 / 13124
6000 / 19686
M / FT

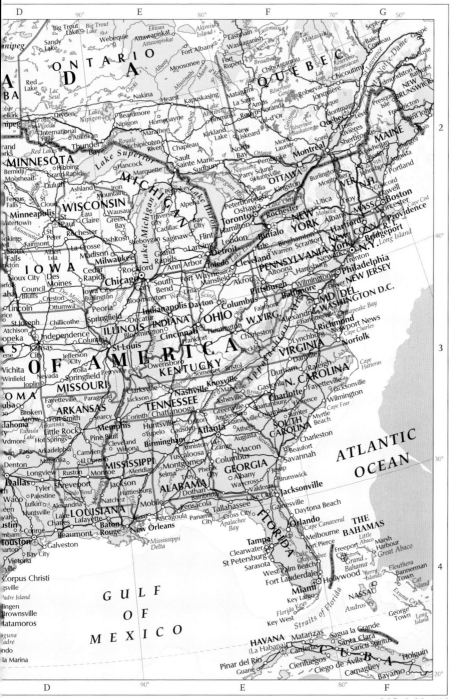

ATLANTIC
OCEAN

GULF
OF
MEXICO

THE
BAHAMAS

Lambert Azimuthal Equal Area Projection

1 : 8 000 000

MILES 0   50   100   150

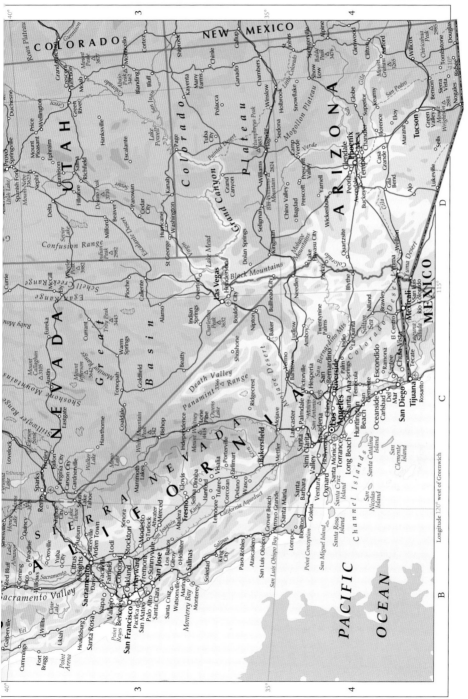

© Collins Bartholomew Ltd

0   100   200 KM

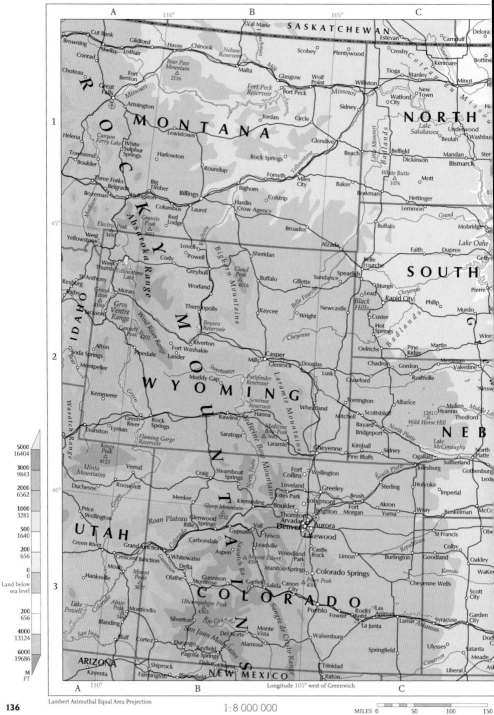

Lambert Azimuthal Equal Area Projection

1 : 8 000 000

MILES 0   50   100   150

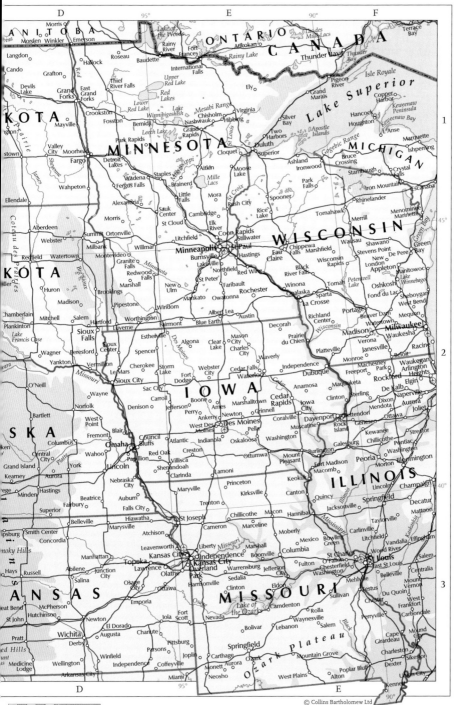

Lambert Azimuthal Equal Area Projection

1 : 8 000 000

MILES 0   50   100   150

Longitude 85° west of Greenwich

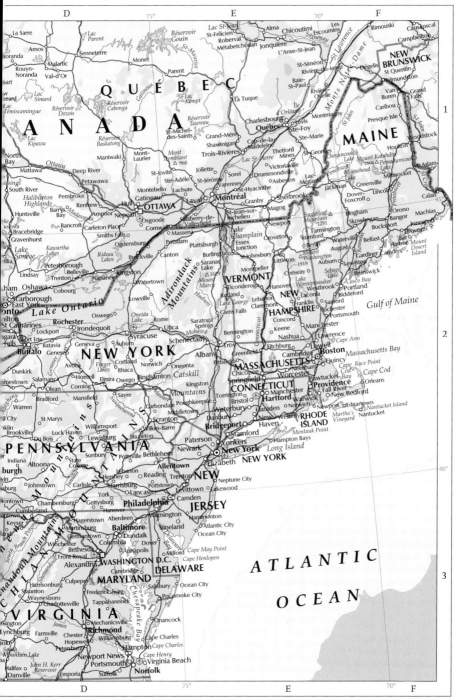

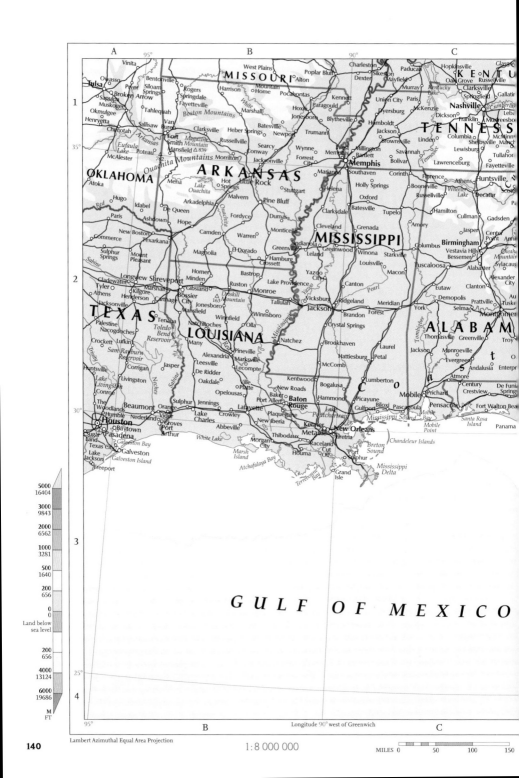

Lambert Azimuthal Equal Area Projection

1:8 000 000

MILES 0   50   100   150

Longitude 90° west of Greenwich

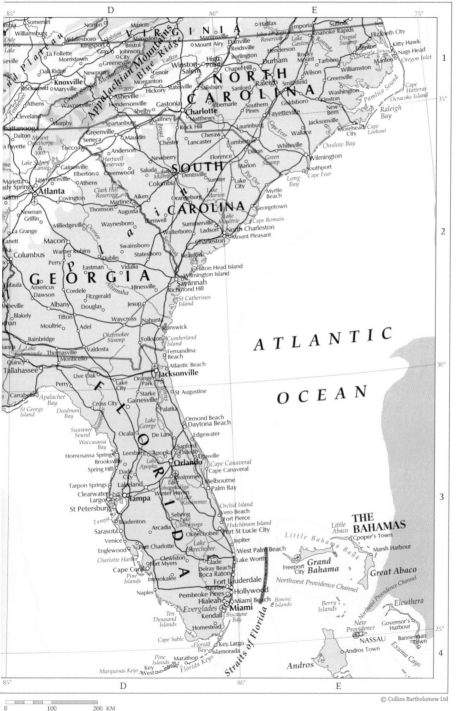

© Collins Bartholomew Ltd

0   100   200 KM

Lambert Azimuthal Equal Area Projection

1 : 8 000 000

MILES 0   50   100   150

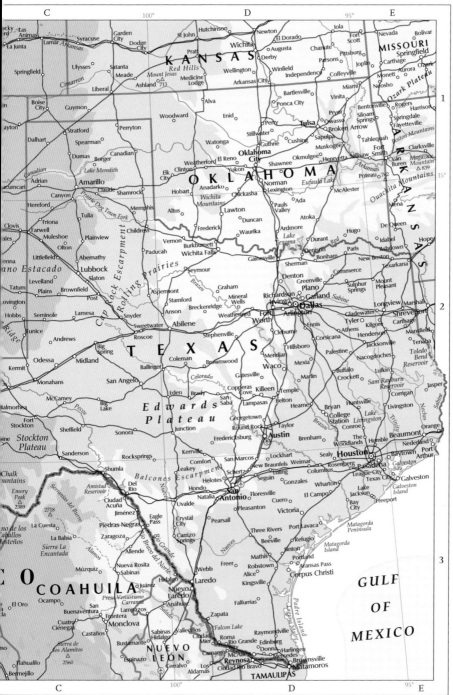

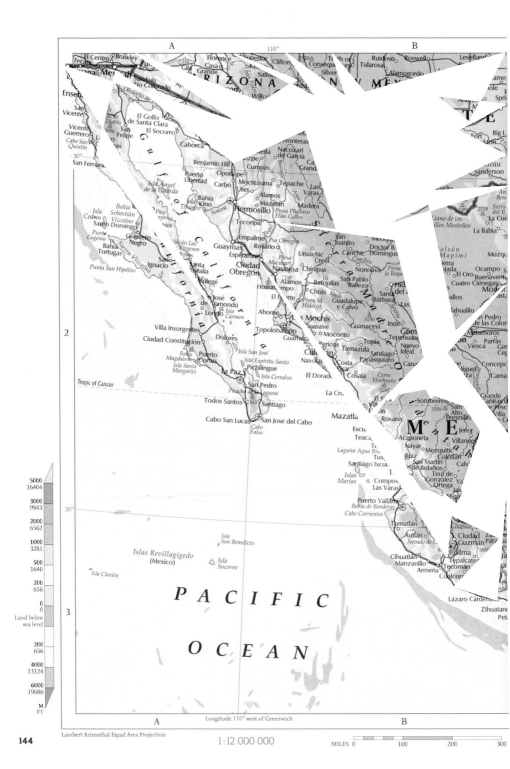

110°

El Centro   Brawley
Tec   Yu
Mey   Ka
San Luis   Gila
Río Colorado   Grande
Ensen   Casa
                        Willcox
San
Vicente
El Golfo
Diablo   de Santa Clara
Vicente   San   El Socorro
Guerrero   Felipe
Cabo San   L
Quintín   as
– 30°
San Fernand

ARIZONA

Florence   Superior   Clifton
                Safford
Río Colorado   Ar
                        Fronteras

Ensenada

Caborca

Benjamín Hill
Puerto   Opodepe
Libertad   Carbó   Moctezuma   Tepache
Isla Ángel   Ures   Alamos
de la Guarda   Bahía   Mazatán
Isla   Kino   Sonora
Tiburón

Isla   Bahía K.   Pico
Cedros   Sebastián   coverha
Santo Domingo   Vizcaíno   1908
Punta   Volcán Las
Eugenia   Bahía   Vírgenes
Tortugas   1996
        Serra Vizcaíno
Punta San Hipólito

Ruidoso   Roswell   Levelland
Consequ   ees   Tularosa   Alamogordo
Silver
                    A   Spri
                                        Big
                        Fort   Lon

Hermosillo
Tecoripa
Presa Plutarco
Elías Calles

San
Juanito

Guaymas   Empalme   Psa Obregón
        Rosario   Uruachic
Esperanza   Creel
Ciudad   Navojoa   Chínipas   Nonoava
Obregón
        Álamos   Batopilas   San Pablo
Múlege   Choix   Ballesa
        Río F.   El Fuerte   Presa M.
        Presa M.
        Ahome   Hidalgo   Bárbara

Doctor B.
Carichi   Dominguez

Presa   Ocampo
la Boqu   El Oro
        Cuatro Ciénegas

Villa Insurgentes
Ciudad Constitución   Dolores
Bahía   Isla San José
Magdalena   Puerto
Isla Santa   Cortés   Navola
Margarita   Isla Espíritu Santo   Pichilingue
        Isla Cerralvo
La Paz   El Dorado

Los Mochis
Guasave
Guamú   Mocorito
Pericos
Tamazula
Costa
ica   Papasquiaro

Tropic of Cancer

San Pedro
Todos Santos   Santiago
Cabo San Lucas   San José del Cabo
Cabo
Falso

La Cru.

Mazatlán   Rosario

Escu.
Teaca
Laguna Agua Br
Santiago Ixcui
Islas   Compos.
Marías   Las Varas

Acaponeta
Nayar
Tux.   Ruiz   Mezquitic
San Martín   Colotlán
de Bolaños
Teul de
González
Ortega

20°

Isla
San Benedicto

Islas Revillagigedo
(Mexico)   Isla
        Socorro
Isla Clarión

Puerto Vallarta
Bahía de Banderas
Cabo Corrientes

Tomatlán

Autlán
Nevado de C

Cihuatlán
Manzanillo   Tecomán
Armería
Coahom

Ciudad
Guzmán   Patz
Colima
Tecalpate

P A C I F I C

O C E A N

Lázaro Cárdena
Zihuatane
Peta

Lambert Azimuthal Equal Area Projection

A   Longitude 110° west of Greenwich   **B**

1:12 000 000   MILES 0   100   200   300

5000
16404
3000
9843
2000
6562
1000
3281
500
1640
200
656
0
0
Land below
sea level
200
656
4000
13124
6000
19686
M
FT

1

Denton
Graham Mineral Wells
Breckenridge Richar...
Abilene Weatherford
ter

Commerce
Magnolia El Dorado Crosset
Bastrop Lake Canton
Yazoo City Pearl
Demopolis Tuskegee
Selma Montgomery
ALABAMA
Greenville Troy
Andalusia Ozark
Enterprise

Stephenville

T A T
Brownwood
Colorado Gatesville
Brady Lampasas
unction Georgetown
Round Rock
icksburg Austin
San M...
New F

X A Kille
College
Conroe
Huntsville
Rayburn R
ckett
Lufkin

Atmore
Mobile
Biloxi Pascagoula
ilport Mobile
Bay

FLORIDA
Crestview Pensacola
Fort
Walton
Beach

30°

S T A
Plaquemine Kenner
rleans
ctor Chandeleur
and Islands

Anto
Uvald

n Port
ena Arthur
ytown
Abbevi
Iberia
Morgan Houma
City
Marsh
Island
Thibodaux
Raceland
Sulphur
Grand
Isle
Mississippi
Delta

Terre Bonne

Cuero El Campo Wharton
Pleasanton
earsall B y City
Beeville Victoria
ort Lavaca
Mathis Sinton
Matagorda
land
as Pass
s Christi
Galveston Bay
Galveston
Freeport

Alice kings

1

2

nde
a Ros
nas
uárez
o Laredo
nua
pazos

ecillos
rias Mier
algo Camargo
Cerralvo
Monterrey Reynos
tillo Cao
Al
Montan
Galea
Salvador
ado Cerro
Nevada
pala
Doctor 3644
Arroyo Jaum
as

ville
n
vnsville
amoros

na Madre

scal

esca

Marina

GULF

O F

M E X I C O

Tropic of Cancer

Arrecife
Alacrán

Yucatan Channel

ma
s
iudad Madero
ampico
o
Laguna de
al Tamiahua
yuca Naranjos
Cerro Azul
Tuxpan
Tihuatlán
ipán Poza Rica
P xtla
ancingo
la

Telchac
Progreso Puerto Dzilam de Bravo
Sisal Motul
Celestun Izamal
Calkiní
Tenabo Te
Campeche Hope

Río
Lagartos
Cabo Catoche
Cancún
Tizimín

Cozumel
Isla
de Cozumel

Bahía
de la Ascensión

20°

2

Bahía
de Campeche
Champotó
Aguad
Ciudad del Carmen
Frontera
malcalco Parais
llaherm
Jon

Banco
Chinchorro

Chetumal
Corozal
Orange Walk
Ambergris
Caye
Belize Turneffe
Islands

BELMOPAN
Dangriga

3

la
Balsas Acatlan
Chilpancingo Tlapa
oyuca de Tierra
Benítez Colorada
Ayutla
apulco
Copala
Pinotepa Nacional
Puerto Escondido
Puerto Ángel

Chilpancingo Tlapa
de Lе
Tlaxiaco
Zacatepec
tepec
San Miguel Miahu
Sola de Vega
Pochutla

Madre del

Juchitán
Salina Arriaga
Cruz Tonalá
Pijijiapan
Gulf of Cuauhtén
Tehuantepec Mapastepec
Huixtla
Tapachula
Ciudad Hidalgo
Retalhuleu

Flores

rtad
1120
BELIZE
Punta Gorda

San Pedro
Puerto Sula
Barrios El Progreso
Santa Rosa
de Copán
HONDURAS
La Paz

3

0 200 400 KM

© Collins Bartholomew Ltd

145

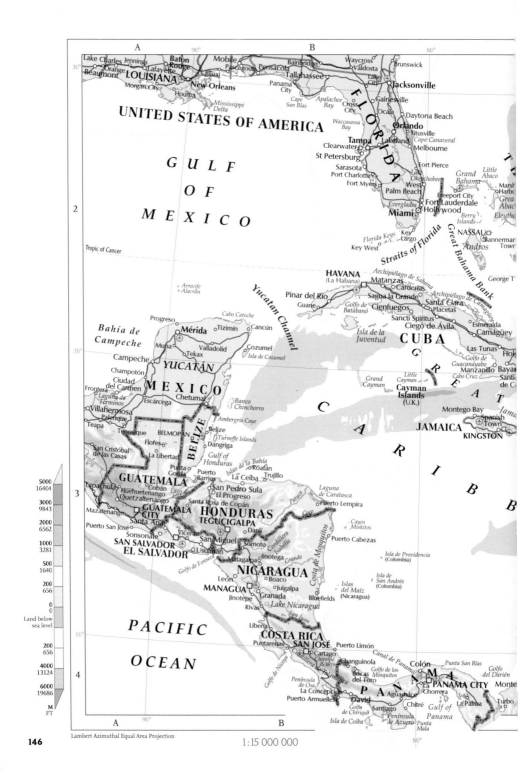

Lambert Azimuthal Equal Area Projection

1:15 000 000

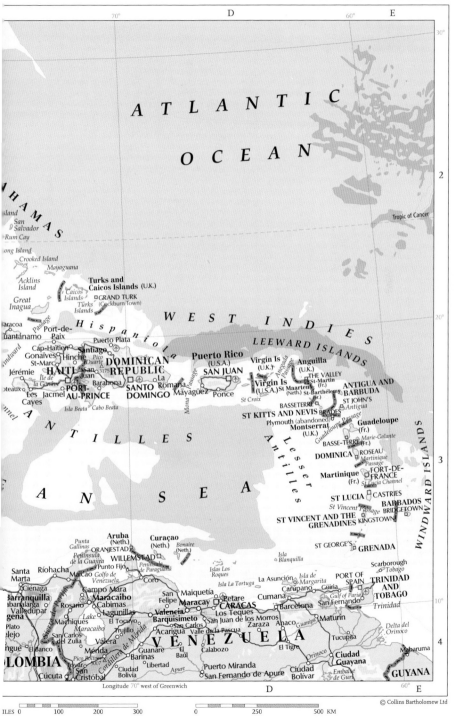

ATLANTIC

OCEAN

Tropic of Cancer

2

H A M A S

island
San
Salvador
Rum Cay

ong Island

Crooked Island

Mayaguana

Acklins
Island

Turks and
Caicos Islands (U.K.)

Great
Inagua

Caicos
Islands
□GRAND TURK
(Cockburn Town)
Türks
Islands

aracoa

W E S T   I N D I E S

20°

Hispaniola

Passage

Port-de-
Paix

uantánamo

Cap-Haïtien

Santiago

Pico
Duarte

Puerto Plata

Puerto Rico
(U.S.A.)

LEEWARD ISLANDS

Gonaïves
Hinche
St-Marc

DOMINICAN
REPUBLIC

Virgin Is
(U.K.)

Anguilla
(U.K.)

Jérémie

HAITI
San
Juan

3175

La

SAN JUAN

Virgin Is
(U.S.A.)

THE VALLEY

St-Martin
(Fr.)

ANTIGUA AND
BARBUDA

Îndward

Île de
la Gonâve

Barahona

Romana

Ponce

St Maarten
(Neth.)

St-Barthélemy

oteaux

Les
Cayes

Jacmel

PORT-
AU-PRINCE

SANTO
DOMINGO

Mayagüez

St Croix

BASSETERRE

ST JOHN'S

Antigua

Isla Beata Cabo Beata

BRADES

Mona

ST KITTS AND NEVIS

hnel

Plymouth (abandoned)

Guadeloupe
(Fr.)

N

T   *I*   *L*   *L*   *E*   *S*

Montserrat
(U.K.)

Guadeloupe Passage

BASSE-TERRE (Fr.)

Marie-Galante

3

Lesser

DOMINICA

ROSEAU

Martinique
Passage

A   *N*

S   *E*   *A*

Antilles

Martinique
(Fr.)

FORT-DE-
FRANCE

St Lucia Channel

ST LUCIA

CASTRIES

St Vincent Passage

BARBADOS

BRIDGETOWN

ST VINCENT AND THE
GRENADINES

KINGSTOWN

WINDWARD ISLANDS

Aruba
(Neth.)

Curaçao
(Neth.)

Bonaire
(Neth.)

Punta
Gallinas

ORANJESTAD

WILLEMSTAD

ST GEORGE'S

GRENADA

Santa
Marta

Peninsula
de la Guajira

Punto Fijo

Peninsula
de Paraguaná

Isla
Blanquilla

Scarborough
Tobago

Ríohacha

Maicao

Golfo de
Venezuela

Coro

Islas Los
Roques

La Asunción

Isla de
Margarita

PORT OF
SPAIN

Ciénaga

Campo Mara

San
Felipe

Maiquetía

Isla La Tortuga

Cumaná

Carúpano

Güiria

TRINIDAD
AND

10°

ibanalarga

Valledupar

gena

Rosario

MARACAIBO

Cabimas

Lagunillas

Maracay

Petare

CARACAS

Barcelona

Gulf of Paria

San Fernando

TOBAGO

Trinidad

Valencia

San Carlos

Los Teques

Plato

Machiques

Lake
Maracaibo

El Tocuyo

Barquisimeto

San Juan de los Morros

Maturin

Delta del
Orinoco

lejo

San Carlos
del Zulia

Trujillo

Acarigua

Valle de la Pascua

Zaraza

Anaco

Guanipa

ngue

El Banco

Valera

Guanare

El
Baúl

V E N E Z U E L A

Tucupita

abaruma

LOMBIA

Mérida

Pico Bolívar

San

Barinas

Libertad

Apure

Calabozo

El Tigre

Orinoco

Ciudad
Guayana

GUYANA

Cúcuta

Cristóbal

Ciudad
Bolivia

Puerto Miranda

San Fernando de Apure

Ciudad
Bolívar

Embalse
de Gurí

Cordillera de Mérida

ILES 0   100   200   300   0   250   500 KM

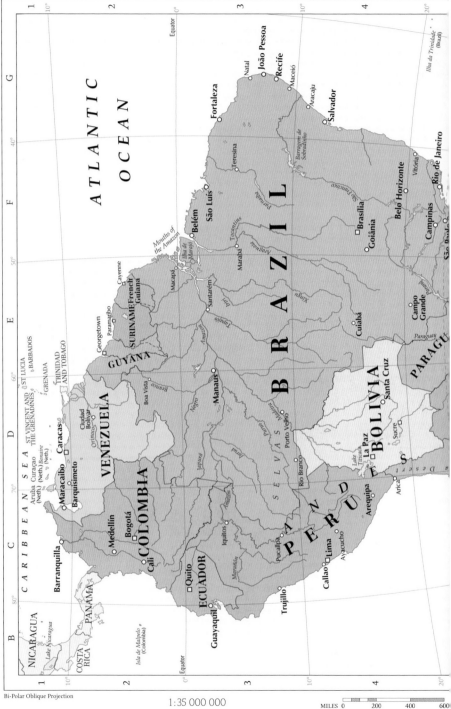

Bi-Polar Oblique Projection

1:35 000 000

MILES 0   200   400   600

ATLANTIC OCEAN

CARIBBEAN SEA

BRAZIL

VENEZUELA

COLOMBIA

GUYANA

SURINAME

French Guiana

ECUADOR

PERÚ

BOLIVIA

PARAGU

NICARAGUA

COSTA RICA

PANAMA

TRINIDAD AND TOBAGO

ST LUCIA
ST VINCENT AND THE GRENADINES
BARBADOS
GRENADA
Aruba (Neth)
Curaçao (Neth.)
Bonaire (Neth)

Ilha da Trinidade (Brazil)

Mouths of the Amazon

Ilha de Marajó

Lake Nicaragua

Isla de Malpelo (Colombia)

Equator

Natal
João Pessoa
Recife
Maceió
Aracaju
Salvador
Fortaleza
Teresina
São Luís
Belém
Rio de Janeiro
Vitória
Campinas
São Paulo
Belo Horizonte
Brasília
Goiânia
Campo Grande
Cuiabá
Santa Cruz
Sucre
La Paz
Manaus
Boa Vista
Georgetown
Paramaribo
Cayenne
Macapá
Santarém
Marabá
Porto Velho
Rio Branco
Arica
Arequipa
Ayacucho
Lima
Callao
Trujillo
Pucallpa
Iquitos
Guayaquil
Quito
Cali
Bogotá
Medellín
Barranquilla
Maracaibo
Barquisimeto
Caracas
Ciudad Bolívar

Barragem de Sobradinho

Lake Titicaca

SELVAS

ANDES

Amazon
Negro
Branco
Orinoco
Paraguay
Marañón
São Francisco
Tocantins
Xingu
Tapajós
Madeira
Paraná
Purus
Juruá
Japurá

Desert

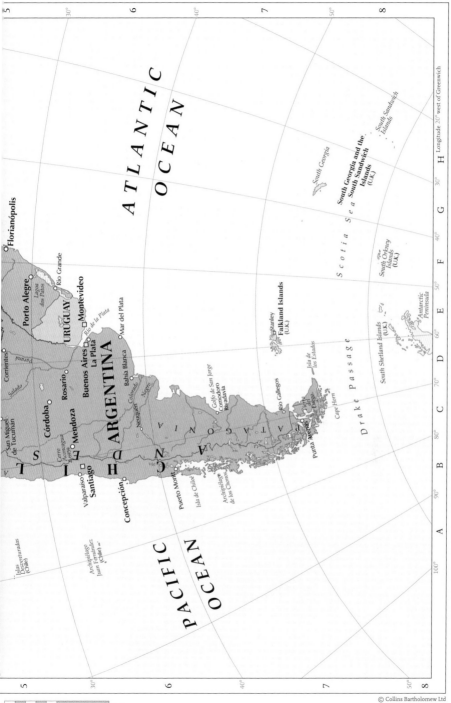

ATLANTIC OCEAN

PACIFIC OCEAN

Drake Passage

Scotia Sea

Florianópolis

Porto Alegre

URUGUAY

Montevideo

ARGENTINA

Buenos Aires

La Plata

Rosario

Córdoba

Mendoza

Santiago

Valparaíso

Concepción

San Miguel de Tucumán

Cerro Aconcagua

Corrientes

Mar del Plata

Bahía Blanca

Puerto Montt

Isla de Chiloé

Archipiélago de los Chonos

Río Grande

Laguna dos Patos

Río de la Plata

Paraná

Salado

Colorado

Neuquén

Negro

PATAGONIA

ANDES

CHILE

Golfo de San Jorge

Comodoro Rivadavia

Río Gallegos

Punta Arenas

Tierra del Fuego

Cape Horn

Isla de los Estados

Stanley

Falkland Islands (U.K.)

Islas Desventuradas (Chile)

Archipiélago Juan Fernández (Chile)

South Georgia

South Georgia and the South Sandwich Islands (U.K.)

South Sandwich Islands

South Orkney Islands (U.K.)

South Shetland Islands (U.K.)

Antarctic Peninsula

Longitude 20° west of Greenwich

0   500   1000 KM

© Collins Bartholomew Ltd

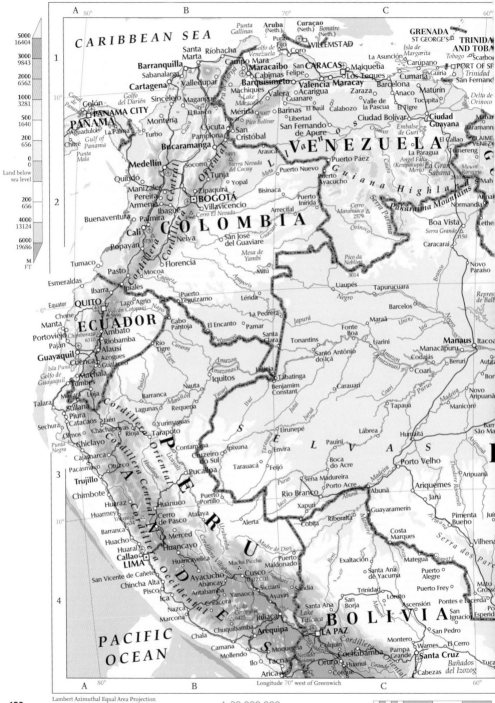

Lambert Azimuthal Equal Area Projection

1:20 000 000

MILES 0   100   200   300   400

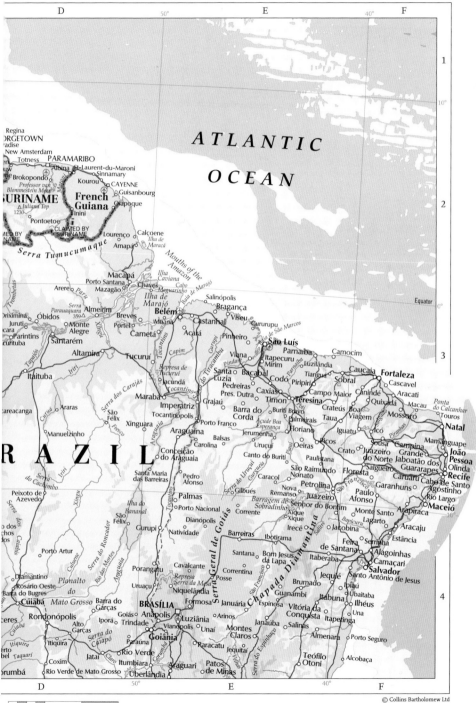

D 50° E 40° F

1

10°

ATLANTIC

Regina
ORGETOWN
adise
New Amsterdam
Totness PARAMARIBO
Albina St-Laurent-du-Maroni
Sinnamary
e Brokopondo
Kourou CAYENNE
Professor van
Blommestein Meer Guisanbourg
SURINAME French
Juliana Top Guiana Oiapoque
1230 Inini
Pontoetoe
CLAIMED BY
SURINAME Lourenço Calçoene
ED BY Ilha de
NAME Amapá Maracá

Serra Tumucumaque

OCEAN

Mouths of the
Amazon

Macapá Ilha
Porto Santana Caviana
Arere Paru Mazagão Chaves Cabo
Magueirinho de Marajó
Ilha de Salinópolis
Parauaquara Almeirim Marajó Bragança
riximina Óbidos 350 Belém Viseu
Juruti Breves Muaná Castanhal Cururupu
cara Monte Portel Acará Pinheiro Ilha de São Marcos
Parintins Alegre Cametá
curituba Santarém

Equator 0°

2

Serra dos Carajás

3

Altamira Tucuruí Capim Viana Parnaíba Camocim
Itaituba Iriri Represa de Santa Bacabal Mirim Luzilândia
Jacundá Tucuruí Luzia Codó Piripiri Tianguá Caucaia Fortaleza
Pedreiras Pres. Dutra Caxias Campo Maior Cascavel
Marabá Grajaú Timon Teresina Crateús Boa Quixadá Aracati
areacanga Araras São Imperatriz Barra do Buriti Bravo Palmeiras Taua Viagem Macau Calcanhar
Felix Tocantinópolis Corda Açude Boa Crateús Mossoró Touros
Manuelzinho Xinguara Porto Franco Esperança Floriano Iguatu Icó Natal
Araguaína Jerumenha Picos Piranhas
Balsas Oeiras Campina
Carolina Uruçuí Crato Juazeiro Grande João
do Serra Conceição Canto do Buriti Paulistana do Norte Jaboatão dos Pessoa
do Cachimbo do Araguaia São Raimundo Floresta Salgueiro Guararapes Recife
Santa Maria Pedro Nonato Petrolina Garanhuns Caruaru Cabo de Santo
Peixoto de das Barreiras Afonso Caracol Gilbués Nova Juazeiro Paulo Agostinho
Azevedo Remanso Senhor do Bonfim Afonso Rio Largo
Palmas Barragem de Monte Santo Maceió
Ilha do Porto Nacional Sobradinho Corrente Xique Lagarto Arapiraca
Bananal Dianópolis Xique Irecê Jacobina Aracaju
São Natividade Barreiras Ibotirama Serrinha Estância
Porto Artur Felix Gurupi Santana Bom Jesus Itaberaba de Santana Alagoinhas
Diamantino Cavalcante da Lapa Camaçari
Rosário Oeste Planalto Correntina Feira Santo Antônio de Jesus
Barra do Bugres do Porangatu Posse Itaberaba Salvador
Cuiabá Mato Grosso Barra do Uruaçu Niquelândia Brumado Jequié Ipiaú
ceres Garças Formosa Januária Guanambi Ilhéus
Rondonópolis Iporá Goiás Anápolis BRASÍLIA Vitória da Itabuna Una
Alto Trindade Luziânia Arinos Conquista Itapetinga
Garças Vianópolis Unaí Janaúba Salinas Almenara Porto Seguro
Itiquira Paranã Goiânia Paracatu Montes Jequitaí Teófilo Alcobaça
Coxim Jataí Itumbiara Araguari Claros Otoni
rumbá Rio Verde de Mato Grosso Uberlândia Patos de Minas

4

10°

Planalto
Mato Grosso

BRAZIL

Rio Verde

D 50° E 40° F

0 200 400 600 KM

151

Lambert Azimuthal Equal Area Projection

1 : 20 000 000

MILES 0   100   200   300

© Collins Bartholomew Ltd

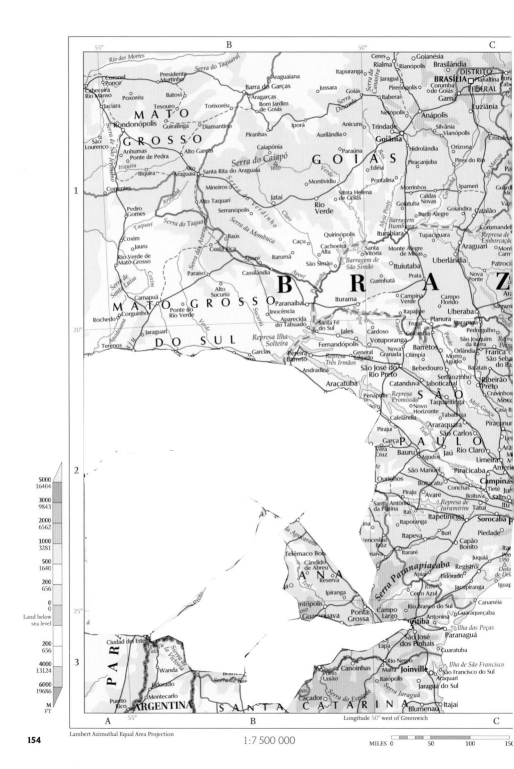

Lambert Azimuthal Equal Area Projection

1:7 500 000

MILES 0   50   100   150

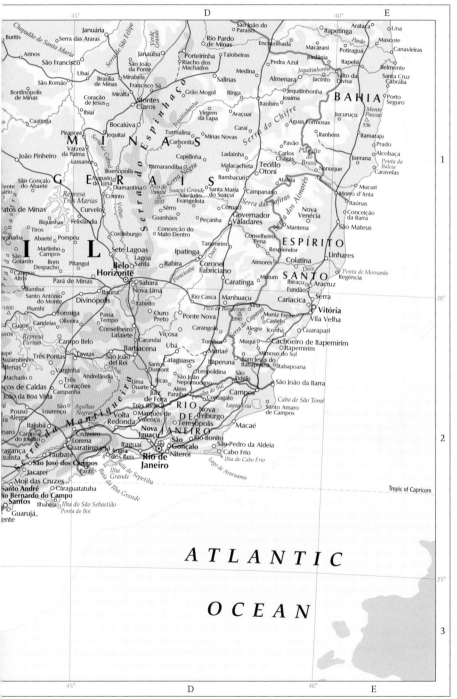

ATLANTIC

OCEAN

Tropic of Capricorn

0   100   200 KM

© Collins Bartholomew Ltd

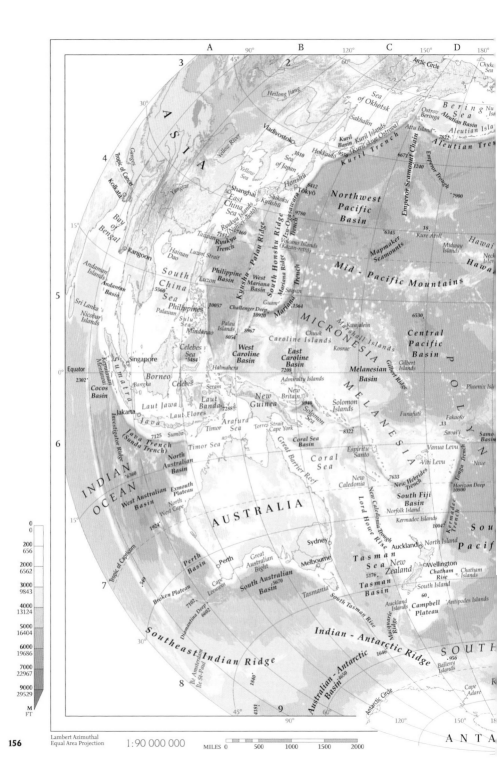

Arctic Circle

Chuku
Sea

3

2

Heilong Jiang

Sea
of Okhotsk

60°

Bering
Sea

Nu
Isl

30°

Vladivostok

Sakhalin

Ostrov
Beringa

Aleutian Basin

Attu Island

Aleutian Isla

A   S   I   A

Kuril
Basin

Kuril Islands
(Kuril'skiye Ostrova)

Aleutian Tre

4

Ganges

Tropic of Cancer

Yellow River

Hokkaido

.3510
9550

Kuril Trench

6671

Emperor Seamount Chain

Emperor Trough

1240

.7900

Kolkata

Yangtze

Yellow
Sea

Sea
of Japan

Honshu

.8412

Tokyo

Northwest

18.

Kure Atoll

15°

Shanghai

Shikoku
Kyushu

Izu-Ogasawara
Trench

Pacific

Kazan-retto)

Midway
Islands

Hawai

East
China
Sea

Ryukyu Islands
(Nansei-shoto)

9780

Basin

.6345

Mapmaker
Seamounts

Neck
Islan

Bay
of
Bengal

Rangoon

Hainan
Dao

Taiwan

7151

7460

Luzon Strait

Ryukyu
Trench

Volcano Islands

South Honshu Ridge

Mariana Ridge

Mid - Pacific Mountains

Hawa

Andaman
Islands

South

Philippine
Basin

West
Mariana
Basin

Saipan

MICRONESIA

6530.

Central
Pacific
Basin

Andaman
Sea

5560

Luzon

Kyushu - Palau Ridge

Sri Lanka

Andaman
Basin

China

Philippines

10057

Guam

Challenger Deep
10920

Mariana Trench

.1564

Kwajalein

P

5

Nicobar
Islands

Sea

Palawan

Mindanao

Sulu
Sea

Palau
Islands

8967

8054

Chuuk

Caroline Islands

Marshall Islands

Kosrae

O

Equator

0°

Singapore

Celebes
Sea

.5484

Halmahera

West
Caroline
Basin

East
Caroline
Basin

7208

Melanesian
Basin

Gilbert
Islands

Gilbert Ridge

L

Y

2302.

Sumatra

Mentawai

Cocos
Basin

Bangka

Borneo

Celebes

Seram

Admiralty Islands

New
Britain

M

E

L

A

N

E

Phoenix Isl

Jakarta

Laut Jawa

Laut
Banda

2288

New
Guinea

8940

Solomon
Islands

Funafuti

Fakaofo
.13

Samo
Basin

Java

Laut Flores

Solomon
Sea

S

I

Savai'i

7125

Sumba

Timor

Arafura
Sea

Torres Strait

Cape York

-8322

Vanua Levu

Niue

Java Trench
(Sunda Trench)

Timor Sea

Coral Sea
Basin

Espiritu
Santo

Viti Levu

Tonga Trench

6

Investigator Ridge

North
Australian
Basin

Great Barrier Reef

Coral
Sea

New
Caledonia

7633

New Hebrides
Trench

Horizon Deep
10800

.6360

INDIAN

West Australian
Basin

Exmouth
Plateau

New Caledonia Trough

South Fiji
Basin

15°

OCEAN

North
West Cape

1924

Lord Howe Rise

Norfolk Island

Kermadec Islands

10047

Sou

Pacif

AUSTRALIA

Sydney

Tasman
Sea New
Zealand

Auckland

North Island

Kermadec Trench

Perth
Basin

Perth

Great
Australian
Bight

Melbourne

5176.

Wellington

Chatham
Rise

Chatham
Islands

Tropic of Capricorn

249

Cape
Leeuwin

South Australian
Basin

5670

Tasmania

Tasman
Basin

South Island

60

7

Broken Plateau

7102

Diamantina Deep
6602

Dampier Deep

South Tasman Rise

Auckland
Islands

Campbell
Plateau

Antipodes Islands

30°

Southeast

Indian

Ridge

Ile Amsterdam
Ile St-Paul

7340

Indian - Antarctic Ridge

1646

Macquarie Ridge

.956

Balleny
Islands

SOUTH

8

4650

Australian - Antarctic
Basin

Antarctic Circle

Cape
Adare

R

4181

9

45°

90°

60°

120°

150°

Lambert Azimuthal
Equal Area Projection

1:90 000 000

MILES 0   500   1000   1500   2000

0
0

200
656

2000
6562

3000
9843

4000
13124

5000
16404

6000
19686

7000
22967

9000
29529

M
FT

A   N   T   A

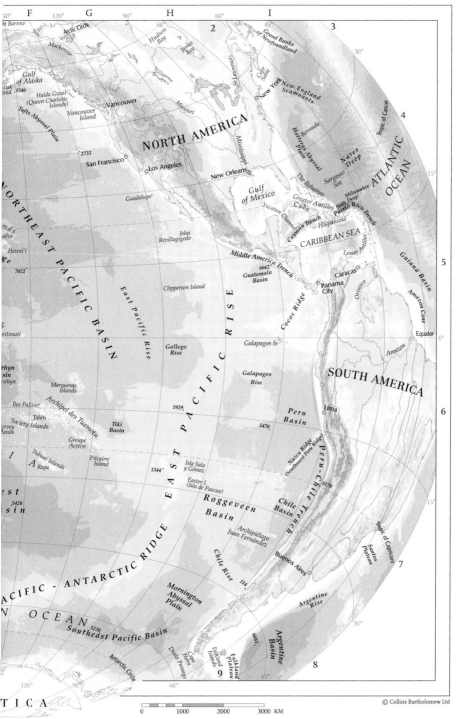

F 120° G 90° H 60° I

at Barrow
Arctic Circle
Mackenzie
Gulf of Alaska
odiak
nd .1546
Haida Gwaii
(Queen Charlotte
Islands)
Vancouver
Island
Tufts Abyssal Plain
Hudson Bay
James Bay
Foxe Basin
Cape Farewell
Grand Banks of Neufoundland
2
3
30°
4
Missouri
Vancouver
NORTH AMERICA
.2733
San Francisco
Los Angeles
New York
New England Seamounts
Bermuda
Hatteras Abyssal Plain
Nares Deep
Tropic of Cancer
15°
Guadalupe
Gulf of Mexico
Yucatan Channel
Greater Antilles
Cuba
Milwaukee Deep 8605
Puerto Rico Trench
Sargasso Sea
The Bahamas
ATLANTIC OCEAN
NORTHEAST PACIFIC BASIN
nds
ahu
Hawai'i
ge
7022
Islas Revillagigedo
Middle America Trench
6662
Guatemala Basin
Cayman Trench
Hispaniola
CARIBBEAN SEA
Lesser Antilles
Guiana Basin
Amazon Cone
5
Clipperton Island
East Pacific Rise
Cocos Ridge
Caracas
Panama City
Orinoco
ritimati
Gallego Rise
Galapagos Is
Galapagos Rise
Amazon
Equator 0°
SOUTH AMERICA
rhyn
sin
arhyn
Marquesas Islands
Iles Palliser
Archipel des Tuamotu
Tahiti
Society Islands
ervey
ands
Groupe Actéon
Tiki Basin
1929
EAST PACIFIC RISE
Peru Basin
5470
Lima
6
I
A
Tubuai Islands
Rapa
Pitcairn Island
1344
Isla Sala y Gómez
Nazca Ridge
(Sidewest Peru Ridge)
Peru-Chile Trench
8170
st
sin
5420
Easter I. (Isla de Pascua)
Roggeveen Basin
Chile Basin
15°
Archipiélago Juan Fernández
Santos Plateau
Tropic of Capricorn
7
PACIFIC - ANTARCTIC RIDGE
Chile Rise
114
Buenos Aires
Argentine Rise
OCEAN
5230
Southeast Pacific Basin
Mornington Abyssal Plain
Cape Horn
Drake Passage
Falkland Islands
Falkland Plateau
6693
Argentine Basin
8
30°
Antarctic Circle
9
120° 90° 60° 45°
TICA

0 1000 2000 3000 KM

157

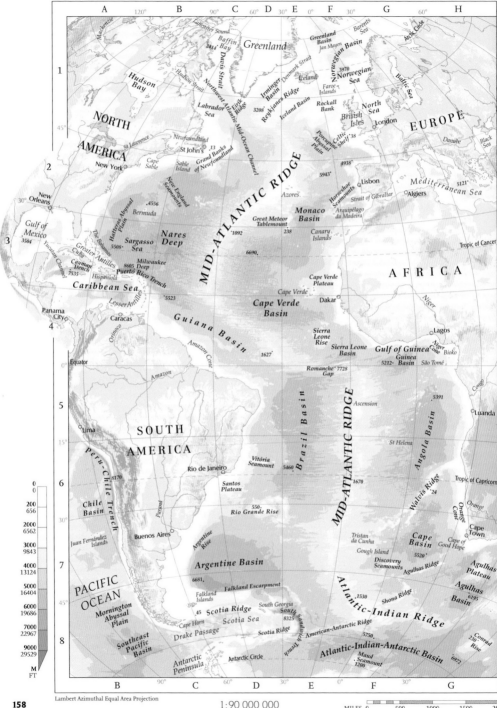

Lambert Azimuthal Equal Area Projection

1:90 000 000

MILES 0   500   1000   1500   200

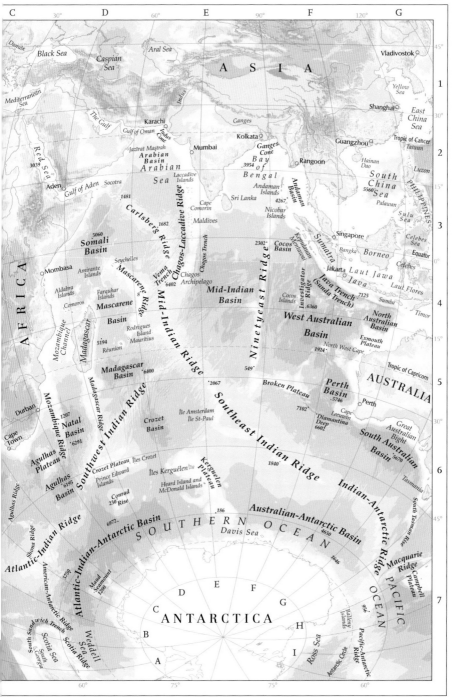

C   30°   D   60°   E   90°   F   120°   G

Danube
Black Sea
Caspian Sea
Aral Sea
Vladivostok
A S I A
45°
Mediterranean Sea
Yellow Sea
1
Shanghai
East China Sea
The Gulf
Karachi
Gulf of Oman
Indus
Ganges
30°
Tropic of Cancer
Guangzhou
Taiwan
Jazīrat Maşīrah
Arabian Basin
Indus Cone
Mumbai
Kolkata
Ganges Cone
Rangoon
Hainan Dao
Luzon
2
3039
Arabian Sea
Ganges 3954
Bay of Bengal
South China Sea
5560
Aden
Gulf of Aden
Socotra
Laccadive Islands
Andaman Islands
Palawan
PHILIPPINES
15°
1481
Sri Lanka
Andaman Basin
Sulu Sea
Carlsberg Ridge
Cape Comorin
4267
Celebes Sea
Somali Basin
5060
1682
Maldives
Nicobar Islands
Singapore
Equator
3
Mombasa
Seychelles
Chagos-Laccadive Ridge
Chagos Trench
2302
Cocos Basin
Sumatra
Bangka
Borneo
Celebes
Amirante Islands
Vema Trench
6402
Chagos Archipelago
Kerguelen Merantau
Jakarta
Laut Jawa
Java
Laut Flores
Aldabra Islands
Farquhar Islands
Mascarene Ridge
Mid-Indian Basin
Cocos Islands
Java Trench (Sunda Trench)
7125
Sumba
Comoros
Mascarene Basin
Investigator Ridge
6360
North Australian Basin
Timor
4
5194
Rodrigues Island
Mauritius
Réunion
Mid-Indian Ridge
Ninetyeast Ridge
West Australian Basin
Exmouth Plateau
15°
Madagascar Basin
6400
549
North West Cape
1924
Tropic of Capricorn
AUSTRALIA
5
Durban
1207
2067
Broken Plateau
Perth Basin
5746
Perth
Mozambique Ridge
Natal Basin
6291
Crozet Basin
Southeast Indian Ridge
7102
Cape Leeuwin
Great Australian Bight
South Australian Basin
5670
30°
Cape Town
Agulhas Plateau
Southwest Indian Ridge
Ile Amsterdam
Ile St-Paul
Diamantina Deep 6602
Agulhas Ridge
Agulhas Basin
6195
Madagascar Ridge
Crozet Plateau
Iles Crozet
1840
Indian-Antarctic Ridge
Tasmania
6
Prince Edward Islands
Îles Kerguélen
Kerguelen Plateau
South Tasman Rise
Shona Ridge
Conrad Rise 230
Heard Island and McDonald Islands
Macquarie Ridge
Atlantic-Indian Ridge
6972
186
Australian-Antarctic Basin
4650
Campbell Plateau
45°
Atlantic-Indian-Antarctic Basin
SOUTHERN OCEAN
Davis Sea
1646
PACIFIC
American-Antarctic Ridge
5759
Maud Seamount 1200
D
E
F
G
956
Balleny Islands
Pacific-Antarctic Ridge
7
C
H
OCEAN
South Sandwich Trench
Scotia Sea
South Georgia
Scotia Ridge
Weddell Sea
B
A
ANTARCTICA
I
Ross Sea
Antarctic Circle

60°   75°   75°   60°

1000   2000   3000 KM   1:90 000 000

159

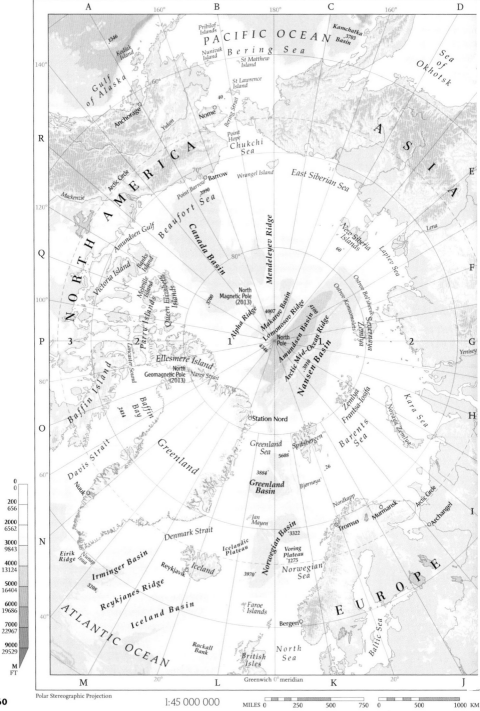

M	FT
0	0
200	656
2000	6562
3000	9843
4000	13124
5000	16404
6000	19686
7000	22967
9000	29529
M	FT

A 160° **B** 180° **C** 160° **D**

PACIFIC OCEAN

Pribilof Islands

Kamchatka 3703 Basin

Bering Sea

Nunivak Island

St Matthew Island

Sea of Okhotsk

Gulf of Alaska

St Lawrence Island

1546

Kodiak Island

60°

R

Anchorage

Yukon

Nome

Bering Strait

40

Point Hope

Chukchi Sea

A S I A

140°

Arctic Circle

Barrow

Wrangel Island

East Siberian Sea

E

70°

Point Barrow

3990

NORTH AMERICA

Mackenzie

Beaufort Sea

Amundsen Gulf

Canada Basin

Banks Island

New Siberia Islands

Lena

120°

Q

Victoria Island

Melville Island

80°

Mendeleyev Ridge

60

Laptev Sea

F

Parry Islands

Queen Elizabeth Islands

3700

North Magnetic Pole (2013)

4007

Makarov Basin

Lomonosov Ridge

2700

Ostrov Bol'shevik

Ostrov Komsomolets

Severnaya Zemlya

Yenisey

100°

P 3 Lancaster Sound **2** **1** Alpha Ridge

North Pole

4343

Amundsen Basin

3910

Arctic Mid-Ocean Ridge

Nansen Basin

2 G

Ellesmere Island

North Geomagnetic Pole (2013)

Nares Strait

Zemlya Frantsa Iosifa

Kara Sea

Baffin Bay

Baffin Island

Novaya Zemlya

H

3414

Station Nord

Spitsbergen

Barents Sea

80°

O

Davis Strait

Greenland

Greenland Sea

5608

Nordkapp

Archangel

Arctic Circle

I

60°

Nuuk

3884

Bjørnøya

.26

Tromsø

Murmansk

Greenland Basin

Jan Mayen

Nordkapp

N

Eirik Ridge

Nuuk hua

Denmark Strait

Icelandic Plateau

Norwegian Basin

3322

Voring Plateau 1275

Bergen

EUROPE

Irminger Basin

Reykjanes Ridge

Reykjavik

Iceland

3970

Norwegian Sea

Baltic Sea

3208

40°

M 20° **L** Greenwich 0° meridian **K** 20° **J**

Iceland Basin

Faroe Islands

Bergen

ATLANTIC OCEAN

Rockall Bank

British Isles

North Sea

Polar Stereographic Projection

1:45 000 000

MILES 0   250   500   750

0   500   1000 KM

INTRODUCTION TO THE INDEX

The index includes all names shown on the maps in the Atlas of the World. Names are referenced by page number and by a grid reference. The grid reference correlates to the alphanumeric values which appear within each map frame. Each entry also includes the country or geographical area in which the feature is located. Entries relating to names appearing on insets are indicated by a small box symbol: □, followed by a grid reference if the inset has its own alphanumeric values.

Name forms are as they appear on the maps, with additional alternative names or name forms included as cross-references which refer the user to the entry for the map form of the name. Names beginning with Mc or Mac are alphabetized exactly as they appear. The terms Saint, Sainte, Sankt, etc, are abbreviated to St, Ste, St, etc, but alphabetized as if in the full form.

Names of physical features beginning with generic geographical terms are permuted – the descriptive term is placed after the main part of the name. For example, Lake Superior is indexed as Superior, Lake; Mount Everest as Everest, Mount. This policy is applied to all languages.

Entries, other than those for towns and cities, include a descriptor indicating the type of geographical feature. Descriptors are not included where the type of feature is implicit in the name itself.

Administrative divisions are included to differentiate entries of the same name and feature type within the one country. In such cases, duplicate names are alphabetized in order of administrative division. Additional qualifiers are also included for names within selected geographical areas.

INDEX ABBREVIATIONS

admin. div.	administrative division	Fr. Polynesia	French Polynesia	Pol.	Poland
Afgh.	Afghanistan	g.	gulf	Port.	Portugal
Alg.	Algeria	Ger.	Germany	prov.	province
Arg.	Argentina	Guat.	Guatemala	pt	point
Austr.	Australia	hd	headland	r.	river
aut. comm.	autonomous community	Hond.	Honduras	r. mouth	river mouth
aut. reg.	autonomous region	i.	island	reg.	region
aut. rep.	autonomous republic	imp. l.	impermanent lake	resr	reservoir
Azer.	Azerbaijan	Indon.	Indonesia	rf	reef
b.	bay	is.	islands	Rus. Fed.	Russian Federation
B.I.O.T.	British Indian Ocean Territory	isth.	isthmus	S.	South
		Kazakh.	Kazakhstan	salt l.	salt lake
Bangl.	Bangladesh	Kyrg.	Kyrgyzstan	sea chan.	sea channel
Bol.	Bolivia	l.	lake	special admin. reg.	special administrative region
Bos.-Herz.	Bosnia Herzegovina	lag.	lagoon		
Bulg.	Bulgaria	Lith.	Lithuania	str.	strait
c.	cape	Lux.	Luxembourg	Switz.	Switzerland
Can.	Canada	Madag.	Madagascar	Tajik.	Tajikistan
C.A.R.	Central African Republic	Maur.	Mauritania	Tanz.	Tanzania
Col.	Colombia	Mex.	Mexico	terr.	territory
Czech Rep.	Czech Republic	Moz.	Mozambique	Thai.	Thailand
Dem. Rep.	Democratic	mt.	mountain	Trin. and Tob.	Trinidad and Tobago
Congo	Republic of the Congo	mts	mountains	Turkm.	Turkmenistan
depr.	depression	mun.	municipality	U.A.E.	United Arab Emirates
des.	desert	N.	North	U.K.	United Kingdom
Dom. Rep.	Dominican Republic	Neth.	Netherlands	Ukr.	Ukraine
Equat.	Equatorial Guinea	Nic.	Nicaragua	union terr.	union territory
Guinea		N.Z.	New Zealand	Uru.	Uruguay
esc.	escarpment	Pak.	Pakistan	U.S.A.	United States of America
est.	estuary	Para.	Paraguay	Uzbek.	Uzbekistan
Eth.	Ethiopia	pen.	peninsula	val.	valley
Fin.	Finland	Phil.	Philippines	Venez.	Venezuela
for.	forest	plat.	plateau	vol.	volcano
Fr. Guiana	French Guiana	P.N.G.	Papua New Guinea	vol. crater	volcanic crater

1

128 B2 100 Mile House Can.

A

93 E4 Aabenraa Denmark
100 C2 Aachen Ger.
93 E4 Aalborg Denmark
102 C2 Aalen Ger.
100 B2 Aalst Belgium
93 I3 Äänekoski Fin.
105 D2 Aarau Switz.
100 B2 Aarschot Belgium
70 A2 Aba China
119 D2 Aba Dem. Rep. Congo
115 C4 Aba Nigeria
81 C2 Ābādān Iran
81 D2 Ābādeh Iran
81 D3 Ābādeh Ṭashk Iran
114 B1 Abadla Alg.
155 C1 Abaeté Brazil
Abagnar Qi China see
Xilinhot
135 E3 Abajo Peak U.S.A.
115 C4 Abakaliki Nigeria
83 H3 Abakan Rus. Fed.
150 B4 Abancay Peru
81 D2 Abarküh Iran
66 D2 Abashiri Japan
66 D2 Abashiri-wan b. Japan
59 D3 Abaya P.N.G.
Abaya, Lake l. Eth. see
Lake Abaya
Ābaya Wenz r. Eth. see
Blue Nile
83 H3 Abaza Rus. Fed.
108 A2 Abbasanta Italy
104 C1 Abbeville France
141 C2 Abbeville AL U.S.A.
140 B3 Abbeville LA U.S.A.
97 B2 Abbeyfeale Ireland
55 R2 Abbot Ice Shelf Antarctica
74 B1 Abbottabad Pak.
115 E3 Abéché Chad
114 B4 Abengourou Côte d'Ivoire
114 C4 Abeokuta Nigeria
99 A3 Aberaeron U.K.
96 C2 Aberchirder U.K.
Abercorn Zambia see
Mbala
99 B4 Aberdare U.K.
99 A3 Aberdaron U.K.
53 D2 Aberdeen Austr.
122 B3 Aberdeen S. Africa
96 C2 Aberdeen U.K.
139 D3 Aberdeen MD U.S.A.
137 D1 Aberdeen SD U.S.A.
134 B1 Aberdeen WA U.S.A.
129 E1 Aberdeen Lake Can.
96 C2 Aberfeldy U.K.
96 B2 Aberfoyle U.K.
99 B4 Abergavenny U.K.
Abergwaun U.K. see
Fishguard
Aberhonddu U.K. see
Brecon
143 C2 Abernathy U.S.A.
134 B2 Abert, Lake U.S.A.
Abertawe U.K. see Swansea
Aberteifi U.K. see Cardigan
99 B4 Abertillery U.K.
99 A3 Aberystwyth U.K.
86 F2 Abez' Rus. Fed.
78 B3 Abhā Saudi Arabia
81 C2 Abhar Iran
Abiad, Bahr el r. Africa see
White Nile
114 B4 Abidjan Côte d'Ivoire
137 D3 Abilene KS U.S.A.
143 D2 Abilene TX U.S.A.
99 C4 Abingdon U.K.
138 C3 Abingdon U.S.A.
91 D3 Abinsk Rus. Fed.
130 B3 Abitibi, Lake Can.
Åbo Fin. see Turku
74 B1 Abohar India
114 B4 Aboisso Côte d'Ivoire
114 C4 Abomey Benin
60 A1 Abongabong, Gunung mt.
Indon.
118 B2 Abong Mbang Cameroon
64 A3 Aborlan Phil.
115 D3 Abou Déia Chad
106 B2 Abrantes Port.
152 B2 Abra Pampa Arg.
142 A3 Abreojos, Punta pt Mex.

116 B2 'Abri Sudan
136 A2 Absaroka Range mts U.S.A.
81 C1 Abşeron Yarımadası pen.
Azer.
78 B3 Abū 'Arīsh Saudi Arabia
116 A2 Abū Ballāş h. Egypt
79 C2 Abu Dhabi U.A.E.
116 B3 Abu Hamed Sudan
116 B3 Abu Haraz Sudan
115 C4 Abuja Nigeria
81 C2 Abū Kamāl Syria
118 C2 Abumombazi
Dem. Rep. Congo
152 B1 Abunã r. Bol./Brazil
150 C3 Abunã Brazil
74 B2 Abu Road India
78 B2 Abū Şādi, Jabal h.
Saudi Arabia
116 B2 Abū Sunbul Egypt
116 A3 Abu Zabad Sudan
Abū Ẓaby U.A.E. see
Abu Dhabi
117 A4 Abyei Sudan
145 B2 Acambaro Mex.
120 B2 Acampamento de Caça do
Mucusso Angola
106 B1 A Cañiza Spain
144 B2 Acaponeta Mex.
145 C3 Acapulco Mex.
151 E3 Acará Brazil
154 A3 Acaray, Represa de resr
Para.
150 C2 Acarigua Venez.
110 B1 Acâş Romania
145 C3 Acatlán Mex.
145 C3 Acayucán Mex.
114 B4 Accra Ghana
98 B3 Accrington U.K.
74 B2 Achalpur India
97 A2 Achill Island Ireland
101 D1 Achim Ger.
96 B2 Achnasheen U.K.
91 D2 Achuyevo Rus. Fed.
111 C3 Acıpayam Turkey
109 C3 Acireale Italy
147 C2 Acklins Island Bahamas
153 A3 Aconcagua, Cerro mt. Arg.
106 B1 A Coruña Spain
108 A2 Acqui Terme Italy
103 D2 Ács Hungary
49 N6 Actéon, Groupe is
Fr. Polynesia
145 C2 Actopán Mex.
143 D2 Ada U.S.A.
Adabazar Turkey see
Adapazarı
79 C2 Adam Oman
111 B3 Adamas Greece
135 B3 Adams Peak U.S.A.
'Adan Yemen see Aden
80 B2 Adana Turkey
111 D2 Adapazarı Turkey
97 B2 Adare Ireland
55 M2 Adare, Cape Antarctica
108 A1 Adda r. Italy
78 B2 Ad Dafinah Saudi Arabia
78 B2 Ad Dahnā' des. Saudi Arabia
79 B2 Ad Dahnā' des. Saudi Arabia
Ad Dammam Saudi Arabia
see Dammam
78 A2 Ad Dār al Ḥamrā'
Saudi Arabia
78 B3 Ad Darb Saudi Arabia
78 B2 Ad Dawādimī Saudi Arabia
Ad Dawḥah Qatar see Doha
Aḑ Diffah plat. Egypt/Libya
see Libyan Plateau
78 B2 Ad Dilam Saudi Arabia
78 B2 Ad Dir'īyah Saudi Arabia
117 B4 Addis Ababa Eth.
81 C2 Ad Dīwānīyah Iraq
141 D2 Adel U.S.A.
52 A2 Adelaide Austr.
55 A3 Adelaide Island Antarctica
50 C1 Adelaide River Austr.
101 D2 Adelebsen Ger.
55 K2 Adélie Land reg. Antarctica
116 C3 Aden Yemen
116 C3 Aden, Gulf of Somalia/
Yemen
100 C2 Adenau Ger.
115 C3 Aderbissinat Niger
79 C2 Adh Dhayd U.A.E.
59 C3 Adi i. Indon.
116 B3 Adī Ark'ay Eth.
108 B1 Adige r. Italy
116 B3 Adigrat Eth.
78 A3 Adi Keyih Eritrea
74 B3 Adilabad India
115 D2 Adīrī Libya

139 E2 Adirondack Mountains
U.S.A.
Ādīs Ābeba Eth. see
Addis Ababa
117 B4 Ādīs Alem Eth.
80 B2 Adiyaman Turkey
110 C1 Adjud Romania
50 B1 Admiralty Gulf Austr.
128 A2 Admiralty Island U.S.A.
59 D3 Admiralty Islands P.N.G.
73 B3 Adoni India
104 B3 Adour r. France
106 C2 Adra Spain
114 B2 Adrar Alg.
138 C2 Adrian MI U.S.A.
143 C1 Adrian TX U.S.A.
108 B2 Adriatic Sea Europe
Adua Eth. see Ādwa
116 B3 Ādwa Eth.
83 K2 Adycha r. Rus. Fed.
91 D3 Adygeysk Rus. Fed.
114 B4 Adzopé Côte d'Ivoire
111 B3 Aegean Sea Greece/Turkey
101 D1 Aerzen Ger.
106 B1 A Estrada Spain
116 B3 Afabet Eritrea
Affreville Alg. see
Khemis Miliana
76 C3 Afghanistan country Asia
78 B2 'Afif Saudi Arabia
136 A2 Afton U.S.A.
80 B2 Afyon Turkey
115 C3 Agadez Niger
114 B1 Agadir Morocco
113 I7 Agalega Islands Mauritius
Agana Guam see Hagåtña
74 B2 Agar India
119 D2 Agaro Eth.
75 D2 Agartala India
81 C2 Ağdam (abandoned) Azer.
105 C3 Agde France
115 D1 Agedabia Libya see Ajdābiyā
104 C3 Agen France
122 A2 Aggeneys S. Africa
111 C3 Agia Varvara Greece
111 B3 Agios Dimitrios Greece
111 C3 Agios Efstratios i. Greece
111 C3 Agios Kirykos Greece
111 C3 Agios Nikolaos Greece
78 A3 Agirwat Hills Sudan
123 C2 Agisanang S. Africa
110 B1 Agnita Romania
74 B2 Agra India
81 C2 Ağrı Turkey
Ağrı Dağı mt. Turkey see
Ararat, Mount
108 B3 Agrigento Italy
111 B3 Agrinio Greece
109 B2 Agropoli Italy
87 E3 Agryz Rus. Fed.
144 B2 Agua Brava, Laguna lag.
Mex.
154 B2 Água Clara Brazil
145 C3 Aguada Mex.
146 B4 Aguadulce Panama
144 B2 Aguanaval r. Mex.
144 B1 Agua Prieta Mex.
144 B2 Aguascalientes Mex.
155 D1 Águas Formosas Brazil
154 C2 Agudos Brazil
106 B1 Águeda Port.
114 C3 Aguelhok Mali
106 C1 Aguilar de Campoo Spain
107 C2 Águilas Spain
144 B3 Aguililla Mex.
122 B3 Agulhas, Cape S. Africa
158 F7 Agulhas Basin
Southern Ocean
155 D2 Agulhas Negras mt. Brazil
158 F7 Agulhas Plateau
Southern Ocean
158 F7 Agulhas Ridge
S. Atlantic Ocean
111 C2 Ağva Turkey
115 C2 Ahaggar plat. Alg.
115 C2 Ahaggar, Tassili oua-n-
plat. Alg.
81 C2 Ahar Iran
100 C1 Ahaus Ger.
81 C2 Ahlat Turkey
100 C2 Ahlen Ger.
74 B2 Ahmadabad India
74 B3 Ahmadnagar India
74 B1 Ahmadpur East Pak.
74 B1 Ahmadpur Sial Pak.
117 C4 Ahmar mts Eth.
Ahmedabad India see
Ahmadabad
Ahmednagar India see
Ahmadnagar

144 B2 Ahome Mex.
81 D3 Ahram Iran
101 E1 Ahrensburg Ger.
104 C2 Ahun France
93 F4 Åhus Sweden
81 C2 Ahvāz Iran
Ahvenanmaa is Fin. see
Åland Islands
122 A2 Ai-Ais Namibia
74 A1 Aībak Afgh.
80 B2 Aigialousa Cyprus
111 B3 Aigio Greece
Aihui China see Heihe
Aijal India see Aizawl
141 D2 Aiken U.S.A.
97 B1 Ailt an Chorráin Ireland
155 D1 Aimorés Brazil
155 D1 Aimorés, Serra dos hills
Brazil
105 D2 Ain r. France
107 E2 Aïn Azel Alg.
115 C1 Aïn Beïda Alg.
114 B2 'Aïn Ben Tili Maur.
107 D2 Aïn Defla Alg.
114 B1 Aïn Sefra Alg.
136 D2 Ainsworth U.S.A.
Aintab Turkey see Gaziar
107 D2 Aïn Taya Alg.
107 D2 Aïn Tédélès Alg.
107 C2 Aïn Temouchent Alg.
115 C3 Aïr, Massif de l' mts Nige
60 A1 Airbangis Indon.
128 C2 Airdrie Can.
96 C3 Airdrie U.K.
104 B3 Aire-sur-l'Adour France
101 E3 Aisch r. Ger.
128 A1 Aishihik Lake Can.
100 A3 Aisne r. France
59 D3 Aitape P.N.G.
137 E1 Aitkin U.S.A.
110 B1 Aiud Romania
105 D3 Aix-en-Provence France
Aix-la-Chapelle Ger. see
Aachen
105 D2 Aix-les-Bains France
75 D2 Aizawl India
88 C2 Aizkraukle Latvia
88 B2 Aizpute Latvia
67 C3 Aizuwakamatsu Japan
105 D3 Ajaccio France
Ajayameru India see Ajme
115 E1 Ajdābiyā Libya
115 C2 Ajjer, Tassili n' plat. Alg.
79 C2 'Ajman U.A.E.
74 B2 Ajmer India
Ajmer-Merwara India see
Ajmer
142 A2 Ajo U.S.A.
77 D2 Akadyr Kazakh.
119 D2 Ak'ak'ī Beseka Eth.
Akamagaseki Japan see
Shimonoseki
54 B2 Akaroa N.Z.
87 E3 Akbulak Rus. Fed.
80 B2 Akçakale Turkey
114 A3 Akchâr reg. Maur.
111 C3 Akdağ mt. Turkey
80 B2 Akdağmadeni Turkey
88 A2 Åkersberga Sweden
118 C2 Aketi Dem. Rep. Congo
81 C1 Akhalkalaki Georgia
81 C1 Akhaltsikhe Georgia
79 C2 Akhdar, Jabal mts Oman
111 C3 Akhisar Turkey
87 D4 Akhtubinsk Rus. Fed.
118 B3 Akiéni Gabon
130 B2 Akimiski Island Can.
66 D3 Akita Japan
114 A3 Akjoujt Maur.
Akkerman Ukr. see
Bilhorod-Dnistrovs'kyy
77 D1 Akkol' Kazakh.
Ak-Mechet Kazakh. see
Kyzylorda
88 B2 Akmenrags pt Latvia
Akmola Kazakh. see Astana
Akmolinsk Kazakh. see
Astana
67 B4 Akō Japan
117 B4 Akobo S. Sudan
74 B2 Akola India
118 B2 Akonolinga Cameroon
78 A3 Akordat Eritrea
131 D1 Akpatok Island Can.
77 D2 Akqi China
92 □A3 Akranes Iceland
136 C2 Akron CO U.S.A.
138 C2 Akron OH U.S.A.
75 D1 Aksai Chin terr. Asia
80 B2 Aksaray Turkey

86 F2 Aksarka Rus. Fed.
76 B1 Aksay Kazakh.
91 D2 Aksay Rus. Fed.
80 B2 Akşehir Turkey
76 C2 Akshiganak Kazakh.
77 E2 Aksu China
116 B3 Āksum Eth.
76 B2 Aktau Kazakh.
76 B1 Aktobe Kazakh.
77 D2 Aktogay Kazakh.
88 C3 Aktsyabrski Belarus
Aktyubinsk Kazakh. see
Aktobe
67 B4 Akune Japan
115 C4 Akure Nigeria
92 □B2 Akureyri Iceland
Akyab Myanmar see Sittwe
111 D2 Akyazı Turkey
77 C2 Akzhaykyn, Ozero salt l.
Kazakh.
140 C2 Alabama r. U.S.A.
140 C2 Alabama state U.S.A.
140 C2 Alabaster U.S.A.
111 C3 Alaçatı Turkey
145 D2 Alacrán, Arrecife rf Mex.
81 C1 Alagir Rus. Fed.
151 F4 Alagoinhas Brazil
107 C1 Alagón Spain
78 B2 Al Aḩmadī Kuwait
77 E2 Alakol', Ozero salt l.
Kazakh.
92 J2 Alakurtti Rus. Fed.
78 B3 Al 'Alayyah Saudi Arabia
81 C2 Al 'Amādiyah Iraq
81 C2 Al 'Amārah Iraq
80 A2 Al 'Āmirīyah Egypt
143 C3 Alamitos, Sierra de los mt.
Mex.
135 C3 Alamo U.S.A.
142 B2 Alamogordo U.S.A.
144 A2 Alamos Sonora Mex.
144 B2 Alamos Sonora Mex.
144 B2 Alamos r. Mex.
136 B3 Alamosa U.S.A.
Åland Islands
93 G3 Åland Islands is Fin.
80 B2 Alanya Turkey
73 B4 Alappuzha India
80 B3 Al 'Aqabah Jordan
78 B2 Al 'Aqiq Saudi Arabia
107 C2 Alarcón, Embalse de resr
Spain
80 B2 Al 'Arīsh Egypt
78 B2 Al Arṭāwīyah Saudi Arabia
61 C2 Alas Indon.
111 C3 Alaşehir Turkey
128 A2 Alaska state U.S.A.
124 D4 Alaska, Gulf of U.S.A.
126 B3 Alaska Peninsula U.S.A.
126 C2 Alaska Range mts U.S.A.
81 C2 Ālāt Azer.
87 D3 Alatyr' Rus. Fed.
150 B3 Alausí Ecuador
93 H3 Alavus Fin.
52 B2 Alawoona Austr.
79 C2 Al 'Ayn U.A.E.
108 A2 Alba Italy
107 C2 Albacete Spain
78 B2 Al Badā'i' Saudi Arabia
78 B2 Al Badī' Saudi Arabia
110 B3 Alba Iulia Romania
109 C2 Albania country Europe
50 A3 Albany Austr.
141 D2 Albany r. Can.
141 D2 Albany U.S.A.
139 E2 Albany NY U.S.A.
134 B2 Albany OR U.S.A.
Al Başrah Iraq see Basra
51 D1 Albatross Bay Austr.
116 A2 Al Bawīṭī Egypt
15 E1 Al Baydā' Libya
78 B3 Al Baydā' Yemen
141 D1 Albemarle U.S.A.
141 E1 Albemarle Sound sea chan.
U.S.A.
108 A2 Albenga Italy
51 C2 Alberga watercourse Austr.
119 D2 Albert, Lake
Dem. Rep. Congo/Uganda
128 C2 Alberta prov. Can.
100 B2 Albert Kanaal canal Belgium
37 E2 Albert Lea U.S.A.
17 B4 Albert Nile r. Sudan/
Uganda
23 C2 Alberton S. Africa
Albertville Dem. Rep.
Congo see Kalemie
105 D2 Albertville France
104 C3 Albi France

151 D2 Albina Suriname
78 A2 Al Bi'r Saudi Arabia
78 B3 Al Birk Saudi Arabia
78 B2 Al Biyāḍh reg. Saudi Arabia
106 C2 Alborán, Isla de i. Spain
106 C2 Alboran Sea Europe
Alborz, Reshteh-ye mts Iran
see Elburz Mountains
107 D2 Albox Spain
106 B2 Albufeira Port.
142 B1 Albuquerque U.S.A.
79 C2 Al Buraymī Oman
115 E1 Al Burdī Libya
53 C3 Albury Austr.
106 B2 Alcácer do Sal Port.
106 C1 Alcalá de Henares Spain
106 C2 Alcalá la Real Spain
108 B3 Alcamo Italy
107 C1 Alcañiz Spain
106 B2 Alcántara Spain
107 C2 Alcantarilla Spain
106 C2 Alcaraz Spain
106 C2 Alcaraz, Sierra de mts Spain
106 C2 Alcaudete Spain
106 C2 Alcázar de San Juan Spain
Alcazarquivir Morocco see
Ksar el Kebir
91 D2 Alchevs'k Ukr.
155 E1 Alcobaça Brazil
107 C2 Alcoy-Alcoi Spain
107 D2 Alcúdia Spain
113 H6 Aldabra Islands Seychelles
142 B3 Aldama Mex.
145 C2 Aldama Mex.
83 J3 Aldan Rus. Fed.
83 J2 Aldan r. Rus. Fed.
99 D3 Aldeburgh U.K.
95 C4 Alderney i. Channel Is
99 C4 Aldershot U.K.
114 A3 Aleg Maur.
155 D2 Alegre Brazil
152 C2 Alegrete Brazil
89 D1 Alekhovshchina Rus. Fed.
89 E2 Aleksandrov Rus. Fed.
Aleksandrovsk Ukr. see
Zaporizhzhya
83 K3 Aleksandrovsk-Sakhalinskiy
Rus. Fed.
82 E1 Aleksandry, Zemlya i.
Rus. Fed.
91 D1 Alekseyevka Rus. Fed.
91 D1 Alekseyevka Rus. Fed.
89 E3 Aleksin Rus. Fed.
109 D2 Aleksinac Serbia
118 B3 Alèmbè Gabon
155 D2 Além Paraíba Brazil
93 F3 Ålen Norway
104 C2 Alençon France
80 B2 Aleppo Syria
150 B4 Alerta Peru
128 B2 Alert Bay Can.
105 C3 Alès France
110 B1 Aleşd Romania
Aleshki Ukr. see
Tsyurupyns'k
108 A2 Alessandria Italy
Alessio Albania see Lezhë
93 E3 Ålesund Norway
156 D2 Aleutian Basin Bering Sea
124 A4 Aleutian Islands U.S.A.
83 L3 Alevina, Mys c. Rus. Fed.
128 A2 Alexander Archipelago is
U.S.A.
122 A2 Alexander Bay S. Africa
140 C2 Alexander City U.S.A.
55 A2 Alexander Island Antarctica
53 C3 Alexandra Austr.
54 A3 Alexandra N.Z.
153 E5 Alexandra, Cape S. Georgia
Alexandra Land i. Rus. Fed.
see Aleksandry, Zemlya
111 B3 Alexandreia Greece
116 A1 Alexandria Egypt
110 C2 Alexandria Romania
123 C3 Alexandria S. Africa
96 B3 Alexandria U.K.
140 B2 Alexandria LA U.S.A.
137 D1 Alexandria MN U.S.A.
139 D3 Alexandria VA U.S.A.
52 A3 Alexandrina, Lake Austr.
111 C2 Alexandroupoli Greece
131 E2 Alexis r. Can.
128 B2 Alexis Creek Can.
77 E1 Aleysk Rus. Fed.
107 C1 Alfaro Spain
81 C3 Al Fāw Iraq
80 B3 Al Fayyūm Egypt
101 D2 Alfeld (Leine) Ger.

155 C2 Alfenas Brazil
96 C2 Alford U.K.
Al Fujayrah U.A.E. see
Fujairah
Al Furāt r. Iraq/Syria see
Euphrates
93 E4 Ålgård Norway
106 B2 Algarve reg. Port.
106 B2 Algeciras Spain
107 C2 Algemesí Spain
78 A3 Algena Eritrea
Alger Alg. see Algiers
114 C2 Algeria country Africa
79 C3 Al Ghaydah Yemen
108 A2 Alghero Italy
116 B2 Al Ghurdaqah Egypt
79 B2 Al Ghwaybiyah Saudi Arabia
115 C1 Algiers Alg.
123 C3 Algoa Bay S. Africa
137 E2 Algona U.S.A.
106 C1 Algorta Spain
Alguieirao Moz. see
Hacufera
81 C2 Al Ḥadithah Iraq
79 C2 Al Ḩajar al Gharbī mts
Oman
107 C2 Alhama de Murcia Spain
80 A2 Al Ḩammām Egypt
78 B2 Al Ḩanākiyah Saudi Arabia
81 C2 Al Ḩasakah Syria
78 B2 Al Ḩawiyah Saudi Arabia
81 C2 Al Ḩayy Iraq
79 C3 Al Ḩibāk des. Saudi Arabia
78 B2 Al Ḩillah Iraq see Hillah
79 B2 Al Ḩinnāh Saudi Arabia
Al Ḩudaydah Yemen see
Hodeidah
79 B2 Al Ḩufūf Saudi Arabia
115 D2 Al Ḩulayq al Kabīr hills
Libya
79 C2 'Alīābād Iran
111 C3 Aliağa Turkey
111 B2 Aliakmonas r. Greece
107 C2 Alicante Spain
143 D3 Alice U.S.A.
109 C3 Alice, Punta pt Italy
51 C2 Alice Springs Austr.
77 D3 Alichur Tajik.
74 B2 Aligarh India
81 C2 Aligudarz Iran
69 E1 Alihe China
118 B3 Alima r. Congo
118 C2 Alindao C.A.R.
111 C3 Aliova r. Turkey
74 B2 Alirajpur India
117 C3 Ali Sabieh Djibouti
78 A1 Al 'Īsāwīyah Saudi Arabia
Al Iskandarīyah Egypt see
Alexandria
116 B1 Al Ismā'īlīyah Egypt
123 C3 Aliwal North S. Africa
115 D2 Al Jaghbūb Libya
78 B2 Al Jahrah Kuwait
115 E2 Al Jawf Libya
115 D1 Al Jawsh Libya
106 B2 Aljezur Port.
Al Jīzah Egypt see Giza
78 B2 Al Jubayl Saudi Arabia
79 B2 Al Jubaylah Saudi Arabia
115 D2 Al Jufrah Libya
79 C2 Al Jumaylīyah Qatar
78 B2 Al Junaynah Saudi Arabia
106 B2 Aljustrel Port.
78 B2 Al Kahfah Saudi Arabia
79 C2 Al Kāmil Oman
80 B2 Al Karak Jordan
79 C2 Al Kāẓimīyah Iraq
79 C2 Al Khābūrah Oman
78 B2 Al Khamāsin Saudi Arabia
116 B2 Al Khārijah Egypt
78 B3 Al Khaṣab Oman
79 C2 Al Khawr Qatar
115 D1 Al Khums Libya
79 B2 Al Khunn Saudi Arabia
79 C2 Al Kidan well Saudi Arabia
79 C2 Al Kir'ānah Qatar
100 B1 Alkmaar Neth.
115 E2 Al Kufrah Libya
81 C2 Al Kūt Iraq
Al Kuwayt Kuwait see
Kuwait
Al Lādhiqīyah Syria see
Latakia
75 C2 Allahabad India
83 K2 Allakh-Yun' Rus. Fed.
78 A2 'Allāqī, Wādī al watercourse
Egypt
139 D2 Allegheny r. U.S.A.

139 C3 Allegheny Mountains
U.S.A.
97 B1 Allen, Lough l. Ireland
145 B2 Allende Mex.
145 B2 Allende Mex.
139 D2 Allentown U.S.A.
Alleppey India see
Alappuzha
101 D1 Aller r. Ger.
136 C2 Alliance NE U.S.A.
138 C2 Alliance OH U.S.A.
78 B2 Al Līth Saudi Arabia
96 C2 Alloa U.K.
131 C3 Alma Can.
Alma-Ata Kazakh. see
Almaty
106 B2 Almada Port.
106 C2 Almadén Spain
Al Madīnah Saudi Arabia
see Medina
80 B2 Al Mafraq Jordan
78 B3 Al Maḩwīt Yemen
78 B2 Al Majma'ah Saudi Arabia
116 B2 Al Maks al Baḩrī Egypt
Al Manāmah Bahrain see
Manama
135 B2 Almanor, Lake U.S.A.
107 C2 Almansa Spain
80 B2 Al Manṣūrah Egypt
79 C2 Al Mariyyah U.A.E.
115 E1 Al Marj Libya
77 D2 Almaty Kazakh.
Al Mawşil Iraq see Mosul
81 C2 Al Mayādin Syria
106 C1 Almazán Spain
151 D3 Almeirim Brazil
100 C1 Almelo Neth.
155 D1 Almenara Brazil
106 B1 Almendra, Embalse de resr
Spain
106 B2 Almendralejo Spain
106 C2 Almería Spain
106 C2 Almería, Golfo de b. Spain
87 E3 Al'met'yevsk Rus. Fed.
78 B2 Al Mindak Saudi Arabia
116 B2 Al Minyā Egypt
79 B2 Al Mish'āb Saudi Arabia
106 B2 Almodôvar Port.
106 B2 Almonte Spain
75 B2 Almora India
79 B2 Al Mubarrez Saudi Arabia
79 C2 Al Mudaybī Oman
80 B3 Al Mudawwarah Jordan
Al Mukallā Yemen see
Mukalla
Al Mukhā Yemen see
Mocha
106 C2 Almuñécar Spain
81 C2 Al Muqdādīyah Iraq
78 A2 Al Musayjid Saudi Arabia
78 A2 Al Muwaylih Saudi Arabia
111 B3 Almyros Greece
96 B2 Alness U.K.
98 C2 Alnwick U.K.
62 A1 Along India
111 B3 Alonnisos i. Greece
59 C3 Alor i. Indon.
59 C3 Alor, Kepulauan is Indon.
60 B1 Alor Star Malaysia
Alost Belgium see Aalst
59 E3 Alotau P.N.G.
86 C2 Alozero (abandoned)
Rus. Fed.
138 C1 Alpena U.S.A.
160 Q1 Alpha Ridge Arctic Ocean
100 B1 Alphen aan den Rijn Neth.
142 B3 Alpine AZ U.S.A.
143 C2 Alpine TX U.S.A.
84 E4 Alps mts Europe
79 B3 Al Qa'āmīyāt reg.
Saudi Arabia
115 D1 Al Qaddāḩīyah Libya
Al Qāhirah Egypt see Cairo
78 B2 Al Qā'īyah Saudi Arabia
81 C2 Al Qāmishlī Syria
80 B2 Al Qaryatayn Syria
79 B2 Al Qaṣab Saudi Arabia
116 A2 Al Qaṣr Egypt
79 B3 Al Qaṭn Yemen
115 D2 Al Qaṭrūn Libya
106 B2 Alqueva, Barragem de resr
Port.
80 B2 Al Qunayṭirah (abandoned)
Syria
78 B3 Al Qunfidhah Saudi Arabia
116 B2 Al Quṣayr Egypt
78 B2 Al Quwārah Saudi Arabia
78 B2 Al Quwayīyah Saudi Arabia
101 D2 Alsfeld Ger.
98 B2 Alston U.K.

92 H2 Alta Norway
92 H2 Altaelva r. Norway
68 B1 Altai Mountains Asia
141 D2 Altamaha r. U.S.A.
151 D3 Altamira Brazil
109 C2 Altamura Italy
144 A1 Altar, Desierto de des. Mex.
77 E2 Altay China
68 C1 Altay Mongolia
105 D2 Altdorf Switz.
107 C2 Altea Spain
92 H1 Alteidet Norway
101 F2 Altenburg Ger.
100 C2 Altenkirchen (Westerwald) Ger.
101 F1 Altentreptow Ger.
111 C3 Altınoluk Turkey
111 D3 Altıntaş Turkey
152 B1 Altiplano plain Bol.
96 B1 Altnaharra U.K.
154 B1 Alto Araguaia Brazil
154 B1 Alto Garças Brazil
121 C2 Alto Ligonha Moz.
121 C2 Alto Molócuè Moz.
99 C4 Alton U.K.
137 E3 Alton U.S.A.
129 E3 Altona Can.
139 D2 Altoona U.S.A.
154 B1 Alto Sucuriú Brazil
154 B1 Alto Taquari Brazil
102 C2 Altötting Ger.
68 C2 Altun Shan mt. China
68 B2 Altun Shan mts China
134 B2 Alturas U.S.A.
143 D2 Altus U.S.A.
88 C2 Alūksne Latvia
78 A2 Al 'Ulā Saudi Arabia
115 D1 Al 'Uqaylah Libya
Al Uqşur Egypt see Luxor
91 C3 Alushta Ukr.
115 E2 Al 'Uwaynāt Libya
143 D1 Alva U.S.A.
145 C3 Alvarado Mex.
93 F3 Älvdalen Sweden
93 F3 Älvdalen val. Sweden
92 H2 Älvsbyn Sweden
78 A2 Al Wajh Saudi Arabia
79 C2 Al Wakrah Qatar
74 B2 Alwar India
81 C2 Al Widyān plat. Iraq/Saudi Arabia
Alxa Youqi China see Ehen Hudag
Alxa Zuoqi China see Bayan Hot
51 C1 Alyangula Austr.
88 B3 Alytus Lith.
136 C1 Alzada U.S.A.
101 D3 Alzey Ger.
50 C2 Amadeus, Lake imp. l. Austr.
127 H2 Amadjuak Lake Can.
106 B2 Amadora Port.
78 B2 Amā'ir Saudi Arabia
67 B4 Amakusa-Shimo-shima i. Japan
93 F4 Åmål Sweden
111 B3 Amaliada Greece
59 D3 Amamapare Indon.
154 A2 Amambaí Brazil
154 B2 Amambaí r. Brazil
154 A2 Amambaí, Serra de hills Brazil/Para.
69 E3 Amami-Ō-shima i. Japan
69 E3 Amami-shotō is Japan
77 C1 Amankel'dy Kazakh.
109 C3 Amantea Italy
151 D2 Amapá Brazil
106 B2 Amareleja Port.
88 B2 Āmari Estonia
143 C1 Amarillo U.S.A.
108 B2 Amaro, Monte mt. Italy
80 B1 Amasya Turkey
150 D2 Amazon r. S. America
151 E2 Amazon, Mouths of the Brazil
Amazonas r. S. America see Amazon
157 I5 Amazon Cone S. Atlantic Ocean
74 B1 Ambala India
121 □D2 Ambalavao Madag.
121 □D2 Ambanja Madag.
104 B3 Ambarès-et-Lagrave France
150 B3 Ambato Ecuador
121 □D3 Ambato Boeny Madag.
121 □D3 Ambato Finandrahana Madag.
121 □D2 Ambatolampy Madag.
121 □D3 Ambatondrazaka Madag.
101 E3 Amberg Ger.

146 B3 Ambergris Caye i. Belize
75 C2 Ambikapur India
121 □D2 Ambilobe Madag.
98 C2 Amble U.K.
98 B2 Ambleside U.K.
121 □D3 Amboasary Madag.
121 □D2 Ambodifotatra Madag.
121 □D2 Ambohidratrimo Madag.
121 □D3 Ambohimahasoa Madag.
Amboina Indon. see Ambon
59 C3 Ambon Indon.
59 C3 Ambon i. Indon.
121 □D3 Ambositra Madag.
121 □D3 Ambovombe Madag.
135 C4 Amboy U.S.A.
Ambre, Cap d' c. Madag. see Bobaomby, Tanjona
120 A1 Ambriz Angola
Ambrizete Angola see N'zeto
86 F2 Amderma Rus. Fed.
Amdo China see Lharigarbo
145 B2 Amealco Mex.
144 B2 Ameca Mex.
100 B1 Ameland i. Neth.
154 C2 Americana Brazil
55 D4 American-Antarctic Ridge S. Atlantic Ocean
134 D2 American Falls U.S.A.
134 D2 American Falls Reservoir U.S.A.
135 D2 American Fork U.S.A.
49 J5 American Samoa terr. S. Pacific Ocean
141 D2 Americus U.S.A.
100 B1 Amersfoort Neth.
55 H2 Amery Ice Shelf Antarctica
137 E2 Ames U.S.A.
111 B3 Amfissa Greece
83 J2 Amga Rus. Fed.
66 C1 Amgu Rus. Fed.
115 C2 Amguid Alg.
83 K3 Amgun' r. Rus. Fed.
131 D3 Amherst Can.
104 C2 Amiens France
79 C3 Amilḥayt, Wādī al r. Oman
73 B3 Amindivi Islands India
122 A1 Aminuis Namibia
74 A2 Amir Chah Pak.
129 D2 Amisk Lake Can.
143 C3 Amistad Reservoir Mex./U.S.A.
98 A3 Amlwch U.K.
80 B2 'Ammān Jordan
127 J2 Ammassalik Greenland
81 D2 Āmol Iran
111 C3 Amorgos i. Greece
140 C2 Amory U.S.A.
130 C3 Amos Can.
Amoy China see Xiamen
121 □D3 Ampanihy Madag.
155 C2 Amparo Brazil
121 □D2 Ampasimanolotra Madag.
107 D1 Amposta Spain
78 B3 'Amrān Yemen
Amraoti India see Amravati
74 B2 Amravati India
74 B2 Amreli India
74 B1 Amritsar India
100 B1 Amstelveen Neth.
100 B1 Amsterdam Neth.
123 D2 Amsterdam S. Africa
156 A8 Amsterdam, Île i. Indian Ocean
103 C2 Amstetten Austria
115 E3 Am Timan Chad
76 B2 Amudar'ya r. Asia
126 F1 Amund Ringnes Island Can.
55 J3 Amundsen, Mount Antarctica
160 H1 Amundsen Basin Arctic Ocean
126 D2 Amundsen Gulf Can.
55 P2 Amundsen Ridges Southern Ocean
55 P2 Amundsen Sea Antarctica
61 C2 Amuntai Indon.
Amur r. China see Heilong Jiang
78 A3 'Amur, Wadi watercourse Sudan
61 D2 Anabanua Indon.
83 I2 Anabar r. Rus. Fed.
83 I2 Anabarskiy Zaliv b. Rus. Fed.
150 C2 Anaco Venez.
134 D2 Anaconda U.S.A.
134 B1 Anacortes U.S.A.
143 D1 Anadarko U.S.A.

80 B1 Anadolu Dağları mts Turkey
83 M2 Anadyr' Rus. Fed.
83 M2 Anadyr' r. Rus. Fed.
81 C2 'Ānah Iraq
145 B2 Anáhuac Mex.
73 B3 Anai Mudi India
121 □D2 Analalava Madag.
121 □D3 Analavelona mts Madag.
60 B1 Anambas, Kepulauan is Indon.
137 E2 Anamosa U.S.A.
80 B2 Anamur Turkey
67 B4 Anan Japan
73 B3 Anantapur India
74 B1 Anantnag India
90 B2 Anan'yiv Ukr.
91 D3 Anapa Rus. Fed.
154 C1 Anápolis Brazil
81 D2 Anār Iran
152 B2 Añatuya Arg.
97 B1 An Baile Thiar Ireland
97 B1 An Bun Beag Ireland
65 B2 Anbyon N. Korea
104 B2 Ancenis France
126 C2 Anchorage U.S.A.
108 B2 Ancona Italy
153 A4 Ancud Chile
Anda China see Daqing
97 A2 An Daingean Ireland
93 E3 Åndalsnes Norway
106 C2 Andalucía aut. comm. Spain
Andalusia aut. comm. Spain see Andalucía
140 C2 Andalusia U.S.A.
159 F3 Andaman Basin Indian Ocean
73 D3 Andaman Islands India
63 A2 Andaman Sea Indian Ocean
121 □D2 Andapa Madag.
100 B2 Andelst Neth.
92 G2 Andenes Norway
100 B2 Andenne Belgium
100 B2 Anderlecht Belgium
105 D2 Andermatt Switz.
126 D2 Anderson r. Can.
126 C2 Anderson AK U.S.A.
138 B2 Anderson IN U.S.A.
141 D2 Anderson SC U.S.A.
148 C3 Andes mts S. America
77 D2 Andijon Uzbek.
121 □D2 Andilamena Madag.
121 □D2 Andilanatoby Madag.
Andizhan Uzbek. see Andijon
74 A1 Andkhōy Afgh.
121 □D2 Andoany Madag.
Andong China see Dandong
65 B2 Andong S. Korea
104 C3 Andorra country Europe
104 C3 Andorra la Vella Andorra
99 C4 Andover U.K.
154 B2 Andradina Brazil
89 D2 Andreapol' Rus. Fed.
155 D2 Andrelândia Brazil
143 C2 Andrews U.S.A.
109 C2 Andria Italy
121 □D3 Androka Madag.
Andropov Rus. Fed. see Rybinsk
146 C2 Andros i. Bahamas
111 B3 Andros i. Greece
111 B3 Andros i. Greece
141 E4 Andros Town Bahamas
73 B3 Andrott i. India
90 B1 Andrushivka Ukr.
92 G2 Andselv Norway
106 C2 Andújar Spain
120 A2 Andulo Angola
114 C3 Anéfis Mali
147 D3 Anegada Passage Virgin Is (U.K.)
114 C4 Aného Togo
107 D1 Aneto mt. Spain
115 D3 Aney Niger
97 B1 An Fál Carrach Ireland
83 H3 Angara r. Rus. Fed.
68 C1 Angarsk Rus. Fed.
93 G3 Änge Sweden
Angel, Salto del waterfall Venez. see Angel Falls
144 A2 Ángel de la Guarda, Isla i. Mex.
64 B2 Angeles Phil.
150 C2 Angel Falls Venez.
93 F4 Ängelholm Sweden
92 G3 Ångermanälven r. Sweden
104 B2 Angers France
129 E1 Angikuni Lake Can.
52 B3 Anglesea Austr.
98 A3 Anglesey i. U.K.

121 C2 Angoche Moz.
79 C2 Angohrān Iran
120 A2 Angola country Africa
138 C2 Angola U.S.A.
158 F6 Angola Basin S. Atlantic Ocean
128 A2 Angoon U.S.A.
104 C2 Angoulême France
155 D2 Angra dos Reis Brazil
77 D2 Angren Uzbek.
147 D3 Anguilla terr. West Indies
75 C2 Angul India
93 F4 Anholt i. Denmark
71 B3 Anhua China
70 B2 Anhui prov. China
154 B1 Anhumas Brazil
Anhwei prov. China see Anhui
154 C1 Aniva, Mys c. Rus. Fed.
66 D1 Aniva, Zaliv b. Rus. Fed.
88 C1 Anjalankoski Fin.
104 B2 Anjou reg. France
65 B2 Anjū N. Korea
70 A2 Ankang China
80 B2 Ankara Turkey
121 □D3 Ankazoabo Madag.
121 □D2 Ankazobe Madag.
137 E2 Ankeny U.S.A.
102 C1 Anklam Ger.
121 □D2 Ankofa mt. Madag.
70 B2 Anlu China
55 G3 Ann, Cape Antarctica
139 E2 Ann, Cape U.S.A.
89 F3 Anna Rus. Fed.
115 C1 Annaba Alg.
101 F2 Annaberg-Buchholtz Ger.
78 B3 An Nābiyah Yemen
80 B2 An Nabk Syria
78 B2 An Nafūd des. Saudi Arabia
150 D2 Annai Guyana
81 C2 An Najaf Iraq
63 B2 Annam Highlands mts Laos/Vietnam
96 C3 Annan U.K.
139 D3 Annapolis U.S.A.
75 C2 Annapurna I mt. Nepal
138 C2 Ann Arbor U.S.A.
150 D2 Anna Regina Guyana
An Nás Ireland see Naas
81 C2 An Nāşirīyah Iraq
115 D1 An Nawfalīyah Libya
105 D2 Annecy France
78 B3 An Nimāş Saudi Arabia
71 A3 Anning China
140 C2 Anniston U.S.A.
105 C2 Annonay France
79 B2 An Nu'ayrīyah Saudi Arab
121 □D2 Anorontany, Tanjona hd Madag.
71 B3 Anpu China
70 B2 Anqing China
65 B2 Ansan S. Korea
101 E3 Ansbach Ger.
70 C1 Anshan China
71 A3 Anshun China
143 D2 Anson U.S.A.
114 C3 Ansongo Mali
96 C2 Anstruther U.K.
150 B4 Antabamba Peru
80 B2 Antakya Turkey
121 □E2 Antalaha Madag.
80 B2 Antalya Turkey
80 B2 Antalya Körfezi g. Turkey
121 □D2 Antananarivo Madag.
55 A2 Antarctic Peninsula Antarctica
96 B2 An Teallach mt. U.K.
106 C2 Antequera Spain
142 B2 Anthony U.S.A.
114 B2 Anti-Atlas mts Morocco
105 D3 Antibes France
131 D2 Anticosti, Île d' i. Can.
131 D2 Antigonish Can.
147 D3 Antigua i. Antigua
Antigua and Barbuda country West Indies
145 C2 Antiguo-Morelos Mex.
111 B3 Antikythira i. Greece
49 I8 Antipodes Islands N.Z.
An t-Ob U.K. see Leverburgh
152 A2 Antofagasta Chile
154 C3 Antonina Brazil
96 C3 Antonine Wall tourist site U.K.
António Enes Moz. see Angoche
97 C1 Antrim U.K.

97 C1 Antrim Hills U.K.
121 □D2 Antsalova Madag.
Antseranana Madag. see
Antsiranana
121 □D2 Antsirabe Madag.
121 □D2 Antsiranana Madag.
121 □D2 Antsohihy Madag.
100 B2 Antwerp Belgium
Antwerpen Belgium see
Antwerp
An Uaimh Ireland see
Navan
74 B2 Anupgarh India
73 C4 Anuradhapura Sri Lanka
Anvers Belgium see Antwerp
51 C3 Anxious Bay Austr.
70 B2 Anyang China
65 B2 Anyang S. Korea
108 B2 Anzio Italy
67 C4 Aoga-shima i. Japan
66 D2 Aomori Japan
54 B2 Aoraki N.Z.
108 A1 Aosta Italy
114 B2 Aoukâr reg. Mali/Maur.
114 C2 Aoulef Alg.
114 A2 Aoussard Western Sahara
115 D2 Aozou Chad
141 D3 Apalachee Bay U.S.A.
150 C3 Apaporis r. Col.
154 D2 Aparecida do Tabuado Brazil
64 B2 Aparri Phil.
86 C2 Apatity Rus. Fed.
144 B3 Apatzingán Mex.
100 B1 Apeldoorn Neth.
100 C1 Apen Ger.
108 A2 Apennines mts Italy
49 J5 Apia Samoa
154 B2 Apiaí Brazil
64 B3 Apo, Mount vol. Phil.
101 E2 Apolda Ger.
52 B3 Apollo Bay Austr.
141 D3 Apopka U.S.A.
141 D3 Apopka, Lake U.S.A.
154 B1 Aporé r. Brazil
154 B1 Aporé Brazil
138 A1 Apostle Islands U.S.A.
80 B2 Apostolos Andreas, Cape Cyprus
91 C2 Apostolove Ukr.
133 F3 Appalachian Mountains U.S.A.
Appennino mts Italy see
Apennines
53 D2 Appin Austr.
100 C1 Appingedam Neth.
98 B2 Appleby-in-Westmorland U.K.
138 B2 Appleton U.S.A.
108 B2 Aprilia Italy
62 A1 Aprunyi India
91 D3 Apsheronsk Rus. Fed.
Apsheronskaya Rus. Fed. see
Apsheronsk
154 B2 Apucarana Brazil
154 B2 Apucarana, Serra do hills Brazil
83 M2 Apuka Rus. Fed.
64 A3 Apurahuan Phil.
147 D4 Apure r. Venez.
78 A2 Aqaba, Gulf of Asia
81 D2 'Aqdā Iran
75 C1 Aqqikkol Hu salt l. China
154 A1 Aquidauana r. Brazil
104 B3 Aquitaine reg. France
75 C2 Ara India
117 A4 Arab, Bahr el watercourse Sudan
Arabian Gulf g. Asia see
The Gulf
78 B2 Arabian Peninsula Asia
56 B4 Arabian Sea Indian Ocean
151 F4 Aracaju Brazil
154 A2 Aracanguy, Montes de hills Para.
151 F3 Aracati Brazil
154 B2 Araçatuba Brazil
155 D1 Aracruz Brazil
155 D1 Araçuaí Brazil
110 B1 Arad Romania
115 E3 Arada Chad
79 C2 'Arādah U.A.E.
156 C6 Arafura Sea Austr./Indon.
154 B1 Aragarças Brazil
107 C1 Aragón aut. comm. Spain
107 C1 Aragón r. Spain
151 E3 Araguaia r. Brazil
151 E3 Araguaiana Brazil
151 E3 Araguaína Brazil
154 C1 Araguari Brazil

115 C2 Arak Alg.
81 C2 Arāk Iran
62 A1 Arakan Yoma mts Myanmar
81 C1 Arak's r. Armenia
76 C2 Aral Sea salt l. Kazakh./Uzbek.
76 C2 Aral'sk Kazakh.
Aral'skoye More salt l. Kazakh./Uzbek. see Aral Sea
106 C2 Aranda de Duero Spain
109 D2 Aranđelovac Serbia
97 B2 Aran Islands Ireland
106 C1 Aranjuez Spain
122 A1 Aranos Namibia
143 D3 Aransas Pass U.S.A.
67 B4 Arao Japan
114 B3 Araouane Mali
151 F3 Arapiraca Brazil
154 B2 Arapongas Brazil
154 C3 Araquari Brazil
78 B1 'Ar'ar Saudi Arabia
154 C2 Araraquara Brazil
151 D3 Araras Brazil
154 C2 Araras Brazil
154 B1 Araras, Serra das hills Brazil
154 B3 Araras, Serra das mts Brazil
81 C2 Ararat Armenia
52 B3 Ararat Austr.
81 C2 Ararat, Mount Turkey
155 D2 Araruama, Lago de lag. Brazil
155 E1 Arataca Brazil
Aratürük China see Yiwu
150 B2 Arauca Col.
154 C1 Araxá Brazil
81 C2 Arbīl Iraq
129 E2 Arborg Can.
96 C2 Arbroath U.K.
74 A2 Arbū-ye Shamālī, Dasht-e des. Afgh.
104 B3 Arcachon France
141 D3 Arcadia U.S.A.
134 B2 Arcata U.S.A.
145 B3 Arcelia Mex.
86 D2 Archangel Rus. Fed.
51 D1 Archer r. Austr.
49 M5 Archipel des Tuamotu is Fr. Polynesia
149 B6 Archipiélago Juan Fernández S. Pacific Ocean
134 D2 Arco Col.
106 B2 Arcos de la Frontera Spain
127 G2 Arctic Bay Can.
Arctic Institute Islands is Rus. Fed. see Arkticheskogo Instituta, Ostrova
160 J1 Arctic Mid-Ocean Ridge Arctic Ocean
160 Arctic Ocean
126 D2 Arctic Red r. Can.
81 C2 Ardabīl Iran
81 C1 Ardahan Turkey
93 E3 Ardalstangen Norway
97 C2 Ardee Ireland
100 B3 Ardennes plat. Belgium
135 B3 Arden Town U.S.A.
81 D2 Ardestān Iran
97 D1 Ardglass U.K.
53 C2 Ardlethan Austr.
143 D2 Ardmore U.S.A.
96 A2 Ardnamurchan, Point of U.K.
52 A2 Ardrossan Austr.
96 B3 Ardrossan U.K.
96 B2 Ardvasar U.K.
135 B3 Arena, Point U.S.A.
93 E4 Arendal Norway
100 B2 Arendonk Belgium
101 E1 Arendsee (Altmark) Ger.
150 B4 Arequipa Peru
151 D3 Arere Brazil
106 C1 Arévalo Spain
108 B2 Arezzo Italy
108 B2 Argenta Italy
104 B2 Argentan France
153 C2 Argentina country S. America
158 D7 Argentine Basin S. Atlantic Ocean
157 I8 Argentine Rise S. Atlantic Ocean
153 A5 Argentino, Lago l. Arg.
104 C2 Argenton-sur-Creuse France
110 C2 Argeș r. Romania
74 A1 Arghandāb Rōd r. Afgh.
111 B3 Argolikos Kolpos b. Greece
111 B3 Argos Greece
111 B3 Argostoli Greece
107 C1 Arguis Spain

69 E1 Argun' r. China/Rus. Fed.
131 D3 Argyle Can.
50 B1 Argyle, Lake Austr.
Argyrokastron Albania see Gjirokastër
Ar Horqin Qi China see Tianshan
93 F4 Århus Denmark
122 A2 Ariamsvlei Namibia
152 A1 Arica Chile
96 A2 Arinagour U.K.
155 C1 Arinos Brazil
150 D4 Aripuanā Brazil
150 C3 Aripuanā r. Brazil
150 C3 Ariquemes Brazil
154 B1 Ariranhá r. Brazil
96 B2 Arisaig U.K.
96 B2 Arisaig, Sound of sea chan. U.K.
104 B3 Arizgoiti Spain
142 A2 Arizona state U.S.A.
144 A1 Arizpe Mex.
78 B2 'Arjah Saudi Arabia
61 C2 Arjasa Indon.
92 G2 Arjeplog Sweden
140 B2 Arkadelphia U.S.A.
77 C1 Arkalyk Kazakh.
140 B2 Arkansas r. U.S.A.
140 B1 Arkansas state U.S.A.
137 D3 Arkansas City U.S.A.
Arkhangel'sk Rus. Fed. see Archangel
97 C2 Arklow Ireland
102 C1 Arkona, Kap c. Ger.
82 G1 Arkticheskogo Instituta, Ostrova is Rus. Fed.
105 C3 Arles France
143 D2 Arlington U.S.A.
138 B2 Arlington Heights U.S.A.
115 C3 Arlit Niger
100 B3 Arlon Belgium
97 C1 Armagh U.K.
116 B2 Armant Egypt
87 D4 Armavir Rus. Fed.
81 C1 Armenia country Asia
150 B2 Armenia Col.
Armenopolis Romania see Gherla
144 B3 Armeria Mex.
53 D2 Armidale Austr.
134 D1 Armington U.S.A.
130 B2 Armstrong Can.
91 C2 Armyans'k Ukr.
Armyanskaya S.S.R. country Asia see Armenia
Arnaoutis, Cape c. Cyprus see Arnauti, Cape
130 C2 Arnaud r. Can.
80 D2 Arnauti, Cape Cyprus
100 B2 Arnhem Neth.
51 C1 Arnhem, Cape Austr.
51 C1 Arnhem Bay Austr.
51 C1 Arnhem Land reg. Austr.
122 B3 Arniston S. Africa
108 B2 Arno r. Italy
52 A2 Arno Bay Austr.
130 C3 Arnprior Can.
101 D2 Arnsberg Ger.
101 E2 Arnstadt Ger.
122 A2 Aroab Namibia
154 B2 Aroeira Brazil
101 D2 Arolsen Ger.
78 A3 Aroma Sudan
108 A1 Arona Italy
144 B2 Aros r. Mex.
Arquipélago dos Açores aut. reg. Port. see Azores
Arrah India see Ara
81 C2 Ar Ramādī Iraq
96 B3 Arran i. U.K.
97 B1 Arranmore Island Ireland
80 B2 Ar Raqqah Syria
105 C1 Arras France
106 C1 Arrasate Spain
78 B2 Ar Rass Saudi Arabia
79 C2 Ar Rayyān Qatar
150 C2 Arrecifal Col.
145 C3 Arriagá Mex.
79 C2 Ar Rimāl reg. Saudi Arabia
Ar Riyāḍ Saudi Arabia see Riyadh
54 A2 Arrowtown N.Z.
135 B3 Arroyo Grande U.S.A.
145 C2 Arroyo Seco Mex.
79 C2 Ar Rustāq Oman
79 C2 Ar Ruţbah Iraq
78 B2 Ar Ruwaydah Saudi Arabia
81 C2 Arsenajān Iran
66 D2 Arsen'yev Rus. Fed.
117 C3 Arta Djibouti

111 B3 Arta Greece
144 B3 Arteaga Mex.
66 B2 Artem Rus. Fed.
91 D2 Artemivs'k Ukr.
104 C2 Artenay France
142 C2 Artesia U.S.A.
51 E2 Arthur Point Austr.
54 B2 Arthur's Pass N.Z.
152 C3 Artigas Uru.
129 D1 Artillery Lake Can.
123 C1 Artisia Botswana
104 C1 Artois reg. France
90 B2 Artsyz Ukr.
Artur de Paiva Angola see Kuvango
77 D3 Artux China
81 C1 Artvin Turkey
59 C3 Aru, Kepulauan is Indon.
119 D2 Arua Uganda
147 D3 Aruba terr. West Indies
75 C2 Arun r. Nepal
62 A1 Arunachal Pradesh state India
119 D3 Arusha Tanz.
136 B3 Arvada U.S.A.
68 C1 Arvayheer Mongolia
129 E1 Arviat Can.
92 G2 Arvidsjaur Sweden
93 F4 Arvika Sweden
108 A2 Arzachena Italy
87 D3 Arzamas Rus. Fed.
107 C2 Arzew Alg.
100 C2 Arzfeld Ger.
Arzila Morocco see Asilah
101 F2 Aš Czech Rep.
115 C4 Asaba Nigeria
74 B1 Asadābād Afgh.
66 D2 Asahi-dake vol. Japan
66 D2 Asahikawa Japan
78 B3 Asalē l. Eth.
75 C2 Asansol India
117 C3 Asayita Eth.
130 C3 Asbestos Can.
122 B2 Asbestos Mountains S. Africa
119 E2 Asbe Teferi Eth.
109 C2 Ascea Italy
152 B1 Ascensión Bol.
113 B6 Ascension i. S. Atlantic Ocean
145 D3 Ascensión, Bahía de la b. Mex.
101 D3 Aschaffenburg Ger.
100 C2 Ascheberg Ger.
101 E2 Aschersleben Ger.
108 B2 Ascoli Piceno Italy
119 D2 Asela Eth.
92 G3 Åsele Sweden
111 B2 Asenovgrad Bulg.
76 B3 Asgabat Turkm.
78 B2 Asharat Saudi Arabia
50 A2 Ashburton watercourse Austr.
54 B2 Ashburton N.Z.
140 B2 Ashdown U.S.A.
141 D1 Asheville U.S.A.
53 D1 Ashford Austr.
97 C2 Ashford Ireland
99 D3 Ashford U.K.
66 D2 Ashibetsu Japan
98 C2 Ashington U.K.
67 B4 Ashizuri-misaki pt Japan
Ashkhabad Turkm. see Asgabat
136 D3 Ashland KS U.S.A.
138 C3 Ashland KY U.S.A.
138 C3 Ashland OH U.S.A.
134 B2 Ashland OR U.S.A.
138 A1 Ashland WI U.S.A.
53 C1 Ashley Austr.
88 C3 Ashmyany Belarus
66 D2 Ashoro Japan
81 C2 Ash Shabakah Iraq
78 B3 Ash Sharawrah Saudi Arabia
Ash Shāriqah U.A.E. see Sharjah
81 C2 Ash Sharqāţ Iraq
78 B3 Ash Shaţrah Iraq
78 B3 Ash Shaykh 'Uthman Yemen
79 B3 Ash Shiḩr Yemen
79 C2 Ash Shināş Oman
78 B2 Ash Shu'aybah Saudi Arabia
78 B2 Ash Shu'bah Saudi Arabia
78 B2 Ash Shubaykīyah Saudi Arabia
78 B2 Ash Shumlūl Saudi Arabia
78 B3 Ash Shuqayq Saudi Arabia
115 D2 Ash Shuwayrif Libya
138 C2 Ashtabula U.S.A.

Ashuanipi Lake

131 D2 Ashuanipi Lake Can.
75 B3 Asifabad India
106 B2 Asilah Morocco
108 A2 Asinara, Golfo dell' b. Italy
108 A2 Asinara, Isola i. Italy
82 G3 Asino Rus. Fed.
88 C3 Asipovichy Belarus
78 B2 'Asīr reg. Saudi Arabia
93 F4 Asker Norway
93 F4 Askim Norway
68 C1 Askiz Rus. Fed.
116 B3 Asmara Eritrea
93 F4 Åsnen l. Sweden
116 B2 Asoteriba, Jebel mt. Sudan
103 D2 Aspang-Markt Austria
136 B3 Aspen U.S.A.
143 C2 Aspermont U.S.A.
54 A2 Aspiring, Mount N.Z.
116 C3 Assab Eritrea
79 B3 Aş Şadārah Yemen
Aş Şaḥrā' al Gharbīyah des. Egypt see Western Desert
Aş Şaḥrā' ash Sharqīyah des. Egypt see Eastern Desert
78 B2 As Salamiyah Saudi Arabia
78 B3 Aş Şalīf Yemen
75 D2 Assam state India
81 C2 As Samāwah Iraq
79 C2 Aş Şanām reg. Saudi Arabia
115 E2 As Sarīr reg. Libya
108 A3 Assemini Italy
100 C1 Assen Neth.
100 B2 Assesse Belgium
115 D1 As Sidrah Libya
129 D3 Assiniboia Can.
128 C2 Assiniboine, Mount Can.
154 B3 Assis Brazil
78 B2 Aş Şubayḩiyah Kuwait
81 C2 As Sulaymānīyah Iraq
78 B2 As Sulaymī Saudi Arabia
78 B2 As Sulayyil Saudi Arabia
78 B2 As Sūq Saudi Arabia
80 B2 As Suwaydā' Syria
79 C2 As Suwayq Oman
As Suways Egypt see Suez
111 B3 Astakos Greece
77 D1 Astana Kazakh.
81 C2 Āstārā Iran
100 B2 Asten Neth.
Asterabad Iran see Gorgān
108 A2 Asti Italy
74 B1 Astor Pak.
106 B1 Astorga Spain
134 B1 Astoria U.S.A.
Astrabad Iran see Gorgān
87 D4 Astrakhan' Rus. Fed.
Astrakhan' Bazar Azer. see Cälilabad
88 C3 Astravyets Belarus
Astrida Rwanda see Butare
106 B1 Asturias aut. comm. Spain
111 C3 Astypalaia i. Greece
152 C2 Asunción Para.
116 B2 Aswān Egypt
116 B2 Asyūţ Egypt
Atacama, Desierto de des. Chile see Atacama Desert
152 B2 Atacama, Puna de plat. Arg.
152 B2 Atacama, Salar de salt flat Chile
152 B3 Atacama Desert des. Chile
114 C4 Atakpamé Togo
111 B3 Atalanti Greece
150 B4 Atalaya Peru
155 D1 Ataléia Brazil
77 C3 Atamyrat Turkm.
78 B3 'Ataq Yemen
114 A2 Atâr Maur.
135 B3 Atascadero U.S.A.
77 D2 Atasu Kazakh.
111 C3 Atavyros mt. Greece
116 B3 Atbara Sudan
116 B3 Atbara r. Sudan
77 C1 Atbasar Kazakh.
140 B3 Atchafalaya Bay U.S.A.
137 D3 Atchison U.S.A.
Ateränsk Kazakh. see Atyrau
108 B2 Aterno r. Italy
108 B2 Atessa Italy
100 A2 Ath Belgium
128 C2 Athabasca Can.
129 C2 Athabasca r. Can.
129 D2 Athabasca, Lake Can.
97 C2 Athboy Ireland
97 B2 Athenry Ireland
111 B3 Athens Greece
140 C2 Athens AL U.S.A.
141 D2 Athens GA U.S.A.

138 C3 Athens OH U.S.A.
141 D1 Athens TN U.S.A.
143 D2 Athens TX U.S.A.
51 D1 Atherton Austr.
Athina Greece see Athens
97 C2 Athlone Ireland
111 B2 Athos mt. Greece
97 C2 Athy Ireland
115 D3 Ati Chad
130 A3 Atikokan Can.
87 D3 Atkarsk Rus. Fed.
141 D2 Atlanta U.S.A.
141 D2 Atlantic U.S.A.
141 D2 Atlantic Beach U.S.A.
139 E3 Atlantic City U.S.A.
55 D3 Atlantic-Indian-Antarctic Basin S. Atlantic Ocean
158 E8 Atlantic-Indian Ridge Southern Ocean
158 Atlantic Ocean
122 A3 Atlantis S. Africa
114 B1 Atlas Mountains Africa
114 C1 Atlas Saharien mts Alg.
128 A2 Atlin Can.
128 A2 Atlin Lake Can.
140 C2 Atmore U.S.A.
152 B2 Atocha Bol.
143 D2 Atoka U.S.A.
75 C2 Atrai r. India
80 B2 Aţ Ţafilah Jordan
78 B2 Aţ Ţā'if Saudi Arabia
63 B2 Attapu Laos
130 B2 Attawapiskat Can.
130 B2 Attawapiskat r. Can.
130 B2 Attawapiskat Lake Can.
100 C2 Attendorn Ger.
100 A3 Attichy France
116 B2 Aţ Ţūr Egypt
78 B3 At Turbah Yemen
135 B3 Atwater U.S.A.
76 B2 Atyrau Kazakh.
105 D3 Aubagne France
105 C3 Aubenas France
126 D2 Aubry Lake Can.
135 B3 Auburn AL U.S.A.
138 B2 Auburn IN U.S.A.
139 E2 Auburn ME U.S.A.
137 D2 Auburn NE U.S.A.
139 D2 Auburn NY U.S.A.
104 C3 Aubusson France
104 C3 Auch France
54 B1 Auckland N.Z.
48 H9 Auckland Islands N.Z.
117 C4 Audo mts Eth.
101 F2 Aue Ger.
101 F2 Auerbach Ger.
102 C2 Augsburg Ger.
50 A3 Augusta Austr.
109 C3 Augusta Italy
137 D3 Augusta GA U.S.A.
139 F2 Augusta KS U.S.A.
139 F2 Augusta ME U.S.A.
155 D1 Augusto de Lima Brazil
51 E1 Augustów Pol.
50 A2 Augustus, Mount Austr.
100 A2 Aulnoye-Aymeries France
Aumale Alg. see Sour el Ghozlane
62 A2 Aunglan Myanmar
122 B2 Auob watercourse Namibia/S. Africa
131 D2 Aupaluk Can.
74 B3 Aurangabad India
104 B2 Auray France
100 C1 Aurich Ger.
154 B1 Aurilândia Brazil
104 C3 Aurillac France
136 C3 Aurora CO U.S.A.
138 B2 Aurora IL U.S.A.
137 E3 Aurora MO U.S.A.
137 D2 Aurora NE U.S.A.
122 A2 Aus Namibia
138 C2 Au Sable r. U.S.A.
137 E2 Austin MN U.S.A.
135 C3 Austin NV U.S.A.
143 D2 Austin TX U.S.A.
Australes, Îles is Fr. Polynesia see Tubuai Islands
50 B2 Australia country Oceania
55 K4 Australian-Antarctic Basin sea feature Southern Ocean
53 C3 Australian Capital Territory admin. div. Austr.
102 C2 Austria country Europe
150 D3 Autazes Brazil
144 B3 Autlán Mex.
105 C2 Autun France
104 C3 Auvergne reg. France

105 C2 Auvergne, Monts d' mts France
105 C2 Auxerre France
105 D2 Auxonne France
105 C2 Avallon France
131 E3 Avalon Peninsula Can.
154 C2 Avaré Brazil
91 D2 Avdiyivka Ukr.
106 B1 Aveiro Port.
109 B2 Avellino Italy
108 B2 Aversa Italy
100 A2 Avesnes-sur-Helpe France
93 G3 Avesta Sweden
104 C3 Aveyron r. France
108 B2 Avezzano Italy
96 C2 Aviemore U.K.
109 C2 Avigliano Italy
105 C3 Avignon France
106 C1 Ávila Spain
106 B1 Avilés Spain
52 B3 Avoca Austr.
139 D2 Avoca U.S.A.
109 C3 Avola Italy
99 B3 Avon r. England U.K.
99 B4 Avon r. England U.K.
99 C3 Avon r. England U.K.
142 A2 Avondale U.S.A.
104 B2 Avranches France
54 B1 Awanui N.Z.
78 B3 Awārik, 'Urūq al des. Saudi Arabia
119 D2 Āwasa Eth.
117 C4 Āwash Eth.
117 C3 Āwash r. Eth.
115 D2 Awbārī Libya
115 D2 Awbārī, Idhān des. Libya
117 C4 Aw Dheegle Somalia
96 B2 Awe, Loch l. U.K.
117 A4 Aweil S. Sudan
115 C4 Awka Nigeria
126 F1 Axel Heiberg Island Can.
114 B4 Axim Ghana
150 B4 Ayacucho Peru
77 E2 Ayagoz Kazakh.
Ayaguz Kazakh. see Ayagoz
68 B2 Ayakkum Hu salt l. China
106 B2 Ayamonte Spain
83 K3 Ayan Rus. Fed.
150 B4 Ayaviri Peru
76 A2 Aybas Kazakh.
91 D2 Aydar r. Ukr.
77 C2 Aydarko'l ko'li l. Uzbek.
111 C3 Aydın Turkey
Ayers Rock h. Austr. see Uluru
83 I2 Aykhal Rus. Fed.
99 C4 Aylesbury U.K.
106 C1 Ayllón Spain
129 D1 Aylmer Lake Can.
117 B4 Ayod S. Sudan
83 M2 Ayon, Ostrov i. Rus. Fed.
114 B3 'Ayoûn el 'Atroûs Maur.
51 D1 Ayr Austr.
96 B3 Ayr U.K.
98 A2 Ayre, Point of Isle of Man
76 C2 Ayteke Bi Kazakh.
110 C2 Aytos Bulg.
63 B2 Ayutthaya Thai.
111 C3 Ayvacık Turkey
111 C3 Ayvalık Turkey
91 C3 Ayya, Mys pt Ukr.
107 C2 Azahar, Costa del coastal area Spain
114 B3 Azaouâd reg. Mali
114 C3 Azaouagh, Vallée de watercourse Mali/Niger
115 D3 Azare Nigeria
77 C2 Azat, Gory h. Kazakh.
Azbine mts Niger see Aïr, Massif de l'
81 C1 Azerbaijan country Asia
Azerbaydzhanskaya S.S.R. country Asia see Azerbaijan
77 C1 Azhibeksor, Ozero salt l. Kazakh.
150 B3 Azogues Ecuador
86 D2 Azopol'ye Rus. Fed.
112 A2 Azores aut. reg. Port.
91 D2 Azov Rus. Fed.
91 D2 Azov, Sea of Rus. Fed./Ukr.
Azraq, Bahr el r. Sudan see Blue Nile
112 A2 Azuaga Spain
146 B4 Azuero, Península de pen. Panama
153 C3 Azul Arg.
105 D3 Azur, Côte d' coastal area France
108 A3 Azzaba Alg.

Aẓ Ẓahrān Saudi Arabia see Dhahran
80 B2 Az Zaqāzīq Egypt
80 B2 Az Zarqā' Jordan
115 D1 Az Zāwiyah Libya
78 B3 Az Zaydīyah Yemen
114 C2 Azzel Matti, Sebkha salt pan Alg.
78 B2 Az Zilfī Saudi Arabia
78 B3 Az Zuqur i. Yemen

B

63 B2 Ba, Sông r. Vietnam
117 C4 Baardheere Somalia
77 C3 Bābā, Kōh-e mts Afgh.
111 C3 Baba Burnu pt Turkey
110 C2 Babadag Romania
111 C2 Babaeski Turkey
116 C3 Bāb al Mandab str. Africa/Asia
61 C2 Babana Indon.
119 C1 Babanusa Sudan
59 C3 Babar i. Indon.
119 D3 Babati Tanz.
89 E2 Babayevo Rus. Fed.
59 C2 Babeldaob i. Palau
128 B2 Babine r. Can.
128 B2 Babine Lake Can.
59 C3 Babo Indon.
81 D2 Bābol Iran
122 A3 Baboon Point S. Africa
88 C3 Babruysk Belarus
Babu China see Hezhou
64 B2 Babuyan i. Phil.
64 B2 Babuyan Channel Phil.
64 B2 Babuyan Islands Phil.
151 E3 Bacabal Brazil
145 D3 Bacalar Mex.
59 C3 Bacan i. Indon.
110 C1 Bacău Romania
52 B3 Bacchus Marsh Austr.
62 B1 Băc Giang Vietnam
77 D3 Bachu China
129 E1 Back r. Can.
109 C1 Bačka Palanka Serbia
109 C1 Bačka Topola Serbia
52 A3 Backstairs Passage Austr.
63 B3 Bac Liêu Vietnam
142 B3 Bacobampo Mex.
64 B2 Bacolod Phil.
130 C2 Bacqueville, Lac l. Can.
Bada China see Xilin
70 A1 Badain Jaran Shamo des. China
106 B2 Badajoz Spain
Badaojiang China see Baishan
75 D2 Badarpur India
101 E2 Bad Berka Ger.
101 D2 Bad Berleburg Ger.
101 E1 Bad Bevensen Ger.
101 D3 Bad Dürkheim Ger.
100 C2 Bad Ems Ger.
103 D2 Baden Austria
102 B2 Baden-Baden Ger.
101 E2 Bad Harzburg Ger.
101 D2 Bad Hersfeld Ger.
102 C2 Bad Hofgastein Austria
101 D2 Bad Homburg vor der Höhe Ger.
101 D1 Bad Iburg Ger.
74 A2 Badin Pak.
Bādiyat ash Shām des. Asia see Syrian Desert
101 E2 Bad Kissingen Ger.
100 C3 Bad Kreuznach Ger.
136 C1 Badlands reg. ND U.S.A.
136 C2 Badlands reg. SD U.S.A.
101 E2 Bad Lauterberg im Harz Ger.
101 D2 Bad Lippspringe Ger.
101 E2 Bad Mergentheim Ger.
101 D2 Bad Nauheim Ger.
100 C2 Bad Neuenahr-Ahrweiler Ger.
101 E2 Bad Neustadt an der Saale Ger.
101 E1 Bad Oldesloe Ger.
101 D2 Bad Pyrmont Ger.
78 A2 Badr Ḩunayn Saudi Arabia
101 D1 Bad Salzdetfurth Ger.
101 E2 Bad Salzuflen Ger.
101 E2 Bad Salzungen Ger.
102 C1 Bad Schwartau Ger.
101 E1 Bad Segeberg Ger.
100 C3 Bad Sobernheim Ger.
73 C4 Badulla Sri Lanka
101 D1 Bad Zwischenahn Ger.

65 A2	Baengnyeong-do *i.* S. Korea	
106 C2	Baeza Spain	
114 A3	Bafatá Guinea-Bissau	
127 H2	Baffin Bay *sea* Can./Greenland	
127 H2	Baffin Island Can.	
118 B2	Bafia Cameroon	
114 A3	Bafing *r.* Africa	
114 A3	Bafoulabé Mali	
118 B2	Bafoussam Cameroon	
81 D2	Bāfq Iran	
80 B1	Bafra Turkey	
79 C2	Bāft Iran	
119 C2	Bafwasende Dem. Rep. Congo	
119 D3	Bagamoyo Tanz.	
	Bagan Datoh Malaysia *see* Bagan Datuk	
60 B1	Bagan Datuk Malaysia	
64 B3	Baganga Phil.	
120 B2	Bagani Namibia	
60 B1	Bagansiapiapi Indon.	
118 B3	Bagata Dem. Rep. Congo	
91 E2	Bagayevskiy Rus. Fed.	
142 A2	Bagdad U.S.A.	
152 C3	Bagé Brazil	
97 C2	Bagenalstown Ireland	
81 C2	Baghdād Iraq	
79 C1	Bāghīn Iran	
77 C3	Baghlān Afgh.	
104 C3	Bagnères-de-Luchon France	
105 C2	Bagnols-sur-Cèze France	
	Bago Myanmar *see* Pegu	
88 B3	Bagrationovsk Rus. Fed.	
	Bagrax China *see* Bohu	
64 B2	Baguio Phil.	
115 C3	Bagzane, Monts *mts* Niger	
79 D2	Bāhā Kālāt Iran	
146 C2	Bahamas, The *country* West Indies	
75 C2	Baharampur India	
	Bahariya Oasis *oasis* Egypt *see* Wāḥāt al Baḥrīyah	
76 B3	Baharly Turkm.	
60 B1	Bahau Malaysia	
74 B2	Bahawalnagar Pak.	
74 B2	Bahawalpur Pak.	
155 E1	Bahia Brazil *state* Brazil	
	Bahia Brazil *see* Salvador	
146 B3	Bahía, Islas de la *is* Hond.	
153 B3	Bahía Blanca Arg.	
144 A2	Bahía Kino Mex.	
152 C2	Bahía Negra Para.	
144 A2	Bahía Tortugas Mex.	
117 B3	Bahir Dar Eth.	
79 C2	Bahlā Oman	
75 C2	Bahraich India	
79 C2	Bahrain *country* Asia	
89 D3	Bahushewsk Belarus	
110 C2	Baia Romania	
120 A2	Baía dos Tigres Angola	
110 B1	Baia Mare Romania	
118 B2	Baïbokoum Chad	
69 E1	Baicheng China	
	Baidoa Somalia *see* Baydhabo	
	Baie-aux-Feuilles Can. *see* Tasiujaq	
131 D3	Baie-Comeau Can.	
	Baie-du-Poste Can. *see* Mistissini	
131 C3	Baie-St-Paul Can.	
131 E3	Baie Verte Can.	
65 B1	Baihe China	
69 D1	Baikal, Lake Rus. Fed.	
	Baile Átha Cliath Ireland *see* Dublin	
110 B2	Băilești Romania	
68 C2	Baima China	
141 D2	Bainbridge U.S.A.	
	Baingoin China *see* Porong	
48 I3	Bairiki Kiribati	
	Bairin Youqi China *see* Daban	
53 C3	Bairnsdale Austr.	
71 A3	Baise China	
65 B1	Baishan *Jilin* China	
65 B1	Baishan *Jilin* China	
65 B1	Baitou Shan *mt.* China/N. Korea	
120 A2	Baixo-Longa Angola	
70 A2	Baiyin China	
116 B3	Baiyuda Desert Sudan	
103 D2	Baja Hungary	
144 A1	Baja California *pen.* Mex.	
61 D2	Bajawa Indon.	
78 B3	Bājil Yemen	
115 D3	Bajoga Nigeria	
109 D2	Bajram Curri Albania	
114 A3	Bakel Senegal	
135 C3	Baker *CA* U.S.A.	
140 B2	Baker *LA* U.S.A.	
136 C1	Baker *MT* U.S.A.	
134 C2	Baker *OR* U.S.A.	
134 B1	Baker, Mount *vol.* U.S.A.	
129 E1	Baker Foreland *hd* Can.	
49 J3	Baker Island *terr.* N. Pacific Ocean	
129 E1	Baker Lake Can.	
129 E1	Baker Lake *l.* Can.	
135 C3	Bakersfield U.S.A.	
	Bakharden Turkm. *see* Baharly	
91 C2	Bakhchysaray Ukr.	
91 C1	Bakhmach Ukr.	
	Bakhmut Ukr. *see* Artemivs'k	
	Bākhtarān Iran *see* Kermānshāh	
	Bakı Azer. *see* Baku	
111 C2	Bakırköy Turkey	
92 □C2	Bakkaflói *b.* Iceland	
118 C2	Bakouma C.A.R.	
81 C1	Baku Azer.	
99 B3	Bala U.K.	
64 A3	Balabac Phil.	
64 A3	Balabac *i.* Phil.	
61 C1	Balabac Strait Malaysia/Phil.	
75 C2	Balaghat India	
60 C2	Balaiberkuak Indon.	
52 A2	Balaklava Austr.	
91 C3	Balaklava Ukr.	
91 D2	Balakliya Ukr.	
87 D3	Balakovo Rus. Fed.	
76 C3	Bālā Murghāb Afgh.	
145 C3	Balancán Mex.	
111 C3	Balanga Dağı *h.* Turkey	
64 B2	Balanga Phil.	
87 D3	Balashov Rus. Fed.	
	Balaton, Lake *l.* Hungary *see* Lake Balaton	
103 D2	Balatonboglár Hungary	
150 D3	Balbina, Represa de *resr* Brazil	
97 C2	Balbriggan Ireland	
52 A2	Balcanoona Austr.	
110 C2	Balchik Bulg.	
54 A3	Balclutha N.Z.	
143 C3	Balcones Escarpment U.S.A.	
129 E2	Baldock Lake Can.	
138 D2	Baldwin U.S.A.	
129 D2	Baldy Mountain *h.* Can.	
142 B2	Baldy Peak U.S.A.	
	Baleares, Islas *is* Spain *see* Balearic Islands	
107 D2	Balearic Islands	
155 E1	Baleia, Ponta da *pt* Brazil	
130 C2	Baleine, Grande Rivière de la *r.* Can.	
131 D2	Baleine, Rivière à la *r.* Can.	
75 C2	Baleshwar India	
108 A2	Balestrieri, Punta *mt.* Italy	
50 B2	Balgo Austr.	
79 B3	Bālḥaf Yemen	
61 C2	Bali *i.* Indon.	
115 D4	Bali Nigeria	
60 A1	Balige Indon.	
75 C2	Baliguda India	
111 C3	Balıkesir Turkey	
61 C2	Balikpapan Indon.	
59 D3	Balimo P.N.G.	
101 E3	Balingen Ger.	
64 B2	Balintang Channel Phil.	
	Bali Sea *sea* Indon. *see* Laut Bali	
78 B2	Baljurshi Saudi Arabia	
76 B3	Balkanabat Turkm.	
110 B2	Balkan Mountains Bulg./Serbia	
77 D2	Balkash Kazakh.	
	Balkash, Lake *see* Balkhash, Lake	
77 C1	Balkashino Kazakh.	
77 D2	Balkhash, Lake Kazakh.	
	Balla Balla Zimbabwe *see* Mbalabala	
96 B2	Ballachulish U.K.	
50 B3	Balladonia Austr.	
97 B2	Ballaghaderreen Ireland	
92 G2	Ballangen Norway	
96 B3	Ballantrae U.K.	
52 B3	Ballarat Austr.	
50 B2	Ballard, Lake *imp. l.* Austr.	
96 C2	Ballater U.K.	
114 B3	Ballé Mali	
55 M3	Balleny Islands Antarctica	
53 D1	Ballina Austr.	
97 B1	Ballina Ireland	
97 B2	Ballinasloe Ireland	
97 B3	Ballineen Ireland	
143 D2	Ballinger U.S.A.	
97 B2	Ballinrobe Ireland	
97 B1	Ballycastle Ireland	
97 C1	Ballycastle U.K.	
97 D1	Ballyclare U.K.	
97 D1	Ballyhaunis Ireland	
97 C1	Ballymena U.K.	
97 C1	Ballymoney U.K.	
97 D1	Ballynahinch U.K.	
97 B1	Ballyshannon Ireland	
95 B2	Ballyvoy U.K.	
52 B3	Balmoral Austr.	
143 C2	Balmorhea U.S.A.	
120 A2	Balombo Angola	
51 D2	Balonne *r.* Austr.	
74 B2	Balotra India	
77 D2	Balpyk Bi Kazakh.	
75 C2	Balrampur India	
52 B2	Balranald Austr.	
110 B2	Balş Romania	
151 E3	Balsas Brazil	
145 C3	Balsas Mex.	
145 B3	Balsas *r.* Mex.	
90 B2	Balta Ukr.	
90 B2	Bălți Moldova	
93 G4	Baltic Sea *g.* Europe	
80 B2	Balṭīm Egypt	
97 B3	Baltimore Ireland	
123 C1	Baltimore S. Africa	
139 D3	Baltimore U.S.A.	
97 C2	Baltinglass Ireland	
88 A3	Baltiysk Rus. Fed.	
75 D2	Balu India	
52 A2	Balumbah Austr.	
88 C2	Balvi Latvia	
77 D2	Balykchy Kyrg.	
87 E4	Balykshi Kazakh.	
79 C2	Bam Iran	
71 A3	Bama China	
51 D1	Bamaga Austr.	
130 A2	Bamaji Lake Can.	
114 B3	Bamako Mali	
118 C2	Bambari C.A.R.	
101 E3	Bamberg Ger.	
119 C2	Bambili Dem. Rep. Congo	
118 B2	Bambio C.A.R.	
119 C2	Bambouti C.A.R.	
155 C2	Bambuí Brazil	
98 C2	Bamburgh U.K.	
118 B2	Bamenda Cameroon	
118 C2	Bamingui C.A.R.	
79 D2	Bampūr Iran	
79 C2	Bampūr *watercourse* Iran	
77 C3	Bāmyān Afgh.	
119 C2	Banalia Dem. Rep. Congo	
71 A3	Banan China	
151 D4	Bananal, Ilha do *i.* Brazil	
74 B2	Banas *r.* India	
111 C3	Banaz Turkey	
62 B2	Ban Ban Laos	
99 C3	Banbridge U.K.	
99 C3	Banbury U.K.	
96 C2	Banchory U.K.	
130 C3	Bancroft Zambia *see* Chililabombwe	
119 C2	Banda Dem. Rep. Congo	
75 C2	Banda India	
59 C3	Banda, Kepulauan *is* Indon.	
59 C3	Banda, Laut *sea* Indon.	
60 A1	Banda Aceh Indon.	
	Bandar Abbas Iran *see* Bandar-e 'Abbās	
	Bandar Maharani Malaysia *see* Muar	
75 D2	Bandarban Bangl.	
79 C2	Bandar-e 'Abbās Iran	
81 C2	Bandar-e Anzalī Iran	
79 C2	Bandar-e Chārak Iran	
81 C2	Bandar-e Emām Khomeynī Iran	
79 C2	Bandar-e Lengeh Iran	
79 C2	Bandar-e Maqām Iran	
81 C2	Bandar-e Pahlavī Iran *see* Bandar-e Anzalī	
	Bandar-e Shāhpūr Iran *see* Bandar-e Emām Khomeynī	
60 B2	Bandar Lampung Indon.	
61 C1	Bandar Seri Begawan Brunei	
	Banda Sea *see* Indon. *see* Banda, Laut	
155 D2	Bandeiras, Pico de *mt.* Brazil	
123 C1	Bandelierkop S. Africa	
144 B2	Banderas, Bahía de *b.* Mex.	
114 B3	Bandiagara Mali	
111 C2	Bandırma Turkey	
	Bandjarmasin Indon. *see* Banjarmasin	
97 B3	Bandon Ireland	
97 B3	Bandon *r.* Ireland	
118 B3	Bandundu Dem. Rep. Congo	
60 B2	Bandung Indon.	
128 C2	Banff Can.	
96 C2	Banff U.K.	
114 B3	Banfora Burkina Faso	
64 B3	Banga Phil.	
73 B3	Bangalore India	
118 C2	Bangassou C.A.R.	
61 D2	Banggai Indon.	
61 D2	Banggai, Kepulauan *is* Indon.	
61 C1	Banggi *i.* Malaysia	
	Banghāzī Libya *see* Benghazi	
60 B2	Bangka *i.* Indon.	
60 B2	Bangka, Selat *sea chan.* Indon.	
61 C2	Bangkalan Indon.	
60 B1	Bangkinang Indon.	
60 B2	Bangko Indon.	
63 B2	Bangkok Thai.	
75 C2	Bangladesh *country* Asia	
63 B2	Ba Ngoi Vietnam	
97 D1	Bangor *Northern Ireland* U.K.	
98 A3	Bangor *Wales* U.K.	
139 F2	Bangor U.S.A.	
63 A2	Bang Saphan Yai Thai.	
64 B2	Bangued Phil.	
118 B2	Bangui C.A.R.	
64 B2	Bangui Phil.	
121 B2	Bangweulu, Lake Zambia	
80 B2	Banhā Egypt	
62 B2	Ban Huai Khon Thai.	
118 C2	Bani C.A.R.	
80 B3	Banī Mazār Egypt	
116 B2	Banī Suwayf Egypt	
115 D1	Banī Walīd Libya	
80 B2	Bāniyās Syria	
109 C2	Banja Luka Bos.-Herz.	
61 C2	Banjarmasin Indon.	
114 A3	Banjul Gambia	
63 A3	Ban Khok Kloi Thai.	
128 A2	Banks Island *B.C.* Can.	
126 D2	Banks Island *N.W.T.* Can.	
48 H5	Banks Islands Vanuatu	
129 E1	Banks Lake Can.	
54 B2	Banks Peninsula N.Z.	
51 D4	Banks Strait Austr.	
75 C2	Bankura India	
62 A1	Banmauk Myanmar	
62 B2	Ban Mouang Laos	
97 C1	Bann *r.* U.K.	
63 A2	Ban Napè Laos	
63 A3	Ban Na San Thai.	
146 C2	Bannerman Town Bahamas	
	Banningville Dem. Rep. Congo *see* Bandundu	
71 A4	Ban Nong Kung Thai.	
74 B2	Bannu Pak.	
62 B2	Ban Phôn-Hông Laos	
103 D2	Banská Bystrica Slovakia	
111 B2	Bansko Bulg.	
74 B2	Banswara India	
62 B2	Ban Taviang Laos	
63 A3	Ban Tha Kham Thai.	
62 A2	Ban Tha Song Yang Thai.	
63 B2	Ban Tôp Laos	
97 B3	Bantry Ireland	
97 B3	Bantry Bay Ireland	
60 A1	Banyak, Pulau-pulau *is* Indon.	
118 B2	Banyo Cameroon	
107 D1	Banyoles Spain	
61 C2	Banyuwangi Indon.	
	Banzyville Dem. Rep.Congo *see* Mobayi-Mbongo	
	Bao'an China *see* Shenzhen	
70 B1	Baochang China	
70 B2	Baoding China	
70 B2	Baoji China	
62 B1	Bao Lac Vietnam	
63 B2	Bao Lôc Vietnam	
66 B1	Baoqing China	
118 B2	Baoro C.A.R.	
62 A1	Baoshan China	
70 B1	Baotou China	
74 B2	Bap India	
81 C2	Ba'qūbah Iraq	
109 C2	Bar Montenegro	
90 B2	Bar Ukr.	
116 B3	Bara Sudan	
117 C4	Baraawe Somalia	

147 C2 Baracoa Cuba
53 C2 Baradine Austr.
147 C3 Barahona Dom. Rep.
116 B3 Baraka *watercourse* Eritrea/Sudan
106 C1 Barakaldo Spain
61 C1 Baram *r.* Malaysia
150 D2 Baramanni Guyana
74 B1 Baramulla India
89 D3 Baran' Belarus
74 B2 Baran India
88 C3 Baranavichy Belarus
116 B2 Baranīs Egypt
90 B1 Baranivka Ukr.
128 A2 Baranof Island U.S.A.
Baranowicze Belarus *see* Baranavichy
59 C3 Barat Daya, Kepulauan *is* Indon.
155 D2 Barbacena Brazil
147 E3 Barbados *country* West Indies
107 D1 Barbastro Spain
106 B2 Barbate Spain
123 D2 Barberton S. Africa
104 B2 Barbezieux-St-Hilaire France
51 D2 Barcaldine Austr.
107 D1 Barcelona Spain
150 C1 Barcelona Venez.
105 D3 Barcelonnette France
150 C3 Barcelos Brazil
114 B4 Barclayville Liberia
51 D2 Barcoo *watercourse* Austr.
Barcoo Creek *watercourse* Austr. *see* Cooper Creek
103 D2 Barcs Hungary
92 □B3 Bárðarbunga *mt.* Iceland
75 C2 Barddhaman India
103 E2 Bardejov Slovakia
Bardera Somalia *see* Baardheere
79 C2 Bardsīr Iran
75 B2 Bareilly India
82 D1 Barents Sea Arctic Ocean
78 A3 Barentu Eritrea
72 C1 Barga China
108 B2 Barga Italy
75 C2 Barh India
52 B3 Barham Austr.
139 F2 Bar Harbor U.S.A.
109 C2 Bari Italy
107 E2 Barika Alg.
74 B1 Barī Kōṭ Afgh.
150 B2 Barinas Venez.
75 C2 Baripada India
75 D2 Barisal Bangl.
60 B2 Barisan, Pegunungan *mts* Indon.
61 C2 Barito *r.* Indon.
79 C2 Barkā Oman
88 C2 Barkava Latvia
74 A2 Barkhan Pak.
51 C1 Barkly Tableland *reg.* Austr.
122 B2 Barkly West S. Africa
68 C2 Barkol China
110 C1 Bârlad Romania
105 D2 Bar-le-Duc France
50 A2 Barlee, Lake *imp. l.* Austr.
109 C2 Barletta Italy
53 C2 Barmedman Austr.
Barmen-Elberfeld Ger. *see* Wuppertal
74 B2 Barmer India
52 B2 Barmera Austr.
99 A3 Barmouth U.K.
101 D1 Barmstedt Ger.
98 C2 Barnard Castle U.K.
53 B2 Barnato Austr.
82 G3 Barnaul Rus. Fed.
127 H2 Barnes Icecap Can.
100 B1 Barneveld Neth.
98 C3 Barnsley U.K.
99 A4 Barnstaple U.K.
99 A4 Barnstaple Bay U.K.
141 D2 Barnwell U.S.A.
Baroda India *see* Vadodara
150 C1 Barquisimeto Venez.
96 A2 Barra *i.* U.K.
53 D2 Barraba Austr.
151 D4 Barra do Bugres Brazil
151 E3 Barra do Corda Brazil
154 B1 Barra do Garças Brazil
150 D3 Barra do São Manuel Brazil
Barraigh *i.* U.K. *see* Barra
150 B4 Barranca Peru
150 B3 Barranca Peru
152 C2 Barranqueras Arg.
150 B1 Barranquilla Col.
105 D3 Barre des Écrins *mt.* France
151 E4 Barreiras Brazil
63 A2 Barren Island India
154 C2 Barretos Brazil
128 C2 Barrhead Can.
130 C3 Barrie Can.
128 B2 Barrière Can.
52 B2 Barrier Range *hills* Austr.
53 D2 Barrington, Mount Austr.
129 D2 Barrington Lake Can.
53 C1 Barringun Austr.
97 C2 Barrow *r.* Ireland
126 B2 Barrow U.S.A.
126 B2 Barrow, Point U.S.A.
51 C2 Barrow Creek Austr.
98 B2 Barrow-in-Furness U.K.
50 A2 Barrow Island Austr.
126 F2 Barrow Strait Can.
99 B4 Barry U.K.
122 B3 Barrydale S. Africa
130 C3 Barrys Bay Can.
74 B2 Barsalpur India
101 D1 Barsinghausen Ger.
135 C4 Barstow U.S.A.
105 C2 Bar-sur-Aube France
102 C1 Barth Ger.
80 B1 Bartın Turkey
51 D1 Bartle Frere, Mount Austr.
143 D1 Bartlesville U.S.A.
137 D2 Bartlett *NE* U.S.A.
140 C1 Bartlett *TN* U.S.A.
98 C3 Barton-upon-Humber U.K.
103 E1 Bartoszyce Pol.
61 C2 Barung *i.* Indon.
69 D1 Baruun-Urt Mongolia
91 D2 Barvinkove Ukr.
53 C2 Barwon *r.* Austr.
88 C3 Barysaw Belarus
118 B2 Basankusu Dem. Rep. Congo
110 C2 Basarabi Romania
64 B1 Basco Phil.
105 D2 Basel Switz.
71 C3 Bashi Channel Phil./Taiwan
91 C2 Bashtanka Ukr.
64 B3 Basilan *i.* Phil.
99 D4 Basildon U.K.
99 C4 Basingstoke U.K.
81 C2 Başkale Turkey
130 C3 Baskatong, Réservoir *resr* Can.
Basle Switz. *see* Basel
118 C2 Basoko Dem. Rep. Congo
81 C2 Basra Iraq
128 C2 Bassano Can.
114 C4 Bassar Togo
63 A2 Bassein Myanmar
147 D3 Basse-Terre Guadeloupe
147 D3 Basseterre St Kitts and Nevis
114 B3 Bassikounou Maur.
114 C4 Bassila Benin
51 D3 Bass Strait Austr.
79 C2 Bastak Iran
101 E2 Bastheim Ger.
75 C2 Basti India
105 D3 Bastia France
100 B2 Bastogne Belgium
140 B2 Bastrop U.S.A.
Basuo China *see* Dongfang
Basutoland *country* Africa *see* Lesotho
118 A2 Bata Equat. Guinea
146 B2 Batabanó, Golfo de *b.* Cuba
83 J2 Batagay Rus. Fed.
154 B2 Bataguassu Brazil
74 B1 Batala India
106 B2 Batalha Port.
64 B1 Batan *i.* Phil.
118 B2 Batangafo C.A.R.
64 B2 Batangas Phil.
60 B2 Batanghari *r.* Indon.
64 B1 Batan Islands Phil.
154 C2 Batatais Brazil
139 D2 Batavia U.S.A.
91 D2 Bataysk Rus. Fed.
130 B3 Batchawana Mountain *h.* Can.
50 C1 Batchelor Austr.
63 B2 Bătdâmbâng Cambodia
118 B3 Batéké, Plateaux Congo
53 D3 Batemans Bay Austr.
140 B1 Batesville *AR* U.S.A.
140 C2 Batesville *MS* U.S.A.
89 D2 Batetskiy Rus. Fed.
99 B4 Bath U.K.
96 C3 Bathgate U.K.
74 B1 Bathinda India
53 C2 Bathurst Austr.
131 D3 Bathurst Can.
Bathurst Gambia *see* Banjul
126 E2 Bathurst Inlet *inlet* Can.
126 E2 Bathurst Inlet (abandoned) Can.
50 C1 Bathurst Island Austr.
126 F1 Bathurst Island Can.
78 B1 Bāṭin, Wādī al *watercourse* Asia
53 C3 Batlow Austr.
81 C2 Batman Turkey
115 C1 Batna Alg.
140 B2 Baton Rouge U.S.A.
144 B2 Batopilas Mex.
118 B2 Batouri Cameroon
154 B1 Batovi Brazil
Batrā' *tourist site* Jordan *see* Petra
92 I1 Båtsfjord Norway
73 C4 Batticaloa Sri Lanka
109 B2 Battipaglia Italy
128 D2 Battle *r.* Can.
138 B2 Battle Creek U.S.A.
135 C2 Battle Mountain U.S.A.
74 B1 Battura Glacier Pak.
117 B4 Batu *mt.* Eth.
60 A2 Batu, Pulau-pulau *is* Indon.
61 D2 Batuata *i.* Indon.
61 D2 Batudaka *i.* Indon.
64 B3 Batulaki Phil.
Batum Georgia *see* Bat'umi
81 C1 Bat'umi Georgia
60 B1 Batu Pahat Malaysia
61 D2 Baubau Indon.
115 C3 Bauchi Nigeria
137 E1 Baudette U.S.A.
Baudouinville Dem. Rep. Congo *see* Moba
104 B2 Baugé France
105 D2 Baume-les-Dames France
154 C2 Bauru Brazil
154 B1 Baús Brazil
88 B2 Bauska Latvia
102 C1 Bautzen Ger.
144 B2 Bavispe *r.* Mex.
87 E3 Bavly Rus. Fed.
62 A1 Bawdwin Myanmar
61 C2 Bawean *i.* Indon.
114 B3 Bawku Ghana
Baxian China *see* Banan
146 C2 Bayamo Cuba
Bayan Gol China *see* Dengkou
68 C1 Bayanhongor Mongolia
70 A2 Bayan Hot China
70 A1 Bayannur China
70 A1 Bayan Obo China
69 D2 Bayan Shutu China
69 D1 Bayan-Uul Mongolia
136 C2 Bayard *NE* U.S.A.
142 B2 Bayard *NM* U.S.A.
64 B3 Bayawan Phil.
81 C1 Bayburt Turkey
138 C2 Bay City *MI* U.S.A.
143 D3 Bay City *TX* U.S.A.
86 F2 Baydaratskaya Guba Rus. Fed.
117 C4 Baydhabo Somalia
102 C2 Bayern *reg.* Ger.
104 B2 Bayeux France
136 B3 Bayfield U.S.A.
78 B3 Bayḥān al Qiṣāb Yemen
Bay Islands *is* Hond. *see* Bahía, Islas de la
81 C2 Bayjī Iraq
Baykal, Ozero *l.* Rus. Fed. *see* Baikal, Lake
Baykal Range *mts* Rus. Fed. *see* Baykal'skiy Khrebet
83 I3 Baykal'skiy Khrebet *mts* Rus. Fed.
76 C2 Baykonyr Kazakh.
87 E3 Baymak Rus. Fed.
64 B2 Bayombong Phil.
104 B3 Bayonne France
76 C3 Bayramaly Turkm.
111 C3 Bayramiç Turkey
101 E3 Bayreuth Ger.
78 B3 Bayt al Faqīh Yemen
143 D3 Baytown U.S.A.
106 C2 Baza Spain
106 C2 Baza, Sierra de *mts* Spain
74 B1 Bāzā'ī Gonbad Afgh.
74 A1 Bāzārak Afgh.
76 A2 Bazardyuzyu, Gora *mt.* Azer./Rus. Fed.
104 B3 Bazas France
74 A2 Bazdar Pak.
70 A2 Bazhong China
79 D2 Bazmān Iran
79 D2 Bazmān, Kūh-e *mt.* Iran
Bé, Nossi *i.* Madag. *see* Bé, Nosy
121 □D2 Bé, Nosy *i.* Madag.
136 C1 Beach U.S.A.
52 B3 Beachport Austr.
99 D4 Beachy Head *hd* U.K.
123 C3 Beacon Bay S. Africa
50 B1 Beagle Gulf Austr.
121 □D2 Bealanana Madag.
97 B1 Béal an Mhuirthead Irela
121 □D3 Beampingaratra *mts* Madag.
134 D2 Bear *r.* U.S.A.
130 B3 Beardmore Can.
Bear Island *i.* Arctic Ocea *see* Bjørnøya
134 E1 Bear Paw Mountain U.S.A
147 C3 Beata, Cabo *c.* Dom. Rep.
147 C3 Beata, Isla *i.* Dom. Rep.
137 D2 Beatrice U.S.A.
135 C3 Beatty U.S.A.
53 D1 Beaudesert Austr.
52 B3 Beaufort Austr.
61 C1 Beaufort Malaysia
141 D2 Beaufort U.S.A.
126 D2 Beaufort Sea Can./U.S.A.
122 B3 Beaufort West S. Africa
96 B2 Beauly U.K.
96 B2 Beauly *r.* U.K.
100 B2 Beaumont Belgium
54 A3 Beaumont N.Z.
143 E2 Beaumont U.S.A.
105 C2 Beaune France
100 B2 Beauraing Belgium
129 E2 Beauséjour Can.
104 C2 Beauvais France
129 D2 Beauval Can.
129 D2 Beaver *r.* Can.
135 D3 Beaver U.S.A.
128 A1 Beaver Creek Can.
138 B2 Beaver Dam U.S.A.
129 E2 Beaver Hill Lake Can.
138 B1 Beaver Island U.S.A.
128 C2 Beaverlodge Can.
74 B2 Beawar India
154 C2 Bebedouro Brazil
101 D2 Bebra Ger.
99 D3 Beccles U.K.
109 D1 Bečej Serbia
106 B1 Becerreá Spain
114 B1 Béchar Alg.
Bechuanaland *country* Afr *see* Botswana
138 C3 Beckley U.S.A.
117 B4 Bedelē Eth.
99 C3 Bedford U.K.
138 B3 Bedford U.S.A.
98 C2 Bedlington U.K.
100 C1 Bedum Neth.
53 C3 Beechworth Austr.
53 D2 Beecroft Peninsula Austr.
101 F1 Beelitz Ger.
53 D1 Beenleigh Austr.
80 B2 Beersheba Israel
Be'ér Sheva' Israel *see* Beersheba
143 D3 Beeville U.S.A.
121 □D2 Befandriana Avaratra Madag.
53 C3 Bega Austr.
107 D1 Begur, Cap de *c.* Spain
81 D2 Behbehān Iran
128 C1 Behchokò Can.
81 D2 Behshahr Iran
69 E1 Bei'an China
71 A3 Beihai China
70 B2 Beijing China
100 C1 Beilen Neth.
118 B2 Béinamar Chad
96 B3 Beinn an Oir *h.* U.K.
96 A2 Beinn Mhòr *h.* U.K.
Beinn na Faoghlia *i.* U.K. *see* Benbecula
121 C2 Beira Moz.
80 B2 Beirut Lebanon
123 C1 Beitbridge Zimbabwe
106 B2 Beja Port.
115 C1 Bejaïa Alg.
106 B1 Béjar Spain
74 A2 Beji *r.* Pak.
103 E2 Békés Hungary
103 E2 Békéscsaba Hungary
121 □D3 Bekily Madag.
114 B4 Bekwai Ghana
75 C2 Bela India
74 A2 Bela Pak.
123 C1 Bela-Bela S. Africa
118 B2 Bélabo Cameroon
109 D2 Bela Crkva Serbia
61 C1 Belaga Malaysia

88 C3 Belarus country Europe
121 C3 Bela Vista Moz.
60 A1 Belawan Indon.
83 M2 Belaya r. Rus. Fed.
Belchatow Pol. see
Belchatów
103 D1 Belchatów Pol.
130 C2 Belcher Islands Can.
87 E3 Belebey Rus. Fed.
117 C4 Beledweyne Somalia
118 B2 Bélel Cameroon
151 E3 Belém Brazil
142 B2 Belen U.S.A.
110 C2 Belene Bulg.
89 E3 Belev Rus. Fed.
97 D1 Belfast U.K.
139 F2 Belfast U.S.A.
136 C1 Belfield U.S.A.
105 D2 Belfort France
73 B3 Belgaum India
Belgian Congo
country Africa see
Congo, Democratic
Republic of the
100 B2 Belgium country Europe
91 D1 Belgorod Rus. Fed.
109 D2 Belgrade Serbia
134 D1 Belgrade U.S.A.
109 C1 Beli Manastir Croatia
60 B2 Belinyu Indon.
60 B2 Belitung i. Indon.
118 B3 Belize Angola
146 B3 Belize Belize
146 B3 Belize country
Central America
83 K1 Bel'kovskiy, Ostrov i.
Rus. Fed.
128 B2 Bella Bella Can.
104 C2 Bellac France
128 B2 Bella Coola Can.
73 B3 Bellary India
53 C1 Bellata Austr.
138 C2 Bellefontaine U.S.A.
136 C2 Belle Fourche U.S.A.
136 C2 Belle Fourche r. U.S.A.
141 D3 Belle Glade U.S.A.
104 B2 Belle-Île i. France
131 E2 Belle Isle i. Can.
131 E2 Belle Isle, Strait of Can.
130 C3 Belleville Can.
138 B3 Belleville Il. U.S.A.
137 D3 Belleville KS U.S.A.
134 D2 Bellevue ID U.S.A.
134 B1 Bellevue WA U.S.A.
Bellin Can. see Kangirsuk
53 D2 Bellingen Austr.
134 B1 Bellingham U.S.A.
55 R2 Bellingshausen Sea
Antarctica
105 D2 Bellinzona Switz.
96 C2 Bell Rock i. U.K.
108 B1 Belluno Italy
122 A3 Bellville S. Africa
53 D2 Belmont Austr.
155 E1 Belmonte Brazil
146 B3 Belmopan Belize
Belmullet Ireland see
Béal an Mhuirthead
69 E1 Belogorsk Rus. Fed.
121 □D3 Beloha Madag.
155 D1 Belo Horizonte Brazil
138 B2 Beloit U.S.A.
86 C2 Belomorsk Rus. Fed.
89 E3 Beloomut Rus. Fed.
91 D3 Belorechensk Rus. Fed.
Belorechensk Rus. Fed.
see Belorechensk
87 E3 Beloretsk Rus. Fed.
Belorussia country Europe
see Belarus
Belorusskaya S.S.R. country
Europe see Belarus
Belostok Pol. see Białystok
121 □D2 Belo Tsiribihina Madag.
86 F2 Beloyarskiy Rus. Fed.
89 E1 Beloye, Ozero l. Rus. Fed.
Beloye More sea Rus. Fed.
see White Sea
89 E1 Belozersk Rus. Fed.
52 A2 Beltana Austr.
143 D2 Belton U.S.A.
Bel'ts' Moldova see Bălţi
Bel'tsy Moldova see Bălţi
97 C1 Belturbet Ireland
77 E2 Belukha, Gora mt. Kazakh./
Rus. Fed.
86 D2 Belush'ye Rus. Fed.
86 C2 Belvidere U.S.A.
51 D2 Belyando r. Austr.
89 D2 Belyy Rus. Fed.

82 F2 Belyy, Ostrov i. Rus. Fed.
101 F1 Belzig Ger.
137 E1 Bemidji U.S.A.
118 C3 Bena Dibele
Dem. Rep. Congo
53 C3 Benalla Austr.
Benares India see Varanasi
115 D1 Ben Arous Tunisia
118 C3 Bena-Sungu
Dem. Rep. Congo
106 B3 Benavente Spain
96 A2 Benbecula i. U.K.
134 B2 Bend U.S.A.
123 C3 Bendearg mt. S. Africa
Bender Moldova see
Tighina
Bendery Moldova see
Tighina
52 B3 Bendigo Austr.
121 C2 Bene Moz.
102 C2 Benešov Czech Rep.
109 B2 Benevento Italy
73 C3 Bengal, Bay of sea
Indian Ocean
Bengaluru India see
Bangalore
70 B2 Bengbu China
115 E1 Benghazi Libya
60 B1 Bengkalis Indon.
60 B1 Bengkayang Indon.
60 B2 Bengkulu Indon.
120 A2 Benguela Angola
Benha Egypt see Banhā
96 B1 Ben Hope h. U.K.
152 B1 Beni r. Bol.
119 C2 Beni Dem. Rep. Congo
114 B1 Beni Abbès Alg.
107 C2 Benidorm Spain
114 B1 Beni Mellal Morocco
114 C3 Benin country Africa
114 C4 Benin, Bight of g. Africa
115 C4 Benin City Nigeria
107 C2 Beni Saf Alg.
Beni Suef Egypt see
Banī Suwayf
153 C3 Benito Juárez Arg.
150 C3 Benjamim Constant Brazil
144 A1 Benjamín Hill Mex.
59 C3 Benjina Indon.
136 C2 Benkelman U.S.A.
96 B2 Ben Lawers mt. U.K.
96 B2 Ben Lomond h. U.K.
96 C2 Ben Macdui mt. U.K.
96 A2 Ben More mt. U.K.
96 B2 Ben More mt. U.K.
54 B2 Benmore, Lake N.Z.
96 B1 Ben More Assynt h. U.K.
128 A2 Bennett Can.
83 K1 Bennetta, Ostrov i.
Rus. Fed.
Bennett Island i. Rus. Fed.
see Bennetta, Ostrov
96 B2 Ben Nevis mt. U.K.
139 E2 Bennington U.S.A.
123 C2 Benoni S. Africa
115 D4 Bénoué Chad
101 D3 Bensheim Ger.
114 B1 Ben Slimane Morocco
142 A2 Benson U.S.A.
61 D2 Benteng Indon.
117 A4 Bentiu S. Sudan
118 B2 Bentonville U.S.A.
140 B2 Benton Harbor U.S.A.
63 B2 Bên Tre Vietnam
115 C4 Benue r. Nigeria
97 B1 Benwee Head hd Ireland
96 B2 Ben Wyvis mt. U.K.
70 C1 Benxi China
Beograd Serbia see Belgrade
52 B2 Beohari India
114 B4 Béoumi Côte d'Ivoire
67 B4 Beppu Japan
109 C2 Berane Montenegro
109 C2 Berat Albania
59 C3 Berau, Teluk b. Indon.
116 B3 Berber Sudan
117 C3 Berbera Somalia
118 B2 Berbérati C.A.R.
104 C1 Berck France
91 D2 Berdyans'k Ukr.
90 B2 Berdychiv Ukr.
59 D3 Bereina P.N.G.
129 E2 Berens River Can.
137 D2 Beresford U.S.A.
91 D2 Berezanskaya Rus. Fed.
90 A2 Berezhany Ukr.
90 C2 Berezivka Ukr.
90 B1 Berezne Ukr.

86 D2 Bereznik Rus. Fed.
86 E3 Berezniki Rus. Fed.
Berezovo Rus. Fed. see
Berezovo
86 F2 Berezovo Rus. Fed.
107 D1 Berga Spain
111 C3 Bergama Turkey
108 A1 Bergamo Italy
102 C1 Bergen Ger.
101 D1 Bergen Ger.
100 B1 Bergen Neth.
93 E3 Bergen Norway
100 B2 Bergen op Zoom Neth.
104 C3 Bergerac France
100 C2 Bergheim (Erft) Ger.
100 C2 Bergisch Gladbach Ger.
122 A1 Bergland Namibia
74 B1 Bhairi Hol mt. Pak.
74 B1 Bhakkar Pak.
75 C2 Bhaktapur Nepal
62 A1 Bhamo Myanmar
75 C3 Bhanjanagar India
74 B2 Bharatpur India
93 G3 Bergsjö Sweden
92 H2 Bergsviken Sweden
Berhampur India see
Baharampur
83 M3 Beringa, Ostrov i. Rus. Fed.
100 B2 Beringen Belgium
124 A4 Bering Sea N. Pacific Ocean
124 B3 Bering Strait Rus. Fed./
U.S.A.
100 C1 Berkel r. Neth.
135 B3 Berkeley U.S.A.
100 B1 Berkhout Neth.
55 B2 Berkner Island Antarctica
110 B2 Berkovitsa Bulg.
92 I1 Berlevåg Norway
101 F1 Berlin Ger.
139 E2 Berlin U.S.A.
101 E2 Berlingerode Ger.
53 D3 Bermagui Austr.
144 B2 Bermejillo Mex.
152 B2 Bermejo Bol.
131 D2 Bermen, Lac l. Can.
125 L6 Bermuda terr.
N. Atlantic Ocean
105 D2 Bern Switz.
101 E2 Bernburg (Saale) Ger.
127 G2 Bernier Bay Can.
50 A2 Bernier Island Austr.
101 D3 Bernkastel-Kues Ger.
121 □D3 Beroroha Madag.
52 B2 Berri Austr.
115 C1 Berriane Alg.
107 D2 Berrouaghia Alg.
53 D2 Berry Austr.
146 C2 Berry Islands Bahamas
122 A2 Berseba Namibia
101 C1 Bersenbrück Ger.
90 B2 Bershad' Ukr.
131 D2 Berté, Lac l. Can.
118 B2 Bertoua Cameroon
150 C3 Beruri Brazil
96 B2 Berwick-upon-Tweed U.K.
91 C2 Beryslav Ukr.
121 □D2 Besalampy Madag.
105 D2 Besançon France
81 D3 Beshneh Iran
129 D2 Besnard Lake Can.
140 C2 Bessemer U.S.A.
76 B2 Besshoky, Gora h. Kazakh.
100 B2 Best Neth.
121 □D2 Betafo Madag.
106 B1 Betanzos Spain
118 B2 Bétaré Oya Cameroon
122 A2 Bethanie Namibia
139 D3 Bethesda U.S.A.
123 C2 Bethlehem S. Africa
139 D2 Bethlehem U.S.A.
123 C3 Bethulie S. Africa
104 C1 Béthune France
121 □D3 Betioky Madag.
77 D2 Betpakdala plain Kazakh.
121 □D3 Betroka Madag.
131 D3 Betsiamites Can.
121 □D2 Betsiboka r. Madag.
66 D2 Betsukai Japan
137 E2 Bettendorf U.S.A.
75 C2 Bettiah India
74 B2 Betul India
74 B2 Betwa r. India
99 B3 Betws-y-coed U.K.
100 C2 Betzdorf Ger.
136 C1 Beulah U.S.A.
98 C1 Beverley U.K.
100 B1 Beverwijk Neth.
99 D4 Bexhill U.K.
111 C2 Beykoz Turkey
114 B4 Beyla Guinea
76 B2 Beyneu Kazakh.
80 B1 Beypazarı Turkey

Beyrouth Lebanon see
Beirut
80 B2 Beyşehir Turkey
80 B2 Beyşehir Gölü l. Turkey
91 D2 Beysug r. Rus. Fed.
91 D2 Beysugskiy Liman lag.
Rus. Fed.
88 C2 Bezhanitsy Rus. Fed.
89 E2 Bezhetsk Rus. Fed.
105 C3 Béziers France
Bhadgaon Nepal see
Bhaktapur
75 C2 Bhadrak India
73 B3 Bhadravati India
75 C2 Bhagalpur India
74 A2 Bhairi Hol mt. Pak.
74 B1 Bhakkar Pak.
75 C2 Bhaktapur Nepal
62 A1 Bhamo Myanmar
75 C3 Bhanjanagar India
74 B2 Bharatpur India
74 B2 Bharuch India
74 B2 Bhavnagar India
75 C3 Bhawanipatna India
123 D2 Bhekuzulu S. Africa
74 B1 Bhera Pak.
74 B2 Bhilwara India
73 B3 Bhima r. India
74 B2 Bhind India
123 C3 Bhisho S. Africa
74 B2 Bhiwani India
123 C3 Bhongweni S. Africa
74 B2 Bhopal India
75 C2 Bhubaneshwar India
Bhubaneswar India see
Bhubaneshwar
74 A2 Bhuj India
74 B2 Bhusawal India
75 D2 Bhutan country Asia
Biafra, Bight of g. Africa see
Benin, Bight of
59 D3 Biak Indon.
59 D3 Biak i. Indon.
103 E1 Biała Podlaska Pol.
103 D1 Białogard Pol.
103 E1 Białystok Pol.
109 C3 Bianco Italy
74 B2 Biaora India
104 B3 Biarritz France
105 D2 Biasca Switz.
66 D2 Bibai Japan
120 A2 Bibala Angola
53 C3 Bibbenluke Austr.
102 B2 Biberach an der Riß Ger.
155 D2 Bicas Brazil
73 B3 Bid India
115 C4 Bida Nigeria
73 B3 Bidar India
139 E2 Biddeford U.S.A.
96 B2 Bidean nam Bian mt. U.K.
99 A4 Bideford U.K.
Bideford Bay b. U.K. see
Barnstaple Bay
Bié Angola see Kuito
120 A2 Bié, Planalto do Angola
101 D2 Biedenkopf Ger.
105 D2 Biel Switz.
101 D1 Bielefeld Ger.
108 A1 Biella Italy
103 D2 Bielsko-Biała Pol.
63 B2 Biên Hoa Vietnam
130 C2 Bienville, Lac l. Can.
100 B3 Bièvre Belgium
118 B3 Bifoun Gabon
111 C2 Biga Turkey
134 D1 Big Belt Mountains U.S.A.
123 D2 Big Bend Swaziland
129 D2 Biggar Can.
96 C3 Biggar U.K.
99 C3 Biggleswade U.K.
134 D1 Big Hole r. U.S.A.
134 E1 Bighorn r. U.S.A.
136 B2 Bighorn r. U.S.A.
136 B2 Bighorn Mountains U.S.A.
143 D2 Big Lake U.S.A.
138 B2 Big Rapids U.S.A.
129 D2 Big River Can.
129 E2 Big Sand Lake Can.
137 D2 Big Sioux r. U.S.A.
143 C2 Big Spring U.S.A.
134 E1 Big Timber U.S.A.
130 B1 Big Trout Lake Can.
130 A1 Big Trout Lake l. Can.
109 C2 Bihać Bos.-Herz.
75 C2 Bihar state India
75 C2 Bihar Sharif India
110 B1 Bihor, Vârful mt. Romania
114 A3 Bijagós, Arquipélago dos is
Guinea-Bissau
73 B3 Bijapur India

Bijär

81 C2 Bijär Iran
109 C2 Bijeljina Bos.-Herz.
109 C2 Bijelo Polje Montenegro
71 A3 Bijie China
74 B2 Bikaner India
66 B1 Bikin Rus. Fed.
66 B1 Bikin r. Rus. Fed.
118 B3 Bikoro Dem. Rep. Congo
79 C2 Bilād Banī Bū 'Alī Oman
75 C2 Bilaspur India
81 C2 Biläsuvar Azer.
90 C2 Bila Tserkva Ukr.
63 A2 Bilauktaung Range mts Myanmar/Thai.
106 C1 Bilbao Spain
109 C2 Bileća Bos.-Herz.
111 C2 Bilecik Turkey
103 E1 Biłgoraj Pol.
119 D3 Bilharamulo Tanz.
90 C2 Bilhorod-Dnistrovs'kyy Ukr.
119 C2 Bili Dem. Rep. Congo
83 M2 Bilibino Rus. Fed.
109 D2 Bilisht Albania
104 B3 Billère France
134 E1 Billings U.S.A.
99 B4 Bill of Portland hd U.K.
142 A1 Bill Williams Mountain U.S.A.
115 D3 Bilma Niger
115 D3 Bilma, Grand Erg de des. Niger
51 E2 Biloela Austr.
91 C2 Bilohirs'k Ukr.
90 B1 Bilohir"ya Ukr.
91 C1 Bilopillya Ukr.
91 D2 Bilovods'k Ukr.
140 C2 Biloxi U.S.A.
51 C2 Bilpa Morea Claypan salt flat Austr.
101 E2 Bilshausen Ger.
115 E3 Biltine Chad
90 C2 Bilyayivka Ukr.
114 C4 Bimbila Ghana
118 B2 Bimbo C.A.R.
141 E3 Bimini Islands Bahamas
74 B2 Bina-Etawa India
59 C3 Binaija, Gunung mt. Indon.
53 C1 Bindle Austr.
118 B3 Bindu Dem. Rep. Congo
121 C2 Bindura Zimbabwe
107 D1 Binéfar Spain
120 B2 Binga Zimbabwe
53 D1 Bingara Austr.
100 C3 Bingen am Rhein Ger.
114 B4 Bingerville Côte d'Ivoire
139 F1 Bingham U.S.A.
139 D2 Binghamton U.S.A.
115 D2 Bin Ghanimah, Jabal hills Libya
81 C2 Bingöl Turkey
62 A1 Bingzhongluo China
60 A1 Binjai Indon.
53 C2 Binnaway Austr.
60 B1 Bintan i. Indon.
60 B2 Bintuhan Indon.
61 C1 Bintulu Malaysia
115 C3 Bin-Yauri Nigeria
70 B2 Binzhou China
109 C2 Biograd na Moru Croatia
118 A2 Bioko i. Equat. Guinea
155 C1 Biquinhas Brazil
115 D2 Birāk Libya
118 C1 Birao C.A.R.
75 C2 Biratnagar Nepal
52 B3 Birchip Austr.
128 C2 Birch Mountains Can.
51 C2 Birdsville Austr.
80 B2 Birecik Turkey
Birendranagar Nepal see Surkhet
60 A1 Bireun Indon.
75 C2 Birganj Nepal
154 B2 Birigüi Brazil
118 C2 Birini C.A.R.
76 B3 Bīrjand Iran
98 B3 Birkenhead U.K.
99 C3 Birmingham U.K.
140 C2 Birmingham U.S.A.
114 A2 Bîr Mogreïn Maur.
115 C3 Birnin-Kebbi Nigeria
115 C3 Birnin Konni Niger
69 E1 Birobidzhan Rus. Fed.
97 C2 Birr Ireland
96 C1 Birsay U.K.
78 A2 Bi'r Shalatayn Egypt
91 D1 Biryuch Rus. Fed.
88 B2 Biržai Lith.
75 B2 Bisalpur India
142 B2 Bisbee U.S.A.

104 A2 Biscay, Bay of sea France/Spain
141 D3 Biscayne Bay U.S.A.
102 C2 Bischofshofen Austria
77 D2 Bishkek Kyrg.
135 C3 Bishop U.S.A.
98 C2 Bishop Auckland U.K.
69 E1 Bishui China
150 C2 Bisinaca Col.
115 C1 Biskra Alg.
64 B3 Bislig Phil.
136 C1 Bismarck U.S.A.
59 D3 Bismarck Archipelago is P.N.G.
59 D3 Bismarck Sea P.N.G.
107 D2 Bissa, Djebel mt. Alg.
114 A3 Bissau Guinea-Bissau
129 E2 Bissett Can.
128 C2 Bistcho Lake Can.
110 B1 Bistrița Romania
110 C1 Bistrița r. Romania
100 C3 Bitburg Ger.
105 D2 Bitche France
115 D3 Bitkine Chad
81 C2 Bitlis Turkey
111 B2 Bitola Macedonia
Bitolj Macedonia see Bitola
109 C2 Bitonto Italy
101 F2 Bitterfeld Ger.
122 A3 Bitterfontein S. Africa
134 D1 Bitterroot r. U.S.A.
134 C1 Bitterroot Range mts U.S.A.
89 E3 Bityug r. Rus. Fed.
115 D3 Biu Nigeria
67 C3 Biwa-ko l. Japan
77 E1 Biysk Rus. Fed.
Bizerta Tunisia see Bizerte
115 C1 Bizerte Tunisia
92 □A2 Bjargtangar hd Iceland
92 G3 Bjästa Sweden
109 C1 Bjelovar Croatia
92 G2 Bjerkvik Norway
Björneborg Fin. see Pori
82 C2 Bjørnøya Arctic Ocean
114 B3 Bla Mali
137 E3 Black r. U.S.A.
51 D2 Blackall Austr.
98 B3 Blackburn U.K.
134 D2 Blackfoot U.S.A.
102 B2 Black Forest mts Ger.
136 C2 Black Hills U.S.A.
96 B2 Black Isle pen. U.K.
129 D2 Black Lake Can.
129 D2 Black Lake l. Can.
99 B4 Black Mountains hills U.K.
142 A1 Black Mountains U.S.A.
98 B3 Blackpool U.K.
142 B2 Black Range mts U.S.A.
62 B1 Black River r. Vietnam
138 A2 Black River Falls U.S.A.
134 C2 Black Rock Desert U.S.A.
138 C3 Blacksburg U.S.A.
80 B1 Black Sea Asia/Europe
131 D3 Blacks Harbour Can.
97 A1 Blacksod Bay Ireland
97 C2 Blackstairs Mountains hills Ireland
114 B4 Black Volta r. Africa
51 D2 Blackwater Austr.
97 C2 Blackwater r. Ireland
128 B1 Blackwater Lake Can.
50 A3 Blackwood r. Austr.
87 D4 Blagodarnyy Rus. Fed.
111 B2 Blagoevgrad Bulg.
69 E1 Blagoveshchensk Rus. Fed.
129 D2 Blaine Lake Can.
137 D2 Blair U.S.A.
96 C2 Blair Atholl U.K.
96 C2 Blairgowrie U.K.
105 D2 Blakely U.S.A.
105 D2 Blanc, Mont mt. France/Italy
153 B3 Blanca, Bahía b. Arg.
52 A1 Blanche, Lake imp. l. Austr.
152 B1 Blanco r. Bol.
134 B2 Blanco, Cape U.S.A.
131 E2 Blanc-Sablon Can.
92 □A2 Blanda r. Iceland
99 B4 Blandford Forum U.K.
135 E3 Blanding U.S.A.
107 D1 Blanes Spain
60 A1 Blangkejeren Indon.
100 A2 Blankenberge Belgium
100 C2 Blankenheim Ger.
100 C2 Blankenrath Ger.
147 D3 Blanquilla, Isla i. Venez.
103 D2 Blansko Czech Rep.
121 C2 Blantyre Malawi
97 B3 Blarney Ireland
98 C2 Blaydon U.K.

53 C2 Blayney Austr.
54 B2 Blenheim N.Z.
115 C1 Blida Alg.
130 B3 Blind River Can.
123 C2 Bloemfontein S. Africa
123 C2 Bloemhof S. Africa
123 C2 Bloemhof Dam S. Africa
104 C2 Blois France
92 □A2 Blönduós Iceland
97 B1 Bloody Foreland pt Ireland
142 B1 Bloomfield U.S.A.
138 B2 Bloomington IL U.S.A.
138 B3 Bloomington IN U.S.A.
102 B2 Bludenz Austria
137 E2 Blue Earth U.S.A.
138 C3 Bluefield U.S.A.
146 B3 Bluefields Nic.
53 C2 Blue Mountains Austr.
134 C1 Blue Mountains U.S.A.
116 B3 Blue Nile r. Eth./Sudan
126 E2 Bluenose Lake Can.
138 C3 Blue Ridge mts U.S.A.
128 C2 Blue River Can.
97 B1 Blue Stack Mountains hills Ireland
54 A3 Bluff N.Z.
135 E3 Bluff U.S.A.
154 C3 Blumenau Brazil
52 A2 Blyth Austr.
98 C2 Blyth U.K.
135 D4 Blythe U.S.A.
140 C1 Blytheville U.S.A.
114 A4 Bo Sierra Leone
64 B2 Boac Phil.
146 B3 Boaco Nic.
151 E3 Boa Esperança, Açude resr Brazil
134 C1 Boardman U.S.A.
123 C1 Boatlaname Botswana
151 F3 Boa Viagem Brazil
150 C2 Boa Vista Brazil
53 C2 Bobadah Austr.
71 B3 Bobai China
121 □D2 Bobaomby, Tanjona c. Madag.
114 B3 Bobo-Dioulasso Burkina Faso
121 B3 Bobonong Botswana
Bobriki Rus. Fed. see Novomoskovsk
89 F3 Bobrov Rus. Fed.
91 C1 Bobrovytsya Ukr.
91 C2 Bobrynets' Ukr.
121 □D3 Boby mt. Madag.
150 C3 Boca do Acre Brazil
155 D1 Bocaiúva Brazil
154 A2 Bocajá Brazil
118 B2 Bocaranga C.A.R.
141 D3 Boca Raton U.S.A.
146 B4 Bocas del Toro Panama
103 E2 Bochnia Pol.
100 B2 Bocholt Belgium
100 C2 Bocholt Ger.
100 C2 Bochum Ger.
101 E1 Bockenem Ger.
110 B1 Bocşa Romania
118 B2 Boda C.A.R.
83 I3 Bodaybo Rus. Fed.
96 C2 Boddam U.K.
115 D3 Bodélé reg. Chad
92 H2 Boden Sweden
Bodensee l. Ger./Switz. see Constance, Lake
99 A4 Bodmin U.K.
99 A4 Bodmin Moor moorland U.K.
92 F2 Bodø Norway
111 C3 Bodrum Turkey
118 C3 Boende Dem. Rep. Congo
63 A2 Bogale Myanmar
140 C2 Bogalusa U.S.A.
114 B3 Bogandé Burkina Faso
118 B2 Bogangolo C.A.R.
80 B2 Bogazlıyan Turkey
68 B2 Bogda Shan mts China
53 D1 Boggabilla Austr.
53 D2 Boggabri Austr.
97 B2 Boggeragh Mountains hills Ireland
Boghari Alg. see Ksar el Boukhari
59 D3 Bogia P.N.G.
100 B3 Bogny-sur-Meuse France
97 C2 Bog of Allen reg. Ireland
53 C3 Bogong, Mount Austr.
60 B2 Bogor Indon.
89 E3 Bogoroditsk Rus. Fed.
150 B2 Bogotá Col.
83 G3 Bogotol Rus. Fed.

Bogoyavlenskoye Rus. Fed. see Pervomayskiy
83 H3 Boguchany Rus. Fed.
91 E2 Boguchar Rus. Fed.
114 A3 Bogué Maur.
70 B2 Bo Hai g. China
100 A3 Bohain-en-Vermandois France
70 B2 Bohai Wan b. China
Bohemian Forest mts Ger. see Böhmer Wald
123 C2 Bohlokong S. Africa
101 F3 Böhmer Wald mts Ger.
91 D1 Bohodukhiv Ukr.
64 B3 Bohol i. Phil.
64 B3 Bohol Sea Phil.
77 E2 Bohu China
155 C2 Boi, Ponta do pt Brazil
123 C2 Boikhutso S. Africa
154 B3 Boi Preto, Serra de r. Brazil
154 B1 Bois r. Brazil
126 D2 Bois, Lac des l. Can.
134 C2 Boise U.S.A.
143 C1 Boise City U.S.A.
129 D3 Boissevain Can.
123 C2 Boitumelong S. Africa
154 C2 Boituva Brazil
101 E1 Boizenburg Ger.
76 B3 Bojnürd Iran
75 C2 Bokaro India
118 B3 Bokatola Dem. Rep. Congo
114 A3 Boké Guinea
118 C3 Bokele Dem. Rep. Congo
93 E4 Boknafjorden sea chan. Norway
115 D3 Bokoro Chad
63 A2 Bokpyin Myanmar
89 D2 Boksitogorsk Rus. Fed.
122 B2 Bokspits Botswana
118 C3 Bokungu Dem. Rep. Congo
115 D3 Bol Chad
114 A3 Bolama Guinea-Bissau
63 B2 Bolávén, Phouphiang plat. Laos
104 C2 Bolbec France
77 E2 Bole China
118 B3 Boleko Dem. Rep. Congo
114 B3 Bolgatanga Ghana
90 B2 Bolhrad Ukr.
66 B1 Boli China
118 B3 Bolia Dem. Rep. Congo
92 H3 Boliden Sweden
62 B2 Bolikhamxai Laos
110 C2 Bolintin-Vale Romania
137 E3 Bolivar MO U.S.A.
140 C1 Bolivar TN U.S.A.
150 B2 Bolívar, Pico mt. Venez.
152 B1 Bolivia country S. America
89 E3 Bolkhov Rus. Fed.
105 C3 Bollène France
93 G3 Bollnäs Sweden
53 C1 Bollon Austr.
101 E2 Bollstedt Ger.
93 F4 Bolmen l. Sweden
118 B3 Bolobo Dem. Rep. Congo
108 B2 Bologna Italy
89 D2 Bologovo Rus. Fed.
89 D2 Bologoye Rus. Fed.
123 C2 Bolokanang S. Africa
118 B2 Bolomba Dem. Rep. Congo
108 B2 Bolsena, Lago di l. Italy
83 H1 Bol'shevik, Ostrov i. Rus. Fed.
86 E2 Bol'shezemel'skaya Tundra lowland Rus. Fed.
66 B2 Bol'shoy Kamen' Rus. Fed.
Bol'shoy Kavkaz mts Asia/Europe see Caucasus
83 K2 Bol'shoy Lyakhovskiy, Ostrov i. Rus. Fed.
Bol'shoy Tokmak Kyrg. see Tokmak
Bol'shoy Tokmak Ukr. see Tokmak
100 B3 Bolsward Neth.
98 B3 Bolton U.K.
80 B1 Bolu Turkey
59 E3 Bolubolu P.N.G.
92 □A2 Bolungarvík Iceland
108 B1 Bolzano Italy
118 B3 Boma Dem. Rep. Congo
53 D2 Bomaderry Austr.
53 C3 Bombala Austr.
Bombay India see Mumbai
155 C1 Bom Despacho Brazil
75 D2 Bomdila India
154 B1 Bom Jardim de Goiás Brazil
151 E4 Bom Jesus da Lapa Brazil

155 D2 Bom Jesus do Itabapoana Brazil
115 D1 Bon, Cap c. Tunisia
147 D3 Bonaire mun. West Indies
134 C1 Bonaparte, Mount U.S.A.
50 B1 Bonaparte Archipelago is Austr.
131 E3 Bonavista Can.
131 E3 Bonavista Bay Can.
118 C2 Bondo Dem. Rep. Congo
114 B4 Bondoukou Côte d'Ivoire
Bône Alg. see Annaba
61 D2 Bonerate, Kepulauan is Indon.
155 C1 Bonfinópolis de Minas Brazil
117 B4 Bonga Eth.
75 D2 Bongaigaon India
118 C2 Bongandanga Dem. Rep. Congo
122 B2 Bongani S. Africa
118 C2 Bongo, Massif des mts C.A.R.
121 □D2 Bongolava mts Madag.
115 D3 Bongor Chad
114 B4 Bongouanou Côte d'Ivoire
63 B2 Bông Sơn Vietnam
143 D2 Bonham U.S.A.
105 D3 Bonifacio France
108 A2 Bonifacio, Strait of France/ Italy
69 F3 Bonin Islands is Japan
100 C2 Bonn Ger.
134 C1 Bonners Ferry U.S.A.
105 D2 Bonneville France
50 A3 Bonnie Rock Austr.
129 C2 Bonnyville Can.
108 A2 Bonorva Italy
53 D1 Bonshaw Austr.
61 C1 Bontang Indon.
114 A4 Bonthe Sierra Leone
64 B2 Bontoc Phil.
61 C2 Bontosunggu Indon.
123 C3 Bontrug S. Africa
53 C1 Boolba Austr.
52 B2 Booligal Austr.
53 C1 Boomi Austr.
53 D1 Boonah Austr.
137 E2 Boone U.S.A.
140 C2 Booneville U.S.A.
137 E3 Boonville U.S.A.
52 B2 Booroorban Austr.
53 C2 Boorowa Austr.
117 C3 Boosaaso Somalia
126 G2 Boothia, Gulf of Can.
126 F2 Boothia Peninsula Can.
118 B3 Booué Gabon
100 C2 Boppard Ger.
144 B2 Boquilla, Presa de la resr Mex.
109 D2 Bor Serbia
117 B4 Bor S. Sudan
80 B2 Bor Turkey
119 E2 Bor, Lagh watercourse Kenya/Somalia
121 □E2 Boraha, Nosy i. Madag.
76 B2 Borankul Kazakh.
93 F4 Borås Sweden
81 D3 Borāzjān Iran
150 D3 Borba Brazil
104 B3 Bordeaux France
126 E1 Borden Island Can.
127 G2 Borden Peninsula Can.
52 B3 Bordertown Austr.
107 D2 Bordj Bou Arréridj Alg.
107 D2 Bordj Bounaama Alg.
114 B2 Bordj Flye Ste-Marie Alg.
115 C1 Bordj Messaouda Alg.
114 C2 Bordj Mokhtar Alg.
Bordj Omer Driss Alg. see
115 C2 Bordj Omer Driss Alg.
Bordj Omer Driss Alg. see
94 B1 Borðoy i. Faroe Is
Borgå Fin. see Porvoo
92 □A3 Borgarnes Iceland
143 C1 Borger U.S.A.
93 G4 Borgholm Sweden
100 B2 Borgloon Belgium
108 A1 Borgosesia Italy
87 D3 Borisoglebsk Rus. Fed.
89 E2 Borisoglebskiy Rus. Fed.
91 D1 Borisovka Rus. Fed.
119 E2 Bo River Post S. Sudan
100 C2 Borken Ger.
92 G2 Borkenes Norway
100 C1 Borkum Ger.
100 C1 Borkum i. Ger.
93 G3 Borlänge Sweden
101 F2 Borna Ger.
100 C1 Borne Neth.

61 C1 Borneo i. Asia
93 F4 Bornholm i. Denmark
111 C3 Bornova Turkey
90 B1 Borodyanka Ukr.
77 E2 Borohoro Shan mts China
114 B3 Boron Mali
89 D2 Borovichi Rus. Fed.
89 E2 Borovsk Rus. Fed.
76 C1 Borovskoy Kazakh.
107 C2 Borriana Spain
51 C1 Borroloola Austr.
110 B1 Borşa Romania
76 B2 Borsakelmas sho'rxogi salt marsh Uzbek.
90 B2 Borshchiv Ukr.
69 D1 Borshchovochnyy Khrebet mts Rus. Fed.
101 E1 Börßum Ger.
Bortala China see Bole
81 C2 Borūjerd Iran
65 B2 Boryeong S. Korea
90 A2 Boryslav Ukr.
90 C1 Boryspil' Ukr.
91 C1 Borzna Ukr.
69 D1 Borzya Rus. Fed.
109 C1 Bosanska Dubica Bos.-Herz.
109 C1 Bosanska Gradiška Bos.-Herz.
109 C2 Bosanska Krupa Bos.-Herz.
109 C2 Bosanski Novi Bos.-Herz.
109 C2 Bosansko Grahovo Bos.-Herz.
123 C2 Boshof S. Africa
109 C2 Bosnia-Herzegovina country Europe
118 B2 Bosobolo Dem. Rep. Congo
111 C2 Bosporus str. Turkey
142 B2 Bosque U.S.A.
118 B2 Bossangoa C.A.R.
118 B2 Bossembélé C.A.R.
140 B2 Bossier City U.S.A.
122 A2 Bossiesvlei Namibia
68 B2 Bosten Hu l. China
99 C3 Boston U.K.
139 E2 Boston U.S.A.
140 B1 Boston Mountains U.S.A.
53 D2 Botany Bay Austr.
120 B3 Boteti r. Botswana
110 B2 Botev mt. Bulg.
80 A1 Botevgrad Bulg.
92 G3 Bothnia, Gulf of Fin./ Sweden
110 C1 Botoşani Romania
70 B2 Botou China
123 C2 Botshabelo S. Africa
120 B3 Botswana country Africa
109 C3 Botte Donato, Monte mt. Italy
136 C1 Bottineau U.S.A.
100 C2 Bottrop Ger.
154 C2 Botucatu Brazil
114 B4 Bouaké Côte d'Ivoire
118 B2 Bouar C.A.R.
114 B1 Bou Arfa Morocco
131 D3 Bouctouche Can.
107 E2 Bougaa Alg.
48 G4 Bougainville Island P.N.G.
Bougie Alg. see Bejaïa
114 B3 Bougouni Mali
100 B3 Bouillon Belgium
107 D2 Bouira Alg.
114 A2 Boujdour Western Sahara
50 B3 Boulder Austr.
136 B2 Boulder CO U.S.A.
134 D1 Boulder MT U.S.A.
135 D3 Boulder City U.S.A.
Boulhaut Morocco see Ben Slimane
51 C2 Boulia Austr.
104 C2 Boulogne-Billancourt France
104 C1 Boulogne-sur-Mer France
118 C2 Bouloula C.A.R.
118 B3 Boumango Gabon
111 C3 Boumerdès Alg.
114 B4 Boundiali Côte d'Ivoire
114 B3 Boundiali Côte d'Ivoire
134 D2 Bountiful U.S.A.
49 I8 Bounty Islands N.Z.
114 B3 Bourem Mali
104 C2 Bourganeuf France
105 D2 Bourg-en-Bresse France
104 C2 Bourges France
Bourgogne reg. France see Burgundy
105 D2 Bourgoin-Jallieu France
53 C2 Bourke Austr.
99 C3 Bourne U.K.

99 C4 Bournemouth U.K.
118 C1 Bourtoutou Chad
115 C1 Bou Saâda Alg.
115 D3 Bousso Chad
100 A2 Boussu Belgium
114 A3 Boutilimit Maur.
128 C3 Bow r. Can.
Bowa China see Muli
51 D2 Bowen Austr.
53 C3 Bowen, Mount Austr.
129 C3 Bow Island Can.
138 B3 Bowling Green KY U.S.A.
137 E3 Bowling Green MO U.S.A.
138 C2 Bowling Green OH U.S.A.
136 C1 Bowman U.S.A.
101 D3 Boxberg Ger.
100 B2 Boxtel Neth.
80 B1 Boyabat Turkey
Boyang China see Poyang
97 B2 Boyle Ireland
97 C2 Boyne r. Ireland
136 B2 Boysen Reservoir U.S.A.
152 B2 Boyuibe Bol.
111 C3 Bozburun Turkey
111 C3 Bozcaada i. Turkey
111 C3 Bozdağ mt. Turkey
111 C3 Boz Dağları mts Turkey
111 C3 Bozdoğan Turkey
134 D1 Bozeman U.S.A.
118 B2 Bozoum C.A.R.
111 D3 Bozüyük Turkey
109 C2 Brač i. Croatia
130 C3 Bracebridge Can.
93 G3 Bräcke Sweden
99 C4 Bracknell U.K.
141 D3 Bradano r. Italy
147 B3 Bradenton U.S.A.
98 C3 Brades Montserrat
139 D2 Bradford U.K.
143 D2 Bradford NE U.S.A.
96 C2 Brady U.S.A.
106 B1 Braemar U.K.
151 E3 Braga Port.
106 B1 Bragança Brazil
155 C2 Bragança Port.
89 D3 Bragança Paulista Brazil
75 D2 Brahin Belarus
75 C3 Brahmanbaria Bangl.
62 A1 Brahmapur India
53 C3 Brahmaputra r. China/India
110 C1 Braidwood Austr.
137 E1 Brăila Romania
99 D4 Brainerd U.S.A.
100 B2 Braintree U.K.
101 D1 Braives Belgium
122 A1 Brake (Unterweser) Ger.
98 B2 Brakwater Namibia
101 D1 Brampton Can.
101 D2 Bramsche Ger.
150 C2 Branco r. Brazil
101 F1 Brandberg mt. Namibia
129 E3 Brandenburg Ger.
140 C2 Brandon Can.
97 A2 Brandon Mountain h. Ireland
122 B3 Brandvlei S. Africa
103 D1 Braniewo Pol.
130 B3 Brantford Can.
53 D2 Branxton Austr.
131 D3 Bras d'Or Lake Can.
155 D1 Brasil, Planalto do plat. Brazil
154 C1 Brasilândia Brazil
154 C1 Brasília Brazil
155 D1 Brasília de Minas Brazil
88 C2 Braslaw Belarus
110 C1 Braşov Romania
103 D2 Bratislava Slovakia
83 H3 Bratsk Rus. Fed.
102 C2 Braunau am Inn Austria
101 E1 Braunschweig Ger.
92 □A2 Brautarholt Iceland
Bravo del Norte, Río r. Mex./U.S.A. see Rio Grande
135 C4 Brawley U.S.A.
97 C2 Bray Ireland
150 D2 Brazil country S. America
158 E6 Brazil Basin S. Atlantic Ocean
143 D3 Brazos r. U.S.A.
118 B3 Brazzaville Congo
109 C2 Brčko Bos.-Herz.
96 C2 Brechin U.K.
100 B2 Brecht Belgium
137 D3 Breckenridge U.S.A.
103 D2 Břeclav Czech Rep.
99 B4 Brecon U.K.
99 B4 Brecon Beacons reg. U.K.
100 B2 Breda Neth.

122 B3 Bredasdorp S. Africa
102 B2 Bregenz Austria
92 H1 Breivikbotn Norway
92 E3 Brekstad Norway
101 D1 Bremen Ger.
101 D1 Bremerhaven Ger.
Bremersdorp Swaziland see Manzini
134 B1 Bremerton U.S.A.
101 D1 Bremervörde Ger.
143 D2 Brenham U.S.A.
105 E2 Brennero Italy
102 C2 Brenner Pass Austria/Italy
99 D4 Brentwood U.K.
108 B1 Brescia Italy
100 A2 Breskens Neth.
105 E2 Bressanone Italy
96 □ Bressay i. U.K.
104 B2 Bressuire France
104 B2 Brest Belarus
104 B2 Brest France
Brest-Litovsk Belarus see Brest
Bretagne reg. France see Brittany
140 C3 Breton Sound b. U.S.A.
151 E3 Breves Brazil
53 C1 Brewarrina Austr.
134 C1 Brewster U.S.A.
89 E2 Breytovo Rus. Fed.
Brezhnev Rus. Fed. see Naberezhnyye Chelny
109 C1 Brezovo Polje plain Croatia
118 C2 Bria C.A.R.
105 D3 Briançon France
90 B2 Briceni Moldova
Brichany Moldova see Briceni
99 B4 Bridgend U.K.
139 E2 Bridgeport CT U.S.A.
136 C2 Bridgeport NE U.S.A.
147 E3 Bridgetown Barbados
131 D3 Bridgewater Can.
99 B3 Bridgnorth U.K.
99 B4 Bridgwater U.K.
99 B4 Bridgwater Bay U.K.
98 C2 Bridlington U.K.
98 C2 Bridlington Bay U.K.
99 B4 Bridport U.K.
105 D2 Brig Switz.
134 D2 Brigham City U.S.A.
53 C3 Bright Austr.
54 B3 Brighton N.Z.
99 C4 Brighton U.K.
136 C3 Brighton CO U.S.A.
138 C2 Brighton MI U.S.A.
105 D3 Brignoles France
114 A3 Brikama Gambia
101 D2 Brilon Ger.
109 C2 Brindisi Italy
91 D2 Brin'kovskaya Rus. Fed.
Brinlack Ireland see Bun na Leaca
53 D1 Brisbane Austr.
99 B4 Bristol U.K.
139 E2 Bristol CT U.S.A.
141 D1 Bristol TN U.S.A.
99 A4 Bristol Channel est. U.K.
128 B2 British Columbia prov. Can.
British Guiana country S. America see Guyana
British Honduras country Central America see Belize
56 C6 British Indian Ocean Territory terr. Indian Ocean
95 B2 British Isles is Europe
British Solomon Islands country S. Pacific Ocean see Solomon Islands
123 C2 Brits S. Africa
122 B3 Britstown S. Africa
104 B2 Brittany reg. France
104 C2 Brive-la-Gaillarde France
106 C1 Briviesca Spain
99 B4 Brixham U.K.
103 D2 Brno Czech Rep.
141 D2 Broad r. U.S.A.
130 C2 Broadback r. Can.
53 C3 Broadford Austr.
96 B2 Broadford U.K.
96 C3 Broad Law h. U.K.
136 B1 Broadus U.S.A.
129 D2 Brochet Can.
129 D2 Brochet, Lac l. Can.
101 E1 Bröckel Ger.
101 E1 Brocken mt. Ger.
126 E1 Brock Island Can.
130 C3 Brockville Can.

Brodeur Peninsula

127 G2	Brodeur Peninsula Can.
96 B3	Brodick U.K.
103 D1	Brodnica Pol.
90 B1	Brody Ukr.
143 D1	Broken Arrow U.S.A.
137 D2	Broken Bow U.S.A.
52 B2	Broken Hill Austr.
	Broken Hill Zambia see
	Kabwe
159 F6	Broken Plateau
	Indian Ocean
151 D2	Brokopondo Suriname
99 B3	Bromsgrove U.K.
93 E4	Brønderslev Denmark
123 C2	Bronkhorstspruit S. Africa
92 F2	Brønnøysund Norway
64 A3	Brooke's Point Phil.
140 B2	Brookhaven U.S.A.
134 B2	Brookings OR U.S.A.
137 D2	Brookings SD U.S.A.
128 C2	Brooks Can.
126 C2	Brooks Range mts U.S.A.
141 D1	Brooksville U.S.A.
139 D2	Brookville U.S.A.
96 B2	Broom, Loch inlet U.K.
50 B1	Broome Austr.
134 B2	Brothers U.S.A.
	Broughton Island Can. see
	Qikiqtarjuaq
90 C1	Brovary Ukr.
143 C2	Brownfield U.S.A.
128 C3	Browning U.S.A.
140 C1	Brownsville TN U.S.A.
143 D3	Brownsville TX U.S.A.
143 D2	Brownwood U.S.A.
92 □A2	Brú Iceland
104 C1	Bruay-la-Bussière France
138 B1	Bruce Crossing U.S.A.
130 B3	Bruce Peninsula Can.
103 D2	Bruck an der Mur Austria
	Bruges Belgium see Brugge
100 A2	Brugge Belgium
62 A1	Bruini India
128 C2	Brûlé Can.
151 E4	Brumado Brazil
93 F3	Brumunddal Norway
61 C1	Brunei country Asia
	Brunei Brunei see
	Bandar Seri Begawan
102 C2	Brunico Italy
	Brünn Czech Rep. see Brno
101 D1	Brunsbüttel Ger.
141 D2	Brunswick GA U.S.A.
139 F2	Brunswick ME U.S.A.
53 D1	Brunswick Heads Austr.
123 D2	Bruntville S. Africa
136 C2	Brush U.S.A.
100 B2	Brussels Belgium
	Bruxelles Belgium see
	Brussels
143 D2	Bryan U.S.A.
89 D3	Bryansk Rus. Fed.
91 D2	Bryukhovetskaya Rus. Fed.
103 D1	Brzeg Pol.
	Brześć nad Bugiem Belarus
	see Brest
114 A3	Buba Guinea-Bissau
111 C3	Buca Turkey
80 B2	Bucak Turkey
150 B2	Bucaramanga Col.
90 B2	Buchach Ukr.
53 C3	Buchan Austr.
114 A4	Buchanan Liberia
110 C2	Bucharest Romania
65 B2	Bucheon S. Korea
101 D1	Bucholz in der Nordheide
	Ger.
110 C1	Bucin, Pasul pass Romania
101 D1	Bückeburg Ger.
142 A2	Buckeye U.S.A.
96 C2	Buckhaven U.K.
96 C2	Buckie U.K.
99 C3	Buckingham U.K.
51 C1	Buckingham Bay Austr.
51 C1	Buckland Tableland reg.
	Austr.
52 A2	Buckleboo Austr.
139 F2	Bucksport U.S.A.
103 D2	Bučovice Czech Rep.
	București Romania see
	Bucharest
89 D3	Buda-Kashalyova Belarus
103 D2	Budapest Hungary
75 B2	Budaun India
108 A2	Buddusò Italy
99 A4	Bude U.K.
87 D4	Budennovsk Rus. Fed.
	Budennoye Rus. Fed. see
	Biryuch
89 D2	Budogoshch' Rus. Fed.

108 A2	Budoni Italy
	Budweis Czech Rep. see
	České Budějovice
118 A2	Buea Cameroon
135 B4	Buellton U.S.A.
150 B2	Buenaventura Col.
144 B2	Buenaventura Mex.
	Buena Vista i. N. Mariana Is
	see Tinian
106 C1	Buendía, Embalse de resr
	Spain
155 D1	Buenópolis Brazil
153 C3	Buenos Aires Arg.
153 A4	Buenos Aires, Lago l. Arg./
	Chile
139 D2	Buffalo NY U.S.A.
136 C1	Buffalo SD U.S.A.
143 D2	Buffalo TX U.S.A.
136 B2	Buffalo WY U.S.A.
129 D2	Buffalo Narrows Can.
121 C3	Buffalo Range Zimbabwe
122 A2	Buffels watercourse S. Africa
123 C1	Buffels Drift S. Africa
110 C2	Buftea Romania
103 E1	Bug r. Pol.
61 C2	Bugel, Tanjung pt Indon.
118 C3	Bugojno Bos.-Herz.
86 D2	Bugrino Rus. Fed.
64 A3	Bugsuk i. Phil.
87 E3	Bugul'ma Rus. Fed.
87 E3	Buguruslan Rus. Fed.
121 C2	Buhera Zimbabwe
110 C1	Buhuşi Romania
99 B3	Builth Wells U.K.
69 D1	Buir Nur l. Mongolia
120 A3	Buitepos Namibia
109 D2	Bujanovac Serbia
119 C3	Bujumbura Burundi
69 D1	Bukachacha Rus. Fed.
120 B2	Bukalo Namibia
119 C3	Bukavu Dem. Rep. Congo
	Bukhara Uzbek.
	see Buxoro
60 B2	Bukittinggi Indon.
119 D3	Bukoba Tanz.
103 D1	Bukowiec h. Pol.
59 C3	Bula Indon.
53 D2	Bulahdelah Austr.
121 B3	Bulawayo Zimbabwe
111 C3	Buldan Turkey
123 D2	Bulembu Swaziland
68 C1	Bulgan Mongolia
110 C2	Bulgaria country Europe
54 B2	Buller r. N.Z.
142 A1	Bullhead City U.S.A.
52 B1	Bulloo watercourse Austr.
52 B1	Bulloo Downs Austr.
122 A1	Büllsport Namibia
61 D2	Bulukumba Indon.
118 B3	Bulungu Dem. Rep. Congo
118 C2	Bumba Dem. Rep. Congo
62 A1	Bumhkang Myanmar
118 B3	Buna Dem. Rep. Congo
	Bun Beg Ireland see
	An Bun Beag
50 A3	Bunbury Austr.
97 C2	Bunclody Ireland
97 C1	Buncrana Ireland
119 D3	Bunda Tanz.
51 E2	Bundaberg Austr.
53 C1	Bundaleer Austr.
53 D2	Bundarra Austr.
74 B2	Bundi India
97 B1	Bundoran Ireland
75 C2	Bundu India
53 C3	Bungendore Austr.
119 D2	Bungoma Kenya
67 B4	Bungo-suidō sea chan.
	Japan
119 D2	Bunia Dem. Rep. Congo
118 C3	Bunianga Dem. Rep. Congo
97 B1	Bun na Leaca Ireland
63 B2	Buôn Ma Thuột Vietnam
119 D3	Bura Kenya
117 C3	Buraan Somalia
	Burang China see Jirang
78 B2	Buraydah Saudi Arabia
101 D2	Burbach Ger.
117 C4	Burco Somalia
100 B1	Burdaard Neth.
111 D3	Burdur Turkey
	Burdwan India see
	Barddhaman
117 B3	Burē Eth.
99 D3	Bure r. U.K.
69 E1	Bureinskiy Khrebet mts
	Rus. Fed.
101 D2	Büren Ger.
74 B1	Burewala Pak.

	Bureya Range mts Rus. Fed.
	see Bureinskiy Khrebet
110 C2	Burgas Bulg.
101 E1	Burg bei Magdeburg Ger.
101 E1	Burgdorf Ger.
101 E1	Burgdorf Ger.
131 E3	Burgeo Can.
123 C3	Burgersdorp S. Africa
123 D1	Burgersfort S. Africa
100 A2	Burgh-Haamstede Neth.
101 F3	Burglengenfeld Ger.
145 C2	Burgos Mex.
106 C1	Burgos Spain
105 C2	Burgundy reg. France
111 C3	Burhaniye Turkey
74 B2	Burhanpur India
75 C2	Burhar-Dhanpuri India
101 D1	Burhave (Butjadingen) Ger.
154 C2	Buri Brazil
60 B2	Buriai Indon.
64 B2	Burias i. Phil.
131 E3	Burin Can.
63 B2	Buriram Thai.
154 C1	Buriti Alegre Brazil
151 E3	Buriti Bravo Brazil
155 C1	Buritis Brazil
107 C2	Burjassot Spain
143 D2	Burkburnett U.S.A.
51 C1	Burketown Austr.
	Burkina country Africa see
	Burkina Faso
114 B3	Burkina Faso country Africa
134 D2	Burley U.S.A.
136 C3	Burlington CO U.S.A.
137 E2	Burlington IA U.S.A.
141 E1	Burlington NC U.S.A.
139 E2	Burlington VT U.S.A.
	Burma country Asia see
	Myanmar
134 B2	Burney U.S.A.
51 D4	Burnie Austr.
98 B3	Burnley U.K.
134 C2	Burns U.S.A.
134 C2	Burns Junction U.S.A.
128 B2	Burns Lake Can.
137 E2	Burnsville U.S.A.
101 F1	Burow Ger.
77 E2	Burqin China
52 A2	Burra Austr.
109 D2	Burrel Albania
97 B2	Burren reg. Ireland
53 C2	Burrendong, Lake Austr.
53 C2	Burren Junction Austr.
53 C2	Burrinjuck Reservoir Austr.
144 B2	Burro, Serranías del mts
	Mex.
111 C2	Bursa Turkey
116 B2	Bür Safājah Egypt
	Bür Sa'īd Egypt see
	Port Said
	Bür Sa'īd Egypt see
	Port Said
	Bür Sudan Sudan see
	Port Sudan
130 C2	Burton, Lac l. Can.
	Burtonport Ireland see
	Ailt an Chorráin
99 C3	Burton upon Trent U.K.
52 B2	Burtundy Austr.
59 C3	Buru i. Indon.
119 C3	Burundi country Africa
119 C3	Bururi Burundi
96 C1	Burwick U.K.
98 B3	Bury U.K.
91 C1	Buryn' Ukr.
99 D3	Bury St Edmunds U.K.
108 B3	Busambra, Rocca mt. Italy
65 B2	Busan S. Korea
118 C3	Busanga Dem. Rep. Congo
81 D3	Büshehr Iran
119 D3	Bushenyi Uganda
	Bushire Iran see Büshehr
118 C2	Businga Dem. Rep. Congo
50 A3	Busselton Austr.
143 C3	Bustamante Mex.
108 A1	Busto Arsizio Italy
118 C2	Buta Dem. Rep. Congo
119 C3	Butare Rwanda
96 B3	Bute i. U.K.
119 C2	Butembo Dem. Rep. Congo
123 C2	Butha-Buthe Lesotho
139 D2	Butler U.S.A.
61 D2	Buton i. Indon.
134 D1	Butte U.S.A.
60 B1	Butterworth Malaysia
96 A1	Butt of Lewis hd U.K.
131 D1	Button Bay Can.
131 D1	Button Islands Can.
64 B3	Butuan Phil.
89 F3	Buturlinovka Rus. Fed.

75 C2	Butwal Nepal
101 D2	Butzbach Ger.
117 C4	Buulobarde Somalia
117 C5	Buur Gaabo Somalia
117 C4	Buurhabaka Somalia
78 A2	Buwāṭah Saudi Arabia
76 C3	Buxoro Uzbek.
101 D1	Buxtehude Ger.
98 C3	Buxton U.K.
89 F2	Buy Rus. Fed.
87 D4	Buynaksk Rus. Fed.
111 C3	Büyükmenderes r. Turkey
65 A1	Buyun Shan mt. China
110 C1	Buzău Romania
110 C1	Buzău r. Romania
121 C2	Búzi Moz.
87 E3	Buzuluk Rus. Fed.
88 C3	Byahoml' Belarus
110 C2	Byala Bulg.
110 C2	Byala Bulg.
88 C3	Byalynichy Belarus
88 D3	Byarezina r. Belarus
88 B3	Byaroza Belarus
88 C3	Byarozawka Belarus
103 D1	Bydgoszcz Pol.
	Byelorussia country Europe
	see Belarus
88 C3	Byerazino Belarus
88 C2	Byeshankovichy Belarus
89 D3	Bykhaw Belarus
127 G2	Bylot Island Can.
53 C2	Byrock Austr.
53 D1	Byron Bay Austr.
83 J2	Bytantay r. Rus. Fed.
103 D1	Bytom Pol.
103 D1	Bytów Pol.

C

120 A2	Caála Angola
154 B2	Caarapó Brazil
155 C1	Caatinga Brazil
144 B2	Caballos Mesteños,
	Llano de los plain Mex.
106 B1	Cabanaquinta Spain
64 B2	Cabanatuan Phil.
117 C3	Cabdul Qaadir Somalia
154 A1	Cabeceira Rio Manso Braz.
154 C1	Cabeceiras Brazil
106 B2	Cabeza del Buey Spain
152 B1	Cabezas Bol.
150 B1	Cabimas Venez.
120 A1	Cabinda Angola
118 B3	Cabinda prov. Angola
151 F3	Cabo de Santo Agostinho
	Brazil
155 D2	Cabo Frio Brazil
155 D2	Cabo Frio, Ilha do i. Brazil
130 C3	Cabonga, Réservoir resr
	Can.
53 D1	Caboolture Austr.
150 B3	Cabo Pantoja Peru
121 C2	Cabora Bassa, Lake resr
	Moz.
144 A1	Caborca Mex.
144 B2	Cabo San Lucas Mex.
131 D3	Cabot Strait Can.
106 C2	Cabra Spain
155 D1	Cabral, Serra do mts Brazil
107 D2	Cabrera, Illa de i. Spain
106 B1	Cabrera, Sierra de la mts
	Spain
129 D2	Cabri Can.
107 C2	Cabriel r. Spain
154 B3	Caçador Brazil
109 D2	Čačak Serbia
108 A2	Caccia, Capo c. Italy
106 B2	Cáceres Port.
151 D4	Cáceres Brazil
106 B2	Cáceres Spain
128 B2	Cache Creek Can.
114 A3	Cacheu Guinea-Bissau
151 D3	Cachimbo, Serra do hills
	Brazil
154 B1	Cachoeira Alta Brazil
155 D2	Cachoeiro de Itapemirim
	Brazil
114 A3	Cacine Guinea-Bissau
120 A2	Cacolo Angola
120 A2	Caconda Angola
154 B1	Caçu Brazil
103 D2	Čadca Slovakia
101 D1	Cadenberge Ger.
145 B2	Cadereyta Mex.
138 B2	Cadillac U.S.A.
64 B2	Cadiz Phil.
106 B2	Cádiz Spain
106 B2	Cádiz, Golfo de g. Spain

128 C2 Cadotte Lake Can.
104 B2 Caen France
Caerdydd U.K. see Cardiff
Caerfyrddin U.K. see Carmarthen
Caergybi U.K. see Holyhead
98 A3 Caernarfon U.K.
99 A3 Caernarfon Bay U.K.
Caernarvon U.K. see Caernarfon
152 B2 Cafayate Arg.
154 C2 Cafelândia Brazil
64 B3 Cagayan de Oro Phil.
64 A3 Cagayan de Tawi-Tawi i. Phil.
108 B2 Cagli Italy
108 A3 Cagliari Italy
108 A3 Cagliari, Golfo di b. Italy
76 B2 Çagyl Turkm.
120 A2 Cahama Angola
97 B3 Caha Mountains hills Ireland
97 A3 Cahermore Ireland
97 C2 Cahir Ireland
97 A3 Cahirsiveen Ireland
Cahora Bassa, Lago de resr Moz. see Cabora Bassa, Lake
97 C2 Cahore Point Ireland
104 C3 Cahors France
90 B2 Cahul Moldova
121 C2 Caia Moz.
151 D4 Caiabis, Serra dos hills Brazil
120 B2 Caianda Angola
154 B1 Caiapó, Serra do mts Brazil
154 B1 Caiapônia Brazil
147 C2 Caicos Islands Turks and Caicos Is
96 C2 Cairngorm Mountains U.K.
96 B3 Cairnryan U.K.
51 D1 Cairns Austr.
116 B1 Cairo Egypt
Caisleán an Bharraigh Ireland see Castlebar
98 C3 Caistor U.K.
120 A2 Caiundo Angola
150 B3 Cajamarca Peru
109 C1 Čakovec Croatia
123 C3 Cala S. Africa
115 C4 Calabar Nigeria
150 C2 Calabozo Venez.
110 B2 Calafat Romania
153 A5 Calafate Arg.
107 C1 Calahorra Spain
120 A2 Calai Angola
104 C1 Calais France
139 F1 Calais U.S.A.
152 B2 Calama Chile
64 A2 Calamian Group is Phil.
107 C1 Calamocha Spain
120 A1 Calandula Angola
60 A1 Calang Indon.
64 B2 Calapan Phil.
110 C2 Călărași Romania
107 C1 Calatayud Spain
64 B2 Calayan i. Phil.
64 B2 Calbayog Phil.
151 F3 Calcanhar, Ponta do pt Brazil
151 D2 Calçoene Brazil
Calcutta India see Kolkata
106 B2 Caldas da Rainha Port.
154 C1 Caldas Novas Brazil
152 A2 Caldera Chile
51 D2 Caldervale Austr.
134 C2 Caldwell U.S.A.
123 C3 Caledon r. Lesotho/S. Africa
122 A3 Caledon S. Africa
153 B4 Caleta Olivia Arg.
98 A2 Calf of Man i. Isle of Man
128 C2 Calgary Can.
150 B2 Cali Col.
Calicut India see Kozhikode
135 D3 Caliente U.S.A.
135 B2 California state U.S.A.
144 A1 California, Gulf of g. Mex.
135 B3 California Aqueduct canal U.S.A.
81 C2 Cälilabad Azer.
122 B3 Calitzdorp S. Africa
145 C2 Calkiní Mex.
52 B1 Callabonna, Lake imp. l. Austr.
135 C3 Callaghan, Mount U.S.A.
97 C2 Callan Ireland
96 B2 Callander U.K.
150 B4 Callao Peru
99 A4 Callington U.K.
108 B3 Caltagirone Italy
108 B3 Caltanissetta Italy
120 A1 Calulo Angola
120 B2 Calunda Angola
120 A2 Caluquembe Angola
117 D3 Caluula Somalia
105 D3 Calvi France
107 D2 Calvià Spain
144 B2 Calvillo Mex.
122 A3 Calvinia S. Africa
109 C2 Calvo, Monte mt. Italy
120 A1 Camabatela Angola
151 F4 Camaçari Brazil
144 B2 Camacho Mex.
120 A2 Camacupa Angola
146 C2 Camagüey Cuba
146 C2 Camagüey, Archipiélago de is Cuba
150 B4 Camana Peru
120 B2 Camanongue Angola
154 B1 Camapuã Brazil
145 C2 Camargo Mex.
63 B3 Ca Mau Vietnam
63 B3 Ca Mau, Mui c. Vietnam
Cambay India see Khambhat
63 B2 Cambodia country Asia
99 A4 Camborne U.K.
105 C1 Cambrai France
99 B3 Cambrian Mountains hills U.K.
138 C2 Cambridge Can.
54 C1 Cambridge N.Z.
99 D3 Cambridge U.K.
139 E2 Cambridge MA U.S.A.
139 D3 Cambridge MD U.S.A.
137 E1 Cambridge MN U.S.A.
138 C2 Cambridge OH U.S.A.
126 E2 Cambridge Bay Can.
131 D2 Cambrien, Lac l. Can.
120 B1 Cambulo Angola
53 D2 Camden Austr.
140 B2 Camden AR U.S.A.
139 F2 Camden ME U.S.A.
139 D3 Camden NJ U.S.A.
137 E3 Camdenton U.S.A.
137 E3 Cameron U.S.A.
118 B2 Cameroon country Africa
118 B2 Cameroon Highlands slope Cameroon/Nigeria
118 A2 Cameroun, Mont vol. Cameroon
151 E3 Cametá Brazil
64 B2 Camiguin i. Phil.
64 B3 Camiguin i. Phil.
152 B2 Camiri Bol.
151 E3 Camocim Brazil
51 C1 Camooweal Austr.
63 A3 Camorta i. India
153 A4 Campana, Isla i. Chile
155 D1 Campanário Brazil
155 C2 Campanha Brazil
122 B2 Campbell S. Africa
54 B2 Campbell, Cape N.Z.
48 H9 Campbell Island N.Z.
156 D9 Campbell Plateau S. Pacific Ocean
128 B2 Campbell River Can.
138 B3 Campbellsville U.S.A.
131 D3 Campbellton Can.
53 D2 Campbelltown Austr.
96 B3 Campbeltown U.K.
145 C3 Campeche Mex.
145 C3 Campeche, Bahía de g. Mex.
52 B3 Camperdown Austr.
110 C1 Câmpina Romania
151 F3 Campina Grande Brazil
154 C2 Campinas Brazil
154 C1 Campina Verde Brazil
108 B2 Campobasso Italy
155 C2 Campo Belo Brazil
154 C1 Campo Florido Brazil
152 B2 Campo Gallo Arg.
154 B2 Campo Grande Brazil
154 C3 Campo Largo Brazil
151 E3 Campo Maior Brazil
106 B2 Campo Maior Port.
150 B1 Campo Mara Venez.
109 C2 Campomarino Italy
154 B2 Campo Mourão Brazil
155 D2 Campos Brazil
155 C1 Campos Altos Brazil
155 C2 Campos do Jordão Brazil
110 C1 Câmpulung Romania
142 A2 Camp Verde U.S.A.
Cam Ranh Vietnam see Ba Ngoi
63 B2 Cam Ranh, Vinh b. Vietnam
Cam Ranh Bay b. Vietnam see Cam Ranh, Vinh
128 C2 Camrose Can.
129 D2 Camsell Portage Can.
111 C2 Çan Turkey
126 F2 Canada country N. America
160 A2 Canada Basin Arctic Ocean
143 C1 Canadian U.S.A.
143 D1 Canadian r. U.S.A.
111 C2 Çanakkale Turkey
144 A1 Cananea Mex.
154 C2 Cananéia Brazil
114 A2 Canary Islands is N. Atlantic Ocean
154 C1 Canastra, Serra da mts Goiás Brazil
155 C1 Canastra, Serra da mts Minas Gerais Brazil
144 B2 Canatlán Mex.
141 D3 Canaveral, Cape U.S.A.
155 E1 Canavieiras Brazil
53 C2 Canbelego Austr.
53 C3 Canberra Austr.
145 D2 Cancún Mex.
111 C3 Çandarlı Turkey
155 C2 Candeias Brazil
145 C3 Candelaria Mex.
154 B2 Cândido de Abreu Brazil
129 D2 Candle Lake Can.
137 D1 Cando U.S.A.
120 A2 Cangamba Angola
106 B1 Cangas Spain
106 B1 Cangas del Narcea Spain
120 B2 Cangombe Angola
152 C3 Canguçu Brazil
70 B2 Cangzhou China
131 D2 Caniapiscau Can.
131 D2 Caniapiscau r. Can.
131 C2 Caniapiscau, Réservoir de resr Can.
Caniçado Moz. see Guija
108 B3 Canicattì Italy
151 F3 Canindé Brazil
144 B2 Cañitas de Felipe Pescador Mex.
80 B1 Çankırı Turkey
128 C2 Canmore Can.
96 A2 Canna i. U.K.
Cannanore India see Kannur
105 D3 Cannes France
99 B3 Cannock U.K.
53 C3 Cann River Austr.
152 C2 Canoas Brazil
129 D2 Canoe Lake Can.
154 B3 Canoinhas Brazil
136 B3 Canon City U.S.A.
52 B2 Canopus Austr.
129 D2 Canora Can.
53 C2 Canowindra Austr.
131 D3 Canso Can.
Cantabrian Mountains mts Spain see Cantábrica, Cordillera
Cantabrian Sea sea Spain see Cantábrico, Mar
106 C1 Cantábrica, Cordillera mts Spain
106 B1 Cantábrico, Mar sea Spain
99 D4 Canterbury U.K.
54 B2 Canterbury Bight b. N.Z.
54 B2 Canterbury Plains N.Z.
63 B2 Cân Thơ Vietnam
151 E3 Canto do Buriti Brazil
Canton China see Guangzhou
137 E2 Canton MO U.S.A.
140 C2 Canton MS U.S.A.
139 D2 Canton NY U.S.A.
138 C2 Canton OH U.S.A.
143 C1 Canyon U.S.A.
134 D1 Canyon Ferry Lake U.S.A.
134 B2 Canyonville U.S.A.
62 B1 Cao Bằng Vietnam
109 C2 Capaccio Italy
154 C2 Capão Bonito Brazil
155 D2 Caparaó, Serra do mts Brazil
139 E1 Cap-de-la-Madeleine Can.
51 D4 Cape Barren Island Austr.
158 F7 Cape Basin S. Atlantic Ocean
52 A3 Cape Borda Austr.
131 D3 Cape Breton Island Can.
141 D3 Cape Canaveral U.S.A.
139 D3 Cape Charles U.S.A.
114 B4 Cape Coast Ghana
139 E2 Cape Cod Bay U.S.A.
141 D3 Cape Coral U.S.A.
127 G2 Cape Dorset Can.
141 E2 Cape Fear r. U.S.A.
137 F3 Cape Girardeau U.S.A.
155 D1 Capelinha Brazil
100 B2 Capelle aan de IJssel Neth.
Capelongo Angola see Kuvango
139 E3 Cape May Point U.S.A.
120 A1 Capenda-Camulemba Angola
122 A3 Cape Town S. Africa
158 E3 Cape Verde country N. Atlantic Ocean
158 D4 Cape Verde Basin N. Atlantic Ocean
51 D1 Cape York Peninsula Austr.
147 C3 Cap-Haïtien Haiti
151 E3 Capim r. Brazil
154 A2 Capitán Bado Para.
58 D1 Capitol Hill N. Mariana Is
154 B2 Capivara, Represa resr Brazil
109 C2 Čapljina Bos.-Herz.
109 B3 Capo d'Orlando Italy
108 A2 Capraia, Isola di i. Italy
108 A2 Caprara, Punta pt Italy
108 B2 Capri, Isola di i. Italy
51 E2 Capricorn Channel Austr.
120 B2 Caprivi Strip reg. Namibia
143 C2 Cap Rock Escarpment U.S.A.
143 C1 Capulin U.S.A.
150 C3 Caquetá r. Col.
110 B2 Caracal Romania
150 C2 Caracarai Brazil
150 C1 Caracas Venez.
151 E3 Caracol Brazil
155 C2 Caraguatatuba Brazil
153 A3 Carahue Chile
155 D1 Caraí Brazil
151 D3 Carajás, Serra dos hills Brazil
155 D2 Carandaí Brazil
155 D2 Carangola Brazil
110 B1 Caransebeş Romania
131 D3 Caraquet Can.
146 B3 Caratasca, Laguna de lag. Hond.
155 D1 Caratinga Brazil
150 C3 Carauari Brazil
107 C2 Caravaca de la Cruz Spain
155 E1 Caravelas Brazil
129 E3 Carberry Can.
144 A2 Carbó Mex.
107 C2 Carbon, Cap c. Alg.
153 B5 Carbón, Laguna del l. Arg.
108 A3 Carbonara, Capo c. Italy
136 B3 Carbondale CO U.S.A.
138 B3 Carbondale IL U.S.A.
139 D2 Carbondale PA U.S.A.
131 E3 Carbonear Can.
155 D1 Carbonita Brazil
107 C2 Carcaixent Spain
104 C3 Carcassonne France
128 A1 Carcross Can.
146 B2 Cárdenas Cuba
145 C3 Cárdenas Mex.
99 B4 Cardiff U.K.
99 A3 Cardigan U.K.
99 A3 Cardigan Bay U.K.
154 C2 Cardoso Brazil
128 C3 Cardston Can.
110 B1 Carei Romania
104 B2 Carentan France
50 B2 Carey, Lake imp. l. Austr.
155 D2 Cariacica Brazil
146 B3 Caribbean Sea N. Atlantic Ocean
128 B2 Cariboo Mountains Can.
139 F1 Caribou U.S.A.
130 B2 Caribou Lake Can.
128 C2 Caribou Mountains Can.
144 B2 Carichic Mex.
100 B3 Carignan France
53 C2 Carinda Austr.
107 C1 Cariñena Spain
130 C3 Carleton Place Can.
123 C2 Carletonville S. Africa
135 C2 Carlin U.S.A.
97 C1 Carlingford Lough inlet Ireland/U.K.
138 B3 Carlinville U.S.A.
98 B2 Carlisle U.K.
139 D2 Carlisle U.S.A.
155 D1 Carlos Chagas Brazil
97 C2 Carlow Ireland
96 A1 Carloway U.K.
135 C4 Carlsbad CA U.S.A.
142 C2 Carlsbad NM U.S.A.
129 D3 Carlyle Can.
128 A1 Carmacks Can.

Carman

129 E3 Carman Can.
99 A4 Carmarthen U.K.
99 A4 Carmarthen Bay U.K.
104 C3 Carmaux France
145 C3 Carmelita Guat.
144 A2 Carmen, Isla i. Mex.
155 C1 Carmo do Paranaíba Brazil
Carmona Angola see Uíge
106 B2 Carmona Spain
104 B2 Carnac France
50 A2 Carnarvon Austr.
122 B3 Carnarvon S. Africa
97 C1 Carndonagh Ireland
129 D3 Carnduff Can.
50 B2 Carnegie, Lake imp. l. Austr.
96 B2 Carn Eige mt. U.K.
55 P2 Carney Island Antarctica
73 D4 Car Nicobar i. India
118 B2 Carnot C.A.R.
52 A2 Carnot, Cape Austr.
96 C2 Carnoustie U.K.
97 C2 Carnsore Point Ireland
151 E3 Carolina Brazil
49 L4 Caroline Island Kiribati
59 D2 Caroline Islands N. Pacific Ocean
122 A2 Carolusberg S. Africa
103 D2 Carpathian Mountains Europe
Carpaţii Meridionali mts Romania see Transylvanian Alps
51 C1 Carpentaria, Gulf of Austr.
105 D3 Carpentras France
108 B2 Carpi Italy
141 D3 Carrabelle U.S.A.
97 B3 Carrantuohill mt. Ireland
108 B2 Carrara Italy
97 D1 Carrickfergus U.K.
97 C2 Carrickmacross Ireland
97 B2 Carrick-on-Shannon Ireland
97 C2 Carrick-on-Suir Ireland
137 D1 Carrington U.S.A.
143 D3 Carrizo Springs U.S.A.
142 B2 Carrizozo U.S.A.
137 E2 Carroll U.S.A.
141 C2 Carrollton U.S.A.
129 D2 Carrot River Can.
135 C3 Carson City U.S.A.
135 C3 Carson Sink l. U.S.A.
Carstensz-top mt. Indon. see Jaya, Puncak
150 B1 Cartagena Col.
107 C2 Cartagena Spain
146 B4 Cartago Costa Rica
54 C2 Carterton N.Z.
137 E3 Carthage MO U.S.A.
143 E2 Carthage TX U.S.A.
131 E2 Cartwright Can.
151 F3 Caruaru Brazil
150 C1 Carúpano Venez.
52 B1 Caryapundy Swamp Austr.
114 B1 Casablanca Morocco
154 C2 Casa Branca Brazil
144 B1 Casa de Janos Mex.
142 A2 Casa Grande U.S.A.
108 A1 Casale Monferrato Italy
109 C2 Casarano Italy
144 B1 Casas Grandes Mex.
134 C2 Cascade U.S.A.
134 B2 Cascade Range mts Can./U.S.A.
106 B2 Cascais Port.
151 F3 Cascavel Brazil
154 B2 Cascavel Brazil
139 F2 Casco Bay U.S.A.
108 B2 Caserta Italy
97 C2 Cashel Ireland
153 B3 Casilda Arg.
53 D1 Casino Austr.
Casnewydd U.K. see Newport
107 C1 Caspe Spain
136 B2 Casper U.S.A.
76 A2 Caspian Lowland Kazakh./Rus. Fed.
81 C1 Caspian Sea Asia/Europe
Cassaigne Alg. see Sidi Ali
154 C2 Cássia Brazil
128 B2 Cassiar Can.
128 A2 Cassiar Mountains Can.
154 B1 Cassilândia Brazil
120 A2 Cassinga Angola
108 B2 Cassino Italy
96 B2 Cassley r. U.K.
151 E3 Castanhal Brazil
152 B3 Castaño r. Arg.
144 B2 Castaños Mex.
104 C3 Casteljaloux France

105 D3 Castellane France
107 C2 Castellón de la Plana Spain
155 D2 Castelo Brazil
106 B2 Castelo Branco Port.
104 C3 Castelsarrasin France
108 B3 Castelvetrano Italy
52 B3 Casterton Austr.
108 B2 Castiglione della Pescaia Italy
106 B2 Castilla, Playa de coastal area Spain
106 C2 Castilla-La Mancha aut. comm. Spain
106 C1 Castilla y León aut. comm. Spain
97 B2 Castlebar Ireland
96 A2 Castlebay U.K.
97 C1 Castleblayney Ireland
97 C1 Castlederg U.K.
96 C3 Castle Douglas U.K.
128 C3 Castlegar Can.
97 B2 Castleisland Ireland
52 B3 Castlemaine Austr.
97 C2 Castlepollard Ireland
97 B2 Castlerea Ireland
53 C2 Castlereagh r. Austr.
136 C3 Castle Rock U.S.A.
128 C2 Castor Can.
104 C3 Castres France
100 B1 Castricum Neth.
147 D3 Castries St Lucia
154 C2 Castro Brazil
153 A4 Castro Chile
106 B2 Castro Verde Port.
109 C3 Castrovillari Italy
150 A3 Catacaos Peru
155 D2 Cataguases Brazil
154 C1 Catalão Brazil
Catalonia aut. comm. Spain see Cataluña
107 D1 Cataluña aut. comm. Spain
152 B2 Catamarca Arg.
64 B2 Catanduanes i. Phil.
154 C2 Catanduva Brazil
154 B3 Catanduvas Brazil
109 C3 Catania Italy
109 C3 Catanzaro Italy
64 B2 Cataraman Phil.
107 C2 Catarroja Spain
64 B2 Catbalogan Phil.
145 C3 Catemaco Mex.
120 A1 Catete Angola
Catherine, Mount mt. Egypt see Kātrīna, Jabal
147 C2 Cat Island Bahamas
130 A2 Cat Lake Can.
145 D2 Catoche, Cabo c. Mex.
139 E2 Catskill Mountains U.S.A.
123 D2 Catuane Moz.
64 B3 Cauayan Phil.
131 D2 Caubvick, Mount Can.
151 F3 Caucaia Brazil
81 C1 Caucasus mts Asia/Europe
100 A2 Caudry France
109 C3 Caulonia Italy
120 A1 Caungula Angola
150 C2 Caura r. Venez.
131 D3 Causapscal Can.
90 B2 Căușeni Moldova
105 D3 Cavaillon France
151 E4 Cavalcante Brazil
114 B4 Cavally r. Côte d'Ivoire/Liberia
97 C2 Cavan Ireland
154 B3 Cavernoso, Serra do mts Brazil
151 D2 Caviana, Ilha i. Brazil
Cawnpore India see Kanpur
151 E3 Caxias Brazil
152 C2 Caxias do Sul Brazil
120 A1 Caxito Angola
151 D2 Cayenne Fr. Guiana
146 B3 Cayman Islands terr. West Indies
158 C3 Cayman Trench Caribbean Sea
117 C4 Caynabo Somalia
120 A1 Cazombo Angola
Ceará Brazil see Fortaleza
Ceatharlach Ireland see Carlow
144 B2 Ceballos Mex.
64 B2 Cebu Phil.
64 B2 Cebu i. Phil.
108 B2 Cecina Italy
137 F2 Cedar r. U.S.A.
135 D3 Cedar City U.S.A.
137 E2 Cedar Falls U.S.A.
129 D2 Cedar Lake Can.

137 E2 Cedar Rapids U.S.A.
144 A2 Cedros, Isla i. Mex.
51 C3 Ceduna Austr.
117 C4 Ceeldheere Somalia
117 C3 Ceerigaabo Somalia
108 B3 Cefalù Italy
145 B2 Celaya Mex.
61 D2 Celebes i. Indon.
156 C5 Celebes Sea Indon./Phil.
145 C2 Celestún Mex.
101 E1 Celle Ger.
95 B3 Celtic Sea Ireland/U.K.
59 D3 Cenderawasih, Teluk b. Indon.
140 C2 Center Point U.S.A.
150 B2 Central, Cordillera mts Col.
150 B4 Central, Cordillera mts Peru
64 B2 Central, Cordillera mts Phil.
Central African Empire country Africa see Central African Republic
118 C2 Central African Republic country Africa
74 A2 Central Brahui Range mts Pak.
137 D2 Central City U.S.A.
138 B3 Centralia IL U.S.A.
134 B1 Centralia WA U.S.A.
74 A2 Central Makran Range mts Pak.
156 D5 Central Pacific Basin Pacific Ocean
134 B2 Central Point U.S.A.
Central Provinces state India see Madhya Pradesh
59 D3 Central Range mts P.N.G.
89 E3 Central Russian Upland hills Rus. Fed.
83 I2 Central Siberian Plateau plat. Rus. Fed.
140 C2 Ceos i. Greece see Kea
111 B3 Cephalonia i. Greece
Ceram i. Indon. see Seram
Ceram Sea sea Indon. see Laut Seram
101 F3 Čerchov mt. Czech Rep.
152 B2 Ceres Arg.
154 C1 Ceres Brazil
122 A3 Ceres S. Africa
105 C3 Céret France
106 C1 Cerezo de Abajo Spain
109 C2 Cerignola Italy
110 C2 Cerigo i. Greece see Kythira
90 C2 Cernavodă Romania
145 C2 Cerralvo Mex.
144 B2 Cerralvo, Isla i. Mex.
145 B2 Cerritos Mex.
154 C2 Cerro Azul Brazil
150 B4 Cerro de Pasco Peru
105 D3 Cervione France
106 B1 Cervo Spain
108 B2 Cesena Italy
108 B2 Cesenatico Italy
88 C2 Cēsis Latvia
102 C2 České Budějovice Czech Rep.
101 F3 Český les mts Czech Rep.
111 C3 Çeşme Turkey
53 D2 Cessnock Austr.
104 B2 Cesson-Sévigné France
104 B3 Cestas France
109 C2 Cetinje Montenegro
106 B2 Ceuta N. Africa
105 C3 Cévennes mts France
Ceylon country Asia see Sri Lanka
79 D2 Chābahār Iran
150 B3 Chachapoyas Peru
89 D3 Chachersk Belarus
63 B2 Chachoengsao Thai.
142 B1 Chaco Boreal reg. Para.
142 B1 Chaco Mesa plat. U.S.A.
115 D3 Chad country Africa
115 D3 Chad, Lake Africa
68 C1 Chadaasan Mongolia
68 C1 Chadan Rus. Fed.
123 C1 Chadibe Botswana
136 C2 Chadron U.S.A.
Chadyr-Lunga Moldova see Ciadir-Lunga
77 D2 Chaek Kyrg.
65 B2 Chaeryŏng N. Korea
74 A2 Chagai Afgh.
77 C3 Chaghcharān Afgh.
89 E2 Chagoda Rus. Fed.
56 I10 Chagos Archipelago is B.I.O.T.

159 E4 Chagos-Laccadive Ridge Indian Ocean
159 E4 Chagos Trench Indian Ocean
75 C2 Chaibasa India
63 B2 Chainat Thai.
63 A3 Chaiya Thai.
63 B2 Chaiyaphum Thai.
152 C3 Chajarí Arg.
119 D3 Chake Chake Tanz.
131 D2 Chakonipau, Lac Can.
150 B4 Chala Peru
121 C2 Chaláua Moz.
131 D3 Chaleur Bay inlet Can.
74 B2 Chalisgaon India
111 C3 Chalki i. Greece
111 B3 Chalkida Greece
143 C3 Chalk Mountains U.S.A.
104 B2 Challans France
134 D2 Challis U.S.A.
105 C2 Châlons-en-Champagne France
Châlons-sur-Marne France see Châlons-en-Champagne
105 C2 Chalon-sur-Saône France
101 F3 Cham Ger.
142 B1 Chama U.S.A.
121 C2 Chama Zambia
74 A1 Chaman Pak.
74 B1 Chamba India
74 B2 Chambal r. India
137 D2 Chamberlain U.S.A.
142 B1 Chambers U.S.A.
139 D3 Chambersburg U.S.A.
105 D2 Chambéry France
121 C2 Chambeshi Zambia
121 B2 Chambeshi r. Zambia
Chamdo China see Qamdo
119 D2 Ch'amo Hāyk' l. Eth.
105 D2 Chamonix-Mont-Blanc France
105 C2 Champagne reg. France
138 B3 Champaign U.S.A.
139 E2 Champlain, Lake Can./U.S.A.
145 C3 Champotón Mex.
Chanak Turkey see Çanakkale
152 A2 Chañaral Chile
Chanda India see Chandrapur
126 C2 Chandalar r. U.S.A.
140 C3 Chandeleur Islands U.S.A.
74 B1 Chandigarh India
131 D3 Chandler Can.
142 A2 Chandler U.S.A.
75 C2 Chandpur Bangl.
75 B3 Chandrapur India
63 B2 Chang, Ko i. Thai.
Chang'an China see Rong'an
121 C3 Changane r. Moz.
121 C2 Changara Moz.
65 B1 Changbai China
65 B1 Changbai Shan mts China/N. Korea
Changchow China see Zhangzhou
Changchow China see Changzhou
69 E2 Changchun China
71 B3 Changde China
65 B2 Ch'angdo N. Korea
70 B2 Changge China
Chang Jiang r. China see Yangtze
Changjiang Kou r. mouth China see Yangtze, Mouth of the
65 B1 Changjin N. Korea
65 B1 Changjin-gang r. N. Korea
Changkiang China see Zhanjiang
Changning China see Xunwu
Ch'ang-pai Shan mts China/N. Korea see Changbai Shan
71 B3 Changsha China
70 C2 Changshu China
Changteh China see Changde
65 B1 Changting Fujian China
66 A2 Changting Heilong. China
65 A1 Changtu China
146 B4 Changuinola Panama
65 B2 Changwon S. Korea
69 E2 Changyŏn N. Korea
70 B2 Changyuan China
70 B2 Changzhi China

70 B2	Changzhou China
11 B3	Chania Greece
95 C4	Channel Islands English Chan.
135 C4	Channel Islands U.S.A.
131 E3	Channel-Port-aux-Basques Can.
106 B1	Chantada Spain
63 B2	Chanthaburi Thai.
104 C2	Chantilly France
137 D3	Chanute U.S.A.
82 G3	Chany, Ozero salt l. Rus. Fed.
70 B2	Chaohu China
71 B3	Chaoyang Guangdong China
	Chaoyang China see Huinan
70 C1	Chaoyang Liaoning China
71 B3	Chaozhou China
74 B2	Chapala, Laguna de l. Mex.
76 B1	Chapayevo Kazakh.
87 D3	Chapayevsk Rus. Fed.
152 C2	Chapecó Brazil
141 E1	Chapel Hill U.S.A.
130 B3	Chapleau Can.
89 E3	Chaplygin Rus. Fed.
91 C2	Chaplynka Ukr.
	Chapra India see Chhapra
145 B2	Charcas Mex.
99 B4	Chard U.K.
	Chardzhev Turkm. see Türkmenabat
	Chardzhou Turkm. see Türkmenabat
104 B2	Charente r. France
118 B1	Chari r. Cameroon/Chad
77 C3	Chārīkār Afgh.
86 E2	Charkayuvom Rus. Fed.
	Charklik China see Ruoqiang
100 B2	Charleroi Belgium
139 D3	Charles, Cape U.S.A.
139 E1	Charlesbourg Can.
141 E1	Charles City U.S.A.
138 B3	Charleston Il. U.S.A.
137 F3	Charleston MO U.S.A.
141 E1	Charleston SC U.S.A.
138 C3	Charleston WV U.S.A.
135 C3	Charleston Peak U.S.A.
51 D2	Charleville Austr.
97 B2	Charleville Ireland
105 C2	Charleville-Mézières France
138 B1	Charlevoix U.S.A.
141 D1	Charlotte U.S.A.
141 D3	Charlotte Harbor b. U.S.A.
139 D3	Charlottesville U.S.A.
131 D3	Charlottetown Can.
52 B3	Charlton Austr.
130 C2	Charlton Island Can.
51 D2	Charters Towers Austr.
104 C2	Chartres France
128 C2	Chase Can.
88 C3	Chashniki Belarus
54 A3	Chaslands Mistake c. N.Z.
65 B1	Chasŏng N. Korea
104 B2	Chassiron, Pointe de pt France
104 C2	Châteaubriant France
104 C2	Château-du-Loir France
104 C2	Châteaudun France
104 C2	Châteaulin France
105 D3	Châteauneuf-les-Martigues France
104 C2	Châteauneuf-sur-Loire France
104 C2	Châteauroux France
105 C2	Château-Thierry France
128 C2	Chateh Can.
100 B2	Châtelet Belgium
104 C2	Châtellerault France
138 C2	Chatham Can.
99 D4	Chatham U.K.
49 J8	Chatham Islands N.Z.
105 C2	Châtillon-sur-Seine France
141 D2	Chattahoochee r. U.S.A.
141 C1	Chattanooga U.S.A.
63 B2	Châu Đôc Vietnam
62 A1	Chauk Myanmar
105 D2	Chaumont France
105 C2	Chauny France
	Chau Phu Vietnam see Châu Đôc
89 D3	Chavusy Belarus
151 E3	Chaves Brazil
106 B1	Chaves Port.
130 C2	Chavigny, Lac l. Can.
89 D3	Chayevo Rus. Fed.
89 E2	Chayevo Rus. Fed.
86 E3	Chaykovskiy Rus. Fed.

140 C2	Cheaha Mountain h. U.S.A.
102 C1	Cheb Czech Rep.
87 D3	Cheboksary Rus. Fed.
138 C1	Cheboygan U.S.A.
140 A1	Checotah U.S.A.
	Chefoo China see Yantai
126 B2	Chefornak U.S.A.
114 B2	Chegga Maur.
121 C2	Chegutu Zimbabwe
134 B1	Chehalis U.S.A.
89 E2	Chekhov Rus. Fed.
	Chekiang prov. China see Zhejiang
134 B1	Chelan, Lake U.S.A.
103 E1	Chełm Pol.
99 D4	Chelmer r. U.K.
103 D1	Chełmno Pol.
99 D4	Chelmsford U.K.
99 B4	Cheltenham U.K.
87 F3	Chelyabinsk Rus. Fed.
101 F2	Chemnitz Ger.
	Chemulpo S. Korea see Incheon
134 B2	Chemult U.S.A.
74 B2	Chenab r. India/Pak.
114 B2	Chenachane Alg.
134 C1	Cheney U.S.A.
	Chengchow China see Zhengzhou
70 B1	Chengde China
70 A2	Chengdu China
71 A3	Chengguan China
	Chengjiang China see Taihe
71 B4	Chengmai China
	Chengshou China see Yingshan
	Chengtu China see Chengdu
70 A2	Chengxian China
	Chengxiang China see Wuxi
	Chengxiang China see Mianning
73 C3	Chennai India
	Chenstokhov Pol. see Częstochowa
71 B3	Chenzhou China
65 B2	Cheonan S. Korea
65 B2	Cheongju S. Korea
65 B2	Cheorwon S. Korea
99 B4	Chepstow U.K.
141 E2	Cheraw U.S.A.
104 B2	Cherbourg-Octeville France
	Cherchen China see Qiemo
89 E3	Cheremisinovo Rus. Fed.
68 C1	Cheremkhovo Rus. Fed.
89 E2	Cherepovets Rus. Fed.
91 C2	Cherkasy Ukr.
87 D4	Cherkessk Rus. Fed.
89 E3	Chern' Rus. Fed.
91 C1	Chernihiv Ukr.
91 D2	Chernivtsi Ukr.
90 B2	Chernivtsi Ukr.
91 C2	Chernobyl' Ukr.
90 B1	Chernyakhiv Ukr.
88 B3	Chernyakhovsk Rus. Fed.
89 E3	Chernyanka Rus. Fed.
69 D1	Chernyshevsk Rus. Fed.
83 I2	Chernyshevskiy Rus. Fed.
	Chernyy Rynok Rus. Fed. see Kochubey
137 D2	Cherokee U.S.A.
83 L2	Cherskiy Rus. Fed.
83 K2	Cherskogo, Khrebet mts Rus. Fed.
91 E2	Chertkovo Rus. Fed.
	Chervonoarmeyskoye Ukr. see Vil'nyans'k
	Chervonoarmiys'k Ukr. see Krasnoarmiys'k
	Chervonoarmiys'k Ukr. see Radyvyliv
90 A1	Chervonohrad Ukr.
88 C3	Chervyen' Belarus
89 D3	Cherykaw Belarus
139 D3	Chesapeake Bay U.S.A.
86 D2	Cheshskaya Guba b. Rus. Fed.
98 B3	Chester U.K.
138 B3	Chester Il. U.S.A.
141 D2	Chester SC U.S.A.
139 D3	Chester VA U.S.A.
98 C3	Chesterfield U.K.
137 E3	Chesterfield U.S.A.
129 E1	Chesterfield Inlet Can.
129 E1	Chesterfield Inlet inlet Can.
139 F1	Chesuncook Lake l. U.S.A.
108 A3	Chétaïbi Alg.
131 D3	Chéticamp Can.
145 D3	Chetumal Mex.

128 B2	Chetwynd Can.
98 B2	Cheviot Hills U.K.
119 D2	Che'w Bahir salt l. Eth.
136 C2	Cheyenne U.S.A.
136 C2	Cheyenne r. U.S.A.
136 C3	Cheyenne Wells U.S.A.
75 C2	Chhapra India
75 B2	Chhatarpur India
75 C2	Chhattisgarh state India
74 B2	Chhindwara India
75 C2	Chhukha Bhutan
62 A2	Chiang Dao Thai.
120 A2	Chiange Angola
62 A2	Chiang Mai Thai.
62 A2	Chiang Rai Thai.
145 C3	Chiapa Mex.
108 A1	Chiavenna Italy
69 F2	Chiba Japan
70 B3	Chibi China
	Chibizovka Rus. Fed. see Zherdevka
121 C3	Chiboma Moz.
130 C3	Chibougamau Can.
123 D1	Chibuto Moz.
75 D1	Chibuzhang Co l. China
138 B2	Chicago U.S.A.
128 A2	Chichagof Island U.S.A.
99 C4	Chichester U.K.
50 A2	Chichester Range mts Austr.
143 D1	Chickasha U.S.A.
106 B2	Chiclana de la Frontera Spain
150 B3	Chiclayo Peru
153 B4	Chico r. Arg.
153 B4	Chico r. Arg.
135 B3	Chico U.S.A.
139 E2	Chicopee U.S.A.
64 B2	Chico Sapocoy, Mount Phil.
131 C3	Chicoutimi Can.
131 D1	Chidley, Cape Can.
63 A3	Chieo Lan, Ang Kep Nam Thai.
108 B2	Chieti Italy
145 C3	Chietla Mex.
70 B1	Chifeng China
155 D1	Chifre, Serra do mts Brazil
121 C2	Chifunde Moz.
145 C3	Chignahuapán Mex.
121 C3	Chigubo Moz.
62 A1	Chigu Co l. China
74 A1	Chihil Abdālān, Kōh-e mts Afgh.
144 B2	Chihuahua Mex.
88 C2	Chikhachevo Rus. Fed.
67 C3	Chikuma-gawa r. Japan
128 B2	Chilanko r. Can.
74 B1	Chilas Pak.
143 C2	Childress U.S.A.
153 A3	Chile country S. America
158 C6	Chile Basin S. Pacific Ocean
152 B2	Chilecito Arg.
157 G8	Chile Rise S. Pacific Ocean
75 C3	Chilika Lake India
121 B2	Chililabombwe Zambia
128 B2	Chilko r. Can.
128 B2	Chilko Lake Can.
153 A3	Chillán Chile
138 B2	Chillicothe Il. U.S.A.
137 E3	Chillicothe MO U.S.A.
138 C3	Chillicothe OH U.S.A.
128 B3	Chilliwack Can.
153 A4	Chiloé, Isla de i. Chile
145 C3	Chilpancingo Mex.
53 C3	Chiltern Austr.
99 C4	Chiltern Hills U.K.
120 B1	Chiluage Angola
71 C3	Chilung Taiwan
119 D3	Chimala Tanz.
121 C2	Chimanimani Zimbabwe
152 B3	Chimbas Arg.
150 B3	Chimborazo mt. Ecuador
150 B3	Chimbote Peru
76 B2	Chimboy Uzbek.
	Chimishliya Moldova see Cimișlia
	Chimkent Kazakh. see Shymkent
121 C2	Chimoio Moz.
77 C3	Chimtargha, Qullai mt. Tajik.
68 C2	China country Asia
145 C2	China Mex.
150 B4	China Alta Peru
128 C2	Chinchaga r. Can.
145 D3	Chinchorro, Banco atoll Mex.
121 C2	Chinde Moz.
65 B3	Chindo S. Korea
65 B2	Chindu China
62 A1	Chindwin r. Myanmar

65 B2	Chinghwa N. Korea
121 B2	Chingola Zambia
120 A2	Chinguar Angola
121 C2	Chinhoyi Zimbabwe
	Chini India see Kalpa
	Chining China see Jining
74 B1	Chiniot Pak.
144 B2	Chinipas Mex.
65 B2	Chinju S. Korea
118 C2	Chinko r. C.A.R.
142 B1	Chinle U.S.A.
	Chinnamp'o N. Korea see Namp'o
67 C3	Chino Japan
135 C4	Chino U.S.A.
104 C2	Chinon France
134 E1	Chinook U.S.A.
142 A2	Chino Valley U.S.A.
77 C2	Chinoz Uzbek.
121 C2	Chinsali Zambia
108 B1	Chioggia Italy
111 C3	Chios Greece
111 C3	Chios i. Greece
121 C2	Chipata Zambia
120 A2	Chipindo Angola
	Chipinga Zimbabwe see Chipinge
121 C3	Chipinge Zimbabwe
73 B3	Chiplun India
99 B4	Chippenham U.K.
138 A2	Chippewa Falls U.S.A.
99 C4	Chipping Norton U.K.
	Chipuriro Zimbabwe see Guruve
145 D3	Chiquimula Guat.
77 C2	Chirchiq Uzbek.
121 C3	Chiredzi Zimbabwe
142 B2	Chiricahua Peak U.S.A.
146 B4	Chiriquí, Golfo de b. Panama
146 B4	Chirripó mt. Costa Rica
121 B2	Chirundu Zimbabwe
130 C2	Chisasibi Can.
137 E1	Chisholm U.S.A.
	Chisimaio Somalia see Kismaayo
90 B2	Chişinău Moldova
87 D3	Chistopol' Rus. Fed.
69 D1	Chita Rus. Fed.
120 A2	Chitado Angola
	Chitaldrug India see Chitradurga
121 C2	Chitambo Zambia
120 B1	Chitato Angola
121 C1	Chitipa Malawi
121 C3	Chitobe Moz.
	Chitor India see Chittaurgarh
66 D2	Chitose Japan
73 B3	Chitradurga India
74 B1	Chitral Pak.
146 B4	Chitré Panama
75 D2	Chittagong Bangl.
74 B2	Chittaurgarh India
73 B3	Chittoor India
	Chittorgarh India see Chittaurgarh
121 B2	Chitungwiza Zimbabwe
120 B2	Chiume Angola
121 C2	Chivhu Zimbabwe
70 B2	Chizhou China
	Chkalov Rus. Fed. see Orenburg
114 C1	Chlef Alg.
107 D2	Chlef, Oued r. Alg.
101 F2	Chodov Czech Rep.
153 B3	Choele Choel Arg.
	Chogori Feng mt. China/Pakistan see K2
48 G4	Choiseul i. Solomon Is.
144 B2	Choix Mex.
102 C1	Chojna Pol.
103 D1	Chojnice Pol.
117 B3	Ch'ok'ē Mountains Eth.
117 B3	Ch'ok'ē Terara mt. Eth.
	Chokue Moz. see Chókwè
83 L2	Chokurdakh Rus. Fed.
121 C3	Chókwè Moz.
104 B2	Cholet France
145 C3	Cholula Mex.
120 B2	Choma Zambia
	Chomo China see Yadong
102 C1	Chomutov Czech Rep.
83 I2	Chona r. Rus. Fed.
63 B2	Chon Buri Thai.
150 A3	Chone Ecuador
	Chong'an China see Wuyishan
65 B1	Ch'ŏngjin N. Korea
65 B2	Chŏngju N. Korea

Chŏngp'yŏng

65 B2 Chŏngp'yŏng N. Korea
70 A3 Chongqing China
70 A2 Chongqing mun. China
121 B2 Chongwe Zambia
71 A3 Chongzuo China
153 A4 Chonos, Archipiélago de los is Chile
154 B3 Chopimzinho Brazil
111 B3 Chora Sfakion Greece
98 B3 Chorley U.K.
91 C2 Chornobay Ukr.
90 C1 Chornobyl' Ukr.
91 C2 Chornomors'ke Ukr.
90 B2 Chortkiv Ukr.
65 B1 Ch'osan N. Korea
67 D3 Chōshi Japan
153 A3 Chos Malal Arg.
103 D1 Choszczno Pol.
134 D1 Choteau U.S.A.
114 A2 Choûm Maur.
69 D1 Choybalsan Mongolia
69 D1 Choyr Mongolia
54 B2 Christchurch N.Z.
99 C4 Christchurch U.K.
127 H2 Christian, Cape Can.
123 C2 Christiana S. Africa
Christianshåb Greenland see Qasigiannguit
54 A2 Christina, Mount N.Z.
58 B3 Christmas Island terr. Indian Ocean
111 B2 Chrysoupoli Greece
Chu Kazakh. see Shu
Chubarovka Ukr. see Polohy
153 B4 Chubut r. Arg.
89 F3 Chuchkovo Rus. Fed.
90 B1 Chudniv Ukr.
89 D2 Chudovo Rus. Fed.
126 C2 Chugach Mountains U.S.A.
67 B4 Chūgoku-sanchi mts Japan
Chuguchak China see Tacheng
66 B2 Chuguyevka Rus. Fed.
91 D2 Chuhuyiv Ukr.
Chukchi Peninsula pen. Rus. Fed. see Chukotskiy Poluostrov
160 J3 Chukchi Sea sea Rus. Fed./U.S.A.
89 F2 Chukhloma Rus. Fed.
83 N2 Chukotskiy Poluostrov pen. Rus. Fed.
Chulaktau Kazakh. see Karatau
135 C4 Chula Vista U.S.A.
82 G3 Chulym Rus. Fed.
152 B2 Chumbicha Arg.
83 K3 Chumikan Rus. Fed.
63 A2 Chumphon Thai.
65 B2 Chuncheon S. Korea
Chungking China see Chongqing
Ch'ungmu S. Korea see Tongyeong
83 H2 Chunya r. Rus. Fed.
119 D3 Chunya Tanz.
150 B4 Chuquibamba Peru
152 B2 Chuquicamata Chile
105 D2 Chur Switz.
62 A1 Churachandpur India
83 J2 Churapcha Rus. Fed.
129 E2 Churchill Can.
129 E2 Churchill r. Man. Can.
131 D2 Churchill r. Nfld. and Lab. Can.
129 E2 Churchill, Cape Can.
131 D2 Churchill Falls Can.
129 D2 Churchill Lake Can.
74 B2 Churu India
63 B2 Chư Sê Vietnam
142 B1 Chuska Mountains U.S.A.
86 E3 Chusovoy Rus. Fed.
131 C3 Chute-des-Passes Can.
48 G3 Chuuk is Micronesia
62 B1 Chuxiong China
91 C2 Chyhyryn Ukr.
Chymyshliya Moldova see Cimişlia
Ciadâr-Lunga Moldova see Ciadîr-Lunga
90 B2 Ciadîr-Lunga Moldova
60 B2 Ciamis Indon.
60 B2 Cianjur Indon.
154 B2 Cianorte Brazil
142 A2 Cibuta, Sierra mt. Mex.
80 B1 Cide Turkey
103 E1 Ciechanów Pol.
146 C2 Ciego de Ávila Cuba
147 C3 Ciénaga Col.
146 B2 Cienfuegos Cuba

107 C2 Cieza Spain
106 C2 Cigüela r. Spain
80 B2 Cihanbeyli Turkey
144 B3 Cihuatlán Mex.
106 C2 Cíjara, Embalse de resr Spain
109 C2 Çikës, Maja e mt. Albania
60 B2 Cilacap Indon.
Cill Airne Ireland see Killarney
Cill Chainnigh Ireland see Kilkenny
143 C1 Cimarron r. U.S.A.
90 B2 Cimişlia Moldova
108 B2 Cimone, Monte mt. Italy
Cîmpina Romania see Câmpina
Cîmpulung Romania see Câmpulung
60 B2 Cina, Tanjung c. Indon.
138 C3 Cincinnati U.S.A.
Cinco de Outubro Angola see Xá-Muteba
111 C3 Çine Turkey
100 B2 Ciney Belgium
134 C2 Cinnabar Mountain U.S.A.
145 C3 Cintalapa Mex.
71 B3 Ciping China
153 B3 Cipolletti Arg.
126 C2 Circle AK U.S.A.
136 B1 Circle MT U.S.A.
60 B2 Cirebon Indon.
99 C4 Cirencester U.K.
108 A1 Ciriè Italy
109 C3 Cirò Marina Italy
110 B1 Cisnădie Romania
109 C2 Čitluk Bos.-Herz.
122 A3 Citrusdal S. Africa
135 B3 Citrus Heights U.S.A.
110 C1 Ciucaş, Vârful mt. Romania
145 B2 Ciudad Acuña Mex.
145 B3 Ciudad Altamirano Mex.
150 C2 Ciudad Bolívar Venez.
147 C4 Ciudad Bolivia Venez.
144 B2 Ciudad Camargo Mex.
144 A2 Ciudad Constitución Mex.
145 C3 Ciudad Cuauhtémoc Mex.
145 C3 Ciudad del Carmen Mex.
144 B2 Ciudad del Este Para.
145 C2 Ciudad Delicias Mex.
144 B2 Ciudad de Valles Mex.
150 C2 Ciudad Guayana Venez.
142 B3 Ciudad Guerrero Mex.
144 B3 Ciudad Guzmán Mex.
145 C3 Ciudad Hidalgo Mex.
145 C3 Ciudad Ixtepec Mex.
145 B2 Ciudad Juárez Mex.
145 C2 Ciudad Madero Mex.
145 C2 Ciudad Mante Mex.
145 C2 Ciudad Mier Mex.
145 B2 Ciudad Obregón Mex.
106 C2 Ciudad Real Spain
145 C2 Ciudad Río Bravo Mex.
106 B1 Ciudad Rodrigo Spain
Ciudad Trujillo Dom. Rep. see Santo Domingo
145 C2 Ciudad Victoria Mex.
107 D1 Ciutadella Spain
108 B1 Cividale del Friuli Italy
108 B2 Civitanova Marche Italy
108 B2 Civitavecchia Italy
104 C2 Civray France
111 C3 Çivril Turkey
70 C2 Cixi China
99 D4 Clacton-on-Sea U.K.
130 C3 Claire, Lake Can.
105 C2 Clamecy France
140 C2 Clanton U.S.A.
122 A3 Clanwilliam S. Africa
97 C2 Clara Ireland
52 A2 Clare Austr.
138 C2 Clare r. U.S.A.
97 A2 Clare Island Ireland
139 E2 Claremont U.S.A.
97 B2 Claremorris Ireland
54 B2 Clarence N.Z.
55 B3 Clarence Island Antarctica
131 E3 Clarenville Can.
128 C2 Claresholm Can.
137 D2 Clarinda U.S.A.
144 A3 Clarión, Isla i. Mex.
123 C3 Clarkebury S. Africa
141 D2 Clark Hill Reservoir U.S.A.
128 C2 Clarksburg U.S.A.
140 B2 Clarksdale U.S.A.
134 C1 Clarks Fork r. U.S.A.
134 C1 Clarkston U.S.A.
140 B1 Clarksville AR U.S.A.
140 C1 Clarksville TN U.S.A.
154 B1 Claro r. Brazil

143 C1 Claude U.S.A.
143 C1 Clayton U.S.A.
97 B3 Clear, Cape Ireland
126 C3 Cleare, Cape U.S.A.
137 E2 Clear Lake U.S.A.
135 B3 Clear Lake l. U.S.A.
128 C2 Clearwater Can.
129 C2 Clearwater r. Can.
141 D3 Clearwater U.S.A.
134 C1 Clearwater r. U.S.A.
143 D2 Cleburne U.S.A.
101 E1 Clenze Ger.
51 D2 Clermont Austr.
105 C2 Clermont-Ferrand France
100 C2 Clervaux Lux.
52 A2 Cleve Austr.
140 B2 Cleveland MS U.S.A.
138 C2 Cleveland OH U.S.A.
141 D1 Cleveland TN U.S.A.
134 D1 Cleveland, Mount U.S.A.
154 B3 Clevelândia Brazil
97 B2 Clew Bay Ireland
141 D3 Clewiston U.S.A.
97 A2 Clifden Ireland
53 D1 Clifton Austr.
142 B2 Clifton U.S.A.
142 B1 Clines Corners U.S.A.
128 B2 Clinton Can.
137 E2 Clinton IA U.S.A.
137 E3 Clinton MO U.S.A.
143 D1 Clinton OK U.S.A.
125 H8 Clipperton, Île terr. N. Pacific Ocean
96 A2 Clisham h. U.K.
98 B3 Clitheroe U.K.
97 B3 Clonakilty Ireland
51 D2 Cloncurry Austr.
97 C1 Clones Ireland
97 C2 Clonmel Ireland
101 D1 Cloppenburg Ger.
137 E1 Cloquet U.S.A.
50 A2 Cloud Break Austr.
136 B2 Cloud Peak U.S.A.
135 C3 Clovis CA U.S.A.
143 C2 Clovis NM U.S.A.
Cluain Meala Ireland see Clonmel
129 D2 Cluff Lake Mine Can.
110 B1 Cluj-Napoca Romania
51 C2 Cluny Austr.
105 D2 Cluses France
54 A3 Clutha r. N.Z.
96 B3 Clyde r. U.K.
96 B3 Clyde, Firth of est. U.K.
127 H2 Clyde River Can.
144 B3 Coalcomán Mex.
128 C3 Coaldale Can.
135 C3 Coaldale U.S.A.
128 B2 Coal River Can.
150 C3 Coari Brazil
150 C3 Coari r. Brazil
141 C2 Coastal Plain U.S.A.
128 B2 Coast Mountains Can.
134 B2 Coast Ranges mts U.S.A.
96 B3 Coatbridge U.K.
139 F1 Coats Island Can.
55 C2 Coats Land reg. Antarctica
145 C3 Coatzacoalcos Mex.
146 A3 Cobán Guat.
53 C2 Cobar Austr.
97 B3 Cobh Ireland
152 B1 Cobija Bol.
Coblenz Ger. see Koblenz
130 C3 Cobourg Can.
50 C1 Cobourg Peninsula Austr.
53 C3 Cobram Austr.
101 E2 Coburg Ger.
100 C2 Cochem Ger.
Cochin India see Kochi
128 C2 Cochrane Alta Can.
130 B3 Cochrane Ont. Can.
153 A4 Cochrane Chile
52 A2 Cockaleechie Austr.
52 B2 Cockburn Austr.
Cockburn Town Turks and Caicos Is see Grand Turk
98 B2 Cockermouth U.K.
50 B3 Cocklebiddy Austr.
122 B3 Cockscomb mt. S. Africa
63 A2 Coco r. Hond./Nic.
125 J9 Coco, Isla de i. N. Pacific Ocean
159 F4 Cocos Basin Indian Ocean
58 A3 Cocos Islands terr. Indian Ocean
144 B2 Cocula Mex.

150 B2 Cocuy, Sierra Nevada del mt. Col.
139 E2 Cod, Cape U.S.A.
150 C3 Codajás Brazil
108 B2 Codigoro Italy
131 D2 Cod Island Can.
151 E3 Codó Brazil
136 B2 Cody U.S.A.
51 D1 Coen Austr.
100 C2 Coesfeld Ger.
113 I6 Coëtivy i. Seychelles
134 C1 Coeur d'Alene U.S.A.
100 C1 Coevorden Neth.
123 C3 Coffee Bay S. Africa
137 D3 Coffeyville U.S.A.
52 A2 Coffin Bay Austr.
53 D2 Coffs Harbour Austr.
123 C3 Cofimvaba S. Africa
104 B2 Cognac France
118 A2 Cogo Equat. Guinea
52 B3 Cohuna Austr.
146 B4 Coiba, Isla de i. Panama
153 A4 Coihaique Chile
73 B3 Coimbatore India
106 B1 Coimbra Port.
152 B1 Coipasa, Salar de salt flat Bol.
52 B3 Colac Austr.
155 D1 Colatina Brazil
136 C3 Colby U.S.A.
99 D4 Colchester U.K.
129 C2 Cold Lake Can.
96 C3 Coldstream U.K.
53 C2 Coleambally Austr.
143 D2 Coleman U.S.A.
52 B3 Coleraine Austr.
97 C1 Coleraine U.K.
123 C3 Colesberg S. Africa
153 A3 Colico Chile
144 B3 Colima Mex.
144 B3 Colima, Nevado de vol. Mex.
96 A2 Coll i. U.K.
53 C1 Collarenebri Austr.
143 D2 College Station U.S.A.
53 C1 Collerina Austr.
50 A3 Collie Austr.
50 B1 Collier Bay Austr.
138 C2 Collingwood Can.
54 B2 Collingwood N.Z.
126 F2 Collinson Peninsula Can.
101 F2 Collmberg h. Ger.
97 B1 Collooney Ireland
105 D2 Colmar France
98 B3 Colne U.K.
100 C2 Cologne Ger.
Colomb-Béchar Alg. see Béchar
154 C2 Colômbia Brazil
150 B2 Colombia country S. America
73 B4 Colombo Sri Lanka
104 C3 Colomiers France
152 C3 Colón Arg.
146 C4 Colón Panama
59 D2 Colonia Micronesia
153 B4 Colonia Las Heras Arg.
109 C3 Colonna, Capo c. Italy
96 A2 Colonsay i. U.K.
153 B3 Colorado r. Arg.
142 A2 Colorado r. Mex./U.S.A.
143 D3 Colorado r. U.S.A.
136 B3 Colorado state U.S.A.
135 C4 Colorado Desert U.S.A.
135 E3 Colorado Plateau U.S.A.
136 C3 Colorado Springs U.S.A.
144 B2 Colotlán Mex.
152 B1 Colquiri Bol.
136 B1 Colstrip U.S.A.
138 B3 Columbia KY U.S.A.
139 D3 Columbia MD U.S.A.
137 E3 Columbia MO U.S.A.
141 D2 Columbia SC U.S.A.
140 C1 Columbia TN U.S.A.
134 B1 Columbia r. U.S.A.
128 C2 Columbia, Mount Can.
134 D1 Columbia Falls U.S.A.
134 C1 Columbia Mountains Can.
134 C1 Columbia Plateau U.S.A.
141 D2 Columbus GA U.S.A.
138 B3 Columbus IN U.S.A.
140 C2 Columbus MS U.S.A.
136 C2 Columbus NE U.S.A.
142 B2 Columbus NM U.S.A.
138 C3 Columbus OH U.S.A.
143 D3 Columbus TX U.S.A.
134 C1 Colville r. U.S.A.
126 B2 Colville r. U.S.A.

26	D2	Colville Lake Can.
98	B3	Colwyn Bay U.K.
08	B2	Comacchio Italy
45	C3	Comalcalco Mex.
10	C1	Comănești Romania
30	C2	Comencho, Lac l. Can.
97	C2	Comeragh Mountains hills Ireland
43	D2	Comfort U.S.A.
75	D2	Comilla Bangl.
08	A2	Comino, Capo c. Italy
45	C3	Comitán de Domínguez Mex.
05	C2	Commentry France
43	B2	Commerce U.S.A.
27	G2	Committee Bay Can.
08	A1	Como Italy
		Como, Lago di l. Italy see Como, Lake
08	A1	Como, Lake l. Italy
53	B4	Comodoro Rivadavia Arg.
45	B2	Comonfort Mex.
21	D2	Comoros country Africa
28	B3	Comox Can.
05	C2	Compiègne France
44	B2	Compostela Mex.
90	B2	Comrat Moldova
14	A4	Conakry Guinea
04	B2	Concarneau France
55	E1	Conceição da Barra Brazil
51	E3	Conceição do Araguaia Brazil
55	D1	Conceição do Mato Dentro Brazil
52	B2	Concepción Arg.
53	A3	Concepción Chile
44	B2	Concepción Mex.
35	B4	Conception, Point U.S.A.
52	C2	Conchas Brazil
42	C1	Conchas Lake U.S.A.
44	B2	Conchos r. Mex.
45	C2	Conchos r. Mex.
35	B3	Concord CA U.S.A.
39	E2	Concord NH U.S.A.
52	C3	Concordia Arg.
22	A2	Concordia S. Africa
37	D3	Concordia U.S.A.
53	C2	Condobolin Austr.
04	C3	Condom France
34	B1	Condon U.S.A.
08	B1	Conegliano Italy
51	E1	Conflict Group is P.N.G.
04	C2	Confolens France
35	D3	Confusion Range mts U.S.A.
75	C2	Congdü China
98	B3	Congleton U.K.
18	B3	Congo country Africa
18	B3	Congo r. Congo/ Dem. Rep. Congo
		Congo (Brazzaville) country Africa see Congo
		Congo (Kinshasa) country Africa see Congo, Democratic Republic of the
18	C3	Congo, Democratic Republic of the country Africa
18	C3	Congo Basin Dem. Rep. Congo
		Congo Free State country Africa see Congo, Democratic Republic of the
29	C2	Conklin Can.
97	B1	Conn, Lough l. Ireland
97	B2	Connacht reg. Ireland
39	E2	Connecticut r. U.S.A.
39	E2	Connecticut state U.S.A.
96	B2	Connel U.K.
97	B2	Connemara reg. Ireland
34	D1	Conrad U.S.A.
59	D7	Conrad Rise Southern Ocean
43	D2	Conroe U.S.A.
55	D2	Conselheiro Lafaiete Brazil
55	D1	Conselheiro Pena Brazil
00	B3	Consenvoye France
98	C2	Consett U.K.
63	B3	Côn Sơn, Đao i. Vietnam
		Constance Ger. see Konstanz
05	D2	Constance, Lake Ger./Switz.
10	C2	Constanța Romania
06	C1	Constantina Spain
15	C1	Constantine Alg.
34	D2	Contact U.S.A.
55	D2	Contagalo Brazil
50	B3	Contamana Peru

153	A5	Contreras, Isla i. Chile
126	E2	Contwoyto Lake Can.
140	B1	Conway AR U.S.A.
139	E2	Conway NH U.S.A.
51	C2	Coober Pedy Austr.
		Cooch Behar India see Koch Bihar
		Cook, Mount mt. N.Z. see Aoraki
141	D2	Cookeville U.S.A.
126	B2	Cook Inlet sea chan. U.S.A.
49	K5	Cook Islands terr. S. Pacific Ocean
131	E2	Cook's Harbour Can.
97	C1	Cookstown U.K.
54	B2	Cook Strait N.Z.
51	D1	Cooktown Austr.
53	C2	Coolabah Austr.
53	C2	Coolamon Austr.
53	D1	Coolangatta Austr.
50	B3	Coolgardie Austr.
53	C3	Cooma Austr.
52	B2	Coombah Austr.
53	C2	Coonabarabran Austr.
52	A3	Coonalpyn Austr.
53	C2	Coonamble Austr.
53	C1	Coongoola Austr.
137	E1	Coon Rapids U.S.A.
52	A1	Cooper Creek watercourse Austr.
141	E3	Cooper's Town Bahamas
134	B2	Coos Bay U.S.A.
53	C2	Cootamundra Austr.
97	C1	Cootehill Ireland
145	C3	Copainalá Mex.
145	C3	Copala Mex.
93	F4	Copenhagen Denmark
109	C2	Copertino Italy
53	D1	Copeton Reservoir Austr.
152	A2	Copiapó Chile
143	D2	Copperas Cove U.S.A.
138	B1	Copper Harbor U.S.A.
		Coppermine Can. see Kugluktuk
126	E2	Coppermine r. Can.
122	B2	Copperton S. Africa
		Coquilhatville Dem. Rep. Congo see Mbandaka
152	A2	Coquimbo Chile
152	A2	Coquimbo, Bahía de b. Chile
110	B2	Corabia Romania
155	D1	Coração de Jesus Brazil
150	B4	Coracora Peru
53	D1	Coraki Austr.
50	A2	Coral Bay Austr.
127	G2	Coral Harbour Can.
156	D7	Coral Sea S. Pacific Ocean
48	G5	Coral Sea Islands Territory terr. Austr.
137	E2	Coralville U.S.A.
52	B3	Corangamite, Lake Austr.
99	C3	Corby U.K.
		Corcaigh Ireland see Cork
135	C3	Corcoran U.S.A.
153	A4	Corcovado, Golfo de sea chan. Chile
141	D2	Cordele U.S.A.
64	B2	Cordilleras Range mts Phil.
155	D1	Cordisburgo Brazil
152	B3	Córdoba Arg.
145	C3	Córdoba Mex.
106	C2	Córdoba Spain
153	B3	Córdoba, Sierras de mts Arg.
126	C2	Cordova U.S.A.
51	D2	Corfield Austr.
111	A3	Corfu i. Greece
154	B1	Corguinho Brazil
106	B1	Coria Spain
106	B2	Coria del Río Spain
53	D2	Coricudgy mt. Austr.
111	B3	Corinth Greece
140	C2	Corinth U.S.A.
		Corinth, Gulf of sea chan. Greece see Gulf of Corinth
155	D1	Corinto Brazil
97	B3	Cork Ireland
111	C2	Çorlu Turkey
154	B2	Cornélio Procópio Brazil
131	E3	Corner Brook Can.
53	C3	Corner Inlet b. Austr.
139	D2	Corning NY U.S.A.
		Corn Islands is Nic. see Maíz, Islas del
108	B2	Corno Grande mt. Italy
130	C3	Cornwall Can.
126	F1	Cornwallis Island Can.

150	C1	Coro Venez.
155	D1	Coroaci Brazil
154	C1	Coromandel Brazil
73	C3	Coromandel Coast India
54	C1	Coromandel Peninsula N.Z.
64	B2	Coron Phil.
129	C2	Coronation Can.
126	E2	Coronation Gulf Can.
55	B3	Coronation Island S. Atlantic Ocean
155	D1	Coronel Fabriciano Brazil
152	C2	Coronel Oviedo Para.
154	B1	Coronel Ponce Brazil
153	B3	Coronel Pringles Arg.
154	A2	Coronel Sapucaia Brazil
153	B3	Coronel Suárez Arg.
150	B4	Coropuna, Nudo mt. Peru
109	D2	Çorovodë Albania
145	D3	Corozal Belize
143	D3	Corpus Christi U.S.A.
152	B1	Corque Bol.
151	E4	Corrente Brazil
154	B1	Correntes Brazil
151	E4	Correntina Brazil
97	B2	Corrib, Lough l. Ireland
152	C2	Corrientes Arg.
153	C3	Corrientes, Cabo c. Arg.
144	B2	Corrientes, Cabo c. Mex.
143	E2	Corrigan U.S.A.
53	C3	Corryong Austr.
		Corse i. France see Corsica
105	D3	Corse, Cap c. France
105	D3	Corsica i. France
143	D2	Corsicana U.S.A.
105	D3	Corte France
106	B2	Cortegana Spain
136	B3	Cortez U.S.A.
108	B1	Cortina d'Ampezzo Italy
139	D2	Cortland U.S.A.
108	B2	Cortona Italy
106	B2	Coruche Port.
		Çoruh r. Turkey see Artvin
80	B1	Çorum Turkey
151	D4	Corumbá Brazil
154	C1	Corumbá r. Brazil
154	C1	Corumbá de Goiás Brazil
		Corunna Spain see A Coruña
134	B2	Corvallis U.S.A.
99	B3	Corwen U.K.
		Cos i. Greece see Kos
144	B2	Cosalá Mex.
145	C3	Cosamaloapan Mex.
109	C3	Cosenza Italy
105	C2	Cosne-Cours-sur-Loire France
152	B3	Cosquín Arg.
107	C2	Costa Blanca coastal area Spain
107	D1	Costa Brava coastal area Spain
106	B2	Costa de la Luz coastal area Spain
106	B2	Costa del Sol coastal area Spain
107	D1	Costa Dorada coastal area Spain
150	C4	Costa Marques Brazil
154	B3	Costa Rica Brazil
146	B3	Costa Rica country Central America
144	B2	Costa Rica Mex.
		Costermansville Dem. Rep. Congo see Bukavu
110	B2	Costeşti Romania
64	B3	Cotabato Phil.
137	D1	Coteau des Prairies reg. U.S.A.
136	C1	Coteau du Missouri reg. U.S.A.
147	C3	Coteaux Haiti
114	B4	Côte d'Ivoire country Africa
		Côte Française de Somalis country Africa see Djibouti
150	B3	Cotopaxi, Volcán vol. Ecuador
99	B4	Cotswold Hills U.K.
134	B2	Cottage Grove U.S.A.
102	C1	Cottbus Ger.
105	D3	Cottian Alps mts France/ Italy
104	B2	Coubre, Pointe de la pt France
52	A3	Couedic, Cape du Austr.
105	C2	Coulommiers France
137	D2	Council Bluffs U.S.A.
88	B2	Courland Lagoon b. Lith./ Rus. Fed.
100	A3	Courmelles France

128	B3	Courtenay Can.
104	B2	Coutances France
104	B2	Coutras France
100	B2	Couvin Belgium
99	C3	Coventry U.K.
106	B1	Covilhã Port.
141	D2	Covington GA U.S.A.
138	C3	Covington KY U.S.A.
138	C3	Covington VA U.S.A.
50	B3	Cowan, Lake imp. l. Austr.
96	C2	Cowdenbeath U.K.
52	A2	Cowell Austr.
53	C3	Cowes Austr.
134	B1	Cowlitz r. U.S.A.
53	C2	Cowra Austr.
154	B1	Coxim Brazil
154	B1	Coxim r. Brazil
75	D2	Cox's Bazar Bangl.
145	B3	Coyuca de Benítez Mex.
75	C1	Cozhê China
145	D2	Cozumel Mex.
145	D2	Cozumel, Isla de i. Mex.
52	A2	Cradock Austr.
123	C3	Cradock S. Africa
136	B2	Craig U.S.A.
102	C2	Crailsheim Ger.
110	B2	Craiova Romania
129	D2	Cranberry Portage Can.
53	C3	Cranbourne Austr.
128	C3	Cranbrook Can.
151	E3	Crateús Brazil
151	F3	Crato Brazil
154	C2	Cravinhos Brazil
136	C2	Crawford U.S.A.
138	B2	Crawfordsville U.S.A.
99	C4	Crawley U.K.
134	D1	Crazy Mountains U.S.A.
129	D2	Cree r. Can.
144	B2	Creel Mex.
129	D2	Cree Lake Can.
129	D2	Creighton Can.
104	C2	Creil France
100	B1	Creil Neth.
108	A1	Crema Italy
108	B1	Cremona Italy
108	B2	Cres i. Croatia
134	B2	Crescent City U.S.A.
53	D2	Crescent Head Austr.
135	E3	Crescent Junction U.S.A.
128	C3	Creston Can.
137	E2	Creston U.S.A.
140	C2	Crestview U.S.A.
111	B3	Crete i. Greece
107	D1	Creus, Cabo de c. Spain
107	C2	Crevillent Spain
98	B3	Crewe U.K.
96	B2	Crianlarich U.K.
152	D2	Criciúma Brazil
96	C2	Crieff U.K.
108	B1	Crikvenica Croatia
91	C2	Crimea pen. Ukr.
101	F2	Crimmitschau Ger.
96	B2	Crinan U.K.
118	B2	Cristal, Monts de mts Equat. Guinea/Gabon
154	C1	Cristalina Brazil
110	B1	Crişul Alb r. Romania
101	E1	Crivitz Ger.
		Crna Gora country Europe see Montenegro
109	C1	Črnomelj Slovenia
97	B2	Croagh Patrick h. Ireland
109	C1	Croatia country Europe
61	C1	Crocker, Banjaran mts Malaysia
143	D2	Crockett U.S.A.
59	C3	Croker Island Austr.
96	B2	Cromarty U.K.
99	D3	Cromer U.K.
54	A3	Cromwell N.Z.
147	C2	Crooked Island Bahamas
137	D1	Crookston U.S.A.
53	C1	Crookwell Austr.
136	C1	Crosby U.S.A.
140	B2	Cross City U.S.A.
140	B2	Crossett U.S.A.
98	B2	Cross Fell h. U.K.
129	E2	Cross Lake Can.
140	C1	Crossville U.S.A.
109	C3	Crotone Italy
139	E1	Crow Agency U.S.A.
99	C4	Crowborough U.K.
140	B2	Crowley U.S.A.
53	C2	Crows Nest Austr.
53	D1	Crowsnest Pass Can.
51	D1	Croydon Austr.
159	D7	Crozet, Îles is Indian Ocean
146	B3	Cruz, Cabo c. Cuba
152	C2	Cruz Alta Brazil

Cruz del Eje

152 B3 Cruz del Eje Arg.
155 D2 Cruzeiro Brazil
150 B3 Cruzeiro do Sul Brazil
52 A2 Crystal Brook Austr.
143 D3 Crystal City U.S.A.
138 B1 Crystal Falls U.S.A.
140 B2 Crystal Springs U.S.A.
103 E2 Csongrád Hungary
103 D2 Csorna Hungary
121 C2 Cuamba Moz.
120 B2 Cuando r. Angola/Zambia
120 A2 Cuangar Angola
118 B3 Cuango r. Angola/
    Dem. Rep. Congo
120 A1 Cuanza r. Angola
144 B2 Cuatro Ciénegas Mex.
144 B2 Cuauhtémoc Mex.
145 C3 Cuautla Mex.
146 B2 Cuba country West Indies
120 A2 Cubal Angola
120 B2 Cubango r. Angola/Namibia
150 B2 Cúcuta Col.
73 B3 Cuddalore India
    Cuddapah India see Kadapa
50 A2 Cue Austr.
106 C1 Cuéllar Spain
120 A2 Cuemba Angola
150 B3 Cuenca Ecuador
107 C1 Cuenca Spain
107 C1 Cuenca, Serranía de mts
    Spain
145 C3 Cuernavaca Mex.
143 D3 Cuero U.S.A.
104 C3 Cugnaux France
151 D4 Cuiabá Brazil
151 D4 Cuiabá r. Brazil
96 A2 Cuillin Sound sea chan. U.K.
120 A1 Cuilo Angola
120 B2 Cuito r. Angola
120 A2 Cuito Cuanavale Angola
60 B1 Cukai Malaysia
64 B2 Culasi Phil.
53 C3 Culcairn Austr.
100 B2 Culemborg Neth.
53 C1 Culgoa r. Austr.
144 B2 Culiacán Mex.
64 A2 Culion i. Phil.
107 C2 Cullera Spain
140 C2 Cullman U.S.A.
97 C1 Cullybackey U.K.
139 D3 Culpeper U.S.A.
151 D4 Culuene r. Brazil
54 B2 Culverden N.Z.
150 C1 Cumaná Venez.
139 D3 Cumberland U.S.A.
138 B3 Cumberland r. U.S.A.
141 D2 Cumberland Island U.S.A.
129 C2 Cumberland Lake Can.
127 H2 Cumberland Peninsula
    Can.
140 C1 Cumberland Plateau U.S.A.
127 H2 Cumberland Sound
    sea chan. Can.
96 C3 Cumbernauld U.K.
135 B3 Cummings U.S.A.
96 B3 Cumnock U.K.
144 B1 Cumpas Mex.
145 C3 Cunduacán Mex.
108 A2 Cuneo Italy
53 C1 Cunnamulla Austr.
108 A1 Cuorgnè Italy
96 C2 Cupar U.K.
110 B2 Ćuprija Serbia
147 D3 Curaçao terr. West Indies
150 B3 Curaray r. Ecuador
153 A3 Curicó Chile
154 C2 Curitiba Brazil
52 A2 Curnamona Austr.
135 C3 Currant U.S.A.
51 D3 Currie Austr.
135 D2 Currie U.S.A.
51 E2 Curtis Island Austr.
151 D3 Curuá r. Brazil
60 B2 Curup Indon.
151 E3 Cururupu Brazil
155 D1 Curvelo Brazil
150 B4 Cusco Peru
97 C1 Cushendun U.K.
143 D1 Cushing U.S.A.
136 C2 Custer U.S.A.
134 D1 Cut Bank U.S.A.
140 B3 Cut Off U.S.A.
75 C2 Cuttack India
120 A2 Cuvelai Angola
101 D1 Cuxhaven Ger.
64 B2 Cuyo Islands Phil.
    Cuzco Peru see Cusco
99 B4 Cwmbrân U.K.
119 C3 Cyangugu Rwanda
111 B3 Cyclades is Greece

129 C3 Cypress Hills Can.
80 B2 Cyprus country Asia
80 B2 Cyprus i. Asia
102 C2 Czech Republic country
    Europe
103 D1 Czersk Pol.
103 D1 Częstochowa Pol.

D

    Đa, Sông r. Vietnam see
    Black River
69 D2 Daban China
103 D2 Dabas Hungary
114 A3 Dabola Guinea
103 D1 Dąbrowa Górnicza Pol.
110 B2 Dăbuleni Romania
    Dacca Bangl. see Dhaka
102 C2 Dachau Ger.
    Dachuan China see Dazhou
141 D3 Dade City U.S.A.
    Dadong China see
    Donggang
    Dadra India see Achalpur
74 B2 Dadra and Nagar Haveli
    union terr. India
74 A2 Dadu Pak.
65 B2 Daegu S. Korea
65 B2 Daejeon S. Korea
65 B3 Daejeong S. Korea
64 B2 Daet Phil.
114 A3 Dagana Senegal
119 D2 Daga Post S. Sudan
88 C2 Dagda Latvia
64 B2 Dagupan Phil.
    Dahalach, Isole is Eritrea
    see Dahlak Archipelago
74 B3 Dahanu India
69 D2 Da Hinggan Ling mts
    China
116 C3 Dahlak Archipelago is
    Eritrea
100 C2 Dahlem Ger.
78 B3 Dahm, Ramlat des.
    Saudi Arabia/Yemen
74 B2 Dahod India
    Dahomey country Africa see
    Benin
    Dahra Senegal see Dara
81 C2 Dahūk Iraq
60 B2 Daik Indon.
106 C2 Daimiel Spain
    Dairen China see Dalian
51 C2 Dajarra Austr.
70 A2 Dajing China
114 A3 Dakar Senegal
117 C4 Daketa Shet' watercourse
    Eth.
114 A2 Dakhla Western Sahara
    Dakhla Oasis oasis Egypt
    see Wāḥāt ad Dākhilah
63 A3 Dakoank India
88 C3 Dakol'ka r. Belarus
    Đakovica Kosovo see
    Gjakovë
109 C1 Đakovo Croatia
120 B2 Dala Angola
70 A1 Dalain Hob China
93 G3 Dalälven r. Sweden
111 C3 Dalaman Turkey
111 C3 Dalaman r. Turkey
68 C2 Dalandzadgad Mongolia
64 B2 Dalanganem Islands Phil.
63 B2 Đa Lat Vietnam
    Dalatando Angola see
    N'dalatando
74 A2 Dalbandin Pak.
96 C3 Dalbeattie U.K.
53 D1 Dalby Austr.
93 E3 Dale Norway
141 C1 Dale Hollow Lake U.S.A.
53 C3 Dalgety Austr.
143 C1 Dalhart U.S.A.
131 D3 Dalhousie Can.
62 B1 Dali China
70 C2 Dalian China
96 C3 Dalkeith U.K.
143 D2 Dallas U.S.A.
128 A2 Dall Island U.S.A.
    Dalmacija reg. Bos.-Herz./
    Croatia see Dalmatia
96 B2 Dalmally U.K.
109 C2 Dalmatia reg. Bos.-Herz./
    Croatia
96 B3 Dalmellington U.K.
66 C2 Dal'negorsk Rus. Fed.
66 B1 Dal'nerechensk Rus. Fed.
    Dalny China see Dalian

114 B4 Daloa Côte d'Ivoire
71 A3 Dalou Shan mts China
51 D2 Dalrymple, Mount Austr.
92 □A3 Dalsmynni Iceland
75 C2 Daltenganj India
141 D2 Dalton U.S.A.
    Daltonganj India see
    Daltenganj
60 B1 Daludalu Indon.
71 B3 Daluo Shan mt. China
92 □B2 Dalvík Iceland
96 B2 Dalwhinnie U.K.
50 C1 Daly r. Austr.
51 C1 Daly Waters Austr.
74 B2 Daman India
74 B2 Daman and Diu union terr.
    India
116 B1 Damanhūr Egypt
59 C3 Damar i. Indon.
118 B2 Damara C.A.R.
80 B2 Damascus Syria
115 D3 Damaturu Nigeria
76 B3 Damāvand, Qolleh-ye mt.
    Iran
120 A1 Damba Angola
118 B1 Damboa Nigeria
81 D2 Dāmghān Iran
    Damietta Egypt see Dumyāṭ
79 C2 Dammam Saudi Arabia
101 D1 Damme Ger.
75 B2 Damoh India
114 B4 Damongo Ghana
50 A2 Dampier Austr.
59 C3 Dampir, Selat sea chan.
    Indon.
75 C2 Damqoq Zangbo r. China
    Damxung China see
    Gongtang
117 C3 Danakil reg. Africa
114 B4 Danané Côte d'Ivoire
63 B2 Đa Năng Vietnam
139 E2 Danbury U.S.A.
70 C1 Dandong China
117 B3 Dangila Eth.
146 B3 Dangriga Belize
70 B2 Dangshan China
89 F2 Danilov Rus. Fed.
89 E2 Danilovskaya
    Vozvyshennost' hills
    Rus. Fed.
70 B2 Danjiangkou China
79 C2 Dank Oman
89 E3 Dankov Rus. Fed.
146 B3 Danlí Hond.
101 E1 Dannenberg (Elbe) Ger.
54 C2 Dannevirke N.Z.
62 B2 Dan Sai Thai.
    Dantu China see Zhenjiang
110 A1 Danube r. Europe
110 C1 Danube Delta Romania/
    Ukr.
138 B2 Danville IL U.S.A.
138 C3 Danville KY U.S.A.
139 D3 Danville VA U.S.A.
    Danxian China see Danzhou
71 A4 Danzhou China
    Danzig, Gulf of g.
    Pol./Rus. Fed. see
    Gdańsk, Gulf of
    Daojiang China see Daoxian
115 D2 Dao Timmi Niger
    Daoud Alg. see Aïn Beïda
114 B4 Daoukro Côte d'Ivoire
71 B3 Daoxian China
64 B3 Dapa Phil.
114 C3 Dapaong Togo
68 C2 Da Qaidam China
69 E1 Daqing China
114 A3 Dara Senegal
80 B2 Dar'ā Syria
81 D3 Dārāb Iran
81 D2 Dārān Iran
    Daravica mt. Kosovo see
    Gjeravicë
75 C2 Darbhanga India
    Dardo China see Kangding
119 D3 Dar es Salaam Tanz.
117 A3 Darfur reg. Sudan
74 B1 Dargai Pak.
54 B1 Dargaville N.Z.
53 C3 Dargo Austr.
68 D1 Darhan Mongolia
150 B1 Darién, Golfo del g. Col.
    Darjeeling India see
    Darjiling
75 C2 Darjiling India
52 B2 Darling r. Austr.
53 C1 Darling Downs hills Austr.
50 A3 Darling Range hills Austr.
98 C2 Darlington U.K.

53 C2 Darlington Point Austr.
103 D1 Darłowo Pol.
101 D3 Darmstadt Ger.
115 E1 Darnah Libya
52 B2 Darnick Austr.
55 H3 Darnley, Cape Antarctica
107 C1 Daroca Spain
99 D4 Dartford U.K.
99 A4 Dartmoor hills U.K.
131 D3 Dartmouth Can.
99 B4 Dartmouth U.K.
59 D3 Daru P.N.G.
59 C2 Daruba Indon.
50 C1 Darwin Austr.
153 C5 Darwin Falkland Is
79 C2 Dārzīn Iran
65 A1 Dashiqiao China
    Dashkhovuz Turkm. see
    Daşoguz
74 A2 Dasht r. Pak.
76 B2 Daşoguz Turkm.
61 C1 Datadian Indon.
111 C3 Datça Turkey
66 D2 Date Japan
71 B3 Datian China
70 B1 Datong China
64 B3 Datu Piang Phil.
74 B1 Daud Khel Pak.
88 B2 Daugava r. Latvia
88 C2 Daugavpils Latvia
100 C2 Daun Ger.
129 D2 Dauphin Can.
129 E2 Dauphin Lake Can.
73 B3 Davangere India
64 B3 Davao Phil.
64 B3 Davao Gulf Phil.
137 E2 Davenport U.S.A.
99 C3 Daventry U.K.
123 C2 Daveyton S. Africa
146 B4 David Panama
129 D2 Davidson Can.
126 F3 Davidson Lake Can.
135 B3 Davis U.S.A.
131 D2 Davis Inlet (abandoned)
    Can.
55 I3 Davis Sea sea Antarctica
160 P3 Davis Strait str. Can./
    Greenland
105 D2 Davos Switz.
88 C3 Davyd-Haradok Belarus
78 A2 Dawmat al Jandal
    Saudi Arabia
79 C3 Dawqah Oman
78 B3 Dawqah Saudi Arabia
126 C2 Dawson Can.
141 D2 Dawson U.S.A.
128 B2 Dawson Creek Can.
128 B2 Dawsons Landing Can.
68 C2 Dawu China
71 C3 Dawu Taiwan
    Dawukou China see
    Shizuishan
79 C2 Dawwah Oman
104 B3 Dax France
    Daxian China see Dazhou
68 C2 Da Xueshan mts China
52 B3 Daylesford Austr.
70 A2 Dayong China
    Zhangjiajie
81 C2 Dayr az Zawr Syria
138 C3 Dayton U.S.A.
141 D3 Daytona Beach U.S.A.
71 B3 Dayu China
    Da Yunhe canal China se
    Jinghang Yunhe
79 C2 Dayyer Iran
70 A2 Dazhou China
122 B3 De Aar S. Africa
141 D3 Deadman Bay U.S.A.
80 B2 Dead Sea salt l. Asia
99 D4 Deal U.K.
71 B3 De'an China
152 B3 Deán Funes Arg.
128 B2 Dease Lake Can.
126 E2 Dease Strait Can.
135 C3 Death Valley depr. U.S.A.
104 C2 Deauville France
61 C1 Debak Malaysia
111 B2 Debar Macedonia
103 E1 Dębica Pol.
103 E1 Dęblin Pol.
114 B3 Débo, Lac l. Mali
103 E2 Debrecen Hungary
117 B3 Debre Markos Eth.
119 D2 Debre Sīna Eth.
117 B3 Debre Tabor Eth.
117 B4 Debre Zeyit Eth.
140 C2 Decatur AL U.S.A.
138 B3 Decatur IL U.S.A.
73 B3 Deccan plat. India

53 D1 Deception Bay Austr.
71 A3 Dechang China
02 C1 Děčín Czech Rep.
37 E2 Decorah U.S.A.
54 C2 Dedo de Deus mt. Brazil
88 C2 Dedovichi Rus. Fed.
21 C2 Dedza Malawi
99 B3 Dee r. England/Wales U.K.
96 C2 Dee r. Scotland U.K.
30 C3 Deep River Can.
53 D1 Deepwater Austr.
31 E3 Deer Lake Can.
34 D1 Deer Lodge U.S.A.
38 C2 Defiance U.S.A.
40 C2 De Funiak Springs U.S.A.
68 C2 Dêgê China
47 C4 Degeh Bur Eth.
39 F1 Dégelis Can.
02 C2 Deggendorf Ger.
91 E2 Degtevo Rus. Fed.
74 B1 Dehra Dun India
75 C2 Dehri India
69 E2 Dehui China
00 A2 Deinze Belgium
40 B1 Dej Romania
38 B2 De Kalb U.S.A.
46 B3 Dekemhare Eritrea
48 C3 Dekese Dem. Rep. Congo
48 B2 Dékoa C.A.R.
41 D3 De Land U.S.A.
35 C3 Delano U.S.A.
35 D3 Delano Peak U.S.A.
48 I3 Delap-Uliga-Djarrit Marshall Is
23 C2 Delareyville S. Africa
29 D2 Delaronde Lake Can.
38 C2 Delaware U.S.A.
39 D3 Delaware r. U.S.A.
39 D3 Delaware state U.S.A.
39 D3 Delaware Bay U.S.A.
53 C3 Delegate Austr.
48 C2 Délembé C.A.R.
05 D2 Delémont Switz.
00 B1 Delft Neth.
00 C1 Delfzijl Neth.
21 D2 Delgado, Cabo c. Moz.
68 C1 Delgerhaan Mongolia
68 C2 Delhi China
74 B2 Delhi India
60 B2 Deli i. Indon.
28 B1 Déljne Can.
Delingha China see Delhi
01 F2 Delitzsch Ger.
07 D2 Dellys Alg.
35 C4 Del Mar U.S.A.
01 D1 Delmenhorst Ger.
09 B1 Delnice Croatia
36 B3 Del Norte U.S.A.
83 L1 De-Longa, Ostrova is Rus. Fed.
De Long Islands is Rus. Fed. see De-Longa, Ostrova
De Long Strait sea chan. Rus. Fed. see Longa, Proliv
29 D3 Deloraine Can.
11 B3 Delphi tourist site Greece
41 D3 Delray Beach U.S.A.
43 C3 Del Rio U.S.A.
36 B3 Delta CO U.S.A.
35 D3 Delta UT U.S.A.
26 C2 Delta Junction U.S.A.
09 D3 Delvinë Albania
06 C1 Demanda, Sierra de la mts Spain
Demavend mt. Iran see Damāvand, Qolleh-ye
48 C2 Demba Dem. Rep. Congo
19 D1 Dembech'a Eth.
17 B4 Dembī Dolo Eth.
Demerara Guyana see Georgetown
91 C3 Demerdzhi mt. Ukr.
89 D2 Demidov Rus. Fed.
42 B2 Deming U.S.A.
11 C3 Demirci Turkey
11 C2 Demirköy Turkey
02 C1 Demmin Ger.
09 D3 Demopolis U.S.A.
60 B2 Dempo, Gunung vol. Indon.
89 D2 Demyansk Rus. Fed.
22 B3 De Naawte S. Africa
98 B3 Denbigh U.K.
00 B1 Den Burg Neth.
62 B2 Den Chai Thai.
60 B2 Dendang Indon.
00 B2 Dendermonde Belgium
00 B2 Dendre r. Belgium

Dengjiabu China see Yujiang
70 A1 Dengkou China
Dengxian China see Dengzhou
70 B2 Dengzhou China
Dengzhou China see Penglai
Den Haag Neth. see The Hague
50 A2 Denham Austr.
100 B1 Den Helder Neth.
107 D2 Dénia Spain
52 B3 Deniliquin Austr.
134 C2 Denio U.S.A.
137 D2 Denison IA U.S.A.
143 D2 Denison TX U.S.A.
111 C3 Denizli Turkey
53 D2 Denman Austr.
50 A3 Denmark Austr.
93 E4 Denmark country Europe
84 B2 Denmark Strait Greenland/Iceland
77 C3 Denov Uzbek.
61 C2 Denpasar Indon.
143 D2 Denton U.S.A.
50 A3 D'Entrecasteaux, Point Austr.
59 E3 D'Entrecasteaux Islands P.N.G.
141 D2 Dentsville U.S.A.
136 B3 Denver U.S.A.
75 C2 Deogarh Odisha India
74 B2 Deogarh Rajasthan India
75 C2 Deoghar India
138 B2 De Pere U.S.A.
83 K2 Deputatskiy Rus. Fed.
62 A1 Dêqên China
140 B2 De Queen U.S.A.
74 A2 Dera Bugti Pak.
74 B1 Dera Ghazi Khan Pak.
74 B1 Dera Ismail Khan Pak.
87 D4 Derbent Rus. Fed.
50 B1 Derby Austr.
99 C3 Derby U.K.
137 D3 Derby U.S.A.
99 D3 Dereham U.K.
97 B2 Derg, Lough l. Ireland
91 D1 Derhachi Ukr.
140 B2 De Ridder U.S.A.
91 D2 Derkul r. Rus. Fed./Ukr.
Derry U.K. see Londonderry
75 B1 Dêrub China
116 B3 Derudeb Sudan
122 B3 De Rust S. Africa
109 C2 Derventa Bos.-Herz.
98 C3 Derwent r. England U.K.
98 C3 Derwent r. England U.K.
98 B2 Derwent Water l. U.K.
77 C1 Derzhavinsk Kazakh.
Derzhavinskiy Kazakh. see Derzhavinsk
152 B1 Desaguadero r. Bol.
49 M5 Désappointement, Îles du is Fr. Polynesia
129 D2 Deschambault Lake Can.
134 B1 Deschutes r. U.S.A.
117 B3 Desē Eth.
153 B4 Deseado Arg.
153 B4 Deseado r. Arg.
142 A2 Desemboque Mex.
137 E2 Des Moines U.S.A.
137 E2 Des Moines r. U.S.A.
91 C1 Desna r. Rus. Fed./Ukr.
89 D3 Desnogorsk Rus. Fed.
101 F2 Dessau Ger.
Dessye Eth. see Desē
128 A1 Destruction Bay Can.
149 C4 Desventuradas, Islas is S. Pacific Ocean
128 C1 Detah Can.
120 B2 Dete Zimbabwe
101 D2 Detmold Ger.
138 C2 Detroit U.S.A.
137 D1 Detroit Lakes U.S.A.
Dett Zimbabwe see Dete
100 B2 Deurne Neth.
110 B1 Deva Romania
100 C1 Deventer Neth.
96 C2 Deveron r. U.K.
103 D2 Devét skal h. Czech Rep.
137 D1 Devils Lake U.S.A.
128 A2 Devils Paw mt. U.S.A.
99 C4 Devizes U.K.
74 B2 Devli India
110 C2 Devnya Bulg.
128 C2 Devon Can.
126 F1 Devon Island Can.
51 D4 Devonport Austr.
74 B2 Dewas India

137 F3 Dexter U.S.A.
70 A2 Deyang China
59 D3 Deyong, Tanjung pt Indon.
81 C2 Dezfūl Iran
70 B2 Dezhou China
79 C2 Dhahran Saudi Arabia
75 D2 Dhaka Bangl.
78 B3 Dhamār Yemen
75 C2 Dhamtari India
75 C2 Dhanbad India
74 B2 Dhandhuka India
75 C2 Dhankuta Nepal
74 B2 Dhar India
75 D2 Dharmanagar India
73 B3 Dharmapuri India
75 C2 Dharmjaygarh India
73 B3 Dharwad India
Dharwar India see Dharwad
74 B2 Dhasa India
75 C2 Dhaulagiri I mt. Nepal
78 B3 Dhubāb Yemen
74 B2 Dhule India
Dhulia India see Dhule
117 C4 Dhuusa Marreeb Somalia
144 A1 Diablo, Picacho del mt. Mex.
142 B2 Diablo Plateau U.S.A.
121 C2 Diaca Moz.
51 C2 Diamantina watercourse Austr.
155 D1 Diamantina Brazil
151 E4 Diamantina, Chapada plat. Brazil
159 F6 Diamantina Deep sea feature Indian Ocean
151 D4 Diamantino Brazil
154 B1 Diamantino Brazil
71 B3 Dianbai China
151 E4 Dianópolis Brazil
114 B4 Dianra Côte d'Ivoire
114 C3 Diapaga Burkina Faso
79 C2 Dibā al Ḩiṣn U.A.E.
79 C2 Ḏibab Oman
118 C3 Dibaya Dem. Rep. Congo
122 B2 Dibeng S. Africa
72 D2 Dibrugarh India
136 C1 Dickinson U.S.A.
140 C1 Dickson U.S.A.
Dicle r. Turkey see Tigris
105 D3 Die France
Diedenhofen France see Thionville
129 D2 Diefenbaker, Lake Can.
Diégo Suarez Madag. see Antsirañana
114 B3 Diéma Mali
101 D2 Diemel r. Ger.
62 B1 Điên Biên Phu Vietnam
62 B2 Diên Châu Vietnam
101 D1 Diepholz Ger.
104 C2 Dieppe France
100 B2 Diest Belgium
115 D3 Diffa Niger
131 D3 Digby Can.
105 D3 Digne-les-Bains France
105 C2 Digoin France
64 B3 Digos Phil.
59 D3 Digul r. Indon.
105 D2 Dijon France
115 D4 Dik Chad
117 C3 Dikhil Djibouti
111 C3 Dikili Turkey
100 A2 Diksmuide Belgium
82 G2 Dikson Rus. Fed.
115 D3 Dikwa Nigeria
117 B4 Dīla Eth.
74 A1 Dīlārām Afgh.
59 C3 Dili East Timor
101 D2 Dillenburg Ger.
117 A3 Dilling Sudan
126 B3 Dillingham U.S.A.
134 D1 Dillon MT U.S.A.
141 E2 Dillon SC U.S.A.
118 C4 Dilolo Dem. Rep. Congo
72 D2 Dimapur India
Dimashq Syria see Damascus
52 B3 Dimboola Austr.
110 C2 Dimitrovgrad Bulg.
87 D3 Dimitrovgrad Rus. Fed.
Dimitrovo Bulg. see Pernik
64 B2 Dinagat i. Phil.
75 C2 Dinajpur Bangl.
104 B2 Dinan France
100 B2 Dinant Belgium
111 D3 Dinar Turkey
81 D2 Dīnār, Kūh-e mt. Iran
104 C2 Dinard France
Dinbych U.K. see Denbigh
73 B3 Dindigul India

118 B1 Dindima Nigeria
123 D1 Dindiza Moz.
101 E2 Dingelstädt Ger.
75 C2 Dinggyê China
Dingle Ireland see An Daingean
97 A2 Dingle Bay Ireland
71 B3 Dingnan China
102 C2 Dingolfing Ger.
114 A3 Dinguiraye Guinea
96 B2 Dingwall U.K.
70 A2 Dingxi China
123 C1 Dinokwe Botswana
91 D2 Dinskaya Rus. Fed.
100 C2 Dinslaken Ger.
135 C3 Dinuba U.S.A.
114 B3 Dioïla Mali
154 B3 Dionísio Cerqueira Brazil
114 A3 Diourbel Senegal
75 D2 Diphu India
64 B3 Dipolog Phil.
74 B1 Dir Pak.
51 D1 Direction, Cape Austr.
117 C4 Dirē Dawa Eth.
120 B2 Dirico Angola
115 D1 Dirj Libya
50 A2 Dirk Hartog Island Austr.
53 C1 Dirranbandi Austr.
78 B3 Ḏirs Saudi Arabia
153 E5 Disappointment, Cape S. Georgia
134 B1 Disappointment, Cape imp. l. Austr.
50 B2 Disappointment, Lake imp. l. Austr.
52 B3 Discovery Bay Austr.
Disko i. Greenland see Qeqertarsuaq
141 E1 Dismal Swamp U.S.A.
99 D3 Diss U.K.
154 C1 Distrito Federal admin. dist. Brazil
108 B3 Dittaino r. Italy
74 B2 Diu India
155 D2 Divinópolis Brazil
87 D4 Divnoye Rus. Fed.
114 B4 Divo Côte d'Ivoire
80 B2 Divriği Turkey
74 A2 Diwana Pak.
138 B2 Dixon U.S.A.
128 A2 Dixon Entrance sea chan. Can./U.S.A.
81 C2 Diyarbakır Turkey
74 A2 Diz Pak.
115 D2 Djado Niger
115 D2 Djado, Plateau du Niger
Djakarta Indon. see Jakarta
118 B3 Djambala Congo
115 C2 Djanet Alg.
115 D3 Djédaa Chad
115 C1 Djelfa Alg.
119 C2 Djéma C.A.R.
114 B3 Djenné Mali
114 B3 Djibo Burkina Faso
117 C3 Djibouti country Africa
117 C3 Djibouti Djibouti
Djidjelli Alg. see Jijel
118 C2 Djolu Dem. Rep. Congo
114 C4 Djougou Benin
118 B2 Djoum Cameroon
115 D3 Djourab, Erg du des. Chad
92 □C3 Djúpivogur Iceland
89 F3 Dmitriyevka Rus. Fed.
89 E3 Dmitriyev-L'govskiy Rus. Fed.
Dmitriyevsk Ukr. see Makiyivka
89 E2 Dmitrov Rus. Fed.
Dmytriyevs'k Ukr. see Makiyivka
Dnepr r. Rus. Fed. see Dnieper
89 D3 Dnieper r. Rus. Fed.
91 C2 Dnieper r. Ukr.
90 B2 Dniester r. Ukr.
Dnipro r. Ukr. see Dnieper
91 C2 Dniprodzerzhyns'k Ukr.
91 D2 Dnipropetrovs'k Ukr.
91 C2 Dniprorudne Ukr.
Dnister r. Ukr. see Dniester
90 B2 Dnistrovs'kyy Lyman lag. Ukr.
88 C2 Dno Rus. Fed.
121 C2 Doa Moz.
115 D4 Doba Chad
88 B2 Dobele Latvia
101 F2 Döbeln Ger.
59 C3 Doberai, Jazirah pen. Indon.
Doberai Peninsula pen. Indon. see Doberai, Jazirah

59 C3 Dobo Indon.
109 C2 Doboj Bos.-Herz.
103 E1 Dobre Miasto Pol.
110 C2 Dobrich Bulg.
89 F3 Dobrinka Rus. Fed.
89 E3 Dobroye Rus. Fed.
89 D3 Dobrush Belarus
86 E3 Dobryanka Rus. Fed.
155 E1 Doce r. Brazil
145 B2 Doctor Arroyo Mex.
144 B2 Doctor Belisario Domínguez Mex.
Doctor Petru Groza Romania see Ştei
111 C3 Dodecanese is Greece see Dodecanese
136 C3 Dodge City U.S.A.
119 D3 Dodoma Tanz.
100 C1 Doesburg Neth.
100 C2 Doetinchem Neth.
59 C3 Dofa Indon.
75 C1 Dogai Coring salt l. China
128 B2 Dog Creek Can.
67 B3 Dōgo i. Japan
115 C3 Dogondoutchi Niger
81 C2 Doğubeyazıt Turkey
79 C2 Doha Qatar
62 A2 Doi Saket Thai.
81 D2 Dokali Iran
100 B1 Dokkum Neth.
88 C3 Dokshytsy Belarus
91 D2 Dokuchayevs'k Ukr.
142 A1 Dolan Springs U.S.A.
130 C3 Dolbeau-Mistassini Can.
104 B2 Dol-de-Bretagne France
105 D2 Dole France
91 D2 Dolgaya Kosa spit Rus. Fed.
99 B3 Dolgellau U.K.
89 E3 Dolgorukovo Rus. Fed.
89 E3 Dolgoye Rus. Fed.
69 F1 Dolinsk Rus. Fed.
103 D2 Dolný Kubín Slovakia
59 D3 Dolok, Pulau i. Indon.
108 B1 Dolomites mts Italy
Dolomiti mts Italy see Dolomites
117 C4 Dolo Odo Eth.
144 A2 Dolores Mex.
126 E2 Dolphin and Union Strait Can.
90 A2 Dolyna Ukr.
102 C2 Domažlice Czech Rep.
93 E3 Dombås Norway
103 D2 Dombóvár Hungary
Dombrovitsa Ukr. see Dubrovytsya
Dombrowa Pol. see Dąbrowa Górnicza
128 B2 Dome Creek Can.
147 D3 Dominica country West Indies
147 C3 Dominican Republic country West Indies
118 C3 Domiongo Dem. Rep. Congo
117 C4 Domo Eth.
89 E2 Domodedovo Rus. Fed.
111 B3 Domokos Greece
61 C2 Dompu Indon.
153 A3 Domuyo, Volcán vol. Arg.
142 B3 Don Mex.
89 E3 Don r. Rus. Fed.
96 C2 Don r. U.K.
97 D1 Donaghadee U.K.
52 B3 Donald Austr.
Donau r. Austria/Ger. see Danube
102 C2 Donauwörth Ger.
106 B2 Don Benito Spain
98 C3 Doncaster U.K.
120 A1 Dondo Angola
121 C2 Dondo Moz.
73 C4 Dondra Head hd Sri Lanka
97 B1 Donegal Ireland
97 B1 Donegal Bay Ireland
91 D2 Donets'k Ukr.
91 D2 Donets'kyy Kryazh hills Rus. Fed./Ukr.
118 B2 Donga Nigeria
50 A2 Dongara Austr.
71 A3 Dongchuan China
65 B2 Dongducheon S. Korea
71 A4 Dongfang China
66 B1 Dongfanghong China
61 C2 Donggala Indon.
65 A2 Donggang China
Donggou China see Donggang

71 B3 Dongguan China
62 B2 Đông Ha Vietnam
65 B2 Donghae S. Korea
Dong Hai sea N. Pacific Ocean see East China Sea
62 B2 Đông Hới Vietnam
116 B3 Dongola Sudan
118 B2 Dongou Congo
Dong Phaya Yen Range mts Thai. see San Khao Phang Hoei
63 B2 Dong Phraya Yen esc. Thai.
Dongping China see Anhua
71 B3 Dongshan China
70 B2 Dongsheng China
70 C2 Dongtai China
71 B3 Dongting Hu l. China
Dong Ujimqin Qi China see Uliastai
71 C3 Dongyang China
70 B2 Dongying China
143 D3 Donna U.S.A.
54 B1 Donnellys Crossing N.Z.
100 C3 Donnersberg h. Ger.
Donostia Spain see San Sebastián
81 D1 Donyztau, Sor dry lake Kazakh.
51 C1 Doomadgee Austr.
138 B2 Door Peninsula U.S.A.
50 B2 Dora, Lake imp. l. Austr.
99 B4 Dorchester U.K.
122 A1 Dordabis Namibia
104 B2 Dordogne r. France
100 B2 Dordrecht Neth.
123 C3 Dordrecht S. Africa
122 A1 Doreenville Namibia
129 D2 Doré Lake Can.
101 D1 Dorfmark Ger.
114 B3 Dori Burkina Faso
122 A3 Doring r. S. Africa
100 C2 Dormagen Ger.
96 B2 Dornoch U.K.
96 B2 Dornoch Firth est. U.K.
89 D3 Doro Mali
89 D3 Dorogobuzh Rus. Fed.
110 C1 Dorohoi Romania
68 C1 Döröö Nuur salt l. Mongolia
92 G3 Dorotea Sweden
50 A2 Dorre Island Austr.
53 D2 Dorrigo Austr.
100 C2 Dorsten Ger.
100 C2 Dortmund Ger.
100 C2 Dortmund-Ems-Kanal canal Ger.
81 C2 Dorūd Iran
153 B4 Dos Bahías, Cabo c. Arg.
101 F1 Dosse r. Ger.
114 C3 Dosso Niger
141 C2 Dothan U.S.A.
101 D1 Dötlingen Ger.
105 C1 Douai France
118 A2 Douala Cameroon
104 B2 Douarnenez France
105 D2 Doubs r. France/Switz.
54 A3 Doubtful Sound N.Z.
114 B3 Douentza Mali
98 A2 Douglas Isle of Man
122 B2 Douglas S. Africa
128 A2 Douglas AK U.S.A.
142 B2 Douglas AZ U.S.A.
141 D2 Douglas GA U.S.A.
136 B2 Douglas WY U.S.A.
71 C3 Douliu Taiwan
104 C1 Doullens France
154 B1 Dourada, Serra hills Brazil
154 B2 Dourados Brazil
154 B2 Dourados r. Brazil
154 B2 Dourados, Serra dos hills Brazil
106 B1 Douro r. Port.
99 D4 Dover U.K.
139 D3 Dover U.S.A.
95 D3 Dover, Strait of France/U.K.
139 F1 Dover-Foxcroft U.S.A.
99 B3 Dovey r. U.K.
121 C2 Dowa Malawi
79 C2 Dowlatābād Iran
79 C2 Dowlatābād Iran
97 D1 Downpatrick U.K.
77 C3 Dowshī Afgh.
67 B3 Dōzen is Japan
130 C3 Dozois, Réservoir resr Can.
114 B2 Drâa, Hamada du plat. Alg.
154 B2 Dracena Brazil
100 C1 Drachten Neth.
110 B2 Drăgănești-Olt Romania

110 B2 Drăgășani Romania
105 D3 Draguignan France
88 C3 Drahichyn Belarus
53 D1 Drake Austr.
123 C2 Drakensberg mts Lesotho/S. Africa
123 C2 Drakensberg mts S. Africa
149 C8 Drake Passage S. Atlantic Ocean
111 B2 Drama Greece
93 F4 Drammen Norway
109 C1 Drava r. Europe
128 C2 Drayton Valley Can.
101 D2 Dreieich Ger.
102 C1 Dresden Ger.
104 C2 Dreux France
100 B1 Driemond Neth.
98 C2 Driffield U.K.
137 D1 Drift Prairie reg. U.S.A.
109 C2 Drina r. Bosnia-Herzegovina/Serbia
140 B2 Driskill Mountain h. U.S.A.
Drissa Belarus see Vyerkhnyadzvinsk
109 C2 Drniš Croatia
110 B2 Drobeta-Turnu Severin Romania
90 B2 Drochia Moldova
101 D1 Drochtersen Ger.
97 C2 Drogheda Ireland
90 A2 Drohobych Ukr.
99 B3 Droitwich Spa U.K.
Drokiya Moldova see Drochia
97 B1 Dromahair Ireland
97 C1 Dromore U.K.
100 B1 Dronten Neth.
74 B1 Drosh Pak.
53 C3 Drouin Austr.
128 C2 Drumheller Can.
138 C1 Drummond Island U.S.A.
131 C3 Drummondville Can.
96 B3 Drummore U.K.
96 B2 Drumnadrochit U.K.
Druskieniki Lith. see Druskininkai
88 B3 Druskininkai Lith.
88 C2 Druya Belarus
91 D2 Druzhkivka Ukr.
88 D2 Druzhnaya Gorka Rus. Fed.
89 D3 Drybin Belarus
130 A3 Dryden Can.
50 B1 Drysdale r. Austr.
147 C3 Duarte, Pico mt. Dom. Rep.
78 A2 Dubā Saudi Arabia
79 C2 Dubai U.A.E.
90 B2 Dubăsari Moldova
129 D1 Dubawnt Lake Can.
Dubayy U.A.E. see Dubai
78 A2 Dubbagh, Jabal ad mt. Saudi Arabia
53 C2 Dubbo Austr.
Dubesar' Moldova see Dubăsari
97 C2 Dublin Ireland
141 D2 Dublin U.S.A.
89 E2 Dubna Rus. Fed.
90 B1 Dubno Ukr.
139 D2 Du Bois U.S.A.
Dubossary Moldova see Dubăsari
114 A4 Dubréka Guinea
109 C2 Dubrovnik Croatia
90 B1 Dubrovytsya Ukr.
89 D3 Dubrowna Belarus
137 E2 Dubuque U.S.A.
63 B2 Đưc Bôn Vietnam
135 D2 Duchesne U.S.A.
129 D2 Duck Bay Can.
101 E2 Duderstadt Ger.
82 G2 Dudinka Rus. Fed.
99 B3 Dudley U.K.
106 B1 Duero r. Spain
48 H4 Duff Islands Solomon Is
131 C2 Duffreboy, Lac l. Can.
96 C2 Dufftown U.K.
109 C2 Dugi Otok i. Croatia
109 C2 Dugi Rat Croatia
100 C2 Duisburg Ger.
123 C3 Dukathole S. Africa
117 B4 Duk Fadiat S. Sudan
79 C2 Dukhān Qatar
89 D2 Dukhovshchina Rus. Fed.
Dukou China see Panzhihua
88 C2 Dūkštas Lith.
68 C2 Dulan China
152 B3 Dulce r. Arg.
142 B1 Dulce U.S.A.
97 C2 Duleek Ireland
100 C2 Dülmen Ger.

110 C2 Dulovo Bulg.
137 E1 Duluth U.S.A.
64 B3 Dumaguete Phil.
60 B1 Dumai Indon.
64 B2 Dumaran i. Phil.
140 B2 Dumas AR U.S.A.
143 C1 Dumas TX U.S.A.
96 B3 Dumbarton U.K.
103 D2 Ďumbier mt. Slovakia
96 C3 Dumfries U.K.
89 E3 Duminichi Rus. Fed.
75 C2 Dumka India
55 L3 Dumont d'Urville Sea Antarctica
116 B1 Dumyât Egypt
Duna r. Hungary see Danube
Dünaburg Latvia see Daugavpils
Dunaj r. Slovakia see Danube
103 D2 Dunakeszi Hungary
97 C2 Dunany Point Ireland
Dunărea r. Romania see Danube
Dunării, Delta delta Romania/Ukr. see Danube Delta
103 D2 Dunaújváros Hungary
Dunav r. Bulg./Croatia/Serbia see Danube
90 B2 Dunayivtsi Ukr.
96 C2 Dunbar U.K.
96 C1 Dunbeath U.K.
128 B3 Duncan Can.
143 D2 Duncan U.S.A.
96 C1 Duncansby Head hd U.K.
88 B2 Dundaga Latvia
97 C1 Dundalk Ireland
139 D3 Dundalk U.S.A.
97 C2 Dundalk Bay Ireland
127 H1 Dundas Greenland
Dún Dealgan Ireland see Dundalk
123 D2 Dundee S. Africa
96 C2 Dundee U.K.
97 D1 Dundrum Bay U.K.
54 B3 Dunedin N.Z.
53 C2 Dunedoo Austr.
96 C2 Dunfermline U.K.
97 C1 Dungannon U.K.
74 B2 Dungarpur India
97 C2 Dungarvan Ireland
99 D4 Dungeness hd U.K.
97 C1 Dungiven U.K.
53 D2 Dungog Austr.
119 C2 Dungu Dem. Rep. Congo
60 B1 Dungun Malaysia
116 B2 Dungunab Sudan
69 E2 Dunhua China
68 C2 Dunhuang China
96 C2 Dunkeld U.K.
Dunkerque France see Dunkirk
104 C1 Dunkirk France
139 D2 Dunkirk U.S.A.
97 C2 Dún Laoghaire Ireland
97 B3 Dunmanway Ireland
97 D1 Dunmurry U.K.
96 C1 Dunnet Head hd U.K.
96 C3 Duns U.K.
134 B2 Dunsmuir U.S.A.
99 C4 Dunstable U.K.
100 B3 Dun-sur-Meuse France
96 A2 Dunvegan U.K.
70 B1 Duolun China
Duperré Alg. see Aïn Defl
110 B2 Dupnitsa Bulg.
136 C1 Dupree U.S.A.
Duque de Bragança Angola see Calandula
138 B3 Du Quoin U.S.A.
50 B1 Durack r. Austr.
105 C3 Durance r. France
144 B2 Durango Mex.
106 C1 Durango Spain
136 B3 Durango U.S.A.
143 D2 Durant U.S.A.
153 C3 Durazno Uru.
Durazzo Albania see Durrë
123 D2 Durban S. Africa
105 C3 Durban-Corbières France
122 A3 Durbanville S. Africa
100 B2 Durbuy Belgium
100 C2 Düren Ger.
75 C2 Durg India
98 C2 Durham U.K.
141 E1 Durham U.S.A.
60 B1 Duri Indon.
109 C2 Durmitor mt. Montenegro

96 B1 Durness U.K.
09 C2 Durrës Albania
97 A3 Dursey Island Ireland
11 C3 Dursunbey Turkey
17 C4 Durüksi Somalia
59 D3 D'Urville, Tanjung pt Indon.
54 B2 D'Urville Island N.Z.
71 A3 Dushan China
77 C3 Dushanbe Tajik.
00 C2 Düsseldorf Ger.
Dutch East Indies country Asia see Indonesia
Dutch Guiana country S. America see Suriname
23 C3 Dutywa S. Africa
17 C4 Duudka, Taagga reg. Somalia
Duvno Bos.-Herz. see Tomislavgrad
71 A3 Duyun China
80 B1 Düzce Turkey
91 D2 Dvorichna Ukr.
74 A2 Dwarka India
23 C1 Dwarsberg S. Africa
34 C1 Dworshak Reservoir U.S.A.
89 D3 Dyat'kovo Rus. Fed.
96 C2 Dyce U.K.
27 H2 Dyer, Cape Can.
40 C1 Dyersburg U.S.A.
Dyersville U.S.A.
Dyfrdwy r. U.K. see Dee
03 D2 Dyje r. Austria/Czech Rep.
03 D1 Dylewska Góra h. Pol.
91 D2 Dymytrov Ukr.
23 C3 Dyoki S. Africa
51 D2 Dysart Austr.
22 B3 Dysselsdorp S. Africa
87 E3 Dyurtyuli Rus. Fed.
69 D2 Dzamin Üüd Mongolia
21 D2 Dzaoudzi Mayotte
91 D2 Dzerzhyns'k Ukr.
Dzhaltyr Kazakh. see Zhaltyr
Dzhambul Kazakh. see Taraz
91 C2 Dzhankoy Ukr.
Dzharkent Kazakh. see Zharkent
Dzhetygara Kazakh. see Zhitikara
Dzhezkazgan Kazakh. see Zhezkazgan
Dzhizak Uzbek. see Jizzax
91 D3 Dzhubga Rus. Fed.
83 K3 Dzhugdzhur, Khrebet mts Rus. Fed.
03 E1 Działdowo Pol.
45 D2 Dzilam de Bravo Mex.
69 D1 Dzuunmod Mongolia
88 C3 Dzyarzhynsk Belarus
88 C3 Dzyatlavichy Belarus

E

31 E2 Eagle r. Can.
34 C1 Eagle Cap mt. U.S.A.
30 A3 Eagle Lake Can.
34 B2 Eagle Lake U.S.A.
43 C3 Eagle Pass U.S.A.
26 C2 Eagle Plain Can.
Eap i. Micronesia see Yap
30 A2 Ear Falls Can.
35 C3 Earlimart U.S.A.
96 C2 Earn r. U.K.
55 J2 East Antarctica reg. Antarctica
East Bengal country Asia see Bangladesh
99 D4 Eastbourne U.K.
59 D2 East Caroline Basin N. Pacific Ocean
69 E2 East China Sea N. Pacific Ocean
54 B1 East Coast Bays N.Z.
East Dereham U.K. see Dereham
29 D3 Eastend Can.
57 G7 Easter Island S. Pacific Ocean
23 C3 Eastern Cape prov. S. Africa
16 B2 Eastern Desert Egypt
73 B3 Eastern Ghats mts India
Eastern Samoa terr. S. Pacific Ocean see American Samoa
Eastern Sayan Mountains mts Rus. Fed. see Vostochnyy Sayan

Eastern Transvaal prov. S. Africa see Mpumalanga
129 E2 Easterville Can.
153 C5 East Falkland i. Falkland Is
100 C1 East Frisian Islands Ger.
135 C3 Eastgate U.S.A.
137 D1 East Grand Forks U.S.A.
96 B3 East Kilbride U.K.
138 C2 East Lansing U.S.A.
99 C4 Eastleigh U.K.
96 C3 East Linton U.K.
138 C2 East Liverpool U.S.A.
123 C3 East London S. Africa
130 C2 Eastmain Can.
130 C2 Eastmain r. Can.
141 D2 Eastman U.S.A.
157 G8 East Pacific Rise N. Pacific Ocean
East Pakistan country Asia see Bangladesh
138 A3 East St Louis U.S.A.
East Sea sea N. Pacific Ocean see Japan, Sea of
83 K2 East Siberian Sea Rus. Fed.
59 C3 East Timor country Asia
53 C2 East Toorale Austr.
139 D2 East York Can.
138 A2 Eau Claire U.S.A.
130 C2 Eau Claire, Lac à l' l. Can.
59 D2 Eauripik atoll Micronesia
145 C2 Ebano Mex.
99 B4 Ebbw Vale U.K.
118 B2 Ebebiyin Equat. Guinea
102 C1 Eberswalde-Finow Ger.
66 D2 Ebetsu Japan
77 E2 Ebinur Hu salt l. China
118 C2 Ebola r. Dem. Rep. Congo
109 C2 Eboli Italy
118 B2 Ebolowa Cameroon
107 D1 Ebro r. Spain
Echeng China see Ezhou
144 A2 Echeverria, Pico mt. Mex.
67 C3 Echizen Japan
129 E2 Echoing r. Can.
100 C3 Echternach Lux.
52 B3 Echuca Austr.
106 B2 Écija Spain
102 B1 Eckernförde Ger.
127 G2 Eclipse Sound sea chan. Can.
150 B3 Ecuador country S. America
116 C3 Ed Eritrea
96 C1 Eday i. U.K.
117 A3 Ed Da'ein Sudan
117 B3 Ed Damazin Sudan
116 B3 Ed Damer Sudan
116 B3 Ed Debba Sudan
116 B3 Ed Dueim Sudan
51 D4 Eddystone Point Austr.
100 B1 Ede Neth.
118 B2 Edéa Cameroon
154 C1 Edéia Brazil
53 C3 Eden Austr.
98 B2 Eden r. U.K.
143 D2 Eden U.S.A.
123 C2 Edenburg S. Africa
97 C2 Edenderry Ireland
52 B3 Edenhope Austr.
141 E1 Edenton U.S.A.
111 B2 Edessa Greece
139 E2 Edgartown U.S.A.
Edge Island i. Svalbard see Edgeøya
82 C1 Edgeøya i. Svalbard
141 D3 Edgeøya Svalbard
138 C2 Edina U.S.A.
143 D3 Edinburg U.S.A.
96 C3 Edinburgh U.K.
90 B2 Edineţ Moldova
111 C2 Edirne Turkey
Edith Ronne Land Antarctica see Ronne Ice Shelf
128 C2 Edmonton Can.
131 D3 Edmundston Can.
111 C3 Edremit Turkey
111 C3 Edremit Körfezi b. Turkey
128 C2 Edson Can.
119 C3 Edward, Lake Dem. Rep. Congo/Uganda
116 A3 Edwardesabad Pak. see Bannu
143 D3 Edwards Plateau U.S.A.
55 O2 Edward VII Peninsula Antarctica
96 C2 Edzell U.K.
100 A3 Eeklo Belgium
135 B2 Eel r. U.S.A.
100 C1 Eemskanaal canal Neth.

100 C1 Eenrum Neth.
138 B3 Effingham U.S.A.
135 C3 Egan Range mts U.S.A.
103 E2 Eger Hungary
93 E4 Egersund Norway
100 B2 Eghezée Belgium
92 ◻C2 Egilsstaðir Iceland
80 B2 Eğirdir Turkey
80 B2 Eğirdir Gölü l. Turkey
104 C2 Égletons France
Egmont, Mount vol. N.Z. see Taranaki, Mount
83 M2 Egvekinot Rus. Fed.
116 A2 Egypt country Africa
70 A2 Ehen Hudag China
106 C1 Eibar Spain
100 C1 Eibergen Neth.
100 C2 Eifel hills Ger.
96 A2 Eigg i. U.K.
73 B4 Eight Degree Channel India/Maldives
50 B1 Eighty Mile Beach Austr.
80 B3 Eilat Israel
101 F2 Eilenburg Ger.
101 D2 Einbeck Ger.
100 B2 Eindhoven Neth.
160 N4 Eirik Ridge N. Atlantic Ocean
150 C3 Eirunepé Brazil
120 B2 Eiseb watercourse Namibia
101 E2 Eisenach Ger.
101 E2 Eisenberg Ger.
102 C1 Eisenhüttenstadt Ger.
103 D2 Eisenstadt Austria
101 E2 Eisleben Lutherstadt Ger.
Eivissa Spain see Ibiza
Eivissa i. Spain see Ibiza
107 C1 Ejea de los Caballeros Spain
121 ◻D3 Ejeda Madag.
Ejin Qi China see Dalain Hob
93 H4 Ekenäs Fin.
93 F4 Eksjö Sweden
122 A2 Eksteenfontein S. Africa
130 B2 Ekwan r. Can.
62 A2 Ela Myanmar
El Aaiún Western Sahara see Laâyoune
123 C2 Elandsdoorn S. Africa
El Araïche Morocco see Larache
El Asnam Alg. see Chlef
111 B3 Elassona Greece
80 B2 Elat Israel see Eilat
108 B2 Elba, Isola d' i. Italy
150 B2 El Banco Col.
142 B2 El Barreal l. Mex.
109 D2 Elbasan Albania
150 C2 El Baúl Venez.
114 C1 El Bayadh Alg.
101 D1 Elbe r. Ger.
136 B3 Elbert, Mount U.S.A.
141 D2 Elberton U.S.A.
104 C2 Elbeuf France
80 B2 Elbistan Turkey
103 D1 Elbląg Pol.
87 D4 El'brus mt. Rus. Fed.
81 C2 Elburz Mountains mts Iran
150 C2 El Callao Venez.
143 D3 El Campo U.S.A.
135 C4 El Centro U.S.A.
152 B1 El Cerro Bol.
107 C2 Elche-Elx Spain
145 C2 El Chichónal vol. Mex.
107 C2 Elda Spain
119 D2 Eldama Ravine Kenya
137 E3 Eldon U.S.A.
154 B3 Eldorado Arg.
154 C2 Eldorado Brazil
144 B2 El Dorado Mex.
140 B2 El Dorado AR U.S.A.
137 D3 El Dorado KS U.S.A.
134 D1 Electric Peak U.S.A.
114 B2 El Eglab plat. Alg.
116 C2 El Ejido Spain
89 E2 Elektrostal' Rus. Fed.
150 B3 El Encanto Col.
146 C2 Eleuthera i. Bahamas
116 A3 El Fasher Sudan
144 B2 El Fuerte Mex.
117 A3 El Fula Sudan
116 A3 El Geneina Sudan
116 B3 El Getaina Sudan
96 C2 Elgin U.K.
138 B2 Elgin U.S.A.
83 K2 El'ginskiy Rus. Fed.
El Gîza Egypt see Giza

115 C1 El Goléa Alg.
144 A1 El Golfo de Santa Clara Mex.
119 D2 Elgon, Mount Kenya/Uganda
108 A3 El Hadjar Alg.
114 A2 El Ḥammâmi reg. Maur.
114 B2 El Ḥank esc. Mali/Maur.
114 A2 El Hierro i. Islas Canarias
145 C2 El Higo Mex.
114 C2 El Homr Alg.
Elichpur India see Achalpur
126 B2 Elim U.S.A.
Élisabethville Dem. Rep. Congo see Lubumbashi
El Iskandariya Egypt see Alexandria
87 D4 Elista Rus. Fed.
139 E2 Elizabeth U.S.A.
141 E1 Elizabeth City U.S.A.
138 B3 Elizabethtown U.S.A.
114 B1 El Jadida Morocco
103 E1 Ełk Pol.
108 A3 El Kala Alg.
88 C2 Elkas kalns h. Latvia
143 D1 Elk City U.S.A.
114 B1 El Kelaâ des Srarhna Morocco
128 C2 Elkford Can.
138 B2 Elkhart U.S.A.
El Khartûm Sudan see Khartoum
110 C2 Elkhovo Bulg.
139 D3 Elkins U.S.A.
128 C3 Elko Can.
134 C2 Elko U.S.A.
129 C2 Elk Point Can.
137 E1 Elk River U.S.A.
126 F1 Ellef Ringnes Island Can.
137 D1 Ellendale U.S.A.
134 B1 Ellensburg U.S.A.
54 B2 Ellesmere, Lake N.Z.
127 G1 Ellesmere Island Can.
98 B3 Ellesmere Port U.K.
126 F2 Ellice r. Can.
Ellice Islands country S. Pacific Ocean see Tuvalu
123 C3 Elliotdale S. Africa
138 C1 Elliot Lake Can.
96 C2 Ellon U.K.
139 F2 Ellsworth U.S.A.
55 R2 Ellsworth Mountains Antarctica
114 B2 El-Mahbas Western Sahara
111 C3 Elmalı Turkey
115 C1 El Meghaïer Alg.
139 D2 Elmira U.S.A.
107 C2 El Moral Spain
101 D1 Elmshorn Ger.
117 A3 El Muglad Sudan
150 B2 El Nevado, Cerro mt. Col.
64 A2 El Nido Phil.
116 B3 El Obeid Sudan
144 B2 El Oro Mex.
115 C1 El Oued Alg.
142 A2 Eloy U.S.A.
El Paso U.S.A. see Derby
142 B2 El Paso U.S.A.
145 C3 El Pinalón, Cerro mt. Guat.
144 B1 El Porvenir Mex.
107 D1 El Prat de Llobregat Spain
146 B3 El Progreso Hond.
106 B2 El Puerto de Santa María Spain
El Qâhira Egypt see Cairo
El Quds Israel/West Bank see Jerusalem
143 D1 El Reno U.S.A.
128 A1 Elsa Can.
145 B2 El Salado Mex.
144 B2 El Salto Mex.
146 B3 El Salvador country Central America
152 B2 El Salvador Chile
145 B2 El Salvador Mex.
142 B3 El Sauz Mex.
144 A1 El Socorro Mex.
142 B3 El Sueco Mex.
El Suweis Egypt see Suez
108 A3 El Tarf Alg.
106 B1 El Teleno mt. Spain
145 C2 El Temascal Mex.
150 C2 El Tigre Venez.
147 D4 El Tocuyo Venez.
73 C3 Eluru India
88 C2 Elva Estonia
106 B2 Elvas Port.
93 F3 Elverum Norway

119 E2 **El Wak** Kenya
99 D3 **Ely** U.K.
137 E1 **Ely** *MN* U.S.A.
135 D3 **Ely** *NV* U.S.A.
123 C2 **eMalahleni** S. Africa
81 D2 **Emāmrūd** Iran
93 G4 **Emån** *r.* Sweden
123 D3 **eManzamtoti** S. Africa
76 B2 **Emba** Kazakh.
123 C2 **Embalenhle** S. Africa
118 B2 **Embondo** Dem. Rep. Congo
154 C1 **Emborcação, Represa de** *resr* Brazil
119 D3 **Embu** Kenya
100 C1 **Emden** Ger.
51 D2 **Emerald** Austr.
129 E3 **Emerson** Can.
111 C3 **Emet** Turkey
123 D2 **eMgwenya** S. Africa
115 D3 **Emi Koussi** *mt.* Chad
145 C3 **Emiliano Zapata** Mex.
110 C2 **Emine, Nos** *pt* Bulg.
80 B2 **Emirdağ** Turkey
123 D2 **eMjindini** S. Africa
123 D2 **eMkhondo** S. Africa
88 B2 **Emmaste** Estonia
100 B1 **Emmeloord** Neth.
100 C2 **Emmelshausen** Ger.
100 C1 **Emmen** Neth.
123 D2 **eMondlo** S. Africa
143 C3 **Emory Peak** U.S.A.
144 A2 **Empalme** Mex.
123 D2 **Empangeni** S. Africa
156 D2 **Emperor Trough** N. Pacific Ocean
108 B2 **Empoli** Italy
111 C3 **Emponas** Greece
137 D3 **Emporia** *KS* U.S.A.
139 D3 **Emporia** *VA* U.S.A.
Empty Quarter *des.* Saudi Arabia *see* **Rub' al Khālī**
100 C1 **Ems** *r.* Ger.
100 C1 **Emsdetten** Ger.
88 C2 **Emumägi** *h.* Estonia
123 C2 **eMzinoni** S. Africa
59 D3 **Enarotali** Indon.
144 B2 **Encarnación** Mex.
152 C2 **Encarnación** Para.
155 D1 **Encruzilhada** Brazil
61 D2 **Ende** Indon.
55 G2 **Enderby Land** *reg.* Antarctica
126 B2 **Endicott Mountains** U.S.A.
50 A2 **Eneabba** Austr.
91 C2 **Enerhodar** Ukr.
111 C2 **Enez** Turkey
97 C2 **Enfield** Ireland
87 D3 **Engel's** Rus. Fed.
60 B2 **Enggano** *i.* Indon.
99 C3 **England** *admin. div.* U.K.
130 C3 **Englehart** Can.
141 D3 **Englewood** U.S.A.
130 A2 **English** *r.* Can.
English Bazar India *see* **Ingraj Bazar**
95 C4 **English Channel** France/ U.K.
143 D1 **Enid** U.S.A.
Enkeldoorn Zimbabwe *see* **Chivhu**
100 B1 **Enkhuizen** Neth.
93 G4 **Enköping** Sweden
108 B3 **Enna** Italy
129 D1 **Ennadai Lake** Can.
116 A3 **En Nahud** Sudan
115 E3 **Ennedi, Massif** *mts* Chad
53 C1 **Enngonia** Austr.
97 B2 **Ennis** Ireland
143 D2 **Ennis** U.S.A.
97 C2 **Enniscorthy** Ireland
97 C2 **Enniskerry** Ireland
97 C1 **Enniskillen** U.K.
97 B2 **Ennistymon** Ireland
102 C2 **Enns** Austria
102 C2 **Enns** *r.* Austria
92 J3 **Eno** Fin.
92 H2 **Enontekiö** Fin.
53 C3 **Ensay** Austr.
100 C1 **Enschede** Neth.
144 A1 **Ensenada** Mex.
70 A2 **Enshi** China
119 D2 **Entebbe** Uganda
128 C1 **Enterprise** Can.
140 C2 **Enterprise** *AL* U.S.A.
134 C1 **Enterprise** *OR* U.S.A.
152 B2 **Entre Ríos** Bol.
106 B2 **Entroncamento** Port.
115 C4 **Enugu** Nigeria
150 B3 **Envira** Brazil
118 B2 **Epéna** Congo
105 C2 **Épernay** France
135 D3 **Ephraim** U.S.A.
134 C1 **Ephrata** U.S.A.
105 D2 **Épinal** France
99 D4 **Epping** U.K.
99 C4 **Epsom** U.K.
118 A2 **Equatorial Guinea** *country* Africa
104 B2 **Équeurdreville-Hainneville** France
101 D3 **Erbach** Ger.
101 F3 **Erbendorf** Ger.
100 C3 **Erbeskopf** *h.* Ger.
81 C2 **Erciş** Turkey
80 B2 **Erciyes Dağı** *mt.* Turkey
103 D2 **Érd** Hungary
Erdaobaihe China *see* **Baihe**
65 B1 **Erdao Jiang** *r.* China
111 C2 **Erdek** Turkey
80 B2 **Erdemli** Turkey
154 B2 **Eré, Campos** *hills* Brazil
55 M2 **Erebus, Mount** *vol.* Antarctica
152 C2 **Erechim** Brazil
69 D1 **Ereentsav** Mongolia
80 B2 **Ereğli** Turkey
80 B1 **Ereğli** Turkey
69 D2 **Erenhot** China
Erevan Armenia *see* **Yerevan**
101 E2 **Erfurt** Ger.
80 B2 **Ergani** Turkey
114 B2 **'Erg Chech** *des.* Alg./Mali
111 C2 **Ergene** *r.* Turkey
78 A3 **Erheib** Sudan
96 B2 **Ericht, Loch** *l.* U.K.
138 C2 **Erie** U.S.A.
138 C2 **Erie, Lake** Can./U.S.A.
66 D2 **Erimo-misaki** *c.* Japan
116 B3 **Eritrea** *country* Africa
101 E3 **Erlangen** Ger.
50 C2 **Erldunda** Austr.
123 C2 **Ermelo** S. Africa
80 B2 **Ermenek** Turkey
111 B3 **Ermoupoli** Greece
73 B4 **Ernakulam** India
97 B1 **Erne** *r.* Ireland/U.K.
73 B3 **Erode** India
100 B2 **Erp** Neth.
114 B1 **Er Rachidia** Morocco
116 B3 **Er Rahad** Sudan
97 B1 **Errigal** *h.* Ireland
97 A1 **Erris Head** *hd* Ireland
48 H5 **Erromango** *i.* Vanuatu
109 D2 **Ersekë** Albania
62 B1 **Ertan Reservoir** China
89 F3 **Ertil'** Rus. Fed.
101 D2 **Erwitte** Ger.
101 F2 **Erzgebirge** *mts* Czech Rep./ Ger.
80 B2 **Erzincan** Turkey
81 C2 **Erzurum** Turkey
66 D2 **Esashi** Japan
93 E4 **Esbjerg** Denmark
135 D3 **Escalante** U.S.A.
135 D3 **Escalante Desert** U.S.A.
144 B2 **Escalón** Mex.
138 B1 **Escanaba** U.S.A.
145 C3 **Escárcega** Mex.
107 C1 **Escatrón** Spain
100 A2 **Escaut** *r.* Belgium/France
101 E1 **Eschede** Ger.
100 B3 **Esch-sur-Alzette** Lux.
101 E2 **Eschwege** Ger.
100 C2 **Eschweiler** Ger.
135 C4 **Escondido** U.S.A.
144 B2 **Escuinapa** Mex.
111 C3 **Eşen** Turkey
81 D2 **Eşfahān** Iran
123 D2 **Eshowe** S. Africa
123 D2 **eSikhaleni** S. Africa
98 B2 **Esk** *r.* U.K.
131 D2 **Esker** Can.
92 □C2 **Eskifjörður** Iceland
93 G4 **Eskilstuna** Sweden
Eskimo Point Can. *see* **Arviat**
111 D3 **Eskişehir** Turkey
81 C2 **Eslāmābād-e Gharb** Iran
111 C3 **Esler Dağı** *mt.* Turkey
111 C3 **Eşme** Turkey
146 C2 **Esmeralda** Cuba
150 B2 **Esmeraldas** Ecuador
107 D2 **Es Mercadal** Spain
79 D2 **Espakeh** Iran
105 C3 **Espalion** France
130 B3 **Espanola** Can.
142 B1 **Espanola** U.S.A.
50 B3 **Esperance** Austr.
153 A5 **Esperanza** Arg.
144 B2 **Esperanza** Mex.
106 B2 **Espichel, Cabo** *c.* Port.
154 B3 **Espigão, Serra do** *mts* Brazil
143 C3 **Espinazo** Mex.
155 D1 **Espinhaço, Serra do** *mts* Brazil
151 E4 **Espinosa** Brazil
Espirito Santo Brazil *see* **Vila Velha**
155 D1 **Espírito Santo** *state* Brazil
48 H5 **Espiritu Santo** *i.* Vanuatu
144 A2 **Espíritu Santo, Isla** *i.* Mex.
93 H3 **Espoo** Fin.
153 A4 **Esquel** Arg.
114 B1 **Essaouira** Morocco
100 C2 **Essen** Ger.
150 D2 **Essequibo** *r.* Guyana
139 E2 **Essex Junction** U.S.A.
114 A2 **Es-Smara** Western Sahara
83 L3 **Esso** Rus. Fed.
106 B1 **Estaca de Bares, Punta de** *pt* Spain
153 B5 **Estados, Isla de los** *i.* Arg.
81 D3 **Eşţahbān** Iran
151 F4 **Estância** Brazil
123 C2 **Estcourt** S. Africa
107 C1 **Estella** Spain
106 B2 **Estepona** Spain
106 C1 **Esteras de Medinaceli** Spain
129 D2 **Esterhazy** Can.
152 B2 **Esteros** Para.
136 B2 **Estes Park** U.S.A.
129 D3 **Estevan** Can.
137 E2 **Estherville** U.S.A.
129 D2 **Eston** Can.
88 C2 **Estonia** *country* Europe
Estonskaya S.S.R. *country* Europe *see* **Estonia**
106 B1 **Estrela, Serra da** *mts* Port.
106 B2 **Estremoz** Port.
52 A1 **Etadunna** Austr.
104 C2 **Étampes** France
104 C1 **Étaples** France
75 B2 **Etawah** India
123 D2 **Ethandakukhanya** S. Africa
122 B2 **E'Thembini** S. Africa
117 B4 **Ethiopia** *country* Africa
Etna, Monte *vol.* Italy *see* **Etna, Mount**
109 C3 **Etna, Mount** *vol.* Italy
93 E4 **Etne** Norway
128 A2 **Etolin Island** U.S.A.
120 A2 **Etosha Pan** *salt pan* Namibia
110 B2 **Etropole** Bulg.
100 C3 **Ettelbruck** Lux.
100 B2 **Etten-Leur** Neth.
107 C1 **Etxarri-Aranatz** Spain
99 D4 **Eu** France
53 C2 **Euabalong** Austr.
Euboea *i.* Greece *see* **Evvoia**
50 B3 **Eucla** Austr.
138 C2 **Euclid** U.S.A.
141 C2 **Eufaula** U.S.A.
143 D1 **Eufaula Lake** *resr* U.S.A.
134 B2 **Eugene** U.S.A.
144 A2 **Eugenia, Punta** *pt* Mex.
53 C1 **Eulo** Austr.
53 C2 **Eumungerie** Austr.
143 C2 **Eunice** U.S.A.
81 C2 **Euphrates** *r.* Asia
134 B2 **Eureka** *CA* U.S.A.
134 C1 **Eureka** *MT* U.S.A.
135 C3 **Eureka** *NV* U.S.A.
52 B2 **Euriowie** Austr.
53 C3 **Euroa** Austr.
106 B2 **Europa Point** Gibraltar
140 C2 **Eutaw** U.S.A.
128 B2 **Eutsuk Lake** Can.
123 C2 **Evander** S. Africa
130 C2 **Evans, Lac** *l.* Can.
53 D1 **Evans Head** Austr.
127 G2 **Evans Strait** Can.
138 B2 **Evanston** *IL* U.S.A.
136 A2 **Evanston** *WY* U.S.A.
138 B3 **Evansville** U.S.A.
Eva Perón Arg. *see* **La Plata**
123 C2 **Evaton** S. Africa
79 C2 **Evaz** Iran
83 L2 **Evensk** Rus. Fed.
50 C2 **Everard Range** *hills* Austr.
75 C2 **Everest, Mount** China/ Nepal
134 B1 **Everett** U.S.A.
100 A2 **Evergem** Belgium
141 D3 **Everglades** *swamp* U.S.A.
140 C2 **Evergreen** U.S.A.
99 C3 **Evesham** U.K.
118 B2 **Evinayong** Equat. Guinea
93 E4 **Evje** Norway
106 B2 **Évora** Port.
69 F1 **Evoron, Ozero** *l.* Rus. Fed
111 B2 **Evosmos** Greece
104 C2 **Évreux** France
111 C2 **Evros** *r.* Greece/Turkey
111 B3 **Evrotas** *r.* Greece
104 C2 **Évry** France
80 B2 **Evrychou** Cyprus
111 B3 **Evvoia** *i.* Greece
119 E2 **Ewaso Ngiro** *r.* Kenya
152 B1 **Exaltación** Bol.
99 B4 **Exe** *r.* U.K.
99 B4 **Exeter** U.K.
99 B4 **Exmoor** *hills* U.K.
50 A2 **Exmouth** Austr.
99 B4 **Exmouth** U.K.
50 A2 **Exmouth Gulf** Austr.
106 B2 **Extremadura** *aut. comm.* Spain
146 C2 **Exuma Cays** *is* Bahamas
118 C3 **Eyangu** Dem. Rep. Congo
119 D3 **Eyasi, Lake** *salt l.* Tanz.
96 C3 **Eyemouth** U.K.
92 □B2 **Eyjafjörður** *inlet* Iceland
117 C4 **Eyl** Somalia
52 A1 **Eyre (North), Lake** *imp. l.* Austr.
52 A2 **Eyre Peninsula** Austr.
94 B1 **Eysturoy** *i.* Faroe Is
123 D2 **Ezakheni** S. Africa
123 C2 **Ezenzeleni** S. Africa
70 B2 **Ezhou** China
86 E2 **Ezhva** Rus. Fed.
111 C3 **Ezine** Turkey

F

102 C1 **Faaborg** Denmark
142 B2 **Fabens** U.S.A.
108 B2 **Fabriano** Italy
115 D3 **Fachi** Niger
114 C3 **Fada-N'Gourma** Burkina Faso
108 B2 **Faenza** Italy
Faeroes *terr.* N. Atlantic Ocean *see* **Faroe Islands**
59 C3 **Fafanlap** Indon.
110 B1 **Făgăraş** Romania
49 J5 **Fagatogo** American Samo
93 E3 **Fagernes** Norway
93 G4 **Fagersta** Sweden
153 B5 **Fagnano, Lago** *l.* Arg./Ch
114 B3 **Faguibine, Lac** *l.* Mali
92 □B3 **Fagurhólsmýri** Iceland
126 C2 **Fairbanks** U.S.A.
137 D2 **Fairbury** U.S.A.
135 B3 **Fairfield** *CA* U.S.A.
138 C3 **Fairfield** *OH* U.S.A.
96 □ **Fair Isle** *i.* U.K.
137 E2 **Fairmont** *MN* U.S.A.
138 C3 **Fairmont** *WV* U.S.A.
128 C2 **Fairview** Can.
128 A2 **Fairweather, Mount** Can./U.S.A.
59 D2 **Fais** *i.* Micronesia
74 B1 **Faisalabad** Pak.
136 C1 **Faith** U.S.A.
77 D3 **Faīzābād** Afgh.
75 C2 **Faizabad** India
156 E6 **Fakaofo** *atoll* Tokelau
99 D3 **Fakenham** U.K.
59 C3 **Fakfak** Indon.
65 A1 **Faku** China
114 A4 **Falaba** Sierra Leone
Falcarragh Ireland *see* **An Fál Carrach**
143 D3 **Falcon Lake** Mex./U.S.A.
Faleshty Moldova *see* **Făleşti**
90 B2 **Făleşti** Moldova
143 D3 **Falfurrias** U.S.A.
128 C2 **Falher** Can.
101 F2 **Falkenberg** Ger.
93 F4 **Falkenberg** Sweden
101 F1 **Falkensee** Ger.
96 C3 **Falkirk** U.K.
158 D8 **Falkland Escarpment** S. Atlantic Ocean
153 C5 **Falkland Islands** *terr.* S. Atlantic Ocean
157 I9 **Falkland Plateau** S. Atlantic Ocean
93 F4 **Falköping** Sweden
101 D1 **Fallingbostel** Ger.
135 C3 **Fallon** U.S.A.
139 E2 **Fall River** U.S.A.

137 D2	Falls City U.S.A.
99 A4	Falmouth U.K.
122 A3	False Bay S. Africa
144 B2	Falso, Cabo c. Mex.
93 F5	Falster i. Denmark
110 C1	Fălticeni Romania
93 G3	Falun Sweden
152 B2	Famailla Arg.
121 □D3	Fandriana Madag.
71 A3	Fangcheng China
	Fangchenggang China see Fangcheng
71 C3	Fangshan Taiwan
66 A1	Fangzheng China
108 B2	Fano Italy
119 C2	Faradje Dem. Rep. Congo
121 □D3	Farafangana Madag.
	Farafra Oasis oasis Egypt see Waḥāt al Farāfirah
76 C3	Farāh Afgh.
114 A3	Faranah Guinea
79 C3	Fararah Oman
78 B3	Farasān, Jazā'ir is Saudi Arabia
59 D2	Faraulep atoll Micronesia
127 J2	Farewell, Cape c. Greenland
54 B2	Farewell, Cape N.Z.
137 D1	Fargo U.S.A.
77 C2	Farg'ona Uzbek.
137 E2	Faribault U.S.A.
130 C2	Faribault, Lac l. Can.
74 B2	Faridabad India
75 C2	Faridpur Bangl.
139 E2	Farmington ME U.S.A.
142 B1	Farmington NM U.S.A.
139 D3	Farmville U.S.A.
99 C4	Farnborough U.K.
128 C2	Farnham, Mount Can.
128 A1	Faro Can.
106 B2	Faro Port.
88 A2	Fårö i. Sweden
147 C3	Faro, Punta pt Col.
106 B1	Faro, Serra do mts Spain
94 B1	Faroe Islands terr. N. Atlantic Ocean
113 I6	Farquhar Group is Seychelles
81 D3	Farrāshband Iran
	Farrukhabad India see Fatehgarh
79 C3	Fartak, Ra's c. Yemen
154 B3	Fartura, Serra da mts Brazil
143 C2	Farwell U.S.A.
79 C2	Fāryāb Iran
79 C2	Fāryāb Iran
81 D3	Fasā Iran
109 C2	Fasano Italy
90 B1	Fastiv Ukr.
119 D2	Fataki Dem. Rep. Congo
75 B2	Fatehgarh India
75 C2	Fatehpur India
114 A3	Fatick Senegal
131 D3	Fatima Port.
123 C2	Fauresmith S. Africa
92 G2	Fauske Norway
92 □A3	Faxaflói b. Iceland
92 G3	Faxälven r. Sweden
115 D3	Faya Chad
140 B1	Fayetteville AR U.S.A.
141 E1	Fayetteville NC U.S.A.
140 C1	Fayetteville TN U.S.A.
74 B1	Fazilka India
114 A2	Fdérik Maur.
141 E2	Fear, Cape U.S.A.
54 C2	Featherston N.Z.
104 C2	Fécamp France
	Federated Malay States country Asia see Malaysia
102 C1	Fehmarn i. Ger.
101 F1	Fehrbellin Ger.
155 D2	Feia, Lagoa lag. Brazil
150 B3	Feijó Brazil
54 C2	Feilding N.Z.
151 F4	Feira de Santana Brazil
107 D2	Felanitx Spain
101 F1	Feldberg Ger.
145 D3	Felipe C. Puerto Mex.
155 D1	Felixlândia Brazil
99 D4	Felixstowe U.K.
101 D2	Felsberg Ger.
108 B1	Feltre Italy
93 F3	Femunden l. Norway
	Fénérive Madag. see Fenoarivo Atsinanana
	Fengcheng China see Fengshan
71 B3	Fengcheng Jiangxi China
65 A1	Fengcheng Liaoning China
62 A1	Fengqing China
71 A3	Fengshan China

	Fengshan China see Fengqing
70 B2	Fengxian Jiangsu China
70 A2	Fengxian Shaanxi China
	Fengxiang China see Lincang
	Fengyi China see Zheng'an
71 C3	Fengyuan Taiwan
70 B1	Fengzhen China
105 D3	Feno, Capo di c. France
121 □D2	Fenoarivo Atsinanana Madag.
70 B2	Fenyang China
91 D2	Feodosiya Ukr.
108 A3	Fer, Cap de c. Alg.
137 D1	Fergus Falls U.S.A.
51 E1	Fergusson Island P.N.G.
109 D2	Ferizaj Kosovo
114 B4	Ferkessédougou Côte d'Ivoire
108 B2	Fermo Italy
131 D2	Fermont Can.
106 B1	Fermoselle Spain
97 B2	Fermoy Ireland
141 D2	Fernandina Beach U.S.A.
154 B2	Fernandópolis Brazil
	Fernando Poó i. Equat. Guinea see Bioko
134 B1	Ferndale U.S.A.
99 C4	Ferndown U.K.
128 C3	Fernie Can.
135 C3	Fernley U.S.A.
97 C2	Ferns Ireland
	Ferozepore India see Firozpur
108 B2	Ferrara Italy
154 B2	Ferreiros Brazil
108 A2	Ferro, Capo c. Italy
106 B1	Ferrol Spain
	Ferryville Tunisia see Menzel Bourguiba
100 B1	Ferwert Neth.
114 B1	Fès Morocco
118 B3	Feshi Dem. Rep. Congo
137 E3	Festus U.S.A.
110 C2	Fetești Romania
97 C2	Fethard Ireland
111 C3	Fethiye Turkey
96 □	Fetlar i. U.K.
130 C2	Feuilles, Rivière aux r. Can.
	Fez Morocco see Fès
121 □D3	Fianarantsoa Madag.
115 D4	Fianga Chad
117 B4	Fichè Eth.
109 C2	Fier Albania
96 C2	Fife Ness pt U.K.
104 C3	Figeac France
106 B1	Figueira da Foz Port.
107 D1	Figueres Spain
114 B1	Figuig Morocco
49 I5	Fiji country S. Pacific Ocean
152 B2	Filadelfia Para.
55 B2	Filchner Ice Shelf Antarctica
98 C2	Filey U.K.
108 B3	Filicudi, Isola i. Italy
114 C3	Filingué Niger
111 B3	Filippiada Greece
93 F4	Filipstad Sweden
92 E3	Fillan Norway
135 D3	Fillmore U.S.A.
119 E2	Filtu Eth.
55 D2	Fimbul Ice Shelf Antarctica
96 C2	Findhorn r. U.K.
138 C2	Findlay U.S.A.
51 D4	Fingal Austr.
139 D2	Finger Lakes U.S.A.
111 D3	Finike Turkey
106 B1	Finisterre, Cape c. Spain
93 I3	Finland country Europe
93 H4	Finland, Gulf of Europe
128 B2	Finlay r. Can.
53 C3	Finley Austr.
134 C1	Finley U.S.A.
101 E2	Finne ridge Ger.
92 H2	Finnmarksvidda reg. Norway
92 G2	Finnsnes Norway
93 G4	Finspång Sweden
97 C1	Fintona U.K.
96 A2	Fionnphort U.K.
111 C3	Fira Greece
	Firat r. Turkey see Euphrates
	Firenze Italy see Florence
105 C2	Firminy France
74 B2	Firozabad India
74 B1	Firozpur India
81 D3	Firūzābād Iran
122 A2	Fish watercourse Namibia

122 B3	Fish r. S. Africa
129 F1	Fisher Strait Can.
99 A4	Fishguard U.K.
105 C2	Fismes France
	Fisterra, Cabo c. Spain see Finisterre, Cape
139 E2	Fitchburg U.S.A.
128 C2	Fitzgerald Can.
141 D2	Fitzgerald U.S.A.
50 B1	Fitzroy Crossing Austr.
	Fiume Croatia see Rijeka
54 A3	Five Rivers N.Z.
108 B2	Fivizzano Italy
119 C3	Fizi Dem. Rep. Congo
93 F3	Fjällnäs Sweden
92 G3	Fjällsjöälven r. Sweden
123 C3	Flagstaff S. Africa
142 A1	Flagstaff U.S.A.
130 C2	Flaherty Island Can.
98 C2	Flamborough Head hd U.K.
101 F1	Fläming hills Ger.
136 B2	Flaming Gorge Reservoir U.S.A.
134 D1	Flathead r. U.S.A.
134 D1	Flathead Lake U.S.A.
51 D1	Flattery, Cape Austr.
134 B1	Flattery, Cape U.S.A.
98 B3	Fleetwood U.K.
93 E4	Flekkefjord Norway
102 B1	Flensburg Ger.
104 B2	Flers France
100 B2	Fleurus Belgium
104 C2	Fleury-les-Aubrais France
51 D1	Flinders r. Austr.
50 A3	Flinders Bay Austr.
51 D3	Flinders Island Austr.
52 A2	Flinders Ranges mts Austr.
129 D2	Flin Flon Can.
98 B3	Flint U.K.
138 C2	Flint U.S.A.
49 L5	Flint Island Kiribati
101 F2	Flöha Ger.
107 D1	Florac France
100 C3	Florange France
108 B2	Florence Italy
140 C2	Florence AL U.S.A.
142 A2	Florence AZ U.S.A.
134 B2	Florence OR U.S.A.
141 E2	Florence SC U.S.A.
150 B2	Florencia Col.
146 B3	Flores Guat.
61 D2	Flores i. Indon.
61 C2	Flores, Laut Indon.
	Floreshty Moldova see Florești
	Flores Sea sea Indon. see Flores, Laut
C2	
151 F3	Floresta Brazil
90 B2	Florești Moldova
143 D3	Floresville U.S.A.
151 E3	Floriano Brazil
152 D2	Florianópolis Brazil
153 C3	Florida Uru.
141 D2	Florida state U.S.A.
141 D4	Florida, Straits of N. Atlantic Ocean
141 D4	Florida Bay U.S.A.
141 D4	Florida Keys is U.S.A.
111 B2	Florina Greece
93 E3	Florø Norway
129 D2	Foam Lake Can.
110 C1	Focșani Romania
71 B3	Fogang China
109 C2	Foggia Italy
131 E3	Fogo Island Can.
104 C3	Foix France
89 D3	Fokino Rus. Fed.
130 B3	Foleyet Can.
108 B2	Foligno Italy
99 D4	Folkestone U.K.
141 D2	Folkston U.S.A.
108 B2	Follonica Italy
129 D2	Fond-du-Lac Can.
129 D2	Fond du Lac r. Can.
138 B2	Fond du Lac U.S.A.
106 B1	Fondevila Spain
108 B2	Fondi Italy
146 B3	Fonseca, Golfo do b. Central America
150 C3	Fonte Boa Brazil
104 B2	Fontenay-le-Comte France
92 □C2	Fontur pt Iceland
53 C2	Forbes Austr.
75 C2	Forbesganj India
101 E3	Forchheim Ger.
93 E3	Førde Norway
53 C1	Fords Bridge Austr.
140 B2	Fordyce U.S.A.

140 C2	Forest U.S.A.
53 C3	Forest Hill Austr.
131 D3	Forestville Can.
96 C2	Forfar U.K.
134 B1	Forks U.S.A.
108 B2	Forlì Italy
107 D2	Formentera i. Spain
107 D2	Formentor, Cap de c. Spain
155 C2	Formiga Brazil
152 C2	Formosa Arg.
	Formosa country Asia see Taiwan
154 C1	Formosa Brazil
	Formosa Strait str. China/Taiwan see Taiwan Strait
	Føroyar terr. N. Atlantic Ocean see Faroe Islands
96 C2	Forres U.K.
50 B3	Forrest Austr.
140 B1	Forrest City U.S.A.
51 D1	Forsayth Austr.
93 H3	Forssa Fin.
53 D2	Forster Austr.
136 B1	Forsyth U.S.A.
74 B2	Fort Abbas Pak.
130 B2	Fort Albany Can.
151 F3	Fortaleza Brazil
	Fort Archambault Chad see Sarh
128 C2	Fort Assiniboine Can.
96 B2	Fort Augustus U.K.
123 C3	Fort Beaufort S. Africa
134 D1	Fort Benton U.S.A.
	Fort Brabant Can. see Tuktoyaktuk
135 B3	Fort Bragg U.S.A.
	Fort Carnot Madag. see Ikongo
	Fort Charlet Alg. see Djanet
	Fort Chimo Can. see Kuujjuaq
129 C2	Fort Chipewyan Can.
136 B2	Fort Collins U.S.A.
	Fort Crampel C.A.R. see Kaga Bandoro
	Fort-Dauphin Madag. see Tôlañaro
147 D3	Fort-de-France Martinique
	Fort de Polignac Alg. see Illizi
137 E2	Fort Dodge U.S.A.
139 D2	Fort Erie Can.
	Fort Flatters Alg. see Bordj Omer Driss
	Fort Foureau Cameroon see Kousséri
130 A3	Fort Frances Can.
	Fort Franklin Can. see Déline
	Fort Gardel Alg. see Zaouatallaz
	Fort George Can. see Chisasibi
126 D2	Fort Good Hope Can.
	Fort Gouraud Maur. see Fdérik
96 C2	Forth r. U.K.
96 C2	Forth, Firth of est. U.K.
	Fort Hall Kenya see Murang'a
	Fort Hertz Myanmar see Putao
152 C2	Fortín Madrejón Para.
	Fort Jameson Zambia see Chipata
	Fort Johnston Malawi see Mangochi
	Fort Lamy Chad see Ndjamena
	Fort Laperrine Alg. see Tamanrasset
141 D3	Fort Lauderdale U.S.A.
128 B1	Fort Liard Can.
128 C2	Fort Mackay Can.
137 E2	Fort Madison U.S.A.
	Fort Manning Malawi see Mchinji
128 C2	Fort McMurray Can.
126 D2	Fort McPherson Can.
136 C2	Fort Morgan U.S.A.
141 D3	Fort Myers U.S.A.
128 B2	Fort Nelson Can.
128 B2	Fort Nelson r. Can.
	Fort Norman Can. see Tulita
140 C2	Fort Payne U.S.A.
136 B1	Fort Peck U.S.A.
136 B1	Fort Peck Reservoir U.S.A.
141 D3	Fort Pierce U.S.A.

Fort Portal

119 D2	Fort Portal Uganda
128 C1	Fort Providence Can.
129 D2	Fort Qu'Appelle Can.
128 C1	Fort Resolution Can.
96 B2	Fortrose U.K.
	Fort Rosebery Zambia see Mansa
	Fort Rousset Congo see Owando
	Fort Rupert Can. see Waskaganish
128 B2	Fort St James Can.
128 B2	Fort St John Can.
	Fort Sandeman Pak. see Zhob
128 C2	Fort Saskatchewan Can.
137 E3	Fort Scott U.S.A.
130 B2	Fort Severn Can.
76 B2	Fort-Shevchenko Kazakh.
128 B1	Fort Simpson Can.
128 C1	Fort Smith Can.
140 B1	Fort Smith U.S.A.
143 C2	Fort Stockton U.S.A.
142 C2	Fort Sumner U.S.A.
	Fort Trinquet Maur. see Bîr Mogreïn
134 B2	Fortuna U.S.A.
131 E3	Fortune Bay Can.
128 C2	Fort Vermilion Can.
	Fort Victoria Zimbabwe see Masvingo
	Fort Walton Beach
140 C2	Fort Walton Beach U.S.A.
136 B2	Fort Washakie U.S.A.
138 B2	Fort Wayne U.S.A.
96 B2	Fort William U.K.
143 D2	Fort Worth U.S.A.
126 C2	Fort Yukon U.S.A.
71 B3	Foshan China
92 F3	Fosna pen. Norway
93 E3	Fosnavåg Norway
92 □B3	Foss Iceland
92 □A2	Fossá Iceland
108 A2	Fossano Italy
137 D1	Fosston U.S.A.
53 C3	Foster Austr.
118 B3	Fougamou Gabon
104 B2	Fougères France
96 □	Foula i. U.K.
99 D4	Foulness Point U.K.
118 B2	Foumban Cameroon
111 C3	Fournoi i. Greece
114 A3	Fouta Djallon reg. Guinea
54 A3	Foveaux Strait N.Z.
136 C3	Fowler U.S.A.
50 C3	Fowlers Bay Austr.
81 C2	Fowman Iran
128 C2	Fox Creek Can.
127 G2	Foxe Basin g. Can.
127 G2	Foxe Channel Can.
127 G2	Foxe Peninsula Can.
54 B2	Fox Glacier N.Z.
128 C2	Fox Lake Can.
128 A1	Fox Mountain Can.
54 C2	Foxton N.Z.
129 D2	Fox Valley Can.
97 C1	Foyle r. Ireland/U.K.
97 C1	Foyle, Lough b. Ireland/U.K.
97 B2	Foynes Ireland
154 B3	Foz de Areia, Represa de resr Brazil
120 A2	Foz do Cunene Angola
154 B3	Foz do Iguaçu Brazil
107 D1	Fraga Spain
100 A2	Frameries Belgium
154 C2	Franca Brazil
109 C2	Francavilla Fontana Italy
104 C2	France country Europe
118 B3	Franceville Gabon
137 D2	Francis Case, Lake U.S.A.
155 D1	Francisco Sá Brazil
120 B3	Francistown Botswana
128 B2	François Lake Can.
100 B1	Franeker Neth.
101 D2	Frankenberg (Eder) Ger.
101 D3	Frankenthal (Pfalz) Ger.
101 E2	Frankenwald mts Ger.
138 C3	Frankfort KY U.S.A.
138 B2	Frankfort MI U.S.A.
	Frankfurt Ger. see Frankfurt am Main
101 D2	Frankfurt am Main Ger.
102 C1	Frankfurt an der Oder Ger.
102 C2	Fränkische Alb hills Ger.
101 E3	Fränkische Schweiz reg. Ger.
139 E2	Franklin NH U.S.A.
139 D2	Franklin PA U.S.A.
140 C1	Franklin TN U.S.A.
126 D2	Franklin Bay Can.
134 C1	Franklin D. Roosevelt Lake U.S.A.
128 B1	Franklin Mountains Can.
126 F2	Franklin Strait Can.
53 C3	Frankston Austr.
82 E1	Frantsa-Iosifa, Zemlya is Rus. Fed.
54 B2	Franz Josef Glacier N.Z.
	Franz Josef Land is Rus. Fed. see Frantsa-Iosifa, Zemlya
108 A3	Frasca, Capo della c. Italy
128 B3	Fraser r. B.C. Can.
131 D2	Fraser r. Nfld. and Lab. Can.
122 B3	Fraserburg S. Africa
96 C2	Fraserburgh U.K.
130 B3	Fraserdale Can.
51 E2	Fraser Island Austr.
128 B2	Fraser Lake Can.
128 B2	Fraser Plateau Can.
153 C3	Fray Bentos Uru.
93 E4	Fredericia Denmark
143 D2	Frederick U.S.A.
143 D2	Fredericksburg TX U.S.A.
139 D3	Fredericksburg VA U.S.A.
128 A2	Frederick Sound sea chan. U.S.A.
131 D3	Fredericton Can.
	Frederikshåb Greenland see Paamiut
93 F4	Frederikshavn Denmark
	Frederikshamn Fin. see Hamina
93 F4	Fredrikstad Norway
138 B2	Freeport IL U.S.A.
143 D3	Freeport TX U.S.A.
146 C2	Freeport City Bahamas
143 D3	Freer U.S.A.
123 C2	Free State prov. S. Africa
114 A4	Freetown Sierra Leone
106 B2	Fregenal de la Sierra Spain
104 B2	Fréhel, Cap c. France
102 B2	Freiburg im Breisgau Ger.
102 C2	Freising Ger.
102 C2	Freistadt Austria
105 D3	Fréjus France
50 A3	Fremantle Austr.
135 B3	Fremont CA U.S.A.
137 D2	Fremont NE U.S.A.
138 C2	Fremont OH U.S.A.
	French Guiana country Africa see Congo
151 D2	French Guiana terr. S. America
	French Guinea country Africa see Guinea
134 E1	Frenchman r. U.S.A.
49 M5	French Polynesia terr. S. Pacific Ocean
	French Somaliland country Africa see Djibouti
	French Sudan country Africa see Mali
	French Territory of the Afars and Issas country Africa see Djibouti
151 D3	Fresco r. Brazil
144 B2	Fresnillo Mex.
135 C3	Fresno U.S.A.
107 D2	Freu, Cap des c. Spain
105 D2	Freyming-Merlebach France
114 A3	Fria Guinea
152 B2	Frías Arg.
102 B2	Friedberg (Hessen) Ger.
102 B2	Friedrichshafen Ger.
101 F1	Friesack Ger.
100 C1	Friesoythe Ger.
143 C2	Friona U.S.A.
136 B3	Frisco U.S.A.
127 H2	Frobisher Bay b. Can. see Iqaluit
101 F2	Frohburg Ger.
87 D4	Frolovo Rus. Fed.
103 D1	Frombork Pol.
99 B4	Frome U.K.
52 A2	Frome, Lake imp. l. Austr.
52 A2	Frome Downs Austr.
100 C2	Fröndenberg Ger.
143 C3	Frontera Mex.
145 C3	Frontera Mex.
144 B1	Fronteras Mex.
139 D3	Front Royal U.S.A.
108 B2	Frosinone Italy
92 E3	Frøya i. Norway
	Frunze Kyrg. see Bishkek
154 C2	Frutal Brazil
105 D2	Frutigen Switz.
103 D2	Frýdek-Místek Czech Rep.
71 B3	Fu'an China
71 C3	Fuding China
106 C1	Fuenlabrada Spain
152 C2	Fuerte Olimpo Para.
114 A2	Fuerteventura i. Islas Canarias
64 B2	Fuga i. Phil.
79 C2	Fujairah U.A.E.
67 C3	Fuji Japan
71 B3	Fujian prov. China
67 C3	Fujinomiya Japan
67 C3	Fuji-san vol. Japan
	Fukien prov. China see Fujian
67 C3	Fukui Japan
67 B4	Fukuoka Japan
67 D3	Fukushima Japan
101 D2	Fulda Ger.
101 D2	Fulda r. Ger.
70 A3	Fuling China
137 E3	Fulton U.S.A.
105 C2	Fumay France
49 I4	Funafuti atoll Tuvalu
114 A1	Funchal Arquipélago da Madeira
155 D1	Fundão Brazil
106 B1	Fundão Port.
131 D3	Fundy, Bay of g. Can.
121 C3	Funhalouro Moz.
70 B2	Funing Jiangsu China
71 A3	Funing Yunnan China
115 C3	Funtua Nigeria
79 C2	Fürgun, Küh-e mt. Iran
89 F2	Furmanov Rus. Fed.
	Furmanovka Kazakh. see Moyynkum
	Furmanovo Kazakh. see Zhalpaktal
155 C2	Furnas, Represa resr Brazil
51 D4	Furneaux Group is Austr.
	Furong China see Wan'an
100 C1	Fürstenau Ger.
101 E3	Fürth Ger.
127 G2	Fury and Hecla Strait Can.
70 C1	Fushun China
65 B1	Fusong China
79 C2	Fuwayrit Qatar
	Fuxian China see Wafangdian
70 A2	Fuxian China
70 C1	Fuxin China
70 B2	Fuyang China
69 E1	Fuyu China
	Fuyu China see Songyuan
68 B1	Fuyun China
71 B3	Fuzhou China
71 B3	Fuzhou China
93 F4	Fyn i. Denmark
96 B3	Fyne, Loch inlet U.K.
	F.Y.R.O.M. country Europe see Macedonia

G

117 C4	Gaalkacyo Somalia
120 A2	Gabela Angola
	Gaberones Botswana see Gaborone
115 D1	Gabès Tunisia
115 D1	Gabès, Golfe de g. Tunisia
118 B3	Gabon country Africa
123 C1	Gaborone Botswana
79 C2	Gābrīk Iran
110 C2	Gabrovo Bulg.
114 A3	Gabú Guinea-Bissau
73 B3	Gadag India
75 C2	Gadchiroli India
101 E1	Gadebusch Ger.
140 C2	Gadsden U.S.A.
118 B2	Gadzi C.A.R.
110 C2	Găești Romania
108 B2	Gaeta Italy
108 B2	Gaeta, Golfo di g. Italy
141 D1	Gaffney U.S.A.
115 C1	Gafsa Tunisia
89 E2	Gagarin Rus. Fed.
114 B4	Gagnoa Côte d'Ivoire
131 D2	Gagnon Can.
	Gago Coutinho Angola see Lumbala N'guimbo
81 C1	Gagra Georgia
122 A2	Gaiab watercourse Namibia
111 C3	Gaïdouronisi i. Greece
104 C3	Gaillac France
	Gaillimh Ireland see Galway
141 D3	Gainesville FL U.S.A.
141 D2	Gainesville GA U.S.A.
143 D2	Gainesville TX U.S.A.
98 C3	Gainsborough U.K.
52 A2	Gairdner, Lake imp. l. Aus[t]
96 B2	Gairloch U.K.
122 B2	Gakarosa mt. S. Africa
119 E3	Galana r. Kenya
103 D2	Galanta Slovakia
	Galápagos, Islas is Ecuado[r] see Galapagos Islands
125 I10	Galapagos Islands is Ecuador
157 G6	Galapagos Rise Pacific Ocean
96 C3	Galashiels U.K.
110 C1	Galaţi Romania
93 E3	Galdhøpiggen mt. Norway
145 B2	Galeana Mex.
128 C2	Galena Bay Can.
138 A2	Galesburg U.S.A.
122 B2	Galeshewe S. Africa
89 F2	Galich Rus. Fed.
106 B1	Galicia aut. comm. Spain
80 B2	Galilee, Sea of l. Israel
78 A3	Gallabat Sudan
140 C1	Gallatin U.S.A.
73 C4	Galle Sri Lanka
157 G6	Gallego Rise Pacific Ocean
150 B1	Gallinas, Punta pt Col.
109 C2	Gallipoli Italy
111 C2	Gallipoli Turkey
92 H2	Gällivare Sweden
142 B1	Gallup U.S.A.
117 C4	Galmudug reg. Somalia
114 A2	Galtat-Zemmour Western Sahara
97 B2	Galtymore mt. Ireland
143 E3	Galveston U.S.A.
143 E3	Galveston Bay U.S.A.
143 E3	Galveston Island U.S.A.
97 B2	Galway Ireland
97 B2	Galway Bay Ireland
154 C1	Gamá Brazil
123 D3	Gamalakhe S. Africa
117 B4	Gambēla Eth.
114 A3	Gambia r. Gambia
114 A3	Gambia, The country Africa
49 N6	Gambier, Îles is Fr. Polynesia
52 A3	Gambier Islands Austr.
131 E3	Gambo Can.
118 B3	Gamboma Congo
118 B3	Gamboma Congo
92 H2	Gammelstaden Sweden
142 B1	Ganado U.S.A.
123 C2	Ga-Nala S. Africa
81 C1	Gäncä Azer.
61 C2	Gandadiwata, Bukit mt. Indon.
118 C3	Gandajika Dem. Rep. Cong[o]
106 B1	Gándara Spain
131 E3	Gander Can.
131 E3	Gander r. Can.
101 D1	Ganderkesee Ger.
107 D1	Gandesa Spain
74 B2	Gandhidham India
74 B2	Gandhinagar India
74 B2	Gandhi Sagar resr India
107 C2	Gandia Spain
	Ganga r. Bangl./India see Ganges
153 B4	Gangán Arg.
74 B2	Ganganagar India
62 A1	Gangaw Myanmar
68 C2	Gangca China
75 C1	Gangdisê Shan mts China
75 D2	Ganges r. Bangl./India
105 C3	Ganges France
75 C2	Ganges, Mouths of the Bangl./India
159 E2	Ganges Cone Indian Ocea[n]
65 B2	Gangneung S. Korea
75 C2	Gangtok India
75 C3	Ganjam India
71 B3	Gan Jiang r. China
71 A3	Ganluo China
105 C2	Gannat France
136 B2	Gannett Peak U.S.A.
122 A3	Gansbaai S. Africa
70 A1	Gansu prov. China
115 D4	Ganye Nigeria
71 B3	Ganzhou China
114 B3	Gao Mali
	Gaoleshan China see Xianfeng
97 B1	Gaoth Dobhair Ireland
114 B3	Gaoua Burkina Faso
114 A3	Gaoual Guinea
71 C3	Gaoxiong Taiwan
70 B2	Gaoyou China

70 B2 Gaoyou Hu l. China
105 D3 Gap France
64 B2 Gapan Phil.
75 C1 Gar China
97 B2 Gara, Lough l. Ireland
76 C3 Garabil Belentligi hills Turkm.
76 B2 Garabogaz Turkm.
76 B2 Garabogazköl Turkm.
76 B2 Garabogazköl Aýlagy b. Turkm.
117 C4 Garacad Somalia
53 C1 Garah Austr.
123 C2 Ga-Rankuwa S. Africa
118 C1 Garar, Plaine de plain Chad
117 C4 Garbaharrey Somalia
135 B2 Garberville U.S.A.
101 D1 Garbsen Ger.
154 C2 Garça Brazil
154 B2 Garcias Brazil
108 B1 Garda, Lake l. Italy
108 A3 Garde, Cap de c. Alg.
101 E1 Gardelegen Ger.
136 C3 Garden City U.S.A.
129 E2 Garden Hill Can.
77 C3 Gardēz Afgh.
139 F2 Gardiner U.S.A.
Gardner atoll Micronesia see Faraulep
135 C2 Gardnerville U.S.A.
136 B3 Garfield U.S.A.
88 B2 Gargždai Lith.
123 C3 Gariep Dam dam S. Africa
122 A3 Garies S. Africa
119 D3 Garissa Kenya
88 B2 Garkalne Latvia
143 D2 Garland U.S.A.
102 C2 Garmisch-Partenkirchen Ger.
52 B2 Garnpung Lake imp. l. Austr.
104 B3 Garonne r. France
117 C4 Garoowe Somalia
74 B2 Garoth India
118 B2 Garoua Cameroon
118 B2 Garoua Boulaï Cameroon
Garqêntang China see Sog
96 B2 Garry r. U.K.
126 F2 Garry Lake Can.
119 E3 Garsen Kenya
76 B2 Garşy Turkm.
122 A2 Garub Namibia
60 B2 Garut Indon.
138 B2 Gary U.S.A.
145 B2 Garza García Mex.
68 C2 Garzê China
Gascogne reg. France see Gascony
Gascony, Golfe de g. France see Gascony, Gulf of
104 B3 Gascony reg. France
104 B3 Gascony, Gulf of France
50 A2 Gascoyne r. Austr.
118 B2 Gashaka Nigeria
115 D3 Gashua Nigeria
59 E3 Gasmata P.N.G.
131 D3 Gaspé Can.
131 D3 Gaspésie, Péninsule de la pen. Can.
141 E1 Gaston, Lake U.S.A.
141 D1 Gastonia U.S.A.
107 C2 Gata, Cabo de c. Spain
88 D2 Gatchina Rus. Fed.
98 C2 Gateshead U.K.
143 D2 Gatesville U.S.A.
139 D1 Gatineau Can.
130 C3 Gatineau r. Can.
Gatooma Zimbabwe see Kadoma
53 D1 Gatton Austr.
129 E2 Gauer Lake Can.
93 E4 Gausta mt. Norway
Gauteng prov. S. Africa
79 C2 Gāvbandī Iran
111 B3 Gavdos i. Greece
93 G3 Gävle Sweden
93 G3 Gävlebukten b. Sweden
89 F2 Gavrilov Posad Rus. Fed.
89 E2 Gavrilov-Yam Rus. Fed.
122 A2 Gawachab Namibia
62 A1 Gawai Myanmar
52 A2 Gawler Austr.
52 A2 Gawler Ranges hills Austr.
75 C2 Gaya India
114 C3 Gaya Niger
114 C3 Gayéri Burkina Faso
138 C1 Gaylord U.S.A.
86 E2 Gayny Rus. Fed.
116 B1 Gaza terr. Asia

80 B2 Gaza Gaza
80 B2 Gaziantep Turkey
76 C2 Gazojak Turkm.
114 B4 Gbarnga Liberia
118 A2 Gboko Nigeria
103 D1 Gdańsk Pol.
88 A3 Gdańsk, Gulf of Pol./Rus. Fed.
88 C2 Gdov Rus. Fed.
103 D1 Gdynia Pol.
116 B3 Gedaref Sudan
101 D2 Gedern Ger.
111 C3 Gediz Turkey
111 C3 Gediz r. Turkey
102 C1 Gedser Denmark
100 B2 Geel Belgium
52 B3 Geelong Austr.
101 E1 Geesthacht Ger.
75 C1 Gê'gyai China
129 D2 Geikie r. Can.
93 E3 Geilo Norway
119 D3 Geita Tanz.
109 C3 Gela Italy
108 B3 Gela, Golfo di g. Italy
91 D3 Gelendzhik Rus. Fed.
Gelibolu Turkey see Gallipoli
100 C2 Gelsenkirchen Ger.
118 B2 Gemena Dem. Rep. Congo
111 C2 Gemlik Turkey
108 B1 Gemona del Friuli Italy
117 C4 Genalē Wenz r. Eth.
81 D3 Genāveh Iran
153 B5 General Acha Arg.
153 B3 General Alvear Arg.
153 C3 General Belgrano Arg.
144 B2 General Cepeda Mex.
General Freire Angola see Muxaluando
General Machado Angola see Camacupa
153 B3 General Pico Arg.
153 B3 General Roca Arg.
154 B2 General Salgado Brazil
64 B3 General Santos Phil.
139 D2 Genesee r. U.S.A.
138 A2 Geneseo IL U.S.A.
139 D2 Geneseo NY U.S.A.
105 D2 Geneva Switz.
139 D2 Geneva U.S.A.
105 D2 Geneva, Lake l. France/Switz.
Genève Switz. see Geneva
106 B2 Genil r. Spain
100 B2 Genk Belgium
53 C3 Genoa Austr.
108 A2 Genoa Italy
108 A2 Genoa, Gulf of g. Italy
Genova Italy see Genoa
Gent Belgium see Ghent
61 C2 Genteng i. Indon.
101 F1 Genthin Ger.
50 A3 Geographe Bay Austr.
65 B2 Geongju S. Korea
131 D2 George r. Can.
122 B3 George S. Africa
52 A3 George, Lake salt flat Austr.
141 D3 George, Lake FL U.S.A.
139 E2 George, Lake NY U.S.A.
146 C2 George Town Bahamas
151 D2 Georgetown Guyana
60 B1 George Town Malaysia
138 C3 Georgetown KY U.S.A.
141 E2 Georgetown SC U.S.A.
143 D2 Georgetown TX U.S.A.
55 L2 George V Land reg. Antarctica
81 C1 Georgia country Asia
141 D2 Georgia state U.S.A.
130 B3 Georgian Bay Can.
51 C2 Georgina watercourse Austr.
Georgiu-Dezh Rus. Fed. see Liski
77 E2 Georgiyevka Kazakh.
87 D4 Georgiyevsk Rus. Fed.
101 F2 Gera Ger.
151 E4 Geral de Goiás, Serra hills Brazil
54 B2 Geraldine N.Z.
50 A2 Geraldton Austr.
80 B1 Gerede Turkey
102 C2 Geretsried Ger.
135 C2 Gerlach U.S.A.
103 E2 Gerlachovský štít mt. Slovakia
139 D3 Germantown U.S.A.
102 C1 Germany country Europe
100 C2 Gerolstein Ger.
101 E3 Gerolzhofen Ger.

53 D2 Gerringong Austr.
101 D2 Gersfeld (Rhön) Ger.
Géryville Alg. see El Bayadh
75 C1 Gêrzê China
117 C4 Gestro Wenz, Wabē r. Eth.
106 C1 Getafe Spain
139 D3 Gettysburg PA U.S.A.
136 D1 Gettysburg SD U.S.A.
55 P2 Getz Ice Shelf Antarctica
111 B2 Gevgelija Macedonia
111 C3 Geyikli Turkey
122 B2 Ghaap Plateau S. Africa
Ghadamés Libya see Ghadāmis
115 C1 Ghadāmis Libya
75 C2 Ghaghara r. India
114 B4 Ghana country Africa
120 B3 Ghanzi Botswana
115 C1 Ghardaïa Alg.
78 A2 Ghārib, Jabal mt. Egypt
115 D2 Gharyān Libya
115 D2 Ghāt Libya
75 C2 Ghatal India
75 B2 Ghaziabad India
75 C2 Ghazipur India
77 C3 Ghaznī Afgh.
78 B2 Ghazzālah Saudi Arabia
100 A2 Ghent Belgium
Gheorghe Gheorghiu-Dej Romania see Oneşti
110 C1 Gheorgheni Romania
110 B1 Gherla Romania
105 D3 Ghisonaccia France
74 B2 Ghotaru India
74 A2 Ghotki Pak.
75 C2 Ghugri r. India
76 C3 Ghūrīān Afgh.
97 C1 Giant's Causeway lava field U.K.
61 C2 Gianyar Indon.
109 C3 Giarre Italy
108 A1 Giaveno Italy
122 A2 Gibeon Namibia
106 B2 Gibraltar Gibraltar
106 B2 Gibraltar, Strait of Morocco/Spain
140 B2 Gibsland U.S.A.
50 B2 Gibson Desert Austr.
68 C1 Gichgeniyn Nuruu mts Mongolia
117 B4 Gīdolē Eth.
105 C2 Gien France
101 D2 Gießen Ger.
101 E1 Gifhorn Ger.
128 C2 Gift Lake Can.
67 C3 Gifu Japan
96 B3 Gigha i. U.K.
76 C2 G'ijduvon Uzbek.
106 B1 Gijón Spain
142 A2 Gila r. U.S.A.
142 A2 Gila Bend U.S.A.
51 D1 Gilbert r. Austr.
48 I4 Gilbert Islands is Kiribati
156 D5 Gilbert Ridge Pacific Ocean
151 E3 Gilbués Brazil
134 D1 Gildford U.S.A.
Gilf Kebir Plateau plat. Egypt see Haḍabat al Jilf al Kabīr
53 C2 Gilgandra Austr.
74 B1 Gilgit Pak.
74 B1 Gilgit r. Pak.
53 C2 Gilgunnia Austr.
129 E2 Gillam Can.
136 B2 Gillette U.S.A.
99 D4 Gillingham U.K.
130 C2 Gilmour Island Can.
135 B3 Gilroy U.S.A.
65 B2 Gimcheon S. Korea
65 B2 Gimhae S. Korea
129 E2 Gimli Can.
64 B3 Gingoog Phil.
117 C4 Gīnīr Eth.
109 C2 Ginosa Italy
109 C2 Gioia del Colle Italy
53 C3 Gippsland reg. Austr.
74 A2 Girdar Dhor r. Pak.
79 D1 Gīrdī Iran
80 B1 Giresun Turkey
Girgenti Italy see Agrigento
53 C2 Girilambone Austr.
76 C3 Girishk Afgh.
Giron Sweden see Kiruna
107 D1 Girona Spain
104 B2 Gironde est. France
53 C2 Girral Austr.
96 B3 Girvan U.K.
54 C1 Gisborne N.Z.
93 F4 Gislaved Sweden

119 C3 Gitarama Rwanda
119 C3 Gitega Burundi
Giuba r. Somalia see Jubba
108 B2 Giulianova Italy
110 C2 Giurgiu Romania
110 C1 Giuvala, Pasul pass Romania
105 C2 Givors France
123 D1 Giyani S. Africa
119 D2 Giyon Eth.
116 B2 Giza Egypt
103 E1 Giżycko Pol.
109 D2 Gjakovë Kosovo
109 D2 Gjeravicë mt. Kosovo
109 D2 Gjilan Kosovo
109 D2 Gjirokastër Albania
126 F2 Gjoa Haven Can.
93 F3 Gjøvik Norway
131 E3 Glace Bay Can.
134 B1 Glacier Peak vol. U.S.A.
143 E2 Gladewater U.S.A.
51 E2 Gladstone Qld Austr.
52 A2 Gladstone S.A. Austr.
92 □A2 Gláma mts Iceland
109 C2 Glamoč Bos.-Herz.
100 C3 Glan r. Ger.
97 B2 Glanaruddery Mountains hills Ireland
96 B3 Glasgow U.K.
138 B3 Glasgow KY U.S.A.
136 B1 Glasgow MT U.S.A.
99 B4 Glastonbury U.K.
101 F2 Glauchau Ger.
86 E3 Glazov Rus. Fed.
89 E3 Glazunovka Rus. Fed.
123 D2 Glencoe S. Africa
96 B2 Glen Coe val. U.K.
142 A2 Glendale AZ U.S.A.
138 B2 Glendale WI U.S.A.
53 D2 Glen Davis Austr.
51 D2 Glenden Austr.
136 C1 Glendive U.S.A.
52 B3 Glenelg r. Austr.
96 B2 Glenfinnan U.K.
53 D1 Glen Innes Austr.
96 B2 Glen More val. U.K.
53 C1 Glenmorgan Austr.
126 C2 Glennallen U.S.A.
134 C2 Glenns Ferry U.S.A.
136 B2 Glenrock U.S.A.
96 C2 Glenrothes U.K.
139 E2 Glens Falls U.S.A.
96 C2 Glen Shee val. U.K.
97 B1 Glenties Ireland
142 B2 Glenwood U.S.A.
136 B3 Glenwood Springs U.S.A.
101 E1 Glinde Ger.
103 D1 Gliwice Pol.
142 A2 Globe U.S.A.
103 D1 Głogów Pol.
92 F2 Glomfjord Norway
93 F4 Glomma r. Norway
53 D2 Gloucester Austr.
99 B4 Gloucester U.K.
131 E3 Glovertown Can.
101 F1 Glöwen Ger.
77 E1 Glubokoye Kazakh.
101 D1 Glückstadt Ger.
103 C2 Gmünd Austria
102 C2 Gmunden Austria
101 D1 Gnarrenburg Ger.
103 D1 Gniezno Pol.
Gnjilane Kosovo see Gjilan
75 D2 Goalpara India
96 B3 Goat Fell h. U.K.
117 C4 Goba Eth.
122 A1 Gobabis Namibia
153 A4 Gobernador Gregores Arg.
152 C2 Gobernador Virasoro Arg.
68 D2 Gobi des. China/Mongolia
67 C4 Gobō Japan
100 C2 Goch Ger.
122 A1 Gochas Namibia
74 B3 Godavari r. India
73 C3 Godavari, Mouths of the India
75 C2 Godda India
117 C4 Godē Eth.
130 B3 Goderich Can.
Godhavn Greenland see Qeqertarsuaq
74 B2 Godhra India
129 E2 Gods r. Can.
129 E2 Gods Lake Can.
Godthåb Greenland see Nuuk
Godwin-Austen, Mount mt. China/Pakistan see K2
Goedgegun Swaziland see Nhlangano

130 C3 **Goéland, Lac au** *l.* Can.
131 D2 **Goélands, Lac aux** *l.* Can.
100 A2 **Goes** Neth.
138 B1 **Gogebic Range** *hills* U.S.A.
88 C1 **Gogland, Ostrov** *i.* Rus. Fed.
Gogra *r.* India *see* **Ghaghara**
119 C2 **Gogrial** S. Sudan
154 C1 **Goiandira** Brazil
154 C1 **Goianésia** Brazil
154 C1 **Goiânia** Brazil
154 B1 **Goiás** Brazil
154 B1 **Goiás** *state* Brazil
154 C1 **Goiatuba** Brazil
154 B2 **Goio-Erê** Brazil
111 C2 **Gökçeada** *i.* Turkey
111 C3 **Gökçedağ** Turkey
121 B2 **Gokwe** Zimbabwe
93 E3 **Gol** Norway
62 A1 **Golaghat** India
111 C2 **Gölcük** Turkey
103 E1 **Gołdap** Pol.
101 F1 **Goldberg** Ger.
Gold Coast *country* Africa *see* **Ghana**
53 D1 **Gold Coast** Austr.
114 B4 **Gold Coast** *coastal area* Ghana
128 C2 **Golden** Can.
54 B2 **Golden Bay** N.Z.
134 B1 **Goldendale** U.S.A.
128 B3 **Golden Hinde** *mt.* Can.
97 B2 **Golden Vale** *lowland* Ireland
135 C3 **Goldfield** U.S.A.
128 B3 **Gold River** Can.
141 E1 **Goldsboro** U.S.A.
103 C1 **Goleniów** Pol.
135 C4 **Goleta** U.S.A.
Golfe du St-Laurent *g.* Can. *see* **St Lawrence, Gulf of**
111 C3 **Gölhisar** Turkey
Gollel Swaziland *see* **Lavumisa**
75 D1 **Golmud** China
81 D2 **Golpāyegān** Iran
96 C2 **Golspie** U.K.
Golyshi Rus. Fed. *see* **Vetluzhskiy**
119 C3 **Goma** Dem. Rep. Congo
75 C2 **Gomati** *r.* India
115 D3 **Gombe** Nigeria
115 D3 **Gombi** Nigeria
Gomel' Belarus *see* **Homyel'**
144 B2 **Gómez Palacio** Mex.
81 D2 **Gomīshān** Iran
75 C1 **Gomo** China
147 C3 **Gonaïves** Haiti
147 C3 **Gonâve, Île de la** *i.* Haiti
81 D2 **Gonbad-e Kāvūs** Iran
74 B2 **Gondal** India
Gondar Eth. *see* **Gonder**
116 B3 **Gonder** Eth.
75 C2 **Gondia** India
111 C2 **Gönen** Turkey
62 A1 **Gonggar** China
70 A3 **Gongga Shan** *mt.* China
68 C2 **Gonghe** China
65 B2 **Gongju** S. Korea
115 D4 **Gongola** *r.* Nigeria
53 C2 **Gongolgon** Austr.
75 D1 **Gongtang** China
123 C3 **Gonubie** S. Africa
145 C2 **Gonzáles** Mex.
143 D3 **Gonzales** U.S.A.
122 A3 **Good Hope, Cape of** S. Africa
134 D2 **Gooding** U.S.A.
136 C3 **Goodland** U.S.A.
53 C1 **Goodooga** Austr.
98 C3 **Goole** U.K.
53 C2 **Goolgowi** Austr.
52 A3 **Goolwa** Austr.
53 D1 **Goondiwindi** Austr.
134 B2 **Goose Lake** U.S.A.
102 B2 **Göppingen** Ger.
75 C2 **Gorakhpur** India
109 C2 **Goražde** Bos.-Herz.
111 C3 **Gördes** Turkey
89 D3 **Gordeyevka** Rus. Fed.
136 C2 **Gordon** U.S.A.
51 D4 **Gordon, Lake** Austr.
115 D4 **Goré** Chad
117 B4 **Gorē** Eth.
54 A3 **Gore** N.Z.
97 C2 **Gorey** Ireland
81 D2 **Gorgān** Iran
81 C1 **Gori** Georgia
100 B2 **Gorinchem** Neth.
108 B1 **Gorizia** Italy
Gor'kiy Rus. Fed. *see* **Nizhniy Novgorod**
89 F2 **Gor'kovskoye Vodokhranilishche** *resr* Rus. Fed.
103 E2 **Gorlice** Pol.
103 C1 **Görlitz** Ger.
Gorna Dzhumaya Bulg. *see* **Blagoevgrad**
109 D2 **Gornji Milanovac** Serbia
109 C2 **Gornji Vakuf** Bos.-Herz.
77 E1 **Gorno-Altaysk** Rus. Fed.
86 F2 **Gornopravdinsk** Rus. Fed.
110 C2 **Gornotrakiyska Nizina** *lowland* Bulg.
66 D1 **Gornozavodsk** Rus. Fed.
77 E1 **Gornyak** Rus. Fed.
59 D3 **Goroka** P.N.G.
52 B3 **Goroke** Austr.
114 B3 **Gorom Gorom** Burkina Faso
121 C2 **Gorongosa** Moz.
61 D1 **Gorontalo** Indon.
89 E3 **Gorshechnoye** Rus. Fed.
97 B2 **Gorumna Island** Ireland
91 D3 **Goryachiy Klyuch** Rus. Fed.
65 B2 **Goryeong** S. Korea
103 D1 **Gorzów Wielkopolski** Pol.
59 E3 **Goschen Strait** P.N.G.
53 D2 **Gosford** Austr.
74 A2 **Goshanak** Pak.
66 D2 **Goshogawara** Japan
101 E2 **Goslar** Ger.
109 C2 **Gospić** Croatia
99 C4 **Gosport** U.K.
111 B2 **Gostivar** Macedonia
Göteborg Sweden *see* **Gothenburg**
101 E2 **Gotha** Ger.
93 F4 **Gothenburg** Sweden
136 C2 **Gothenburg** U.S.A.
93 G4 **Gotland** *i.* Sweden
111 B2 **Gotse Delchev** Bulg.
93 G4 **Gotska Sandön** *i.* Sweden
67 B4 **Gōtsu** Japan
101 D2 **Göttingen** Ger.
128 B2 **Gott Peak** Can.
Gottwaldow Czech Rep. *see* **Zlín**
Gotval'd Ukr. *see* **Zmiyiv**
100 B1 **Gouda** Neth.
114 A3 **Goudiri** Senegal
115 D3 **Goudoumaria** Niger
158 E7 **Gough Island** S. Atlantic Ocean
130 C3 **Gouin, Réservoir** *resr* Can.
53 C2 **Goulburn** Austr.
53 D2 **Goulburn** *r. N.S.W.* Austr.
53 C3 **Goulburn** *r. Vic.* Austr.
114 B3 **Goundam** Mali
107 D2 **Gouraya** Alg.
114 B3 **Gourcy** Burkina Faso
104 C3 **Gourdon** France
115 D3 **Gouré** Niger
122 B3 **Gourits** *r.* S. Africa
114 B3 **Gourma-Rharous** Mali
53 C3 **Gourock Range** *mts* Austr.
155 D1 **Governador Valadares** Brazil
141 E3 **Governor's Harbour** Bahamas
68 C2 **Govĭ Altayn Nuruu** *mts* Mongolia
75 C2 **Govind Ballash Pant Sagar** *resr* India
99 A3 **Gower** *pen.* U.K.
152 C2 **Goya** Arg.
81 C1 **Göyçay** Azer.
80 B1 **Göynük** Turkey
115 E3 **Goz-Beïda** Chad
75 C1 **Gozha Co** *salt l.* China
122 B3 **Graaff-Reinet** S. Africa
122 A3 **Graafwater** S. Africa
101 E1 **Grabow** Ger.
109 C2 **Gračac** Croatia
87 E3 **Grachevka** Rus. Fed.
109 C2 **Gradačac** Bos.-Herz.
104 B3 **Gradignan** France
101 F2 **Gräfenhainichen** Ger.
53 D1 **Grafton** Austr.
137 D1 **Grafton** U.S.A.
143 D2 **Graham** U.S.A.
142 B2 **Graham, Mount** U.S.A.
Graham Bell Island *i.* Rus. Fed. *see* **Greem-Bell, Ostrov**
128 A2 **Graham Island** Can.
55 A3 **Graham Land** *pen.* Antarctica
123 C3 **Grahamstown** S. Africa
97 C2 **Graiguenamanagh** Ireland
151 E3 **Grajaú** Brazil
103 E1 **Grajewo** Pol.
111 B2 **Grammos** *mt.* Greece
96 B2 **Grampian Mountains** U.K.
146 B3 **Granada** Nic.
106 C2 **Granada** Spain
97 C2 **Granard** Ireland
139 E1 **Granby** Can.
114 A2 **Gran Canaria** *i.* Islas Canarias
152 B3 **Gran Chaco** *reg.* Arg./Para.
136 F2 **Grand** *r. MO* U.S.A.
136 C1 **Grand** *r. SD* U.S.A.
146 C2 **Grand Bahama** *i.* Bahamas
131 E3 **Grand Bank** Can.
158 D2 **Grand Banks of Newfoundland** N. Atlantic Ocean
Grand Canal *canal* China *see* **Jinghang Yunhe**
Grand Canary *i.* Islas Canarias *see* **Gran Canaria**
142 A1 **Grand Canyon** U.S.A.
142 A1 **Grand Canyon** *gorge* U.S.A.
146 B3 **Grand Cayman** *i.* Cayman Is
129 C2 **Grand Centre** Can.
134 C1 **Grand Coulee** U.S.A.
152 B1 **Grande** *r.* Bol.
154 B2 **Grande** *r.* Brazil
146 B3 **Grande** *r.* Nic.
153 B5 **Grande, Bahía** *b.* Arg.
155 D2 **Grande, Ilha** *i.* Brazil
150 C2 **Grande, Serra** *mt.* Brazil
128 C2 **Grande Cache** Can.
Grande Comore *i.* Comoros *see* **Ngazidja**
128 C2 **Grande Prairie** Can.
114 B1 **Grand Erg Occidental** *des.* Alg.
115 C2 **Grand Erg Oriental** *des.* Alg.
131 D3 **Grande-Rivière** Can.
152 B3 **Grandes, Salinas** *salt flat* Arg.
131 D3 **Grand Falls** Can.
131 E3 **Grand Falls-Windsor** Can.
128 C3 **Grand Forks** Can.
137 D1 **Grand Forks** U.S.A.
138 B2 **Grand Haven** U.S.A.
128 C1 **Grandin, Lac** *l.* Can.
137 D2 **Grand Island** U.S.A.
140 B3 **Grand Isle** U.S.A.
136 B3 **Grand Junction** U.S.A.
114 B4 **Grand-Lahou** Côte d'Ivoire
131 D3 **Grand Lake** *N.B.* Can.
131 E3 **Grand Lake** *Nfld. and Lab.* Can.
137 E1 **Grand Marais** U.S.A.
130 C3 **Grand-Mère** Can.
106 B2 **Grândola** Port.
129 E2 **Grand Rapids** Can.
138 B2 **Grand Rapids** *MI* U.S.A.
137 E1 **Grand Rapids** *MN* U.S.A.
136 A2 **Grand Teton** *mt.* U.S.A.
147 C2 **Grand Turk** Turks and Caicos Is
129 D2 **Grandview** Can.
134 C1 **Grandview** U.S.A.
134 C1 **Grangeville** U.S.A.
128 B2 **Granisle** Can.
137 D2 **Granite Falls** U.S.A.
134 E1 **Granite Peak** *MT* U.S.A.
134 C2 **Granite Peak** *NV* U.S.A.
108 B3 **Granitola, Capo** *c.* Italy
93 F4 **Gränna** Sweden
101 F1 **Gransee** Ger.
99 C3 **Grantham** U.K.
96 C2 **Grantown-on-Spey** U.K.
142 B1 **Grants** U.S.A.
134 B2 **Grants Pass** U.S.A.
104 B2 **Granville** France
129 D2 **Granville Lake** Can.
155 D1 **Grão Mogol** Brazil
123 D1 **Graskop** S. Africa
105 D3 **Grasse** France
50 B3 **Grass Patch** Austr.
107 D1 **Graus** Spain
92 F2 **Gravdal** Norway
104 B2 **Grave, Pointe de** *pt* France
129 D3 **Gravelbourg** Can.
130 C3 **Gravenhurst** Can.
53 D1 **Gravesend** Austr.
99 D4 **Gravesend** U.K.
105 D2 **Gray** France
141 D1 **Gray** U.S.A.
138 C2 **Grayling** U.S.A.
99 D4 **Grays** U.K.
103 D2 **Graz** Austria
146 C2 **Great Abaco** *i.* Bahamas
50 B3 **Great Australian Bight** *g.* Austr.
146 C2 **Great Bahama Bank** Bahamas
54 C1 **Great Barrier Island** N.Z.
51 D1 **Great Barrier Reef** Austr.
135 C3 **Great Basin** U.S.A.
128 C1 **Great Bear Lake** Can.
93 F4 **Great Belt** *sea chan.* Denmark
137 D3 **Great Bend** U.S.A.
95 C3 **Great Britain** *i.* Europe
63 A2 **Great Coco Island** Cocos
53 B3 **Great Dividing Range** *mts* Austr.
Great Eastern Erg *des.* Alg. *see* **Grand Erg Oriental**
146 B2 **Greater Antilles** *is* Caribbean Sea
Greater Khingan Mountains *mts* China *see* **Da Hinggan Ling**
58 A3 **Greater Sunda Islands** Indon.
134 D1 **Great Falls** U.S.A.
123 C3 **Great Fish** *r.* S. Africa
123 C3 **Great Fish Point** S. Africa
147 C2 **Great Inagua** *i.* Bahamas
122 B3 **Great Karoo** *plat.* S. Africa
123 C3 **Great Kei** *r.* S. Africa
99 B3 **Great Malvern** U.K.
122 A2 **Great Namaqualand** *reg.* Namibia
73 D4 **Great Nicobar** *i.* India
98 B3 **Great Ormes Head** *hd* U.K.
99 D3 **Great Ouse** *r.* U.K.
136 C2 **Great Plains** U.S.A.
119 D3 **Great Rift Valley** Africa
119 D3 **Great Ruaha** *r.* Tanz.
134 D2 **Great Salt Lake** U.S.A.
134 D2 **Great Salt Lake Desert** U.S.A.
116 A2 **Great Sand Sea** *des.* Egypt/Libya
50 B2 **Great Sandy Desert** Austr.
128 C1 **Great Slave Lake** Can.
141 D1 **Great Smoky Mountains** U.S.A.
99 A4 **Great Torrington** U.K.
50 B2 **Great Victoria Desert** Aus
70 B1 **Great Wall** *tourist site* China
Great Western Erg *des.* Al *see* **Grand Erg Occidental**
99 D3 **Great Yarmouth** U.K.
Grebenkovskiy Ukr. *see* **Hrebinka**
106 B1 **Gredos, Sierra de** *mts* Spain
111 B3 **Greece** *country* Europe
136 C2 **Greeley** U.S.A.
82 F1 **Greem-Bell, Ostrov** *i.* Rus. Fed.
138 F1 **Green** *r.* Can.
138 B3 **Green** *r. KY* U.S.A.
136 B3 **Green** *r. WY* U.S.A.
138 B2 **Green Bay** U.S.A.
138 B1 **Green Bay** *b.* U.S.A.
138 C3 **Greenbrier** *r.* U.S.A.
138 B3 **Greencastle** U.S.A.
141 D1 **Greeneville** U.S.A.
139 E2 **Greenfield** U.S.A.
129 D2 **Green Lake** Can.
127 J2 **Greenland** *terr.* N. Americ
160 L2 **Greenland Basin** Arctic Ocean
160 R2 **Greenland Sea** *sea* Greenland/Svalbard
96 B3 **Greenock** U.K.
97 C1 **Greenore** Ireland
135 D3 **Green River** *UT* U.S.A.
136 B2 **Green River** *WY* U.S.A.
138 B3 **Greensburg** *IN* U.S.A.
139 D2 **Greensburg** *PA* U.S.A.
141 E2 **Green Swamp** U.S.A.
142 A2 **Green Valley** U.S.A.
114 B4 **Greenville** Liberia
140 C2 **Greenville** *AL* U.S.A.
139 F1 **Greenville** *ME* U.S.A.
140 B2 **Greenville** *MS* U.S.A.
141 E1 **Greenville** *NC* U.S.A.
141 D2 **Greenville** *SC* U.S.A.
143 D2 **Greenville** *TX* U.S.A.
53 D2 **Greenwell Point** Austr.
131 D3 **Greenwood** Can.
140 B2 **Greenwood** *MS* U.S.A.
141 D2 **Greenwood** *SC* U.S.A.
50 B2 **Gregory, Lake** *imp. l.* Aust
51 D1 **Gregory Range** *hills* Austr.
102 C1 **Greifswald** Ger.
101 F2 **Greiz** Ger.
93 F4 **Grenaa** Denmark
140 C2 **Grenada** U.S.A.

147 D3 Grenada *country* West Indies
104 C3 Grenade France
93 F4 Grenen *spit* Denmark
53 C2 Grenfell Austr.
129 D2 Grenfell Can.
105 D2 Grenoble France
51 D1 Grenville, Cape Austr.
134 B1 Gresham U.S.A.
140 B3 Gretna *LA* U.S.A.
139 D3 Gretna *VA* U.S.A.
100 C1 Greven Ger.
111 B2 Grevena Greece
100 C2 Grevenbroich Ger.
101 E1 Grevesmühlen Ger.
136 B2 Greybull U.S.A.
128 A1 Grey Hunter Peak Can.
131 E2 Grey Islands Can.
54 B2 Greymouth N.Z.
52 B1 Grey Range *hills* Austr.
97 C2 Greystones Ireland
123 D2 Greytown S. Africa
91 E1 Gribanovskiy Rus. Fed.
118 B2 Gribingui *r.* C.A.R.
101 D3 Griesheim Ger.
141 D2 Griffin U.S.A.
53 C2 Griffith U.S.A.
88 C3 Grigiškės Lith.
118 C2 Grimari C.A.R.
101 F2 Grimma Ger.
102 C1 Grimmen Ger.
98 C3 Grimsby U.K.
128 C2 Grimshaw Can.
92 □B2 Grímsstaðir Iceland
93 E4 Grimstad Norway
92 □A3 Grindavík Iceland
93 E4 Grindsted Denmark
137 E2 Grinnell U.S.A.
123 C3 Griqualand East *reg.* S. Africa
122 B2 Griqualand West *reg.* S. Africa
127 G1 Grise Fiord Can.
Grishino Ukr. *see* Krasnoarmiys'k
99 D4 Gris Nez, Cap *c.* France
96 C1 Gritley U.K.
123 C2 Groblersdal S. Africa
122 B2 Groblershoop S. Africa
Grodno Belarus *see* Hrodna
103 D1 Grodzisk Wielkopolski Pol.
92 □A3 Gróf Iceland
104 B2 Groix, Île de *i.* France
100 C1 Gronau (Westfalen) Ger.
92 F3 Grong Norway
100 C1 Groningen Neth.
122 B3 Groot Brakrivier S. Africa
122 B2 Grootdrink S. Africa
51 C1 Groote Eylandt *i.* Austr.
120 A2 Grootfontein Namibia
122 A2 Groot Karas Berg *plat.* Namibia
122 B3 Groot Swartberge *mts* S. Africa
122 B2 Grootvloer *salt pan* S. Africa
123 C3 Groot Winterberg *mt.* S. Africa
101 D2 Großenlüder Ger.
101 E2 Großer Beerberg *h.* Ger.
102 C2 Großer Rachel *mt.* Ger.
103 C2 Grosser Speikkogel *mt.* Austria
108 B2 Grosseto Italy
101 D3 Groß-Gerau Ger.
102 C2 Großglockner *mt.* Austria
101 E2 Groß-Hesepe Ger.
101 E2 Großlohra Ger.
122 A1 Gross Ums Namibia
136 A2 Gros Ventre Range *mts* U.S.A.
131 E3 Groswater Bay Can.
130 B3 Groundhog *r.* Can.
135 B3 Grover Beach U.S.A.
140 B3 Groves U.S.A.
139 E2 Groveton U.S.A.
87 D4 Groznyy Rus. Fed.
109 C1 Grubišno Polje Croatia
Grudovo Bulg. *see* Sredets
103 D1 Grudziądz Pol.
122 A1 Grünau Namibia
92 □A3 Grundarfjörður Iceland
Gruzinskaya S.S.R. *country* Asia *see* Georgia
89 E3 Gryazi Rus. Fed.
89 F2 Gryazovets Rus. Fed.
103 D1 Gryfice Pol.
102 C1 Gryfino Pol.
153 E5 Grytviken S. Georgia

146 C2 Guacanayabo, Golfo de *b.* Cuba
144 B2 Guadalajara Mex.
106 C1 Guadalajara Spain
106 C1 Guadalajara *reg.* Spain
48 H4 Guadalcanal *i.* Solomon Is
107 C1 Guadalope *r.* Spain
106 B2 Guadalquivir *r.* Spain
132 B4 Guadalupe *i.* Mex.
106 B2 Guadalupe, Sierra de *mts* Spain
142 C2 Guadalupe Peak U.S.A.
144 B2 Guadalupe Victoria Mex.
144 B2 Guadalupe y Calvo Mex.
106 C1 Guadarrama, Sierra de *mts* Spain
147 D3 Guadeloupe *terr.* West Indies
147 D3 Guadeloupe Passage Caribbean Sea
106 B2 Guadiana *r.* Port./Spain
106 C2 Guadix Spain
154 B2 Guaíra Brazil
147 C3 Guajira, Península de la *pen.* Col.
150 B3 Gualaceo Ecuador
59 D2 Guam *terr.* N. Pacific Ocean
144 B2 Guamúchil Mex.
144 B2 Guanacevi Mex.
151 E4 Guanambi Brazil
150 C2 Guanare Venez.
146 B2 Guane Cuba
70 A2 Guang'an China
71 B3 Guangchang China
71 B3 Guangdong *prov.* China
Guanghua China *see* Laohekou
71 A3 Guangxi Zhuangzu Zizhiqu *aut. reg.* China
70 A2 Guangyuan China
71 B3 Guangzhou China
155 D1 Guanhães Brazil
147 D4 Guanipa *r.* Venez.
71 A3 Guanling China
65 A1 Guanshui China
Guansuo China *see* Guanling
147 C2 Guantánamo Cuba
155 C2 Guapé Brazil
150 C4 Guaporé *r.* Bol./Brazil
154 B3 Guarapari Brazil
154 B3 Guarapuava Brazil
155 C3 Guaraqueçaba Brazil
154 C3 Guaratinguetá Brazil
154 C3 Guaratuba Brazil
106 B1 Guarda Port.
Guardafui, Cape *c.* Somalia *see* Gwardafuy, Gees
154 C1 Guarda Mor Brazil
106 C1 Guardo Spain
155 C2 Guarujá Brazil
144 B2 Guasave Mex.
146 A3 Guatemala *country* Central America
Guatemala Guat. *see* Guatemala City
157 G5 Guatemala Basin N. Pacific Ocean
146 A3 Guatemala City Guat.
150 C2 Guaviare *r.* Col.
155 C2 Guaxupé Brazil
150 B3 Guayaquil Ecuador
150 A3 Guayaquil, Golfo de *g.* Ecuador
152 B1 Guayaramerín Bol.
144 A2 Guaymas Mex.
68 C2 Guazhou China
117 B3 Guba Eth.
86 E1 Guba Dolgaya Rus. Fed.
108 B2 Gubbio Italy
89 E3 Gubkin Rus. Fed.
73 C3 Gudivada India
105 D2 Guebwiller France
Guecho Spain *see* Algorta
114 C1 Guelb er Richât *h.* Maur.
114 A3 Guelma Alg.
115 C1 Guelma Alg.
114 A1 Guelmim Morocco
130 B3 Guelph Can.
145 C2 Guémez Mex.
104 C2 Guérande France
115 E2 Guerende Libya
95 C4 Guernsey *i.* Channel Is
95 C4 Guernsey *terr.* Channel Is
144 A2 Guerrero Negro Mex.
131 D2 Guers, Lac *l.* Can.
158 D4 Guiana Basin N. Atlantic Ocean
150 C2 Guiana Highlands *mts* Guyana/Venez.

Guichi China *see* Chizhou
118 B2 Guider Cameroon
108 B2 Guidonia-Montecelio Italy
71 A3 Guigang China
100 A3 Guignicourt France
123 D1 Guija Moz.
99 C4 Guildford U.K.
71 B3 Guilin China
130 C2 Guillaume-Delisle, Lac *l.* Can.
106 B1 Guimarães Port.
114 A3 Guinea *country* Africa
114 A3 Guinea, Gulf of Africa
114 A3 Guinea-Bissau *country* Africa
104 B2 Guingamp France
104 B2 Guipavas France
154 B1 Guiratinga Brazil
150 C1 Güiria Venez.
151 D2 Guisanbourg Fr. Guiana
98 C2 Guisborough U.K.
100 A3 Guise France
64 B2 Guiuan Phil.
71 A3 Guiyang China
71 A3 Guizhou *prov.* China
104 B3 Gujan-Mestras France
74 B2 Gujarat *state* India
Gujerat *state* India *see* Gujarat
74 B1 Gujranwala Pak.
74 B1 Gujrat Pak.
91 D2 Gukovo Rus. Fed.
74 A2 Gulang China
53 C2 Gulargambone Austr.
73 B3 Gulbarga India
88 C2 Gulbene Latvia
Gulf of California *g.* Mex. *see* California, Gulf of
Gulf of Chihli *g.* China *see* Bo Hai
111 B3 Gulf of Corinth *sea chan.* Greece
77 C2 Gulfport U.S.A.
79 C2 Gulf, The Asia
53 C2 Gulgong Austr.
69 E1 Gulian China
77 C2 Gulistan Uzbek.
Gulja China *see* Yining
75 C2 Gull Lake Can.
111 C3 Güllük Turkey
Gulü China *see* Xincai
119 D2 Gulu Uganda
120 B2 Gumare Botswana
76 B3 Gumdag Turkm.
65 B2 Gumel Nigeria
65 B2 Gumi S. Korea
75 C2 Gumla India
100 C2 Gummersbach Ger.
87 C4 Gümüşhane Turkey
74 B2 Guna India
53 C3 Gundagai Austr.
111 C3 Güney Turkey
118 B3 Gungu Dem. Rep. Congo
129 E2 Gunisao *r.* Can.
53 D2 Gunnedah Austr.
136 B3 Gunnison *CO* U.S.A.
135 D3 Gunnison *UT* U.S.A.
136 B3 Gunnison *r.* U.S.A.
73 B3 Gunsan S. Korea
73 B3 Guntakal India
73 B3 Guntur India
51 C1 Gununa Austr.
60 A1 Gunungsitoli Indon.
60 A1 Gunungtua Indon.
102 C2 Günzburg Ger.
102 C2 Gunzenhausen Ger.
70 B2 Guojiaba China
Guoluezhen China *see* Lingbao
74 B2 Gurgaon India
74 B2 Gurgueia *r.* Brazil
151 E3 Guri, Embalse de *resr* Venez.
154 C1 Gurinhatã Brazil
74 B2 Guro Moz.
151 E4 Gurupi Brazil
151 E3 Gurupi *r.* Brazil
74 B2 Guru Sikhar *mt.* India
121 C2 Guruve Zimbabwe
Gur'yev Kazakh. *see* Atyrau
115 C3 Gusau Nigeria
88 B3 Gusev Rus. Fed.
65 B2 Gushan China
70 B2 Gushi China
116 C3 Gusinoozersk Rus. Fed.
109 B2 Guspini Italy
128 A2 Gustavus U.S.A.
101 F1 Güstrow Ger.
101 D2 Gütersloh Ger.
143 D1 Guthrie U.S.A.
152 B1 Gutiérrez Bol.
121 C2 Gutu Mupandawana Zimbabwe
75 D2 Guwahati India
150 D2 Guyana *country* S. America
Guyi China *see* Sanjiang
143 C1 Guymon U.S.A.
53 D2 Guyra Austr.
70 A2 Guyuan China
Guzhou China *see* Rongjiang
144 B1 Guzmán Mex.
77 C3 G'uzor Uzbek.
88 B3 Gvardeysk Rus. Fed.
74 A2 Gwadar Pak.
Gwadar Pak. *see* Gwadar
128 A2 Gwaii Haanas Can.
121 B3 Gwalior India
121 B3 Gwanda Zimbabwe
65 B2 Gwangju S. Korea
117 D3 Gwardafuy, Gees *c.* Somalia
97 B1 Gweebarra Bay Ireland
Gweedore Ireland *see* Gaoth Dobhair
Gwelo Zimbabwe *see* Gweru
121 B2 Gweru Zimbabwe
115 D3 Gwoza Nigeria
53 D2 Gwydir *r.* Austr.
Gya'gya China *see* Saga
Gyandzha Azer. *see* Gäncä
Gyangkar China *see* Dinggyê
75 C2 Gyangzê China
75 C1 Gyaring Co *l.* China
68 C2 Gyaring Hu *l.* China
86 G1 Gydan Peninsula *pen.* Rus. Fed.
Gydanskiy Poluostrov *pen.* Rus. Fed. *see* Gydan Peninsula
Gyêgu China *see* Yushu
65 B2 Gyeonggi-man *b.* S. Korea
Gyixong China *see* Gonggar
51 E2 Gympie Austr.
62 A2 Gyobingauk Myanmar
103 D2 Gyöngyös Hungary
103 D2 Győr Hungary
136 B3 Gypsum U.S.A.
129 E2 Gypsumville Can.
131 D2 Gyrfalcon Islands Can.
111 B3 Gytheio Greece
103 E2 Gyula Hungary
81 C1 Gyumri Armenia
116 A2 Gzhatsk Rus. Fed. *see* Gagarin

H

88 B3 Haapsalu Estonia
100 B1 Haarlem Neth.
122 B3 Haarlem S. Africa
101 C2 Haarstrang *ridge* Ger.
54 A2 Haast N.Z.
74 A2 Hab *r.* Pak.
79 C2 Ḩabarūt Oman
78 B3 Ḩabbān Yemen
81 C2 Ḩabbānīyah, Hawr al *l.* Iraq
70 B1 Habirag China
67 C4 Hachijō-jima *i.* Japan
66 D2 Hachinohe Japan
67 C3 Hachiōji Japan
81 C1 Hacıqabul Azer.
121 C3 Hacufera Moz.
116 A2 Ḩaḍabat al Jilf al Kabīr Egypt
79 C2 Ḩadd, Ra's al *pt* Oman
96 C3 Haddington U.K.
115 D3 Hadejia Nigeria
93 E4 Haderslev Denmark
78 B2 Ḩādhah Saudi Arabia
79 B3 Ḩaḍramawt, Wādī *watercourse* Yemen
98 B2 Hadrian's Wall *tourist site* U.K.
91 C1 Hadyach Ukr.
65 B2 Haeju N. Korea
65 B2 Haeju-man *b.* N. Korea
65 B2 Haenam S. Korea
78 B2 Ḩafar al Bāṭin Saudi Arabia
74 B1 Hafizabad Pak.
75 D2 Haflong India
92 □A3 Hafnarfjörður Iceland
117 A3 Hagar Nish Plateau Eritrea/Sudan

Hagåtña

59 D2 **Hagåtña** Guam
101 F1 **Hagelberg** *h.* Ger.
100 C2 **Hagen** Ger.
101 E1 **Hagenow** Ger.
128 B2 **Hagensborg** Can.
139 D3 **Hagerstown** U.S.A.
93 F3 **Hagfors** Sweden
134 D1 **Haggin, Mount** U.S.A.
67 B4 **Hagi** Japan
62 B1 **Ha Giang** Vietnam
97 B2 **Hag's Head** *hd* Ireland
104 B2 **Hague, Cap de la** *c.* France
69 F3 **Hahajima-rettō** *is* Japan
119 D3 **Hai** Tanz.
70 C1 **Haicheng** China *see* Haifeng
128 A2 **Haida Gwaii** Can.
62 B1 **Hai Dương** Vietnam
80 B2 **Haifa** Israel
71 B3 **Haifeng** China
71 B3 **Haikang** China *see* Leizhou
78 B2 **Haikou** China
Ḥāʼil Saudi Arabia
Hailar China *see* Hulun Buir
Hailong China *see* Meihekou
92 H2 **Hailuoto** *i.* Fin.
71 A4 **Hainan** *prov.* China
69 D3 **Hainan Dao** *i.* China
128 A2 **Haines** U.S.A.
128 A1 **Haines Junction** Can.
101 E2 **Hainich** *ridge* Ger.
101 E2 **Hainleite** *ridge* Ger.
Haiphong Vietnam *see* Hai Phong
62 B1 **Hai Phong** Vietnam
147 C3 **Haiti** *country* West Indies
116 B3 **Haiya** Sudan
103 E2 **Hajdúböszörmény** Hungary
103 E2 **Hajdúszoboszló** Hungary
67 C3 **Hajiki-zaki** *pt* Japan
78 B3 **Hajjah** Yemen
81 D3 **Ḥājjiābād** Iran
103 E1 **Hajnówka** Pol.
62 A1 **Haka** Myanmar
81 C2 **Hakkâri** Turkey
66 D2 **Hakodate** Japan
122 B2 **Hakseen Pan** *salt pan* S. Africa
Ḥalab Syria *see* Aleppo
78 B2 **Ḥalabān** Saudi Arabia
81 C2 **Ḥalabja** Iraq
116 B2 **Halaib** Sudan
78 A2 **Halaib Triangle** *terr.* Egypt/Sudan
79 C3 **Ḥalāniyāt, Juzur al** *is* Oman
78 A2 **Ḥalat ʻAmmār** Saudi Arabia
Halban Mongolia *see* Tsetserleg
101 E2 **Halberstadt** Ger.
64 B2 **Halcon, Mount** Phil.
93 F4 **Halden** Norway
101 E1 **Haldensleben** Ger.
75 B2 **Haldwani** India
79 C2 **Hāleh** Iran
54 A3 **Halfmoon Bay** N.Z.
139 D1 **Haliburton Highlands** *hills* Can.
131 D3 **Halifax** Can.
98 C3 **Halifax** U.K.
139 D3 **Halifax** U.S.A.
65 B3 **Halla-san** *mt.* S. Korea
127 G2 **Hall Beach** Can.
100 B2 **Halle** Belgium
101 E2 **Halle (Saale)** Ger.
102 C2 **Hallein** Austria
101 E2 **Halle-Neustadt** Ger.
48 G3 **Hall Islands** Micronesia
137 D1 **Hallock** U.S.A.
127 H2 **Hall Peninsula** Can.
50 B1 **Halls Creek** Austr.
59 C2 **Halmahera** *i.* Indon.
93 F4 **Halmstad** Sweden
62 B1 **Ha Long** Vietnam
Hälsingborg Sweden *see* Helsingborg
100 B2 **Halsteren** Neth.
98 B2 **Haltwhistle** U.K.
67 B4 **Hamada** Japan
81 C2 **Hamadān** Iran
80 B2 **Ḥamāh** Syria
67 C4 **Hamamatsu** Japan
93 F3 **Hamar** Norway
116 B2 **Ḥamāṭah, Jabal** *mt.* Egypt
73 C4 **Hambantota** Sri Lanka
101 D1 **Hamburg** Ger.
123 C3 **Hamburg** S. Africa
140 B2 **Hamburg** U.S.A.

78 A2 **Ḥamḍ, Wādī al** *watercourse* Saudi Arabia
78 B3 **Ḥamḍah** Saudi Arabia
139 E2 **Hamden** U.S.A.
93 H3 **Hämeenlinna** Fin.
101 D1 **Hameln** Ger.
50 A2 **Hamersley Range** *mts* Austr.
65 B1 **Hamgyŏng-sanmaek** *mts* N. Korea
65 B2 **Hamhŭng** N. Korea
68 C2 **Hami** China
116 B2 **Hamid** Sudan
52 B3 **Hamilton** Austr.
130 C3 **Hamilton** Bermuda
Hamilton *r.* Can. *see* Churchill
54 C1 **Hamilton** N.Z.
96 B3 **Hamilton** U.K.
140 C2 **Hamilton** *AL* U.S.A.
134 D1 **Hamilton** *MT* U.S.A.
138 C3 **Hamilton** *OH* U.S.A.
115 E1 **Hamīm, Wādī al** *watercourse* Libya
93 I3 **Hamina** Fin.
100 C2 **Hamm** Ger.
115 D1 **Hammamet, Golfe de** *g.* Tunisia
81 C2 **Ḥammār, Hawr al** *imp. l.* Iraq
101 D2 **Hammelburg** Ger.
92 G3 **Hammerdal** Sweden
92 H1 **Hammerfest** Norway
140 B2 **Hammond** U.S.A.
139 E3 **Hammonton** U.S.A.
139 D3 **Hampton** U.S.A.
139 E2 **Hampton Bays** U.S.A.
115 D2 **Ḥamrāʼ, Al Ḥamādah al** *plat.* Libya
78 A2 **Ḥanak** Saudi Arabia
66 D3 **Hanamaki** Japan
101 D2 **Hanau** Ger.
69 D2 **Hanbogd** Mongolia
70 B2 **Hancheng** China
138 B1 **Hancock** U.S.A.
70 B2 **Handan** China
119 D3 **Handeni** Tanz.
135 C3 **Hanford** U.S.A.
68 C1 **Hangayn Nuruu** *mts* Mongolia
Hangzhou China *see* Hangzhou
Hanggin Houqi China *see* Xamba
Hangö Fin. *see* Hanko
70 C2 **Hangzhou** China
70 C2 **Hangzhou Wan** *b.* China
79 B2 **Ḥanīdh** Saudi Arabia
Hanjia China *see* Pengshui
Hanjiang China *see* Yangzhou
93 H4 **Hanko** Fin.
135 D3 **Hanksville** U.S.A.
54 B2 **Hanmer Springs** N.Z.
128 C2 **Hanna** Can.
136 B2 **Hanna** U.S.A.
137 E3 **Hannibal** U.S.A.
101 D1 **Hannover** Ger.
101 D2 **Hannoversch Münden** Ger.
93 F4 **Hanöbukten** *b.* Sweden
62 B1 **Ha Nôi** Vietnam
Hanoi Vietnam *see* Ha Nôi
130 B3 **Hanover** Can.
Hanover *see* Hannover
122 B3 **Hanover** S. Africa
139 E2 **Hanover** *NH* U.S.A.
139 D3 **Hanover** *PA* U.S.A.
92 G2 **Hansnes** Norway
93 E4 **Hanstholm** Denmark
88 C3 **Hantsavichy** Belarus
75 C2 **Hanumana** India
74 B2 **Hanumangarh** India
70 A2 **Hanzhong** China
49 M5 **Hao** *atoll* Fr. Polynesia
75 C2 **Haora** India
92 H2 **Haparanda** Sweden
100 B2 **Hapert** Neth.
131 D2 **Happy Valley-Goose Bay** Can.
78 B2 **Ḥaql** Saudi Arabia
79 B2 **Ḥaraḍh** Saudi Arabia
88 C2 **Haradok** Belarus
78 B3 **Harajā** Saudi Arabia
79 C3 **Harare** Zimbabwe
79 C3 **Harāsīs, Jiddat al** *des.* Oman
69 D1 **Har-Ayrag** Mongolia
115 D3 **Haraze-Mangueigne** Chad
114 A4 **Harbel** Liberia
69 E1 **Harbin** China

138 C2 **Harbor Beach** U.S.A.
131 E3 **Harbour Breton** Can.
74 B2 **Harda** India
92 E4 **Hardangerfjorden** *sea chan.* Norway
61 C1 **Harden, Bukit** *mt.* Indon.
100 C1 **Hardenberg** Neth.
100 B1 **Harderwijk** Neth.
122 A3 **Hardeveld** *mts* S. Africa
134 E1 **Hardin** U.S.A.
128 C1 **Hardisty Lake** Can.
93 E3 **Hareid** Norway
100 C1 **Haren (Ems)** Ger.
117 C4 **Härer** Eth.
117 C4 **Hargeysa** Somalia
110 C1 **Harghita-Mădăraş, Vârful** *mt.* Romania
68 C2 **Har Hu** *l.* China
88 B2 **Hari kurk** *sea chan.* Estonia
74 B1 **Haripur** Pak.
74 A1 **Hari Röd** *r.* Afgh./Iran
110 C1 **Hârlău** Romania
100 B1 **Harlingen** Neth.
143 D3 **Harlingen** U.S.A.
99 D4 **Harlow** U.K.
134 E1 **Harlowton** U.S.A.
134 C2 **Harney Basin** U.S.A.
134 C2 **Harney Lake** U.S.A.
93 G3 **Härnösand** Sweden
69 E1 **Har Nur** China
68 C1 **Har Nuur** *l.* Mongolia
96 □ **Haroldswick** U.K.
114 B4 **Harper** Liberia
101 D1 **Harpstedt** Ger.
130 C2 **Harricana, Rivière d'** *r.* Can.
53 D2 **Harrington** Austr.
131 E2 **Harrington Harbour** Can.
96 A2 **Harris** *reg.* U.K.
96 A2 **Harris, Sound of** *sea chan.* U.K.
138 B3 **Harrisburg** *IL* U.S.A.
134 B2 **Harrisburg** *OR* U.S.A.
139 D2 **Harrisburg** *PA* U.S.A.
123 C2 **Harrismith** S. Africa
140 B1 **Harrison** U.S.A.
131 E2 **Harrison, Cape** Can.
126 B2 **Harrison Bay** U.S.A.
139 D3 **Harrisonburg** U.S.A.
128 B3 **Harrison Lake** Can.
137 E3 **Harrisonville** U.S.A.
98 C3 **Harrogate** U.K.
110 C2 **Hârşova** Romania
92 G2 **Harstad** Norway
122 B2 **Hartbees** *watercourse* S. Africa
103 D2 **Hartberg** Austria
139 E2 **Hartford** *CT* U.S.A.
137 D2 **Hartford** *SD* U.S.A.
99 A4 **Hartland** U.K.
98 C2 **Hartlepool** U.K.
Hartley Zimbabwe *see* Chegutu
128 B2 **Hartley Bay** Can.
123 B2 **Harts** *r.* S. Africa
141 D2 **Hartwell Reservoir** U.S.A.
68 C1 **Har Us Nuur** *l.* Mongolia
136 C1 **Harvey** U.S.A.
99 D4 **Harwich** U.K.
74 B2 **Haryana** *state* India
101 E2 **Harz** *hills* Ger.
101 E2 **Harzgerode** Ger.
78 B2 **Ḥasan, Jabal** *h.* Saudi Arabia
80 B2 **Hasan Dağı** *mts* Turkey
99 C4 **Haslemere** U.K.
73 B3 **Hassan** India
100 B2 **Hasselt** Belgium
101 E2 **Haßfurt** Ger.
115 C2 **Hassi Bel Guebbour** Alg.
115 C1 **Hassi Messaoud** Alg.
93 F4 **Hässleholm** Sweden
100 B2 **Hastière-Lavaux** Belgium
53 C3 **Hastings** *Vic.* Austr.
54 C1 **Hastings** N.Z.
99 D4 **Hastings** U.K.
137 D2 **Hastings** *MN* U.S.A.
137 D2 **Hastings** *NE* U.S.A.
Hatay Turkey *see* Antakya
142 B2 **Hatch** U.S.A.
129 D2 **Hatchet Lake** Can.
110 B1 **Haţeg** Romania
52 B2 **Hatfield** Austr.
68 C1 **Hatgal** Mongolia
62 B2 **Ha Tinh** Vietnam
52 B2 **Hattah** Austr.
141 E1 **Hatteras, Cape** U.S.A.
157 H3 **Hatteras Abyssal Plain** S. Atlantic Ocean
140 C2 **Hattiesburg** U.S.A.
100 C2 **Hattingen** Ger.

63 B3 **Hat Yai** Thai.
117 C4 **Haud** *reg.* Eth.
93 E4 **Haugesund** Norway
93 E4 **Haukeligrend** Norway
92 I2 **Haukipudas** Fin.
54 C1 **Hauraki Gulf** N.Z.
54 A3 **Hauroko, Lake** N.Z.
114 B1 **Haut Atlas** *mts* Morocco
131 D3 **Hauterive** Can.
Haute-Volta *country* Africa *see* Burkina Faso
114 B1 **Hauts Plateaux** Alg.
146 B2 **Havana** Cuba
99 C4 **Havant** U.K.
101 E1 **Havel** *r.* Ger.
101 F1 **Havelberg** Ger.
54 B2 **Havelock** N.Z.
Havelock Swaziland *see* Bulembu
54 C1 **Havelock North** N.Z.
99 A4 **Haverfordwest** U.K.
100 C2 **Havixbeck** Ger.
103 D2 **Havlíčkův Brod** Czech Rep.
92 H1 **Havøysund** Norway
111 C3 **Havran** Turkey
134 E1 **Havre** U.S.A.
131 D3 **Havre-Aubert** Can.
131 D3 **Havre-St-Pierre** Can.
49 L2 **Hawaiʻi** *i.* U.S.A.
156 E4 **Hawaiʻian Islands** *is* N. Pacific Ocean
78 B2 **Ḥawallī** Kuwait
98 B3 **Hawarden** U.K.
54 A2 **Hawea, Lake** N.Z.
54 B1 **Hawera** N.Z.
98 B2 **Hawes** U.K.
96 C3 **Hawick** U.K.
54 C1 **Hawke Bay** N.Z.
52 A2 **Hawker** Austr.
52 B1 **Hawkers Gate** Austr.
122 A3 **Hawston** S. Africa
135 C3 **Hawthorne** U.S.A.
52 B2 **Hay** Austr.
128 C1 **Hay** *r.* Can.
100 C3 **Hayange** France
134 C1 **Hayden** U.S.A.
129 E2 **Hayes** *r. Man.* Can.
126 F2 **Hayes** *r. Nunavut* Can.
79 C3 **Hayma'** Oman
77 C2 **Hayotboshi togʻi** *mt.* Uzbe.
111 C2 **Hayrabolu** Turkey
128 C1 **Hay River** Can.
137 D3 **Hays** U.S.A.
78 B3 **Ḥays** Yemen
90 B2 **Haysyn** Ukr.
135 B3 **Hayward** U.S.A.
99 C4 **Haywards Heath** U.K.
81 D2 **Hazar** Turkm.
74 A1 **Hazārajāt** *reg.* Afgh.
138 C3 **Hazard** U.S.A.
75 C2 **Hazaribagh** India
75 C2 **Hazaribagh Range** *mts* India
104 C1 **Hazebrouck** France
128 B2 **Hazelton** Can.
139 D2 **Hazleton** U.S.A.
78 B3 **Hazm al Jawf** Yemen
135 B3 **Healdsburg** U.S.A.
53 C3 **Healesville** Austr.
159 E7 **Heard Island** Indian Ocean
143 D2 **Hearne** U.S.A.
130 B3 **Hearst** Can.
55 A3 **Hearst Island** Antarctica
70 B2 **Hebei** *prov.* China
53 C1 **Hebel** Austr.
140 B1 **Heber Springs** U.S.A.
131 D2 **Hebron** Can.
128 A2 **Hecate Strait** Can.
71 A3 **Hechi** China
100 B2 **Hechtel** Belgium
54 C2 **Hector, Mount** N.Z.
93 F3 **Hede** Sweden
100 C1 **Heerde** Neth.
100 B1 **Heerenveen** Neth.
100 B1 **Heerhugowaard** Neth.
100 B2 **Heerlen** Neth.
Hefa Israel *see* Haifa
70 B2 **Hefei** China
70 B3 **Hefeng** China
70 B1 **Hegang** China
119 D1 **Heiban** Sudan
102 B1 **Heide** Ger.
122 A1 **Heide** Namibia
101 D3 **Heidelberg** Ger.
122 B3 **Heidelberg** S. Africa
69 E1 **Heihe** China
102 B2 **Heilbronn** Ger.
66 A1 **Heilongjiang** *prov.* China
69 E1 **Heilong Jiang** *r.* China
93 I3 **Heinola** Fin.

		Hejaz reg. Saudi Arabia see Hijaz
92	A3	Hekla vol. Iceland
92	F3	Helagsfjället mt. Sweden
70	A2	Helan Shan mts China
140	B2	Helena AR U.S.A.
134	D1	Helena MT U.S.A.
96	B2	Helensburgh U.K.
102	B1	Helgoland i. Ger.
102	B1	Helgoländer Bucht g. Ger.
		Heligoland i. Ger. see Helgoland
		Heligoland Bight g. Ger. see Helgoländer Bucht
92	□A3	Hella Iceland
100	B2	Hellevoetsluis Neth.
107	C2	Hellín Spain
		Hell-Ville Madag. see Andoany
76	C3	Helmand r. Afgh.
101	E2	Helmbrechts Ger.
122	A2	Helmeringhausen Namibia
100	B2	Helmond Neth.
96	C1	Helmsdale U.K.
96	C1	Helmsdale r. U.K.
98	C2	Helmsley U.K.
101	E1	Helmstedt Ger.
65	B1	Helong China
143	D3	Helotes U.S.A.
93	F4	Helsingborg Sweden
		Helsingfors Fin. see Helsinki
93	F4	Helsingør Denmark
93	H3	Helsinki Fin.
99	A4	Helston U.K.
97	C2	Helvick Head hd Ireland
99	C4	Hemel Hempstead U.K.
101	D1	Hemmoor Ger.
92	F2	Hemnesberget Norway
70	B2	Henan prov. China
111	D2	Hendek Turkey
138	B3	Henderson KY U.S.A.
141	E1	Henderson NC U.S.A.
135	D3	Henderson NV U.S.A.
143	E2	Henderson TX U.S.A.
49	O6	Henderson Island Pitcairn Is
141	D1	Hendersonville U.S.A.
99	C4	Hendon U.K.
62	A1	Hengduan Shan mts China
100	C1	Hengelo Neth.
		Hengnan China see Hengyang
71	B3	Hengshan China
70	B2	Hengshui China
71	A3	Hengxian China
71	B3	Hengyang China
		Hengzhou China see Hengxian
91	C2	Heniches'k Ukr.
139	D3	Henlopen, Cape U.S.A.
100	C2	Hennef (Sieg) Ger.
130	B2	Henrietta Maria, Cape Can.
		Henrique de Carvalho Angola see Saurimo
139	D3	Henry, Cape U.S.A.
143	D1	Henryetta U.S.A.
127	H2	Henry Kater, Cape Can.
101	D1	Henstedt-Ulzburg Ger.
120	A3	Hentiesbaai Namibia
101	D3	Heppenheim (Bergstraße) Ger.
71	B3	Hepu China
76	C3	Herāt Afgh.
129	D2	Herbert Can.
54	C2	Herbertville N.Z.
101	D2	Herbstein Ger.
109	C2	Herceg-Novi Montenegro
99	B3	Hereford U.K.
143	C2	Hereford U.S.A.
101	D1	Herford Ger.
100	C2	Herkenbosch Neth.
96	□	Herma Ness hd U.K.
122	A3	Hermanus S. Africa
53	C2	Hermidale Austr.
134	C1	Hermiston U.S.A.
59	D3	Hermit Islands P.N.G.
144	A2	Hermosillo Mex.
154	B3	Hernandarias Para.
100	C2	Herne Ger.
93	E4	Herning Denmark
104	B3	Hérouville-St-Clair France
106	B2	Herrera del Duque Spain
139	D2	Hershey U.S.A.
99	C4	Hertford U.K.
123	C2	Hertzogville S. Africa
51	E2	Hervey Bay Austr.
101	F2	Herzberg Ger.
101	E3	Herzogenaurach Ger.
71	A3	Heshan China
135	C4	Hesperia U.S.A.
128	A1	Hess r. Can.
101	E1	Hessen Ger.
101	D2	Hessisch Lichtenau Ger.
136	C1	Hettinger U.S.A.
101	E2	Hettstedt Ger.
98	B2	Hexham U.K.
81	C2	Heydarābād Iran
98	B2	Heysham U.K.
71	B3	Heyuan China
52	B3	Heywood Austr.
70	B2	Heze China
71	B3	Hezhou China
141	D3	Hialeah U.S.A.
137	D3	Hiawatha U.S.A.
137	E1	Hibbing U.S.A.
141	D1	Hickory U.S.A.
54	C1	Hicks Bay N.Z.
66	D2	Hidaka-sanmyaku mts Japan
143	D3	Hidalgo Mex.
145	C2	Hidalgo Mex.
144	B2	Hidalgo del Parral Mex.
154	C1	Hidrolândia Brazil
67	A4	Higashi-suidō str. Japan
		High Atlas mts Morocco see Haut Atlas
134	B2	High Desert U.S.A.
128	C2	High Level Can.
141	E1	High Point U.S.A.
128	C2	High Prairie Can.
128	C2	High River Can.
129	D2	Highrock Lake Can.
		High Tatras mts Pol./Slovakia see Tatry
99	C4	High Wycombe U.K.
88	B2	Hiiumaa i. Estonia
78	A2	Hijaz reg. Saudi Arabia
54	C1	Hikurangi mt. N.Z.
101	E2	Hildburghausen Ger.
100	C2	Hilden Ger.
101	E2	Hilders Ger.
101	D1	Hildesheim Ger.
81	C2	Hillah Iraq
100	B1	Hillegom Neth.
100	C2	Hillesheim Ger.
138	C3	Hillsboro OH U.S.A.
143	D2	Hillsboro TX U.S.A.
53	C2	Hillston Austr.
96	□	Hillswick U.K.
49	L2	Hilo U.S.A.
141	D2	Hilton Head Island U.S.A.
100	B1	Hilversum Neth.
74	B1	Himachal Pradesh state India
68	B2	Himalaya mts Asia
74	B2	Himatnagar India
67	B4	Himeji Japan
123	C2	Himeville S. Africa
67	C3	Himi Japan
		Ḥimṣ Syria see Homs
147	C3	Hinche Haiti
51	D1	Hinchinbrook Island Austr.
52	B3	Hindmarsh, Lake dry lake Austr.
74	A1	Hindu Kush mts Afgh./Pak.
73	B3	Hindupur India
134	C2	Hines U.S.A.
141	D2	Hinesville U.S.A.
74	B2	Hinganghat India
81	C2	Hınıs Turkey
92	G2	Hinnøya i. Norway
106	B2	Hinojosa del Duque Spain
100	C1	Hinte Ger.
62	A2	Hinthada Myanmar
128	C2	Hinton Can.
75	C2	Hirakud Reservoir India
		Hîrlău Romania see Hârlău
66	D2	Hiroo Japan
66	D2	Hirosaki Japan
67	B4	Hiroshima Japan
101	E3	Hirschaid Ger.
101	E2	Hirschberg Ger.
105	C2	Hirson France
		Hîrşova Romania see Hârşova
93	E4	Hirtshals Denmark
74	B2	Hisar India
147	C2	Hispaniola i. Caribbean Sea
81	C2	Ḩīt Iraq
67	D3	Hitachi Japan
67	D3	Hitachinaka Japan
92	E3	Hitra i. Norway
49	N4	Hiva Oa i. Fr. Polynesia
93	G4	Hjälmaren l. Sweden
129	D1	Hjalmar Lake Can.
93	F4	Hjørring Denmark
123	D2	Hlabisa S. Africa
92	□B2	Hlíð Iceland
91	C2	Hlobyne Ukr.
123	C2	Hlohlowane S. Africa
123	C2	Hlotse Lesotho
91	C1	Hlukhiv Ukr.
88	C3	Hlusk Belarus
88	C2	Hlybokaye Belarus
114	C4	Ho Ghana
62	B2	Hoa Binh Vietnam
122	A1	Hoachanas Namibia
120	A2	Hoanib watercourse Namibia
120	A2	Hoarusib watercourse Namibia
51	D4	Hobart Austr.
143	D1	Hobart U.S.A.
143	C2	Hobbs U.S.A.
93	E4	Hobro Denmark
117	C4	Hobyo Somalia
		Hồ Chi Minh Vietnam see Ho Chi Minh City
63	B2	Ho Chi Minh City Vietnam
114	B3	Hôd reg. Maur.
117	D3	Hodda mt. Somalia
78	B3	Hodeidah Yemen
103	E2	Hódmezővásárhely Hungary
		Hoek van Holland Neth. see Hook of Holland
65	B2	Hoeyang N. Korea
101	E2	Hof Ger.
101	E2	Hofheim in Unterfranken Ger.
92	□B3	Höfn Iceland
92	□A2	Höfn Iceland
92	□B3	Hofsjökull Iceland
67	B4	Hōfu Japan
93	G4	Högsby Sweden
93	E3	Høgste Breakulen mt. Norway
101	D2	Hohe Rhön mts Ger.
100	C2	Hohe Venn moorland Belgium
70	B1	Hohhot China
75	C1	Hoh Xil Shan mts China
63	B2	Hôi An Vietnam
119	D2	Hoima Uganda
75	D2	Hojai India
54	B2	Hokitika N.Z.
66	D2	Hokkaidō i. Japan
91	C2	Hola Prystan' Ukr.
128	B2	Holberg Can.
53	C3	Holbrook Austr.
142	A2	Holbrook U.S.A.
137	D2	Holdrege U.S.A.
146	C2	Holguín Cuba
92	□B2	Hóll Iceland
103	D2	Hollabrunn Austria
138	B2	Holland U.S.A.
		Hollandia Indon. see Jayapura
135	B3	Hollister U.S.A.
103	E2	Hollóháza Hungary
93	I3	Hollola Fin.
100	B1	Hollum Neth.
140	B1	Holly Springs U.S.A.
134	C4	Hollywood U.S.A.
141	D3	Hollywood U.S.A.
92	F2	Holm Norway
		Holman Can. see Ulukhaktok
92	H3	Holmsund Sweden
122	A2	Holoog Namibia
93	E4	Holstebro Denmark
141	D1	Holston r. U.S.A.
98	A3	Holyhead U.K.
98	C2	Holy Island England U.K.
98	A3	Holy Island Wales U.K.
136	C2	Holyoke U.S.A.
		Holy See Europe see Vatican City
101	D2	Holzminden Ger.
62	A1	Homalin Myanmar
101	D2	Homberg (Efze) Ger.
114	B3	Hombori Mali
100	C3	Homburg Ger.
127	H2	Home Bay Can.
140	B2	Homer U.S.A.
141	D3	Homestead U.S.A.
92	F3	Hommelvik Norway
141	D3	Homosassa Springs U.S.A.
80	B2	Homs Syria
89	D3	Homyel' Belarus
		Honan prov. China see Henan
122	A3	Hondeklipbaai S. Africa
145	D3	Hondo r. Belize/Mex.
142	B2	Hondo NM U.S.A.
143	D3	Hondo TX U.S.A.
146	B3	Honduras country Central America
146	B3	Honduras, Gulf of Belize/Hond.
93	F3	Hønefoss Norway
135	B2	Honey Lake U.S.A.
104	C2	Honfleur France
70	B3	Honghu China
71	A3	Hongjiang China
71	B3	Hong Kong China
71	B3	Hong Kong aut. reg. China
		Hongqizhen China see Wuzhishan
131	D3	Honguedo, Détroit d' sea chan. Can.
65	B1	Hongwŏn N. Korea
70	B2	Hongze Hu l. China
48	H4	Honiara Solomon Is
99	B4	Honiton U.K.
92	I1	Honningsvåg Norway
49	L1	Honolulu U.S.A.
67	B3	Honshū i. Japan
134	B1	Hood, Mount vol. U.S.A.
50	A3	Hood Point Austr.
134	B1	Hood River U.S.A.
100	B2	Hoogerheide Neth.
100	C1	Hoogeveen Neth.
100	C1	Hoogezand-Sappemeer Neth.
100	C2	Hoog-Keppel Neth.
100	B2	Hook of Holland Neth.
128	A2	Hoonah U.S.A.
123	C2	Hoopstad S. Africa
100	B1	Hoorn Neth.
49	J5	Hoorn, Îles de is Wallis and Futuna Is
128	B3	Hope Can.
140	B2	Hope U.S.A.
83	N2	Hope, Point U.S.A.
131	D2	Hopedale Can.
		Hopei prov. China see Hebei
145	D3	Hopelchén Mex.
131	D2	Hope Mountains Can.
		Hopes Advance Bay Can. see Aupaluk
52	B3	Hopetoun Austr.
122	B2	Hopetown S. Africa
139	D3	Hopewell U.S.A.
130	C2	Hopewell Islands Can.
50	B2	Hopkins, Lake imp. l. Austr.
138	B3	Hopkinsville U.S.A.
134	B1	Hoquiam U.S.A.
81	C1	Horasan Turkey
93	F4	Hörby Sweden
89	D3	Horki Belarus
91	D2	Horlivka Ukr.
79	D2	Hormak Iran
79	C2	Hormuz, Strait of Iran/Oman
103	D2	Horn Austria
92	□A2	Horn c. Iceland
153	B5	Horn, Cape Chile
139	D2	Hornell U.S.A.
130	B3	Hornepayne Can.
		Hornos, Cabo de c. Chile see Horn, Cape
53	D2	Hornsby Austr.
98	C3	Hornsea U.K.
90	B2	Horodenka Ukr.
91	C1	Horodnya Ukr.
90	B2	Horodok Ukr.
90	A2	Horodok Ukr.
90	A1	Horokhiv Ukr.
		Horqin Youyi Qianqi China see Ulanhot
131	E2	Horse Islands Can.
52	B3	Horsham Austr.
99	C4	Horsham U.K.
93	F4	Horten Norway
126	E2	Horton r. Can.
117	B4	Hosaʼina Eth.
74	A2	Hoshab Pak.
74	B1	Hoshiarpur India
142	B1	Hosta Butte mt. U.S.A.
75	C1	Hotan China
122	B2	Hotazel S. Africa
92	G3	Hoting Sweden
140	B2	Hot Springs AR U.S.A.
136	C2	Hot Springs SD U.S.A.
		Hot Springs U.S.A. see Truth or Consequences
128	C1	Hottah Lake Can.
62	B1	Houayxay Laos
100	B2	Houffalize Belgium
138	B1	Houghton U.S.A.
139	F1	Houlton U.S.A.
70	B2	Houma China
140	B3	Houma U.S.A.
128	B2	Houston Can.
143	D3	Houston U.S.A.
50	A2	Houtman Abrolhos is Austr.

122 B3 Houwater S. Africa
68 C1 Hovd Mongolia
99 C4 Hove U.K.
90 A2 Hoverla, Hora mt. Ukr.
68 C1 Hövsgöl Nuur l. Mongolia
116 A3 Howar, Wadi watercourse Sudan
98 C3 Howden U.K.
53 C3 Howe, Cape Austr.
123 D2 Howick S. Africa
49 J3 Howland Island terr. N. Pacific Ocean
53 C3 Howlong Austr.
　 Howrah India see Haora
140 B1 Hoxie U.S.A.
101 D2 Höxter Ger.
96 C1 Hoy i. U.K.
93 E3 Høyanger Norway
102 C1 Hoyerswerda Ger.
62 A2 Hpapun Myanmar
103 D1 Hradec Králové Czech Rep.
109 C2 Hrasnica Bos.-Herz.
92 □B2 Hraun Iceland
91 C1 Hrebinka Ukr.
88 B3 Hrodna Belarus
62 A1 Hsi-hseng Myanmar
62 A1 Hsipaw Myanmar
70 A2 Huachi China
150 B4 Huacho Peru
70 B1 Huade China
65 B1 Huadian China
70 B2 Huai'an China
70 B2 Huaibei China
70 B2 Huai He r. China
71 A3 Huaihua China
70 B2 Huainan China
70 B2 Huaiyang China
145 C3 Huajuápan de León Mex.
59 C3 Huaki Indon.
71 C3 Hualian Taiwan
150 B3 Huallaga r. Peru
120 A2 Huambo Angola
150 B4 Huancavelica Peru
150 B4 Huancayo Peru
　 Huangcaoba China see Xingyi
70 B2 Huangchuan China
　 Huang Hai sea N. Pacific Ocean see Yellow Sea
　 Huang He r. China see Yellow River
71 A4 Huangliu China
70 B3 Huangshan China
70 B2 Huangshi China
70 A2 Huangtu Gaoyuan plat. China
71 C3 Huangyan China
70 A2 Huangyuan China
65 B1 Huanren China
150 B3 Huánuco Peru
152 B1 Huanuni Bol.
150 B4 Huaral Peru
150 B3 Huaraz Peru
150 B4 Huarmey Peru
152 A2 Huasco Chile
152 A2 Huasco r. Chile
144 B2 Huatabampo Mex.
145 C3 Huatusco Mex.
71 A3 Huayuan China
70 B2 Hubei prov. China
73 B3 Hubli India
100 C2 Hückelhoven Ger.
99 C3 Hucknall U.K.
98 C3 Huddersfield U.K.
93 G3 Hudiksvall Sweden
139 E2 Hudson r. U.S.A.
129 D2 Hudson Bay Can.
127 G3 Hudson Bay sea Can.
128 B2 Hudson's Hope Can.
127 H2 Hudson Strait Can.
63 B2 Huê Vietnam
146 A3 Huehuetenango Guat.
144 B2 Huehueto, Cerro mt. Mex.
145 C2 Huejutla Mex.
106 B2 Huelva Spain
107 C2 Huércal-Overa Spain
107 C1 Huesca Spain
106 C2 Huéscar Spain
51 D2 Hughenden Austr.
50 B3 Hughes (abandoned) Austr.
75 C2 Hugli r. mouth India
143 D2 Hugo U.S.A.
　 Huhehot China see Hohhot
122 B2 Huhudi S. Africa
122 A2 Huib-Hoch Plateau Namibia
71 B3 Huichang China
　 Huicheng China see Huilai
65 B1 Huich'ŏn N. Korea

120 A2 Huíla, Planalto da Angola
71 B3 Huilai China
71 A3 Huili China
70 B2 Huimin China
69 E2 Huinan China
　 Huinan China see Nanhui
93 H3 Huittinen Fin.
145 C3 Huixtla Mex.
71 A3 Huize China
78 B2 Hujr Saudi Arabia
122 B1 Hukuntsi Botswana
78 B2 Ḩulayfah Saudi Arabia
66 B1 Hulin China
130 C3 Hull Can.
70 C1 Huludao China
69 D1 Hulun Buir China
69 D1 Hulun Nur l. China
91 D2 Hulyaypole Ukr.
69 E1 Huma China
150 C3 Humaitá Brazil
122 B3 Humansdorp S. Africa
98 C2 Humber est. U.K.
143 D3 Humble U.S.A.
129 D2 Humboldt Can.
135 C2 Humboldt NV U.S.A.
140 C1 Humboldt TN U.S.A.
135 C2 Humboldt r. U.S.A.
103 E2 Humenné Slovakia
53 C3 Hume Reservoir Austr.
142 A1 Humphreys Peak U.S.A.
115 D2 Hūn Libya
92 □A2 Húnaflói b. Iceland
71 B3 Hunan prov. China
65 C1 Hunchun China
110 B1 Hunedoara Romania
101 D2 Hünfeld Ger.
103 D2 Hungary country Europe
52 B1 Hungerford Austr.
65 B2 Hüngnam N. Korea
65 A1 Hun He r. China
　 Hunjiang China see Baishan
99 D3 Hunstanton U.K.
101 D1 Hunte r. Ger.
48 I6 Hunter Island S. Pacific Ocean
51 D4 Hunter Islands Austr.
99 C3 Huntingdon U.K.
138 B2 Huntington IN U.S.A.
138 C3 Huntington WV U.S.A.
135 C4 Huntington Beach U.S.A.
54 C1 Huntly N.Z.
96 C2 Huntly U.K.
130 C3 Huntsville Can.
140 C2 Huntsville AL U.S.A.
143 D2 Huntsville TX U.S.A.
　 Hunyani r. Moz./Zimbabwe see Manyame
59 D3 Huon Peninsula P.N.G.
154 B2 Huoxian China see Huozhou
70 B2 Huozhou China
　 Hupeh prov. China see Hubei
　 Hurghada Egypt see Al Ghurdaqah
137 D2 Huron U.S.A.
138 C2 Huron, Lake Can./U.S.A.
135 D3 Hurricane U.S.A.
100 C2 Hürth Ger.
92 □B2 Húsavík Iceland
110 C1 Huşi Romania
126 B2 Huslia U.S.A.
78 B3 Ḩusn Äl 'Abr Yemen
102 B1 Husum Ger.
68 C1 Hutag-Öndör Mongolia
60 A1 Hutanopan Indon.
137 D3 Hutchinson U.S.A.
141 D3 Hutchinson Island U.S.A.
70 B2 Hutuo He r. China
100 B2 Huy Belgium
70 C2 Huzhou China
92 □C3 Hvalnes Iceland
92 □B3 Hvannadalshnúkur vol. Iceland
109 C2 Hvar Croatia
109 C2 Hvar i. Croatia
91 C2 Hvardiys'ke Ukr.
120 B2 Hwange Zimbabwe
　 Hwang Ho r. China see Yellow River
136 C2 Hyannis U.S.A.
68 C1 Hyargas Nuur salt l. Mongolia
50 A3 Hyden Austr.
73 B3 Hyderabad India
74 A2 Hyderabad Pak.
　 Hydra i. Greece see Ydra
105 D3 Hyères France
105 D3 Hyères, Îles d' is France

65 B1 Hyesan N. Korea
128 B2 Hyland Post Can.
67 B3 Hyōno-sen mt. Japan
99 D4 Hythe U.K.
67 B4 Hyūga Japan
93 H3 Hyvinkää Fin.

I

114 B2 Iabès, Erg des. Alg.
150 C3 Iaco r. Brazil
110 C2 Ialomiţa r. Romania
110 C1 Ianca Romania
110 C1 Iaşi Romania
64 A2 Iba Phil.
115 C4 Ibadan Nigeria
150 B2 Ibagué Col.
154 B2 Ibaiti Brazil
150 B2 Ibarra Ecuador
78 B3 Ibb Yemen
100 C1 Ibbenbüren Ger.
115 C4 Ibi Nigeria
107 C2 Ibi Spain
155 C1 Ibiá Brazil
155 D1 Ibiaí Brazil
155 D1 Ibiraçu Brazil
107 D2 Ibiza Spain
107 D2 Ibiza i. Spain
151 E4 Ibotirama Brazil
79 C2 Ibrā' Oman
79 C2 Ibrī Oman
150 B4 Ica Peru
　 Icaria i. Greece see Ikaria
　 İçel Turkey see Mersin
92 □B2 Iceland country Europe
160 M4 Iceland Basin N. Atlantic Ocean
160 L3 Icelandic Plateau N. Atlantic Ocean
66 D3 Ichinoseki Japan
91 C1 Ichnya Ukr.
65 B2 Ich'ŏn N. Korea
151 F3 Icó Brazil
155 D2 Iconha Brazil
143 E2 Idabel U.S.A.
115 C4 Idah Nigeria
134 D2 Idaho state U.S.A.
134 D2 Idaho Falls U.S.A.
100 C3 Idar-Oberstein Ger.
68 C1 Ideriyn Gol r. Mongolia
116 B2 Idfū Egypt
　 Idi Amin Dada, Lake l. Dem. Rep. Congo/Uganda see Edward, Lake
118 B3 Idiofa Dem. Rep. Congo
80 B2 Idlib Syria
88 B2 Iecava Latvia
154 B2 Iepê Brazil
100 A2 Ieper Belgium
111 C3 Ierapetra Greece
119 D3 Ifakara Tanz.
121 □D3 Ifanadiana Madag.
115 C4 Ife Nigeria
114 C3 Ifôghas, Adrar des hills Mali
118 C3 Ifumo Dem. Rep. Congo
61 C1 Igan Malaysia
119 D2 Iganga Uganda
154 C2 Igarapava Brazil
82 G2 Igarka Rus. Fed.
74 B3 Igatpuri India
81 C2 Iğdır Turkey
108 A3 Iglesias Italy
127 G2 Igloolik Can.
　 Igluligaarjuk Can. see Chesterfield Inlet
130 A3 Ignace Can.
88 C2 Ignalina Lith.
111 C2 İğneada Turkey
110 C2 İğneada Burnu pt Turkey
111 B3 Igoumenitsa Greece
86 E3 Igra Rus. Fed.
86 F2 Igrim Rus. Fed.
154 B3 Iguaçu r. Brazil
154 B3 Iguaçu Falls Arg./Brazil
145 C3 Iguala Mex.
154 C2 Igualada Spain
154 C2 Iguape Brazil
154 B2 Iguatemi Brazil
154 B2 Iguatemi r. Brazil
151 F3 Iguatu Brazil
118 A3 Iguéla Gabon
114 C2 Iguidi, Erg des. Alg./Maur.
119 D3 Igunga Tanz.
121 □D2 Iharaña Madag.
121 □D3 Ihosy Madag.
92 I2 Iijoki r. Fin.
92 I3 Iisalmi Fin.
67 B4 Iizuka Japan

115 C4 Ijebu-Ode Nigeria
100 B1 IJmuiden Neth.
100 B1 IJssel r. Neth.
100 B1 IJsselmeer l. Neth.
152 C2 Ijuí Brazil
　 Ikaahuk Can. see Sachs Harbour
123 C2 Ikageleng S. Africa
123 C2 Ikageng S. Africa
111 C3 Ikaria i. Greece
118 C3 Ikela Dem. Rep. Congo
110 B2 Ikhtiman Bulg.
67 A4 Iki-shima i. Japan
118 A2 Ikom Nigeria
121 □D3 Ikongo Madag.
65 B2 Iksan S. Korea
119 D3 Ikungu Tanz.
114 C2 Ilaferh, Oued watercourse Alg.
64 B2 Ilagan Phil.
81 C2 Īlām Iran
75 C2 Ilam Nepal
103 D1 Iława Pol.
79 C2 Ilazārān, Küh-e mt. Iran
129 D2 Île-à-la-Crosse Can.
129 D2 Île-à-la-Crosse, Lac l. Can
118 C3 Ilebo Dem. Rep. Congo
119 D2 Ilemi Triangle terr. Kenya
119 D2 Ileret Kenya
99 D4 Ilford U.K.
99 A4 Ilfracombe U.K.
155 C2 Ilhabela Brazil
155 D2 Ilha Grande, Baía da b. Brazil
154 B2 Ilha Grande, Represa resr Brazil
154 B2 Ilha Solteíra, Represa resr Brazil
106 B1 Ílhavo Port.
151 F4 Ilhéus Brazil
　 Ili Kazakh. see Kapshagay
126 B3 Iliamna Lake U.S.A.
64 B3 Iligan Phil.
　 Iliysk Kazakh. see Kapshagay
98 C3 Ilkley U.K.
152 A3 Illapel Chile
90 C2 Illichivs'k Ukr.
138 A3 Illinois r. U.S.A.
138 B3 Illinois state U.S.A.
90 B2 Illintsi Ukr.
115 C2 Illizi Alg.
89 D2 Il'men', Ozero l. Rus. Fed.
101 E2 Ilmenau Ger.
150 B4 Ilo Peru
64 B2 Iloilo Phil.
92 J3 Ilomantsi Fin.
115 C4 Ilorin Nigeria
87 D4 Ilovlya Rus. Fed.
53 D1 Iluka Austr.
127 I2 Ilulissat Greenland
　 Iman Rus. Fed. see Dal'nerechensk
66 B1 Iman r. Rus. Fed.
67 A4 Imari Japan
93 I3 Imatra Fin.
　 imeni Petra Stuchki Latvia see Aizkraukle
117 C4 İmi Eth.
65 B2 Imjin-gang r. N. Korea/S. Korea
141 D3 Immokalee U.S.A.
108 B2 Imola Italy
151 E3 Imperatriz Brazil
108 A2 Imperia Italy
136 C2 Imperial U.S.A.
118 B2 Impfondo Congo
72 D2 Imphal India
111 C2 İmroz Turkey
67 C3 Ina Japan
150 C4 Inambari r. Peru
115 C2 In Aménas Alg.
115 C2 In Amguel Alg.
54 B2 Inangahua Junction N.Z.
59 C3 Inanwatan Indon.
92 I2 Inari Fin.
92 I2 Inarijärvi l. Fin.
67 D3 Inawashiro-ko l. Japan
80 B1 İnce Burun pt Turkey
65 B2 Incheon S. Korea
121 C2 Inchope Moz.
123 D2 Incomati r. Moz.
116 B3 Inda Silasē Eth.
144 B2 Indé Mex.
135 C3 Independence CA U.S.A.
137 E2 Independence IA U.S.A.
137 D3 Independence KS U.S.A.
137 E3 Independence MO U.S.A.
134 C2 Independence Mountains U.S.A.

76 B2	Inderbor Kazakh.
72 B2	India country Asia
139 D2	Indiana U.S.A.
138 B2	Indiana state U.S.A.
138 B3	Indianapolis U.S.A.
129 D2	Indian Head Can.
159	Indian Ocean
137 E2	Indianola IA U.S.A.
140 B2	Indianola MS U.S.A.
135 D3	Indian Peak U.S.A.
135 C3	Indian Springs U.S.A.
86 D2	Indiga Rus. Fed.
83 K2	Indigirka r. Rus. Fed.
109 D1	Inđija Serbia
135 C4	Indio U.S.A.
58 B3	Indonesia country Asia
74 B2	Indore India
60 B2	Indramayu, Tanjung pt Indon.
	Indrapura, Gunung vol. Indon. see Kerinci, Gunung
75 C3	Indravati r. India
104 C2	Indre r. France
74 A2	Indus r. China/Pak.
74 A2	Indus, Mouths of the Pak.
159 E2	Indus Cone Indian Ocean
80 B1	Inebolu Turkey
111 C2	Inegöl Turkey
	Infantes Spain see Villanueva de los Infantes
144 B3	Infiernillo, Presa resr Mex.
51 D1	Ingham Austr.
53 D1	Inglewood Austr.
102 C2	Ingolstadt Ger.
75 C2	Ingraj Bazar India
123 D2	Ingwavuma S. Africa
120 B2	Ingwe Zambia
123 D2	Inhaca Moz.
121 C3	Inhambane Moz.
121 C2	Inhaminga Moz.
151 D2	Inini Fr. Guiana
	Inis Ireland see Ennis
97 A2	Inishbofin i. Ireland
97 B2	Inishmore i. Ireland
97 C1	Inishowen pen. Ireland
54 B2	Inland Kaikoura Range mts N.Z.
	Inland Sea sea Japan see Seto-naikai
102 C2	Inn r. Europe
127 H1	Innaanganeq c. Greenland
54 A2	Innamincka Austr.
70 A1	Inner Mongolia aut. reg. China
96 B2	Inner Sound sea chan. U.K.
51 D1	Innisfail Austr.
102 C2	Innsbruck Austria
97 C2	Inny r. Ireland
154 B1	Inocência Brazil
118 B3	Inongo Dem. Rep. Congo
111 D3	İnönü Turkey
	Inoucdjouac Can. see Inukjuak
103 D1	Inowrocław Pol.
114 C2	In Salah Alg.
62 A2	Insein Myanmar
110 C2	Însurăţei Romania
86 F2	Inta Rus. Fed.
105 D2	Interlaken Switz.
137 E1	International Falls U.S.A.
63 A2	Interview Island India
130 C2	Inukjuak Can.
126 D2	Inuvik Can.
96 B2	Inveraray U.K.
96 C2	Inverbervie U.K.
54 A3	Invercargill N.Z.
53 D1	Inverell Austr.
96 B2	Invergordon U.K.
128 C2	Invermere Can.
131 D3	Inverness Can.
96 B2	Inverness U.K.
141 D3	Inverness U.S.A.
159 F4	Investigator Ridge Indian Ocean
52 A3	Investigator Strait Austr.
77 E1	Inya Rus. Fed.
	Inyanga Zimbabwe see Nyanga
119 D3	Inyonga Tanz.
87 D3	Inza Rus. Fed.
111 B3	Ioannina Greece
137 D3	Iola U.S.A.
96 A2	Iona i. U.K.
111 C3	Ionian Islands Greece
109 C3	Ionian Sea Greece/Italy
	Ionioi Nisoi is Greece see Ionian Islands
111 C3	Ios i. Greece
137 E2	Iowa state U.S.A.
137 E2	Iowa City U.S.A.
154 C1	Ipameri Brazil
155 D1	Ipatinga Brazil
81 C1	Ipatovo Rus. Fed.
123 C2	Ipelegeng S. Africa
150 B2	Ipiales Col.
151 F4	Ipiaú Brazil
154 B3	Ipiranga Brazil
150 B3	Ipixuna Brazil
60 B1	Ipoh Malaysia
154 B1	Iporá Brazil
118 C2	Ippy C.A.R.
111 C2	İpsala Turkey
53 D1	Ipswich Austr.
99 D3	Ipswich U.K.
127 H2	Iqaluit Can.
152 A2	Iquique Chile
150 B3	Iquitos Peru
	Irakleio Greece see Iraklion
111 C3	Iraklion Greece
61 C1	Iran country Asia
	Iran, Pegunungan mts Indon.
79 D2	Īrānshahr Iran
144 B2	Irapuato Mex.
81 C2	Iraq country Asia
154 B3	Irati Brazil
88 B2	Irbe Strait Estonia/Latvia
80 B2	Irbid Jordan
86 F3	Irbit Rus. Fed.
151 E4	Irecê Brazil
97 C2	Ireland country Europe
118 C3	Irema Dem. Rep. Congo
	Iri S. Korea see Iksan
115 E3	Iriba Chad
114 B3	Irîgui reg. Mali/Maur.
119 D3	Iringa Tanz.
151 D3	Iriri r. Brazil
	Irish Free State country Europe see Ireland
95 B3	Irish Sea Ireland/U.K.
68 C1	Irkutsk Rus. Fed.
160 M4	Irminger Basin N. Atlantic Ocean
139 D2	Irondequoit U.S.A.
110 B2	Iron Gates gorge Romania/ Serbia
52 A2	Iron Knob Austr.
138 B1	Iron Mountain U.S.A.
138 A1	Ironton U.S.A.
138 B1	Ironwood U.S.A.
130 B3	Iroquois Falls Can.
64 B2	Irosin Phil.
67 C4	Irō-zaki pt Japan
90 C1	Irpin' Ukr.
62 A2	Irrawaddy r. Myanmar
63 A2	Irrawaddy, Mouths of the Myanmar
86 F2	Irtysh r. Kazakh./Rus. Fed.
107 C1	Irun Spain
96 B3	Irvine U.K.
143 D2	Irving U.S.A.
64 B3	Isabela Phil.
146 B3	Isabelia, Cordillera mts Nic.
92 □A2	Ísafjarðardjúp est. Iceland
92 □A2	Ísafjörður Iceland
67 B4	Isahaya Japan
102 C2	Isar r. Ger.
96 □	Isbister U.K.
108 B2	Ischia, Isola d' i. Italy
67 C4	Ise Japan
118 C2	Isengi Dem. Rep. Congo
105 C3	Isère r. France
100 C2	Iserlohn Ger.
101 D1	Isernhagen Ger.
67 C4	Ise-wan b. Japan
114 C4	Iseyin Nigeria
	Isfahan Iran see Eşfahān
66 D2	Ishikari-wan b. Japan
82 F3	Ishim Rus. Fed.
67 D3	Ishinomaki Japan
67 D3	Ishioka Japan
67 B4	Ishizuchi-san mt. Japan
74 B1	Ishkoshim Tajik.
138 B1	Ishpeming U.S.A.
111 C2	Işıklar Dağı mts Turkey
111 C3	Işıklı Turkey
123 D2	Isipingo S. Africa
119 C2	Isiro Dem. Rep. Congo
80 B2	İskenderun Turkey
82 G3	İskitim Rus. Fed.
110 B2	Iskŭr r. Bulg.
117 D3	Iskushuban Somalia
128 A2	Iskut r. Can.
74 B1	Islamabad Pak.
141 D4	Islamorada U.S.A.
52 A2	Island Lagoon imp. l. Austr.
129 E2	Island Lake Can.
54 B1	Islands, Bay of N.Z.
	Islas Canarias is N. Atlantic Ocean see Canary Islands
96 A3	Islay i. U.K.
98 A2	Isle of Man i. Irish Sea
	Ismail Ukr. see Izmayil
77 D3	Ismoili Somoní, Qullai mt. Tajik.
116 B2	Isnā Egypt
121 □D3	Isoanala Madag.
121 C2	Isoka Zambia
93 H3	Isokyrö Fin.
109 C3	Isola di Capo Rizzuto Italy
	Ispahan Iran see Eşfahān
80 B2	Isparta Turkey
110 C2	Isperikh Bulg.
	Ispisar Tajik. see Khüjand
80 B2	Israel country Asia
50 B3	Israelite Bay Austr.
105 C2	Issoire France
	Issyk-Kul' Kyrg. see Balykchy
111 C2	İstanbul Turkey
	İstanbul Boğazı str. Turkey see Bosporus
103 D2	Isten dombja h. Hungary
111 B3	Istiaia Greece
141 D3	Istokpoga, Lake U.S.A.
	Istra pen. Croatia see Istria
105 C3	Istres France
108 B1	Istria pen. Croatia
155 D2	Itabapoana Brazil
151 E4	Itaberaba Brazil
154 C1	Itaberaí Brazil
155 D1	Itabira Brazil
155 D2	Itabirito Brazil
151 F4	Itabuna Brazil
150 D3	Itacoatiara Brazil
155 D2	Itaguaí Brazil
154 B2	Itaguajé Brazil
154 C2	Itaí Brazil
154 C3	Itaiópolis Brazil
154 B3	Itaipu, Represa de resr Brazil
151 D3	Itaituba Brazil
155 C2	Itajaí Brazil
155 C2	Itajubá Brazil
108 B2	Italy country Europe
155 E1	Itamaraju Brazil
155 D1	Itamarandiba Brazil
155 D1	Itambacuri Brazil
155 D1	Itambé, Pico de mt. Brazil
75 D2	Itanagar India
154 C2	Itanhaém Brazil
155 D1	Itaobím Brazil
154 C1	Itapajipe Brazil
151 E3	Itapebi Brazil
151 E3	Itapecuru Mirim Brazil
155 D1	Itapemirim Brazil
155 D2	Itaperuna Brazil
154 B2	Itapetinga Brazil
154 C2	Itapetininga Brazil
154 B2	Itapeva Brazil
151 F4	Itapicuru r. Brazil
154 C2	Itapira Brazil
154 C2	Itaporanga Brazil
154 C2	Itapuranga Brazil
154 C2	Itararé Brazil
74 B2	Itarsi India
154 B1	Itarumã Brazil
150 D2	Itaúna Brazil
155 E1	Itaúnas Brazil
64 B1	Itbayat i. Phil.
139 D2	Ithaca U.S.A.
101 D1	Ith Hils ridge Ger.
118 C2	Itimbiri r. Dem. Rep. Congo
155 D1	Itinga Brazil
154 B1	Itiquira Brazil
154 A1	Itiquira r. Brazil
67 C4	Itō Japan
154 C2	Itu Brazil
150 B3	Itui r. Brazil
154 C1	Ituiutaba Brazil
119 C3	Itula Dem. Rep. Congo
154 C1	Itumbiara Brazil
154 C1	Itumbiara, Barragem resr Brazil
154 B1	Iturama Brazil
101 D1	Itzehoe Ger.
83 N2	Iul'tin Rus. Fed.
154 B1	Ivaí r. Brazil
92 I2	Ivalo Rus. Fed.
88 C3	Ivalo Fin.
88 C3	Ivanava Belarus
109 C1	Ivanec Croatia
	Ivangrad Montenegro see Berane
52 B2	Ivanhoe Austr.
90 B1	Ivankiv Ukr.
90 A2	Ivano-Frankivs'k Ukr.
89 F2	Ivanovo Rus. Fed.
88 C3	Ivatsevichy Belarus
111 C2	Ivaylovgrad Bulg.
86 F2	Ivdel' Rus. Fed.
110 C1	Iveşti Romania
154 B2	Ivinheima Brazil
154 B2	Ivinheima r. Brazil
127 I2	Ivittuut Greenland
	Iviza i. Spain see Ibiza
89 E3	Ivnya Rus. Fed.
	Ivory Coast country Africa see Côte d'Ivoire
108 A1	Ivrea Italy
111 C3	İvrindi Turkey
	Ivugivik Can. see Ivujivik
127 G2	Ivujivik Can.
67 D3	Iwaki Japan
67 B4	Iwakuni Japan
66 D2	Iwamizawa Japan
66 D2	Iwanai Japan
66 D3	Iwate-san vol. Japan
115 C4	Iwye Belarus
113 A3	Ixopo S. Africa
144 B2	Ixtlán Mex.
146 B3	Izabal, Lago de l. Guat.
145 D2	Izamal Mex.
81 C1	Izberbash Rus. Fed.
86 E3	Izhevsk Rus. Fed.
	Izhma Rus. Fed. see Sosnogorsk
86 E2	Izhma Rus. Fed.
89 E3	Izmalkovo Rus. Fed.
90 B2	İzmayil Ukr.
111 C3	İzmir Turkey
111 C2	İzmit Turkey
111 C2	İznik Gölü l. Turkey
152 B1	Izozog, Bañados del swamp Bol.
67 B3	Izumo Japan
156 C4	Izu-Ogasawara Trench N. Pacific Ocean
67 C4	Izu-shotō is Japan
90 B1	Izyaslav Ukr.
91 D2	Izyum Ukr.

J

106 C2	Jabalón r. Spain
75 B2	Jabalpur India
74 A1	Jabal us Sarāj Afgh.
50 C1	Jabiru Austr.
109 C2	Jablanica Bos.-Herz.
151 F3	Jaboatão dos Guararapes Brazil
154 C2	Jaboticabal Brazil
107 C1	Jaca Spain
145 C2	Jacala Mex.
151 D3	Jacareacanga Brazil
155 C2	Jacareí Brazil
154 B1	Jaciara Brazil
155 D1	Jacinto Brazil
139 E1	Jackman U.S.A.
140 C2	Jackson AL U.S.A.
138 C2	Jackson MI U.S.A.
140 B2	Jackson MS U.S.A.
140 C1	Jackson TN U.S.A.
136 A2	Jackson WY U.S.A.
54 A2	Jackson Head hd N.Z.
140 B2	Jacksonville AL U.S.A.
141 D3	Jacksonville FL U.S.A.
138 A3	Jacksonville IL U.S.A.
141 E2	Jacksonville NC U.S.A.
143 D2	Jacksonville TX U.S.A.
147 C3	Jacmel Haiti
74 A2	Jacobabad Pak.
151 E4	Jacobina Brazil
131 D3	Jacques-Cartier, Mont mt. Can.
151 E3	Jacundá Brazil
154 C2	Jacupiranga Brazil
101 D1	Jadebusen b. Ger.
	Jadotville Dem. Rep. Congo see Likasi
109 C2	Jadovnik mt. Bos.-Herz.
150 B3	Jaén Peru
106 C2	Jaén Spain
	Jaffa Israel see Tel Aviv-Yafo
52 A3	Jaffa, Cape Austr.
73 B4	Jaffna Sri Lanka
75 C3	Jagdalpur India
123 C2	Jagersfontein S. Africa
79 C2	Jaghin Iran
154 C2	Jaguarão Brazil
154 C2	Jaguariaíva Brazil
74 B2	Jahrom Iran
74 B2	Jaipur India
74 B2	Jaisalmer India

74 B2 Jaitgarh h. India
75 C2 Jajarkot Nepal
109 C2 Jajce Bos.-Herz.
Jajnagar state India see Odisha
60 B2 Jakarta Indon.
128 A1 Jakes Corner Can.
92 G2 Jäkkvik Sweden
Jakobshavn Greenland see Ilulissat
92 H3 Jakobstad Fin.
77 D3 Jalālābād Afgh.
77 D2 Jalal-Abad Kyrg.
74 B1 Jalandhar India
Jalapa Mex. see Xalapa
154 B2 Jales Brazil
74 B2 Jalgaon India
115 D4 Jalingo Nigeria
74 B3 Jalna India
74 B2 Jalore India
144 B2 Jalostotitlán Mex.
144 B2 Jalpa Mex.
75 C2 Jalpaiguri India
145 C2 Jalpan Mex.
115 E2 Jālū Libya
146 C3 Jamaica country West Indies
146 C3 Jamaica Channel Haiti/Jamaica
75 C2 Jamalpur Bangl.
60 B2 Jambi Indon.
137 D2 James r. N. Dakota/S. Dakota U.S.A.
139 D3 James r. VA U.S.A.
130 B2 James Bay Can.
52 A2 Jamestown Austr.
123 C3 Jamestown S. Africa
Jamestown Can. see Wawa
137 D1 Jamestown ND U.S.A.
139 D2 Jamestown NY U.S.A.
74 B1 Jammu India
74 B1 Jammu and Kashmir state India
74 B2 Jamnagar India
74 B2 Jampur Pak.
93 I3 Jämsä Fin.
78 A2 Jamsah Egypt
75 C2 Jamshedpur India
75 C2 Jamuna r. Bangl.
75 C2 Janakpur Nepal
155 D1 Janaúba Brazil
81 D2 Jandaq Iran
114 A2 Jandía h. Islas Canarias
135 B2 Janesville CA U.S.A.
138 B2 Janesville WI U.S.A.
65 B3 Jangheung S. Korea
154 A2 Jango Brazil
65 B2 Jangseong S. Korea
80 B2 Janīn West Bank
114 A3 Janjanbureh Gambia
82 A2 Jan Mayen terr. Arctic Ocean
81 D1 Jaña Turkm.
122 B3 Jansenville S. Africa
155 D1 Januária Brazil
74 B2 Jaora India
67 C3 Japan country Asia
66 B3 Japan, Sea of N. Pacific Ocean
150 C3 Japurá r. Brazil
154 C1 Jaraguá Brazil
154 C3 Jaraguá, Serra mts Brazil
154 C3 Jaraguá do Sul Brazil
154 B2 Jaraguari Brazil
70 A2 Jarantai China
152 C2 Jardim Brazil
103 D1 Jarocin Pol.
103 E1 Jarosław Pol.
92 F3 Järpen Sweden
150 C4 Jarú Brazil
Jarud China see Lubei
88 C1 Järvenpää Fin.
49 K4 Jarvis Island terr. S. Pacific Ocean
93 G3 Järvsö Sweden
79 C2 Jāsk Iran
103 E2 Jasło Pol.
55 A3 Jason Peninsula Antarctica
128 C2 Jasper Can.
140 C2 Jasper AL U.S.A.
138 B3 Jasper IN U.S.A.
143 E2 Jasper TX U.S.A.
Jassy Romania see Iași
103 D2 Jastrzębie-Zdrój Pol.
103 D2 Jászberény Hungary
154 B1 Jataí Brazil
74 A2 Jati India
154 C2 Jaú Brazil
150 C3 Jaú r. Brazil
145 C2 Jaumave Mex.
75 C2 Jaunpur India

154 B1 Jauru Brazil
61 B2 Java i. Indon.
107 C1 Javalambre, Sierra de mts Spain
Java Sea sea Indon. see Laut Jawa
156 B6 Java Trench Indian Ocean
Jawa i. Indon. see Java
117 C4 Jawhar Somalia
103 D1 Jawor Pol.
103 D1 Jaworzno Pol.
59 D3 Jaya, Puncak mt. Indon.
59 D3 Jayapura Indon.
79 C2 Jazīrat Maşīrah i. Oman
79 C2 Jaz Mūrīān, Hāmūn-e imp. l. Iran
128 B1 Jean Marie River Can.
131 D2 Jeannin, Lac l. Can.
81 D2 Jebel Turkm.
Jebel, Bahr el r. Africa see White Nile
116 A3 Jebel Abyad Plateau Sudan
65 B2 Jecheon S. Korea
96 C3 Jedburgh U.K.
78 A2 Jeddah Saudi Arabia
108 A3 Jedeida Tunisia
101 E1 Jeetze r. Ger.
137 E2 Jefferson U.S.A.
135 C3 Jefferson, Mount U.S.A.
137 E3 Jefferson City U.S.A.
122 B3 Jeffreys Bay S. Africa
65 B3 Jeju S. Korea
65 B3 Jeju-do i. S. Korea
65 B3 Jeju-haehyeop sea chan. S. Korea
88 C2 Jēkabpils Latvia
103 D1 Jelenia Góra Pol.
88 B2 Jelgava Latvia
60 B1 Jemaja i. Indon.
61 C2 Jember Indon.
101 E2 Jena Ger.
Jengish Chokusu mt. China/Kyrg. see Pobeda Peak
140 B2 Jennings U.S.A.
65 B2 Jeongeup S. Korea
65 B2 Jeonju S. Korea
151 E4 Jequié Brazil
155 D1 Jequitaí Brazil
155 D1 Jequitinhonha Brazil
155 E1 Jequitinhonha r. Brazil
147 C3 Jérémie Haiti
144 B2 Jerez Mex.
106 B2 Jerez de la Frontera Spain
109 D3 Jergucat Albania
115 C1 Jerid, Chott el salt l. Tunisia
134 D2 Jerome U.S.A.
95 C4 Jersey i. Channel Is
95 C4 Jersey terr. Channel Is
151 E3 Jerumenha Brazil
80 B2 Jerusalem Israel/West Bank
53 D3 Jervis Bay Territory admin. div. Austr.
108 B1 Jesenice Slovenia
108 B2 Jesi Italy
Jesselton Malaysia see Kota Kinabalu
101 F2 Jessen Ger.
75 C2 Jessore Bangl.
141 D2 Jesup U.S.A.
137 D3 Jesus, Mount h. U.S.A.
145 C3 Jesús Carranza Mex.
100 C1 Jever Ger.
109 C2 Jezercë, Maja mt. Albania
74 B2 Jhalawar India
74 B1 Jhang Pak.
74 B2 Jhansi India
75 C2 Jharkhand state India
75 C2 Jharsuguda India
74 B1 Jhelum India/Pak.
74 B1 Jhelum r. India/Pak.
74 B2 Jhunjhunun India
70 C2 Jiading China
69 E1 Jiamusi China
71 B3 Ji'an Jiangxi China
65 B1 Ji'an Jilin China
Jianchang China see Nancheng
62 A1 Jianchuan China
70 B3 Jiande China
Jiangling China see Jingzhou
71 B3 Jiangmen China
70 B2 Jiangsu prov. China
71 B3 Jiangxi prov. China
70 A2 Jiangyou China
70 B2 Jianli China
70 B2 Jianqiao China
71 B3 Jianyang Fujian China
70 A2 Jianyang Sichuan China
69 E2 Jiaohe China

Jiaojiang China see Taizhou
70 C2 Jiaozhou China
70 B2 Jiaozuo China
Jiashan China see Mingguang
70 C2 Jiaxing China
71 C3 Jiayi Taiwan
68 C2 Jiayuguan China
Jiddah Saudi Arabia see Jeddah
92 G2 Jiehkkevárri mt. Norway
70 B2 Jiexiu China
70 A2 Jigzhi China
103 D2 Jihlava Czech Rep.
115 C1 Jijel Alg.
117 C4 Jijiga Eth.
117 C4 Jilib Somalia
69 E2 Jilin China
65 B1 Jilin prov. China
65 A1 Jilin Hada Ling mts China
117 B4 Jīma Eth.
110 B1 Jimbolia Romania
144 B2 Jiménez S. Korea
143 C3 Jiménez Mex.
70 B2 Jinan China
70 A2 Jinchang China
70 B2 Jincheng China
Jinchuan China see Jinchang
70 B2 Jindabyne Austr.
65 B3 Jin-do i. S. Korea
103 D2 Jindřichův Hradec Czech Rep.
Jin'e China see Longchang
62 B1 Jingbian China
70 C2 Jingdezhen China
62 B1 Jingdong China
70 A2 Jingellic Austr.
Jinggangshan China see Ciping
70 B2 Jinghong Yunhe canal China
62 B1 Jinghong China
70 B2 Jingmen China
70 A2 Jingning China
70 B2 Jingsha China see Jingzhou
70 B2 Jingtai China
71 A3 Jingxi China
Jingxian China see Jingzhou
65 B1 Jingyu China
70 A2 Jingyuan China
70 B2 Jingzhou Hubei China
70 B2 Jingzhou Hubei China
71 A3 Jingzhou Hunan China
65 B2 Jinhae S. Korea
71 B3 Jinhua China
70 B2 Jining China
119 D2 Jinja Uganda
117 B4 Jinka Eth.
71 B3 Jinmen Taiwan
146 B3 Jinotega Nic.
146 B3 Jinotepe Nic.
71 A3 Jinping China
Jinsha Jiang r. China see Yangtze
70 B3 Jinshi China
Jinshi China see Xinning
Jinxi China see Lianshan
70 B2 Jinzhong China
70 C1 Jinzhou China
150 C3 Ji-Paraná r. Brazil
75 C1 Jirang China
65 B2 Jiri-san mt. S. Korea
116 B2 Jirjā Egypt
79 C2 Jīroft Iran
79 C2 Jīrwān Saudi Arabia
71 A3 Jishou China
110 B2 Jiu r. Romania
70 A2 Jiuding Shan mt. China
62 A1 Jiuhe China
70 B3 Jiujiang China
Jiulian China see Mojiang
79 D2 Jiwani Pak.
66 B1 Jixi China
78 B3 Jīzān Saudi Arabia
77 C2 Jizzax Uzbek.
155 D1 Joaíma Brazil
João Belo Moz. see Xai-Xai
151 F3 João Pessoa Brazil
155 C1 João Pinheiro Brazil
74 B2 Jodhpur India
92 I3 Joensuu Fin.
67 C3 Jōetsu Japan
121 C3 Jofane Moz.
88 C2 Jõgeva Estonia
Jogjakarta Indon. see Yogyakarta
123 C2 Johannesburg S. Africa
134 C2 John Day U.S.A.
134 B1 John Day r. U.S.A.

128 C2 John D'Or Prairie Can.
141 E1 John H. Kerr Reservoir U.S.A.
96 C1 John o' Groats U.K.
141 D1 Johnson City U.S.A.
128 A1 Johnson's Crossing Can.
50 B3 Johnston, Lake imp. l. Aus
49 J2 Johnston Atoll N. Pacific Ocean
96 B3 Johnstone U.K.
Johnstone Lake l. Can. see Old Wives Lake
139 D2 Johnstown U.S.A.
60 B1 Johor Bahru Malaysia
88 C2 Jõhvi Estonia
154 C3 Joinville Brazil
105 D2 Joinville France
55 B3 Joinville Island Antarctica
92 G2 Jokkmokk Sweden
92 □B2 Jökulsá á Fjöllum r. Icelan
138 B2 Joliet U.S.A.
130 C3 Joliette Can.
64 B3 Jolo Phil.
64 B3 Jolo i. Phil.
61 C2 Jombang Indon.
75 C2 Jomsom Nepal
88 B2 Jonava Lith.
140 B1 Jonesboro AR U.S.A.
140 B2 Jonesboro LA U.S.A.
139 F2 Jonesport U.S.A.
127 G1 Jones Sound sea chan. Can
93 F4 Jönköping Sweden
131 C3 Jonquière Can.
145 C3 Jonuta Mex.
137 E3 Joplin U.S.A.
80 B2 Jordan country Asia
80 B2 Jordan r. Asia
134 E1 Jordan U.S.A.
155 D1 Jordânia Brazil
134 C2 Jordan Valley U.S.A.
72 D2 Jorhat India
101 D1 Jork Ger.
93 E4 Jørpeland Norway
115 C4 Jos Nigeria
145 C3 José Cardel Mex.
131 D2 Joseph, Lac l. Can.
50 B1 Joseph Bonaparte Gulf Austr.
115 C4 Jos Plateau Nigeria
93 E3 Jotunheimen mts Norway
122 B3 Joubertina S. Africa
123 C2 Jouberton S. Africa
104 C2 Joué-lès-Tours France
93 I3 Joutseno Fin.
134 B1 Juan de Fuca Strait Can./U.S.A.
Juanshui China see Tongcheng
145 B2 Juárez Mex.
144 A1 Juárez, Sierra de mts Mex
151 E3 Juazeiro Brazil
151 F3 Juazeiro do Norte Brazil
117 B4 Juba S. Sudan
117 C5 Jubba r. Somalia
78 B2 Jubbah Saudi Arabia
Jubbulpore India see Jabalpur
145 C3 Juchitán Mex.
155 E1 Jucuruçu Brazil
102 C2 Judenburg Austria
155 E1 Juerana Brazil
101 D2 Jühnde Ger.
146 B3 Juigalpa Nic.
150 D4 Juína Brazil
100 C1 Juist i. Ger.
155 D2 Juiz de Fora Brazil
136 C2 Julesburg U.S.A.
150 B4 Juliaca Peru
Julianatop mt. Indon. see Mandala, Puncak
151 D2 Juliana Top mt. Suriname
Jullundur India see Jalandhar
107 C2 Jumilla Spain
75 C2 Jumla Nepal
Jumna r. India see Yamun
74 B2 Junagadh India
143 D3 Junction U.S.A.
137 D3 Junction City U.S.A.
154 C2 Jundiaí Brazil
128 A2 Juneau U.S.A.
53 C2 Junee Austr.
105 D2 Jungfrau mt. Switz.
139 D2 Juniata r. U.S.A.
153 B3 Junín Arg.
92 G3 Junsele Sweden
134 C2 Juntura U.S.A.
Junxi China see Datian
Junxian China see Danjiangkou

54 B2 Jupía, Represa *resr* Brazil
41 D3 Jupiter U.S.A.
54 C2 Juquiá Brazil
17 A4 Jur *r.* S. Sudan
05 D2 Jura *mts* France/Switz.
96 B2 Jura *i.* U.K.
96 B3 Jura, Sound of *sea chan.* U.K.
88 B2 Jurbarkas Lith.
88 B2 Jūrmala Latvia
50 C3 Jurua *r.* Brazil
50 D3 Juruena *r.* Brazil
54 C2 Jurumirim, Represa de *resr* Brazil
51 D3 Juruti Brazil
54 B1 Jussara Brazil
50 C3 Jutaí *r.* Brazil
01 F2 Jüterbog Ger.
54 B2 Juti Brazil
45 D3 Jutiapa Guat.
93 E4 Jutland *pen.* Denmark
46 B2 Juventud, Isla de la *i.* Cuba
70 B2 Juxian China
81 D3 Jūyom Iran
22 B1 Jwaneng Botswana
   Jylland *pen.* Denmark *see* Jutland
93 I3 Jyväskylä Fin.

K

74 B1 K2 *mt.* China/Pakistan
   Kaakhka Turkm. *see* Kaka
92 I2 Kaamanen Fin.
61 D2 Kabaena *i.* Indon.
19 C3 Kabalo Dem. Rep. Congo
19 C3 Kabambare Dem. Rep. Congo
19 C3 Kabare Dem. Rep. Congo
19 C3 Kabemba Dem. Rep. Congo
30 B3 Kabinakagami Lake Can.
18 C3 Kabinda Dem. Rep. Congo
18 B2 Kabo C.A.R.
20 B2 Kabompo Zambia
19 C3 Kabongo Dem. Rep. Congo
77 C3 Kābul Afgh.
64 B3 Kaburuang *i.* Indon.
21 B2 Kabwe Zambia
09 D2 Kaçanik Kosovo
74 A2 Kachchh, Gulf of India
74 B2 Kachchh, Rann of *marsh* India
83 I3 Kachug Rus. Fed.
81 C1 Kaçkar Dağı *mt.* Turkey
73 B3 Kadapa India
11 C2 Kadıköy Turkey
52 A2 Kadina Austr.
14 B3 Kadiolo Mali
   Kadiyevka Ukr. *see* Stakhanov
73 B3 Kadmat *atoll* India
89 F2 Kadnikov Rus. Fed.
21 B2 Kadoma Zimbabwe
63 A2 Kadonkani Myanmar
17 A3 Kadugli Sudan
15 C3 Kaduna Nigeria
89 E2 Kaduy Rus. Fed.
86 E2 Kadzherom Rus. Fed.
14 A3 Kaédi Maur.
18 B1 Kaélé Cameroon
65 B2 Kaesŏng N. Korea
18 C3 Kafakumba Dem. Rep. Congo
14 A3 Kaffrine Senegal
80 B2 Kafr ash Shaykh Egypt
21 B2 Kafue Zambia
20 B2 Kafue *r.* Zambia
67 C3 Kaga Japan
18 B2 Kaga Bandoro C.A.R.
91 E2 Kagal'nitskaya Rus. Fed.
   Kaganovich Pervyye Ukr. *see* Polis'ke
60 A2 Kagologolo Indon.
67 B4 Kagoshima Japan
   Kagul Moldova *see* Cahul
19 C3 Kahama Tanz.
90 C2 Kaharlyk Ukr.
61 C2 Kahayan *r.* Indon.
18 B3 Kahemba Dem. Rep. Congo
01 E2 Kahla Ger.
79 C2 Kahnūj Iran
92 H2 Kahperusvaarat *mts* Fin.
80 B2 Kahramanmaraş Turkey
79 C2 Kahūrak Iran
59 C3 Kai, Kepulauan *is* Indon.
15 C4 Kaiama Nigeria
54 B2 Kaiapoi N.Z.
59 C3 Kai Besar *i.* Indon.

70 B2 Kaifeng China
   Kaihua China *see* Wenshan
122 B2 Kaiingveld *reg.* S. Africa
59 C3 Kai Kecil *i.* Indon.
54 B2 Kaikoura N.Z.
114 A4 Kailahun Sierra Leone
   Kailas Range *mts* China *see* Gangdisê Shan
71 A3 Kaili China
59 C3 Kaimana Indon.
54 C1 Kaimanawa Mountains N.Z.
72 C2 Kaimur Range *hills* India
88 B2 Käina Estonia
67 C4 Kainan Japan
115 C3 Kainji Reservoir Nigeria
54 B1 Kaipara Harbour N.Z.
74 B2 Kairana India
115 D1 Kairouan Tunisia
100 C3 Kaiserslautern Ger.
55 I2 Kaiser Wilhelm II Land *reg.* Antarctica
54 B1 Kaitaia N.Z.
54 C1 Kaitawa N.Z.
   Kaitong China *see* Tongyu
59 C3 Kaiwatu Indon.
65 A1 Kaiyuan *Liaoning* China
71 A3 Kaiyuan *Yunnan* China
92 I3 Kajaani Fin.
51 D2 Kajabbi Austr.
53 C1 Kajarabie, Lake Austr.
76 B3 Kaka Turkm.
122 B2 Kakamas S. Africa
119 D2 Kakamega Kenya
114 A4 Kakata Liberia
91 C1 Kakhovka Ukr.
91 C2 Kakhovs'ke Vodoskhovyshche *resr* Ukr.
   Kakhul Moldova *see* Cahul
73 C3 Kakinada India
128 C1 Kakisa Can.
67 B4 Kakogawa Japan
119 C3 Kakoswa Dem. Rep. Congo
126 C2 Kaktovik U.S.A.
   Kalaallit Nunaat *terr.* N. America *see* Greenland
59 C3 Kalabahi Indon.
120 B2 Kalabo Zambia
91 E1 Kalach Rus. Fed.
119 D2 Kalacha Dida Kenya
87 D4 Kalach-na-Donu Rus. Fed.
62 A1 Kaladan *r.* India/Myanmar
120 B3 Kalahari Desert Africa
92 H3 Kalajoki Fin.
123 C1 Kalamare Botswana
111 B3 Kalamaria Greece
111 B3 Kalamata Greece
138 B2 Kalamazoo U.S.A.
111 B3 Kalampaka Greece
88 B2 Kalana Estonia
91 C2 Kalanchak Ukr.
115 E2 Kalanshiyū ar Ramlī al Kabīr, Sarīr *des.* Libya
61 D2 Kalao *i.* Indon.
61 D2 Kalaotoa *i.* Indon.
63 B2 Kalasin Thai.
79 C2 Kalāt Iran
74 A2 Kalat Pak.
50 A2 Kalbarri Austr.
   Kalburgi India *see* Gulbarga
111 C3 Kale Turkey
80 B1 Kalecik Turkey
118 C3 Kalema Dem. Rep. Congo
119 C3 Kalemie Dem. Rep. Congo
62 A1 Kalemyo Myanmar
86 C2 Kalevala Rus. Fed.
   Kalgan China *see* Zhangjiakou
50 B3 Kalgoorlie Austr.
109 C2 Kali Croatia
110 C2 Kaliakra, Nos *pt* Bulg.
60 A2 Kaliet Indon.
119 C3 Kalima Dem. Rep. Congo
61 C2 Kalimantan *reg.* Indon.
   Kalinin Rus. Fed. *see* Tver'
88 B3 Kaliningrad Rus. Fed.
91 D2 Kalininskaya Rus. Fed.
88 C3 Kalinkavichy Belarus
134 D1 Kalispell U.S.A.
103 D1 Kalisz Pol.
91 E2 Kalitva *r.* Rus. Fed.
92 H2 Kalix Sweden
92 H2 Kalixälven *r.* Sweden
111 C3 Kalkan Turkey
120 A3 Kalkfeld Namibia
100 C2 Kall Ger.
92 I3 Kallavesi *l.* Fin.
92 F3 Kallsjön *l.* Sweden
93 G4 Kalmar Sweden
93 G4 Kalmarsund *sea chan.* Sweden

73 C4 Kalmunai Sri Lanka
119 C3 Kalole Dem. Rep. Congo
120 B2 Kalomo Zambia
128 B2 Kalone Peak Can.
74 B1 Kalpa India
73 B3 Kalpeni *atoll* India
75 B2 Kalpi India
126 B2 Kaltag U.S.A.
101 D1 Kaltenkirchen Ger.
118 B2 Kaltungo Nigeria
89 E3 Kaluga Rus. Fed.
93 F4 Kalundborg Denmark
74 B1 Kalur Kot Pak.
90 A2 Kalush Ukr.
74 B3 Kalyan India
89 E2 Kalyazin Rus. Fed.
111 C3 Kalymnos Greece
111 C3 Kalymnos *i.* Greece
119 C3 Kama Dem. Rep. Congo
62 A2 Kama Myanmar
86 E3 Kama *r.* Rus. Fed.
66 D3 Kamaishi Japan
80 B2 Kaman Turkey
120 A2 Kamanjab Namibia
78 B3 Kamarān *i.* Yemen
   Kamaran Island *i.* Yemen *see* Kamarān
74 A2 Kamarod Pak.
50 B3 Kambalda Austr.
119 C4 Kambove Dem. Rep. Congo
160 C4 Kamchatka Basin Bering Sea
83 L3 Kamchatka Peninsula Rus. Fed.
110 C2 Kamchiya *r.* Bulg.
108 B2 Kamenjak, Rt *pt* Croatia
86 D2 Kamenka Rus. Fed.
87 D3 Kamenka Rus. Fed.
66 C2 Kamenka Rus. Fed.
91 D1 Kamenka Rus. Fed.
   Kamenka-Strumilovskaya Ukr. *see* Kam"yanka-Buz'ka
91 E3 Kamennomostskiy Rus. Fed.
91 E2 Kamenolomni Rus. Fed.
   Kamenongue Angola *see* Camanongue
83 M2 Kamenskoye Rus. Fed.
   Kamenskoye Ukr. *see* Dniprodzerzhyns'k
91 E2 Kamensk-Shakhtinskiy Rus. Fed.
86 F3 Kamensk-Ural'skiy Rus. Fed.
89 F2 Kameshkovo Rus. Fed.
72 C1 Kamet *mt.* China/India
75 B1 Kamet *mt.* China/India
122 A3 Kamiesberge *mts* S. Africa
122 A3 Kamieskroon S. Africa
129 D1 Kamilukuak Lake Can.
119 C3 Kamina Dem. Rep. Congo
129 E1 Kaminak Lake Can.
90 A1 Kamin'-Kashyrs'kyy Ukr.
119 C3 Kamituga Dem. Rep. Congo
128 B2 Kamloops Can.
54 B1 Kamo N.Z.
116 B3 Kamob Sanha Sudan
119 D2 Kamonia Dem. Rep. Congo
119 D2 Kampala Uganda
60 B1 Kampar *r.* Indon.
60 B1 Kampar Malaysia
100 B1 Kampen Neth.
119 C3 Kampene Dem. Rep. Congo
63 A2 Kamphaeng Phet Thai.
63 B2 Kâmpóng Cham Cambodia
63 B2 Kâmpóng Chhnǎng Cambodia
   Kâmpóng Saôm Cambodia *see* Sihanoukville
63 B2 Kâmpóng Spœ Cambodia
63 B2 Kâmpôt Cambodia
   Kampuchea *country* Asia *see* Cambodia
129 D2 Kamsack Can.
86 E3 Kamskoye Vodokhranilishche *resr* Rus. Fed.
117 C4 Kamsuuma Somalia
90 B2 Kam"yanets'-Podil's'kyy Ukr.
90 A1 Kam"yanka-Buz'ka Ukr.
88 B3 Kamyanyets Belarus
91 D2 Kamyshevatskaya Rus. Fed.
87 D3 Kamyshin Rus. Fed.
135 D3 Kanab U.S.A.
118 C3 Kananga Dem. Rep. Congo
87 D3 Kanash Rus. Fed.
138 C3 Kanawha *r.* U.S.A.
67 C3 Kanazawa Japan
62 A1 Kanbalu Myanmar

63 A2 Kanchanaburi Thai.
73 B3 Kanchipuram India
77 C3 Kandahār Afgh.
86 C2 Kandalaksha Rus. Fed.
61 C2 Kandangan Indon.
74 A2 Kandh Kot Pak.
114 C3 Kandi Benin
74 A2 Kandiaro Pak.
74 B2 Kandla India
53 C2 Kandos Austr.
121 D2 Kandreho Madag.
73 C4 Kandy Sri Lanka
76 B2 Kandyagash Kazakh.
127 H1 Kane Bassin *b.* Greenland
91 D2 Kanevskaya Rus. Fed.
122 B1 Kang Botswana
127 I2 Kangaatsiaq Greenland
114 B3 Kangaba Mali
80 B2 Kangal Turkey
79 C2 Kangān Iran
60 B1 Kangar Malaysia
52 A3 Kangaroo Island Austr.
93 H3 Kangasala Fin.
81 C2 Kangāvar Iran
75 C2 Kangchenjunga *mt.* India/Nepal
70 A2 Kangding China
65 B2 Kangdong N. Korea
61 C2 Kangean, Kepulauan *is* Indon.
119 D2 Kangen *r.* S. Sudan
127 J2 Kangeq *c.* Greenland
127 I2 Kangerlussuaq *inlet* Greenland
127 J2 Kangerlussuaq *inlet* Greenland
127 I2 Kangersuatsiaq Greenland
65 B1 Kanggye N. Korea
131 D2 Kangiqsualujjuaq Can.
127 H2 Kangiqsujuaq Can.
131 C1 Kangirsuk Can.
75 C2 Kangmar China
65 A1 Kangping China
72 D2 Kangto *mt.* China/India
62 A1 Kani Myanmar
118 C3 Kaniama Dem. Rep. Congo
61 C1 Kanibongan Malaysia
86 D2 Kanin, Poluostrov *pen.* Rus. Fed.
86 D2 Kanin Nos Rus. Fed.
86 D2 Kanin Nos, Mys *c.* Rus. Fed.
91 C2 Kaniv Ukr.
52 B3 Kaniva Austr.
93 H3 Kankaanpää Fin.
138 B2 Kankakee U.S.A.
114 B3 Kankan Guinea
75 C2 Kanker India
73 B3 Kannur India
115 C3 Kano Nigeria
122 B3 Kanonpunt *pt* S. Africa
67 B4 Kanoya Japan
75 C2 Kanpur India
136 C3 Kansas *r.* U.S.A.
137 D3 Kansas *state* U.S.A.
137 E3 Kansas City *KS* U.S.A.
137 E3 Kansas City *MO* U.S.A.
83 H3 Kansk Rus. Fed.
   Kansu *prov.* China *see* Gansu
63 B2 Kantaralak Thai.
114 C3 Kantchari Burkina Faso
49 J4 Kanton *atoll* Kiribati
97 B2 Kanturk Ireland
123 D2 Kanyamazane S. Africa
123 C1 Kanye Botswana
120 A2 Kaokoveld *plat.* Namibia
114 A3 Kaolack Senegal
120 B2 Kaoma Zambia
118 C3 Kapanga Dem. Rep. Congo
88 C3 Kapatkyevichy Belarus
100 B2 Kapellen Belgium
121 B2 Kapiri Mposhi Zambia
127 I2 Kapisillit Greenland
130 B2 Kapiskau *r.* Can.
63 A3 Kapoe Thai.
5 B4 Kapoeta S. Sudan
103 D2 Kaposvár Hungary
102 B1 Kappeln Ger.
65 B1 Kapsan N. Korea
76 B2 Kapshagay Kazakh.
77 D2 Kapshagay, Vodokhranilishche *resr* Kazakh.
   Kapsukas Lith. *see* Marijampolė
61 B2 Kapuas *r.* Indon.
52 A2 Kapunda Austr.
130 B3 Kapuskasing Can.

Kaputar

53 D2 **Kaputar** mt. Austr.
103 D2 **Kapuvár** Hungary
88 C3 **Kapyl'** Belarus
77 D3 **Kaqung** China
114 C4 **Kara** Togo
111 C3 **Kara Ada** i. Turkey
77 D2 **Kara-Balta** Kyrg.
76 C1 **Karabalyk** Kazakh.
81 D1 **Karabaur, Uval** hills Kazakh./Uzbek.
Kara-Bogaz-Gol Turkm. see **Garabogazköl**
80 B1 **Karabük** Turkey
76 C2 **Karabutak** Kazakh.
111 C2 **Karacabey** Turkey
111 C2 **Karaçaköy** Turkey
81 C1 **Karachayevsk** Rus. Fed.
89 D3 **Karachev** Rus. Fed.
74 A2 **Karachi** Pak.
77 D2 **Karagandy** Kazakh.
77 D2 **Karagayly** Kazakh.
83 L3 **Karaginskiy Zaliv** b. Rus. Fed.
81 D2 **Karaj** Iran
Kara-Kala Turkm. see **Magtymguly**
64 B3 **Karakelong** i. Indon.
Karaklis Armenia see **Vanadzor**
77 D2 **Kara-Köl** Kyrg.
77 D2 **Karakol** Kyrg.
74 B1 **Karakoram Range** mts Asia
117 B3 **Kara K'orē** Eth.
Karakum, Peski des. Kazakh. see **Karakum Desert**
76 B2 **Karakum Desert** des. Kazakh.
76 C3 **Karakum Desert** Turkm.
Karakumy, Peski des. Turkm. see **Karakum Desert**
80 B2 **Karaman** Turkey
77 E2 **Karamay** China
54 B2 **Karamea** N.Z.
54 B2 **Karamea Bight** b. N.Z.
80 B2 **Karapınar** Turkey
122 A2 **Karasburg** Namibia
86 F1 **Kara Sea** Rus. Fed.
92 I2 **Karasjok** Norway
Kara Strait str. Rus. Fed. see **Karskiye Vorota, Proliv**
111 D2 **Karasu** Turkey
Karasubazar Ukr. see **Bilohirs'k**
77 D1 **Karasuk** Rus. Fed.
77 D2 **Karatau** Kazakh.
77 C2 **Karatau, Khrebet** mts Kazakh.
86 F2 **Karatayka** Rus. Fed.
67 A4 **Karatsu** Japan
111 B3 **Karavas** Greece
60 B2 **Karawang** Indon.
81 C2 **Karbalā'** Iraq
81 D2 **Karbüsh, Küh-e** mt. Iran
103 E2 **Karcag** Hungary
Kardeljevo Croatia see **Ploče**
111 B3 **Karditsa** Greece
88 B2 **Kärdla** Estonia
122 B3 **Kareeberge** mts S. Africa
75 B2 **Kareli** India
88 C3 **Karelichy** Belarus
92 H2 **Karesuando** Sweden
Karghalik China see **Yecheng**
74 B1 **Kargil** India
Kargilik China see **Yecheng**
86 C2 **Kargopol'** Rus. Fed.
118 B1 **Kari** Nigeria
121 B2 **Kariba** Zimbabwe
121 B2 **Kariba, Lake** resr Zambia/Zimbabwe
60 B2 **Karimata, Pulau-pulau** is Indon.
60 B2 **Karimata, Selat** str. Indon.
73 B3 **Karimnagar** India
61 C2 **Karimunjawa, Pulau-pulau** is Indon.
91 C2 **Karkinits'ka Zatoka** g. Ukr.
91 D2 **Karlivka** Ukr.
Karl-Marx-Stadt Ger. see **Chemnitz**
109 C1 **Karlovac** Croatia
102 C1 **Karlovy Vary** Czech Rep.
Karlsburg Romania see **Alba Iulia**
93 F4 **Karlshamn** Sweden
93 F4 **Karlskoga** Sweden
93 G4 **Karlskrona** Sweden
102 B2 **Karlsruhe** Ger.
93 F4 **Karlstad** Sweden
101 D3 **Karlstadt** Ger.

89 D3 **Karma** Belarus
93 E4 **Karmøy** i. Norway
74 B2 **Karnal** India
75 D2 **Karnaphuli Reservoir** Bangl.
110 C2 **Karnobat** Bulg.
74 A2 **Karodi** Pak.
121 B2 **Karoi** Zimbabwe
121 C1 **Karonga** Malawi
116 B3 **Karora** Eritrea
111 C3 **Karpathos** Greece
111 C3 **Karpathos** i. Greece
111 B3 **Karpenisi** Greece
Karpilovka Belarus see **Aktsyabrski**
86 D2 **Karpogory** Rus. Fed.
50 A2 **Karratha** Austr.
81 C1 **Kars** Turkey
88 C2 **Kärsava** Latvia
Karshi Uzbek. see **Qarshi**
111 C3 **Karşıyaka** Turkey
86 E2 **Karskiye Vorota, Proliv** str. Rus. Fed.
Karskoye More sea Rus. Fed. see **Kara Sea**
101 E1 **Karstädt** Ger.
111 C2 **Kartal** Turkey
87 F3 **Kartaly** Rus. Fed.
81 C2 **Kārūn, Rūd-e** r. Iran
73 B3 **Karwar** India
83 I3 **Karymskoye** Rus. Fed.
111 C3 **Karystos** Greece
111 C3 **Kaş** Turkey
130 B2 **Kasabonika Lake** Can.
118 C3 **Kasaï, Plateau du** Dem. Rep. Congo
118 C4 **Kasaji** Dem. Rep. Congo
121 C2 **Kasama** Zambia
120 B2 **Kasane** Botswana
118 B3 **Kasangulu** Dem. Rep. Congo
73 B3 **Kasaragod** India
129 D1 **Kasba Lake** Can.
120 B2 **Kasempa** Zambia
119 C4 **Kasenga** Dem. Rep. Congo
119 C3 **Kasese** Dem. Rep. Congo
119 D2 **Kasese** Uganda
Kasevo Rus. Fed. see **Neftekamsk**
81 D2 **Kāshān** Iran
77 D3 **Kashi** China
Kashgar China see **Kashi**
67 D3 **Kashima-nada** b. Japan
89 E2 **Kashin** Rus. Fed.
89 E3 **Kashira** Rus. Fed.
89 E3 **Kashirskoye** Rus. Fed.
67 C3 **Kashiwazaki** Japan
76 B3 **Kāshmar** Iran
74 B1 **Kashmir** reg. Asia
74 A2 **Kashmore** Pak.
119 C3 **Kashyukulu** Dem. Rep. Congo
89 F3 **Kasimov** Rus. Fed.
138 B3 **Kaskaskia** r. U.S.A.
93 H3 **Kaskinen** Fin.
119 C3 **Kasongo** Dem. Rep. Congo
118 B3 **Kasongo-Lunda** Dem. Rep. Congo
111 C3 **Kasos** i. Greece
Kaspiyskiy Rus. Fed. see **Lagan'**
116 B3 **Kassala** Sudan
101 D2 **Kassel** Ger.
115 C1 **Kasserine** Tunisia
80 B1 **Kastamonu** Turkey
Kastellorizon i. Greece see **Megisti**
111 B2 **Kastoria** Greece
89 D3 **Kastsyukovichy** Belarus
119 D3 **Kasulu** Tanz.
121 C2 **Kasungu** Malawi
139 F1 **Katahdin, Mount** U.S.A.
118 C3 **Katako-Kombe** Dem. Rep. Congo
119 D2 **Katakwi** Uganda
50 A3 **Katanning** Austr.
63 A3 **Katchall** i. India
111 B2 **Katerini** Greece
119 D3 **Katesh** Tanz.
128 A2 **Kate's Needle** mt. Can./U.S.A.
121 C2 **Katete** Zambia
62 A1 **Katha** Myanmar
50 C1 **Katherine** Austr.
50 C1 **Katherine** r. Austr.
74 B2 **Kathiawar** pen. India
75 C2 **Kathmandu** Nepal
122 B2 **Kathu** S. Africa
74 B1 **Kathua** India
114 B3 **Kati** Mali

75 C2 **Katihar** India
54 C1 **Katikati** N.Z.
123 C3 **Katikati** S. Africa
120 B2 **Katima Mulilo** Namibia
114 B4 **Katiola** Côte d'Ivoire
123 C2 **Katlehong** S. Africa
Katmandu Nepal see **Kathmandu**
111 B3 **Kato Achaïa** Greece
119 C3 **Katompi** Dem. Rep. Congo
53 D2 **Katoomba** Austr.
103 D1 **Katowice** Pol.
80 B3 **Kātrīnā, Jabal** mt. Egypt
93 G4 **Katrineholm** Sweden
115 C3 **Katsina** Nigeria
115 C4 **Katsina-Ala** Nigeria
67 D3 **Katsuura** Japan
77 C3 **Kattaqo'rg'on** Uzbek.
93 F4 **Kattegat** str. Denmark/Sweden
100 B1 **Katwijk aan Zee** Neth.
101 D3 **Katzenbuckel** h. Ger.
49 L1 **Kaua'i** i. U.S.A.
93 H3 **Kauhajoki** Fin.
88 B3 **Kaunas** Lith.
115 C3 **Kaura-Namoda** Nigeria
Kaushany Moldova see **Căuşeni**
92 H2 **Kautokeino** Norway
109 D2 **Kavadarci** Macedonia
109 C2 **Kavajë** Albania
111 B2 **Kavala** Greece
66 C2 **Kavalerovo** Rus. Fed.
73 C3 **Kavali** India
73 B3 **Kavaratti** atoll India
110 C2 **Kavarna** Bulg.
59 E3 **Kavieng** P.N.G.
81 D2 **Kavīr, Dasht-e** des. Iran
67 C3 **Kawagoe** Japan
54 B1 **Kawakawa** N.Z.
121 B1 **Kawambwa** Zambia
67 C3 **Kawanishi** Japan
130 C3 **Kawartha Lakes** Can.
67 C3 **Kawasaki** Japan
54 C1 **Kawerau** N.Z.
63 A2 **Kawkareik** Myanmar
62 A1 **Kawlin** Myanmar
63 A2 **Kawmapyin** Myanmar
116 B2 **Kawm Umbū** Egypt
63 A2 **Kawthaung** Myanmar
Kaxgar China see **Kashi**
77 D3 **Kaxgar He** r. China
114 B3 **Kaya** Burkina Faso
111 C3 **Kayacı Dağı** h. Turkey
121 C1 **Kayambi** Zambia
61 C1 **Kayan** r. Indon.
136 B2 **Kaycee** U.S.A.
Kaydanovo Belarus see **Dzyarzhynsk**
142 A1 **Kayenta** U.S.A.
114 A3 **Kayes** Mali
77 D2 **Kaynar** Kazakh.
80 B2 **Kayseri** Turkey
134 D2 **Kaysville** U.S.A.
60 B2 **Kayuagung** Indon.
Kazakhskaya S.S.R. country Asia see **Kazakhstan**
76 B2 **Kazakhskiy Zaliv** b. Kazakh.
76 C2 **Kazakhstan** country Asia
Kazakhstan Kazakh. see **Aksay**
87 D3 **Kazan'** Rus. Fed.
Kazandzhik Turkm. see **Bereket**
110 C2 **Kazanlŭk** Bulg.
Kazan-rettō is Japan see **Volcano Islands**
76 A2 **Kazbek** mt. Georgia/Rus. Fed.
81 D3 **Kāzerūn** Iran
103 E2 **Kazincbarcika** Hungary
118 C3 **Kazumba** Dem. Rep. Congo
66 D2 **Kazuno** Japan
83 J2 **Kazym-Mys** Rus. Fed.
111 B3 **Kea** i. Greece
97 C1 **Keady** U.K.
137 D2 **Kearney** U.S.A.
142 A2 **Kearny** U.S.A.
115 C1 **Kebili** Tunisia
116 A3 **Kebkabiya** Sudan
92 G2 **Kebnekaise** mt. Sweden
117 C4 **K'ebri Dehar** Eth.
60 B2 **Kebumen** Indon.
128 B2 **Kechika** r. Can.
111 D3 **Keçiborlu** Turkey
103 D2 **Kecskemét** Hungary
88 B2 **Kėdainiai** Lith.
114 A3 **Kédougou** Senegal
103 D1 **Kędzierzyn-Koźle** Pol.
128 B1 **Keele** r. Can.

128 A1 **Keele Peak** Can.
Keelung Taiwan see **Chilung**
139 E2 **Keene** U.S.A.
122 A2 **Keetmanshoop** Namibia
129 E3 **Keewatin** Can.
Kefallonia i. Greece see **Cephalonia**
59 C3 **Kefamenanu** Indon.
92 □A3 **Keflavík** Iceland
77 D2 **Kegen** Kazakh.
128 C2 **Keg River** Can.
88 C2 **Kehra** Estonia
62 A1 **Kehsi Mansam** Myanmar
98 C3 **Keighley** U.K.
88 B2 **Keila** Estonia
122 B2 **Keimoes** S. Africa
92 I3 **Keitele** l. Fin.
52 B3 **Keith** Austr.
96 C2 **Keith** U.K.
128 B1 **Keith Arm** b. Can.
134 B2 **Keizer** U.S.A.
103 E2 **Kékes** mt. Hungary
117 C4 **K'elafo** Eth.
Kelang Malaysia see **Klang**
92 J2 **Kelesuayv, Gora** h. Rus. Fed.
102 C2 **Kelheim** Ger.
76 C3 **Kelif Uzboýy** marsh Turkm.
80 B1 **Kelkit** r. Turkey
128 B1 **Keller Lake** Can.
134 C1 **Kellogg** U.S.A.
92 I2 **Kelloselkä** Fin.
97 C2 **Kells** Ireland
88 B2 **Kelmė** Lith.
115 D4 **Kélo** Chad
128 C3 **Kelowna** Can.
96 C3 **Kelso** U.K.
134 B1 **Kelso** U.S.A.
60 B1 **Keluang** Malaysia
129 D2 **Kelvington** Can.
86 C2 **Kem'** Rus. Fed.
Ke Macina Mali see **Macina**
128 B2 **Kemano (abandoned)** Can.
118 C2 **Kembé** C.A.R.
111 C3 **Kemer** Turkey
82 G3 **Kemerovo** Rus. Fed.
92 H2 **Kemi** Fin.
92 I2 **Kemijärvi** Fin.
92 I2 **Kemijärvi** l. Fin.
92 I2 **Kemijoki** r. Fin.
136 A2 **Kemmerer** U.S.A.
92 I3 **Kempele** Fin.
55 D2 **Kemp Land** reg. Antarctica
55 A2 **Kemp Peninsula** Antarctica
53 D2 **Kempsey** Austr.
130 C3 **Kempt, Lac** l. Can.
102 C2 **Kempten (Allgäu)** Ger.
123 C2 **Kempton Park** S. Africa
61 C2 **Kemujan** i. Indon.
126 B2 **Kenai** U.S.A.
129 D2 **Kenaston** Can.
98 B2 **Kendal** U.K.
141 D3 **Kendall** U.S.A.
61 D2 **Kendari** Indon.
60 C2 **Kendawangan** Indon.
115 D3 **Kendégué** Chad
114 A4 **Kenema** Sierra Leone
118 B3 **Kenge** Dem. Rep. Congo
62 A1 **Kengtung** Myanmar
122 B2 **Kenhardt** S. Africa
114 B1 **Kenitra** Morocco
97 B3 **Kenmare** Ireland
136 C1 **Kenmare** U.S.A.
97 A3 **Kenmare River** inlet Ireland
100 C3 **Kenn** Ger.
143 C2 **Kenna** U.S.A.
139 F2 **Kennebec** r. U.S.A.
Kennedy, Cape c. U.S.A. see **Canaveral, Cape**
140 B3 **Kennet** U.S.A.
99 C4 **Kennet** r. U.K.
137 E3 **Kennett** U.S.A.
134 C1 **Kennewick** U.S.A.
130 A3 **Kenora** Can.
138 B2 **Kenosha** U.S.A.
142 C2 **Kent** U.S.A.
77 C2 **Kentau** Kazakh.
138 B3 **Kentucky** r. U.S.A.
138 C3 **Kentucky** state U.S.A.
138 B3 **Kentucky Lake** U.S.A.
140 B2 **Kentwood** U.S.A.
119 D2 **Kenya** country Africa
119 D3 **Kenya, Mount** Kenya
60 B1 **Kenyir, Tasik** resr Malaysia
137 E2 **Keokuk** U.S.A.
75 C2 **Keonjhar** India
111 C3 **Kepsut** Turkey
52 B3 **Kerang** Austr.
91 D2 **Kerch** Ukr.
59 D3 **Kerema** P.N.G.

28 C3 Keremeos Can.
16 B3 Keren Eritrea
81 C2 Kerend Iran
    Kerepakupai Merú *waterfall*
    Venez. *see* Angel Falls
59 E7 Kerguélen, Îles *is*
    Indian Ocean
59 E7 Kerguelen Plateau
    Indian Ocean
19 D3 Kericho Kenya
54 B1 Kerikeri N.Z.
60 B2 Kerinci, Gunung *vol.* Indon.
    Kerintji *vol.* Indon. *see*
    Kerinci, Gunung
00 C2 Kerkrade Neth.
11 A3 Kerkyra Greece
    Kerkyra *i.* Greece *see* Corfu
16 B3 Kerma Sudan
49 J7 Kermadec Islands
    S. Pacific Ocean
79 C1 Kermän Iran
81 C2 Kermänshäh Iran
    Kermine Uzbek. *see* Navoiy
43 C2 Kermit U.S.A.
35 C3 Kern *r.* U.S.A.
14 B4 Kérouané Guinea
00 C2 Kerpen Ger.
29 D2 Kerrobert Can.
43 D2 Kerrville U.S.A.
97 B2 Kerry Head *hd* Ireland
    Keryneia Cyprus *see* Kyrenia
66 D3 Kesagami Lake Can.
11 C2 Keşan Turkey
66 D3 Kesennuma Japan
74 B2 Keshod India
00 C2 Kessel Neth.
98 B2 Keswick U.K.
03 D2 Keszthely Hungary
82 G3 Ket' *r.* Rus. Fed.
60 C2 Ketapang Indon.
28 A2 Ketchikan U.S.A.
34 D2 Ketchum U.S.A.
14 B4 Kete Krachi Ghana
18 B2 Kétté Cameroon
99 C3 Kettering U.K.
38 C3 Kettering U.S.A.
34 C1 Kettle River Range *mts*
    U.S.A.
93 H3 Keuruu Fin.
00 C2 Kevelaer Ger.
48 B2 Kewanee U.S.A.
48 B1 Keweenaw Bay U.S.A.
38 B1 Keweenaw Peninsula U.S.A.
41 D3 Key Largo U.S.A.
74 B1 Keylong India
99 B4 Keynsham U.K.
39 D3 Keyser U.S.A.
41 D4 Key West U.S.A.
23 C2 Kgotsong S. Africa
69 F1 Khabarovsk Rus. Fed.
91 D3 Khadyzhensk Rus. Fed.
75 D2 Khagrachari Bangl.
74 A2 Khairpur Pak.
22 B1 Khakhea Botswana
76 B3 Khalīlābād Iran
86 F2 Khal'mer"yu Rus. Fed.
68 C1 Khamar-Daban, Khrebet
    mts Rus. Fed.
74 B2 Khambhat India
74 B3 Khambhat, Gulf of India
74 B2 Khamgaon India
79 C2 Khamīr Iran
78 B3 Khamis Mushayt
    Saudi Arabia
78 B3 Khamr Yemen
77 C3 Khānābād Afgh.
74 B2 Khandwa India
83 K2 Khandyga Rus. Fed.
74 B1 Khanewal Pak.
    Khan Hung Vietnam *see*
    Soc Trăng
83 J3 Khani Rus. Fed.
66 B2 Khanka, Lake China/
    Rus. Fed.
45 C2 Khannfoussa *h.* Alg.
74 B2 Khanpur Pak.
77 D2 Khantau Kazakh.
83 H2 Khantayskoye, Ozero *l.*
    Rus. Fed.
86 F2 Khanty-Mansiysk Rus. Fed.
63 A3 Khao Chum Thong Thai.
63 A2 Khao Laem, Ang Kep Nam
    Thai.
74 B1 Khaplu Pak.
87 D4 Kharabali Rus. Fed.
74 B2 Kharagpur India
79 C2 Khārān *r.* Iran
    Kharga Oasis *oasis* Egypt
    see Wāḩat al Khārijah
74 B2 Khargon India

91 D2 Kharkiv Ukr.
    Khar'kov Ukr. *see* Kharkiv
111 C2 Kharmanli Bulg.
89 F2 Kharovsk Rus. Fed.
116 B3 Khartoum Sudan
87 D4 Khasavyurt Rus. Fed.
79 D2 Khāsh Iran
78 A3 Khashm el Girba Sudan
78 A3 Khashm el Girba Dam
    Sudan
81 C1 Khashuri Georgia
75 D2 Khasi Hills India
111 C2 Khaskovo Bulg.
83 H2 Khatanga Rus. Fed.
123 C3 Khayamnandi S. Africa
78 A2 Khaybar Saudi Arabia
122 A3 Khayelitsha S. Africa
107 D2 Khemis Miliana Alg.
63 B2 Khemmarat Thai.
115 C1 Khenchela Alg.
81 D3 Kherämeh Iran
91 C2 Kherson Ukr.
83 H2 Kheta *r.* Rus. Fed.
69 D1 Khilok Rus. Fed.
89 E2 Khimki Rus. Fed.
    Khipro Pak.
89 E3 Khlevnoye Rus. Fed.
63 B2 Khlung Thai.
90 B2 Khmel'nyts'kyy Ukr.
    Khmer Republic *country*
    Asia *see* Cambodia
    Khodzheyli Uzbek. *see*
    Xo'jayli
89 E3 Khokhol'skiy Rus. Fed.
74 B2 Khokhropar Pak.
89 D2 Kholm Rus. Fed.
89 D2 Kholm-Zhirkovskiy
    Rus. Fed.
122 A1 Khomas Highland *hills*
    Namibia
89 E3 Khomutovo Rus. Fed.
79 C2 Khonj Iran
63 B2 Khon Kaen Thai.
62 A1 Khonsa India
83 K2 Khonuu Rus. Fed.
86 E2 Khorey-Ver Rus. Fed.
69 D1 Khorinsk Rus. Fed.
120 A3 Khorixas Namibia
66 B2 Khorol Rus. Fed.
91 C2 Khorol Ukr.
81 C2 Khorramābād Iran
81 C2 Khorramshahr Iran
77 D3 Khorugh Tajik.
86 F2 Khoshgort Rus. Fed.
77 C3 Khŏst Afgh.
    Khotan China *see* Hotan
90 B2 Khotyn Ukr.
114 B1 Khouribga Morocco
88 C3 Khoyniki Belarus
62 A1 Khreum Myanmar
76 B1 Khromtau Kazakh.
    Khrushchev Ukr. *see*
    Svitlovods'k
90 B2 Khrystynivka Ukr.
123 B1 Khudumelapye Botswana
77 C2 Khūjand Tajik.
63 B2 Khu Khan Thai.
78 A2 Khulays Saudi Arabia
75 C2 Khulna Bangl.
    Khŭninshahr Iran *see*
    Khorramshahr
79 B2 Khurays Saudi Arabia
74 B1 Khushab Pak.
90 A2 Khust Ukr.
123 C2 Khutsong S. Africa
74 A2 Khuzdar Pak.
81 D2 Khvänsär Iran
81 D3 Khvormüj Iran
81 C2 Khvoy Iran
89 D2 Khvoynaya Rus. Fed.
77 D3 Khyber Pass Afgh./Pak.
53 D2 Kiama Austr.
64 B3 Kiamba Phil.
119 C3 Kiambi Dem. Rep. Congo
    Kiangsi *prov.* China *see*
    Jiangxi
    Kiangsu *prov.* China *see*
    Jiangsu
119 D3 Kibaha Tanz.
119 D3 Kibaya Tanz.
119 D3 Kibiti Tanz.
119 C3 Kibombo Dem. Rep. Congo
119 D3 Kibondo Tanz.
119 D2 Kibre Mengist Eth.
119 D3 Kibungo Rwanda
111 B2 Kičevo Macedonia
114 C3 Kidal Mali
99 B3 Kidderminster U.K.
114 A3 Kidira Senegal
74 B1 Kidmang India

54 C1 Kidnappers, Cape N.Z.
102 C1 Kiel Ger.
103 E1 Kielce Pol.
98 B2 Kielder Water *resr* U.K.
119 C4 Kienge Dem. Rep. Congo
90 C1 Kiev Ukr.
114 A3 Kiffa Maur.
119 D3 Kigali Rwanda
119 C3 Kigoma Tanz.
88 B2 Kihnu *i.* Estonia
92 I2 Kiiminki Fin.
67 B4 Kii-suidō *sea chan.* Japan
109 D1 Kikinda Serbia
119 C3 Kikondja Dem. Rep. Congo
59 D3 Kikori P.N.G.
59 D3 Kikori *r.* P.N.G.
118 B3 Kikwit Dem. Rep. Congo
65 B1 Kilchu N. Korea
97 C2 Kilcock Ireland
97 C2 Kildare Ireland
118 B3 Kilembe Dem. Rep. Congo
143 E2 Kilgore U.S.A.
119 D3 Kilifi Kenya
119 D3 Kilimanjaro *vol.* Tanz.
119 D3 Kilindoni Tanz.
73 C4 Kilinochchi Sri Lanka
80 B2 Kilis Turkey
90 B2 Kiliya Ukr.
97 B2 Kilkee Ireland
97 D1 Kilkeel U.K.
97 C2 Kilkenny Ireland
111 B2 Kilkis Greece
97 B1 Killala Ireland
97 B1 Killala Bay Ireland
97 B2 Killaloe Ireland
128 C2 Killam Can.
97 B2 Killarney Ireland
143 D2 Killeen U.S.A.
96 B2 Killin U.K.
131 D1 Killiniq Can.
97 B2 Killorglin Ireland
97 B1 Killybegs Ireland
96 B3 Kilmarnock U.K.
53 B3 Kilmore Austr.
119 D3 Kilosa Tanz.
97 B2 Kilrush Ireland
119 C3 Kilwa Dem. Rep. Congo
119 D3 Kilwa Masoko Tanz.
119 D3 Kimambi Tanz.
52 A2 Kimba Austr.
136 C2 Kimball U.S.A.
59 E3 Kimbe P.N.G.
128 C3 Kimberley Can.
122 B2 Kimberley S. Africa
50 B1 Kimberley Plateau Austr.
65 B1 Kimch'aek N. Korea
127 H2 Kimmirut Can.
89 E3 Kimovsk Rus. Fed.
118 C3 Kimpanga
    Dem. Rep. Congo
118 B3 Kimpese Dem. Rep. Congo
89 E2 Kimry Rus. Fed.
61 C1 Kinabalu, Gunung *mt.*
    Malaysia
128 C2 Kinbasket Lake Can.
96 C1 Kinbrace U.K.
130 B3 Kincardine Can.
62 A1 Kinchang Myanmar
119 C3 Kinda Dem. Rep. Congo
98 C3 Kinder Scout *h.* U.K.
129 D2 Kindersley Can.
114 A3 Kindia Guinea
119 C3 Kindu Dem. Rep. Congo
89 F2 Kineshma Rus. Fed.
118 B3 Kingandu Dem. Rep. Congo
51 E2 Kingaroy Austr.
135 B3 King City U.S.A.
130 C2 King George Islands Can.
88 C2 Kingisepp Rus. Fed.
51 D3 King Island Austr.
    Kingisseppa Estonia *see*
    Kuressaare
50 B1 King Leopold Ranges *hills*
    Austr.
142 A1 Kingman U.S.A.
135 B3 Kings *r.* U.S.A.
52 A3 Kingscote Austr.
97 C2 Kingscourt Ireland
99 D3 King's Lynn U.K.
51 D3 King Sound *b.* Austr.
134 D2 Kings Peak U.S.A.
141 D1 Kingsport U.S.A.
51 D4 Kingston Austr.
130 C3 Kingston Can.
146 C3 Kingston Jamaica
139 E2 Kingston U.S.A.
52 A3 Kingston South East Austr.
98 C3 Kingston upon Hull U.K.
147 D3 Kingstown St Vincent
143 D3 Kingsville U.S.A.

99 B4 Kingswood U.K.
96 B2 Kingussie U.K.
126 F2 King William Island Can.
123 C3 King William's Town
    S. Africa
67 D3 Kinka-san *i.* Japan
96 B2 Kinlochleven U.K.
93 F4 Kinna Sweden
    Kinneret, Yam *l.* Israel *see*
    Galilee, Sea of
97 B3 Kinsale Ireland
118 B3 Kinshasa Dem. Rep. Congo
141 E1 Kinston U.S.A.
88 B2 Kintai Lith.
114 B4 Kintampo Ghana
96 C2 Kintore U.K.
96 B3 Kintyre *pen.* U.K.
62 A1 Kin-U Myanmar
119 D3 Kiomboi Tanz.
130 C3 Kipawa, Lac *l.* Can.
119 D3 Kipembawe Tanz.
119 D3 Kipengere Range *mts* Tanz.
129 D2 Kipling Can.
    Kipling Station Can. *see*
    Kipling
119 C4 Kipushi Dem. Rep. Congo
119 C4 Kipushia Dem. Rep. Congo
101 D2 Kirchhain Ger.
101 D3 Kirchheim-Bolanden Ger.
83 I3 Kirenga *r.* Rus. Fed.
83 I3 Kirensk Rus. Fed.
89 E3 Kireyevsk Rus. Fed.
    Kirghizia *country* Asia *see*
    Kyrgyzstan
77 D2 Kirghiz Range *mts* Kazakh./
    Kyrg.
    Kirgizskaya S.S.R. *country*
    Asia *see* Kyrgyzstan
49 J4 Kiribati *country*
    Pacific Ocean
80 B2 Kırıkkale Turkey
89 E2 Kirillov Rus. Fed.
    Kirin China *see* Jilin
    Kirin *prov.* China *see* Jilin
    Kirinyaga *mt.* Kenya *see*
    Kenya, Mount
89 D2 Kirishi Rus. Fed.
48 L3 Kiritimati *atoll* Kiribati
111 C3 Kırkağaç Turkey
98 B3 Kirkby U.K.
98 B2 Kirkby Stephen U.K.
96 C2 Kirkcaldy U.K.
96 B3 Kirkcudbright U.K.
92 J2 Kirkenes Norway
88 B1 Kirkkonummi Fin.
130 B3 Kirkland Lake Can.
111 C2 Kırklareli Turkey
137 E2 Kirksville U.S.A.
81 C2 Kirkūk Iraq
96 C1 Kirkwall U.K.
    Kirov Kazakh. *see* Balpyk Bi
86 D3 Kirov Rus. Fed.
86 D3 Kirov Rus. Fed.
    Kirovabad Azer. *see* Gäncä
    Kirovakan Armenia *see*
    Vanadzor
86 E3 Kirovo-Chepetsk Rus. Fed.
    Kirovo-Chepetskiy Rus. Fed.
    see Kirovo-Chepetsk
91 C2 Kirovohrad Ukr.
86 C2 Kirovsk Rus. Fed.
91 D2 Kirovs'ke Ukr.
    Kirovskiy Kazakh. *see*
    Balpyk Bi
66 B1 Kirovskiy Rus. Fed.
96 C2 Kirriemuir U.K.
86 E3 Kirs Rus. Fed.
87 D3 Kirsanov Rus. Fed.
80 B2 Kırşehir Turkey
74 A2 Kirthar Range *mts* Pak.
92 H2 Kiruna Sweden
67 C3 Kiryū Japan
89 E2 Kirzhach Rus. Fed.
119 D3 Kisaki Tanz.
119 C2 Kisangani Dem. Rep. Congo
118 B3 Kisantu Dem. Rep. Congo
60 A1 Kisaran Indon.
82 G3 Kiselevsk Rus. Fed.
75 C2 Kishanganj India
115 C4 Kishi Nigeria
    Kishinev Moldova *see*
    Chișinău
67 C4 Kishiwada Japan
77 D1 Kishkenekol' Kazakh.
75 D2 Kishoreganj Bangl.
74 B1 Kishtwar India
103 D2 Kiskőrös Hungary
103 D2 Kiskunhalas Hungary

87 D4 Kislovodsk Rus. Fed.
117 C5 Kismaayo Somalia
Kismayu Somalia see Kismaayo
119 C3 Kisoro Uganda
111 B3 Kissamos Greece
114 A4 Kissidougou Guinea
141 D3 Kissimmee U.S.A.
141 D3 Kissimmee, Lake U.S.A.
129 D2 Kississing Lake Can.
Kistna r. India see Krishna
119 D3 Kisumu Kenya
103 E2 Kisvárda Hungary
Kisykkamys Kazakh. see Zhanakala
114 B3 Kita Mali
67 D3 Kitaibaraki Japan
66 D3 Kitakami Japan
66 D3 Kitakami-gawa r. Japan
67 B4 Kita-Kyūshū Japan
119 D2 Kitale Kenya
66 D2 Kitami Japan
130 B3 Kitchener Can.
93 J3 Kitee Fin.
119 D2 Kitgum Uganda
128 B2 Kitimat Can.
118 B3 Kitona Dem. Rep. Congo
92 H2 Kittilä Fin.
141 E1 Kitty Hawk U.S.A.
119 D3 Kitunda Tanz.
128 B2 Kitwanga Can.
121 B2 Kitwe Zambia
102 C2 Kitzbühel Austria
101 E3 Kitzingen Ger.
59 D3 Kiunga P.N.G.
92 I3 Kiuruvesi Fin.
92 I2 Kivalo ridge Fin.
90 B1 Kivertsi Ukr.
88 C2 Kiviõli Estonia
91 D2 Kivsharivka Ukr.
119 C3 Kivu, Lac Dem. Rep. Congo/ Rwanda
111 C2 Kıyıköy Turkey
86 E3 Kizel Rus. Fed.
111 C3 Kızılca Dağ mt. Turkey
80 B1 Kızılırmak r. Turkey
87 D4 Kizlyar Rus. Fed.
Kizyl-Arbat Turkm. see Serdar
92 I1 Kjøllefjord Norway
92 G2 Kjøpsvik Norway
102 C1 Kladno Czech Rep.
102 C2 Klagenfurt Austria
88 B2 Klaipėda Lith.
94 B1 Klaksvík Faroe Is
134 B2 Klamath r. U.S.A.
134 B2 Klamath Falls U.S.A.
134 B2 Klamath Mountains U.S.A.
60 B1 Klang Malaysia
102 C2 Klatovy Czech Rep.
122 A3 Klawer S. Africa
128 A2 Klawock U.S.A.
128 B2 Kleena Kleene Can.
122 B2 Kleinbegin S. Africa
122 A2 Klein Karas Namibia
122 A2 Kleinsee S. Africa
123 C2 Klerksdorp S. Africa
90 B1 Klesiv Ukr.
89 D3 Kletnya Rus. Fed.
100 C2 Kleve Ger.
88 C3 Klichaw Belarus
89 D3 Klimavichy Belarus
89 D3 Klimovo Rus. Fed.
89 E2 Klimovsk Rus. Fed.
89 E2 Klin Rus. Fed.
101 F2 Klingenthal Ger.
101 F2 Klínovec mt. Czech Rep.
93 G4 Klintehamn Sweden
89 D3 Klintsy Rus. Fed.
109 C2 Ključ Bos.-Herz.
103 D1 Kłodzko Pol.
100 C1 Kloosterhaar Neth.
103 D2 Klosterneuburg Austria
101 E1 Klötze (Altmark) Ger.
128 A1 Kluane Lake Can.
Kluang Malaysia see Keluang
103 D1 Kluczbork Pol.
Klukhori Rus. Fed. see Karachayevsk
128 A2 Klukwan U.S.A.
89 F2 Klyaz'ma r. Rus. Fed.
88 C3 Klyetsk Belarus
83 L3 Klyuchi Rus. Fed.
98 C2 Knaresborough U.K.
93 F3 Knästen h. Sweden
129 E2 Knee Lake Can.
101 E1 Knesebeck Ger.
101 E3 Knetzgau Ger.
109 C2 Knin Croatia

103 C2 Knittelfeld Austria
109 D2 Knjaževac Serbia
Knob Lake Can. see Schefferville
97 B3 Knockboy h. Ireland
100 A2 Knokke-Heist Belgium
141 D1 Knoxville U.S.A.
127 H1 Knud Rasmussen Land reg. Greenland
122 B3 Knysna S. Africa
60 B2 Koba Indon.
76 B1 Kobda Kazakh.
67 C4 Kōbe Japan
København Denmark see Copenhagen
100 C2 Koblenz Ger.
59 C3 Kobroõr i. Indon.
88 B3 Kobryn Belarus
Kocaeli Turkey see İzmit
111 B2 Kočani Macedonia
111 C2 Kocasu r. Turkey
109 B1 Kočevje Slovenia
75 C2 Koch Bihar India
89 F3 Kochetovka Rus. Fed.
73 B4 Kochi India
67 B4 Kōchi Japan
87 D4 Kochubey Rus. Fed.
75 C2 Kodarma India
126 B3 Kodiak U.S.A.
126 B3 Kodiak Island U.S.A.
123 C1 Kodibeleng Botswana
117 B4 Kodok S. Sudan
90 B2 Kodyma Ukr.
111 C2 Kodzhaele mt. Bulg./Greece
122 A2 Koës Namibia
122 C2 Koffiefontein S. Africa
114 B4 Koforidua Ghana
67 C3 Kōfu Japan
131 D2 Kogaluk r. Can.
117 B5 Kogelo Kenya
114 B3 Kogoni Mali
74 B1 Kohat Pak.
72 D2 Kohima India
88 C2 Kohtla-Järve Estonia
128 A1 Koidern Can.
114 A4 Koidu-Sefadu Sierra Leone
Kokand Uzbek. see Qo'qon
88 B2 Kōkar Fin.
Kokchetav Kazakh. see Kokshetau
122 A2 Kokerboom Namibia
88 C3 Kokhanava Belarus
89 F2 Kokhma Rus. Fed.
92 H3 Kokkola Fin.
88 C2 Koknese Latvia
138 B2 Kokomo U.S.A.
122 B1 Kokong Botswana
123 C2 Kokosi S. Africa
77 E2 Kokpekty Kazakh.
77 C1 Kokshetau Kazakh.
131 D2 Koksoak r. Can.
123 C3 Kokstad S. Africa
Koktokay China see Fuyun
61 D2 Kolaka Indon.
86 C2 Kola Peninsula Rus. Fed.
92 H2 Kolari Fin.
Kolarovgrad Bulg. see Shumen
114 A3 Kolda Senegal
93 E4 Kolding Denmark
119 C2 Kole Dem. Rep. Congo
107 D2 Koléa Alg.
86 D2 Kolguyev, Ostrov i. Rus. Fed.
73 B3 Kolhapur India
88 B2 Kolkasrags pt Latvia
75 C2 Kolkata India
73 B4 Kollam India
100 C1 Kollum Neth.
Köln Ger. see Cologne
103 D1 Koło Pol.
103 D1 Kołobrzeg Pol.
114 B3 Kolokani Mali
89 E2 Kolomna Rus. Fed.
90 B2 Kolomyya Ukr.
114 B3 Kolondiéba Mali
61 D2 Kolonedale Indon.
122 B2 Kolonkwaneng Botswana
82 G3 Kolpashevo Rus. Fed.
89 E3 Kolpny Rus. Fed.
Kol'skiy Poluostrov pen. Rus. Fed. see Kola Peninsula
78 B3 Koluli Eritrea
92 F3 Kolvereid Norway
119 C4 Kolwezi Dem. Rep. Congo
83 L2 Kolyma r. Rus. Fed.
Kolyma Lowland lowland Rus. Fed. see Kolymskaya Nizmennost'

Kolyma Range mts Rus. Fed. see Kolymskiy, Khrebet
83 L2 Kolymskaya Nizmennost' lowland Rus. Fed.
83 M2 Kolymskiy, Khrebet mts Rus. Fed.
122 A2 Komaggas S. Africa
67 C3 Komaki Japan
83 M3 Komandorskiye Ostrova is Rus. Fed.
103 D2 Komárno Slovakia
123 D2 Komati r. S. Africa/ Swaziland
123 D2 Komatipoort S. Africa
67 C3 Komatsu Japan
120 A2 Kombat Namibia
119 C3 Kombe Dem. Rep. Congo
Komintern Ukr. see Marhanets'
90 C2 Kominternivs'ke Ukr.
109 C2 Komiža Croatia
103 D2 Komló Hungary
Kommunarsk Ukr. see Alchevs'k
119 C3 Komono Congo
111 C2 Komotini Greece
Kompong Som Cambodia see Sihanoukville
122 B3 Komsberg mts S. Africa
83 H1 Komsomolets, Ostrov i. Rus. Fed.
89 F2 Komsomol'sk Rus. Fed.
91 C2 Komsomol's'k Ukr.
83 M2 Komsomol'skiy Rus. Fed.
Komsomol'skiy Rus. Fed. see Yugorsk
87 D4 Komsomol'skiy Rus. Fed.
83 K3 Komsomol'sk-na-Amure Rus. Fed.
89 E2 Konakovo Rus. Fed.
75 C3 Kondagaon India
Kondinskoye Rus. Fed. see Oktyabr'skoye
119 D3 Kondoa Tanz.
86 F2 Kondol' Rus. Fed.
86 C2 Kondopoga Rus. Fed.
89 E3 Kondrovo Rus. Fed.
127 J2 Kong Christian IX Land reg. Greenland
127 K2 Kong Christian X Land reg. Greenland
127 J2 Kong Frederik VI Kyst coastal area Greenland
119 C3 Kongolo Dem. Rep. Congo
93 E4 Kongsberg Norway
93 F3 Kongsvinger Norway
77 D3 Kongur Shan mt. China
100 C2 Königswinter Ger.
103 D1 Konin Pol.
109 C2 Konjic Bos.-Herz.
122 A2 Konkiep watercourse Namibia
86 D2 Konosha Rus. Fed.
91 C1 Konotop Ukr.
103 E1 Końskie Pol.
Konstantinograd Ukr. see Krasnohrad
102 B2 Konstanz Ger.
115 C3 Kontagora Nigeria
63 B2 Kon Tum Vietnam
63 B2 Kon Tum, Cao Nguyên Vietnam
80 B2 Konya Turkey
77 D2 Konyrat Kazakh.
100 C3 Konz Ger.
86 E3 Konzhakovskiy Kamen', Gora mt. Rus. Fed.
134 C1 Kooskia U.S.A.
128 C3 Kootenay Lake Can.
53 D2 Kootingal Austr.
122 B3 Kootjieskolk S. Africa
92 □B2 Kópasker Iceland
108 B1 Koper Slovenia
93 G4 Köping Sweden
123 C1 Kopong Botswana
93 G4 Kopparberg Sweden
109 C1 Koprivnica Croatia
89 F3 Korablino Rus. Fed.
73 C3 Koraput India
75 C2 Korba India
101 D2 Korbach Ger.
109 D2 Korçë Albania
109 C2 Korčula Croatia
109 C2 Korčula i. Croatia
70 C2 Korea Bay g. China/N. Korea
65 B1 Korea, North country Asia
65 B2 Korea, South country Asia

65 B3 Korea Strait Japan/S. Kor
89 D3 Korenevo Rus. Fed.
91 D2 Korenovsk Rus. Fed.
Korenovskaya Rus. Fed. see Korenovsk
90 B1 Korets' Ukr.
111 C2 Körfez Turkey
114 B4 Korhogo Côte d'Ivoire
Korinthos Greece see Corinth
103 D2 Kőris-hegy h. Hungary
109 D2 Koritnik mt. Albania/Kose
Koritsa Albania see Korçë
67 D3 Kōriyama Japan
87 F3 Korkino Rus. Fed.
111 D3 Korkuteli Turkey
77 E2 Korla China
103 D2 Körmend Hungary
49 I5 Koro i. Fiji
114 B3 Koro Mali
131 D2 Koroc r. Can.
91 D1 Korocha Rus. Fed.
119 D3 Korogwe Tanz.
59 C2 Koror Palau
103 E2 Körös r. Hungary
90 B1 Korosten' Ukr.
90 B1 Korostyshiv Ukr.
115 D3 Koro Toro Chad
93 H3 Korpo Fin.
66 D1 Korsakov Rus. Fed.
91 C2 Korsun'-Shevchenkivs'ky Ukr.
103 E1 Korsze Pol.
116 B3 Korti Sudan
100 A2 Kortrijk Belgium
83 L3 Koryakskaya, Sopka vol. Rus. Fed.
83 M2 Koryakskoye Nagor'ye mt Rus. Fed.
86 D2 Koryazhma Rus. Fed.
91 C1 Koryukivka Ukr.
111 C3 Kos Greece
111 C3 Kos i. Greece
91 D2 Kosa Biryuchyy Ostriv i. Ukr.
65 B2 Kosan N. Korea
103 D1 Kościan Pol.
Kosciusko, Mount mt. Austr. see Kosciuszko, Mount
53 C3 Kosciuszko, Mount Austr.
77 E2 Kosh-Agach Rus. Fed.
67 A4 Koshikijima-rettō is Japa
103 E2 Košice Slovakia
92 H2 Koskullskulle Sweden
65 B2 Kosong N. Korea
109 D2 Kosovo country Europe
Kosovska Mitrovica Koso see Mitrovicë
48 H3 Kosrae atoll Micronesia
114 B4 Kossou, Lac de l. Côte d'Ivoire
76 C1 Kostanay Kazakh.
110 B2 Kostenets Bulg.
123 C2 Koster S. Africa
116 B3 Kosti Sudan
92 J3 Kostomuksha Rus. Fed.
90 B1 Kostopil' Ukr.
89 F2 Kostroma Rus. Fed.
89 F2 Kostroma r. Rus. Fed.
102 C1 Kostrzyn Pol.
91 D2 Kostyantynivka Ukr.
103 D1 Koszalin Pol.
103 D2 Kőszeg Hungary
74 B2 Kota India
60 B2 Kotaagung Indon.
61 C2 Kotabaru Indon.
61 C1 Kota Belud Malaysia
60 B1 Kota Bharu Malaysia
60 B2 Kota Kinabalu Malaysia
75 C3 Kotaparh India
61 C1 Kota Samarahan Malaysia
86 D3 Kotel'nich Rus. Fed.
87 D4 Kotel'nikovo Rus. Fed.
83 K1 Kotel'nyy, Ostrov i. Rus. Fed.
101 D2 Köthen (Anhalt) Ger.
119 D2 Kotido Uganda
93 I3 Kotka Fin.
86 D2 Kotlas Rus. Fed.
126 B2 Kotlik U.S.A.
109 C2 Kotor Varoš Bos.-Herz.
87 D3 Kotovo Rus. Fed.
91 E1 Kotovsk Rus. Fed.
90 B2 Kotovs'k Ukr.
73 C3 Kottagudem India
115 D4 Kotto r. C.A.R.
83 H2 Kotuy r. Rus. Fed.

26 B2 Kotzebue U.S.A.
26 B2 Kotzebue Sound sea chan.
U.S.A.
14 A3 Koubia Guinea
00 A2 Koudekerke Neth.
14 B3 Koudougou Burkina Faso
22 B3 Kougaberge mts S. Africa
14 B3 Koulamoutou Gabon
18 B3 Koulikoro Mali
18 B2 Koum Cameroon
18 B2 Koumra Chad
14 A3 Koundâra Guinea
Kounradskiy Kazakh. see
Konyrat
61 D2 Kourou Fr. Guiana
14 B3 Kouroussa Guinea
15 D3 Kousséri Cameroon
14 B3 Koutiala Mali
93 I3 Kouvola Fin.
09 D1 Kovačica Serbia
92 J2 Kovdor Rus. Fed.
90 A1 Kovel' Ukr.
Kovno Lith. see Kaunas
89 F2 Kovrov Rus. Fed.
51 D1 Kowanyama Austr.
54 B2 Kowhitirangi N.Z.
Koyamutthoor India see
Coimbatore
11 C3 Köyceğiz Turkey
86 D2 Koyda Rus. Fed.
26 B2 Koyukuk r. U.S.A.
90 C1 Kozelets' Ukr.
89 E3 Kozel'sk Rus. Fed.
73 B3 Kozhikode India
90 B2 Kozyatyn Ukr.
63 A2 Kra, Isthmus of Myanmar/
Thai.
63 A3 Krabi Thai.
63 A2 Kra Buri Thai.
63 B2 Krâchéh Cambodia
93 E4 Kragerø Norway
00 B1 Kraggenburg Neth.
09 D2 Kragujevac Serbia
02 B2 Krakatau i. Indon.
03 D1 Kraków Pol.
91 D2 Kraljevo Serbia
91 D2 Kramators'k Ukr.
93 G3 Kramfors Sweden
11 B3 Kranidi Greece
02 C2 Kranj Slovenia
23 D2 Kranskop S. Africa
86 E1 Krasino Rus. Fed.
88 C2 Krāslava Latvia
01 D3 Kraslice Czech Rep.
89 D3 Krasnapollye Belarus
89 D3 Krasnaya Gora Rus. Fed.
89 F2 Krasnaya Gorbatka
Rus. Fed.
Krasnoarmeysk Kazakh. see
Taiynsha
87 D2 Krasnoarmeysk Rus. Fed.
Krasnoarmeyskaya
Rus. Fed. see Poltavskaya
91 D2 Krasnoarmiys'k Ukr.
91 D2 Krasnoborsk Rus. Fed.
91 D2 Krasnodar Rus. Fed.
91 D2 Krasnodarskoye
Vodokhranilishche resr
Rus. Fed.
91 D2 Krasnodon Ukr.
92 D2 Krasnogorodsk Rus. Fed.
91 D2 Krasnohrad Ukr.
91 D2 Krasnohvardiys'ke Ukr.
86 E3 Krasnokamsk Rus. Fed.
91 D2 Krasnomasskiy Rus. Fed.
91 C2 Krasnoperekops'k Ukr.
87 D3 Krasnoslobodsk Rus. Fed.
86 F3 Krasnotur'insk Rus. Fed.
86 F4 Krasnoufimsk Rus. Fed.
86 E2 Krasnovishersk Rus. Fed.
Krasnovodsk Turkm. see
Türkmenbaşy
83 H3 Krasnoyarsk Rus. Fed.
89 E3 Krasnoye Rus. Fed.
83 M2 Krasnoye, Ozero l. Rus. Fed.
89 F2 Krasnoye-na-Volge
Rus. Fed.
03 E1 Krasnystaw Pol.
89 D3 Krasnyy Rus. Fed.
Krasnyy Kholm Rus. Fed.
91 D2 Krasnyy Luch Ukr.
91 E2 Krasnyy Sulin Rus. Fed.
90 B2 Krasyliv Ukr.
Kraulshavn Greenland see
Nuussuaq
00 C2 Krefeld Ger.

91 C2 Kremenchuk Ukr.
91 C2 Kremenchuts'ke
Vodoskhovyshche resr Ukr.
90 B1 Kremenets' Ukr.
103 D2 Křemešník h. Czech Rep.
Kremges Ukr. see
Svitlovods'k
91 D2 Kreminna Ukr.
136 B2 Kremmling U.S.A.
103 D2 Krems an der Donau
Austria
89 D2 Kresttsy Rus. Fed.
88 B2 Kretinga Lith.
100 C2 Kreuzau Ger.
101 C2 Kreuztal Ger.
118 A2 Kribi Cameroon
111 B3 Krikellos Greece
66 D1 Kril'on, Mys c. Rus. Fed.
111 B3 Krios, Akrotirio pt Greece
73 C3 Krishna r. India
73 C3 Krishna, Mouths of the
India
75 C2 Krishnanagar India
93 E4 Kristiansand Norway
93 F4 Kristianstad Sweden
92 E3 Kristiansund Norway
93 F4 Kristinehamn Sweden
Kristinopol' Ukr. see
Chervonohrad
Kriti i. Greece see Crete
111 C3 Kritiko Pelagos sea Greece
110 B2 Kriva Palanka Macedonia
Krivoy Rog Ukr. see
Kryvyy Rih
109 C1 Križevci Croatia
108 B1 Krk i. Croatia
92 F3 Krokom Sweden
91 C1 Krolevets' Ukr.
89 E3 Kromy Rus. Fed.
101 E2 Kronach Ger.
63 B2 Krŏng Kaôh Kŏng
Cambodia
127 J2 Kronprins Frederik Bjerge
nunatak Greenland
123 C2 Kroonstad S. Africa
91 E2 Kropotkin Rus. Fed.
103 E2 Krosno Pol.
103 D1 Krotoszyn Pol.
60 B2 Krui Indon.
122 B3 Kruisfontein S. Africa
109 C2 Krujë Albania
111 C2 Krumovgrad Bulg.
Krung Thep Thai. see
Bangkok
88 C3 Krupki Belarus
109 D2 Kruševac Serbia
101 E2 Krušné hory mts Czech Rep.
128 A2 Kruzof Island U.S.A.
89 D3 Krychaw Belarus
91 D2 Krylovskaya Rus. Fed.
91 D3 Krymsk Rus. Fed.
Krymskaya Rus. Fed. see
Krymsk
Kryms'kyy Pivostriv pen.
Ukr. see Crimea
Krystynopol Ukr. see
Chervonohrad
91 C2 Kryvyy Rih Ukr.
90 B2 Kryzhopil' Ukr.
114 B2 Ksabi Alg.
107 D2 Ksar el Boukhari Alg.
114 B1 Ksar el Kebir Morocco
Ksar-es-Souk Morocco see
Er Rachidia
89 E3 Kshenskiy Rus. Fed.
78 B2 Kũ', Jabal al h. Saudi Arabia
61 C1 Kuala Belait Brunei
Kuala Dungun Malaysia see
Dungun
60 B1 Kuala Kangsar Malaysia
60 B1 Kuala Kerai Malaysia
60 B1 Kuala Lipis Malaysia
60 B1 Kuala Lumpur Malaysia
61 C2 Kualapembuang Indon.
60 B1 Kuala Terengganu Malaysia
61 C2 Kualatungal Indon.
61 C1 Kuamut Malaysia
65 A1 Kuandian China
60 B1 Kuantan Malaysia
91 D2 Kuban' r. Rus. Fed.
89 E2 Kubenskoye, Ozero l.
Rus. Fed.
110 C2 Kubrat Bulg.
60 B2 Kubu Indon.
61 C1 Kubuang Indon.
60 C1 Kuching Malaysia
109 C2 Kuçovë Albania
61 C1 Kudat Malaysia
61 C2 Kudus Indon.
102 C2 Kufstein Austria

127 G2 Kugaaruk Can.
91 D2 Kugey Rus. Fed.
126 E2 Kugluktuk Can.
126 D2 Kugmallit Bay Can.
92 I3 Kuhmo Fin.
79 C2 Kührän, Küh-e mt. Iran
122 A1 Kuis Namibia
Kuitin China see Kuytun
120 A2 Kuito Angola
92 I2 Kuivaniemi Fin.
65 B2 Kujang N. Korea
66 D2 Kuji Japan
67 B4 Kujū-san vol. Japan
109 D2 Kukës Albania
76 B3 Kükürtli Turkm.
111 C3 Kula Turkey
75 D2 Kula Kangri mt. Bhutan/
China
76 B2 Kulandy Kazakh.
88 B2 Kuldīga Latvia
Kuldja China see Yining
122 B1 Kule Botswana
101 E2 Kulmbach Ger.
77 C3 Külob Tajik.
76 B2 Kul'sary Kazakh.
111 D3 Kulübe Tepe mt. Turkey
77 D1 Kulunda Rus. Fed.
77 D1 Kulundinskoye, Ozero salt l.
Rus. Fed.
127 J2 Kulusuk Greenland
67 C3 Kumagaya Japan
67 B4 Kumamoto Japan
67 C4 Kumano Japan
110 B2 Kumanovo Macedonia
114 B4 Kumasi Ghana
118 A2 Kumba Cameroon
Kum-Dag Turkm. see
Gumdag
78 B2 Kumdah Saudi Arabia
87 E3 Kumertau Rus. Fed.
119 D2 Kumi Uganda
111 C3 Kumkale Turkey
93 G4 Kumla Sweden
115 D3 Kumo Nigeria
62 A1 Kumon Range mts
Myanmar
Kumphawapi Thai.
Kumul China see Hami
120 A2 Kunene r. Angola/Namibia
77 D2 Kungei Alatau mts Kazakh./
Kyrg.
93 F4 Kungsbacka Sweden
118 B3 Kungu Dem. Rep. Congo
86 E3 Kungur Rus. Fed.
62 A1 Kunhing Myanmar
62 A1 Kunlong Myanmar
75 B1 Kunlun Shan mts China
71 A3 Kunming China
50 B1 Kununurra Austr.
101 D3 Künzelsau Ger.
92 I3 Kuopio Fin.
109 C1 Kupa r. Croatia/Slovenia
59 C3 Kupang Indon.
88 B2 Kupiškis Lith.
111 C2 Küplü Turkey
128 A2 Kupreanof Island U.S.A.
91 D2 Kup"yans'k Ukr.
77 E2 Kuqa China
81 C2 Kür r. Azer.
67 B4 Kurashiki Japan
75 C2 Kurasia India
67 B3 Kurayoshi Japan
89 E3 Kurchatov Rus. Fed.
111 C2 Kürdzhali Bulg.
67 B4 Kure Japan
57 T7 Kure Atoll U.S.A.
88 B2 Kuressaare Estonia
87 F3 Kurgan Rus. Fed.
Kuria Muria
Islands is Oman see
Ḩalāniyāt, Juzur al
93 H3 Kurikka Fin.
156 C2 Kuril Basin Sea of Okhotsk
69 F1 Kuril Islands is Rus. Fed.
69 F1 Kuril'sk Rus. Fed.
Kuril'skiye Ostrova is
Rus. Fed. see Kuril Islands
156 C3 Kuril Trench
N. Pacific Ocean
Kurmashkino Kazakh. see
Kurshim
117 B3 Kurmuk Sudan
73 B3 Kurnool India
53 D2 Kurri Kurri Austr.
88 B2 Kuršėnai Lith.
78 B2 Kursh, Jabal mt.
Saudi Arabia
77 E2 Kurshim Kazakh.
89 E3 Kursk Rus. Fed.

109 D2 Kuršumlija Serbia
122 B2 Kuruman S. Africa
122 B2 Kuruman watercourse
S. Africa
67 B4 Kurume Japan
83 I3 Kurumkan Rus. Fed.
73 C4 Kurunegala Sri Lanka
81 D1 Kuryk Kazakh.
111 C3 Kuşadası Turkey
111 C3 Kuşadası, Gulf of b. Turkey
111 C2 Kuş Gölü l. Turkey
91 D2 Kushchevskaya Rus. Fed.
66 D2 Kushiro Japan
Kushka Turkm. see
Serhetabat
75 C2 Kushtia Bangl.
126 B2 Kuskokwim r. U.S.A.
126 B2 Kuskokwim Mountains
U.S.A.
76 C1 Kusmuryn Kazakh.
66 D2 Kussharo-ko l. Japan
Kustanay Kazakh. see
Kostanay
63 B2 Kut, Ko i. Thai.
111 C3 Kütahya Turkey
81 C1 Kutaisi Georgia
Kutaraja Indon. see
Banda Aceh
Kutch, Gulf of g. India see
Kachchh, Gulf of
109 C1 Kutjevo Croatia
103 D1 Kutno Pol.
118 B3 Kutu Dem. Rep. Congo
116 A3 Kutum Sudan
126 E2 Kuujjua r. Can.
131 D2 Kuujjuaq Can.
130 C2 Kuujjuarapik Can.
92 I2 Kuusamo Fin.
120 A2 Kuvango Angola
89 D2 Kuvshinovo Rus. Fed.
78 B2 Kuwait country Asia
78 B2 Kuwait Kuwait
82 G3 Kuybyshev Rus. Fed.
Kuybyshev Rus. Fed. see
Samara
91 D2 Kuybysheve Ukr.
Kuybyshevka-Vostochnaya
Rus. Fed. see Belogorsk
87 D3 Kuybyshevskoye
Vodokhranilishche resr
Rus. Fed.
77 E2 Kuytun China
111 C3 Kuyucak Turkey
87 D3 Kuznetsk Rus. Fed.
90 B1 Kuznetsovs'k Ukr.
86 C2 Kuzomen' Rus. Fed.
92 H1 Kvalsund Norway
123 D2 KwaDukuza S. Africa
123 D2 KwaMashu S. Africa
61 D1 Kwandang Indon.
Kwangchow China see
Guangzhou
Kwangju S. Korea see
Gwangju
Kwangtung prov. China see
Guangdong
65 B1 Kwanmo-bong mt. N. Korea
123 C3 KwaNobuhle S. Africa
123 C3 KwaNojoli S. Africa
122 B3 KwaNonzame S. Africa
115 C3 Kwatarkwashi Nigeria
123 C3 Kwatinidubu S. Africa
122 B3 KwaZamukuhle S. Africa
123 C2 KwaZamukucinga S. Africa
123 C2 KwaZanele S. Africa
123 D2 KwaZulu-Natal prov.
S. Africa
Kweichow prov. China see
Guizhou
Kweiyang China see
Guiyang
121 B2 Kwekwe Zimbabwe
118 B3 Kwenge r. Dem. Rep. Congo
123 C3 Kwezi-Naledi S. Africa
103 D1 Kwidzyn Pol.
59 D3 Kwikila P.N.G.
118 B3 Kwilu r. Angola/
Dem. Rep. Congo
59 C3 Kwoka mt. Indon.
118 B2 Kyabé Chad
53 C3 Kyabram Austr.
62 A2 Kyaikto Myanmar
63 A2 Kya-in Seikkyi Myanmar
68 D1 Kyakhta Rus. Fed.
52 A2 Kyancutta Austr.
62 A1 Kyaukpadaung Myanmar
62 A2 Kyaukpyu Myanmar
62 A1 Kyaukse Myanmar
62 A2 Kyebogyi Myanmar
62 A2 Kyeintali Myanmar
Kyiv Ukr. see Kiev

90 C1 **Kyivs'ke Vodoskhovyshche** *resr* Ukr.
Kyklades *is* Greece *see* **Cyclades**
129 D2 **Kyle** Can.
96 B2 **Kyle of Lochalsh** U.K.
100 C3 **Kyll** *r.* Ger.
111 B3 **Kyllini** *mt.* Greece
111 B3 **Kymi** Greece
52 B3 **Kyneton** Austr.
119 D2 **Kyoga, Lake** Uganda
53 D1 **Kyogle** Austr.
67 C4 **Kyōto** Japan
111 B3 **Kyparissia** Greece
111 B3 **Kyparissiakos Kolpos** *b.* Greece
111 B3 **Kyra Panagia** *i.* Greece
80 B2 **Kyrenia** Cyprus
77 D2 **Kyrgyzstan** *country* Asia
101 F1 **Kyritz** Ger.
93 H3 **Kyrönjoki** *r.* Fin.
86 E2 **Kyrta** Rus. Fed.
86 D2 **Kyssa** Rus. Fed.
83 J2 **Kytalyktakh** Rus. Fed.
111 B3 **Kythira** *i.* Greece
111 B3 **Kythnos** *i.* Greece
128 B2 **Kyuquot** Can.
67 B4 **Kyūshū** *i.* Japan
110 B2 **Kyustendil** Bulg.
62 A2 **Kywebwe** Myanmar
92 H3 **Kyyjärvi** Fin.
68 C1 **Kyzyl** Rus. Fed.
76 C2 **Kyzylkum Desert** Kazakh./Uzbek.
77 C2 **Kyzylorda** Kazakh.
Kzyl-Orda Kazakh. *see* **Kyzylorda**
Kzyltu Kazakh. *see* **Kishkenekol'**

L

145 C3 **La Angostura, Presa de** *resr* Mex.
117 C4 **Laascaanood** Somalia
117 C3 **Laasgoray** Somalia
150 C1 **La Asunción** Venez.
114 A2 **Laâyoune** Western Sahara
87 D4 **Laba** *r.* Rus. Fed.
144 B2 **La Babia** Mex.
61 D2 **Labala** Indon.
152 B2 **La Banda** Arg.
61 C1 **Labang** Malaysia
104 B2 **La Baule-Escoublac** France
102 C1 **Labe** *r.* Czech Rep.
114 A3 **Labé** Guinea
128 A1 **Laberge, Lake** Can.
128 C2 **La Biche, Lac** *l.* Can.
108 B1 **Labin** Croatia
87 D4 **Labinsk** Rus. Fed.
64 B2 **Labo** Phil.
104 B3 **Labouheyre** France
153 B3 **Laboulaye** Arg.
131 D2 **Labrador** *reg.* Can.
131 D2 **Labrador City** Can.
127 I2 **Labrador Sea** Can./Greenland
150 C3 **Lábrea** Brazil
61 C1 **Labuan** Malaysia
61 C2 **Labuhanbajo** Indon.
60 B1 **Labuhanbilik** Indon.
59 C3 **Labuna** Indon.
63 A2 **Labutta** Myanmar
86 F2 **Labytnangi** Rus. Fed.
109 C2 **Laç** Albania
107 D2 **La Cabaneta** Spain
La Calle Alg. *see* **El Kala**
105 C2 **La Capelle** France
73 B3 **Laccadive Islands** India
129 E2 **Lac du Bonnet** Can.
146 B3 **La Ceiba** Hond.
52 A3 **Lacepede Bay** Austr.
134 B1 **Lacey** U.S.A.
105 D2 **La Chaux-de-Fonds** Switz.
53 B2 **Lachlan** *r.* Austr.
146 C4 **La Chorrera** Panama
139 E1 **Lachute** Can.
105 D3 **La Ciotat** France
128 C2 **Lac La Biche** Can.
Lac la Martre Can. *see* **Whatì**
139 E1 **Lac-Mégantic** Can.
128 C2 **Lacombe** Can.
146 B4 **La Concepción** Panama
145 C3 **La Concordia** Mex.
108 A3 **Laconi** Italy
139 E2 **Laconia** U.S.A.
128 C2 **La Crete** Can.

138 A2 **La Crosse** U.S.A.
144 B2 **La Cruz** Mex.
144 B2 **La Cuesta** Mex.
155 D1 **Ladainha** Brazil
74 B1 **Ladakh Range** *mts* India/Pak.
122 B3 **Ladismith** S. Africa
79 D2 **Lādīz** Iran
89 D1 **Ladoga, Lake** Rus. Fed.
Ladozhskoye Ozero *l.* Rus. Fed. *see* **Ladoga, Lake**
141 D2 **Ladson** U.S.A.
123 C3 **Lady Grey** S. Africa
128 B3 **Ladysmith** Can.
123 C2 **Ladysmith** S. Africa
59 D3 **Lae** P.N.G.
143 C3 **La Encantada, Sierra** *mts* Mex.
152 B2 **La Esmeralda** Bol.
93 F4 **Læsø** *i.* Denmark
Lafayette Alg. *see* **Bougaa**
141 C2 **La Fayette** U.S.A.
138 B2 **Lafayette** *IN* U.S.A.
140 B2 **Lafayette** *LA* U.S.A.
115 C4 **Lafia** Nigeria
104 B2 **La Flèche** France
141 D1 **La Follette** U.S.A.
130 C2 **Laforge** Can.
79 C2 **Läft** Iran
108 A3 **Lagan'** Rus. Fed.
87 D4 **Lagan** *r.* U.K.
151 F4 **Lagarto** Brazil
118 B2 **Lagdo, Lac de** *l.* Cameroon
115 C1 **Laghouat** Alg.
155 D1 **Lagoa Santa** Brazil
114 A2 **La Gomera** *i.* Islas Canarias
114 C4 **Lagos** Nigeria
106 B2 **Lagos** Port.
134 C1 **La Grande** U.S.A.
130 C2 **La Grande 3, Réservoir** *resr* Can.
130 C2 **La Grande 4, Réservoir** *resr* Can.
50 B1 **La Grange** Austr.
141 C2 **La Grange** U.S.A.
150 C2 **La Gran Sabana** *plat.* Venez.
152 D2 **Laguna** Brazil
144 A2 **Laguna, Picacho de la** *mt.* Mex.
150 B3 **Lagunas** Peru
147 C3 **Lagunillas** Venez.
La Habana Cuba *see* **Havana**
61 C1 **Lahad Datu** Malaysia
60 B2 **Lahat** Indon.
78 B3 **Laḥij** Yemen
81 C2 **Lāhījān** Iran
100 C2 **Lahnstein** Ger.
74 B1 **Lahore** Pak.
74 A2 **Lahri** Pak.
93 I3 **Lahti** Fin.
115 D4 **Laï** Chad
53 D1 **Laidley** Austr.
104 C2 **L'Aigle** France
111 B3 **Laimos, Akrotirio** *pt* Greece
122 B3 **Laingsburg** S. Africa
92 H2 **Lainioälven** *r.* Sweden
96 B1 **Lairg** U.K.
108 B1 **Laives** Italy
70 B1 **Laiwu** China
70 B2 **Laiyang** China
70 B2 **Laiyuan** China
70 B2 **Laizhou** China
70 B2 **Laizhou Wan** *b.* China
50 C1 **Lajamanu** Austr.
152 C2 **Lajes** Brazil
144 B2 **La Junta** Mex.
136 C3 **La Junta** U.S.A.
103 D2 **Lake Abaya** *l.* Eth.
53 C2 **Lake Balaton** *l.* Hungary
53 C2 **Lake Cargelligo** Austr.
53 D2 **Lake Cathie** Austr.
138 B2 **Lake Charles** U.S.A.
141 D2 **Lake City** *FL* U.S.A.
141 E2 **Lake City** *SC* U.S.A.
128 B3 **Lake Cowichan** Can.
Lake Harbour Can. *see* **Kimmirut**
143 A2 **Lake Havasu City** U.S.A.
143 D3 **Lake Jackson** U.S.A.
50 A3 **Lake King** Austr.
141 D3 **Lakeland** U.S.A.
128 C2 **Lake Louise** Can.
134 B1 **Lake Oswego** U.S.A.
54 B2 **Lake Paringa** N.Z.
140 B2 **Lake Providence** U.S.A.
53 C3 **Lakes Entrance** Austr.
117 B3 **Lake Tana** *l.* Eth.
134 B2 **Lakeview** U.S.A.
137 E2 **Lakeville** U.S.A.

136 B3 **Lakewood** *CO* U.S.A.
139 E2 **Lakewood** *NJ* U.S.A.
141 D3 **Lake Worth** U.S.A.
74 A2 **Lakhpat** India
74 B1 **Lakki Marwat** Pak.
111 B3 **Lakonikos Kolpos** *b.* Greece
114 B4 **Lakota** Côte d'Ivoire
92 H1 **Lakselv** Norway
107 C1 **L'Alcora** Spain
146 A3 **La Libertad** Guat.
106 B1 **Lalín** Spain
106 B2 **La Línea de la Concepción** Spain
74 B2 **Lalitpur** India
Lalitpur Nepal *see* **Patan**
129 D2 **La Loche** Can.
100 B2 **La Louvière** Belgium
108 A2 **La Maddalena** Italy
61 C1 **Lamag** Malaysia
La Manche *str.* France/U.K. *see* **English Channel**
136 C3 **Lamar** U.S.A.
79 C2 **Lamard** Iran
108 A3 **La Marmora, Punta** *mt.* Italy
128 C1 **La Martre, Lac** *l.* Can.
104 B2 **Lamballe** France
118 B3 **Lambaréné** Gabon
122 A3 **Lambert's Bay** S. Africa
92 □A2 **Lambeyri** Iceland
106 B1 **Lamego** Port.
131 D3 **Lamèque, Île** *i.* Can.
150 A3 **La Merced** Peru
52 B3 **Lameroo** Austr.
143 C2 **Lamesa** U.S.A.
111 B3 **Lamia** Greece
137 E2 **Lamoni** U.S.A.
62 A2 **Lampang** Thai.
143 D2 **Lampasas** U.S.A.
145 B2 **Lampazos** Mex.
99 A3 **Lampeter** U.K.
62 A2 **Lamphun** Thai.
119 E3 **Lamu** Kenya
105 D3 **La Mure** France
62 A2 **Lan, Loi** *mt.* Myanmar/Thai.
96 C3 **Lanark** U.K.
63 A2 **Lanbi Kyun** *i.* Myanmar
62 A1 **Lancang** China
Lancang Jiang *r.* China *see* **Mekong**
98 B2 **Lancaster** U.K.
135 C4 **Lancaster** *CA* U.S.A.
138 C3 **Lancaster** *OH* U.S.A.
139 D2 **Lancaster** *PA* U.S.A.
141 D2 **Lancaster** *SC* U.S.A.
127 G2 **Lancaster Sound** *str.* Can.
50 A3 **Lancelin** Austr.
Lanchow China *see* **Lanzhou**
102 C2 **Landeck** Austria
136 B2 **Lander** U.S.A.
99 A4 **Land's End** *pt* U.K.
102 C2 **Landshut** Ger.
93 F4 **Landskrona** Sweden
141 C2 **Lanett** U.S.A.
122 B2 **Langberg** *mts* S. Africa
137 D1 **Langdon** U.S.A.
93 F4 **Langeland** *i.* Denmark
101 D1 **Langeoog** Ger.
100 C1 **Langeoog** *i.* Ger.
82 G2 **Langepas** Rus. Fed.
70 B2 **Langfang** China
101 D2 **Langgöns** Ger.
92 □A3 **Langjökull** Iceland
60 A1 **Langkawi** *i.* Malaysia
105 C3 **Langogne** France
104 B3 **Langon** France
105 D2 **Langres** France
60 A1 **Langsa** Indon.
62 B1 **Lang Sơn** Vietnam
101 D1 **Langwedel** Ger.
70 A2 **Langzhong** China
153 B2 **Lanín, Volcán** *vol.* Arg./Chile
81 C2 **Länkäran** Azer.
104 B2 **Lannion** France
139 E1 **L'Anse** U.S.A.
139 E1 **L'Anse-St-Jean** Can.
138 B2 **Lansing** U.S.A.
71 B3 **Lanxi** China
117 B4 **Lanya** S. Sudan
114 A2 **Lanzarote** *i.* Islas Canarias
70 A2 **Lanzhou** China
64 B1 **Laoag** Phil.
62 B1 **Lao Cai** Vietnam
70 B2 **Laohekou** China
70 B2 **Laojunmiao** China

65 B1 **Laoling** China
65 B1 **Lao Ling** *mts* China
68 C2 **Lao Mangnai** China
105 C2 **Laon** France
62 B2 **Laos** *country* Asia
65 B1 **Laotougou** China
66 A2 **Laoye Ling** *mts* China
154 C3 **Lapa** Brazil
114 A2 **La Palma** *i.* Islas Canarias
146 C4 **La Palma** Panama
150 C2 **La Paragua** Venez.
152 B1 **La Paz** Bol.
145 D3 **La Paz** Hond.
144 A2 **La Paz** Mex.
150 C3 **La Pedrera** Col.
138 C2 **Lapeer** U.S.A.
66 D1 **La Pérouse Strait** Japan/Rus. Fed.
145 C2 **La Pesca** Mex.
144 B2 **La Piedad** Mex.
153 C3 **La Plata** Arg.
153 C3 **La Plata, Río de** *sea chan.* Arg./Uru.
106 B1 **La Pola** Spain
106 B1 **La Pola Siero** Spain
92 H3 **Lappajärvi** *l.* Fin.
93 I3 **Lappeenranta** Fin.
92 G2 **Lappland** *reg.* Europe
111 C2 **Lâpseki** Turkey
Laptevo Rus. Fed. *see* **Yasnogorsk**
83 J1 **Laptev Sea** Rus. Fed.
Laptevykh, More *sea* Rus. Fed. *see* **Laptev Sea**
93 H3 **Lapua** Fin.
152 B2 **La Quiaca** Arg.
108 B2 **L'Aquila** Italy
135 C4 **La Quinta** U.S.A.
79 C2 **Lār** Iran
114 B1 **Larache** Morocco
136 B2 **Laramie** U.S.A.
136 B2 **Laramie Mountains** U.S.A.
154 B3 **Laranjeiras do Sul** Brazil
61 D2 **Larantuka** Indon.
59 C3 **Larat** *i.* Indon.
107 D2 **Larba** Alg.
L'Ardenne, Plateau de *plateau* Belgium *see* **Ardennes**
106 C1 **Laredo** Spain
143 D3 **Laredo** U.S.A.
141 D3 **Largo** U.S.A.
96 B3 **Largs** U.K.
115 D1 **L'Ariana** Tunisia
152 B2 **La Rioja** Arg.
111 B3 **Larisa** Greece
74 A2 **Larkana** Pak.
80 B2 **Larnaca** Cyprus
Larnaka Cyprus *see* **Larnaca**
97 D1 **Larne** U.K.
100 B2 **La Roche-en-Ardenne** Belgium
104 B2 **La Rochelle** France
104 B2 **La Roche-sur-Yon** France
107 C2 **La Roda** Spain
147 D3 **La Romana** Dom. Rep.
129 D2 **La Ronge** Can.
129 D2 **La Ronge, Lac** *l.* Can.
50 C1 **Larrimah** Austr.
55 A3 **Larsen Ice Shelf** Antarctica
93 F4 **Larvik** Norway
136 C3 **Las Animas** U.S.A.
Las Anod Somalia *see* **Laascaanood**
130 C3 **La Sarre** Can.
142 B2 **Las Cruces** U.S.A.
152 A2 **La Serena** Chile
107 D1 **La Seu d'Urgell** Spain
153 C3 **Las Flores** Arg.
153 B3 **Las Heras** Arg.
62 A1 **Lashio** Myanmar
76 C3 **Lashkar Gāh** Afgh.
109 C3 **La Sila** *reg.* Italy
152 B2 **Las Lomitas** Arg.
106 B2 **Las Marismas** *marsh* Spain
144 B2 **Las Nieves** Mex.
114 A2 **Las Palmas de Gran Canaria** Islas Canarias
108 A2 **La Spezia** Italy
153 C3 **Las Piedras** Uru.
153 B3 **Las Plumas** Arg.
155 D1 **Lassance** Brazil
129 D2 **Last Mountain Lake** Can.
152 B3 **Las Tórtolas, Cerro** *mt.* Arg./Chile
118 B3 **Lastoursville** Gabon
109 C2 **Lastovo** *i.* Croatia
144 A2 **Las Tres Vírgenes, Volcán** *vol.* Mex.
146 C2 **Las Tunas** Cuba
144 B2 **Las Varas** Mex.

44	B2	Las Varas Mex.
142	B1	Las Vegas NM U.S.A.
135	C3	Las Vegas NV U.S.A.
131	E2	La Tabatière Can.
80	B2	Latakia Syria
104	B3	La Teste-de-Buch France
108	B2	Latina Italy
147	D3	La Tortuga, Isla i. Venez.
64	B2	La Trinidad Phil.
89	E2	Latskoye Rus. Fed.
130	C1	La Tuque Can.
88	B2	Latvia country Europe
		Latviyskaya S.S.R. country
		Europe see Latvia
59	E3	Lau P.N.G.
102	C1	Lauchhammer Ger.
106	C1	Laudio Spain
101	E3	Lauf an der Pegnitz Ger.
105	D2	Laufen Switz.
92	□A3	Laugarás Iceland
127	H1	Lauge Koch Kyst reg. Greenland
142	C1	Laughlin Peak U.S.A.
51	D4	Launceston Austr.
99	A4	Launceston U.K.
62	A1	Launggyaung Myanmar
153	A4	La Unión Chile
51	D1	Laura Austr.
140	C2	Laurel MS U.S.A.
134	E1	Laurel MT U.S.A.
96	C2	Laurencekirk U.K.
109	C2	Lauria Italy
141	E2	Laurinburg U.S.A.
105	D2	Lausanne Switz.
60	B1	Laut i. Indon.
61	C1	Laut i. Indon.
61	C2	Laut Bali Indon.
101	D2	Lautersbach (Hessen) Ger.
60	C2	Laut Jawa Indon.
61	C2	Laut Kecil, Kepulauan is Indon.
59	C3	Laut Maluku Indon.
61	D2	Laut Sawu Indon.
59	C3	Laut Seram Indon.
100	C1	Lauwersmeer l. Neth.
139	E1	Laval Can.
104	B2	Laval France
107	C2	La Vall d'Uixó Spain
81	D2	Lāvar Meydān salt marsh Iran
50	B2	Laverton Austr.
155	D2	Lavras Brazil
123	D2	Lavumisa Swaziland
61	A1	Lawas Malaysia
78	B3	Lawdar Yemen
62	A1	Lawksawk Myanmar
114	B3	Lawra Ghana
138	B3	Lawrence IN U.S.A.
137	D3	Lawrence KS U.S.A.
139	E2	Lawrence MA U.S.A.
140	C1	Lawrenceburg U.S.A.
141	D2	Lawrenceville U.S.A.
143	D2	Lawton U.S.A.
78	A2	Lawz, Jabal al mt. Saudi Arabia
93	F4	Laxá Sweden
122	B2	Laxey S. Africa
78	B2	Laylá Saudi Arabia
74	B1	Layyah Pak.
109	D2	Lazarevac Serbia
144	A1	Lázaro Cárdenas Mex.
144	B3	Lázaro Cárdenas Mex.
88	B3	Lazdijai Lith.
66	B2	Lazo Rus. Fed.
83	K2	Lazo Rus. Fed.
136	C2	Lead U.S.A.
129	D2	Leader Can.
136	B3	Leadville U.S.A.
		Leaf Bay Can. see Tasiujaq
129	D2	Leaf Rapids Can.
143	D3	League City U.S.A.
143	D3	Leakey U.S.A.
99	C3	Leamington Spa, Royal U.K.
97	B2	Leane, Lough l. Ireland
137	E3	Leavenworth U.S.A.
64	B3	Lebak Phil.
80	B2	Lebanon country Asia
137	E3	Lebanon MO U.S.A.
139	E2	Lebanon NH U.S.A.
134	B2	Lebanon OR U.S.A.
139	D2	Lebanon PA U.S.A.
140	C1	Lebanon TN U.S.A.
89	E3	Lebedyan' Rus. Fed.
91	C1	Lebedyn Ukr.
53	C3	Lebel-sur-Quévillon Can.
104	C2	Le Blanc France
103	D1	Lębork Pol.
123	C1	Lebowakgomo S. Africa
106	B2	Lebrija Spain

153	A3	Lebu Chile
104	C3	Le Bugue France
109	C2	Lecce Italy
108	A1	Lecco Italy
102	C2	Lech r. Austria/Ger.
111	B3	Lechaina Greece
71	B3	Lechang China
102	B1	Leck Ger.
140	B2	Lecompte Phil.
105	C2	Le Creusot France
104	C3	Lectoure France
106	B1	Ledesma Spain
104	C2	Le Dorat France
128	C2	Leduc Can.
97	B3	Lee r. Ireland
137	E1	Leech Lake U.S.A.
98	C1	Leeds U.K.
99	B3	Leek U.K.
97	B2	Leenane Ireland
100	C1	Leer (Ostfriesland) Ger.
141	D3	Leesburg U.S.A.
140	B2	Leesville U.S.A.
53	C2	Leeton Austr.
122	B3	Leeu-Gamka S. Africa
100	B1	Leeuwarden Neth.
50	A3	Leeuwin, Cape Austr.
147	D3	Leeward Islands Caribbean Sea
111	B3	Lefkada Greece
111	B3	Lefkada i. Greece
		Lefkosia Cyprus see Nicosia
64	B2	Legazpi Phil.
103	E1	Legionowo Pol.
108	B1	Legnago Italy
103	D1	Legnica Pol.
74	B1	Leh India
104	C2	Le Havre France
101	E1	Lehre Ger.
122	B1	Lehututu Botswana
103	D2	Leibnitz Austria
99	C3	Leicester U.K.
51	C1	Leichhardt r. Austr.
51	D2	Leichhardt Range mts Austr.
100	B1	Leiden Neth.
100	A2	Leie r. Belgium
52	A2	Leigh Creek Austr.
97	C2	Leighlinbridge Ireland
99	C4	Leighton Buzzard U.K.
93	E3	Leikanger Norway
101	D1	Leine r. Ger.
50	B2	Leinster Austr.
97	C2	Leinster reg. Ireland
97	C2	Leinster, Mount h. Ireland
101	F2	Leipzig Ger.
92	F2	Leiranger Norway
106	B2	Leiria Port.
93	E4	Leirvik Norway
97	C2	Leixlip Ireland
71	B3	Leiyang China
71	B3	Leizhou China
71	A3	Leizhou Bandao pen. China
118	B3	Lékana Congo
122	A2	Lekkersing S. Africa
140	B2	Leland U.S.A.
		Leli China see Tianlin
100	B1	Lelystad Neth.
153	B5	Le Maire, Estrecho de sea chan. Arg.
		Léman, Lac l. France/Switz. see Geneva, Lake
104	C2	Le Mans France
137	D2	Le Mars U.S.A.
102	B2	Lemberg mt. Ger.
154	C2	Leme Brazil
151	C3	Lemesos Cyprus see Limassol
101	D1	Lemförde Ger.
127	H2	Lemieux Islands Can.
136	C1	Lemmon U.S.A.
135	C3	Lemoore U.S.A.
131	D2	Le Moyne, Lac l. Can.
62	A1	Lemro r. Myanmar
109	C2	Le Murge hills Italy
83	J2	Lena r. Rus. Fed.
100	C1	Lengerich Ger.
70	A2	Lenglong Ling mts China
71	B3	Lengshuijiang China

		Leninsk Kazakh. see Baykonyr
89	E3	Leninskiy Rus. Fed.
53	D1	Lennox Head Austr.
141	D1	Lenoir U.S.A.
100	A2	Lens Belgium
105	C1	Lens France
83	I2	Lensk Rus. Fed.
103	D2	Lenti Hungary
109	C3	Lentini Italy
114	B3	Léo Burkina Faso
103	D2	Leoben Austria
		Leodhais, Eilean i. U.K. see Lewis, Isle of
99	B3	Leominster U.K.
144	B2	León Mex.
146	B3	León Nic.
106	B1	León Spain
122	A1	Leonardville Namibia
108	B3	Leonforte Italy
53	C3	Leongatha Austr.
50	B2	Leonora Austr.
		Léopold II, Lac l. Dem. Rep. Congo see Mai-Ndombe, Lac
155	D2	Leopoldina Brazil
		Léopoldville Dem. Rep. Congo see Kinshasa
90	B2	Leova Moldova
		Leovo Moldova see Leova
123	C1	Lephalale S. Africa
123	C1	Lephepe Botswana
123	C3	Lephoi S. Africa
71	B3	Leping China
77	D2	Lepsi Kazakh.
105	C2	Le Puy-en-Velay France
123	C1	Lerala Botswana
123	C2	Leratswana S. Africa
150	B3	Lérida Col.
		Lérida Spain see Lleida
106	C1	Lerma Spain
111	C3	Leros i. Greece
130	C2	Le Roy, Lac l. Can.
93	F4	Lerum Sweden
96	□	Lerwick U.K.
111	C3	Lesbos i. Greece
147	C3	Les Cayes Haiti
104	C3	Les Escaldes Andorra
139	F1	Les Escoumins Can.
70	A3	Leshan China
104	B2	Les Herbiers France
86	D2	Leshukonskoye Rus. Fed.
		Leskhimstroy Ukr. see Syeverodonets'k
109	D2	Leskovac Serbia
104	B2	Lesneven France
		Lesnoy Rus. Fed. see Umba
89	E2	Lesnoye Rus. Fed.
83	H3	Lesosibirsk Rus. Fed.
123	C2	Lesotho country Africa
66	B1	Lesozavodsk Rus. Fed.
104	B2	Les Sables-d'Olonne France
147	D3	Lesser Antilles is Caribbean Sea
81	C1	Lesser Caucasus mts Asia
		Lesser Khingan Mountains mts China see Xiao Hinggan Ling
		Lesser Slave Lake Can.
58	B3	Lesser Sunda Islands Indon.
105	C3	Les Vans France
		Lesvos i. Greece see Lesbos
103	D1	Leszno Pol.
123	D1	Letaba S. Africa
99	C4	Letchworth Garden City U.K.
128	C3	Lethbridge Can.
150	D2	Lethem Guyana
59	C3	Leti, Kepulauan is Indon.
150	C3	Leticia Col.
120	B3	Letlhakane Botswana
123	C1	Letlhakeng Botswana
104	C1	Le Touquet-Paris-Plage France
99	D4	Le Tréport France
123	D1	Letsitele S. Africa
63	A2	Letsok-aw Kyun i. Myanmar
123	C2	Letsopa S. Africa
97	C1	Letterkenny Ireland
105	D2	Leucate, Étang de l. France
100	B2	Leuven Belgium
92	F3	Levanger Norway
143	C2	Levelland U.S.A.
50	B1	Lévêque, Cape Austr.
96	A2	Leverburgh U.K.

100	C2	Leverkusen Ger.
103	D2	Levice Slovakia
54	C2	Levin N.Z.
131	C3	Lévis Can.
139	E2	Levittown U.S.A.
89	E3	Lev Tolstoy Rus. Fed.
99	D4	Lewes U.K.
96	A1	Lewis, Isle of i. U.K.
139	D2	Lewisburg PA U.S.A.
140	C1	Lewisburg TN U.S.A.
138	C3	Lewisburg WV U.S.A.
134	D1	Lewis Range mts U.S.A.
134	C1	Lewiston ID U.S.A.
139	E2	Lewiston ME U.S.A.
134	E1	Lewistown U.S.A.
138	C3	Lexington KY U.S.A.
136	D2	Lexington NE U.S.A.
143	D1	Lexington OK U.S.A.
139	D3	Lexington VA U.S.A.
64	B2	Leyte i. Phil.
109	C2	Lezhë Albania
89	F2	Lezhnevo Rus. Fed.
89	E3	L'gov Rus. Fed.
75	D1	Lharigarbo China
75	D2	Lhasa China
75	C2	Lhazê China
60	A1	Lhokseumawe Indon.
62	A1	Lhünzê China
111	B3	Liakoura mt. Greece
67	B3	Liancourt Rocks is N. Pacific Ocean
		Liangzhou China see Wuwei
70	B2	Liangzi Hu l. China
		Lianhe China see Qianjiang
71	B3	Lianhua China
71	B3	Lianjiang China
		Lianran China see Anning
70	C1	Lianshan China
		Liantang China see Nanchang
		Lianxian China see Lianzhou
		Lianzhou China
70	B2	Lianzhou China
71	B3	Lianzhou China
		Lianzhou China see Hepu
70	B2	Liaocheng China
70	C1	Liaodong Bandao pen. China
70	C1	Liaodong Wan b. China
65	A1	Liao He r. China
70	C1	Liaoning prov. China
71	C1	Liaoyang China
65	A1	Liaoyuan China
128	B1	Liard r. Can.
134	C1	Libby U.S.A.
118	B2	Libenge Dem. Rep. Congo
136	C3	Liberal U.S.A.
103	D1	Liberec Czech Rep.
114	B4	Liberia country Africa
146	B3	Liberia Costa Rica
150	C2	Libertad Venez.
137	E3	Liberty U.S.A.
100	B3	Libin Belgium
64	B2	Libmanan Phil.
71	A3	Libo China
123	C3	Libode S. Africa
119	E2	Liboi Kenya
104	B3	Libourne France
100	B3	Libramont Belgium
142	A3	Sierra mts Mex.
118	A2	Libreville Gabon
115	D2	Libya country Africa
115	D2	Libyan Desert Egypt/Libya
116	A1	Libyan Plateau Egypt/Libya
108	B3	Licata Italy
		Licheng China see Lipu
99	C3	Lichfield U.K.
121	C2	Lichinga Moz.
101	E2	Lichte Ger.
123	C2	Lichtenburg S. Africa
101	E2	Lichtenfels Ger.
88	C3	Lida Belarus
93	F4	Lidköping Sweden
50	C2	Liebig, Mount Austr.
105	C2	Liechtenstein country Europe
100	B2	Liège Belgium
92	J3	Lieksa Fin.
119	C2	Lienart Dem. Rep. Congo
118	B3	Lién Nghia Vietnam
102	C2	Lienz Austria
88	B2	Liepāja Latvia
100	B2	Lier Belgium
102	C2	Liezen Austria
97	C2	Liffey r. Ireland
97	C1	Lifford Ireland
53	C1	Lightning Ridge Austr.
121	C2	Ligonha r. Moz.
105	D3	Ligurian Sea France/Italy
119	C4	Likasi Dem. Rep. Congo

118 C2 **Likati** Dem. Rep. Congo
128 B2 **Likely** Can.
Likhachevo Ukr. *see* **Pervomays'kyy**
Likhachyovo Ukr. *see* **Pervomays'kyy**
89 E2 **Likhoslavl'** Rus. Fed.
60 B1 **Liku** Indon.
105 D3 **L'Île-Rousse** France
71 B3 **Liling** China
93 F4 **Lilla Edet** Sweden
100 B2 **Lille** Belgium
105 C1 **Lille** France
Lille Bælt *sea chan.* Denmark *see* **Little Belt**
93 F3 **Lillehammer** Norway
93 F4 **Lillestrøm** Norway
128 B2 **Lillooet** Can.
121 C2 **Lilongwe** Malawi
64 B3 **Liloy** Phil.
150 B4 **Lima** Peru
138 C2 **Lima** U.S.A.
155 D2 **Lima Duarte** Brazil
79 C2 **Līmah** Oman
80 B2 **Limassol** Cyprus
97 C1 **Limavady** U.K.
153 B3 **Limay** *r.* Arg.
101 F2 **Limbach-Oberfrohna** Ger.
88 B2 **Limbaži** Latvia
118 A2 **Limbe** Cameroon
101 D2 **Limburg an der Lahn** Ger.
122 B2 **Lime Acres** S. Africa
154 C2 **Limeira** Brazil
97 B2 **Limerick** Ireland
93 E4 **Limfjorden** *sea chan.* Denmark
92 I3 **Liminka** Fin.
111 C3 **Limnos** *i.* Greece
104 C2 **Limoges** France
136 C3 **Limon** U.S.A.
104 C2 **Limousin, Plateaux du** France
104 C3 **Limoux** France
123 C1 **Limpopo** *prov.* S. Africa
121 C3 **Limpopo** *r.* S. Africa/Zimbabwe
64 A2 **Linapacan** *i.* Phil.
153 A3 **Linares** Chile
145 C2 **Linares** Mex.
106 C2 **Linares** Spain
108 A3 **Linas, Monte** *mt.* Italy
62 B1 **Lincang** China
Linchuan China *see* **Fuzhou**
98 C3 **Lincoln** U.K.
138 B2 **Lincoln** *IL* U.S.A.
139 F1 **Lincoln** *ME* U.S.A.
137 D2 **Lincoln** *NE* U.S.A.
134 B2 **Lincoln City** U.S.A.
151 D2 **Linden** Guyana
140 C1 **Linden** U.S.A.
119 C2 **Lindi** *r.* Dem. Rep. Congo
119 D3 **Lindi** Tanz.
Lindisfarne *i.* U.K. *see* **Holy Island**
111 C3 **Lindos** Greece
130 C3 **Lindsay** Can.
48 K3 **Line Islands** Kiribati
70 B2 **Linfen** China
64 B2 **Lingayen** Phil.
70 B2 **Lingbao** China
Lingcheng China *see* **Lingshan**
Lingcheng China *see* **Lingshui**
123 C3 **Lingelethu** S. Africa
123 C3 **Lingelihle** S. Africa
100 C1 **Lingen (Ems)** Ger.
60 B2 **Lingga** *i.* Indon.
60 B2 **Lingga, Kepulauan** *is* Indon.
71 B3 **Lingling** China
71 A3 **Lingshan** China
71 A4 **Lingshui** China
114 A3 **Linguère** Senegal
63 B2 **Linh, Ngok** *mt.* Vietnam
155 D1 **Linhares** Brazil
Linjiang China *see* **Shanghang**
65 B1 **Linjiang** China
93 G4 **Linköping** Sweden
66 B1 **Linkou** China
71 B3 **Linli** China
96 B2 **Linnhe, Loch** *inlet* U.K.
70 B2 **Linqing** China
154 C2 **Lins** Brazil
136 C1 **Linton** U.S.A.
69 D2 **Linxi** China
70 A2 **Linxia** China
70 B2 **Linyi** *Shandong* China
70 B2 **Linyi** *Shandong* China
70 B2 **Linying** China
102 C2 **Linz** Austria
105 C3 **Lion, Golfe du** *g.* France
Lions, Gulf of *g.* France *see* **Lion, Golfe du**
109 B3 **Lipari** Italy
108 B3 **Lipari, Isole** *is* Italy
89 E3 **Lipetsk** Rus. Fed.
110 B1 **Lipova** Romania
101 D2 **Lippstadt** Ger.
53 C3 **Liptrap, Cape** Austr.
71 B3 **Lipu** China
119 D2 **Lira** Uganda
108 B2 **Liri** *r.* Italy
76 C1 **Lisakovsk** Kazakh.
118 C2 **Lisala** Dem. Rep. Congo
Lisboa Port. *see* **Lisbon**
106 B2 **Lisbon** Port.
97 C1 **Lisburn** U.K.
97 B2 **Liscannor Bay** Ireland
97 B2 **Lisdoonvarna** Ireland
Lishi China *see* **Dingnan**
71 B3 **Lishui** China
104 C2 **Lisieux** France
99 A4 **Liskeard** U.K.
89 E3 **Liski** Rus. Fed.
53 D1 **Lismore** Austr.
97 C2 **Lismore** Ireland
97 C1 **Lisnaskea** U.K.
97 B2 **Listowel** Ireland
71 A3 **Litang** *Guangxi* China
68 C2 **Litang** *Sichuan* China
138 B3 **Litchfield** *IL* U.S.A.
137 E1 **Litchfield** *MN* U.S.A.
53 D2 **Lithgow** Austr.
111 B3 **Lithino, Akrotirio** *pt* Greece
88 B2 **Lithuania** *country* Europe
111 B2 **Litochoro** Greece
102 C1 **Litoměřice** Czech Rep.
Litovskaya S.S.R. *country* Europe *see* **Lithuania**
146 C2 **Little Abaco** *i.* Bahamas
73 D3 **Little Andaman** *i.* India
141 E3 **Little Bahama Bank** *sea feature* Bahamas
93 E4 **Little Belt** *sea chan.* Denmark
146 B3 **Little Cayman** *i.* Cayman Is
142 A1 **Little Colorado** *r.* U.S.A.
138 C1 **Little Current** Can.
137 E1 **Little Falls** U.S.A.
143 C2 **Littlefield** U.S.A.
99 C4 **Littlehampton** U.K.
122 A2 **Little Karas Berg** *plat.* Namibia
122 B3 **Little Karoo** *plat.* S. Africa
96 A2 **Little Minch** *sea chan.* U.K.
136 C1 **Little Missouri** *r.* U.S.A.
73 D4 **Little Nicobar** *i.* India
140 B2 **Little Rock** U.S.A.
139 E2 **Littleton** U.S.A.
121 C2 **Litunde** Moz.
90 B2 **Lityn** Ukr.
Liuchow China *see* **Liuzhou**
70 B2 **Liujiachang** China
71 A3 **Liupanshui** China
121 C2 **Liupo** Moz.
71 A3 **Liuzhi** China
71 A3 **Liuzhou** China
111 B3 **Livadeia** Greece
88 C2 **Līvāni** Latvia
141 D2 **Live Oak** U.S.A.
50 B1 **Liveringa** Austr.
142 C2 **Livermore, Mount** U.S.A.
53 D2 **Liverpool** Austr.
131 D3 **Liverpool** Can.
98 B3 **Liverpool** U.K.
127 G2 **Liverpool, Cape** Can.
53 C2 **Liverpool Range** *mts* Austr.
96 C3 **Livingston** U.K.
134 D1 **Livingston** *MT* U.S.A.
143 E2 **Livingston** *TX* U.S.A.
143 D2 **Livingston, Lake** U.S.A.
120 B2 **Livingstone** Zambia
55 A3 **Livingston Island** Antarctica
109 C2 **Livno** Bos.-Herz.
89 E3 **Livny** Rus. Fed.
138 C2 **Livonia** U.S.A.
108 B2 **Livorno** Italy
119 D3 **Liwale** Tanz.
99 A5 **Lizard Point** U.K.
108 B1 **Ljubljana** Slovenia
93 G3 **Ljungan** *r.* Sweden
93 F4 **Ljungby** Sweden
93 G3 **Ljusdal** Sweden
93 G3 **Ljusnan** *r.* Sweden
99 B4 **Llandeilo** U.K.
99 B4 **Llandovery** U.K.
99 B3 **Llandrindod Wells** U.K.
98 B3 **Llandudno** U.K.
99 A4 **Llanelli** U.K.
106 C1 **Llanes** Spain
98 A3 **Llangefni** U.K.
99 B3 **Llangollen** U.K.
106 B1 **Llangréu** Spain
99 B3 **Llangurig** U.K.
143 C2 **Llano Estacado** *plain* U.S.A.
150 C2 **Llanos** *plain* Col./Venez.
107 D1 **Lleida** Spain
99 A3 **Lleyn Peninsula** U.K.
107 C2 **Llíria** Spain
128 B2 **Lloyd George, Mount** Can.
129 D2 **Lloyd Lake** Can.
129 C2 **Lloydminster** Can.
152 B2 **Llullaillaco, Volcán** *vol.* Chile
154 B2 **Loanda** Brazil
123 C2 **Lobatse** Botswana
103 D1 **Łobez** Pol.
120 A2 **Lobito** Angola
101 F1 **Loburg** Ger.
96 B2 **Lochaber** *reg.* U.K.
96 B2 **Lochaline** U.K.
Loch Baghasdail U.K. *see* **Lochboisdale**
96 A2 **Lochboisdale** U.K.
104 C2 **Loches** France
96 B2 **Lochgilphead** U.K.
96 B1 **Lochinver** U.K.
96 A2 **Lochmaddy** U.K.
96 C2 **Lochnagar** *mt.* U.K.
Loch nam Madadh U.K. *see* **Lochmaddy**
96 B3 **Lochranza** U.K.
52 A2 **Lock** Austr.
96 C3 **Lockerbie** U.K.
53 C3 **Lockhart** Austr.
143 D3 **Lockhart** U.S.A.
51 D1 **Lockhart River** Austr.
139 D2 **Lock Haven** U.S.A.
139 D2 **Lockport** U.S.A.
63 B2 **Lôc Ninh** Vietnam
105 C3 **Lodève** France
86 C2 **Lodeynoye Pole** Rus. Fed.
74 B2 **Lodhran** Pak.
108 A1 **Lodi** Italy
135 B3 **Lodi** U.S.A.
92 F2 **Løding** Norway
92 G2 **Lødingen** Norway
118 C3 **Lodja** Dem. Rep. Congo
119 D2 **Lodwar** Kenya
103 D1 **Łódź** Pol.
62 B2 **Loei** Thai.
122 A3 **Loeriesfontein** S. Africa
92 F2 **Lofoten** *is* Norway
134 D2 **Logan** U.S.A.
128 A1 **Logan, Mount** Can.
138 B2 **Logansport** U.S.A.
108 B1 **Logatec** Slovenia
115 D3 **Logone** *r.* Africa
106 C1 **Logroño** Spain
93 H3 **Lohja** Fin.
101 D1 **Löhne** Ger.
101 D1 **Lohne (Oldenburg)** Ger.
62 A2 **Loikaw** Myanmar
93 H3 **Loimaa** Fin.
104 B2 **Loire** *r.* France
150 B3 **Loja** Ecuador
106 C2 **Loja** Spain
92 I2 **Lokan tekojärvi** *resr* Fin.
74 B1 **Lōkär** Afgh.
100 B2 **Lokeren** Belgium
122 B1 **Lokgwabe** Botswana
91 C1 **Lokhvytsya** Ukr.
119 D2 **Lokichar** Kenya
119 D2 **Lokichokio** Kenya
93 E4 **Løkken** Denmark
89 D2 **Loknya** Rus. Fed.
115 C4 **Lokoja** Nigeria
89 D3 **Lokot'** Rus. Fed.
88 C2 **Loksa** Estonia
127 H2 **Loks Land** *i.* Can.
114 B4 **Lola** Guinea
93 F5 **Lolland** *i.* Denmark
119 D3 **Lollondo** Tanz.
118 C2 **Lolo** Dem. Rep. Congo
122 B2 **Lolwane** S. Africa
110 B2 **Lom** Bulg.
93 E3 **Lom** Norway
119 C2 **Lomami** *r.* Dem. Rep. Congo
153 C3 **Lomas de Zamora** Arg.
50 B1 **Lombardina** Austr.
61 C2 **Lombok** *i.* Indon.
61 C2 **Lombok, Selat** *sea chan.* Indon.
114 C4 **Lomé** Togo
118 C3 **Lomela** *r.* Dem. Rep. Congo
100 B2 **Lommel** Belgium
96 B2 **Lomond, Loch** *l.* U.K.
88 C2 **Lomonosov** Rus. Fed.
160 A1 **Lomonosov Ridge** Arctic Ocean
61 C2 **Lompobattang, Gunung** Indon.
135 B4 **Lompoc** U.S.A.
63 B2 **Lom Sak** Thai.
103 E1 **Łomża** Pol.
130 B3 **London** Can.
99 C4 **London** U.K.
138 C3 **London** U.S.A.
97 C1 **Londonderry** U.K.
50 B1 **Londonderry, Cape** Austr.
154 B2 **Londrina** Brazil
135 C3 **Lone Pine** U.S.A.
83 M2 **Longa, Proliv** *sea chan.* Rus. Fed.
61 C1 **Long Akah** Malaysia
141 E2 **Long Bay** U.S.A.
135 C4 **Long Beach** U.S.A.
71 A3 **Longchang** China
99 C3 **Long Eaton** U.K.
97 C2 **Longford** Ireland
96 C1 **Longhope** U.K.
119 D3 **Longido** Tanz.
61 C2 **Longiram** Indon.
147 C2 **Long Island** Bahamas
130 C2 **Long Island** Can.
59 D3 **Long Island** P.N.G.
139 E2 **Long Island** U.S.A.
130 B3 **Longlac** Can.
130 B3 **Long Lake** Can.
71 A3 **Longli** China
71 A3 **Longming** China
136 B2 **Longmont** U.S.A.
70 A2 **Longnan** China
Longping China *see* **Luodian**
138 C2 **Long Point** Can.
71 B3 **Longquan** China
131 E3 **Long Range Mountains** Can.
51 D2 **Longreach** Austr.
Longshan China *see* **Long**
99 D3 **Long Stratton** U.K.
98 B2 **Longtown** U.K.
105 D2 **Longuyon** France
143 E2 **Longview** *TX* U.S.A.
134 B1 **Longview** *WA* U.S.A.
61 C1 **Longwai** Indon.
70 A2 **Longxi** China
Longxian China *see* **Wengyuan**
71 B3 **Longxi Shan** *mt.* China
63 B2 **Long Xuyên** Vietnam
71 B3 **Longyan** China
82 C1 **Longyearbyen** Svalbard
108 B1 **Lonigo** Italy
100 C1 **Löningen** Ger.
105 D2 **Lons-le-Saunier** France
141 E2 **Lookout, Cape** U.S.A.
119 D3 **Loolmalasin** *vol. crater* Tanz.
50 B3 **Loongana** Austr.
97 B2 **Loop Head** *hd* Ireland
Lopasnya Rus. Fed. *see* **Chekhov**
63 B2 **Lop Buri** Thai.
64 B2 **Lopez** Phil.
118 A3 **Lopez, Cap** *c.* Gabon
68 C2 **Lop Nur** *salt flat* China
118 B2 **Lopori** *r.* Dem. Rep. Congo
92 H1 **Lopphavet** *b.* Norway
74 A2 **Lora, Hāmūn-i-** *dry lake* Afgh./Pak.
106 B2 **Lora del Río** Spain
138 C2 **Lorain** U.S.A.
74 A1 **Loralai** Pak.
107 C2 **Lorca** Spain
51 E3 **Lord Howe Island** Austr.
142 B2 **Lordsburg** U.S.A.
155 C2 **Lorena** Brazil
59 D3 **Lorengau** P.N.G.
59 D3 **Lorentz** *r.* Indon.
152 B1 **Loreto** Bol.
144 A2 **Loreto** Mex.
104 B2 **Lorient** France
96 B2 **Lorn, Firth of** *est.* U.K.
52 B3 **Lorne** Austr.
105 D2 **Lorraine** *reg.* France
142 B1 **Los Alamos** U.S.A.
143 D3 **Los Aldamas** Mex.
153 A3 **Los Ángeles** Chile
135 C4 **Los Angeles** U.S.A.
135 B3 **Los Banos** U.S.A.
152 B2 **Los Blancos** Arg.
89 F3 **Losevo** Rus. Fed.
108 B2 **Lošinj** *i.* Croatia
144 B2 **Los Mochis** Mex.
118 B2 **Losombo** Dem. Rep. Cong

106 B2 Los Pedroches plat. Spain
147 D3 Los Roques, Islas is Venez.
96 C2 Lossiemouth U.K.
150 C1 Los Teques Venez.
59 E3 Losuia P.N.G.
152 A3 Los Vilos Chile
104 C3 Lot r. France
96 C1 Loth U.K.
134 D1 Lothair U.S.A.
Lothringen reg. France see
Lorraine
119 D2 Lotikipi Plain Kenya/Sudan
118 C3 Loto Dem. Rep. Congo
89 E2 Lotoshino Rus. Fed.
62 B1 Louangnamtha Laos
62 B2 Louangphabang Laos
118 B3 Loubomo Congo
104 B2 Loudéac France
71 B3 Loudi China
118 B3 Loudima Congo
114 A3 Louga Senegal
99 C3 Loughborough U.K.
97 B2 Loughrea Ireland
105 D2 Louhans France
97 B2 Louisburgh Ireland
51 E1 Louisiade Archipelago is
P.N.G.
140 B2 Louisiana state U.S.A.
138 B3 Louisville KY U.S.A.
140 C2 Louisville MS U.S.A.
86 C2 Loukhi Rus. Fed.
118 B3 Loukoléla Congo
106 B2 Loulé Port.
118 A2 Loum Cameroon
130 C2 Loups Marins, Lacs des
lakes Can.
104 B3 Lourdes France
151 D2 Lourenço Brazil
Lourenço Marques Moz. see
Maputo
106 B1 Lousã Port.
53 C2 Louth Austr.
98 C3 Louth U.K.
Louvain Belgium see
Leuven
122 A1 Louwater-Suid Namibia
89 D2 Lovat' r. Rus. Fed.
110 B2 Lovech Bulg.
136 B2 Loveland U.S.A.
136 B2 Lovell U.S.A.
135 C2 Lovelock U.S.A.
88 C1 Loviisa Fin.
143 C2 Lovington U.S.A.
86 C2 Lovozero Rus. Fed.
119 C3 Lowa Dem. Rep. Congo
139 E2 Lowell U.S.A.
119 D2 Lowelli S. Sudan
128 C3 Lower Arrow Lake Can.
Lower California pen. Mex.
see Baja California
54 B2 Lower Hutt N.Z.
97 C1 Lower Lough Erne l. U.K.
128 B2 Lower Post Can.
137 E1 Lower Red Lake U.S.A.
Lower Tunguska r. Rus. Fed.
see Nizhnyaya Tunguska
99 D3 Lowestoft U.K.
103 D1 Łowicz Pol.
139 D2 Lowville U.S.A.
52 B2 Loxton Austr.
Loyang China see Luoyang
48 H6 Loyauté, Îles
New Caledonia
89 D3 Loyew Belarus
92 F2 Løypskardtinden mt.
Norway
109 C2 Loznica Serbia
91 D2 Lozova Ukr.
120 B2 Luacano Angola
70 B2 Lu'an China
120 A1 Luanda Angola
63 B3 Luang, Thale lag. Thai.
121 C2 Luangwa r. Zambia
121 B2 Luanshya Zambia
Luao Angola see Luau
106 B1 Luarca Spain
120 B2 Luau Angola
103 E1 Lubaczów Pol.
103 D1 Lubań Pol.
64 A2 Lubāna Latv.
120 A2 Lubango Angola
119 C3 Lubao Dem. Rep. Congo
103 E1 Lubartów Pol.
101 D1 Lübbecke Ger.
102 C1 Lübben Ger.
143 C2 Lübbock U.S.A.
101 E1 Lübeck Ger.
69 E2 Lubei China
76 B1 Lubenka Kazakh.
119 C3 Lubero Dem. Rep. Congo

103 D1 Lubin Pol.
103 E1 Lublin Pol.
91 C1 Lubny Ukr.
61 C1 Lubok Antu Malaysia
101 E1 Lübow Ger.
101 E1 Lübtheen Ger.
119 C3 Lubudi Dem. Rep. Congo
60 B2 Lubuklinggau Indon.
119 C4 Lubumbashi
Dem. Rep. Congo
120 B2 Lubungu Zambia
119 C3 Lubutu Dem. Rep. Congo
120 A1 Lucala Angola
97 C2 Lucan Ireland
120 B1 Lucapa Angola
108 B2 Lucca Italy
96 B3 Luce Bay U.K.
154 B2 Lucélia Brazil
64 B2 Lucena Phil.
106 C2 Lucena Spain
103 D2 Lučenec Slovakia
109 C2 Lucera Italy
105 D2 Lucerne Switz.
66 B1 Luchegorsk Rus. Fed.
101 E1 Lüchow Ger.
120 A2 Lucira Angola
Łuck Ukr. see Luts'k
101 F1 Luckenwalde Ger.
122 B2 Luckhoff S. Africa
75 C2 Lucknow India
120 A1 Lucunga Angola
120 B2 Lucusse Angola
Lüda China see Dalian
100 C2 Lüdenscheid Ger.
101 E1 Lüder Ger.
120 A3 Lüderitz Namibia
119 D4 Ludewa Tanz.
74 B1 Ludhiana India
99 B3 Ludington U.S.A.
99 B3 Ludlow U.K.
135 C4 Ludlow U.S.A.
110 C2 Ludogorie reg. Bulg.
93 G3 Ludvika Sweden
102 B2 Ludwigsburg Ger.
101 F1 Ludwigsfelde Ger.
101 D3 Ludwigshafen am Rhein
Ger.
101 E1 Ludwigslust Ger.
88 C2 Ludza Latvia
118 C3 Luebo Dem. Rep. Congo
120 A2 Luena Angola
70 A2 Lüeyang China
71 B3 Lufeng China
119 C3 Lufira r. Dem. Rep. Congo
143 E2 Lufkin U.S.A.
88 C2 Luga Rus. Fed.
88 C2 Luga r. Rus. Fed.
105 D2 Lugano Switz.
121 C2 Lugenda r. Moz.
97 C2 Lugnaquilla h. Ireland
106 B1 Lugo Spain
110 B1 Lugoj Romania
91 D2 Luhans'k Ukr.
119 D3 Luhombero Tanz.
90 B1 Luhyny Ukr.
120 B2 Luiana Angola
Luichow Peninsula pen.
China see Leizhou Bandao
118 C3 Luilaka r. Dem. Rep. Congo
Luimneach Ireland see
Limerick
105 D2 Luino Italy
92 I2 Luiro r. Fin.
118 C3 Luiza Dem. Rep. Congo
70 B2 Lujiang China
Lukapa Angola see Lucapa
109 C2 Lukavac Bos.-Herz.
118 B3 Lukenie r. Dem. Rep. Congo
142 A2 Lukeville U.S.A.
89 E3 Lukhovitsy Rus. Fed.
Lukou China see Zhuzhou
103 E1 Łuków Pol.
120 B2 Lukulu Zambia
92 H2 Luleå Sweden
92 H2 Luleälven r. Sweden
111 C2 Lüleburgaz Turkey
70 B2 Lüliang Shan mts China
143 D3 Luling U.S.A.
Luluabourg
Dem. Rep. Congo see
Kananga
61 C2 Lumajang Indon.
75 C1 Lumajangdong Co salt l.
China
Lumbala Angola see
Lumbala Kaquengue
Lumbala Angola see
Lumbala N'guimbo
120 B2 Lumbala Kaquengue
Angola

120 B2 Lumbala N'guimbo Angola
140 C2 Lumberton MS U.S.A.
141 E2 Lumberton NC U.S.A.
61 C1 Lumbis Indon.
106 B1 Lumbrales Spain
63 B2 Lumphät Cambodia
129 D2 Lumsden Can.
54 A3 Lumsden N.Z.
93 F4 Lund Sweden
121 C2 Lundazi Zambia
99 A4 Lundy i. U.K.
101 E1 Lüneburg Ger.
101 E1 Lüneburger Heide reg. Ger.
100 C2 Lünen Ger.
105 D2 Lunéville France
120 B2 Lunga r. Zambia
114 A4 Lungi Sierra Leone
Lungleh India see Lunglei
75 D2 Lunglei India
120 B2 Lungwebungu r. Zambia
74 B2 Luni r. India
88 C3 Luninyets Belarus
104 C3 L'Union France
114 A4 Lunsar Sierra Leone
77 E2 Luntai China
71 A3 Luodian China
71 B3 Luoding China
70 B2 Luohe China
70 B2 Luoyang China
118 B3 Luozi Dem. Rep. Congo
121 B2 Lupane Zimbabwe
110 B1 Lupeni Romania
121 C2 Lupilichi Moz.
101 F2 Luppa Ger.
95 B3 Lurgan U.K.
Luring China see Gêrzê
121 D2 Lúrio Moz.
121 D2 Lurio r. Moz.
92 F2 Lurøy Norway
121 B2 Lusaka Zambia
118 C3 Lusambo Dem. Rep. Congo
109 C2 Lushnjë Albania
70 C2 Lüshun China
123 C3 Lusikisiki S. Africa
136 C2 Lusk U.S.A.
Luso Angola see Luena
76 B3 Lūt, Kavīr-e des. Iran
99 C4 Luton U.K.
61 C1 Lutong Malaysia
129 C1 Łutselk'e Can.
90 B1 Luts'k Ukr.
55 F3 Lützow-Holm Bay
Antarctica
122 B2 Lutzputs S. Africa
122 A3 Lutzville S. Africa
117 C4 Luuq Somalia
137 D2 Luverne U.S.A.
119 C3 Luvua r. Dem. Rep. Congo
123 D1 Luvuvhu r. S. Africa
119 D3 Luwego r. Tanz.
119 D2 Luwero Uganda
61 D2 Luwuk Indon.
100 C3 Luxembourg country Europe
100 C3 Luxembourg Lux.
105 D2 Luxeuil-les-Bains France
123 C3 Luxolweni S. Africa
116 B2 Luxor Egypt
100 B2 Luyksgestel Neth.
86 D2 Luza Rus. Fed.
Luzern Switz. see Lucerne
62 B1 Luzhai China
71 A3 Luzhou China
154 C1 Luziânia Brazil
151 E3 Luzilândia Brazil
64 B1 Luzon i. Phil.
64 B1 Luzon Strait Phil./Taiwan
109 C2 Luzzi Italy
90 A2 L'viv Ukr.
L'vov Ukr. see L'viv
Lwów Ukr. see L'viv
88 C3 Lyakhavichy Belarus
89 D2 Lychkova Rus. Fed.
92 G3 Lycksele Sweden
55 C2 Lyddan Island Antarctica
88 C3 Lyel'chytsy Belarus
88 C2 Lyepyel' Belarus
121 U.S.A. Lyman U.S.A.
99 B4 Lyme Bay U.K.
99 B4 Lyme Regis U.K.
139 D3 Lynchburg U.S.A.
52 A2 Lyndhurst Austr.
129 D2 Lynn Lake Can.
129 D1 Lynx Lake Can.
105 C2 Lyon France

Lyons France see Lyon
89 D2 Lyozna Belarus
103 E1 Łysica h. Pol.
86 E3 Lys'va Rus. Fed.
91 D2 Lysychans'k Ukr.
87 D3 Lysyye Gory Rus. Fed.
88 B3 Lytham St Anne's U.K.
88 C3 Lyuban' Belarus
90 C2 Lyubashivka Ukr.
89 E2 Lyubertsy Rus. Fed.
90 B1 Lyubeshiv Ukr.
89 F2 Lyubim Rus. Fed.
90 A1 Lyuboml' Ukr.
91 D2 Lyubotyn Ukr.
89 D2 Lyubytino Rus. Fed.
89 D3 Lyudinovo Rus. Fed.

M

80 B2 Ma'ān Jordan
70 B2 Ma'anshan China
88 C2 Maardu Estonia
78 B3 Ma'āriḍ, Banī des.
Saudi Arabia
80 B2 Ma'arrat an Nu'mān Syria
100 B1 Maarssen Neth.
100 B2 Maas r. Neth.
100 B2 Maaseik Belgium
64 B2 Maasin Phil.
100 B2 Maastricht Neth.
121 C3 Mabalane Moz.
78 B3 Ma'bar Yemen
150 D2 Mabaruma Guyana
98 D3 Mablethorpe U.K.
123 C2 Mabopane S. Africa
121 C3 Mabote Moz.
122 B2 Mabule Botswana
122 B1 Mabutsane Botswana
155 D2 Macaé Brazil
121 C2 Macaloge Moz.
126 F2 MacAlpine Lake Can.
71 B3 Macao aut. reg. China
151 D2 Macapá Brazil
150 B3 Macará Ecuador
155 D1 Macarani Brazil
121 C3 Macarretane Moz.
Macassar Indon. see
Makassar
Macassar Strait str. Indon.
see Makassar, Selat
121 C2 Macatanja Moz.
151 F3 Macau Brazil
98 B3 Macclesfield U.K.
50 B2 Macdonald, Lake imp. l.
Austr.
50 C2 Macdonnell Ranges mts
Austr.
130 A2 MacDowell Lake Can.
96 C2 Macduff U.K.
106 B1 Macedo de Cavaleiros Port.
52 B3 Macedon mt. Austr.
111 B2 Macedonia country Europe
151 F3 Maceió Brazil
108 B2 Macerata Italy
52 A2 Macfarlane, Lake imp. l.
Austr.
97 B3 Macgillycuddy's Reeks mts
Ireland
74 A2 Mach Pak.
155 C2 Machado Brazil
121 C3 Machaíla Moz.
119 D3 Machakos Kenya
150 B3 Machala Ecuador
68 C2 Machali China
121 C3 Machanga Moz.
Machaze Moz. see Chitobe
70 B2 Macheng China
138 B2 Machesney Park U.S.A.
139 F2 Machias U.S.A.
73 C3 Machilipatnam India
121 C2 Machinga Malawi
150 B4 Machiques Venez.
150 B4 Machu Picchu tourist site
Peru
99 B3 Machynlleth U.K.
123 D2 Macia Moz.
Macias Nguema i.
Equat. Guinea see Bioko
110 C1 Măcin Romania
114 B3 Macina Mali
53 D1 Macintyre r. Austr.
51 D2 Mackay Austr.
50 B2 Mackay, Lake imp. l. Austr.
128 C1 MacKay Lake Can.
128 B1 Mackenzie Can.
128 B2 Mackenzie r. Can.
128 A1 Mackenzie r. Can.
Mackenzie Guyana see
Linden

Mackenzie *atoll* Micronesia *see* Ulithi
55 H3 Mackenzie Bay Antarctica
126 C2 Mackenzie Bay Can.
126 E1 Mackenzie King Island Can.
128 A1 Mackenzie Mountains Can.
Mackillop, Lake *imp. l. Austr. see* Yamma Yamma, Lake
129 D2 Macklin Can.
53 D2 Macksville Austr.
53 D1 Maclean Austr.
123 C3 Maclear S. Africa
50 A2 MacLeod, Lake *dry lake* Austr.
138 A2 Macomb U.S.A.
108 A2 Macomer Italy
121 D2 Macomia Moz.
105 C2 Mâcon France
141 D2 Macon *GA* U.S.A.
137 E3 Macon *MO* U.S.A.
140 C2 Macon *MS* U.S.A.
53 C2 Macquarie *r.* Austr.
48 G9 Macquarie Island S. Pacific Ocean
53 C2 Macquarie Marshes Austr.
53 C2 Macquarie Mountain Austr.
156 D9 Macquarie Ridge S. Pacific Ocean
55 H2 Mac. Robertson Land *reg.* Antarctica
97 B3 Macroom Ireland
52 A1 Macumba *watercourse* Austr.
145 C3 Macuspana Mex.
144 B2 Macuzari, Presa *resr* Mex.
123 D2 Madadeni S. Africa
121 □D3 Madagascar *country* Africa
159 D5 Madagascar Ridge Indian Ocean
115 D2 Madama Niger
111 B2 Madan Bulg.
59 D3 Madang P.N.G.
139 D1 Madawaska *r.* Can.
62 A1 Madaya Myanmar
150 D3 Madeira *r.* Brazil
114 A1 Madeira *terr.* N. Atlantic Ocean
131 D3 Madeleine, Îles de la *is* Can.
99 B3 Madeley U.K.
144 B2 Madera Mex.
135 B3 Madera U.S.A.
73 B3 Madgaon India
74 B2 Madhya Pradesh *state* India
123 C2 Madibogo S. Africa
118 B3 Madingou Congo
121 □D2 Madirovalo Madag.
138 B3 Madison *IN* U.S.A.
137 D2 Madison *SD* U.S.A.
138 B2 Madison *WI* U.S.A.
138 C3 Madison *WV* U.S.A.
134 D1 Madison *r.* U.S.A.
138 B3 Madisonville U.S.A.
61 C2 Madiun Indon.
119 D2 Mado Gashi Kenya
88 C2 Madona Latvia
78 A2 Madrakah Saudi Arabia
79 C3 Madrakah, Ra's *c.* Oman
Madras India *see* Chennai
134 B2 Madras U.S.A.
145 C2 Madre, Laguna *lag.* Mex.
143 D3 Madre, Laguna *lag.* U.S.A.
150 C4 Madre de Dios *r.* Peru
145 B3 Madre del Sur, Sierra *mts* Mex.
144 B2 Madre Occidental, Sierra *mts* Mex.
145 B2 Madre Oriental, Sierra *mts* Mex.
106 C1 Madrid Spain
106 C2 Madridejos Spain
61 C2 Madura *i.* Indon.
61 C2 Madura, Selat *sea chan.* Indon.
73 B4 Madurai India
121 B2 Madzivazvido Zimbabwe
67 C3 Maebashi Japan
62 A2 Mae Hong Son Thai.
62 A1 Mae Sai Thai.
62 A2 Mae Sariang Thai.
99 B4 Maesteg U.K.
62 A2 Mae Suai Thai.
121 □D2 Maevatanana Madag.
123 C2 Mafadi *mt.* Lesotho/S. Africa
Mafeking S. Africa *see* Mahikeng
123 C2 Mafeteng Lesotho
53 C3 Maffra Austr.
119 D3 Mafia Island Tanz.

119 D3 Mafinga Tanz.
154 C3 Mafra Brazil
83 L3 Magadan Rus. Fed.
Magallanes Chile *see* Punta Arenas
Magallanes, Estrecho de *sea chan.* Chile *see* Magellan, Strait of
150 B2 Magangué Col.
140 B1 Magazine Mountain *h.* U.S.A.
114 A4 Magburaka Sierra Leone
69 E1 Magdagachi Rus. Fed.
144 A1 Magdalena Mex.
142 B2 Magdalena U.S.A.
144 A2 Magdalena, Bahía *b.* Mex.
101 E1 Magdeburg Ger.
153 A5 Magellan, Strait of *sea chan.* Chile
Maggiore, Lago *l.* Italy *see* Maggiore, Lake
108 A1 Maggiore, Lake *l.* Italy
116 B2 Maghāghah Egypt
97 C1 Magherafelt U.K.
87 E3 Magnitogorsk Rus. Fed.
140 B2 Magnolia U.S.A.
121 C2 Magoé Moz.
130 C3 Magog Can.
131 D2 Magpie, Lac *l.* Can.
114 A3 Magta' Lahjar Maur.
81 D2 Magtymguly Turkm.
119 D3 Magu Tanz.
151 E3 Maguarinho, Cabo *c.* Brazil
123 D2 Magude Moz.
62 A1 Māgura, Dealul *h.* Moldova
81 C2 Mahābād Iran
74 B2 Mahajan India
121 □D2 Mahajanga Madag.
61 C2 Mahakam *r.* Indon.
123 C1 Mahalapye Botswana
121 □D2 Mahalevona Madag.
75 C2 Mahanadi *r.* India
121 □D2 Mahanoro Madag.
74 B3 Maharashtra *state* India
63 B2 Maha Sarakham Thai.
121 □D2 Mahavavy *r.* Madag.
68 B3 Mahbubnagar India
78 B2 Mahd adh Dhahab Saudi Arabia
107 D2 Mahdia Alg.
150 D2 Mahdia Guyana
113 I6 Mahé *i.* Seychelles
75 C3 Mahendragiri *mt.* India
119 D3 Mahenge Tanz.
54 B3 Maheno N.Z.
74 B2 Mahesana India
74 B2 Mahi *r.* India
54 C1 Mahia Peninsula N.Z.
123 C2 Mahikeng S. Africa
89 D3 Mahilyow Belarus
Mahón Spain *see* Maó
114 B3 Mahou Mali
Mahsana India *see* Mahesana
74 B2 Mahuva India
111 C2 Mahya Dağı *mt.* Turkey
106 B1 Maia Port.
Maiaia Moz. *see* Nacala
147 C3 Maicao Col.
74 A1 Maidan Shahr Afgh.
129 D2 Maidstone Can.
99 D4 Maidstone U.K.
115 D3 Maiduguri Nigeria
75 D2 Maijdi Bangl.
75 C2 Mailani India
76 C3 Maïmanah Afgh.
101 D2 Main *r.* Ger.
118 B3 Mai-Ndombe, Lac *l.* Dem. Rep. Congo
101 E3 Main-Donau-Kanal *canal* Ger.
139 F1 Maine *state* U.S.A.
131 D3 Maine, Gulf of Can./U.S.A.
62 A1 Maingkwan Myanmar
96 C1 Mainland *i. Scotland* U.K.
96 □ Mainland *i. Scotland* U.K.
121 □D2 Maintirano Madag.
101 D2 Mainz Ger.
150 C1 Maiquetía Venez.
120 B3 Maitengwe Botswana
53 D2 Maitland *N.S.W.* Austr.
52 A2 Maitland *S.A.* Austr.
146 B3 Maíz, Islas del *is* Nic.
67 C3 Maizuru Japan
119 D2 Majene Indon.
117 B4 Maji Eth.
107 D2 Majorca *i.* Spain
Majunga Madag. *see* Mahajanga

123 C2 Majwemasweu S. Africa
118 B3 Makabana Congo
61 C2 Makale Indon.
119 C3 Makamba Burundi
77 E2 Makanshy Kazakh.
118 B2 Makanza Dem. Rep. Congo
90 B1 Makariv Ukr.
69 F1 Makarov Rus. Fed.
160 B1 Makarov Basin Arctic Ocean
109 C2 Makarska Croatia
61 C2 Makassar Indon.
61 C2 Makassar, Selat Indon.
76 B2 Makat Kazakh.
119 D3 Makatapora Tanz.
123 D2 Makatini Flats *lowland* S. Africa
114 A4 Makeni Sierra Leone
120 B3 Makgadikgadi *depr.* Botswana
87 D4 Makhachkala Rus. Fed.
123 C1 Makhado S. Africa
76 B2 Makhambet Kazakh.
119 D3 Makindu Kenya
77 D1 Makinsk Kazakh.
91 D2 Makiyivka Ukr.
Makkah Saudi Arabia *see* Mecca
131 E2 Makkovik Can.
103 E2 Makó Hungary
118 B2 Makokou Gabon
119 D3 Makongolosi Tanz.
122 B2 Makopong Botswana
119 C2 Makoro Dem. Rep. Congo
118 B3 Makoua Congo
111 B3 Makrakomi Greece
79 D2 Makran *reg.* Iran/Pak.
74 A2 Makran Coast Range *mts* Pak.
89 E2 Maksatikha Rus. Fed.
81 C2 Mākū Iran
62 A1 Makum India
67 B4 Makurazaki Japan
115 C4 Makurdi Nigeria
92 G2 Malá Sweden
146 B4 Mala, Punta *pt* Panama
73 B3 Malabar Coast India
118 A2 Malabo Equat. Guinea
155 D1 Malacacheta Brazil
Malacca Malaysia *see* Melaka
60 A1 Malacca, Strait of Indon./Malaysia
134 D2 Malad City U.S.A.
88 C3 Maladzyechna Belarus
106 C2 Málaga Spain
Malagasy Republic *country* Africa *see* Madagascar
121 □D2 Malaimbandy Madag.
97 B1 Málainn Mhóir Ireland
48 H4 Malaita *i.* Solomon Is
117 B4 Malakal S. Sudan
48 H5 Malakula *i.* Vanuatu
61 D2 Malamala Indon.
61 C2 Malang Indon.
Malange Angola *see* Malanje
120 A1 Malanje Angola
93 G4 Mälaren *l.* Sweden
153 B3 Malargüe Arg.
130 C3 Malartic Can.
88 B3 Malaryta Belarus
121 C2 Malawi *country* Africa
Malawi, Lake *l.* Africa *see* Nyasa, Lake
89 D2 Malaya Vishera Rus. Fed.
64 B3 Malaybalay Phil.
81 C2 Malāyer Iran
60 B1 Malaysia *country* Asia
81 C2 Malazgirt Turkey
103 D1 Malbork Pol.
101 F1 Malchin Ger.
100 A2 Maldegem Belgium
48 L4 Malden Island Kiribati
56 C5 Maldives *country* Indian Ocean
99 D4 Maldon U.K.
56 19 Male Maldives
111 B3 Maleas, Akrotirio *pt* Greece
103 D2 Malé Karpaty *hills* Slovakia
119 C3 Malela Dem. Rep. Congo
116 A3 Malha Sudan
134 C2 Malheur Lake U.S.A.
114 B3 Mali *country* Africa
114 A3 Mali Guinea
59 C3 Maliana East Timor
58 C3 Malili Indon.
97 C1 Malin Ireland
119 E3 Malindi Kenya

97 C1 Malin Head *hd* Ireland
Malin More Ireland *see* Málainn Mhóir
111 C2 Malkara Turkey
88 C3 Mal'kavichy Belarus
110 C2 Malko Tŭrnovo Bulg.
53 C3 Mallacoota Austr.
53 C3 Mallacoota Inlet *b.* Austr.
96 B2 Mallaig U.K.
116 B2 Mallawī Egypt
129 E1 Mallery Lake Can.
Mallorca *i.* Spain *see* Majorca
97 B2 Mallow Ireland
92 F3 Malm Norway
92 H2 Malmberget Sweden
100 C2 Malmédy Belgium
122 A3 Malmesbury S. Africa
93 F4 Malmö Sweden
71 A3 Malong China
118 C4 Malonga Dem. Rep. Congo
86 C2 Maloshuyka Rus. Fed.
93 E3 Måløy Norway
89 E2 Maloyaroslavets Rus. Fed.
89 E2 Maloye Borisovo Rus. Fed.
86 D2 Malozemel'skaya Tundra *lowland* Rus. Fed.
125 J9 Malpelo, Isla de *i.* N. Pacific Ocean
84 F5 Malta *country* Europe
88 C2 Malta Latvia
134 E1 Malta U.S.A.
122 A1 Maltahöhe Namibia
98 C2 Malton U.K.
Maluku *is* Indon. *see* Moluccas
92 G3 Malung Sweden
123 C2 Maluti Mountains Lesotho
73 B3 Malvan India
140 B2 Malvern U.S.A.
117 B4 Malwal S. Sudan
90 B1 Malyn Ukr.
83 L2 Malyy Anyuy *r.* Rus. Fed.
Malyy Kavkaz *mts* Asia *see* Lesser Caucasus
83 K2 Malyy Lyakhovskiy, Ostrov *i.* Rus. Fed.
123 C2 Mamafubedu S. Africa
151 F3 Mamanguape Brazil
64 B3 Mambajao Phil.
119 C2 Mambasa Dem. Rep. Congo
118 B2 Mambéré *r.* C.A.R.
64 B2 Mamburao Phil.
123 C2 Mamelodi S. Africa
118 A2 Mamfe Cameroon
135 C3 Mammoth Lakes U.S.A.
88 A3 Mamonovo Rus. Fed.
150 C4 Mamoré *r.* Bol./Brazil
114 A3 Mamou Guinea
114 B4 Mampong Ghana
61 C2 Mamuju Indon.
114 B4 Man Côte d'Ivoire
98 A2 Man, Isle of *i.* Irish Sea
150 C3 Manacapuru Brazil
107 D2 Manacor Spain
59 C2 Manado Indon.
146 B3 Managua Nic.
121 □D3 Manakara Madag.
78 B3 Manākhah Yemen
79 C2 Manama Bahrain
59 D3 Manam Island P.N.G.
121 □D3 Mananara *r.* Madag.
121 □D2 Mananara Avaratra Madag.
121 □D3 Mananjary Madag.
114 A3 Manantali, Lac de *l.* Mali
54 A3 Manapouri, Lake N.Z.
77 E2 Manas Hu *l.* China
75 C2 Manaslu *mt.* Nepal
Manastir Macedonia *see* Bitola
59 C3 Manatuto East Timor
62 A2 Man-aung Kyun Myanmar
150 C3 Manaus Brazil
80 B2 Manavgat Turkey
116 A3 Manawashei Sudan
98 B3 Manchester U.K.
139 E2 Manchester *CT* U.S.A.
139 E2 Manchester *NH* U.S.A.
140 C1 Manchester *TN* U.S.A.
81 D3 Mand, Rūd-e *r.* Iran
117 A4 Manda, Jebel *mt.* S. Sudan
121 □D3 Mandabe Madag.
93 E4 Mandal Norway
59 D3 Mandala, Puncak *mt.* Indon.
62 A1 Mandalay Myanmar
68 D1 Mandalgovĭ Mongolia
136 C1 Mandan U.S.A.
118 B1 Mandara Mountains Cameroon/Nigeria

108 A3	Mandas Italy
119 E2	Mandera Kenya
100 C2	Manderscheid Ger.
74 B1	Mandi India
114 B3	Mandiana Guinea
	Mandidzuzure Zimbabwe see Chimanimani
75 C2	Mandla India
121 □D2	Mandritsara Madag.
74 B2	Mandsaur India
50 A3	Mandurah Austr.
73 B3	Mandya India
108 B1	Manerbio Italy
90 B1	Manevychi Ukr.
109 C2	Manfredonia Italy
109 C2	Manfredonia, Golfo di g. Italy
114 B3	Manga Burkina Faso
118 B3	Mangai Dem. Rep. Congo
49 L6	Mangaia i. Cook Is
54 C1	Mangakino N.Z.
110 C2	Mangalia Romania
73 B3	Mangalore India
123 C2	Mangaung S. Africa
60 B2	Mangchi Indon.
	Mangghyshlaq Kazakh. see Mangistau
76 B2	Mangistau Kazakh.
61 C1	Mangkalihat, Tanjung pt Indon.
121 C2	Mangochi Malawi
121 □D3	Mangoky r. Madag.
54 B1	Mangole i. Indon.
54 B1	Mangonui N.Z.
62 A1	Mangshi China
106 B1	Mangualde Port.
154 B3	Mangueirinha Brazil
69 E1	Mangui China
	Mangyshlak Kazakh. see Mangistau
137 D3	Manhattan U.S.A.
121 C3	Manhica Moz.
155 D2	Manhuaçu Brazil
121 □D2	Mania r. Madag.
108 B1	Maniago Italy
121 C2	Maniamba Moz.
150 C3	Manicoré Brazil
131 D3	Manicouagan r. Can.
131 D2	Manicouagan, Petit Lac l. Can.
131 D2	Manicouagan, Réservoir resr Can.
79 B2	Manifah Saudi Arabia
49 K5	Manihiki atoll Cook Is
64 B2	Manila Phil.
53 D2	Manilla Austr.
	Manipur India see Imphal
111 C3	Manisa Turkey
107 C2	Manises Spain
138 B1	Manistee U.S.A.
138 B1	Manistique U.S.A.
129 E2	Manitoba prov. Can.
129 E2	Manitoba, Lake Can.
138 B1	Manitou Islands U.S.A.
130 B3	Manitoulin Island Can.
136 C3	Manitou Springs U.S.A.
130 B3	Manitouwadge Can.
130 C3	Manitowoc U.S.A.
130 C3	Maniwaki Can.
150 B2	Manizales Col.
121 □D3	Manja Madag.
121 C2	Manjacaze Moz.
137 E2	Mankato U.S.A.
114 B4	Mankono Côte d'Ivoire
129 D3	Mankota Can.
73 C4	Mankulam Sri Lanka
74 B2	Manmad India
52 A2	Mannahill Austr.
52 B4	Mannar Sri Lanka
73 B4	Mannar, Gulf of India/Sri Lanka
101 D3	Mannheim Ger.
128 C2	Manning Can.
52 A2	Mannum Austr.
129 C2	Mannville Can.
59 C3	Manokwari Indon.
119 C3	Manono Dem. Rep. Congo
63 A2	Manoron Myanmar
105 D3	Manosque France
131 C2	Manouane, Lac l. Can.
65 B1	Manp'o N. Korea
107 D1	Manresa Spain
121 B2	Mansa Zambia
127 F2	Mansel Island Can.
92 I2	Mansel'ya ridge Fin./Rus. Fed.
53 C3	Mansfield Austr.
98 C3	Mansfield U.K.
140 B1	Mansfield AR U.S.A.
140 B2	Mansfield LA U.S.A.
138 C2	Mansfield OH U.S.A.
139 D2	Mansfield PA U.S.A.
150 A3	Manta Ecuador
64 A3	Mantalingajan, Mount Phil.
155 D1	Mantena Brazil
141 E1	Manteo U.S.A.
104 C2	Mantes-la-Jolie France
155 C2	Mantiqueira, Serra de mts Brazil
	Mantova Italy see Mantua
88 C1	Mäntsälä Fin.
108 B1	Mantua Italy
86 D3	Manturovo Rus. Fed.
151 D3	Manuelzinho Brazil
61 D2	Manui i. Indon.
54 B1	Manukau N.Z.
59 D3	Manus Island P.N.G.
140 B2	Many U.S.A.
121 C2	Manyame r. Moz./Zimbabwe
119 D3	Manyara, Lake salt l. Tanz.
	Manyas Gölü l. Turkey see Kuş Gölü
142 B1	Many Farms U.S.A.
119 D3	Manyoni Tanz.
106 C2	Manzanares Spain
146 C2	Manzanillo Cuba
144 B3	Manzanillo Mex.
119 C3	Manzanza Dem. Rep. Congo
69 D1	Manzhouli China
123 D2	Manzini Swaziland
115 D3	Mao Chad
107 D2	Maó Spain
59 D3	Maoke, Pegunungan mts Indon.
123 C2	Maokeng S. Africa
65 A1	Maokui Shan mt. China
70 A2	Maomao Shan mt. China
71 B3	Maoming China
121 C3	Mapai Moz.
75 C1	Mapam Yumco l. China
61 D2	Mapane Indon.
145 C3	Mapastepec Mex.
144 B2	Mapimí Mex.
144 B2	Mapimí, Bolsón de des. Mex.
121 C3	Mapinhane Moz.
129 D3	Maple Creek Can.
156 D4	Mapmaker Seamounts N. Pacific Ocean
59 D3	Maprik P.N.G.
123 D1	Mapulanguene Moz.
121 C3	Maputo Moz.
123 D2	Maputo r. Moz./S. Africa
123 C2	Maputsoe Lesotho
114 A2	Maqteïr reg. Maur.
120 A1	Maquela do Zombo Angola
153 B4	Maquinchao Arg.
137 E2	Maquoketa U.S.A.
123 C1	Mara S. Africa
150 C3	Maraã Brazil
151 E3	Marabá Brazil
151 D2	Maracá, Ilha de i. Brazil
150 B1	Maracaibo Venez.
	Maracaibo, Lago de lake Venez. see Maracaibo, Lake
150 B2	Maracaibo, Lake inlet Venez.
154 A2	Maracaju Brazil
154 A2	Maracaju, Serra de hills Brazil
150 C1	Maracay Venez.
115 D2	Marādah Libya
115 C3	Maradi Niger
81 C2	Marāgheh Iran
150 C1	Marahuaca, Cerro mt. Venez.
151 E3	Marajó, Baía de est. Brazil
151 D3	Marajó, Ilha de i. Brazil
79 C2	Marākī Iran
119 D2	Maralal Kenya
50 C3	Maralinga Austr.
	Maralwexi China see Bachu
142 A2	Marana U.S.A.
81 C2	Marand Iran
	Marandellas Zimbabwe see Marondera
150 B3	Marañón r. Peru
110 C2	Mărăşeşti Romania
130 B3	Marathon Can.
141 D4	Marathon U.S.A.
106 C2	Marbella Spain
50 A2	Marble Bar Austr.
123 C1	Marble Hall S. Africa
123 D3	Marburg S. Africa
101 D2	Marburg an der Lahn Ger.
103 D2	Marcali Hungary
137 E3	Marceline U.S.A.
99 D3	March U.K.
100 B2	Marche-en-Famenne Belgium
106 B2	Marchena Spain
152 B3	Mar Chiquita, Laguna l. Arg.
150 B4	Marcona Peru
139 E2	Marcy, Mount U.S.A.
74 B1	Mardan Pak.
153 C3	Mar del Plata Arg.
81 C2	Mardin Turkey
96 B2	Maree, Loch l. U.K.
51 D1	Mareeba Austr.
108 B3	Marettimo, Isola i. Italy
89 D2	Marevo Rus. Fed.
142 C2	Marfa U.S.A.
	Margao India see Madgaon
50 A3	Margaret River Austr.
150 C1	Margarita, Isla de i. Venez.
123 D3	Margate S. Africa
99 D4	Margate U.K.
62 A1	Margherita India
	Margherita, Lake l. Eth. see Lake Abaya
119 C2	Margherita Peak mt. Dem. Rep. Congo/Uganda
76 C3	Märgö, Dasht-e des. Afgh.
91 C2	Marhanets' Ukr.
62 A1	Mari Myanmar
152 B2	María Elena Chile
156 C5	Mariana Trench N. Pacific Ocean
	Mariánica, Cordillera mts Spain see Morena, Sierra
140 B2	Marianna AR U.S.A.
141 C2	Marianna FL U.S.A.
102 C2	Mariánské Lázně Czech Rep.
144 B2	Marías, Islas is Mex.
78 B3	Ma'rib Yemen
109 C1	Maribor Slovenia
	Maricourt Can. see Kangiqsualujjuaq
119 C2	Maridi S. Sudan
117 A4	Maridi watercourse S. Sudan
55 P2	Marie Byrd Land reg. Antarctica
147 D3	Marie-Galante i. Guadeloupe
93 G3	Mariehamn Fin.
122 A1	Mariental Namibia
93 F4	Mariestad Sweden
141 D2	Marietta GA U.S.A.
138 C3	Marietta OH U.S.A.
105 D3	Marignane France
83 K3	Marii, Mys pt Rus. Fed.
88 B3	Marijampolė Lith.
154 C2	Marília Brazil
106 B1	Marín Spain
135 B3	Marina U.S.A.
109 C3	Marina di Gioiosa Ionica Italy
88 C3	Mar"ina Horka Belarus
138 B1	Marinette U.S.A.
154 B2	Maringá Brazil
106 B2	Marinha Grande Port.
138 B2	Marion IN U.S.A.
138 C2	Marion OH U.S.A.
141 E2	Marion SC U.S.A.
138 C3	Marion VA U.S.A.
141 D2	Marion, Lake U.S.A.
52 A3	Marion Bay Austr.
152 B2	Mariscal José Félix Estigarribia Para.
105 D3	Maritime Alps mts France/Italy
110 C2	Maritsa r. Bulg.
91 D2	Mariupol' Ukr.
117 C4	Marka Somalia
68 C3	Markam China
100 B1	Markermeer l. Neth.
98 C3	Market Rasen U.K.
98 C3	Market Weighton U.K.
83 I2	Markha r. Rus. Fed.
139 D2	Markham Can.
91 C2	Markivka Ukr.
118 B2	Markounda C.A.R.
140 B2	Marksville U.S.A.
101 D3	Marktheidenfeld Ger.
101 F2	Marktredwitz Ger.
100 C2	Marl Ger.
51 C1	Marla Austr.
100 A3	Marle France
143 D2	Marlin U.S.A.
51 D2	Marlo Austr.
104 C3	Marmande France
111 C2	Marmara, Sea of g. Turkey
	Marmara Denizi g. Turkey see Marmara, Sea of
111 C3	Marmaris Turkey
105 C2	Marne r. France
105 C2	Marne-la-Vallée France
118 B2	Maro Chad
121 □D2	Maroantsetra Madag.
54 B1	Marokopa N.Z.
101 E2	Maroldsweisach Ger.
121 □D2	Maromokotro mt. Madag.
121 C2	Marondera Zimbabwe
151 D2	Maroni r. Fr. Guiana
51 E2	Maroochydore Austr.
61 C2	Maros Indon.
49 M6	Marotiri is Fr. Polynesia
118 B1	Maroua Cameroon
121 □D2	Marovoay Madag.
49 N4	Marquesas Islands Fr. Polynesia
141 D4	Marquesas Keys is U.S.A.
155 D2	Marquês de Valença Brazil
138 B1	Marquette U.S.A.
116 A3	Marra, Jebel mt. Sudan
116 A3	Marra, Jebel Sudan
123 D2	Marracuene Moz.
114 B1	Marrakech Morocco
	Marrakesh Morocco see Marrakech
52 A1	Marree Austr.
121 C2	Marromeu Moz.
121 C2	Marrupa Moz.
116 B2	Marsá al 'Alam Egypt
115 D1	Marsá al Burayqah Libya
119 D2	Marsabit Kenya
78 A2	Marsa Delwein Sudan
108 B3	Marsala Italy
116 A1	Marsá Maṭrūḥ Egypt
101 D2	Marsberg Ger.
108 B2	Marsciano Italy
53 C2	Marsden Austr.
100 B1	Marsdiep sea chan. Neth.
105 D3	Marseille France
	Marseilles France see Marseille
140 B1	Marshall AR U.S.A.
137 D2	Marshall MN U.S.A.
137 E3	Marshall MO U.S.A.
143 E2	Marshall TX U.S.A.
48 H2	Marshall Islands country N. Pacific Ocean
137 E2	Marshalltown U.S.A.
138 A2	Marshfield U.S.A.
146 C2	Marsh Harbour Bahamas
140 B3	Marsh Island U.S.A.
93 G4	Märsta Sweden
	Martaban, Gulf of g. Myanmar see Mottama, Gulf of
61 C2	Martapura Indon.
60 B2	Martapura Indon.
139 E2	Martha's Vineyard i. U.S.A.
105 D2	Martigny Switz.
103 D2	Martin Slovakia
136 C2	Martin U.S.A.
145 C2	Martínez Mex.
141 D2	Martínez U.S.A.
155 C1	Martinho Campos Brazil
147 D3	Martinique terr. West Indies
147 D3	Martinique Passage Dominica/Martinique
139 D3	Martinsburg U.S.A.
138 C3	Martinsville U.S.A.
76 B1	Martok Kazakh.
54 C2	Marton N.Z.
107 D1	Martorell Spain
106 C2	Martos Spain
81 D2	Marv Dasht Iran
105 C3	Marvejols France
76 C3	Mary Turkm.
51 E2	Maryborough Austr.
122 A2	Marydale S. Africa
139 D3	Maryland state U.S.A.
98 B2	Maryport U.K.
137 D3	Marysville U.S.A.
137 E2	Maryville MO U.S.A.
141 D1	Maryville TN U.S.A.
119 D3	Masai Steppe plain Tanz.
119 D3	Masaka Uganda
61 C2	Masalembu Besar i. Indon.
61 D2	Masamba Indon.
65 B2	Masan S. Korea
119 D4	Masasi Tanz.
64 B2	Masbate Phil.
64 B2	Masbate i. Phil.
107 D2	Mascara Alg.
155 E1	Mascote Brazil
123 C2	Maseru Lesotho
	Mashaba Zimbabwe see Mashava
121 C3	Mashava Zimbabwe

76 B3 **Mashhad** Iran
123 D2 **Mashishing** S. Africa
74 A2 **Mashkel, Hamun-i-** *salt flat* Pak.
123 C3 **Masibambane** S. Africa
79 C3 **Masīlah, Wādī al** *watercourse* Yemen
123 C2 **Masilo** S. Africa
118 B3 **Masi-Manimba** Dem. Rep. Congo
119 D2 **Masindi** Uganda
122 B3 **Masinyusane** S. Africa
Masira, Gulf of *b.* Oman *see* **Maşīrah, Khalīj**
79 C3 **Maşīrah, Khalīj** Oman
81 C2 **Masjed-e Soleymān** Iran
97 B2 **Mask, Lough** *l.* Ireland
121 ☐E2 **Masoala, Tanjona** *c.* Madag.
137 E2 **Mason City** U.S.A.
Masqaţ Oman *see* **Muscat**
108 B2 **Massa** Italy
139 E2 **Massachusetts** *state* U.S.A.
139 E2 **Massachusetts Bay** U.S.A.
115 D3 **Massaguet** Chad
115 D3 **Massakory** Chad
121 C3 **Massangena** Moz.
120 A1 **Massango** Angola
116 B3 **Massawa** Eritrea
139 E2 **Massena** U.S.A.
115 D3 **Massenya** Chad
128 A2 **Masset** Can.
105 C2 **Massif Central** *mts* France
138 C2 **Massillon** U.S.A.
121 C3 **Massinga** Moz.
123 D1 **Massingir** Moz.
78 A2 **Mastābah** Saudi Arabia
54 C2 **Masterton** N.Z.
74 B1 **Mastuj** Pak.
74 A2 **Mastung** Pak.
78 A2 **Mastūrah** Saudi Arabia
88 B3 **Masty** Belarus
67 B4 **Masuda** Japan
Masuku Gabon *see* **Franceville**
121 C3 **Masvingo** Zimbabwe
119 D3 **Maswa** Tanz.
69 D1 **Matad** Mongolia
118 B3 **Matadi** Dem. Rep. Congo
146 B3 **Matagalpa** Nic.
130 C3 **Matagami** Can.
130 C3 **Matagami, Lac** *l.* Can.
143 D3 **Matagorda Island** U.S.A.
143 D3 **Matagorda Peninsula** U.S.A.
120 A2 **Matala** Angola
114 A3 **Matam** Senegal
54 C1 **Matamata** N.Z.
144 B2 **Matamoros** Mex.
145 C2 **Matamoros** Mex.
119 D3 **Matandu** *r.* Tanz.
131 D3 **Matane** Can.
146 B2 **Matanzas** Cuba
Matapan, Cape *c.* Greece *see* **Tainaro, Akra**
73 C4 **Matara** Sri Lanka
61 C2 **Mataram** Indon.
50 C1 **Mataranka** Austr.
107 D1 **Mataró** Spain
123 C3 **Matatiele** S. Africa
54 A3 **Mataura** N.Z.
54 A3 **Mataura** *r.* N.Z.
54 C1 **Matawai** N.Z.
152 B1 **Mategua** Bol.
145 B2 **Matehuala** Mex.
109 C2 **Matera** Italy
108 A3 **Mateur** Tunisia
129 E2 **Matheson Island** Can.
143 D3 **Mathis** U.S.A.
74 B2 **Mathura** India
64 B3 **Mati** Phil.
145 C3 **Matías Romero** Mex.
98 C3 **Matlock** U.K.
150 D4 **Mato Grosso** Brazil
154 B1 **Mato Grosso** *state* Brazil
154 B1 **Mato Grosso do Sul** *state* Brazil
123 D2 **Matola** Moz.
106 B1 **Matosinhos** Port.
Matou China *see* **Pingguo**
79 C2 **Maţraḩ** Oman
67 B3 **Matsue** Japan
66 D2 **Matsumae** Japan
67 C3 **Matsumoto** Japan
67 C4 **Matsusaka** Japan
67 B4 **Matsuyama** Japan
130 B2 **Mattagami** *r.* Can.
130 C3 **Mattawa** Can.
105 D2 **Matterhorn** *mt.* Italy/Switz.
134 C2 **Matterhorn** *mt.* U.S.A.
141 D1 **Matthews** U.S.A.
79 C2 **Maţţī, Sabkhat** *salt pan* Saudi Arabia
138 B3 **Mattoon** U.S.A.
150 C2 **Maturín** Venez.
91 D2 **Matveyev Kurgan** Rus. Fed.
123 C2 **Matwabeng** S. Africa
105 C1 **Maubeuge** France
104 C3 **Maubourguet** France
55 E3 **Maud Seamount** *sea feature* S. Atlantic Ocean
49 L1 **Maui** *i.* U.S.A.
141 D2 **Mauldin** U.S.A.
138 C2 **Maumee** *r.* U.S.A.
61 D2 **Maumere** Indon.
120 B2 **Maun** Botswana
75 C2 **Maunath Bhanjan** India
123 C1 **Maunatlala** Botswana
62 A1 **Maungdaw** Myanmar
75 B2 **Mau Ranipur** India
50 C2 **Maurice, Lake** *imp. l.* Austr.
114 A3 **Mauritania** *country* Africa
113 I8 **Mauritius** *country* Indian Ocean
120 B2 **Mavinga** Angola
123 C3 **Mavuya** S. Africa
78 B2 **Māwān, Khashm** *mt.* Saudi Arabia
118 B3 **Mawanga** Dem. Rep. Congo
71 B3 **Mawei** China
62 A1 **Mawkmai** Myanmar
62 A1 **Mawlaik** Myanmar
63 A2 **Mawlamyaing** Myanmar
78 B2 **Mawqaq** Saudi Arabia
55 L3 **Mawson Peninsula** Antarctica
78 B3 **Mawza'** Yemen
108 A3 **Maxia, Punta** *mt.* Italy
121 C3 **Maxixe** Moz.
54 B1 **Maxwell** N.Z.
83 J2 **Maya** *r.* Rus. Fed.
147 C2 **Mayaguana** *i.* Bahamas
147 D3 **Mayagüez** Puerto Rico
81 D2 **Mayāmey** Iran
96 B3 **Maybole** U.K.
116 B3 **Maych'ew** Eth.
117 C3 **Maydh** Somalia
100 C2 **Mayen** Ger.
104 B2 **Mayenne** France
104 B2 **Mayenne** *r.* France
128 C2 **Mayerthorpe** Can.
138 B3 **Mayfield** U.S.A.
91 E3 **Maykop** Rus. Fed.
128 A1 **Mayo** Can.
118 B3 **Mayoko** Congo
Mayo Landing Can. *see* **Mayo**
64 B2 **Mayon** *vol.* Phil.
121 D2 **Mayotte** *terr.* Africa
83 J3 **Mayskiy** Rus. Fed.
138 C3 **Maysville** U.S.A.
118 B3 **Mayumba** Gabon
137 D1 **Mayville** U.S.A.
120 B2 **Mazabuka** Zambia
Mazagan Morocco *see* **El Jadida**
151 D3 **Mazagão** Brazil
104 C3 **Mazamet** France
74 B1 **Mazar** China
108 B3 **Mazara del Vallo** Italy
77 C3 **Mazār-e Sharīf** Afgh.
107 C2 **Mazarrón** Spain
107 C2 **Mazarrón, Golfo de** *b.* Spain
144 A2 **Mazatán** Mex.
146 A3 **Mazatenango** Guat.
144 B2 **Mazatlán** Mex.
88 B2 **Mažeikiai** Lith.
88 B2 **Mazirbe** Latvia
103 E1 **Mazowiecka, Nizina** *lowland* Pol.
71 C3 **Mazu Dao** *i.* Taiwan
121 B3 **Mazunga** Zimbabwe
103 E1 **Mazurskie, Pojezierze** *reg.* Pol.
88 C3 **Mazyr** Belarus
123 D2 **Mbabane** Swaziland
118 B2 **Mbaïki** C.A.R.
118 B2 **Mbakaou, Lac de** *l.* Cameroon
121 C1 **Mbala** Zambia
121 B3 **Mbalabala** Zimbabwe
119 D2 **Mbale** Uganda
118 B2 **Mbalmayo** Cameroon
118 B3 **Mbandaka** Dem. Rep. Congo
118 B2 **Mbandjok** Cameroon
118 A2 **Mbanga** Cameroon
120 A1 **M'banza Congo** Angola
118 B3 **Mbanza-Ngungu** Dem. Rep. Congo
119 D3 **Mbarara** Uganda
118 B2 **Mbé** Cameroon
119 D3 **Mbeya** Tanz.
119 D4 **Mbinga** Tanz.
119 D3 **Mbizi Mountains** Tanz.
119 C2 **Mboki** C.A.R.
123 D2 **Mbombela** S. Africa
118 B2 **Mbomo** Congo
118 B2 **Mbouda** Cameroon
114 A3 **Mbour** Senegal
114 A3 **Mbout** Maur.
118 C3 **Mbuji-Mayi** Dem. Rep. Congo
119 D3 **Mbuyuni** Tanz.
139 F1 **McAdam** Can.
143 D2 **McAlester** U.S.A.
143 D3 **McAllen** U.S.A.
128 B2 **McBride** Can.
134 C2 **McCall** U.S.A.
143 C2 **McCamey** U.S.A.
126 F2 **McClintock Channel** Can.
126 E2 **McClure Strait** Can.
140 B2 **McComb** U.S.A.
136 C2 **McConaughy, Lake** U.S.A.
136 C2 **McCook** U.S.A.
134 C2 **McDermitt** U.S.A.
159 E7 **McDonald Islands** Indian Ocean
134 D1 **McDonald Peak** U.S.A.
135 D3 **McGill** U.S.A.
126 B2 **McGrath** U.S.A.
134 D1 **McGuire, Mount** U.S.A.
121 C2 **Mchinji** Malawi
140 C1 **McKenzie** U.S.A.
51 D2 **McKinlay** Austr.
134 B2 **McKinleyville** U.S.A.
128 C2 **McLennan** Can.
128 B2 **McLeod Lake** Can.
134 B1 **McMinnville** *OR* U.S.A.
140 C1 **McMinnville** *TN* U.S.A.
137 D3 **McPherson** U.S.A.
128 C1 **McTavish Arm** *b.* Can.
123 C3 **Mdantsane** S. Africa
107 E2 **M'Doukal** Alg.
135 D3 **Mead, Lake** *resr* U.S.A.
136 C3 **Meade** U.S.A.
129 D2 **Meadow Lake** Can.
138 C2 **Meadville** U.S.A.
66 D2 **Meaken-dake** *vol.* Japan
106 B1 **Mealhada** Port.
131 E2 **Mealy Mountains** Can.
128 C2 **Meander River** Can.
78 A2 **Mecca** Saudi Arabia
139 D3 **Mechanicsville** U.S.A.
100 B2 **Mechelen** Belgium
100 B2 **Mechelen** Neth.
100 C2 **Mechernich** Ger.
100 C2 **Meckenheim** Ger.
101 E1 **Mecklenburgische Seenplatte** *reg.* Ger.
106 B1 **Meda** Port.
60 A1 **Medan** Indon.
153 B4 **Medanosa, Punta** *pt* Arg.
73 C4 **Medawachchiya** Sri Lanka
107 D2 **Médéa** Alg.
150 B2 **Medellín** Col.
115 D1 **Medenine** Tunisia
134 B2 **Medford** U.S.A.
110 C2 **Medgidia** Romania
110 B1 **Mediaş** Romania
136 B2 **Medicine Bow Mountains** U.S.A.
136 B2 **Medicine Bow Peak** U.S.A.
129 C2 **Medicine Hat** Can.
137 D3 **Medicine Lodge** U.S.A.
155 D1 **Medina** Brazil
78 A2 **Medina** Saudi Arabia
106 C1 **Medinaceli** Spain
106 C1 **Medina del Campo** Spain
106 B1 **Medina de Rioseco** Spain
75 C2 **Medinipur** India
84 E5 **Mediterranean Sea**
129 C2 **Medley** Can.
87 E3 **Mednogorsk** Rus. Fed.
62 A1 **Mêdog** China
88 B2 **Medvėgalio kalnas** *h.* Lith.
83 L2 **Medvezh'i, Ostrova** *is* Rus. Fed.
86 C2 **Medvezh'yegorsk** Rus. Fed.
50 A2 **Meekatharra** Austr.
136 B2 **Meeker** U.S.A.
74 B2 **Meerut** India
100 A2 **Meetkerke** Belgium
119 D2 **Mēga** Eth.
60 B2 **Mega** *i.* Indon.
119 D2 **Mega Escarpment** Eth./Kenya
111 B3 **Megalopoli** Greece
75 D2 **Meghalaya** *state* India
75 C2 **Meghasani** *mt.* India
111 C3 **Megisti** *i.* Greece
92 I1 **Mehamn** Norway
50 A2 **Meharry, Mount** Austr.
137 E3 **Mehlville** U.S.A.
79 C2 **Mehrān** *watercourse* Iran
74 B1 **Mehtar Lām** Afgh.
119 D3 **Meia Meia** Tanz.
154 C1 **Meia Ponte** *r.* Brazil
118 B2 **Meiganga** Cameroon
65 B1 **Meihekou** China
Meijiang China *see* **Ningdu**
62 A1 **Meiktila** Myanmar
101 E2 **Meiningen** Ger.
102 C1 **Meißen** Ger.
Meixian China *see* **Meizhou**
71 B3 **Meizhou** China
152 B2 **Mejicana** *mt.* Arg.
152 A2 **Mejillones** Chile
118 B2 **Mékambo** Gabon
116 B3 **Mek'elē** Eth.
114 C2 **Mekerrhane, Sebkha** *salt pan* Alg.
114 B1 **Meknès** Morocco
63 B2 **Mekong** *r.* Asia
63 B3 **Mekong, Mouths of the** Vietnam
60 B1 **Melaka** Malaysia
156 D6 **Melanesia** *is* Pacific Ocean
156 D5 **Melanesian Basin** Pacific Ocean
53 B3 **Melbourne** Austr.
141 D3 **Melbourne** U.S.A.
108 A2 **Mele, Capo** *c.* Italy
Melekess Rus. Fed. *see* **Dimitrovgrad**
89 F2 **Melenki** Rus. Fed.
131 C2 **Mélèzes, Rivière aux** *r.* Can.
115 D3 **Melfi** Chad
109 C2 **Melfi** Italy
129 D2 **Melfort** Can.
92 F3 **Melhus** Norway
106 B1 **Melide** Spain
114 B1 **Melilla** N. Africa
129 D3 **Melita** Can.
91 D2 **Melitopol'** Ukr.
119 D2 **Melka Guba** Eth.
101 D1 **Melle** Ger.
93 F4 **Mellerud** Sweden
101 E2 **Mellrichstadt** Ger.
101 D1 **Mellum** *i.* Ger.
123 D2 **Melmoth** S. Africa
152 C3 **Melo** Uru.
115 C1 **Melrhir, Chott** *salt l.* Alg.
96 C3 **Melrose** U.K.
Melsetter Zimbabwe *see* **Chimanimani**
52 B3 **Melton** Austr.
99 C3 **Melton Mowbray** U.K.
105 C2 **Melun** France
129 D2 **Melville** Can.
51 D1 **Melville, Cape** Austr.
131 E2 **Melville, Lake** Can.
50 C1 **Melville Island** Austr.
126 E1 **Melville Island** Can.
127 G2 **Melville Peninsula** Can.
61 C2 **Memboro** Indon.
102 C2 **Memmingen** Ger.
60 B1 **Mempawah** Indon.
80 B3 **Memphis** *tourist site* Egypt
140 B1 **Memphis** *TN* U.S.A.
143 C2 **Memphis** *TX* U.S.A.
91 C1 **Mena** Ukr.
140 B2 **Mena** U.S.A.
121 ☐D3 **Menabe** *mts* Madag.
114 C3 **Ménaka** Mali
Mènam Khong *r.* Laos/Thai. *see* **Mekong**
105 C3 **Mende** France
117 B4 **Mendebo** Eth.
116 B3 **Mendefera** Eritrea
160 B1 **Mendeleyev Ridge** Arctic Ocean
145 C2 **Méndez** Mex.
117 B4 **Mendī** Eth.
59 D3 **Mendi** P.N.G.
99 B4 **Mendip Hills** U.K.
138 B2 **Mendota** U.S.A.
153 B3 **Mendoza** Arg.
111 C3 **Menemen** Turkey
100 A2 **Menen** Belgium
70 B2 **Mengcheng** China
60 B2 **Menggala** Indon.
71 A3 **Mengzi** China
131 D2 **Menihek** Can.
52 B2 **Menindee** Austr.
52 B2 **Menindee Lake** Austr.
52 A3 **Meningie** Austr.
104 C2 **Mennecy** France
138 B1 **Menominee** U.S.A.
120 A2 **Menongue** Angola

Menorca i. Spain see Minorca
61 C1 Mensalong Indon.
60 A2 Mentawai, Kepulauan is Indon.
60 B2 Mentok Indon.
61 C1 Menyapa, Gunung mt. Indon.
108 A3 Menzel Bourguiba Tunisia
50 B2 Menzies Austr.
144 B2 Meoqui Mex.
100 C1 Meppel Neth.
100 C1 Meppen Ger.
121 C3 Mepuze Moz.
123 C2 Meqheleng S. Africa
138 B2 Mequon U.S.A.
108 B1 Merano Italy
59 D3 Merauke Indon.
52 B2 Merbein Austr.
Merca Somalia see Marka
135 B3 Merced U.S.A.
152 C2 Mercedes Arg.
143 D3 Mercedes U.S.A.
127 H2 Mercy, Cape Can.
143 C1 Meredith, Lake U.S.A.
91 D2 Merefa Ukr.
116 A3 Merga Oasis Sudan
63 A2 Mergui Archipelago is Myanmar
110 C2 Meriç r. Greece/Turkey
111 C2 Meriç Turkey
145 D2 Mérida Mex.
106 B2 Mérida Spain
150 B2 Mérida Venez.
147 C4 Mérida, Cordillera de mts Venez.
134 C2 Meridian ID U.S.A.
140 C2 Meridian MS U.S.A.
143 D2 Meridian TX U.S.A.
104 B3 Mérignac France
93 H3 Merikarvia Fin.
53 C3 Merimbula Austr.
88 B3 Merkinė Lith.
116 B3 Merowe Sudan
50 A3 Merredin Austr.
96 B3 Merrick h. U.K.
138 B1 Merrill U.S.A.
138 B2 Merrillville U.S.A.
136 C2 Merriman U.S.A.
128 B2 Merritt Can.
53 C2 Merrygoen Austr.
116 C3 Mersa Fatma Eritrea
100 C3 Mersch Lux.
101 E2 Merseburg (Saale) Ger.
98 B2 Mersey r. U.K.
92 B2 Mersin Turkey
60 B1 Mersing Malaysia
99 D4 Mers-les-Bains France
74 B2 Merta India
99 B4 Merthyr Tydfil U.K.
119 D2 Merti Plateau Kenya
106 B2 Mértola Port.
76 B2 Mertvyy Kultuk, Sor dry lake Kazakh.
119 D3 Meru vol. Tanz.
122 B3 Merweville S. Africa
80 B1 Merzifon Turkey
100 C3 Merzig Ger.
142 A2 Mesa AZ U.S.A.
142 C2 Mesa NM U.S.A.
137 E1 Mesabi Range hills U.S.A.
109 C2 Mesagne Italy
142 B2 Mescalero U.S.A.
143 D2 Mescalero Ridge U.S.A.
101 D2 Meschede Ger.
89 E3 Meshchovsk Rus. Fed.
Meshed Iran see Mashhad
91 E2 Meshkovskaya Rus. Fed.
142 B2 Mesilla U.S.A.
111 B2 Mesimeri Greece
111 B3 Mesolongi Greece
115 C1 Messaad Alg.
121 D2 Messalo r. Moz.
109 C3 Messina, Strait of str. Italy
Messina, Stretta di str. Italy see Messina, Strait of
111 B3 Messini Greece
111 B3 Messiniakos Kolpos g. Greece
111 B2 Mesta r. Bulg.
Mesta r. Greece see Nestos
150 B2 Meta r. Col./Venez.
130 C3 Métabetchouan Can.
127 H2 Meta Incognita Peninsula Can.
140 B3 Metairie U.S.A.
152 B2 Metán Arg.
111 B3 Methoni Greece
109 C2 Metković Croatia

121 C2 Metoro Moz.
60 B2 Metro Indon.
100 C3 Mettlach Ger.
135 C3 Mettler U.S.A.
117 B4 Metu Eth.
105 D2 Metz France
100 B2 Meuse r. Belgium/France
105 C2 Meuse, Côtes de ridge France
143 D2 Mexia U.S.A.
144 A1 Mexicali Mex.
144 B2 Mexico country Central America
México Mex. see Mexico City
137 E3 Mexico U.S.A.
125 I7 Mexico, Gulf of Mex./U.S.A.
145 C3 Mexico City Mex.
81 D2 Meybod Iran
101 F1 Meyenburg Ger.
83 M2 Meynypil'gyno Rus. Fed.
86 D2 Mezen' Rus. Fed.
86 D2 Mezen' r. Rus. Fed.
86 E1 Mezhdusharskiy, Ostrov i. Rus. Fed.
103 E2 Mezőtúr Hungary
132 C4 Mezquital r. Mex.
144 B2 Mezquitic Mex.
88 C2 Mežvidi Latvia
121 C2 Mfuwe Zambia
89 D3 Mglin Rus. Fed.
123 D2 Mhlume Swaziland
74 B2 Mhow India
145 C3 Miahuatlán Mex.
106 B2 Miajadas Spain
141 D3 Miami FL U.S.A.
143 E1 Miami OK U.S.A.
141 D3 Miami Beach U.S.A.
81 C2 Miāndowāb Iran
121 □D2 Miandrivazo Madag.
81 C2 Miāneh Iran
71 A3 Mianning China
74 B1 Mianwali Pak.
Mianyang China see Xiantao
70 A2 Mianyang China
121 □D2 Miarinarivo Madag.
87 F3 Miass Rus. Fed.
103 D1 Miastko Pol.
128 C2 Mica Creek Can.
103 E2 Michalovce Slovakia
138 B1 Michigan state U.S.A.
138 B2 Michigan, Lake U.S.A.
138 B2 Michigan City U.S.A.
138 B1 Michipicoten Bay Can.
130 B3 Michipicoten Island Can.
130 B3 Michipicoten River Can.
89 E3 Michurin Bulg. see Tsarevo
89 F3 Michurinsk Rus. Fed.
156 D5 Micronesia is Pacific Ocean
48 G3 Micronesia, Federated States of country N. Pacific Ocean
158 E6 Mid-Atlantic Ridge Atlantic Ocean
100 A2 Middelburg Neth.
123 C3 Middelburg E. Cape S. Africa
123 C2 Middelburg Mpumalanga S. Africa
100 B2 Middelharnis Neth.
134 B2 Middle Alkali Lake U.S.A.
73 D3 Middle Andaman i. India
Middle Congo country Africa see Congo
136 D2 Middle Loup r. U.S.A.
138 C3 Middlesboro U.S.A.
98 C2 Middlesbrough U.K.
139 E2 Middletown NY U.S.A.
138 C3 Middletown OH U.S.A.
78 B3 Midi Yemen
130 C3 Midland Can.
138 C2 Midland MI U.S.A.
143 C2 Midland TX U.S.A.
97 B3 Midleton Ireland
Midnapore India see Medinipur
94 B1 Miðvágur Faroe Is
Midway Oman see Thamarit
57 T7 Midway Islands terr. N. Pacific Ocean
109 D2 Midzhur mt. Bulg./Serbia
103 E1 Mielec Pol.
110 C1 Miercurea-Ciuc Romania
106 B1 Mieres del Camín Spain
101 E1 Mieste Ger.
145 C3 Miguel Alemán, Presa resr Mex.
144 B2 Miguel Auza Mex.

144 B2 Miguel Hidalgo, Presa resr Mex.
63 A2 Migyaunglaung Myanmar
67 B4 Mihara Japan
89 E3 Mikhaylov Rus. Fed.
Mikhaylovgrad Bulg. see Montana
66 B2 Mikhaylovka Rus. Fed.
87 D3 Mikhaylovka Rus. Fed. see Kimovsk
77 D1 Mikhaylovskoye Rus. Fed.
93 I3 Mikkeli Fin.
86 E2 Mikun' Rus. Fed.
67 C3 Mikuni-sanmyaku mts Japan
67 C4 Mikura-jima i. Japan
73 B4 Miladhunmadulu Maldives
108 A1 Milan Italy
121 C2 Milange Moz.
Milano Italy see Milan
111 C3 Milas Turkey
137 D1 Milbank U.S.A.
99 D3 Mildenhall U.K.
52 B2 Mildura Austr.
71 A3 Mile China
136 B1 Miles City U.S.A.
139 D3 Milford DE U.S.A.
135 D3 Milford UT U.S.A.
99 A4 Milford Haven U.K.
54 A2 Milford Sound N.Z.
Milh, Bahr al r. Iraq see Razāzah, Buhayrat ar
107 D2 Miliana Alg.
50 C1 Milikapiti Austr.
51 C1 Milingimbi Austr.
134 E1 Milk r. U.S.A.
116 B3 Milk, Wadi el watercourse Sudan
83 L3 Mil'kovo Rus. Fed.
128 C3 Milk River Can.
105 C3 Millau France
141 D2 Milledgeville U.S.A.
137 E1 Mille Lacs lakes U.S.A.
130 A3 Mille Lacs, Lac des l. Can.
Millennium Island atoll Kiribati see Caroline Island
137 D2 Miller U.S.A.
91 E2 Millerovo Rus. Fed.
52 A2 Millers Creek Austr.
96 B3 Milleur Point U.K.
52 B3 Millicent Austr.
140 C1 Millington U.S.A.
139 F1 Millinocket U.S.A.
55 J3 Mill Island Antarctica
53 D1 Millmerran Austr.
98 B2 Millom U.K.
136 B2 Mills U.S.A.
128 C1 Mills Lake Can.
111 B3 Milos i. Greece
89 E3 Miloslavskoye Rus. Fed.
91 E2 Milove Ukr.
52 B1 Milparinka Austr.
54 A3 Milton N.Z.
99 C3 Milton Keynes U.K.
138 B2 Milwaukee U.S.A.
158 C3 Milwaukee Deep sea feature Caribbean Sea
104 B3 Mimizan France
118 B3 Mimongo Gabon
155 D2 Mimoso do Sul Brazil
81 D2 Mīnāb Iran
61 D1 Minahasa, Semenanjung pen. Indon.
Minahassa Peninsula pen. Indon. see Minahasa, Semenanjung
79 C2 Mīnā' Jabal 'Alī U.A.E.
Minaker Can. see Prophet River
60 B1 Minas Indon.
153 C3 Minas Uru.
79 B2 Mīnā' Sa'ūd Kuwait
155 D1 Minas Gerais state Brazil
155 D1 Minas Novas Brazil
145 C3 Minatitlán Mex.
62 A1 Minbu Myanmar
64 B3 Mindanao i. Phil.
52 B2 Mindarie Austr.
101 D1 Minden Ger.
140 B2 Minden LA U.S.A.
137 D2 Minden NE U.S.A.
64 B2 Mindoro i. Phil.
64 A2 Mindoro Strait Phil.
118 B3 Mindouli Congo
99 B4 Minehead U.K.
154 B1 Mineiros Brazil
143 D2 Mineral Wells U.S.A.
75 C1 Minfeng China
119 C4 Minga Dem. Rep. Congo

81 C1 Mingäçevir Azer.
131 D2 Mingan Can.
52 B2 Mingary Austr.
70 B2 Mingguang China
62 A1 Mingin Myanmar
107 C2 Minglanilla Spain
119 D4 Mingoyo Tanz.
69 E1 Mingshui China
96 A2 Mingulay i. U.K.
71 B3 Mingxi China
Mingzhou China see Suide
70 A2 Minhe China
73 B4 Minicoy atoll India
114 B3 Minignan Côte d'Ivoire
50 A2 Minilya Austr.
131 D2 Minipi Lake Can.
130 A2 Miniss Lake Can.
52 A2 Minlaton Austr.
115 C4 Minna Nigeria
137 E2 Minneapolis U.S.A.
129 E2 Minnedosa Can.
137 E2 Minnesota r. U.S.A.
137 E1 Minnesota state U.S.A.
106 B1 Miño r. Port./Spain
107 D1 Minorca i. Spain
136 C1 Minot U.S.A.
88 C3 Minsk Belarus
103 E1 Mińsk Mazowiecki Pol.
96 D2 Mintlaw U.K.
131 D3 Minto Can.
130 C2 Minto, Lac l. Can.
68 C1 Minusinsk Rus. Fed.
62 A1 Minutang India
70 A2 Minxian China
155 D1 Mirabela Brazil
155 D1 Miralta Brazil
131 D2 Miramichi Can.
111 C3 Mirampellou, Kolpos b. Greece
152 C2 Miranda Brazil
152 C1 Miranda r. Brazil
Miranda Moz. see Macaloge
106 C1 Miranda de Ebro Spain
106 B1 Mirandela Port.
154 B2 Mirandópolis Brazil
154 B2 Mirante, Serra do hills Brazil
79 C3 Mirbāt Oman
61 C1 Miri Malaysia
153 C3 Mirim, Lagoa l. Brazil/Uru.
79 D2 Mīrjāveh Iran
89 D3 Mirnyy Rus. Fed.
83 I2 Mirnyy Rus. Fed.
101 F1 Mirow Ger.
74 A2 Mirpur Khas Pak.
Mirtoan Sea sea Greece see Mirtoö Pelagos
111 B3 Mirtoö Pelagos sea Greece
65 B2 Miryang S. Korea
Mirzachirla Turkm. see Murzechirla
Mirzachul Uzbek. see Guliston
75 C2 Mirzapur India
77 E3 Misalay China
70 B1 Mishan China
51 E1 Misima Island P.N.G.
146 B3 Miskitos, Cayos is Nic.
103 E2 Miskolc Hungary
59 C3 Misoöl i. Indon.
115 D1 Mişrātah Libya
130 B2 Missinaibi r. Can.
130 B3 Missinaibi Lake Can.
128 B3 Mission Can.
130 B2 Missisa Lake Can.
140 C3 Mississippi r. U.S.A.
140 C2 Mississippi state U.S.A.
140 C2 Mississippi Delta U.S.A.
140 C2 Mississippi Sound sea chan. U.S.A.
Missolonghi Greece see Mesolongi
134 D1 Missoula U.S.A.
137 E3 Missouri r. U.S.A.
137 E3 Missouri state U.S.A.
130 C2 Mistassibi r. Can.
130 C2 Mistastin, Lac l. Can.
103 D2 Mistelbach Austria
131 D2 Mistinibi, Lac l. Can.
130 C2 Mistissini Can.
52 A2 Mitchell Austr.
51 C1 Mitchell r. Austr.
137 D2 Mitchell NE U.S.A.
137 D2 Mitchell SD U.S.A.
97 B2 Mitchelstown Ireland
74 A2 Mithi Pak.
67 D3 Mito Japan
119 D3 Mitole Tanz.
109 D2 Mitrovicë Kosovo

Mittagong

53 D2 Mittagong Austr.
101 E2 Mittelhausen Ger.
101 D1 Mittellandkanal canal Ger.
101 F3 Mitterteich Ger.
Mittimatalik Can. see Pond Inlet
150 B2 Mitú Col.
119 C4 Mitumba, Chaîne des mts Dem. Rep. Congo
119 C3 Mitumba, Monts mts Dem. Rep. Congo
119 C3 Mitwaba Dem. Rep. Congo
118 B2 Mitzic Gabon
78 B2 Miyah, Wādī al watercourse Saudi Arabia
67 C4 Miyake-jima i. Japan
66 D3 Miyako Japan
67 B4 Miyakonojō Japan
76 B2 Miyaly Kazakh.
Miyang China see Mile
67 B4 Miyazaki Japan
67 C3 Miyazu Japan
115 D1 Mizdah Libya
97 B3 Mizen Head hd Ireland
90 A2 Mizhhirr"ya Ukr.
Mizo Hills state India see Mizoram
75 D2 Mizoram state India
93 G4 Mjölby Sweden
93 F3 Mjøsa l. Norway
119 D3 Mkomazi Tanz.
103 C1 Mladá Boleslav Czech Rep.
109 D2 Mladenovac Serbia
103 E1 Mława Pol.
109 C2 Mljet i. Croatia
123 C3 Mlungisi S. Africa
90 B1 Mlyniv Ukr.
123 C2 Mmabatho S. Africa
123 C2 Mmathethe Botswana
93 E3 Mo Norway
135 E3 Moab U.S.A.
123 D2 Moamba Moz.
54 B2 Moana N.Z.
118 B3 Moanda Gabon
97 C2 Moate Ireland
119 C3 Moba Dem. Rep. Congo
118 C2 Mobayi-Mbongo Dem. Rep. Congo
137 E3 Moberly U.S.A.
140 C2 Mobile U.S.A.
140 C2 Mobile Bay U.S.A.
140 C2 Mobile Point U.S.A.
136 C1 Mobridge U.S.A.
Mobutu, Lake l. Dem. Rep. Congo/Uganda see Albert, Lake
Mobutu Sese Seko, Lake l. Dem. Rep. Congo/Uganda see Albert, Lake
121 C2 Moçambicano, Planalto plat. Moz.
121 D2 Moçambique Moz.
Moçâmedes Angola see Namibe
62 B1 Mộc Châu Vietnam
78 B3 Mocha Yemen
123 C1 Mochudi Botswana
121 D2 Mocimboa da Praia Moz.
101 D3 Möckmühl Ger.
150 B2 Mocoa Col.
154 C2 Mococa Brazil
144 B2 Mocorito Mex.
144 B1 Moctezuma Mex.
145 B2 Moctezuma Mex.
144 B2 Moctezuma Mex.
121 C2 Mocuba Moz.
105 D2 Modane France
122 B2 Modder r. S. Africa
108 B2 Modena Italy
135 B3 Modesto U.S.A.
109 B3 Modica Italy
123 C1 Modimolle S. Africa
123 D1 Modjadjiskloof S. Africa
53 C3 Moe Austr.
Moero, Lake l. Dem. Rep. Congo/Zambia see Mweru, Lake
100 C2 Moers Ger.
96 C3 Moffat U.K.
117 C4 Mogadishu Somalia
Mogador Morocco see Essaouira
106 B1 Mogadouro, Serra de mts Port.
123 C1 Mogalakwena r. S. Africa
62 A1 Mogaung Myanmar
Mogilev Belarus see Mahilyow
154 C2 Mogi-Mirim Brazil
83 I3 Mogocha Rus. Fed.

123 C1 Mogoditshane Botswana
62 A1 Mogok Myanmar
142 A2 Mogollon Plateau U.S.A.
103 D2 Mohács Hungary
123 C3 Mohale's Hoek Lesotho
74 B1 Mohali India
107 D2 Mohammadia Alg.
142 A2 Mohave Mountains U.S.A.
139 E2 Mohawk r. U.S.A.
62 A1 Mohnyin Myanmar
119 D3 Mohoro Tanz.
90 B2 Mohyliv-Podil's'kyy Ukr.
123 C1 Moijabana Botswana
110 C1 Moineşti Romania
Mointy Kazakh. see Moyynty
92 F2 Mo i Rana Norway
88 C2 Mõisaküla Estonia
104 C3 Moissac France
135 C3 Mojave U.S.A.
135 C3 Mojave Desert U.S.A.
62 B1 Mojiang China
155 C2 Moji das Cruzes Brazil
154 C2 Moji-Guaçu r. Brazil
109 C2 Mojkovac Montenegro
54 B1 Mokau N.Z.
123 C2 Mokhotlong Lesotho
83 J2 Mokhsogollokh Rus. Fed.
118 B1 Mokolo Cameroon
123 C1 Mokopane S. Africa
65 B3 Mokpo S. Korea
109 C2 Mola di Bari Italy
145 C2 Molango Mex.
Moldavia country Europe see Moldova
Moldavskaya S.S.R. country Europe see Moldova
93 E3 Molde Norway
90 B2 Moldova country Europe
110 B2 Moldova Nouă Romania
110 B1 Moldoveanu, Vârful mt. Romania
110 B1 Moldovei, Podişul plat. Romania
90 B2 Moldovei Centrale, Podişul plat. Moldova
123 C1 Molepolole Botswana
88 C2 Molėtai Lith.
109 C2 Molfetta Italy
Molière Alg. see Bordj Bounaama
107 C1 Molina de Aragón Spain
107 C2 Molina de Segura Spain
119 D3 Moliro Dem. Rep. Congo
150 B4 Mollendo Peru
93 F4 Mölnlycke Sweden
91 D2 Molochna r. Ukr.
89 E2 Molokovo Rus. Fed.
53 C2 Molong Austr.
122 B2 Molopo watercourse Botswana/S. Africa
Molotov Rus. Fed. see Perm'
Molotovsk Rus. Fed. see Severodvinsk
Molotovsk Rus. Fed. see Nolinsk
118 B2 Moloundou Cameroon
59 C3 Moluccas is Indon.
Molucca Sea sea Indon. see Laut Maluku
52 B2 Momba Austr.
119 D3 Mombasa Kenya
154 B1 Mombuca, Serra da hills Brazil
111 C2 Momchilgrad Bulg.
93 F4 Møn i. Denmark
105 D3 Monaco country Europe
96 B2 Monadhliath Mountains U.K.
97 C1 Monaghan Ireland
143 C2 Monahans U.S.A.
147 D3 Mona Passage Dom. Rep./ Puerto Rico
120 A1 Mona Quimbundo Angola
Monastir Macedonia see Bitola
89 D3 Monastyrshchina Rus. Fed.
90 B2 Monastyryshche Ukr.
118 B2 Monatélé Cameroon
66 D2 Monbetsu Japan
108 A1 Moncalieri Italy
107 C1 Moncayo mt. Spain
92 J2 Monchegorsk Rus. Fed.
100 C2 Mönchengladbach Ger.
144 B2 Monclova Mex.
131 D3 Moncton Can.
106 B1 Mondego r. Port.
118 C2 Mondjamboli Dem. Rep. Congo
108 A2 Mondovì Italy
111 B3 Monemvasia Greece

66 D1 Moneron, Ostrov i. Rus. Fed.
139 D1 Monet Can.
137 E3 Monett U.S.A.
108 B1 Monfalcone Italy
106 B1 Monforte de Lemos Spain
119 D2 Mongbwalu Dem. Rep. Congo
62 B1 Mông Cai Vietnam
62 A1 Mong Hang Myanmar
Monghyr India see Munger
75 C2 Mongla Bangl.
62 B1 Mong Lin Myanmar
62 A1 Mong Nawng Myanmar
115 D3 Mongo Chad
68 C1 Mongolia country Asia
74 B1 Mongora Pak.
62 A1 Mong Pawk Myanmar
62 A1 Mong Ping Myanmar
120 B2 Mongu Zambia
135 C3 Monitor Range mts U.S.A.
103 E1 Mońki Pol.
99 B4 Monmouth U.K.
114 C4 Mono r. Benin/Togo
135 C3 Mono Lake U.S.A.
109 C2 Monopoli Italy
107 C1 Monreal del Campo Spain
140 B2 Monroe LA U.S.A.
138 C2 Monroe MI U.S.A.
138 B2 Monroe WI U.S.A.
140 C2 Monroeville U.S.A.
114 A4 Monrovia Liberia
100 A2 Mons Belgium
155 E1 Monsarás, Ponta de pt Brazil
100 C2 Montabaur Ger.
122 B3 Montagu S. Africa
109 C3 Montalto mt. Italy
110 B2 Montana Bulg.
134 E1 Montana state U.S.A.
104 C3 Montargis France
104 C3 Montauban France
139 E2 Montauk Point U.S.A.
123 C2 Mont-aux-Sources mt. Lesotho
105 C2 Montbard France
105 D2 Montbéliard France
105 D2 Mont Blanc mt. France/Italy
105 C2 Montbrison France
100 B3 Montcornet France
104 B3 Mont-de-Marsan France
104 C2 Montdidier France
151 D3 Monte Alegre Brazil
154 C1 Monte Alegre de Minas Brazil
139 E1 Montebello Italy
153 B5 Montecarlo Arg.
105 D3 Monte-Carlo Monaco
154 C1 Monte Carmelo Brazil
152 C3 Monte Caseros Arg.
123 C1 Monte Cristo S. Africa
108 B2 Montecristo, Isola di i. Italy
146 C3 Montego Bay Jamaica
105 C3 Montélimar France
109 C2 Montella Italy
145 C2 Montemorelos Mex.
104 B2 Montendre France
109 C2 Montenegro country Europe
121 C2 Montepuez Moz.
108 B2 Montepulciano Italy
135 B3 Monterey U.S.A.
135 B3 Monterey Bay U.S.A.
150 B2 Montería Col.
152 B1 Montero Bol.
145 B2 Monterrey Mex.
109 C2 Montesano sulla Marcellana Italy
109 C2 Monte Sant'Angelo Italy
151 F4 Monte Santo Brazil
108 A2 Monte Santu, Capo di c. Italy
155 D1 Montes Claros Brazil
153 C3 Montevideo Uru.
137 D2 Montevideo U.S.A.
136 B3 Monte Vista U.S.A.
140 C2 Montgomery U.S.A.
100 B3 Monthermé France
105 D2 Monthey Switz.
140 B2 Monticello AR U.S.A.
141 D3 Monticello FL U.S.A.
135 E3 Monticello UT U.S.A.
104 C2 Montignac France
100 B2 Montignies-le-Tilleul Belgium
105 D2 Montigny-le-Roi France
106 B2 Montijo Port.
106 B2 Montijo Spain
106 C2 Montilla Spain
154 B1 Montividiu Brazil
131 D3 Mont-Joli Can.

130 C3 Mont-Laurier Can.
104 C2 Montluçon France
131 C3 Montmagny Can.
104 C2 Montmorillon France
51 E2 Monto Austr.
134 D2 Montpelier ID U.S.A.
139 E2 Montpelier VT U.S.A.
105 C3 Montpellier France
130 C3 Montréal Can.
129 D2 Montreal Lake Can.
129 D2 Montreal Lake l. Can.
99 D4 Montreuil France
105 D2 Montreux Switz.
96 C2 Montrose U.K.
136 B3 Montrose U.S.A.
147 D3 Montserrat terr. West Indi(es)
62 A1 Monywa Myanmar
108 A1 Monza Italy
107 D1 Monzón Spain
123 C1 Mookane Botswana
123 C1 Mookgophong S. Africa
52 A1 Moolawatana Austr.
52 B1 Moomba Austr.
53 D1 Moonie Austr.
53 C1 Moonie r. Austr.
52 A2 Moonta Austr.
50 A3 Moora Austr.
50 A2 Moore, Lake imp. l. Austr.
137 D1 Moorhead U.S.A.
53 C3 Mooroopna Austr.
122 A3 Moorreesburg S. Africa
130 B2 Moose r. Can.
130 B2 Moose Factory Can.
139 F1 Moosehead Lake U.S.A.
129 D2 Moose Jaw Can.
137 E1 Moose Lake U.S.A.
129 D2 Moosomin Can.
130 B2 Moosonee Can.
52 B2 Mootwingee Austr.
123 C1 Mopane S. Africa
114 B3 Mopti Mali
150 B4 Moquegua Peru
103 D2 Mór Hungary
118 B2 Mora Cameroon
93 F3 Mora Sweden
137 E1 Mora U.S.A.
74 B2 Moradabad India
121 □D2 Morafenobe Madag.
121 □D2 Moramanga Madag.
136 A2 Moran U.S.A.
51 D2 Moranbah Austr.
103 D2 Morava r. Europe
96 B2 Moray Firth b. U.K.
100 C3 Morbach Ger.
74 B2 Morbi India
93 G4 Mörbylånga Sweden
104 B3 Morcenx France
69 E1 Mordaga China
129 E3 Morden Can.
89 F3 Mordovo Rus. Fed.
123 C1 Morebeng S. Africa
98 B2 Morecambe U.K.
98 B2 Morecambe Bay U.K.
53 C1 Moree Austr.
59 D3 Morehead P.N.G.
138 C3 Morehead U.S.A.
141 E2 Morehead City U.S.A.
145 B3 Morelia Mex.
107 C1 Morella Spain
74 B2 Morena India
106 B2 Morena, Sierra mts Spain
110 C2 Moreni Romania
142 A3 Moreno Mex.
128 A2 Moresby, Mount Can.
53 D1 Moreton Island Austr.
52 A2 Morgan Austr.
140 B3 Morgan City U.S.A.
141 D1 Morganton U.S.A.
139 D3 Morgantown U.S.A.
105 D2 Morges Switz.
77 C3 Morghāb, Daryā-ye r. Afgh.
68 C2 Mori China
66 D2 Mori Japan
53 C1 Moriarty's Range hills Aus(tr.)
128 B2 Morice Lake Can.
66 D3 Morioka Japan
53 D2 Morisset Austr.
104 B2 Morlaix France
98 C3 Morley U.K.
157 G9 Mornington Abyssal Plain S. Atlantic Ocean
51 C1 Mornington Island Austr.
59 D3 Morobe P.N.G.
114 B1 Morocco country Africa
119 D3 Morogoro Tanz.
64 B3 Moro Gulf Phil.
122 B2 Morokweng S. Africa
121 □D3 Morombe Madag.
68 C1 Mörön Mongolia
121 □D3 Morondava Madag.

06 B2 Morón de la Frontera Spain
21 D2 Moroni Comoros
59 C2 Morotai i. Indon.
19 D2 Moroto Uganda
98 C2 Morpeth U.K.
86 F2 Morrasale Rus. Fed.
40 B1 Morrilton U.S.A.
54 C1 Morrinhos Brazil
54 C1 Morrinsville N.Z.
29 E3 Morris Can.
37 D1 Morris U.S.A.
41 D1 Morristown U.S.A.
54 C2 Morro Agudo Brazil
55 E1 Morro d'Anta Brazil
55 K3 Morse, Cape Antarctica
 Morshanka Rus. Fed. see
 Morshansk
87 D3 Morshansk Rus. Fed.
51 C3 Mortes, Rio das r. Brazil
52 B3 Mortlake Austr.
48 G3 Mortlock Islands
 Micronesia
38 B2 Morton U.S.A.
53 C2 Morundah Austr.
53 D3 Moruya Austr.
96 B2 Morvern reg. U.K.
 Morvi India see Morbi
53 C3 Morwell Austr.
01 D3 Mosbach Ger.
89 E2 Moscow Rus. Fed.
34 C1 Moscow U.S.A.
00 C2 Mosel r. France
22 B2 Moselebe watercourse
 Botswana
05 D2 Moselle r. France
34 C1 Moses Lake U.S.A.
92 ⊓A3 Mosfellsbær Iceland
54 B3 Mosgiel N.Z.
88 C2 Moshchnyy, Ostrov i.
 Rus. Fed.
89 D2 Moshenskoye Rus. Fed.
19 D3 Moshi Tanz.
92 F2 Mosjøen Norway
 Moskva Rus. Fed. see
 Moscow
89 E2 Moskva r. Rus. Fed.
03 D2 Mosonmagyaróvár
 Hungary
46 B3 Mosquitos, Costa de
 coastal area Nic.
46 B4 Mosquitos, Golfo de los b.
 Panama
93 F4 Moss Norway
 Mossâmedes Angola see
 Namibe
22 B3 Mossel Bay S. Africa
22 B3 Mossel Bay b. S. Africa
18 B3 Mossendjo Congo
52 B2 Mossgiel Austr.
52 D1 Mossman Austr.
51 F3 Mossoró Brazil
53 D2 Moss Vale Austr.
02 C1 Most Czech Rep.
14 C1 Mostaganem Alg.
09 C2 Mostar Bos.-Herz.
52 C3 Mostardas Brazil
06 C1 Móstoles Spain
81 C2 Mosul Iraq
45 D3 Motagua r. Guat.
93 G4 Motala Sweden
23 C2 Motetema S. Africa
96 C3 Motherwell U.K.
75 C2 Motihari India
07 C2 Motilla del Palancar Spain
22 B1 Motokwe Botswana
06 C2 Motril Spain
10 B2 Motru Romania
36 C1 Mott U.S.A.
63 A2 Mottama Myanmar
63 A2 Mottama, Gulf of Myanmar
54 B2 Motueka N.Z.
45 D2 Motul Mex.
49 L5 Motu One atoll
 Fr. Polynesia
14 A3 Moudjéria Maur.
11 C3 Moudros Greece
18 B3 Mouila Gabon
52 B3 Moulamein Austr.
05 C2 Moulins France
41 D2 Moultrie U.S.A.
41 E2 Moultrie, Lake U.S.A.
38 B3 Mound City U.S.A.
15 D4 Moundou Chad
38 C3 Moundsville U.S.A.
37 E3 Mountain Grove U.S.A.
40 B1 Mountain Home AR U.S.A.
34 C2 Mountain Home ID U.S.A.
41 D1 Mount Airy U.S.A.
23 C3 Mount Ayliff S. Africa
52 A3 Mount Barker Austr.

53 C3 Mount Beauty Austr.
97 B2 Mountbellew Ireland
121 C2 Mount Darwin Zimbabwe
139 F2 Mount Desert Island U.S.A.
123 C3 Mount Fletcher S. Africa
123 C3 Mount Frere S. Africa
52 B3 Mount Gambier Austr.
59 D3 Mount Hagen P.N.G.
53 C2 Mount Hope Austr.
51 C2 Mount Isa Austr.
52 A3 Mount Lofty Range mts
 Austr.
50 A2 Mount Magnet Austr.
52 B2 Mount Manara Austr.
54 C1 Mount Maunganui N.Z.
97 C2 Mountmellick Ireland
111 B2 Mount Olympus mt. Greece
137 E2 Mount Pleasant IA U.S.A.
138 C2 Mount Pleasant MI U.S.A.
141 E2 Mount Pleasant SC U.S.A.
143 E2 Mount Pleasant TX U.S.A.
135 D3 Mount Pleasant UT U.S.A.
99 A4 Mount's Bay U.K.
134 B2 Mount Shasta U.S.A.
54 B2 Mount Somers N.Z.
138 B3 Mount Vernon IL U.S.A.
138 C2 Mount Vernon OH U.S.A.
134 B1 Mount Vernon WA U.S.A.
51 D2 Moura Austr.
106 B2 Moura Port.
115 E3 Mourdi, Dépression du
 depr. Chad
97 C1 Mourne Mountains hills
 U.K.
100 A2 Mouscron Belgium
115 D3 Moussoro Chad
61 D1 Moutong Indon.
115 C2 Mouydir, Monts du plat.
 Alg.
100 B3 Mouzon France
97 B1 Moy r. Ireland
117 B4 Moyale Eth.
 Moyen Congo country
 Africa see Congo
123 C3 Moyeni Lesotho
76 B2 Mo'ynoq Uzbek.
119 D2 Moyo Uganda
77 D2 Moyynkum Kazakh.
77 D2 Moyynty Kazakh.
121 C3 Mozambique country Africa
113 G8 Mozambique Channel
 Africa
81 C1 Mozdok Rus. Fed.
89 E2 Mozhaysk Rus. Fed.
119 D3 Mpanda Tanz.
121 C2 Mpika Zambia
118 B2 Mpoko r. C.A.R.
121 C1 Mporokoso Zambia
123 C2 Mpumalanga prov. S. Africa
119 D3 Mpwapwa Tanz.
62 A1 Mrauk-U Myanmar
115 D1 M'Saken Tunisia
119 D3 Msambweni Kenya
119 D3 Msata Tanz.
88 C2 Mshinskaya Rus. Fed.
115 C1 M'Sila Alg.
89 D2 Msta r. Rus. Fed.
89 D2 Mstinskiy Most Rus. Fed.
89 D3 Mstsislaw Belarus
123 C3 Mthatha S. Africa
 Mtoko Zimbabwe see
 Mutoko
89 E3 Mtsensk Rus. Fed.
123 D2 Mtubatuba S. Africa
119 E4 Mtwara Tanz.
151 E3 Muana Brazil
118 B3 Muanda Dem. Rep. Congo
62 B1 Muang Hiam Laos
62 B2 Muang Hinboun Laos
63 B2 Muang Không Laos
63 B2 Muang Khôngxédôn Laos
62 B1 Muang Ngoy Laos
62 B2 Muang Pakbeng Laos
63 B2 Muang Phalan Laos
62 B1 Muang Sing Laos
62 B2 Muang Vangviang Laos
60 B1 Muar Malaysia
60 B2 Muarabungo Indon.
60 B2 Muaradua Indon.
61 C2 Muaralaung Indon.
60 A2 Muarasiberut Indon.
60 B2 Muaratembesi Indon.
61 C2 Muarateweh Indon.
 Muara Tuang Malaysia see
 Kota Samarahan
119 D2 Mubende Uganda
115 D3 Mubi Nigeria
120 A2 Muconda Angola
120 A2 Mucope Angola
121 C2 Mucubela Moz.

155 E1 Mucuri Brazil
155 E1 Mucuri r. Brazil
66 A2 Mudanjiang China
66 A1 Mudan Jiang r. China
111 C2 Mudanya Turkey
136 B2 Muddy Gap U.S.A.
101 E1 Müden (Örtze) Ger.
53 C2 Mudgee Austr.
63 A2 Mudon Myanmar
80 B1 Mudurnu Turkey
121 C2 Mueda Moz.
121 B2 Mufulira Zambia
120 B2 Mufumbwe Zambia
111 C3 Muğla Turkey
116 B2 Muhammad Qol Sudan
 Muhammarah Iran see
 Khorramshahr
101 F2 Mühlberg Ger.
101 E2 Mühlhausen (Thüringen)
 Ger.
88 B2 Muhu i. Estonia
96 B3 Muirkirk U.K.
121 C2 Muite Moz.
65 B2 Mukačevo Ukr. see
 Mukacheve
90 A2 Mukacheve Ukr.
61 C1 Mukah Malaysia
79 B3 Mukalla Yemen
63 B2 Mukdahan Thai.
 Mukden China see
 Shenyang
 Mukhtuya Rus. Fed. see
 Lensk
50 A3 Mukinbudin Austr.
60 B2 Mukomuko Indon.
121 C2 Mulanje, Mount Malawi
101 F2 Mulde r. Ger.
119 D3 Muleba Tanz.
144 A2 Mulegé Mex.
143 C2 Muleshoe U.S.A.
106 C2 Mulhacén mt. Spain
100 C2 Mülheim an der Ruhr Ger.
105 D2 Mulhouse France
62 B1 Muli China
66 B2 Muling China
66 B1 Muling He r. China
96 B2 Mull i. U.K.
53 C2 Mullaley Austr.
136 C2 Mullen U.S.A.
61 C1 Muller, Pegunungan mts
 Indon.
50 A2 Mullewa Austr.
97 C2 Mullingar Ireland
96 B3 Mull of Galloway c. U.K.
96 B3 Mull of Kintyre hd U.K.
96 A3 Mull of Oa hd U.K.
53 D1 Mullumbimby Austr.
120 B2 Mulobezi Zambia
74 B1 Multan Pak.
86 F2 Mulym'ya Rus. Fed.
74 B3 Mumbai India
120 B2 Mumbeji Zambia
120 B2 Mumbwa Zambia
61 D2 Muna i. Indon.
145 D2 Muna Mex.
101 E2 Münchberg Ger.
 München Ger. see Munich
 München-Gladbach Ger. see
 Mönchengladbach
138 C2 Muncie U.S.A.
50 B3 Mundrabilla Austr.
138 B3 Munfordville U.S.A.
119 C2 Mungbere
 Dem. Rep. Congo
75 C2 Munger India
52 A1 Mungeranie Austr.
53 C1 Mungindi Austr.
102 C2 Munich Ger.
155 D2 Muniz Freire Brazil
101 E1 Münster Ger.
100 C2 Münster Ger.
97 B2 Munster reg. Ireland
100 C2 Münsterland reg. Ger.
62 B1 Mường Nhe Vietnam
92 H2 Muonio Fin.
92 H2 Muonioälven r. Fin./Sweden
 Muqdisho Somalia see
 Mogadishu
155 D2 Muqui Brazil
103 D2 Mur r. Austria
67 C3 Murakami Japan
119 C3 Muramvya Burundi
119 D3 Murashi Rus. Fed.
81 B2 Murat r. Turkey
111 C2 Muratlı Turkey
67 B4 Murayama Japan
50 A2 Murchison watercourse
 Austr.

107 C2 Murcia Spain
107 C2 Murcia aut. comm. Spain
136 C2 Murdo U.S.A.
131 D3 Murdochville Can.
111 C2 Mürefte Turkey
110 B1 Mureşul r. Romania
104 C3 Muret France
140 C1 Murfreesboro U.S.A.
77 D3 Murghob Tajik.
155 D2 Muriaé Brazil
120 B1 Muriege Angola
101 F1 Müritz l. Ger.
92 J2 Murmansk Rus. Fed.
86 C2 Murmanskiy Bereg
 coastal area Rus. Fed.
87 D3 Murom Rus. Fed.
66 D2 Muroran Japan
106 B1 Muros Spain
67 B4 Muroto Japan
67 B4 Muroto-zaki pt Japan
141 D1 Murphy U.S.A.
53 C1 Murra Murra Austr.
52 A3 Murray r. Austr.
128 B2 Murray r. Can.
138 B3 Murray U.S.A.
59 D3 Murray, Lake P.N.G.
141 D2 Murray, Lake U.S.A.
52 A3 Murray Bridge Austr.
122 B3 Murraysburg S. Africa
52 B3 Murrayville Austr.
52 B2 Murrumbidgee r. Austr.
53 C2 Murrumburrah Austr.
121 C2 Murrupula Moz.
53 D2 Murrurundi Austr.
109 C1 Murska Sobota Slovenia
54 C1 Murupara N.Z.
49 N6 Mururoa atoll
 Fr. Polynesia
75 C2 Murwara India
53 D1 Murwillumbah Austr.
76 C3 Murzechirla Turkm.
115 D2 Murzūq Libya
115 D2 Murzuq, Idhān des. Libya
103 D2 Mürzzuschlag Austria
81 C2 Muş Turkey
110 B2 Musala mt. Bulg.
65 B1 Musan N. Korea
78 B3 Musaymir Yemen
79 C2 Muscat Oman
 Muscat and Oman country
 Asia see Oman
137 E2 Muscatine U.S.A.
50 C2 Musgrave Ranges mts
 Austr.
118 B3 Mushie Dem. Rep. Congo
60 B2 Musi r. Indon.
123 D1 Musina S. Africa
138 B2 Muskegon r. U.S.A.
138 B2 Muskegon U.S.A.
138 C3 Muskingum r. U.S.A.
143 D1 Muskogee U.S.A.
139 D1 Muskoka, Lake Can.
128 B2 Muskwa r. Can.
74 A1 Muslimbagh Pak.
116 B3 Musmar Sudan
119 D3 Musoma Tanz.
59 D3 Mussau Island P.N.G.
96 C3 Musselburgh U.K.
117 C4 Mustahil Eth.
88 B2 Mustjala Estonia
53 D2 Muswellbrook Austr.
116 A2 Müt Egypt
121 C2 Mutare Zimbabwe
121 C2 Mutoko Zimbabwe
66 D2 Mutsu Japan
66 D2 Mutsu-wan b. Japan
121 C2 Mutuali Moz.
155 D1 Mutum Brazil
92 I2 Muurola Fin.
70 A2 Mu Us Shadi des. China
120 A1 Muxaluando Angola
86 C2 Muyezerskiy Rus. Fed.
119 C3 Muyinga Burundi
74 B1 Muzaffargarh Pak.
75 C2 Muzaffarpur India
74 B2 Muzamane Moz.
155 E1 Muzambinho Brazil
144 B2 Múzquiz Mex.
75 D1 Muz Tag mt. China
117 C4 Mvolo S. Sudan
119 C3 Mwanza Dem. Rep. Congo
119 D3 Mwanza Tanz.
118 B3 Mweka Dem. Rep. Congo
118 C3 Mwene-Ditu
 Dem. Rep. Congo
121 C3 Mwenezi Zimbabwe
121 C3 Mwenezi r. Zimbabwe
119 C3 Mweru, Lake
 Dem. Rep. Congo/Zambia

Mweru Wantipa, Lake

121 B1	Mweru Wantipa, Lake Zambia	121 C2	Naiopué Moz.	65 B1	Nangnim-sanmaek *mts* N. Korea	146 C2	Nassau Bahamas

121 B1 Mweru Wantipa, Lake
 Zambia
118 C3 Mwimba Dem. Rep. Congo
120 B2 Mwinilunga Zambia
88 C3 Myadzyel Belarus
62 A2 Myanaung Myanmar
62 A1 Myanmar *country* Asia
63 A2 Myaungmya Myanmar
63 A2 Myeik Myanmar
 Myeik Kyunzu *is* Myanmar
 see Mergui Archipelago
62 A1 Myingyan Myanmar
62 A1 Myitkyina Myanmar
90 A2 Mykolayiv Ukr.
91 C2 Mykolayiv Ukr.
111 C3 Mykonos Greece
111 C3 Mykonos *i.* Greece
86 E2 Myla Rus. Fed.
75 D2 Mymensingh Bangl.
67 C3 Myōkō, Japan
65 B1 Myŏnggan N. Korea
88 C2 Myory Belarus
92 ☐B3 Mýrdalsjökull Iceland
92 G2 Myre Norway
91 C2 Myrhorod Ukr.
111 C3 Myrina Greece
90 C2 Myronivka Ukr.
141 E2 Myrtle Beach U.S.A.
53 C3 Myrtleford Austr.
134 B2 Myrtle Point U.S.A.
83 H1 Mys Chelyuskin Rus. Fed.
89 E2 Myshkin Rus. Fed.
 Myshkino Rus. Fed. *see*
 Myshkin
103 C1 Myślibórz Pol.
73 B3 Mysore India
83 N2 Mys Shmidta Rus. Fed.
 Mysuru India *see* Mysore
63 B2 My Tho Vietnam
111 C3 Mytilini Greece
89 E3 Mytishchi Rus. Fed.
123 C3 Mzamomhle S. Africa
121 C2 Mzimba Malawi
121 C2 Mzuzu Malawi

N

101 F3 Naab *r.* Ger.
100 B1 Naarden Neth.
97 C2 Naas Ireland
122 A2 Nababeep S. Africa
87 E3 Naberezhnyye Chelny
 Rus. Fed.
59 D3 Nabire Indon.
80 B2 Näblus West Bank
121 D2 Nacala Moz.
119 D4 Nachingwea Tanz.
103 D1 Náchod Czech Rep.
73 D3 Nachuge India
143 E2 Nacogdoches U.S.A.
144 B1 Nacozari de García Mex.
 Nada China *see* Danzhou
74 B2 Nadiad India
90 A2 Nadvirna Ukr.
86 C2 Nadvoitsy Rus. Fed.
86 G2 Nadym Rus. Fed.
93 F4 Næstved Denmark
111 B3 Nafpaktos Greece
111 B3 Nafplio Greece
115 D1 Nafūsah, Jabal *hills* Libya
78 B2 Nafy Saudi Arabia
64 B2 Naga Phil.
130 B2 Nagagami *r.* Can.
67 C3 Nagano Japan
67 C3 Nagaoka Japan
75 D2 Nagaon India
74 B1 Nagar India
74 B2 Nagar Parkar Pak.
67 A4 Nagasaki Japan
67 B4 Nagato Japan
74 B2 Nagaur India
73 B4 Nagercoil India
74 A2 Nagha Kalat Pak.
74 B2 Nagina India
67 C3 Nagoya Japan
75 B2 Nagpur India
75 D1 Nagqu China
141 E1 Nags Head U.S.A.
82 E1 Nagurskoye Rus. Fed.
103 D2 Nagyatád Hungary
103 D2 Nagykanizsa Hungary
128 B1 Nahanni Butte Can.
81 C2 Nahāvand Iran
101 E1 Nahrendorf Ger.
153 A4 Nahuel Huapí, Lago *l.* Arg.
141 D2 Nahunta U.S.A.
131 D2 Nain Can.
81 D2 Nā'īn Iran

121 C2 Naiopué Moz.
96 C2 Nairn U.K.
119 D3 Nairobi Kenya
 Naissus Serbia *see* Niš
119 D3 Naivasha Kenya
81 D2 Najafābād Iran
78 B2 Najd *reg.* Saudi Arabia
106 C1 Nájera Spain
65 C1 Najin N. Korea
78 B3 Najrān Saudi Arabia
119 D2 Nakasongola Uganda
67 C3 Nakatsugawa Japan
78 A3 Nakfa Eritrea
66 B2 Nakhodka Rus. Fed.
63 B2 Nakhon Nayok Thai.
63 B2 Nakhon Pathom Thai.
62 B2 Nakhon Phanom Thai.
63 B2 Nakhon Ratchasima Thai.
63 B2 Nakhon Sawan Thai.
63 A3 Nakhon Si Thammarat
 Thai.
 Nakhrachi Rus. Fed. *see*
 Kondinskoye
130 B2 Nakina Can.
126 B3 Naknek U.S.A.
121 C1 Nakonde Zambia
93 F5 Nakskov Denmark
119 D3 Nakuru Kenya
128 C2 Nakusp Can.
75 D2 Nalbari India
87 D4 Nal'chik Rus. Fed.
115 D1 Nālūt Libya
123 D2 Namaacha Moz.
123 C2 Namahadi S. Africa
81 D2 Namak, Daryācheh-ye
 imp. l. Iran
76 B3 Namak, Kavīr-e *salt flat*
 Iran
79 C1 Namakzar-e Shadad *salt flat*
 Iran
77 D2 Namangan Uzbek.
119 D3 Namanyere Tanz.
122 A2 Namaqualand *reg.* S. Africa
51 E2 Nambour Austr.
53 D2 Nambucca Heads Austr.
63 B3 Năm Căn Vietnam
75 D1 Nam Co *salt l.* China
62 B1 Nam Đinh Vietnam
121 C2 Namialo Moz.
120 A3 Namib Desert Namibia
120 A2 Namibe Angola
120 A3 Namibia *country* Africa
72 D2 Namjagbarwa Feng *mt.*
 China
59 C3 Namlea Indon.
62 B2 Nam Ngum Reservoir Laos
53 C2 Namoi *r.* Austr.
134 C2 Nampa U.S.A.
114 B3 Nampala Mali
65 B2 Namp'o N. Korea
121 C2 Nampula Moz.
72 D2 Namrup India
92 F3 Namsang Myanmar
92 F3 Namsos Norway
92 F3 Namsskogan Norway
62 A1 Nam Tok Thai.
83 J2 Namtsy Rus. Fed.
62 A1 Namtu Myanmar
121 C2 Namuno Moz.
100 B2 Namur Belgium
120 B2 Namwala Zambia
65 B2 Namwon S. Korea
62 A1 Namya Ra Myanmar
62 B2 Nan Thai.
128 B3 Nanaimo Can.
71 B3 Nan'an China
122 A1 Nananib Plateau Namibia
 Nan'ao China *see* Dayu
67 C3 Nanao Japan
71 B3 Nanchang *Jiangxi* China
71 B3 Nanchang *Jiangxi* China
71 B3 Nancheng China
70 A2 Nanchong China
63 A3 Nancowry *i.* India
105 D2 Nancy France
75 C1 Nanda Devi *mt.* India
71 A3 Nandan China
74 B3 Nanded India
 Nander India *see* Nanded
53 D2 Nandewar Range *mts*
 Austr.
74 B2 Nandurbar India
73 B3 Nandyal India
71 B3 Nanfeng China
62 A1 Nang China
118 B2 Nanga Eboko Cameroon
61 C2 Nangahpinoh Indon.
77 D3 Nanga Parbat *mt.* Pak.
61 C2 Nangatayap Indon.
63 A2 Nangin Myanmar

65 B1 Nangnim-sanmaek *mts*
 N. Korea
70 B2 Nangong China
119 D3 Nangulangwa Tanz.
70 C2 Nanhui China
70 B2 Nanjing China
 Nanking China *see* Nanjing
67 B4 Nankoku Japan
120 A2 Nankova Angola
70 B2 Nanle China
71 B3 Nan Ling *mts* China
71 A3 Nanning China
127 I2 Nanortalik Greenland
71 A3 Nanpan Jiang *r.* China
75 C2 Nanpara India
71 B3 Nanping China
 Nanpu China *see* Pucheng
 Nansei-shotō *is* Japan *see*
 Ryukyu Islands
160 I1 Nansen Basin Arctic Ocean
65 F1 Nansen Sound *sea chan.*
 Can.
104 B2 Nantes France
70 C2 Nantong China
139 E2 Nantucket U.S.A.
139 F2 Nantucket Island U.S.A.
99 B3 Nantwich U.K.
49 I4 Nanumea *atoll* Tuvalu
155 D1 Nanuque Brazil
64 B3 Nanusa, Kepulauan *is*
 Indon.
71 B3 Nanxiong China
70 B2 Nanyang China
119 D3 Nanyuki Kenya
70 B2 Nanzhang China
 Nanzhao China *see* Zhao'an
107 D2 Nao, Cabo de la *c.* Spain
131 C2 Naococane, Lac *l.* Can.
71 B3 Naozhou Dao *i.* China
135 B3 Napa U.S.A.
126 E2 Napaktulik Lake Can.
127 I2 Napanee Can.
127 I2 Napasoq Greenland
137 F2 Naperville U.S.A.
54 C1 Napier N.Z.
108 B2 Naples Italy
141 D3 Naples U.S.A.
150 B3 Napo *r.* Ecuador/Peru
 Napoli Italy *see* Naples
 Napug China *see* Gê'gyai
114 B3 Nara Mali
88 C3 Narach Belarus
52 B3 Naracoorte Austr.
53 C2 Naradhan Austr.
145 C2 Naranjos Mex.
63 B3 Narathiwat Thai.
74 B3 Narayangaon India
105 C3 Narbonne France
63 A2 Narcondam Island India
127 H1 Nares Strait Can./
 Greenland
103 E1 Narew *r.* Pol.
122 A1 Narib Namibia
87 D4 Narimanov Rus. Fed.
67 D3 Narita Japan
74 B2 Narmada *r.* India
74 B2 Narnaul India
108 B2 Narni Italy
86 F2 Narodnaya, Gora *mt.*
 Rus. Fed.
90 B1 Narodychi Ukr.
89 E2 Naro-Fominsk Rus. Fed.
53 D3 Narooma Austr.
88 C3 Narowlya Belarus
93 H3 Närpes Fin.
53 C2 Narrabri Austr.
53 C2 Narrandera Austr.
53 C2 Narromine Austr.
57 B4 Naruto Japan
88 C2 Narva Estonia
88 C2 Narva Bay Estonia/Rus. Fed.
92 G2 Narvik Norway
88 C2 Narvskoye
 Vodokhranilishche *resr*
 Estonia/Rus. Fed.
86 E2 Nar'yan-Mar Rus. Fed.
77 D2 Naryn Kyrg.
77 D2 Naryn *r.* Kyrg.
74 B2 Nashik India
139 E2 Nashua U.S.A.
140 C1 Nashville U.S.A.
137 E1 Nashwauk U.S.A.
109 C1 Našice Croatia
49 I5 Nasinu Fiji
117 B4 Nasir *r.* S. Sudan
 Nasirabad Bangl. *see*
 Mymensingh
119 C4 Nasondoye
 Dem. Rep. Congo
76 B3 Naşrābād Iran
128 B2 Nass *r.* Can.

146 C2 Nassau Bahamas
116 B2 Nasser, Lake *resr* Egypt
93 F4 Nässjö Sweden
130 C2 Nastapoca *r.* Can.
130 C2 Nastapoka Islands Can.
67 D3 Nasushiobara Japan
89 D2 Nasva Rus. Fed.
120 B3 Nata Botswana
151 F3 Natal Brazil
60 A1 Natal Indon.
 Natal *prov.* S. Africa *see*
 KwaZulu-Natal
159 D6 Natal Basin Indian Ocea...
143 D3 Natalia U.S.A.
131 D2 Natashquan Can.
131 D2 Natashquan *r.* Can.
140 B2 Natchez U.S.A.
140 B2 Natchitoches U.S.A.
53 C3 Nathalia Austr.
107 D1 Nati, Punta *pt* Spain
114 C3 Natitingou Benin
151 E4 Natividade Brazil
67 D3 Natori Japan
119 D3 Natron, Lake *salt l.* Tanz...
60 B1 Natuna, Kepulauan *is*
 Indon.
60 B1 Natuna Besar *i.* Indon.
122 A1 Nauchas Namibia
101 F1 Nauen Ger.
64 B2 Naujan Phil.
88 B2 Naujoji Akmenė Lith.
74 A2 Naukot Pak.
101 E2 Naumburg (Saale) Ger.
48 H34 Nauru *country*
 S. Pacific Ocean
150 B3 Nauta Peru
145 C2 Nautla Mex.
88 C3 Navahrudak Belarus
106 B2 Navalmoral de la Mata
 Spain
106 B2 Navalvillar de Pela Spain
97 C2 Navan Ireland
 Navangar India *see*
 Jamnagar
88 C2 Navapolatsk Belarus
83 M2 Navarin, Mys *c.* Rus. Fed.
153 B5 Navarino, Isla *i.* Chile
107 C1 Navarra *aut. comm.* Spain
 Navarre *aut. comm.* Spain
 see Navarra
96 B1 Naver *r.* U.K.
73 B3 Navi Mumbai India
89 D3 Navlya Rus. Fed.
110 C2 Năvodari Romania
77 C2 Navoiy Uzbek.
144 B2 Navojoa Mex.
144 B2 Navolato Mex.
74 A2 Nawabshah Pak.
75 C2 Nawada India
62 A1 Nawnghkio Myanmar
62 A1 Nawngleng Myanmar
81 C2 Naxçıvan Azer.
111 C3 Naxos Greece
111 C3 Naxos *i.* Greece
144 B2 Nayar Mex.
62 A2 Nayoro Japan
 Nay Pyi Taw Myanmar
 Nazareth Israel *see* Nazer...
144 B2 Nazas Mex.
144 B2 Nazas *r.* Mex.
150 B4 Nazca Peru
157 H7 Nazca Ridge
 S. Pacific Ocean
80 B2 Nazerat Israel
111 C3 Nazilli Turkey
117 B4 Nazrēt Eth.
79 C2 Nazwá Oman
121 B2 Nchelenge Zambia
122 B1 Ncojane Botswana
120 A1 N'dalatando Angola
118 C2 Ndélé C.A.R.
118 B3 Ndendé Gabon
115 D3 Ndjamena Chad
118 A3 Ndogo, Lagune *lag.* Gabon
121 B2 Ndola Zambia
97 C1 Neagh, Lough *l.* U.K.
50 C2 Neale, Lake *imp. l.* Austr.
111 B3 Neapoli Greece
111 B2 Nea Roda Greece
99 B4 Neath U.K.
119 D2 Nebbi Uganda
119 D3 Nebine Creek *r.* Austr.
150 C2 Neblina, Pico da *mt.* Brazil
135 D3 Nebo, Mount U.S.A.
89 D2 Nebolchi Rus. Fed.
136 C2 Nebraska *state* U.S.A.
137 E2 Nebraska City U.S.A.
108 B3 Nebrodi, Monti *mts* Italy
143 E3 Neches *r.* U.S.A.
156 E4 Necker Island U.S.A.

53 C3 Necochea Arg.
43 E3 Nederland U.S.A.
00 B2 Neder Rijn r. Neth.
30 C2 Nedlouc, Lac l. Can.
39 E2 Needham U.S.A.
35 D4 Needles U.S.A.
74 B2 Neemuch India
29 E2 Neepawa Can.
87 E3 Neftekamsk Rus. Fed.
82 F2 Nefteyugansk Rus. Fed.
08 A3 Nefza Tunisia
20 A1 Negage Angola
17 B4 Negēlē Eth.
09 D2 Negotin Serbia
11 B2 Negotino Macedonia
50 A3 Negra, Punta pt Peru
55 D1 Negra, Serra mts Brazil
63 A2 Negrais, Cape Myanmar
53 B4 Negro r. Arg.
50 D3 Negro r. S. America
52 C3 Negro r. Uru.
06 B2 Negro, Cabo c. Morocco
64 B3 Negros i. Phil.
79 D1 Nehbandān Iran
69 E1 Nehe China
70 A3 Neijiang China
29 D2 Neilburg Can.
50 B2 Neiva Col.
29 E2 Nejanilini Lake Can.
Nejd reg. Saudi Arabia see Najd
17 B4 Nek'emtē Eth.
89 F2 Nekrasovskoye Rus. Fed.
89 D2 Nelidovo Rus. Fed.
73 B3 Nellore India
28 C3 Nelson Can.
29 E2 Nelson r. Can.
54 B2 Nelson N.Z.
52 B3 Nelson, Cape Austr.
59 C2 Nelson Bay Austr.
29 E2 Nelson House Can.
34 E1 Nelson Reservoir U.S.A.
14 B3 Néma Maur.
88 B2 Neman Rus. Fed.
05 C2 Nemours France
66 D2 Nemuro Japan
90 B2 Nemyriv Ukr.
97 B2 Nenagh Ireland
99 D3 Nene r. U.K.
69 E1 Nenjiang China
37 E3 Neosho U.S.A.
75 C2 Nepal country Asia
75 C2 Nepalganj Nepal
39 D1 Nepean Can.
35 D3 Nephi U.S.A.
97 B1 Nephin h. Ireland
97 B1 Nephin Beg Range hills Ireland
31 D3 Nepisiguit r. Can.
19 C2 Nepoko r. Dem. Rep. Congo
39 E2 Neptune City U.S.A.
08 B2 Nera r. Italy
04 C3 Nérac France
53 D1 Nerang Austr.
69 D1 Nerchinsk Rus. Fed.
89 F2 Nerekhta Rus. Fed.
09 C2 Neretva r. Bos.-Herz./ Croatia
20 B2 Neriquinha Angola
88 B3 Neris r. Lith.
89 E2 Nerl' r. Rus. Fed.
86 F2 Nerokhi Rus. Fed.
54 C1 Nerópolis Brazil
83 J3 Neryungri Rus. Fed.
92 DC2 Neskaupstaður Iceland
96 B1 Ness, Loch l. U.K.
36 D3 Ness City U.S.A.
Nesterov Ukr. see Zhovkva
11 B2 Nestos r. Greece
00 B1 Netherlands country Europe
47 D3 Netherlands Antilles terr. West Indies
27 H2 Nettilling Lake Can.
01 F1 Neubrandenburg Ger.
05 D2 Neuchâtel Switz.
05 D2 Neuchâtel, Lac de l. Switz.
00 C2 Neuerburg Ger.
00 B3 Neufchâteau Belgium
05 D2 Neufchâteau France
04 C2 Neufchâtel-en-Bray France
01 D2 Neuhof Ger.
01 E3 Neumarkt in der Oberpfalz Ger.
02 B1 Neumünster Ger.
01 D2 Neunkirchen Ger.
53 B4 Neuquén Arg.
53 B3 Neuquén r. Arg.
01 F1 Neuruppin Ger.
00 C2 Neuss Ger.

101 D1 Neustadt am Rübenberge Ger.
101 E3 Neustadt an der Aisch Ger.
Neustadt an der Hardt Ger. see Neustadt an der Weinstraße
101 D3 Neustadt an der Weinstraße Ger.
101 E1 Neustadt-Glewe Ger.
101 F1 Neustrelitz Ger.
99 D5 Neuville-lès-Dieppe France
100 C2 Neuwied Ger.
137 E3 Nevada U.S.A.
135 C3 Nevada state U.S.A.
106 C2 Nevada, Sierra mts Spain
135 B2 Nevada, Sierra mts U.S.A.
88 C2 Nevaišių kalnas h. Lith.
88 C2 Nevel' Rus. Fed.
105 C2 Nevers France
53 C2 Nevertire Austr.
109 C2 Nevesinje Bos.-Herz.
87 D4 Nevinnomyssk Rus. Fed.
80 B2 Nevşehir Turkey
99 C4 New Addington U.K.
128 B2 New Aiyansh Can.
119 D4 Newala Tanz.
138 B3 New Albany U.S.A.
151 D2 New Amsterdam Guyana
139 E2 Newark NJ U.S.A.
138 C2 Newark OH U.S.A.
99 C3 Newark-on-Trent U.K.
139 E2 New Bedford U.S.A.
141 E1 New Bern U.S.A.
138 B1 Newberry MI U.S.A.
141 D2 Newberry SC U.S.A.
143 E2 New Boston U.S.A.
143 D3 New Braunfels U.S.A.
97 C2 Newbridge Ireland
59 D3 New Britain i. P.N.G.
131 D3 New Brunswick prov. Can.
99 C4 Newbury U.K.
64 A2 New Busuanga Phil.
48 H6 New Caledonia terr. S. Pacific Ocean
156 D7 New Caledonia Trough Tasman Sea
53 D2 Newcastle Austr.
123 C2 Newcastle S. Africa
97 D1 Newcastle Austr.
138 C3 Newcastle U.S.A.
99 B3 Newcastle-under-Lyme U.K.
98 C2 Newcastle upon Tyne U.K.
51 C1 Newcastle Waters Austr.
97 B2 Newcastle West Ireland
74 B2 New Delhi India
128 C3 New Denver Can.
53 D2 New England Range mts Austr.
157 H3 New England Seamounts N. Atlantic Ocean
131 E3 Newfoundland i. Can.
131 E2 Newfoundland and Labrador prov. Can.
96 B3 New Galloway U.K.
48 G4 New Georgia Islands Solomon Is
131 D3 New Glasgow Can.
59 D3 New Guinea i. Indon./ P.N.G.
78 A3 New Halfa Sudan
139 E2 New Hampshire state U.S.A.
59 E3 New Hanover i. P.N.G.
139 E2 New Haven U.S.A.
128 B2 New Hazelton Can.
New Hebrides country S. Pacific Ocean see Vanuatu
156 D7 New Hebrides Trench S. Pacific Ocean
140 B2 New Iberia U.S.A.
59 E3 New Ireland i. P.N.G.
139 E3 New Jersey state U.S.A.
130 C3 New Liskeard Can.
138 B2 New London U.S.A.
50 A2 Newman Austr.
54 C2 Newman N.Z.
97 B2 Newmarket Ireland
99 D3 Newmarket U.K.
97 B2 Newmarket-on-Fergus Ireland
142 B2 New Mexico state U.S.A.
141 D2 Newnan U.S.A.
140 B3 New Orleans U.S.A.
138 B3 New Philadelphia U.S.A.
54 B1 New Plymouth N.Z.
99 C4 Newport England U.K.
99 B4 Newport Wales U.K.

140 B1 Newport AR U.S.A.
134 B2 Newport OR U.S.A.
139 E2 Newport RI U.S.A.
141 D1 Newport TN U.S.A.
139 E2 Newport VT U.S.A.
134 C1 Newport WA U.S.A.
139 D3 Newport News U.S.A.
141 E3 New Providence i. Bahamas
99 A4 Newquay U.K.
140 B2 New Roads U.S.A.
97 C2 New Ross Ireland
97 C1 Newry U.K.
83 K1 New Siberia Islands Rus. Fed.
52 B2 New South Wales state Austr.
137 E2 Newton IA U.S.A.
137 D3 Newton KS U.S.A.
99 B4 Newton Abbot U.K.
98 C2 Newton Aycliffe U.K.
96 B3 Newton Mearns U.K.
96 B2 Newtonmore U.K.
96 B3 Newton Stewart U.K.
97 B2 Newtown Ireland
99 B3 Newtown U.K.
136 C1 New Town U.S.A.
97 D1 Newtownabbey U.K.
97 D1 Newtownards U.K.
Newtownbarry Ireland see Bunclody
97 C1 Newtownbutler U.K.
96 C3 Newtown St Boswells U.K.
97 C1 Newtownstewart U.K.
137 E2 New Ulm U.S.A.
139 E2 New York U.S.A.
139 D2 New York state U.S.A.
54 B2 New Zealand country Oceania
86 D3 Neya Rus. Fed.
81 D3 Neyrīz Iran
76 B3 Neyshābūr Iran
145 C3 Nezahualcóyotl, Presa resr Mex.
60 B1 Ngabang Indon.
118 B3 Ngabé Congo
75 C2 Ngamring China
119 D2 Ngangala S. Sudan
75 C1 Nganglong Ringco salt l. China
75 C1 Nganglong Kangri mt. China
75 C1 Nganglong Kangri mts China
75 C1 Ngangzê Co l. China
62 B1 Ngân Sơn Vietnam
62 A2 Ngao Thai.
118 B2 Ngaoundal Cameroon
118 B2 Ngaoundéré Cameroon
54 C1 Ngaruawahia N.Z.
62 A2 Ngathainggyaung Myanmar
121 D2 Ngazidja i. Comoros
Ngiva Angola see Ondjiva
118 B3 Ngo Congo
115 D4 Ngol Bembo Nigeria
68 C2 Ngoring Hu l. China
115 D3 Ngourti Niger
115 D3 Nguigmi Niger
59 D2 Ngulu atoll Micronesia
61 C2 Ngunut Indon.
Ngunza Angola see Sumbe
Ngunza-Kabolu Angola see Sumbe
115 D3 Nguru Nigeria
123 C2 Ngwathe S. Africa
123 D2 Ngwelezana S. Africa
121 C2 Nhamalabué Moz.
120 A2 N'harea Angola
63 B2 Nha Trang Vietnam
52 B3 Nhill Austr.
123 D2 Nhlangano Swaziland
51 C1 Nhulunbuy Austr.
139 D2 Niagara Falls Can.
114 C3 Niamey Niger
119 C2 Niangara Dem. Rep. Congo
114 B3 Niangay, Lac l. Mali
119 C2 Nia-Nia Dem. Rep. Congo
60 A1 Nias i. Indon.
88 B2 Nīca Latvia
146 B3 Nicaragua country Central America
146 B3 Nicaragua, Lago de l. Nic. see Nicaragua, Lake
146 B3 Nicaragua, Lake l. Nic.
109 C3 Nicastro Italy
105 D3 Nice France
73 D4 Nicobar Islands India
80 B2 Nicosia Cyprus
146 B4 Nicoya, Golfo de b. Costa Rica

88 B2 Nida Lith.
103 E1 Nidzica Pol.
102 B1 Niebüll Ger.
101 D2 Niederaula Ger.
118 B2 Niefang Equat. Guinea
101 F1 Niemegk Ger.
101 D1 Nienburg (Weser) Ger.
103 C1 Niesky Ger.
100 B1 Nieuwegein Neth.
100 B1 Nieuwe-Niedorp Neth.
151 D2 Nieuw Nickerie Suriname
122 A3 Nieuwoudtville S. Africa
100 A2 Nieuwpoort Belgium
80 B2 Niğde Turkey
115 C3 Niger country Africa
115 C4 Niger r. Africa
115 C4 Niger, Mouths of the Nigeria
115 C4 Nigeria country Africa
130 B3 Nighthawk Lake Can.
111 B2 Nigrita Greece
67 C3 Niigata Japan
67 B4 Niihama Japan
67 C4 Nii-jima i. Japan
67 B4 Niimi Japan
67 C3 Niitsu Japan
107 C2 Níjar Spain
100 B1 Nijkerk Neth.
100 B2 Nijmegen Neth.
100 C1 Nijverdal Neth.
92 J2 Nikel' Rus. Fed.
Nikolayev Ukr. see Mykolayiv
87 D3 Nikolayevsk Rus. Fed.
Nikolayevskiy Rus. Fed. see Nikolayevsk
86 D3 Nikol'sk Rus. Fed.
Nikol'skiy Kazakh. see Satpayev
83 M3 Nikol'skoye Rus. Fed. see Sheksna
91 C2 Nikopol' Ukr.
80 B1 Niksar Turkey
79 D2 Nīkshahr Iran
109 C2 Nikšić Montenegro
Nîl, Bahr el r. Africa see Nile
135 C4 Niland U.S.A.
116 B1 Nile r. Africa
138 B2 Niles U.S.A.
74 A1 Nīlī Afgh.
Nimach India see Neemuch
105 C3 Nîmes France
53 C3 Nimmitabel Austr.
117 B4 Nimule S. Sudan
53 C1 Nindigully Austr.
73 B4 Nine Degree Channel India
53 C3 Ninety Mile Beach Austr.
54 B1 Ninety Mile Beach N.Z.
70 C2 Ningbo China
71 B3 Ningde China
71 B3 Ningdu China
70 B2 Ningguo China
71 C3 Ninghai China
Ningjiang China see Songyuan
68 C2 Ningjing Shan mts China
70 A2 Ningxia Huizu Zizhiqu aut. reg. China
70 B2 Ningyang China
62 B1 Ninh Binh Vietnam
63 B2 Ninh Hoa Vietnam
66 D2 Ninohe Japan
137 D2 Niobrara r. U.S.A.
62 A1 Nioku India
114 B3 Niono Mali
114 B3 Nioro Mali
104 B2 Niort France
129 D2 Nipawin Can.
130 B3 Nipigon Can.
130 B3 Nipigon, Lake Can.
131 D2 Nipishish Lake Can.
130 C3 Nipissing, Lake Can.
135 C3 Nipton U.S.A.
151 E4 Niquelândia Brazil
74 B3 Nirmal India
109 D2 Niš Serbia
109 D2 Nišava r. Serbia
108 B3 Niscemi Italy
67 B4 Nishino-omote Japan
90 B2 Nisporeni Moldova
155 D2 Niterói Brazil
96 C3 Nith r. U.K.
103 D2 Nitra Slovakia
49 K5 Niue terr. S. Pacific Ocean
92 H3 Nivala Fin.
100 B2 Nivelles Belgium

73 B3 **Nizamabad** India
87 E3 **Nizhnekamsk** Rus. Fed.
87 E3 **Nizhnekamskoye Vodokhranilishche** resr Rus. Fed.
83 H3 **Nizhneudinsk** Rus. Fed.
82 G2 **Nizhnevartovsk** Rus. Fed.
Nizhnevolzhsk Rus. Fed. see Narimanov
83 K2 **Nizhneyansk** Rus. Fed.
Nizhniye Kresty Rus. Fed. see Cherskiy
Nizhniye Ustriki Pol. see Ustrzyki Dolne
89 F3 **Nizhniy Kislyay** Rus. Fed.
87 D3 **Nizhniy Lomov** Rus. Fed.
87 D3 **Nizhniy Novgorod** Rus. Fed.
86 E2 **Nizhniy Odes** Rus. Fed.
86 E3 **Nizhniy Tagil** Rus. Fed.
83 G2 **Nizhnyaya Tunguska** r. Rus. Fed.
86 E3 **Nizhnyaya Tura** Rus. Fed.
91 C1 **Nizhyn** Ukr.
119 D3 **Njinjo** Tanz.
119 D3 **Njombe** Tanz.
118 B2 **Nkambe** Cameroon
119 D4 **Nkhata Bay** Malawi
121 C2 **Nkhotakota** Malawi
118 A3 **Nkomi, Lagune** lag. Gabon
119 D3 **Nkondwe** Tanz.
118 A2 **Nkongsamba** Cameroon
123 C3 **Nkululeko** S. Africa
123 C3 **Nkwenkwezi** S. Africa
67 B4 **Nobeoka** Japan
138 B2 **Noblesville** U.S.A.
52 B1 **Noccundra** Austr.
144 A1 **Nogales** Mex.
142 A2 **Nogales** U.S.A.
104 C2 **Nogent-le-Rotrou** France
83 H2 **Noginsk** Rus. Fed.
89 E2 **Noginsk** Rus. Fed.
83 K3 **Nogliki** Rus. Fed.
74 B2 **Nohar** India
100 C3 **Nohfelden** Ger.
104 B2 **Noires, Montagnes** hills France
104 B2 **Noirmoutier, Île de** i. France
104 B2 **Noirmoutier-en-l'Île** France
67 C4 **Nojima-zaki** c. Japan
74 B2 **Nokha** India
93 H3 **Nokia** Fin.
74 A2 **Nok Kundi** Pak.
118 B2 **Nola** C.A.R.
86 D3 **Nolinsk** Rus. Fed.
126 A2 **Nome** U.S.A.
123 C3 **Nomonde** S. Africa
123 D2 **Nondweni** S. Africa
Nonghui China see Guang'an
62 B2 **Nong Khai** Thai.
75 D2 **Nongstoin** India
52 A2 **Nonning** Austr.
144 B2 **Nonoava** Mex.
65 B2 **Nonsan** S. Korea
63 B2 **Nonthaburi** Thai.
122 B3 **Nonzwakazi** S. Africa
100 B1 **Noordwijk-Binnen** Neth.
77 C3 **Norak** Tajik.
82 C1 **Nordaustlandet** i. Svalbard
128 C2 **Nordegg** Can.
100 C1 **Norden** Ger.
83 H1 **Nordenshel'da, Arkhipelag** is Rus. Fed.
Nordenskjold Archipelago is Rus. Fed. see Nordenshel'da, Arkhipelag
100 C1 **Norderney** Ger.
100 C1 **Norderney** i. Ger.
101 E1 **Norderstedt** Ger.
93 E3 **Nordfjordeid** Norway
Nordfriesische Inseln is Ger. see North Frisian Islands
101 E2 **Nordhausen** Ger.
101 D1 **Nordholz** Ger.
100 C1 **Nordhorn** Ger.
Nordkapp c. Norway see North Cape
92 F3 **Nordli** Norway
102 C2 **Nördlingen** Ger.
92 G3 **Nordmaling** Sweden
94 B1 **Norðoyar** is Faroe Is
97 C2 **Nore** r. Ireland
88 B3 **Noreikiškės** Lith.
137 D2 **Norfolk** NE U.S.A.
139 D3 **Norfolk** VA U.S.A.

48 H6 **Norfolk Island** terr. S. Pacific Ocean
93 E3 **Norheimsund** Norway
82 G2 **Noril'sk** Rus. Fed.
75 C2 **Norkyung** China
143 D1 **Norman** U.S.A.
Normandes, Îles is English Chan. see Channel Islands
150 D2 **Normandia** Brazil
Normandie reg. France see Normandy
104 B2 **Normandy** reg. France
51 D1 **Normanton** Austr.
128 B1 **Norman Wells** Can.
93 G4 **Norrköping** Sweden
93 G4 **Norrtälje** Sweden
50 B3 **Norseman** Austr.
92 G3 **Norsjö** Sweden
55 M2 **North, Cape** Antarctica
98 C2 **Northallerton** U.K.
50 A3 **Northam** Austr.
50 A2 **Northampton** Austr.
99 C3 **Northampton** U.K.
73 D3 **North Andaman** i. India
159 F4 **North Australian Basin** Indian Ocean
129 D2 **North Battleford** Can.
130 C3 **North Bay** Can.
130 C2 **North Belcher Islands** Can.
96 C2 **North Berwick** U.K.
North Borneo state Malaysia see Sabah
92 I1 **North Cape** c. Norway
54 B1 **North Cape** N.Z.
130 A2 **North Caribou Lake** Can.
141 E1 **North Carolina** state U.S.A.
130 B3 **North Channel** lake channel Can.
96 A3 **North Channel** U.K.
141 E2 **North Charleston** U.S.A.
128 B3 **North Cowichan** Can.
136 C1 **North Dakota** state U.S.A.
99 C4 **North Downs** hills U.K.
157 E3 **Northeast Pacific Basin** N. Pacific Ocean
141 E3 **Northeast Providence Channel** Bahamas
101 D2 **Northeim** Ger.
122 A2 **Northern Cape** prov. S. Africa
91 D2 **Northern Donets** r. Rus. Fed./Ukr.
Northern Dvina r. Rus. Fed. see Severnaya Dvina
129 E2 **Northern Indian Lake** Can.
97 C1 **Northern Ireland** prov. U.K.
59 D1 **Northern Mariana Islands** terr. N. Pacific Ocean
Northern Rhodesia country Africa see Zambia
50 C1 **Northern Territory** admin. div. Austr.
Northern Transvaal prov. S. Africa see Limpopo
96 C2 **North Esk** r. U.K.
137 E2 **Northfield** U.S.A.
99 D4 **North Foreland** c. U.K.
102 B1 **North Frisian Islands** is Ger.
160 P2 **North Geomagnetic Pole**
54 B1 **North Island** N.Z.
129 E2 **North Knife Lake** Can.
65 B1 **North Korea** country Asia
72 D2 **North Lakhimpur** India
North Land is Rus. Fed. see Severnaya Zemlya
160 N1 **North Magnetic Pole**
128 B1 **North Nahanni** r. Can.
136 C2 **North Platte** U.S.A.
136 C2 **North Platte** r. U.S.A.
96 C1 **North Ronaldsay** i. U.K.
129 D2 **North Saskatchewan** r. Can.
94 D2 **North Sea** Europe
63 A2 **North Sentinel Island** India
130 A2 **North Spirit Lake** Can.
53 D1 **North Stradbroke Island** Austr.
131 D3 **North Sydney** Can.
54 B1 **North Taranaki Bight** b. N.Z.
130 C2 **North Twin Island** Can.
98 B2 **North Tyne** r. U.K.
96 C1 **North Uist** i. U.K.
131 D3 **Northumberland Strait** Can.
99 D3 **North Walsham** U.K.
123 C2 **North West** prov. S. Africa

158 D1 **Northwest Atlantic Mid-Ocean Channel** sea chan. N. Atlantic Ocean
50 A2 **North West Cape** Austr.
156 D3 **Northwest Pacific Basin** N. Pacific Ocean
141 E3 **Northwest Providence Channel** Bahamas
131 E2 **North West River** Can.
128 B1 **Northwest Territories** admin. div. Can.
98 C2 **North York Moors** moorland U.K.
138 C3 **Norton** U.S.A.
121 C2 **Norton** Zimbabwe
Norton de Matos Angola see Balombo
126 B2 **Norton Sound** sea chan. U.S.A.
55 D2 **Norvegia, Cape** Antarctica
138 C2 **Norwalk** U.S.A.
93 F3 **Norway** country Europe
129 E2 **Norway House** Can.
160 L3 **Norwegian Basin** N. Atlantic Ocean
92 E2 **Norwegian Sea** N. Atlantic Ocean
99 D3 **Norwich** U.K.
139 E2 **Norwich** CT U.S.A.
139 D2 **Norwich** NY U.S.A.
66 D2 **Noshiro** Japan
91 C1 **Nosivka** Ukr.
122 B2 **Nosop** watercourse Africa
86 E2 **Nosovaya** Rus. Fed.
79 C2 **Noşratābād** Iran
122 A1 **Nossob** watercourse Africa
103 D1 **Noteć** r. Pol.
93 E4 **Notodden** Norway
67 C3 **Noto-hantō** pen. Japan
131 D3 **Notre-Dame, Monts** mts Can.
131 E3 **Notre Dame Bay** Can.
130 C2 **Nottaway** r. Can.
99 C3 **Nottingham** U.K.
114 A2 **Nouâdhibou** Maur.
114 A3 **Nouakchott** Maur.
114 A3 **Nouâmghâr** Maur.
63 B2 **Nouei** Vietnam
48 H6 **Nouméa** New Caledonia
114 B3 **Nouna** Burkina Faso
122 B3 **Noupoort** S. Africa
Nouveau-Comptoir Can. see Wemindji
Nouvelle Anvers Dem. Rep. Congo see Makanza
Nouvelles Hébrides country S. Pacific Ocean see Vanuatu
Nova Chaves Angola see Muconda
154 B2 **Nova Esperança** Brazil
Nova Freixa Moz. see Cuamba
155 D2 **Nova Friburgo** Brazil
109 C1 **Nova Gradiška** Croatia
154 C2 **Nova Granada** Brazil
155 D2 **Nova Iguaçu** Brazil
91 C2 **Nova Kakhovka** Ukr.
155 D1 **Nova Lima** Brazil
Nova Lisboa Angola see Huambo
154 B2 **Nova Londrina** Brazil
91 C2 **Nova Odesa** Ukr.
154 C1 **Nova Ponte** Brazil
108 A1 **Novara** Italy
154 B2 **Nova Remanso** Brazil
131 D3 **Nova Scotia** prov. Can.
155 D1 **Nova Venécia** Brazil
83 K1 **Novaya Sibir', Ostrov** i. Rus. Fed.
86 E1 **Novaya Zemlya** is Rus. Fed.
107 C2 **Novelda** Spain
103 D2 **Nové Zámky** Slovakia
Novgorod Rus. Fed. see Velikiy Novgorod
91 C1 **Novhorod-Sivers'kyy** Ukr.
110 B2 **Novi Iskŭr** Bulg.
66 D1 **Novikovo** Rus. Fed.
108 A2 **Novi Ligure** Italy
109 D2 **Novi Pazar** Serbia
109 C1 **Novi Sad** Serbia
Novoalekseyevka Kazakh. see Kobda
87 D3 **Novoanninskiy** Rus. Fed.
150 C3 **Novo Aripuanã** Brazil
91 D2 **Novoazovs'k** Ukr.
91 E2 **Novocherkassk** Rus. Fed.
89 D2 **Novodugino** Rus. Fed.
86 D2 **Novodvinsk** Rus. Fed.

Novoekonomicheskoye Ukr. see Dymytrov
152 C2 **Novo Hamburgo** Brazil
154 C2 **Novo Horizonte** Brazil
90 B1 **Novohrad-Volyns'kyy** Ukr.
Novokazalinsk Kazakh. se Ayteke Bi
91 E1 **Novokhopersk** Rus. Fed.
68 B1 **Novokuznetsk** Rus. Fed.
109 C1 **Novo mesto** Slovenia
91 D3 **Novomikhaylovskiy** Rus. Fed.
89 E3 **Novomoskovsk** Rus. Fed.
91 D2 **Novomoskovs'k** Ukr.
91 C2 **Novomyrhorod** Ukr.
Novonikolayevsk Rus. Fed see Novosibirsk
91 C2 **Novooleksiyivka** Ukr.
150 C2 **Novo Paraíso** Brazil
91 E2 **Novopokrovskaya** Rus. Fe
91 D2 **Novopskov** Ukr.
Novo Redondo Angola se Sumbe
91 D3 **Novorossiysk** Rus. Fed.
88 C2 **Novorzhev** Rus. Fed.
87 E3 **Novosergiyevka** Rus. Fed.
91 D2 **Novoshakhtinsk** Rus. Fed
82 G3 **Novosibirsk** Rus. Fed.
Novosibirskiye Ostrova is Rus. Fed. see New Siberia Islands
89 E3 **Novosil'** Rus. Fed.
89 D2 **Novosokol'niki** Rus. Fed.
91 C2 **Novotroyits'ke** Ukr.
91 C2 **Novoukrayinka** Ukr.
90 A1 **Novovolyns'k** Ukr.
89 E3 **Novovoronezh** Rus. Fed.
Novovoronezhskiy Rus. F see Novovoronezh
89 D3 **Novozybkov** Rus. Fed.
103 D2 **Nový Jíčin** Czech Rep.
86 E2 **Novyy Bor** Rus. Fed.
91 C2 **Novyy Buh** Ukr.
Novyy Donbass Ukr. see Dymytrov
Novyye Petushki Rus. Fed see Petushki
Novyy Margelan Uzbek. see Farg'ona
89 E2 **Novyy Nekouz** Rus. Fed.
91 D1 **Novyy Oskol** Rus. Fed.
86 G2 **Novyy Port** Rus. Fed.
86 G2 **Novyy Urengoy** Rus. Fed.
69 E1 **Novyy Urgal** Rus. Fed.
Novyy Uzen' Kazakh. see Zhanaozen
103 D1 **Nowogard** Pol.
Noworadomsk Pol. see Radomsko
53 D2 **Nowra** Austr.
81 D2 **Nowshahr** Iran
74 B1 **Nowshera** Pak.
103 E2 **Nowy Sącz** Pol.
103 E2 **Nowy Targ** Pol.
82 G2 **Noyabr'sk** Rus. Fed.
105 C2 **Noyon** France
68 C2 **Noyon** Mongolia
121 C2 **Nsanje** Malawi
121 B2 **Nsombo** Zambia
118 B3 **Ntandembele** Dem. Rep. Congo
123 C2 **Ntha** S. Africa
111 B3 **Ntoro, Kavo** pt Greece
118 A2 **Ntoum** Gabon
119 D3 **Ntungamo** Uganda
Nuanetsi r. Zimbabwe see Mwenezi
79 C2 **Nu'aym** reg. Oman
116 B2 **Nubian Desert** Sudan
143 D3 **Nueces** r. U.S.A.
129 E1 **Nueltin Lake** Can.
150 B2 **Nueva Loja** Ecuador
153 A4 **Nueva Lubecka** Arg.
145 B2 **Nueva Rosita** Mex.
144 B1 **Nuevo Casas Grandes** M
144 B2 **Nuevo Ideal** Mex.
145 C2 **Nuevo Laredo** Mex.
117 C4 **Nugaal** watercourse Soma
117 C4 **Nugaaleed, Dooxo** val. Somalia
105 C2 **Nuits-St-Georges** France
Nu Jiang r. China/Myanm see Salween
49 J6 **Nuku'alofa** Tonga
49 M4 **Nuku Hiva** i. Fr. Polynesia
48 G4 **Nukumanu Islands** P.N.G
76 B2 **Nukus** Uzbek.
50 B2 **Nullagine** Austr.
50 C3 **Nullarbor** Austr.

50 B3 Nullarbor Plain Austr.
15 D4 Numan Nigeria
67 C3 Numazu Japan
51 C1 Numbulwar Austr.
93 E3 Numedal val. Norway
59 C3 Numfoor i. Indon.
53 C3 Numurkah Austr.
Nunap Isua c. Greenland see Farewell, Cape
27 G2 Nunavik reg. Can.
29 E1 Nunavut admin. div. Can.
99 C3 Nuneaton U.K.
26 A3 Nunivak Island U.S.A.
06 B1 Nuñomoral Spain
08 A2 Nuoro Italy
78 B2 Nuqrah Saudi Arabia
77 C1 Nura r. Kazakh.
01 E3 Nuremberg Ger.
52 A2 Nuriootpa Austr.
92 I3 Nurmes Fin.
Nürnberg Ger. see Nuremberg
53 C2 Nurri, Mount h. Austr.
62 A1 Nu Shan mts China
74 A2 Nushki Pak.
27 I2 Nuuk Greenland
27 I2 Nuussuaq Greenland
27 I2 Nuussuaq pen. Greenland
80 B3 Nuwaybi' al Muzayyinah Egypt
22 A3 Nuwerus S. Africa
22 B3 Nuweveldberge mts S. Africa
86 F2 Nyagan' Rus. Fed.
75 D1 Nyainqêntanglha Feng mt. China
75 D2 Nyainqêntanglha Shan mts China
Nyakh Rus. Fed. see Nyagan'
47 A3 Nyala Sudan
49 D4 Nyamtumbo Tanz.
Nyande Zimbabwe see Masvingo
86 D2 Nyandoma Rus. Fed.
18 B3 Nyanga Congo
18 B3 Nyanga r. Gabon
21 C2 Nyanga Zimbabwe
Nyang'oma Kenya see Kogelo
21 C2 Nyasa, Lake Africa
Nyasaland country Africa see Malawi
88 C3 Nyasvizh Belarus
62 A2 Nyaunglebin Myanmar
93 F4 Nyborg Denmark
92 I1 Nyborg Norway
93 G4 Nybro Sweden
Nyenchen Tanglha Range mts China see Nyainqêntanglha Shan
49 D3 Nyeri Kenya
68 C3 Nyingchi China
03 E2 Nyíregyháza Hungary
93 F5 Nykøbing Denmark
93 G4 Nyköping Sweden
53 C2 Nymagee Austr.
93 G4 Nynäshamn Sweden
53 C2 Nyngan Austr.
88 B3 Nyoman r. Belarus/Lith.
05 D3 Nyons France
86 E2 Nyrob Rus. Fed.
03 D1 Nysa Pol.
26 C2 Nyssa U.S.A.
49 C3 Nyunzu Dem. Rep. Congo
83 I2 Nyurba Rus. Fed.
91 C2 Nyzhni Sirohozy Ukr.
91 C2 Nyzhn'ohirs'kyy Ukr.
18 B3 Nzambi Congo
19 D3 Nzega Tanz.
14 B4 Nzérékoré Guinea
20 A1 N'zeto Angola

O

86 C2 Oahe, Lake U.S.A.
49 L1 O'ahu i. U.S.A.
52 B2 Oakbank Austr.
40 B2 Oakdale U.S.A.
53 D1 Oakey Austr.
38 B3 Oak Grove U.S.A.
99 C3 Oakham U.K.
34 B1 Oak Harbor U.S.A.
38 C3 Oak Hill U.S.A.
35 B3 Oakland CA U.S.A.
38 B2 Oakland MD U.S.A.
38 B2 Oak Lawn U.S.A.
36 C3 Oakley U.S.A.

50 B2 Oakover r. Austr.
134 B2 Oakridge U.S.A.
141 D1 Oak Ridge U.S.A.
54 B3 Oamaru N.Z.
64 B2 Oas Phil.
145 C3 Oaxaca Mex.
86 F2 Ob' r. Rus. Fed.
Ob, Gulf of sea chan. Rus. Fed. see Obskaya Guba
88 C2 Obal' Belarus
118 B2 Obala Cameroon
96 B2 Oban U.K.
106 B1 O Barco Spain
Obbia Somalia see Hobyo
136 C3 Oberlin U.S.A.
53 C2 Oberon Austr.
101 F3 Oberviechtach Ger.
59 C3 Obi i. Indon.
151 D3 Óbidos Brazil
66 D2 Obihiro Japan
69 E1 Obluch'ye Rus. Fed.
89 E2 Obninsk Rus. Fed.
119 C2 Obo C.A.R.
117 C3 Obock Djibouti
103 D1 Oborniki Pol.
118 B3 Obouya Congo
89 E3 Oboyan' Rus. Fed.
86 D2 Obozerskiy Rus. Fed.
144 B2 Obregón, Presa resr Mex.
109 D2 Obrenovac Serbia
134 B2 O'Brien U.S.A.
87 E3 Obshchiy Syrt hills Kazakh./ Rus. Fed.
86 G2 Obskaya Guba sea chan. Rus. Fed.
114 B4 Obuasi Ghana
90 C1 Obukhiv Ukr.
86 D2 Ob''yachevo Rus. Fed.
141 D3 Ocala U.S.A.
144 B2 Ocampo Mex.
106 C2 Ocaña Spain
150 B2 Occidental, Cordillera mts Col.
150 B4 Occidental, Cordillera mts Peru
139 D3 Ocean City MD U.S.A.
139 E3 Ocean City NJ U.S.A.
128 B2 Ocean Falls Can.
135 C4 Oceanside U.S.A.
91 C2 Ochakiv Ukr.
86 E3 Ocher Rus. Fed.
101 E3 Ochsenfurt Ger.
110 B1 Ocna Mures Romania
90 B2 Ocnița Moldova
141 D2 Oconee r. U.S.A.
145 C3 Ocosingo Mex.
141 E1 Ocracoke Island U.S.A.
October Revolution Island i. Rus. Fed. see Oktyabr'skoy Revolyutsii, Ostrov
59 C3 Ocussi enclave East Timor
116 B3 Oda, Jebel mt. Sudan
66 D2 Ōdate Japan
67 C3 Odawara Japan
93 E3 Odda Norway
106 B2 Odemira Port.
111 C3 Ödemiş Turkey
93 F4 Odense Denmark
101 D3 Odenwald reg. Ger.
102 C1 Oderbucht b. Ger.
Odesa Ukr. see Odessa
90 C2 Odessa Ukr.
143 C2 Odessa U.S.A.
114 B4 Odienné Côte d'Ivoire
75 C2 Odisha state India
89 E3 Odoyev Rus. Fed.
102 C1 Odra r. Ger./Pol.
103 D1 Odra r. Ger./Pol.
151 E3 Oeiras Brazil
101 D2 Oelde Ger.
136 C2 Oelrichs U.S.A.
101 F2 Oelsnitz Ger.
100 B1 Oenkerk Neth.
137 E3 O'Fallon U.S.A.
109 C2 Ofanto r. Italy
101 D2 Offenbach am Main Ger.
102 B2 Offenburg Ger.
66 C3 Oga Japan
117 C4 Ogaden reg. Eth.
66 C3 Oga-hantō pen. Japan
67 C3 Ōgaki Japan
136 C2 Ogallala U.S.A.
Ogasawara-shotō is Japan see Bonin Islands
115 C4 Ogbomosho Nigeria
134 D2 Ogden U.S.A.
139 D2 Ogdensburg U.S.A.
126 C2 Ogilvie r. Can.
126 C2 Ogilvie Mountains Can.

141 D2 Oglethorpe, Mount U.S.A.
130 B2 Ogoki r. Can.
130 B2 Ogoki Reservoir Can.
88 B2 Ogre Latvia
109 C1 Ogulin Croatia
81 D2 Ogurjaly Adasy i. Turkm.
115 C2 Ohanet Alg.
138 B3 Ohio r. U.S.A.
138 C2 Ohio state U.S.A.
101 E2 Ohrdruf Ger.
101 F2 Ohře r. Czech Rep.
111 B2 Ohrid Macedonia
151 D2 Oiapoque Brazil
139 D2 Oil City U.S.A.
100 A3 Oise r. France
67 B4 Ōita Japan
144 B2 Ojinaga Mex.
152 B2 Ojos del Salado, Nevado mt. Arg./Chile
89 F2 Oka r. Rus. Fed.
120 A3 Okahandja Namibia
120 A3 Okakarara Namibia
128 C3 Okanagan Falls Can.
128 C3 Okanagan Lake Can.
134 C1 Okanogan U.S.A.
134 C1 Okanogan r. U.S.A.
74 B1 Okara Pak.
120 B2 Okavango r. Africa
120 B2 Okavango Delta swamp Botswana
67 C3 Okaya Japan
67 B4 Okayama Japan
67 C4 Okazaki Japan
141 D3 Okeechobee U.S.A.
141 D3 Okeechobee, Lake U.S.A.
141 D2 Okefenokee Swamp U.S.A.
99 A4 Okehampton U.K.
115 C4 Okene Nigeria
101 E1 Oker r. Ger.
74 A2 Okha India
83 K3 Okha Rus. Fed.
75 C2 Okhaldhunga Nepal
83 K3 Okhota r. Rus. Fed.
83 K3 Okhotsk Rus. Fed.
83 K3 Okhotsk, Sea of Japan/ Rus. Fed.
Okhotskoye More sea Japan/Rus. Fed. see Okhotsk, Sea of
91 C1 Okhtyrka Ukr.
69 E3 Okinawa i. Japan
67 B3 Okinoshima Japan
67 B4 Oki-shotō is Japan
143 D1 Oklahoma state U.S.A.
143 D1 Oklahoma City U.S.A.
143 D1 Okmulgee U.S.A.
Oknitsa Moldova see Ocnița
78 A2 Oko, Wadi watercourse Sudan
118 B3 Okondja Gabon
128 C2 Okotoks Can.
89 D3 Okovskiy Les for. Rus. Fed.
118 B3 Okoyo Congo
92 H1 Øksfjord Norway
62 A2 Oktwin Myanmar
Oktyabr' Kazakh. see Kandyagash
Oktyabr'sk Kazakh. see Kandyagash
86 D2 Oktyabr'skiy Rus. Fed.
83 L3 Oktyabr'skiy Rus. Fed.
87 E3 Oktyabr'skiy Rus. Fed.
86 F2 Oktyabr'skoye Rus. Fed.
83 H1 Oktyabr'skoy Revolyutsii, Ostrov i. Rus. Fed.
89 D2 Okulovka Rus. Fed.
66 C2 Okushiri-tō i. Japan
92 □B2 Ólafsfjörður Iceland
92 □A3 Ólafsvík Iceland
88 B2 Olaine Latvia
93 G4 Öland i. Sweden
52 B2 Olary Austr.
136 B3 Olathe CO U.S.A.
137 E3 Olathe KS U.S.A.
153 B3 Olavarría Arg.
103 D1 Oława Pol.
108 A2 Olbia Italy
126 C2 Old Crow Can.
101 D1 Oldenburg Ger.
102 C1 Oldenburg in Holstein Ger.
100 C1 Oldenzaal Neth.
98 B3 Oldham U.K.
97 B3 Old Head of Kinsale hd Ireland
96 C2 Oldmeldrum U.K.
128 C2 Olds Can.
129 D2 Old Wives Lake Can.
98 C2 Oldham U.K.
103 E1 Olecko Pol.
83 J2 Olekminsk Rus. Fed.

91 C2 Oleksandrivka Ukr.
Oleksandrivs'k Ukr. see Zaporizhzhya
91 C2 Oleksandriya Ukr.
86 C2 Olenegorsk Rus. Fed.
83 I2 Olenek Rus. Fed.
83 I2 Olenek r. Rus. Fed.
89 D2 Olenino Rus. Fed.
Olenivs'ki Kar''yery Ukr. see Dokuchayevs'k
Olenya Rus. Fed. see Olenegorsk
Oleshky Ukr. see Tsyurupyns'k
103 D1 Olesno Pol.
90 B1 Olevs'k Ukr.
106 B2 Olhão Port.
123 D1 Olifants r. Moz./S. Africa
122 A2 Olifants watercourse Namibia
123 D1 Olifants S. Africa
122 A3 Olifants r. S. Africa
122 B2 Olifantshoek S. Africa
154 C2 Olímpia Brazil
151 F3 Olinda Brazil
123 C1 Oliphants Drift Botswana
107 C2 Oliva Spain
155 D2 Oliveira Brazil
Olivença Moz. see Lupilichi
106 B2 Olivenza Spain
140 B2 Olla U.S.A.
152 B2 Ollagüe Chile
77 C2 Olmaliq Uzbek.
106 C1 Olmedo Spain
105 D3 Olmeto France
150 B3 Olmos Peru
138 B3 Olney U.S.A.
103 D2 Olomouc Czech Rep.
86 C2 Olonets Rus. Fed.
64 B2 Olongapo Phil.
104 B3 Oloron-Ste-Marie France
107 D1 Olot Spain
69 D1 Olovyannaya Rus. Fed.
83 L2 Oloy r. Rus. Fed.
100 C2 Olpe Ger.
103 E1 Olsztyn Pol.
110 B2 Olt r. Romania
110 C2 Oltenița Romania
143 C2 Olton U.S.A.
81 C1 Oltu Turkey
Ol'viopol' Ukr. see Pervomays'k
111 B3 Olympia tourist site Greece
134 B1 Olympia U.S.A.
Olympus, Mount mt. Greece see Mount Olympus
134 B1 Olympus, Mount U.S.A.
83 M3 Olyutorskiy, Mys c. Rus. Fed.
66 D2 Ōma Japan
97 C1 Omagh U.K.
137 D2 Omaha U.S.A.
79 C2 Oman country Asia
79 C2 Oman, Gulf of Asia
54 A2 Omarama N.Z.
120 A3 Omaruru Namibia
120 B2 Omatako watercourse Namibia
122 B2 Omaweneno Botswana
116 B3 Omdurman Sudan
53 C3 Omeo Austr.
145 C3 Ometepec Mex.
78 A3 Om Hajēr Eritrea
81 C2 Omīdīyeh Iran
128 B2 Omineca Mountains Can.
120 A3 Omitara Namibia
67 C3 Ōmiya Japan
100 C1 Ommen Neth.
83 L2 Omolon r. Rus. Fed.
100 B3 Omont France
82 F3 Omsk Rus. Fed.
83 L2 Omsukchan Rus. Fed.
110 C1 Omu, Vârful mt. Romania
67 A4 Ōmura Japan
138 A2 Onalaska U.S.A.
139 D3 Onancock U.S.A.
138 C1 Onaping Lake Can.
131 C3 Onatchiway, Lac l. Can.
63 A2 Onbingwin Myanmar
122 B3 Oncócua Angola
122 B3 Onderstedorings S. Africa
120 A2 Ondjiva Angola
69 D1 Öndörhaan Mongolia
86 C2 One Botswana
86 C2 Onega Rus. Fed.
86 C2 Onega r. Rus. Fed.
86 C2 Onega, Lake Rus. Fed.
137 D2 O'Neill U.S.A.
139 D2 Oneonta U.S.A.

211

110 C1 **Onești** Romania
Onezhskoye Ozero *l.* Rus. Fed. *see* **Onega, Lake**
122 B2 **Ongers** *watercourse* S. Africa
65 B2 **Ongjin** N. Korea
73 C3 **Ongole** India
121 □D3 **Onilahy** *r.* Madag.
115 C4 **Onitsha** Nigeria
120 A3 **Onjati Mountain** Namibia
67 C3 **Ōno** Japan
156 D6 **Onotoa** *atoll* Kiribati
122 A2 **Onseepkans** S. Africa
50 A2 **Onslow** Austr.
141 E2 **Onslow Bay** U.S.A.
130 A2 **Ontario** *prov.* Can.
134 C2 **Ontario** U.S.A.
139 D2 **Ontario, Lake** Can./U.S.A.
107 C2 **Ontinyent** Spain
51 C2 **Oodnadatta** Austr.
Oostende Belgium *see* **Ostend**
100 B2 **Oosterhout** Neth.
100 A2 **Oosterschelde** *est.* Neth.
100 B1 **Oost-Vlieland** Neth.
128 B2 **Ootsa Lake** Can.
128 B2 **Ootsa Lake** *l.* Can.
118 C3 **Opala** Dem. Rep. Congo
130 C2 **Opataca, Lac** *l.* Can.
103 D2 **Opava** Czech Rep.
141 C2 **Opelika** U.S.A.
140 B2 **Opelousas** U.S.A.
117 A4 **Opienge** Dem. Rep. Congo
130 C2 **Opinaca, Réservoir** *resr* Can.
131 D2 **Opiscotéo, Lac** *l.* Can.
88 C2 **Opochka** Rus. Fed.
144 A2 **Opodepe** Mex.
103 D1 **Opole** Pol.
106 B1 **Oporto** Port.
54 C1 **Opotiki** N.Z.
93 E3 **Oppdal** Norway
134 C1 **Opportunity** U.S.A.
54 B1 **Opunake** N.Z.
120 A2 **Opuwo** Namibia
110 B1 **Oradea** Romania
Orahovac Kosovo *see* **Rahovec**
114 B1 **Oran** Alg.
152 B2 **Orán** Arg.
65 B1 **Ŏrang** N. Korea
53 C2 **Orange** Austr.
105 C3 **Orange** France
122 A2 **Orange** *r.* Namibia/S. Africa
143 E2 **Orange** U.S.A.
141 D2 **Orangeburg** U.S.A.
Orange Free State *prov.* S. Africa *see* **Free State**
141 D2 **Orange Park** U.S.A.
138 C2 **Orangeville** Can.
145 D3 **Orange Walk** Belize
101 F1 **Oranienburg** Ger.
122 A2 **Oranjemund** Namibia
147 C3 **Oranjestad** Aruba
120 B3 **Orapa** Botswana
110 B1 **Orăștie** Romania
Orașul Stalin Romania *see* **Brașov**
108 B2 **Orbetello** Italy
53 C3 **Orbost** Austr.
141 D3 **Orchid Island** U.S.A.
111 B3 **Orchomenos** Greece
50 B1 **Ord, Mount** *h.* Austr.
106 B1 **Ordes** Spain
80 B1 **Ordu** Turkey
Ordzhonikidze Rus. Fed. *see* **Vladikavkaz**
91 C2 **Ordzhonikidze** Ukr.
93 G4 **Örebro** Sweden
134 B2 **Oregon** *state* U.S.A.
134 B1 **Oregon City** U.S.A.
141 E1 **Oregon Inlet** U.S.A.
87 C3 **Orekhovo-Zuyevo** Rus. Fed.
89 E3 **Orel** Rus. Fed.
83 K3 **Orel', Ozero** *l.* Rus. Fed.
135 D2 **Orem** U.S.A.
111 C3 **Ören** Turkey
87 E3 **Orenburg** Rus. Fed.
54 A3 **Orepuki** N.Z.
111 C2 **Orestiada** Greece
93 F4 **Öresund** *str.* Denmark/Sweden
Oretana, Cordillera *mts* Spain *see* **Toledo, Montes de**
99 D3 **Orford Ness** *hd* U.K.
89 F2 **Orgtrud** Rus. Fed.
111 C3 **Orhaneli** Turkey
111 C2 **Orhangazi** Turkey
68 D1 **Orhon Gol** *r.* Mongolia
152 B1 **Oriental, Cordillera** *mts* Bol.
150 B2 **Oriental, Cordillera** *mts* Col.
150 B4 **Oriental, Cordillera** *mts* Peru
107 C2 **Orihuela** Spain
91 D2 **Orikhiv** Ukr.
130 C3 **Orillia** Can.
93 I3 **Orimattila** Fin.
150 C2 **Orinoco** *r.* Col./Venez.
150 C2 **Orinoco, Delta del** Venez.
Orissa *state* India *see* **Odisha**
88 B2 **Orissaare** Estonia
108 A3 **Oristano** Italy
93 I3 **Orivesi** *l.* Fin.
151 D3 **Oriximiná** Brazil
145 C3 **Orizaba** Mex.
145 C3 **Orizaba, Pico de** *vol.* Mex.
154 C1 **Orizona** Brazil
92 E3 **Orkanger** Norway
93 F4 **Örkelljunga** Sweden
93 E3 **Orkla** *r.* Norway
96 C1 **Orkney Islands** U.K.
154 C2 **Orlândia** Brazil
141 D3 **Orlando** U.S.A.
104 C2 **Orléans** France
139 F2 **Orleans** U.S.A.
139 E1 **Orléans, Île d'** *i.* Can.
Orléansville Alg. *see* **Chlef**
74 A2 **Ormara** Pak.
64 B2 **Ormoc** Phil.
141 D3 **Ormond Beach** U.S.A.
98 B3 **Ormskirk** U.K.
104 B2 **Orne** *r.* France
92 F2 **Ørnes** Norway
92 G3 **Örnsköldsvik** Sweden
114 B3 **Orodara** Burkina Faso
134 C1 **Orofino** U.S.A.
139 F2 **Orono** U.S.A.
Oroqen Zizhiqi China *see* **Alihe**
64 B3 **Oroquieta** Phil.
108 A2 **Orosei** Italy
108 A2 **Orosei, Golfo di** *b.* Italy
103 E2 **Orosháza** Hungary
135 B3 **Oroville** U.S.A.
52 A2 **Orroroo** Austr.
93 F3 **Orsa** Sweden
89 D3 **Orsha** Belarus
87 E3 **Orsk** Rus. Fed.
110 B2 **Orșova** Romania
93 E3 **Ørsta** Norway
106 B1 **Ortegal, Cabo** *c.* Spain
104 B3 **Orthez** France
106 B1 **Ortigueira** Spain
108 B1 **Ortles** *mt.* Italy
108 B2 **Ortona** Italy
137 D1 **Ortonville** U.S.A.
83 J2 **Orulgan, Khrebet** *mts* Rus. Fed.
Orūmīyeh Iran *see* **Urmia**
Orūmīyeh, Daryācheh-ye *salt l.* Iran *see* **Urmia, Lake**
152 B1 **Oruro** Bol.
104 B2 **Orvault** France
108 B2 **Orvieto** Italy
93 F3 **Os** Norway
146 B4 **Osa, Península de** *pen.* Costa Rica
137 E3 **Osage** *r.* U.S.A.
137 D3 **Osage City** U.S.A.
67 C4 **Ōsaka** Japan
77 D1 **Osakarovka** Kazakh.
66 D3 **Ōsaki** Japan
101 E1 **Oschersleben (Bode)** Ger.
108 A2 **Oschiri** Italy
138 C2 **Oscoda** U.S.A.
89 E3 **Osetr** *r.* Rus. Fed.
139 D1 **Osgoode** Can.
77 D2 **Osh** Kyrg.
120 A2 **Oshakati** Namibia
130 C3 **Oshawa** Can.
120 A2 **Oshikango** Namibia
66 C2 **Ō-shima** *i.* Japan
67 C4 **Ō-shima** *i.* Japan
138 B2 **Oshkosh** U.S.A.
81 C2 **Oshnovīyeh** Iran
115 C4 **Oshogbo** Nigeria
118 B3 **Oshwe** Dem. Rep. Congo
109 C1 **Osijek** Croatia
128 B2 **Osilinka** *r.* Can.
108 B2 **Osimo** Italy
Osipenko Ukr. *see* **Berdyans'k**
123 D2 **oSizweni** S. Africa
137 E2 **Oskaloosa** U.S.A.
93 G4 **Oskarshamn** Sweden
89 E3 **Oskol** *r.* Rus. Fed.
93 F4 **Oslo** Norway
93 F4 **Oslofjorden** *sea chan.* Norway
80 B1 **Osmancık** Turkey
111 C2 **Osmaneli** Turkey
80 B2 **Osmaniye** Turkey
88 C2 **Os'mino** Rus. Fed.
101 D1 **Osnabrück** Ger.
153 A4 **Osorno** Chile
106 C1 **Osorno** Spain
128 C3 **Osoyoos** Can.
100 B2 **Oss** Neth.
51 D4 **Ossa, Mount** Austr.
83 L3 **Ossora** Rus. Fed.
89 D2 **Ostashkov** Rus. Fed.
101 D1 **Oste** *r.* Ger.
100 A2 **Ostend** Belgium
101 E1 **Osterburg (Altmark)** Ger.
93 F3 **Österdalälven** *r.* Sweden
101 D1 **Osterholz-Scharmbeck** Ger.
101 E2 **Osterode am Harz** Ger.
92 F3 **Östersund** Sweden
Ostfriesische Inseln *is* Ger. *see* **East Frisian Islands**
100 C1 **Ostfriesland** *reg.* Ger.
93 G3 **Östhammar** Sweden
103 D2 **Ostrava** Czech Rep.
103 D1 **Ostróda** Pol.
89 E3 **Ostrogozhsk** Rus. Fed.
90 B1 **Ostroh** Ukr.
103 E1 **Ostrołęka** Pol.
101 F2 **Ostrov** Czech Rep.
88 C2 **Ostrov** Rus. Fed.
Ostrovets Pol. *see* **Ostrowiec Świętokrzyski**
86 C2 **Ostrovnoy** Rus. Fed.
89 F2 **Ostrovskoye** Rus. Fed.
103 E1 **Ostrowiec Świętokrzyski** Pol.
103 E1 **Ostrów Mazowiecka** Pol.
Ostrowo Pol. *see* **Ostrów Wielkopolski**
103 D1 **Ostrów Wielkopolski** Pol.
109 C2 **Ostuni** Italy
110 B2 **Osŭm** *r.* Bulg.
67 B4 **Ōsumi-kaikyō** *sea chan.* Japan
67 B4 **Ōsumi-shotō** *is* Japan
106 B2 **Osuna** Spain
139 D2 **Oswego** U.S.A.
99 B3 **Oswestry** U.K.
67 C3 **Ōta** Japan
54 B3 **Otago Peninsula** N.Z.
54 C2 **Otaki** N.Z.
77 D2 **Otar** Kazakh.
66 D2 **Otaru** Japan
120 A2 **Otavi** Namibia
67 D3 **Ōtawara** Japan
92 G2 **Oteren** Norway
134 C1 **Othello** U.S.A.
123 D2 **oThongathi** S. Africa
120 A3 **Otjiwarongo** Namibia
109 C2 **Otočac** Croatia
Otog Qi China *see* **Ulan**
117 B3 **Otoro, Jebel** *mt.* Sudan
Otpor Rus. Fed. *see* **Zabaykal'sk**
93 E4 **Otra** *r.* Norway
109 C2 **Otranto, Strait of** Albania/Italy
67 C3 **Ōtsu** Japan
93 E3 **Otta** Norway
130 C3 **Ottawa** Can.
130 C3 **Ottawa** *r.* Can.
138 B2 **Ottawa** *IL* U.S.A.
137 D3 **Ottawa** *KS* U.S.A.
130 B2 **Ottawa Islands** Can.
98 B2 **Otterburn** U.K.
130 B2 **Otter Rapids** Can.
100 B2 **Ottignies** Belgium
137 E2 **Ottumwa** U.S.A.
150 B3 **Otuzco** Peru
52 B3 **Otway, Cape** Austr.
140 B2 **Ouachita** *r.* U.S.A.
140 B2 **Ouachita, Lake** U.S.A.
140 B2 **Ouachita Mountains** U.S.A.
118 C2 **Ouadda** C.A.R.
115 D3 **Ouaddaï** *reg.* Chad
114 B3 **Ouagadougou** Burkina Faso
114 B3 **Ouahigouya** Burkina Faso
114 B3 **Oualâta** Maur.
114 B3 **Oualâta, Ḍhar** *hills* Maur.
118 C2 **Ouanda Djallé** C.A.R.
114 B2 **Ouarâne** *reg.* Maur.
115 C1 **Ouargla** Alg.
114 B1 **Ouarzazate** Morocco
100 A2 **Oudenaarde** Belgium
122 B3 **Oudtshoorn** S. Africa
107 C2 **Oued Tlélat** Alg.
114 B1 **Oued Zem** Morocco
104 A2 **Ouessant, Île d'** *i.* France
118 B2 **Ouesso** Congo
97 B2 **Oughterard** Ireland
118 B2 **Ouham** *r.* C.A.R./Chad
114 B1 **Oujda** Morocco
92 H3 **Oulainen** Fin.
107 D2 **Ouled Farès** Alg.
92 I2 **Oulu** Fin.
92 I3 **Oulujärvi** *l.* Fin.
108 A1 **Oulx** Italy
115 E3 **Oum-Chalouba** Chad
115 D3 **Oum-Hadjer** Chad
115 E3 **Ounianga Kébir** Chad
100 B2 **Oupeye** Belgium
100 C3 **Our** *r.* Ger./Lux.
106 B1 **Ourense** Spain
154 C2 **Ourinhos** Brazil
155 D2 **Ouro Preto** Brazil
100 B2 **Ourthe** *r.* Belgium
98 C3 **Ouse** *r.* U.K.
Outaouais, Rivière des *r.* Can. *see* **Ottawa**
131 D3 **Outardes, Rivière aux** *r.* Can.
131 D2 **Outardes Quatre, Réserv** *resr* Can.
96 A2 **Outer Hebrides** *is* U.K.
Outer Mongolia *country* Asia *see* **Mongolia**
120 A3 **Outjo** Namibia
129 D2 **Outlook** Can.
92 I3 **Outokumpu** Fin.
52 B3 **Ouyen** Austr.
108 A2 **Ovace, Punta d'** *mt.* Fran
152 A3 **Ovalle** Chile
106 B1 **Ovar** Port.
92 H2 **Överkalix** Sweden
137 E3 **Overland Park** U.S.A.
135 D3 **Overton** U.S.A.
92 H2 **Övertorneå** Sweden
106 B1 **Oviedo** Spain
141 D3 **Oviedo** U.S.A.
88 B2 **Ovišrags** *hd* Latvia
93 E3 **Øvre Årdal** Norway
93 F3 **Øvre Rendal** Norway
90 B1 **Ovruch** Ukr.
118 B3 **Owando** Congo
67 C4 **Owase** Japan
143 D1 **Owasso** U.S.A.
137 E2 **Owatonna** U.S.A.
139 D2 **Owego** U.S.A.
138 B3 **Owensboro** U.S.A.
135 C3 **Owens Lake** U.S.A.
130 B3 **Owen Sound** Can.
51 D1 **Owen Stanley Range** *mts* P.N.G.
115 C4 **Owerri** Nigeria
115 C4 **Owo** Nigeria
138 C2 **Owosso** U.S.A.
134 C2 **Owyhee** U.S.A.
134 C2 **Owyhee** *r.* U.S.A.
129 D3 **Oxbow** Can.
54 B2 **Oxford** N.Z.
99 C4 **Oxford** U.K.
140 C2 **Oxford** U.S.A.
129 E2 **Oxford Lake** Can.
145 D2 **Oxkutzcab** Mex.
52 B2 **Oxley** Austr.
97 B1 **Ox Mountains** *hills* Irelar
135 C4 **Oxnard** U.S.A.
67 C3 **Oyama** Japan
118 B2 **Oyem** Gabon
129 C2 **Oyen** Can.
105 D2 **Oyonnax** France
77 C2 **Oyoqquduq** Uzbek.
64 B3 **Ozamis** Phil.
140 C2 **Ozark** *AL* U.S.A.
137 E3 **Ozark** *MO* U.S.A.
137 E3 **Ozark Plateau** U.S.A.
137 E3 **Ozarks, Lake of the** U.S.A
83 L3 **Ozernovskiy** Rus. Fed.
88 B3 **Ozersk** Rus. Fed.
89 E3 **Ozery** Rus. Fed.
87 D3 **Ozinki** Rus. Fed.

P

127 I2 **Paamiut** Greenland
122 A3 **Paarl** S. Africa
103 D1 **Pabianice** Pol.
75 C2 **Pabna** Bangl.
88 C3 **Pabradė** Lith.
74 A2 **Pab Range** *mts* Pak.
150 B3 **Pacasmayo** Peru
142 B2 **Pacheco** Mex.
109 C3 **Pachino** Italy
145 C2 **Pachuca** Mex.
135 B3 **Pacifica** U.S.A.

57	E9	Pacific-Antarctic Ridge
		S. Pacific Ocean
56		Pacific Ocean
61	C2	Pacitan Indon.
52	B2	Packsaddle Austr.
03	D1	Paczków Pol.
60	D2	Padang Indon.
60	B1	Padang Endau Malaysia
60	B2	Padangpanjang Indon.
60	A1	Padangsidimpuan Indon.
01	D2	Paderborn Ger.
		Padova Italy see Padua
43	D3	Padre Island U.S.A.
99	A4	Padstow U.K.
52	B3	Padthaway Austr.
08	B1	Padua Italy
38	B3	Paducah KY U.S.A.
43	C2	Paducah TX U.S.A.
65	B1	Paegam N. Korea
		Paektu-san mt.
		China/N. Korea see
		Baitou Shan
54	C1	Paeroa N.Z.
		Pafos Cyprus see Paphos
09	C2	Pag Croatia
09	B2	Pag i. Croatia
64	B3	Pagadian Phil.
60	B2	Pagai Selatan i. Indon.
60	B2	Pagai Utara i. Indon.
59	D1	Pagan i. N. Mariana Is
61	C2	Pagatan Indon.
42	A1	Page U.S.A.
88	B2	Pagégiai Lith.
53	E5	Paget, Mount S. Georgia
36	B3	Pagosa Springs U.S.A.
88	C2	Paide Estonia
99	B4	Paignton U.K.
93	I3	Päijänne l. Fin.
75	C2	Paikü Co l. China
60	B2	Painan Indon.
38	C2	Painesville U.S.A.
42	A1	Painted Desert U.S.A.
		Paint Hills Can. see
		Wemindji
96	B3	Paisley U.K.
92	H2	Pajala Sweden
50	A3	Paján Ecuador
50	C2	Pakaraima Mountains mts
		S. America
50	D2	Pakaraima Mountains
		S. America
74	A2	Pakistan country Asia
62	A1	Pakokku Myanmar
88	B2	Pakruojis Lith.
03	D2	Paks Hungary
30	A2	Pakwash Lake Can.
62	B2	Pakxan Laos
63	B2	Pakxé Laos
15	D4	Pala Chad
60	B2	Palabuhanratu, Teluk b.
		Indon.
11	C3	Palaikastro Greece
11	B3	Palaiochora Greece
73	B3	Palakkad India
22	B1	Palamakoloi Botswana
07	D1	Palamós Spain
83	L3	Palana Rus. Fed.
64	B2	Palanan Phil.
61	C2	Palangkaraya Indon.
74	B2	Palanpur India
23	C1	Palapye Botswana
83	L2	Palatka Rus. Fed.
41	D3	Palatka U.S.A.
59	C2	Palau country
		N. Pacific Ocean
63	A2	Palaw Myanmar
64	A3	Palawan i. Phil.
64	A3	Palawan Passage str. Phil.
88	B2	Paldiski Estonia
89	F3	Palekh Rus. Fed.
60	B2	Palembang Indon.
06	C1	Palencia Spain
45	C3	Palenque Mex.
08	B3	Palermo Italy
43	D2	Palestine U.S.A.
62	A1	Paletwa Myanmar
		Palghat India see Palakkad
74	B2	Pali India
48	G4	Palikir Micronesia
09	C2	Palinuro, Capo c. Italy
11	B3	Paliouri, Akrotirio pt Greece
00	B3	Paliseul Belgium
92	I3	Paljakka h. Fin.
89	F2	Palkino Rus. Fed.
73	B4	Palk Strait India/Sri Lanka
		Palla Bianca mt. Austria/
		Italy see Weißkugel
54	C2	Palliser, Cape N.Z.
57	F7	Palliser, Îles is Fr. Polynesia
06	B2	Palma del Río Spain

107	D2	Palma de Mallorca Spain
154	B3	Palmas Brazil
151	E4	Palmas Brazil
154	B3	Palmas, Campos de hills
		Brazil
114	B4	Palmas, Cape Liberia
141	D3	Palm Bay U.S.A.
135	C4	Palmdale U.S.A.
154	C3	Palmeira Brazil
151	E3	Palmeirais Brazil
126	C2	Palmer U.S.A.
55	A2	Palmer Land reg. Antarctica
49	K5	Palmerston atoll Cook Is
54	C2	Palmerston North N.Z.
109	C3	Palmi Italy
145	C2	Palmillas Mex.
150	B2	Palmira Col.
154	B2	Palmital Brazil
135	C4	Palm Springs U.S.A.
		Palmyra Syria see Tadmur
49	K3	Palmyra Atoll
		N. Pacific Ocean
135	B3	Palo Alto U.S.A.
117	B3	Paloich S. Sudan
145	C3	Palomares Mex.
61	D2	Palopo Indon.
107	C2	Palos, Cabo de c. Spain
92	I3	Paltamo Fin.
61	C2	Palu Indon.
83	M2	Palyavaam r. Rus. Fed.
150	B3	Pamar Col.
121	C3	Pambarra Moz.
104	C3	Pamiers France
77	D3	Pamir mts Asia
141	E1	Pamlico Sound sea chan.
		U.S.A.
152	B1	Pampa Grande Bol.
153	B3	Pampas reg. Arg.
150	B2	Pamplona Col.
107	C1	Pamplona Spain
111	D2	Pamukova Turkey
60	B2	Panaitan i. Indon.
73	B3	Panaji India
146	B4	Panama country
		Central America
		Panamá Panama see
		Panamá City
		Panamá, Canal de canal
		Panama
		Panamá see Panama, Gulf of
146	C4	Panama, Gulf of g. Panama
		Panama Canal
		canal Panama see
		Panamá, Canal de
146	C4	Panama City Panama
140	C2	Panama City U.S.A.
135	C3	Panamint Range mts U.S.A.
60	B1	Panarik Indon.
64	B2	Panay i. Phil.
109	D2	Pančevo Serbia
64	B2	Pandan Phil.
64	B1	Pandan Phil.
75	C2	Pandharpur India
73	B3	Pandharpur India
88	B2	Panevėžys Lith.
		Panfilov Kazakh. see
		Zharkent
61	C2	Pangkalanbuun Indon.
60	A1	Pangkalansusu Indon.
60	B2	Pangkalpinang Indon.
61	D2	Pangkalsiang, Tanjung pt
		Indon.
64	A3	Panglima Sugala Phil.
127	H2	Pangnirtung Can.
86	G2	Pangody Rus. Fed.
89	F3	Panino Rus. Fed.
74	B2	Panipat India
74	A2	Panjgur Pak.
		Panjim India see Panaji
118	C3	Pankshin Nigeria
65	C1	Pan Ling mts China
75	C2	Panna India
50	A2	Pannawonica Austr.
154	B2	Panorama Brazil
65	B1	Panshi China
152	C1	Pantanal reg. Brazil
145	C2	Pánuco Mex.
145	C2	Pánuco r. Mex.
62	B1	Panzhihua China
109	C3	Paola Italy
118	B2	Paoua C.A.R.
63	B2	Paôy Pêt Cambodia
03	D2	Pápa Hungary
54	B1	Papakura N.Z.
145	C2	Papantla Mex.
96	□	Papa Stour i. U.K.
54	B1	Papatoetoe N.Z.
49	M5	Papeete Fr. Polynesia
100	C1	Papenburg Ger.

80	B2	Paphos Cyprus
137	D2	Papillion U.S.A.
59	D3	Papua reg. Indon.
59	D3	Papua, Gulf of P.N.G.
59	D3	Papua New Guinea country
		Oceania
89	F3	Para r. Rus. Fed.
50	A2	Paraburdoo Austr.
154	C1	Paracatu Brazil
155	C1	Paracatu r. Brazil
52	A2	Parachilna Austr.
109	D2	Paraćin Serbia
155	D1	Pará de Minas Brazil
135	B3	Paradise U.S.A.
151	D2	Paradise Guyana
135	B3	Paradise U.S.A.
140	B1	Paragould U.S.A.
151	D3	Paraguai r. Brazil
147	D3	Paraguaná, Península de
		pen. Venez.
152	C2	Paraguay r. Arg./Para.
152	C2	Paraguay country S. America
155	D2	Paraíba do Sul r. Brazil
154	B1	Paraíso Brazil
145	C3	Paraíso Mex.
114	C4	Parakou Benin
52	A2	Parakylia Austr.
151	D2	Paramaribo Suriname
83	L3	Paramushir, Ostrov i.
		Rus. Fed.
152	B3	Paraná Arg.
154	B2	Paraná state Brazil
154	A3	Paraná r. S. America
154	C1	Paraná, Serra do hills Brazil
154	B2	Paranaguá Brazil
154	B1	Paranaíba Brazil
154	B2	Paranaíba r. Brazil
154	B2	Paranapanema r. Brazil
154	B2	Paranapiacaba, Serra mts
		Brazil
154	B2	Paranavaí Brazil
90	A2	Parângul Mare, Vârful mt.
		Romania
54	B2	Paraparaumu N.Z.
155	D2	Parati Brazil
52	A2	Paratoo Austr.
151	D3	Parauaquara, Serra h. Brazil
154	B1	Paraúna Brazil
105	C2	Paray-le-Monial France
74	B2	Parbati r. India
73	B3	Parbhani India
101	E1	Parchim Ger.
103	E1	Parczew Pol.
155	C1	Pardo r. Brazil
154	B2	Pardo r. Brazil
154	C2	Pardo r. Brazil
103	D1	Pardubice Czech Rep.
152	C1	Parecis, Serra dos hills
		Brazil
130	C3	Parent Can.
130	C3	Parent, Lac l. Can.
61	C2	Parepare Indon.
89	D2	Parfino Rus. Fed.
111	B3	Parga Greece
109	C3	Parghelia Italy
147	D3	Paria, Gulf of
		Trin. and Tob./Venez.
150	C2	Parima, Serra mts Brazil
151	D3	Parintins Brazil
104	C2	Paris France
140	C1	Paris TN U.S.A.
143	D2	Paris TX U.S.A.
93	H3	Parkano Fin.
142	A2	Parker U.S.A.
138	C3	Parkersburg U.S.A.
53	C2	Parkes Austr.
138	A1	Park Falls U.S.A.
134	B1	Parkland U.S.A.
137	D1	Park Rapids U.S.A.
106	C1	Parla Spain
108	B2	Parma Italy
134	C2	Parma U.S.A.
151	E3	Parnaíba Brazil
151	E3	Parnaíba r. Brazil
54	B2	Parnassus N.Z.
		Parnassus, Mount mt.
		Greece see Liakoura
111	B3	Parnonas mts Greece
88	B2	Pärnu Estonia
65	B2	Paro-ho l. S. Korea
52	B2	Paroo watercourse Austr.
		Paropamisus mts Afgh. see
		Sefīd Kūh, Selseleh-ye
111	C3	Paros i. Greece
135	D3	Parowan U.S.A.
153	A3	Parral Chile
53	D2	Parramatta Austr.
144	B2	Parras Mex.
151	D3	Pārsābād Iran
99	B3	Parsberg Ger.
126	C2	Parry, Cape Can.
126	E1	Parry Islands Can.
130	B3	Parry Sound Can.

137	D3	Parsons U.S.A.
108	B3	Partanna Italy
101	D2	Partenstein Ger.
104	B2	Parthenay France
108	B3	Partinico Italy
66	B2	Partizansk Rus. Fed.
97	B2	Partry Mountains hills
		Ireland
151	D3	Paru r. Brazil
131	E3	Pasadena Can.
135	C4	Pasadena CA U.S.A.
143	D3	Pasadena TX U.S.A.
62	A2	Pasawng Myanmar
140	C2	Pascagoula U.S.A.
110	C1	Pașcani Romania
134	C1	Pasco U.S.A.
155	E1	Pascoal, Monte h. Brazil
		Pascua, Isla de i.
		S. Pacific Ocean see
		Easter Island
		Pas de Calais str. France/
		U.K. see Dover, Strait of
102	C2	Pasewalk Ger.
129	D2	Pasfield Lake Can.
89	D1	Pasha Rus. Fed.
64	B2	Pasig Phil.
60	B1	Pasir Putih Malaysia
103	D1	Pasłęk Pol.
74	A2	Pasni Pak.
153	A4	Paso Río Mayo Arg.
135	B3	Paso Robles U.S.A.
97	B3	Passage West Ireland
155	D2	Passa Tempo Brazil
102	C2	Passau Ger.
152	C2	Passo Fundo Brazil
155	C2	Passos Brazil
88	C2	Pastavy Belarus
150	B3	Pastaza r. Peru
150	B2	Pasto Col.
74	B1	Pasu Pak.
61	C2	Pasuruan Indon.
88	B2	Pasvalys Lith.
103	D2	Pásztó Hungary
153	A5	Patagonia reg. Arg.
54	B2	Patan Nepal
54	B1	Patea N.Z.
139	E2	Paterson U.S.A.
74	B1	Pathankot India
		Pathein Myanmar see
		Bassein
136	B2	Pathfinder Reservoir U.S.A.
61	C2	Pati Indon.
74	B1	Patiala India
62	A1	Patkai Bum mts India/
		Myanmar
111	C3	Patmos i. Greece
75	C2	Patna India
81	C2	Patnos Turkey
154	B3	Pato Branco Brazil
152	C3	Patos, Lagoa dos l. Brazil
155	C1	Patos de Minas Brazil
152	B3	Patquía Arg.
		Patra Greece see Patras
111	B3	Patras Greece
75	C2	Patratu India
154	C1	Patrocínio Brazil
63	B3	Pattani Thai.
63	B2	Pattaya Thai.
128	C2	Pattullo, Mount Can.
129	D2	Patuanak Can.
146	B3	Patuca r. Hond.
144	B3	Pátzcuaro Mex.
104	B3	Pau France
104	B3	Pau, Gave de r. France
104	B2	Pauillac France
150	C3	Pauini Brazil
62	A1	Pauk Myanmar
126	D2	Paulatuk Can.
		Paulis Dem. Rep. Congo see
		Isiro
151	E3	Paulistana Brazil
151	F3	Paulo Afonso Brazil
123	D2	Paulpietersburg S. Africa
143	D2	Pauls Valley U.S.A.
62	A2	Paungde Myanmar
155	D1	Pavão Brazil
108	A1	Pavia Italy
88	B2	Pāvilosta Latvia
110	C2	Pavlikeni Bulg.
77	D1	Pavlodar Kazakh.
91	D2	Pavlohrad Ukr.
91	E1	Pavlovsk Rus. Fed.
91	D2	Pavlovskaya Rus. Fed.
139	E2	Pawtucket U.S.A.
111	B3	Paxoi i. Greece
60	B2	Payakumbuh Indon.
134	C2	Payette U.S.A.
134	C2	Payette r. U.S.A.
86	F2	Pay-Khoy, Khrebet hills
		Rus. Fed.

Payne

Payne Can. see Kangirsuk
130 C2 Payne, Lac l. Can.
152 C3 Paysandú Uru.
81 C1 Pazar Turkey
110 B2 Pazardzhik Bulg.
111 C3 Pazarköy Turkey
108 B1 Pazin Croatia
63 A2 Pe Myanmar
128 C2 Peace r. Can.
128 C2 Peace River Can.
53 C2 Peak Hill N.S.W. Austr.
50 A2 Peak Hill W.A. Austr.
135 E3 Peale, Mount U.S.A.
140 C2 Pearl r. U.S.A.
71 B3 Pearl River r. China
143 D3 Pearsall U.S.A.
126 F1 Peary Channel Can.
121 C2 Pebane Moz.
Peć Kosovo see Pejë
155 D1 Peçanha Brazil
154 C3 Peças, Ilha das i. Brazil
92 J2 Pechenga Rus. Fed.
86 E2 Pechora Rus. Fed.
86 E2 Pechora r. Rus. Fed.
Pechora Sea Rus. Fed.
see Pechorskoye More
86 E2 Pechorskoye More sea
Rus. Fed.
88 C2 Pechory Rus. Fed.
142 B1 Pecos NM U.S.A.
143 C2 Pecos TX U.S.A.
143 C3 Pecos r. U.S.A.
103 D2 Pécs Hungary
142 B3 Pedernales Mex.
155 D1 Pedra Azul Brazil
154 C2 Pedregulho Brazil
151 E3 Pedreiras Brazil
73 C4 Pedro, Point Sri Lanka
151 E3 Pedro Afonso Brazil
152 B2 Pedro de Valdivia Chile
154 B1 Pedro Gomes Brazil
152 C2 Pedro Juan Caballero Para.
106 B1 Pedroso Port.
96 C3 Peebles U.K.
141 E2 Pee Dee r. U.S.A.
126 D2 Peel r. Can.
98 A2 Peel Isle of Man
128 C2 Peerless Lake Can.
54 B2 Pegasus Bay N.Z.
101 E3 Pegnitz Ger.
62 A2 Pegu Myanmar
62 A2 Pegu Yoma mts Myanmar
153 B3 Pehuajó Arg.
101 E1 Peine Ger.
88 C2 Peipus, Lake Estonia/
Rus. Fed.
Peiraias Greece see Piraeus
154 B2 Peixe r. Brazil
155 C2 Peixoto, Represa resr Brazil
151 D4 Peixoto de Azevedo Brazil
109 D2 Pejë Kosovo
123 C2 Peka Lesotho
60 B2 Pekalongan Indon.
60 B1 Pekan Malaysia
60 B1 Pekanbaru Indon.
Peking China see Beijing
130 B3 Pelee Island Can.
61 D2 Peleng i. Indon.
103 D2 Pelhřimov Czech Rep.
92 I2 Pelkosenniemi Fin.
122 A2 Pella S. Africa
137 E2 Pella U.S.A.
59 D3 Pelleluhu Islands P.N.G.
92 H2 Pello Fin.
128 A1 Pelly r. Can.
Pelly Bay Can. see Kugaaruk
128 A1 Pelly Mountains Can.
152 C3 Pelotas Brazil
152 C2 Pelotas, Rio das r. Brazil
139 F1 Pemadumcook Lake U.S.A.
60 B1 Pemangkat Indon.
60 A1 Pematangsiantar Indon.
121 D2 Pemba Moz.
120 B2 Pemba Zambia
119 D3 Pemba Island Tanz.
128 B2 Pemberton Can.
137 D1 Pembina r. Can.
137 D1 Pembina r. Can./U.S.A.
130 C3 Pembroke Can.
99 A4 Pembroke U.K.
141 D3 Pembroke Pines U.S.A.
106 C1 Peñalara mt. Spain
154 B2 Penápolis Brazil
106 B1 Peñaranda de Bracamonte
Spain
107 C1 Peñarroya mt. Spain
106 B2 Peñarroya-Pueblonuevo
Spain
106 B1 Peñas, Cabo de c. Spain
153 A4 Penas, Golfo de g. Chile

111 C2 Pendik Turkey
134 C1 Pendleton U.S.A.
128 B2 Pendleton Bay Can.
134 C1 Pend Oreille Lake U.S.A.
Penfro U.K. see Pembroke
74 B3 Penganga r. India
118 C3 Penge Dem. Rep. Congo
123 D1 Penge S. Africa
70 C2 Penglai China
71 A3 Pengshui China
106 B2 Peniche Port.
96 C3 Penicuik U.K.
60 B1 Peninsular Malaysia pen.
Malaysia
108 B2 Penne Italy
52 A3 Penneshaw Austr.
98 B2 Pennines hills U.K.
139 D2 Pennsylvania state U.S.A.
127 H2 Penny Icecap Can.
89 D2 Peno Rus. Fed.
139 F2 Penobscot r. U.S.A.
52 B3 Penola Austr.
50 C3 Penong Austr.
157 E6 Penrhyn Basin
S. Pacific Ocean
53 D2 Penrith Austr.
98 B2 Penrith U.K.
140 C2 Pensacola U.S.A.
55 B1 Pensacola Mountains
Antarctica
61 C1 Pensiangan Malaysia
128 C3 Penticton Can.
96 C1 Pentland Firth sea chan.
U.K.
99 B3 Penygadair h. U.K.
87 D3 Penza Rus. Fed.
99 A4 Penzance U.K.
83 L2 Penzhinskaya Guba b.
Rus. Fed.
142 A2 Peoria AZ U.S.A.
138 B2 Peoria IL U.S.A.
107 C1 Perales del Alfambra Spain
111 B3 Perama Greece
131 D3 Percé Can.
50 B2 Percival Lakes imp. l. Austr.
51 E2 Percy Isles Austr.
107 D1 Perdido, Monte mt. Spain
154 C1 Perdizes Brazil
86 F2 Peregrebnoye Rus. Fed.
150 B2 Pereira Col.
154 B2 Pereira Barreto Brazil
Pereira de Eça Angola see
Ondjiva
90 A2 Peremyshlyany Ukr.
89 E2 Pereslavl'-Zalesskiy
Rus. Fed.
91 C1 Pereyaslav-Khmel'nyts'kyy
Ukr.
153 B3 Pergamino Arg.
92 H3 Perhonjoki r. Fin.
131 C2 Péribonka, Lac l. Can.
152 B2 Perico Arg.
144 B2 Pericos Mex.
104 C2 Périgueux France
150 B2 Perijá, Sierra de mts Venez.
111 B3 Peristeri Greece
153 A4 Perito Moreno Arg.
101 E1 Perleberg Ger.
86 E3 Perm' Rus. Fed.
109 D2 Përmet Albania
Pernambuco Brazil see
Recife
52 A2 Pernatty Lagoon imp. l.
Austr.
110 B2 Pernik Bulg.
Pernov Estonia see Pärnu
105 C2 Péronne France
145 C3 Perote Mex.
104 C3 Perpignan France
99 A4 Perranporth U.K.
Perrégaux Alg. see
Mohammadia
141 D2 Perry FL U.S.A.
141 D2 Perry GA U.S.A.
137 E2 Perry IA U.S.A.
143 D1 Perry OK U.S.A.
138 C2 Perrysburg U.S.A.
143 C1 Perryton U.S.A.
137 F3 Perryville U.S.A.
Pershotravnevoye Ukr. see
Pershotravens'k
99 B3 Pershore U.K.
91 D2 Pershotravens'k Ukr.
Persia country Asia see Iran
Persian Gulf g. Asia see
The Gulf
50 A3 Perth Austr.
96 C2 Perth U.K.
159 F5 Perth Basin Indian Ocean
86 C2 Pertominsk Rus. Fed.

105 D3 Pertuis France
108 A2 Pertusato, Capo c. France
150 B3 Peru country S. America
138 B2 Peru U.S.A.
157 H6 Peru Basin S. Pacific Ocean
157 H7 Peru-Chile Trench
S. Pacific Ocean
108 B2 Perugia Italy
154 C2 Peruíbe Brazil
100 A2 Péruwelz Belgium
90 C2 Pervomays'k Ukr.
91 C2 Pervomays'ke Ukr.
Pervomayskiy Rus. Fed. see
Novodvinsk
89 F3 Pervomayskiy Rus. Fed.
91 D2 Pervomays'kyy Ukr.
108 B2 Pesaro Italy
108 B2 Pescara Italy
108 B2 Pescara r. Italy
74 B1 Peshawar Pak.
109 D2 Peshkopi Albania
109 C1 Pesnica Slovenia
104 B3 Pessac France
89 E2 Pestovo Rus. Fed.
140 C2 Petal U.S.A.
100 B3 Pétange Lux.
147 D3 Petare Venez.
144 B3 Petatlán Mex.
121 C2 Petauke Zambia
130 C3 Petawawa Can.
138 B2 Petenwell Lake U.S.A.
52 A2 Peterborough Austr.
130 C3 Peterborough Can.
99 C3 Peterborough U.K.
96 D2 Peterhead U.K.
55 R3 Peter I Island Antarctica
129 E1 Peter Lake Can.
50 B2 Petermann Ranges mts
Austr.
129 D2 Peter Pond Lake Can.
128 A2 Petersburg AK U.S.A.
139 D3 Petersburg VA U.S.A.
101 D1 Petershagen Ger.
Peter the Great
Bay b. Rus. Fed. see
Petitjean Morocco see
Sidi Kacem
131 E2 Petit Mécatina r. Can.
145 D2 Peto Mex.
138 C1 Petoskey U.S.A.
80 B2 Petra tourist site Jordan
66 B2 Petra Velikogo, Zaliv b.
Rus. Fed.
111 B2 Petrich Bulg.
Petroaleksandrovsk Uzbek.
see To'rtko'l
88 C2 Petrodvorets Rus. Fed.
Petrokov Pol. see
Piotrków Trybunalski
151 E3 Petrolina Brazil
83 L3 Petropavlovsk-Kamchatskiy
Rus. Fed.
77 C1 Petropavlovskoye Kazakh.
110 B1 Petroşani Romania
Petrovskoye Rus. Fed. see
Svetlograd
89 F3 Petrovskoye Rus. Fed.
89 E2 Petrovskoye Rus. Fed.
69 D1 Petrovsk-Zabaykal'skiy
Rus. Fed.
86 C2 Petrozavodsk Rus. Fed.
123 C2 Petrusburg S. Africa
123 C2 Petrus Steyn S. Africa
122 B3 Petrusville S. Africa
Petsamo Rus. Fed. see
Pechenga
87 F3 Petukhovo Rus. Fed.
89 E2 Petushki Rus. Fed.
60 A1 Peureula Indon.
83 M2 Pevek Rus. Fed.
102 B2 Pforzheim Ger.
102 C2 Pfunds Austria
101 D3 Pfungstadt Ger.
123 C1 Phagameng S. Africa
123 C2 Phahameng S. Africa
123 D1 Phalaborwa S. Africa
74 B2 Phalodi India
63 A3 Phangnga Thai.
62 B1 Phăng Xi Păng mt. Vietnam
63 B2 Phan Rang-Thap Cham
Vietnam
63 B2 Phan Thiết Vietnam
63 B3 Phatthalung Thai.
62 A2 Phayao Thai.
129 D2 Phelps Lake Can.
141 C2 Phenix City U.S.A.
63 A2 Phet Buri Thai.
63 B2 Phetchabun Thai.
63 B2 Phichit Thai.

139 D3 Philadelphia U.S.A.
136 C2 Philip U.S.A.
Philip Atoll atoll Micronesia
see Sorol
Philippeville Alg. see Skikda
100 B2 Philippeville Belgium
51 C2 Philippi, Lake imp. l. Austr.
100 A2 Philippine Neth.
156 C4 Philippine Basin
N. Pacific Ocean
Philippines country Asia
64 B2 Philippine Sea
N. Pacific Ocean
126 C2 Philip Smith Mountains
U.S.A.
122 B3 Philipstown S. Africa
53 C3 Phillip Island Austr.
137 D3 Phillipsburg U.S.A.
63 B2 Phimun Mangsahan Thai.
123 C2 Phiritona S. Africa
63 B2 Phitsanulok Thai.
63 B2 Phnom Penh Cambodia
Phnum Pénh Cambodia see
Phnom Penh
142 A2 Phoenix U.S.A.
49 J4 Phoenix Islands Kiribati
63 B2 Phon Thai.
62 B2 Phong Nha Vietnam
62 B1 Phôngsali Laos
62 B2 Phong Thô Vietnam
62 B2 Phônsavan Laos
62 B2 Phônsavan mt. Laos
62 B2 Phrae Thai.
Phu Cuong Vietnam see
Thu Dâu Môt
120 B3 Phuduhudu Botswana
63 A3 Phuket Thai.
63 B2 Phumĭ Kâmpóng Trâbêk
Cambodia
62 A2 Phumĭphon, Khuan Thai.
63 B2 Phumĭ Sâmraông
Cambodia
63 B2 Phu Quôc, Đao i. Vietnam
123 C2 Phuthaditjhaba S. Africa
Phu Vinh Vietnam see
Tra Vinh
62 A2 Phyu Myanmar
108 A1 Piacenza Italy
108 B2 Pianosa, Isola i. Italy
110 C1 Piatra Neamţ Romania
151 E3 Piauí r. Brazil
108 B1 Piave r. Italy
117 B4 Pibor r. S. Sudan
117 B4 Pibor Post S. Sudan
Picardie France see Picardy
104 C2 Picardy reg. France
140 C2 Picayune U.S.A.
152 B2 Pichanal Arg.
144 A2 Pichilingue Mex.
98 C2 Pickering U.K.
130 A2 Pickle Lake Can.
151 E3 Picos Brazil
153 B4 Pico Truncado Arg.
53 D2 Picton Austr.
54 B2 Picton N.Z.
73 C4 Pidurutalagala mt. Sri Lanka
154 C2 Piedade Brazil
145 C3 Piedras Negras Guat.
145 B2 Piedras Negras Mex.
93 I3 Pieksämäki Fin.
92 I3 Pielinen l. Fin.
136 C2 Pierre U.S.A.
105 C3 Pierrelatte France
123 D2 Pietermaritzburg S. Africa
Pietarsaari Fin. see
Jakobstad
Pietersburg S. Africa see
Polokwane
110 B1 Pietrosa mt. Romania
110 C1 Pietrosu, Vârful mt.
Romania
128 C2 Pigeon Lake Can.
137 F1 Pigeon River U.S.A.
153 B3 Pigüé Arg.
93 I3 Pihlajavesi l. Fin.
92 I3 Pihtipudas Fin.
145 C3 Pijijiapan Mex.
89 D2 Pikalevo Rus. Fed.
130 A2 Pikangikum Can.
136 C3 Pikes Peak U.S.A.
122 A3 Piketberg S. Africa
138 C3 Pikeville U.S.A.
103 D1 Piła Pol.
153 C3 Pilar Arg.
152 C2 Pilar Para.
75 C2 Pilibhit India
53 C2 Pilliga Austr.
154 C1 Pilões, Serra dos mts Brazil
150 C4 Pimenta Bueno Brazil

88 C3 Pina r. Belarus
153 C3 Pinamar Arg.
60 B1 Pinang i. Malaysia
80 B2 Pınarbaşı Turkey
146 B2 Pinar del Río Cuba
64 B2 Pinatubo, Mount vol. Phil.
103 E1 Pińczów Pol.
151 E3 Pindaré r. Brazil
Pindos mts Greece see
Pindus Mountains
150 A3 Pindus Mountains mts
Greece
140 B2 Pine Bluff U.S.A.
136 C2 Pine Bluffs U.S.A.
50 C1 Pine Creek Austr.
136 B2 Pinedale U.S.A.
86 D2 Pinega Rus. Fed.
129 D2 Pinehouse Lake Can.
111 B3 Pineios r. Greece
141 D3 Pine Islands FL U.S.A.
141 D4 Pine Islands FL U.S.A.
128 C1 Pine Point (abandoned)
Can.
136 C2 Pine Ridge U.S.A.
108 A2 Pinerolo Italy
Pines, Isle of i. Cuba see
Juventud, Isla de la
123 D2 Pinetown S. Africa
140 B2 Pineville U.S.A.
70 B2 Pingdingshan China
70 B2 Pingdu China
71 B3 Pingguo China
71 B3 Pingjiang China
70 A2 Pingliang China
70 B1 Pingquan China
71 C3 P'ingtung Taiwan
Pingxi China see Yuping
71 A3 Pingxiang Guangxi China
71 B3 Pingxiang Jiangxi China
70 B2 Pingyin China
155 C2 Pinhal Brazil
151 E3 Pinheiro Brazil
52 B3 Pinnaroo Austr.
101 D1 Pinneberg Ger.
Pinos, Isla de i. Cuba see
Juventud, Isla de la
145 C3 Pinotepa Nacional Mex.
48 H6 Pins, Île des i.
New Caledonia
88 C3 Pinsk Belarus
152 B2 Pinto Arg.
153 D3 Pioche U.S.A.
119 C3 Piodi Dem. Rep. Congo
108 B2 Piombino Italy
86 F2 Pionerskiy Rus. Fed.
103 E1 Pionki Pol.
103 D1 Piotrków Trybunalski Pol.
137 D2 Pipestone U.S.A.
131 C3 Pipmuacan, Réservoir resr
Can.
154 B2 Piquiri r. Brazil
154 C1 Piracanjuba Brazil
154 C2 Piracicaba Brazil
155 D1 Piracicaba r. Brazil
154 C2 Piraçununga Brazil
111 B3 Piraeus Greece
154 C2 Piraí do Sul Brazil
154 C2 Piraju Brazil
154 C2 Pirajuí Brazil
154 B1 Piranhas Brazil
151 F3 Piranhas r. Brazil
155 D1 Pirapora Brazil
154 C1 Pirenópolis Brazil
154 C1 Pires do Rio Brazil
Pirineos mts Europe see
Pyrenees
151 E3 Piripiri Brazil
109 D2 Pirot Serbia
59 C3 Piru Indon.
108 B2 Pisa Italy
152 A1 Pisagua Chile
150 B4 Pisco Peru
102 C2 Písek Czech Rep.
79 D2 Pishin Iran
74 A1 Pishin Pak.
145 D2 Pisté Mex.
109 C2 Pisticci Italy
108 B2 Pistoia Italy
106 C1 Pisuerga r. Spain
134 B2 Pit r. U.S.A.
114 A3 Pita Guinea
155 D1 Pitanga Brazil
155 D1 Pitangui Brazil
49 O6 Pitcairn Island Pitcairn Is
49 O6 Pitcairn Islands terr.
S. Pacific Ocean
92 H2 Piteå Sweden
92 H2 Piteälven r. Sweden
110 B2 Pitești Romania
75 C2 Pithoragarh India

142 A2 Pitiquito Mex.
86 C2 Pitkyaranta Rus. Fed.
96 C2 Pitlochry U.K.
128 B2 Pitt Island Can.
137 E3 Pittsburg U.S.A.
138 D2 Pittsburgh U.S.A.
139 E2 Pittsfield U.S.A.
53 D1 Pittsworth Austr.
155 C2 Piumhí Brazil
150 A3 Piura Peru
90 C2 Pivdennyy Buh r. Ukr.
131 E3 Placentia Can.
135 B3 Placerville U.S.A.
146 C2 Placetas Cuba
143 C2 Plains U.S.A.
143 C2 Plainview U.S.A.
60 B2 Plaju Indon.
61 C2 Plampang Indon.
154 C1 Planaltina Brazil
137 D2 Plankinton U.S.A.
143 D2 Plano U.S.A.
154 C2 Planura Brazil
140 B2 Plaquemine U.S.A.
106 B1 Plasencia Spain
109 C1 Plaški Croatia
147 C4 Plato Col.
137 D2 Platte r. U.S.A.
138 A2 Platteville U.S.A.
139 E2 Plattsburgh U.S.A.
101 F1 Plau Ger.
101 F2 Plauen Ger.
101 F1 Plauer See l. Ger.
89 E3 Plavsk Rus. Fed.
129 E2 Playgreen Lake Can.
63 B2 Plây Ku Vietnam
153 B3 Plaza Huincul Arg.
143 D3 Pleasanton U.S.A.
54 B2 Pleasant Point N.Z.
138 B3 Pleasure Ridge Park U.S.A.
104 C2 Pleaux France
130 B2 Pledger Lake Can.
54 C1 Plenty, Bay of g. N.Z.
136 C1 Plentywood U.S.A.
86 D2 Plesetsk Rus. Fed.
131 C2 Plétipi, Lac l. Can.
100 C2 Plettenberg Ger.
122 B3 Plettenberg Bay S. Africa
110 B2 Pleven Bulg.
Plevna Bulg. see Pleven
109 C2 Pljevlja Montenegro
108 A2 Ploaghe Italy
109 C2 Ploče Croatia
103 D1 Płock Pol.
109 C2 Pločno mt. Bos.-Herz.
104 B2 Ploemeur France
Ploești Romania see Ploiești
111 B2 Ploiești Romania
89 F2 Ploskoye Rus. Fed.
104 B2 Plouzané France
110 B2 Plovdiv Bulg.
Plozk Pol. see Płock
121 B3 Plumtree Zimbabwe
88 B3 Plungė Lith.
144 B2 Plutarco Elías Calles, Presa
resr Mex.
88 C3 Plyeshchanitsy Belarus
99 A4 Plymouth U.K.
138 B2 Plymouth U.S.A.
147 D3 Plymouth (abandoned)
Montserrat
99 B3 Plynlimon h. U.K.
88 C2 Plyussa Rus. Fed.
102 C2 Plzeň Czech Rep.
114 B3 Pô Burkina Faso
108 B1 Po r. Italy
61 C1 Po, Tanjung pt Malaysia
77 E2 Pobeda Peak China/Kyrg.
Pobedy, Pik mt. China/Kyrg.
see Pobeda Peak
140 B1 Pocahontas U.S.A.
134 D2 Pocatello U.S.A.
90 B1 Pochayiv Ukr.
89 D3 Pochep Rus. Fed.
89 D3 Pochinok Rus. Fed.
145 C3 Pochutla Mex.
139 D3 Pocomoke City U.S.A.
155 C2 Poços de Caldas Brazil
89 D2 Poddor'ye Rus. Fed.
91 D1 Podgorenskiy Rus. Fed.
109 C2 Podgorica Montenegro
82 G3 Podgornoye Rus. Fed.
83 H2 Podkamennaya Tunguska r.
Rus. Fed.
89 E2 Podol'sk Rus. Fed.
109 D2 Podujevë Kosovo
Podujevo Kosovo see
Podujevë
122 A3 Pofadder S. Africa
89 D3 Pogar Rus. Fed.
105 E3 Poggibonsi Italy

109 D2 Pogradec Albania
66 B2 Pogranichnyy Rus. Fed.
Po Hai g. China see Bo Hai
65 B2 Pohang S. Korea
48 G3 Pohnpei atoll Micronesia
90 B2 Pohrebyshche Ukr.
110 B2 Poiana Mare Romania
118 C3 Poie Dem. Rep. Congo
118 B3 Pointe-Noire Congo
126 A2 Point Hope U.S.A.
128 C1 Point Lake Can.
138 C3 Point Pleasant U.S.A.
104 C2 Poitiers France
104 C2 Poitou, Plaines et Seuil du
plain France
74 B2 Pokaran India
53 C1 Pokataroo Austr.
75 C2 Pokhara Nepal
119 C2 Poko Dem. Rep. Congo
83 J2 Pokrovsk Rus. Fed.
91 D2 Pokrovskoye Rus. Fed.
Pola Croatia see Pula
142 A1 Polacca U.S.A.
103 D1 Poland country Europe
55 C1 Polar Plateau Antarctica
80 B2 Polatlı Turkey
88 C2 Polatsk Belarus
61 C2 Polewali Indon.
118 B2 Poli Cameroon
102 C1 Police Pol.
109 C2 Policoro Italy
105 D2 Poligny France
64 B2 Polillo Islands Phil.
80 B2 Polis Cyprus
90 B1 Polis'ke Ukr.
109 C3 Polistena Italy
103 D1 Polkowice Pol.
107 D2 Pollença Spain
109 C3 Pollino, Monte mt. Italy
86 F2 Polnovat Rus. Fed.
91 D2 Polohy Ukr.
123 C1 Polokwane S. Africa
123 D1 Polokwane r. S. Africa
90 B1 Polonne Ukr.
134 D1 Polson U.S.A.
91 C2 Poltava Ukr.
66 B2 Poltavka Rus. Fed.
91 D2 Poltavskaya Rus. Fed.
88 C2 Põltsamaa Estonia
88 C2 Põlva Estonia
Polyanovgrad Bulg. see
Karnobat
92 J2 Polyarnyy Rus. Fed.
92 J2 Polyarnyye Zori Rus. Fed.
111 B2 Polygyros Greece
Polykastro Greece
156 E6 Polynesia is Pacific Ocean
106 B2 Pombal Port.
102 C1 Pomeranian Bay b. Ger./Pol.
108 B2 Pomezia Italy
92 I2 Pomokaira reg. Fin.
110 C2 Pomorie Bulg.
Pomorska, Zatoka b. Ger./
Pol. see Pomeranian Bay
155 D1 Pompéu Brazil
143 D1 Ponca City U.S.A.
147 D3 Ponce Puerto Rico
Pondicherry India see
Puducherry
127 G2 Pond Inlet Can.
Ponds Bay Can. see
Pond Inlet
106 B1 Ponferrada Spain
117 A4 Pongo watercourse S. Sudan
123 D2 Pongola r. S. Africa
123 D2 Pongolapoort Dam dam
S. Africa
154 B3 Ponta Grossa Brazil
154 C1 Pontalina Brazil
105 D2 Pont-à-Mousson France
154 A2 Ponta Porã Brazil
105 D2 Pontarlier France
102 B2 Pontcharra France
140 B2 Pontchartrain, Lake
U.S.A.
154 B1 Ponte de Pedra Brazil
106 B2 Ponte de Sor Port.
154 B1 Ponte do Rio Verde Brazil
98 C3 Pontefract U.K.
129 D3 Ponteix Can.
155 D2 Ponte Nova Brazil
150 D4 Pontes e Lacerda Brazil
106 B1 Pontevedra Spain
Ponthierville
Dem. Rep. Congo see
Ubundu
138 B2 Pontiac IL U.S.A.
138 C2 Pontiac MI U.S.A.
60 B2 Pontianak Indon.

Pontine Islands is Italy see
Ponziane, Isole
104 B2 Pontivy France
151 D2 Pontoetoe Suriname
104 C2 Pontoise France
129 E2 Ponton Can.
99 B4 Pontypool U.K.
99 B4 Pontypridd U.K.
108 B2 Ponziane, Isole is Italy
99 C4 Poole U.K.
Poona India see Pune
52 B2 Pooncarie Austr.
152 B1 Poopó, Lago de l. Bol.
150 B2 Popayán Col.
83 I2 Popigay r. Rus. Fed.
52 B2 Popiltah Austr.
129 E2 Poplar r. Can.
137 E3 Poplar Bluff U.S.A.
118 B3 Popokabaka
Dem. Rep. Congo
Popovichskaya Rus. Fed. see
Kalininskaya
110 C2 Popovo Bulg.
103 E2 Poprad Slovakia
151 E4 Porangatu Brazil
74 A2 Porbandar India
126 C2 Porcupine r. Can./U.S.A.
108 B1 Pordenone Italy
108 B1 Poreč Croatia
154 B2 Porecatu Brazil
114 C3 Porga Benin
93 H3 Pori Fin.
54 B2 Porirua N.Z.
88 C2 Porkhov Rus. Fed.
104 B2 Pornic France
83 K3 Poronaysk Rus. Fed.
75 D1 Porong China
111 B3 Poros Greece
92 I1 Porsangerfjorden sea chan.
Norway
93 E4 Porsgrunn Norway
111 D3 Porsuk r. Turkey
97 C1 Portadown U.K.
97 D1 Portaferry U.K.
138 B2 Portage U.S.A.
129 E3 Portage la Prairie Can.
128 B3 Port Alberni Can.
106 B2 Portalegre Port.
143 C2 Portales U.S.A.
128 A2 Port Alexander U.S.A.
128 B2 Port Alice U.S.A.
140 B2 Port Allen U.S.A.
134 B1 Port Angeles U.S.A.
97 C2 Portarlington Ireland
51 D4 Port Arthur Austr.
Port Arthur China see
Lüshunkou
143 E3 Port Arthur U.S.A.
96 A3 Port Askaig U.K.
52 A2 Port Augusta Austr.
147 C3 Port-au-Prince Haiti
131 E2 Port aux Choix Can.
122 B3 Port Beaufort S. Africa
73 D3 Port Blair India
52 B3 Port Campbell Austr.
131 D2 Port-Cartier Can.
54 B3 Port Chalmers N.Z.
141 D3 Port Charlotte U.S.A.
147 C3 Port-de-Paix Haiti
128 A2 Port Edward Can.
155 D1 Porteirinha Brazil
151 D3 Portel Brazil
130 B3 Port Elgin Can.
123 C3 Port Elizabeth S. Africa
96 A3 Port Ellen U.K.
98 A2 Port Erin Isle of Man
122 A3 Porterville S. Africa
135 C3 Porterville U.S.A.
Port Étienne Maur. see
Nouâdhibou
52 B3 Port Fairy Austr.
54 C1 Port Fitzroy N.Z.
Port Francqui
Dem. Rep. Congo see Ilebo
118 A3 Port-Gentil Gabon
115 C4 Port Harcourt Nigeria
128 B2 Port Hardy Can.
Port Harrison Can. see
Inukjuak
131 D3 Port Hawkesbury Can.
99 B4 Porthcawl U.K.
50 A2 Port Hedland Austr.
Port Herald Malawi see
Nsanje
99 A3 Porthmadog U.K.
131 E2 Port Hope Simpson Can.
138 C2 Port Huron U.S.A.
106 B2 Portimão Port.
Port Keats Austr. see
Wadeye

Port Láirge

Port Láirge Ireland see Waterford
53 C2 Portland N.S.W. Austr.
52 B3 Portland Vic. Austr.
139 E2 Portland ME U.S.A.
134 B1 Portland OR U.S.A.
143 D3 Portland TX U.S.A.
99 B4 Portland, Isle of pen. U.K.
128 A2 Portland Canal inlet Can.
97 C2 Portlaoise Ireland
143 D3 Port Lavaca U.S.A.
52 A2 Port Lincoln Austr.
114 A4 Port Loko Sierra Leone
113 I8 Port Louis Mauritius
Port-Lyautrey Morocco see Kenitra
52 B3 Port MacDonnell Austr.
53 D2 Port Macquarie Austr.
Portmadoc U.K. see Porthmadog
128 B2 Port McNeill Can.
131 D3 Port-Menier Can.
59 D3 Port Moresby P.N.G.
96 A3 Portnahaven U.K.
Port Nis U.K. see Port of Ness
122 A2 Port Nolloth S. Africa
Port-Nouveau-Québec Can. see Kangiqsualujjuaq
Porto Port. see Oporto
150 C3 Porto Acre Brazil
154 B2 Porto Alegre Brazil
152 C3 Porto Alegre Brazil
Porto Alexandre Angola see Tombua
Porto Amélia Moz. see Pemba
151 D4 Porto Artur Brazil
154 B2 Porto Camargo Brazil
151 D4 Porto dos Gaúchos Óbidos Brazil
150 D4 Porto Esperidião Brazil
108 B2 Portoferraio Italy
96 A1 Port of Ness U.K.
151 E3 Porto Franco Brazil
147 D3 Port of Spain Trin. and Tob.
108 B1 Portogruaro Italy
108 B2 Portomaggiore Italy
154 B2 Porto Mendes Brazil
152 C2 Porto Murtinho Brazil
151 E4 Porto Nacional Brazil
114 C4 Porto-Novo Benin
154 B2 Porto Primavera, Represa resr Brazil
134 B2 Port Orford U.S.A.
151 D3 Porto Santana Brazil
154 B2 Porto São José Brazil
108 A3 Portoscuso Italy
155 E1 Porto Seguro Brazil
108 B2 Porto Tolle Italy
108 A2 Porto Torres Italy
154 B3 Porto União Brazil
105 D3 Porto-Vecchio France
150 C3 Porto Velho Brazil
150 A3 Portoviejo Ecuador
96 A3 Portpatrick U.K.
52 B3 Port Phillip Bay Austr.
52 A2 Port Pirie Austr.
96 A2 Portree U.K.
128 B3 Port Renfrew Can.
97 C1 Portrush U.K.
116 B1 Port Said Egypt
141 C3 Port St Joe U.S.A.
123 C3 Port St Johns S. Africa
141 D3 Port St Lucie City U.S.A.
123 D3 Port Shepstone S. Africa
99 C4 Portsmouth U.K.
139 E2 Portsmouth NH U.S.A.
138 C3 Portsmouth OH U.S.A.
139 D3 Portsmouth VA U.S.A.
153 B5 Port Stephens Falkland Is
97 C1 Portstewart U.K.
116 B3 Port Sudan Sudan
140 C3 Port Sulphur U.S.A.
99 B4 Port Talbot U.K.
134 B1 Port Townsend U.S.A.
106 B2 Portugal country Europe
Portugália Angola see Chitato
Portuguese Guinea country Africa see Guinea-Bissau
Portuguese Timor country Asia see East Timor
Portuguese West Africa country Africa see Angola
97 B2 Portumna Ireland
105 C3 Port-Vendres France
48 H5 Port Vila Vanuatu
52 A2 Port Wakefield Austr.
50 B1 Port Warrender Austr.
153 A5 Porvenir Chile
93 I3 Porvoo Fin.
152 C2 Posadas Arg.
89 E2 Poshekhon'ye Rus. Fed.
Poshekhon'ye-Volodarsk Rus. Fed. see Poshekhon'ye
92 I2 Posio Fin.
61 D2 Poso Indon.
151 E4 Posse Brazil
101 E2 Pößneck Ger.
143 C2 Post U.S.A.
Poste-de-la-Baleine Can. see Kuujjuarapik
114 C2 Poste Weygand Alg.
122 B2 Postmasburg S. Africa
108 B1 Postojna Slovenia
Postysheve Ukr. see Krasnoarmiys'k
109 C2 Posušje Bos.-Herz.
123 C2 Potchefstroom S. Africa
143 E1 Poteau U.S.A.
109 C2 Potenza Italy
108 B2 Potenza r. Italy
151 E3 Poti r. Brazil
81 C1 Poti Georgia
155 E1 Potiraguá Brazil
115 D3 Potiskum Nigeria
139 D3 Potomac, South Branch r. U.S.A.
152 B1 Potosí Bol.
64 B2 Pototan Phil.
142 C3 Potrero del Llano Mex.
101 F1 Potsdam Ger.
139 E2 Potsdam U.S.A.
139 D2 Pottstown U.S.A.
139 D2 Pottsville U.S.A.
131 E3 Pouch Cove Can.
101 F3 Poughkeepsie U.S.A.
98 B3 Poulton-le-Fylde U.K.
155 C2 Pouso Alegre Brazil
63 B2 Poŭthĭsăt Cambodia
103 D2 Považská Bystrica Slovakia
109 C2 Povlen mt. Serbia
106 B1 Póvoa de Varzim Port.
136 B2 Powell U.S.A.
135 D3 Powell, Lake resr U.S.A.
128 B3 Powell River Can.
154 B1 Poxoréu Brazil
71 B3 Poyang China
71 B3 Poyang Hu l. China
109 D2 Požarevac Serbia
145 C2 Poza Rica Mex.
109 C1 Požega Croatia
109 D2 Požega Serbia
79 D2 Pozm Tīāb Iran
103 D1 Poznań Pol.
106 C2 Pozoblanco Spain
152 C2 Pozo Colorado Para.
109 B3 Pozzallo Italy
108 B2 Pozzuoli Italy
60 B2 Prabumulih Indon.
102 C2 Prachatice Czech Rep.
63 A2 Prachuap Khiri Khan Thai.
103 D1 Praděd mt. Czech Rep.
104 C3 Prades France
155 E1 Prado Brazil
102 C1 Prague Czech Rep.
Praha Czech Rep. see Prague
143 C2 Prairie Dog Town Fork r. U.S.A.
138 A2 Prairie du Chien U.S.A.
60 A1 Prapat Indon.
154 C1 Prata Brazil
108 B2 Prato Italy
137 D3 Pratt U.S.A.
140 C2 Prattville U.S.A.
61 C2 Praya Indon.
63 B2 Preăh Vihéar Cambodia
89 F2 Prechistoye Rus. Fed.
129 D2 Preeceville Can.
88 C2 Preiļi Latvia
53 C2 Premer Austr.
105 C2 Prémery France
101 F1 Premnitz Ger.
102 C1 Prenzlau Ger.
66 B2 Preobrazheniye Rus. Fed.
63 A2 Preparis Island Cocos Is
63 A2 Preparis North Channel Cocos Is
63 A2 Preparis South Channel Cocos Is
103 D2 Přerov Czech Rep.
142 A2 Prescott U.S.A.
142 A2 Prescott Valley U.S.A.
109 D2 Preševo Serbia
152 B2 Presidencia Roque Sáenz Peña Arg.
151 E3 Presidente Dutra Brazil
154 B2 Presidente Epitácio Brazil
64 B3 Presidente Manuel A Roxas Phil.
154 B1 Presidente Murtinho Brazil
155 C1 Presidente Olegário Brazil
154 B2 Presidente Prudente Brazil
142 C3 Presidio U.S.A.
103 E2 Prešov Slovakia
111 B2 Prespa, Lake Europe
139 F1 Presque Isle U.S.A.
Pressburg Slovakia see Bratislava
101 F2 Pressel Ger.
98 B3 Preston U.K.
134 D2 Preston U.S.A.
96 B3 Prestwick U.K.
155 C1 Preto r. Brazil
123 C2 Pretoria S. Africa
Pretoria-Witwatersrand-Vereeniging prov. S. Africa see Gauteng
111 B3 Preveza Greece
63 B2 Prey Vêng Cambodia
124 A4 Pribilof Islands U.S.A.
109 C2 Priboj Serbia
135 D3 Price U.S.A.
140 C2 Prichard U.S.A.
106 C2 Priego de Córdoba Spain
88 B3 Prienai Lith.
122 B2 Prieska S. Africa
103 D2 Prievidza Slovakia
109 C2 Prijedor Bos.-Herz.
109 C2 Prijepolje Serbia
Prikaspiyskaya Nizmennost' lowland Kazakh./Rus. Fed. see Caspian Lowland
111 B2 Prilep Macedonia
91 D2 Primorsko-Akhtarsk Rus. Fed.
129 D2 Primrose Lake Can.
129 D2 Prince Albert Can.
122 B3 Prince Albert S. Africa
126 E2 Prince Albert Peninsula Can.
122 B3 Prince Albert Road S. Africa
126 D2 Prince Alfred, Cape Can.
122 A3 Prince Alfred Hamlet S. Africa
127 G2 Prince Charles Island Can.
55 H2 Prince Charles Mountains Antarctica
131 D3 Prince Edward Island prov. Can.
128 B2 Prince George Can.
51 D1 Prince of Wales Island Austr.
126 F2 Prince of Wales Island Can.
128 A2 Prince of Wales Island U.S.A.
126 E2 Prince of Wales Strait Can.
126 E1 Prince Patrick Island Can.
126 F2 Prince Regent Inlet sea chan. Can.
128 A2 Prince Rupert Can.
51 D1 Princess Charlotte Bay Austr.
55 H2 Princess Elizabeth Land reg. Antarctica
128 B2 Princess Royal Island Can.
128 B3 Princeton Can.
138 B3 Princeton IN U.S.A.
137 E2 Princeton MO U.S.A.
134 B2 Prineville U.S.A.
86 C2 Priozersk Rus. Fed.
Pripet r. Belarus see Prypyats'
90 A1 Pripet Marshes Belarus/Ukr.
109 D2 Prishtinë Kosovo
Priština Kosovo see Prishtinë
101 F1 Pritzwalk Ger.
105 C3 Privas France
109 C2 Privlaka Croatia
89 F2 Privolzhsk Rus. Fed.
109 D2 Prizren Kosovo
87 D4 Proletarskoye Vodokhranilishche l. Rus. Fed.
Prome Myanmar see Pyè
154 C2 Promissão Brazil
154 C2 Promissão, Represa resr Brazil
126 D3 Prophet r. Can.
128 B2 Prophet River Can.
51 D2 Proserpine Austr.
Proskurov Ukr. see Khmel'nyts'kyy
103 D2 Prostějov Czech Rep.
89 E3 Protvino Rus. Fed.
110 C2 Provadiya Bulg.
105 D3 Provence reg. France
139 E2 Providence U.S.A.
146 B3 Providencia, Isla de i. Caribbean Sea
83 N2 Provideniya Rus. Fed.
105 C2 Provins France
135 D2 Provo U.S.A.
129 C2 Provost Can.
91 C2 Prubiynyy, Mys pt Ukr.
154 B3 Prudentópolis Brazil
126 C2 Prudhoe Bay U.S.A.
100 C2 Prüm Ger.
105 D3 Prunelli-di-Fiumorbo France
103 D1 Pruszcz Gdański Pol.
103 E1 Pruszków Pol.
110 C1 Prut r. Europe
88 B3 Pruzhany Belarus
91 D2 Pryazovs'ke Ukr.
91 C1 Pryluky Ukr.
91 D2 Prymors'k Ukr.
91 D2 Prymors'kyy Ukr.
143 D1 Pryor U.S.A.
90 B1 Prypyats' r. Belarus/Ukr.
103 E2 Przemyśl Pol.
103 E1 Przeworsk Pol.
Przheval'sk Kyrg. see Karakol
111 C3 Psara i. Greece
81 C1 Psebay Rus. Fed.
91 D3 Pshekha r. Rus. Fed.
88 C2 Pskov Rus. Fed.
88 C2 Pskov, Lake Estonia/Rus. Fed.
111 B2 Ptolemaïda Greece
109 C1 Ptuj Slovenia
150 B3 Pucallpa Peru
71 B3 Pucheng China
103 D1 Puck Pol.
92 I2 Pudasjärvi Fin.
70 C2 Pudong China
86 C2 Pudozh Rus. Fed.
Puducherri India see Puducherry
145 B3 Puducherry India
145 C3 Puebla Mex.
136 C3 Pueblo U.S.A.
153 B3 Puelén Arg.
153 A3 Puente Alto Chile
106 C2 Puente Genil Spain
152 A4 Puerto Aisén Chile
152 B2 Puerto Alegre Bol.
145 C3 Puerto Ángel Mex.
146 B4 Puerto Armuelles Panama
150 C2 Puerto Ayacucho Venez.
146 B3 Puerto Barrios Guat.
146 B3 Puerto Cabezas Nic.
153 A4 Puerto Cisnes Chile
144 A2 Puerto Cortés Mex.
145 C3 Puerto Escondido Mex.
152 B1 Puerto Frey Bol.
150 C2 Puerto Inírida Col.
152 C1 Puerto Isabel Bol.
150 B3 Puerto Leguizamo Col.
146 B3 Puerto Lempira Hond.
144 A2 Puerto Libertad Mex.
146 B3 Puerto Limón Costa Rica
106 C2 Puertollano Spain
153 B4 Puerto Madryn Arg.
150 C4 Puerto Maldonado Peru
Puerto México Mex. see Coatzacoalcos
147 D4 Puerto Miranda Venez.
153 A4 Puerto Montt Chile
153 A5 Puerto Natales Chile
150 C2 Puerto Nuevo Col.
150 C2 Puerto Páez Venez.
144 A1 Puerto Peñasco Mex.
152 C2 Puerto Pinasco Para.
147 C3 Puerto Plata Dom. Rep.
150 B3 Puerto Portillo Peru
Puerto Presidente Stroessner Para. see Ciudad del Este
64 A3 Puerto Princesa Phil.
154 A3 Puerto Rico Arg.
147 D3 Puerto Rico terr. West Indies
158 C3 Puerto Rico Trench Caribbean Sea
146 A3 Puerto San José Guat.
153 B5 Puerto Santa Cruz Arg.

152 C2 Puerto Sastre Para.
144 B2 Puerto Vallarta Mex.
87 D3 Pugachev Rus. Fed.
74 B2 Pugal India
54 B2 Pukaki, Lake N.Z.
49 K5 Pukapuka atoll Cook Is
129 D2 Pukatawagan Can.
65 B1 Pukchin N. Korea
65 B1 Pukch'ŏng N. Korea
109 C2 Pukë Albania
65 B1 Pukekohe N.Z.
65 B1 Puksubaek-san mt. N. Korea
Pula China see Nyingchi
108 B2 Pula Croatia
108 A3 Pula Italy
152 B2 Pulacayo Bol.
103 E1 Puławy Pol.
77 C3 Pul-e Khumrī Afgh.
92 I3 Pulkkila Fin.
134 C1 Pullman U.S.A.
64 B2 Pulog, Mount Phil.
64 B3 Pulutan Indon.
150 A3 Puná, Isla i. Ecuador
54 B2 Punakaiki N.Z.
123 D1 Punda Maria S. Africa
73 B3 Pune India
65 B1 P'ungsan N. Korea
121 C2 Púngué r. Moz.
119 C3 Punia Dem. Rep. Congo
74 B1 Punjab state India
153 B3 Punta Alta Arg.
153 A5 Punta Arenas Chile
153 C3 Punta del Este Uru.
146 B3 Punta Gorda Belize
146 B3 Puntarenas Costa Rica
117 C4 Puntland reg. Somalia
150 B1 Punto Fijo Venez.
92 I3 Puolanka Fin.
Puqi China see Chibi
82 G2 Pur r. Rus. Fed.
75 C3 Puri India
100 B1 Purmerend Neth.
Purnea India see Purnia
75 C2 Purnia India
150 C3 Purus r. Brazil/Peru
61 B2 Purwakarta Indon.
61 C2 Purwodadi Indon.
65 B1 Puryŏng N. Korea
74 B3 Pusad India
89 E3 Pushchino Rus. Fed.
Pushkin Azer. see Biläsuvar
89 E2 Pushkino Rus. Fed.
88 C2 Pushkinskiye Gory Rus. Fed.
88 C2 Pustoshka Rus. Fed.
62 A1 Pusur Bangl.
71 B3 Putian China
Puting China see De'an
61 C2 Puting, Tanjung pt Indon.
101 F1 Putlitz Ger.
60 B1 Putrajaya Malaysia
122 B2 Putsonderwater S. Africa
102 C1 Puttgarden Ger.
150 B3 Putumayo r. Col.
61 C1 Putusibau Indon.
90 B2 Putyla Ukr.
91 C1 Putyvl' Ukr.
93 I3 Puula i. Fin.
130 C1 Puvirnituq Can.
70 B2 Puyang China
104 C3 Puylaurens France
119 C3 Pweto Dem. Rep. Congo
99 A3 Pwllheli U.K.
92 J2 Pyaozerskiy Rus. Fed.
63 A2 Pyapon Myanmar
63 G2 Pyasina r. Rus. Fed.
87 D4 Pyatigorsk Rus. Fed.
91 C2 P''yatykhatky Ukr.
62 A2 Pyè Myanmar
65 B2 Pyeongtaek S. Korea
88 C3 Pyetrykaw Belarus
93 H3 Pyhäjärvi l. Fin.
92 H3 Pyhäjoki r. Fin.
92 I3 Pyhäsalmi Fin.
62 A1 Pyingaing Myanmar
62 A2 Pyinmana Myanmar
62 A1 Pyin-U-Lwin Myanmar
65 B2 Pyŏksŏng N. Korea
65 B2 P'yŏnggang N. Korea
65 B2 P'yŏngsan N. Korea
65 B2 P'yŏngsong N. Korea
65 B2 P'yŏngyang N. Korea
135 C2 Pyramid Lake U.S.A.
80 B3 Pyramids of Giza tourist site Egypt
107 D1 Pyrenees mts Europe
111 B3 Pyrgetos Greece
111 C3 Pyrgi Greece

111 B3 Pyrgos Greece
91 C1 Pyryatyn Ukr.
103 C1 Pyrzyce Pol.
88 C2 Pytalovo Rus. Fed.
111 B3 Pyxaria mt. Greece

Q

Qaanaaq Greenland see Thule
Qabqa China see Gonghe
123 C3 Qacha's Nek Lesotho
76 B3 Qā'en Iran
Qahremānshahr Iran see Kermānshāh
68 C2 Qaidam Pendi basin China
76 C3 Qal'ah-ye Now Afgh.
79 C2 Qalamat Abū Shafrah Saudi Arabia
77 C3 Qalāt Afgh.
78 A2 Qal'at al Azlam Saudi Arabia
78 A2 Qal'at al Mu'azzam Saudi Arabia
78 B2 Qal'at Bīshah Saudi Arabia
129 E1 Qamanirjuaq Lake Can.
Qamanittuaq Can. see Baker Lake
79 C3 Qamar, Ghubbat al b. Yemen
68 C2 Qamdo China
78 B3 Qam Hadīl Saudi Arabia
127 I2 Qaqortoq Greenland
80 A3 Qārah Egypt
Qarkilik China see Ruoqiang
77 C3 Qarqan China see Qiemo
78 B2 Qaryat al Ulyā Saudi Arabia
127 I2 Qasigiannguit Greenland
116 A2 Qaşr al Farāfirah Egypt
79 D2 Qaşr-e Qand Iran
81 C2 Qaşr-e Shīrīn Iran
127 I2 Qassimiut Greenland
78 B3 Qa'tabah Yemen
79 C2 Qatar country Asia
116 A2 Qattâra Depression Egypt
81 C1 Qazax Azer.
81 C2 Qazvīn Iran
Qena Egypt see Qinā
127 I2 Qeqertarsuaq Greenland
127 I2 Qeqertarsuaq i. Greenland
127 I2 Qeqertarsuatsiaat Greenland
127 I2 Qeqertarsuup Tunua b. Greenland
79 C2 Qeshm Iran
74 A1 Qeysār, Kūh-e mt. Afgh.
70 B3 Qiandao Hu resr China
70 A3 Qianjiang Chongqing China
70 B2 Qianjiang Hubei China
69 E1 Qianjin China
65 A1 Qian Shan mts China
70 B2 Qianshangjie China
71 A3 Qianxi China
70 C2 Qidong China
77 E3 Qiemo China
71 A3 Qijiang China
68 C2 Qijiaojing China
127 H2 Qikiqtarjuaq Can.
74 A2 Qila Ladgasht Pak.
68 C2 Qilian Shan mts China
127 J2 Qillak i. Greenland
70 B3 Qimen China
127 H1 Qimusseriarsuaq b. Greenland
116 B2 Qinā Egypt
72 C2 Qincheng China see Nanfeng
70 A2 Qingcheng China
70 C2 Qingdao China
68 C2 Qinghai Hu salt l. China
68 C2 Qinghai Nanshan mts China
Qingjiang China see Huai'an
Qingjiang China see Zhangshu
70 B2 Qingshuihe China
70 A2 Qingtongxia China
70 A2 Qingyang China
71 B3 Qingyuan Guangdong China
Qingyuan China see Yizhou
65 A1 Qingyuan Liaoning China

Qingzang Gaoyuan plat. China see Tibet, Plateau of
70 B2 Qingzhou China
70 B2 Qinhuangdao China
Qinjiang China see Shicheng
70 A2 Qin Ling mts China
70 B2 Qinyang China
71 A3 Qinzhou China
71 B4 Qionghai China
70 A2 Qionglai Shan mts China
71 B4 Qiongshan China
71 A4 Qiongzhong China
69 E1 Qiqihar China
81 D3 Qīr Iran
Qishan China see Qimen
79 C3 Qishn Yemen
66 B1 Qitaihe China
70 B2 Qixian Henan China
70 B2 Qixian Shanxi China
Qogir Feng mt. China/Pakistan see K2
81 D2 Qom Iran
Qomishēh Iran see Shāhrezā
Qomolangma Feng mt. China/Nepal see Everest, Mount
76 B2 Qo'ng'irot Uzbek.
77 D2 Qo'qon Uzbek.
76 B2 Qoraqalpog'iston Uzbek.
80 B2 Qornet es Saouda mt. Lebanon
81 C2 Qorveh Iran
79 C2 Qotbābād Iran
101 C1 Quakenbrück Ger.
53 C2 Quambone Austr.
63 B2 Quang Ngai Vietnam
63 B2 Quang Tri Vietnam
62 B1 Quan Hoa Vietnam
Quan Long Vietnam see Ca Mau
Quan Phu Quoc i. Vietnam see Phu Quốc, Đao
71 B3 Quanzhou Fujian China
71 B3 Quanzhou Guangxi China
108 A3 Quartu Sant'Elena Italy
142 A2 Quartzsite U.S.A.
81 C1 Quba Azer.
76 B3 Qūchān Iran
53 C3 Queanbeyan Austr.
131 C3 Québec Can.
131 C2 Québec prov. Can.
101 E2 Quedlinburg Ger.
Queen Adelaide Islands is Chile see Reina Adelaida, Archipiélago de la
128 C2 Queen Charlotte Can.
128 B2 Queen Charlotte Sound sea chan. Can.
128 B2 Queen Charlotte Strait Can.
126 E1 Queen Elizabeth Islands Can.
55 I2 Queen Mary Land reg. Antarctica
126 F2 Queen Maud Gulf Can.
55 E2 Queen Maud Land reg. Antarctica
55 P1 Queen Maud Mountains Antarctica
52 B3 Queenscliff Austr.
52 B1 Queensland state Austr.
51 D4 Queenstown Austr.
54 A3 Queenstown N.Z.
Queenstown Ireland see Cobh
123 C3 Queenstown S. Africa
121 C2 Quelimane Moz.
153 A4 Quellón Chile
Quelpart Island i. S. Korea see Jeju-do
142 B2 Quemado U.S.A.
Que Que Zimbabwe see Kwekwe
154 B2 Queréncia do Norte Brazil
145 B2 Querétaro Mex.
101 E2 Querfurt Ger.
128 B2 Quesnel Can.
128 B2 Quesnel Lake Can.
74 A1 Quetta Pak.
146 A3 Quetzaltenango Guat.
64 B2 Quezon Phil.
64 B1 Quezon City Phil.
120 A2 Quibala Angola
150 B2 Quibdó Col.
104 B2 Quiberon France
120 A2 Quilengues Angola

104 C3 Quillan France
153 C3 Quilmes Arg.
Quilon India see Kollam
51 D2 Quilpie Austr.
153 A3 Quilpué Chile
120 A1 Quimbele Angola
152 B2 Quimili Arg.
104 B2 Quimper France
104 B2 Quimperlé France
135 B3 Quincy CA U.S.A.
141 D2 Quincy FL U.S.A.
137 E3 Quincy IL U.S.A.
139 E2 Quincy MA U.S.A.
107 C1 Quinto Spain
121 D2 Quionga Moz.
120 A2 Quirima Angola
53 D2 Quirindi Austr.
154 B1 Quirinópolis Brazil
131 D3 Quispamsis Can.
120 A2 Quitapa Angola
150 B3 Quito Ecuador
151 F3 Quixadá Brazil
71 A3 Qujing China
75 D1 Qumar He r. China
123 C3 Qumrha S. Africa
115 E1 Qunayyin, Sabkhat al salt marsh Libya
129 E1 Quoich r. Can.
52 A2 Quorn Austr.
79 C2 Qurayat Oman
77 C3 Qŭrghonteppa Tajik.
Quxar China see Lhazê
70 C2 Quyang China see Jingzhou
63 B2 Quy Nhơn Vietnam
71 B3 Quzhou China
Qyteti Stalin Albania see Kuçovë
Qyzyltū Kazakh. see Kishkenekol'

R

103 D2 Raab r. Austria
92 H3 Raahe Fin.
100 C1 Raalte Neth.
61 C2 Raas i. Indon.
61 C2 Raba Indon.
114 B1 Rabat Morocco
59 E3 Rabaul P.N.G.
50 C2 Rabbit Flat Austr.
78 A2 Rābigh Saudi Arabia
103 D2 Råbnița Moldova see Rîbnița
Rabyānah, Ramlat des. Libya see Rebiana Sand Sea
131 E3 Race, Cape Can.
140 B3 Raceland U.S.A.
139 E2 Race Point U.S.A.
63 B3 Rach Gia Vietnam
103 D1 Racibórz Pol.
138 B2 Racine U.S.A.
78 B3 Radā' Yemen
110 C1 Rădăuți Romania
138 B3 Radcliff U.S.A.
74 B2 Radhanpur India
130 C2 Radisson Can.
103 E1 Radom Pol.
103 D1 Radomsko Pol.
90 B1 Radomyshl' Ukr.
111 B2 Radoviš Macedonia
88 B2 Radviliškis Lith.
78 A2 Radwá, Jabal mt. Saudi Arabia
90 B1 Radyvyliv Ukr.
75 C2 Rae Bareli India
100 C2 Raeren Belgium
54 C1 Raethi N.Z.
78 A2 Rāf h. Saudi Arabia
152 B3 Rafaela Arg.
118 C2 Rafaï C.A.R.
78 B2 Rafḥa' Saudi Arabia
79 C1 Rafsanjān Iran
119 C2 Raga S. Sudan
64 B3 Ragang, Mount vol. Phil.
109 B3 Ragusa Italy
61 D2 Raha Indon.
88 D3 Rahachow Belarus
78 A3 Rahad r. Sudan
74 B2 Rahimyar Khan Pak.
109 D2 Rahovec Kosovo
73 C2 Raichur India
75 C2 Raigarh India
128 C2 Rainbow Lake Can.
134 B1 Rainier, Mount vol. U.S.A.
130 A2 Rainy Lake Can./U.S.A.
129 E3 Rainy River Can.
75 C2 Raipur India

93 H3 Raisio Fin.
88 C2 Raja Estonia
73 C3 Rajahmundry India
61 C1 Rajang *r.* Malaysia
74 B2 Rajanpur Pak.
73 B4 Rajapalayam India
74 B2 Rajasthan *state* India
74 B2 Rajasthan Canal *canal* India
74 B2 Rajgarh India
60 B2 Rajik Indon.
74 B2 Rajkot India
75 C2 Raj Nandgaon India
74 B2 Rajpur India
75 C2 Rajshahi Bangl.
54 B2 Rakaia *r.* N.Z.
74 B1 Rakaposhi *mt.* Pak.
90 A2 Rakhiv Ukr.
91 D1 Rakitnoye Rus. Fed.
88 C2 Rakke Estonia
88 C2 Rakvere Estonia
141 E1 Raleigh U.S.A.
141 E2 Raleigh Bay U.S.A.
48 H3 Ralik Chain *is* Marshall Is
75 C2 Ramanuj Ganj India
123 C2 Ramatlabama S. Africa
104 C2 Rambouillet France
119 D2 Ramciel S. Sudan
99 A4 Rame Head *hd* U.K.
97 C1 Ramelton Ireland
89 E2 Rameshki Rus. Fed.
74 B2 Ramgarh India
81 C2 Rämhormoz Iran
110 C1 Râmnicu Sărat Romania
110 B1 Râmnicu Vâlcea Romania
89 E3 Ramon' Rus. Fed.
135 C4 Ramona U.S.A.
123 C1 Ramotswa Botswana
75 B2 Rampur India
62 A2 Ramree Island Myanmar
98 A2 Ramsey Isle of Man
130 B3 Ramsey Lake Can.
99 D4 Ramsgate U.K.
81 C2 Rämshir Iran
62 A1 Ramsing India
75 C2 Ranaghat India
61 C1 Ranau Malaysia
153 A3 Rancagua Chile
154 B2 Rancharia Brazil
75 C2 Ranchi India
93 F4 Randers Denmark
63 B3 Rangae Thai.
75 D2 Rangapara India
54 B2 Rangiora N.Z.
49 M5 Rangiroa *atoll* Fr. Polynesia
54 C1 Rangitaiki *r.* N.Z.
60 B2 Rangkasbitung Indon.
63 A2 Rangoon Myanmar
75 C2 Rangpur Bangl.
129 E1 Rankin Inlet Can.
53 C2 Rankin's Springs Austr.
Rankovićevo Serbia *see* Kraljevo
96 B2 Rannoch Moor *moorland* U.K.
63 A3 Ranong Thai.
59 C3 Ransiki Indon.
61 C2 Rantaupanjang Indon.
60 A1 Rantauprapat Indon.
61 C2 Rantepang Indon.
92 I2 Ranua Fin.
93 E4 Ranum Denmark
78 B2 Ranyah, Wādī *watercourse* Saudi Arabia
49 J6 Raoul Island Kermadec Is
49 M6 Rapa *i.* Fr. Polynesia
108 A2 Rapallo Italy
136 C2 Rapid City U.S.A.
88 B2 Rapla Estonia
74 A2 Rapur India
49 L6 Rarotonga *i.* Cook Is
153 B4 Rasa, Punta *pt* Arg.
Ras al Khaimah U.A.E. *see* Ra's al Khaymah
79 C2 Ra's al Khaymah U.A.E.
116 B3 Ras Dejen *mt.* Eth.
88 B2 Raseiniai Lith.
116 B2 Ra's Ghārib Egypt
81 C2 Rasht Iran
74 A2 Ras Koh *mt.* Pak.
110 C1 Râşnov Romania
88 C2 Rasony Belarus
79 C2 Ra's Şirāb Oman
108 B3 Rass Jebel Tunisia
87 D3 Rasskazovo Rus. Fed.
79 C2 Ras Tannūrah Saudi Arabia
101 D1 Rastede Ger.
48 I2 Ratak Chain *is* Marshall Is
93 F3 Rätan Sweden
123 C2 Ratanda S. Africa
74 B2 Ratangarh India

63 A2 Rat Buri Thai.
62 A1 Rathedaung Myanmar
101 F1 Rathenow Ger.
97 C2 Rathfriland U.K.
97 C1 Rathlin Island U.K.
Rathluirc Ireland *see* Charleville
100 C2 Ratingen Ger.
74 B2 Ratlam India
73 B3 Ratnagiri India
73 C4 Ratnapura Sri Lanka
90 A1 Ratne Ukr.
142 C1 Raton U.S.A.
96 D2 Rattray Head *hd* U.K.
93 G3 Rättvik Sweden
101 E1 Ratzeburg Ger.
78 B2 Raudhatain Kuwait
92 □B2 Raufarhöfn Iceland
54 C1 Raukumara Range *mts* N.Z.
93 H3 Rauma Fin.
88 C2 Rauna Latvia
61 C2 Raung, Gunung *vol.* Indon.
66 D2 Raurkela India
90 B2 Răut *r.* Moldova
134 D1 Ravalli U.S.A.
81 C2 Ravänsar Iran
108 B2 Ravenna Italy
102 B2 Ravensburg Ger.
50 B3 Ravensthorpe Austr.
74 B1 Ravi *r.* Pak.
74 B1 Rawalpindi Pak.
103 D1 Rawicz Pol.
50 B3 Rawlinna Austr.
136 B2 Rawlins U.S.A.
153 B4 Rawson Arg.
75 C3 Rayagada India
69 E1 Raychikhinsk Rus. Fed.
78 B3 Raydah Yemen
87 E3 Rayevskiy Rus. Fed.
134 B1 Raymond U.S.A.
53 D2 Raymond Terrace Austr.
143 D3 Raymondville U.S.A.
129 D2 Raymore Can.
145 C2 Rayón Mex.
63 B2 Rayong Thai.
78 A2 Rayyis Saudi Arabia
104 B2 Raz, Pointe du *pt* France
81 C2 Razāzah, Buḩayrat ar *l.* Iraq
110 C2 Razgrad Bulg.
110 C2 Razim, Lacul *lag.* Romania
111 B2 Razlog Bulg.
Raz"yezd 3km Rus. Fed. *see* Novyy Urgal
104 B2 Ré, Île de *i.* France
99 C4 Reading U.K.
138 C3 Reading OH U.S.A.
139 D2 Reading PA U.S.A.
123 C2 Reagile S. Africa
115 E2 Rebiana Sand Sea *des.* Libya
66 D1 Rebun-tō *i.* Japan
50 B3 Recherche, Archipelago of the *is* Austr.
89 D3 Rechytsa Belarus
151 F3 Recife Brazil
123 C3 Recife, Cape S. Africa
100 C2 Recklinghausen Ger.
152 C2 Reconquista Arg.
137 D1 Red *r.* Can./U.S.A.
140 B2 Red *r.* U.S.A.
141 C1 Redange Lux.
141 C1 Red Bank U.S.A.
Red Basin *basin* China *see* Sichuan Pendi
131 E2 Red Bay Can.
135 B2 Red Bluff U.S.A.
98 C2 Redcar U.K.
129 C2 Redcliff Can.
121 B2 Redcliffe Zimbabwe
52 B2 Red Cliffs Austr.
128 C2 Red Deer Can.
126 E3 Red Deer *r.* Can.
129 D2 Red Deer Lake Can.
135 B2 Redding U.S.A.
99 C3 Redditch U.K.
137 D2 Redfield U.S.A.
137 D3 Red Hills U.S.A.
130 A2 Red Lake Can.
130 A2 Red Lake *l.* Can.
137 E1 Red Lakes U.S.A.
134 E1 Red Lodge U.S.A.
135 B2 Redmond U.S.A.
137 D2 Red Oak U.S.A.
104 B2 Redon France
106 B1 Redondela Spain
106 B2 Redondo Port.
78 A2 Red Sea Africa/Asia
128 B1 Redstone *r.* Can.
100 B1 Reduzum Neth.
128 C2 Redwater Can.

137 E2 Red Wing U.S.A.
137 D2 Redwood Falls U.S.A.
97 C2 Ree, Lough *l.* Ireland
134 B2 Reedsport U.S.A.
54 B2 Reefton N.Z.
143 D3 Refugio U.S.A.
102 C2 Regen Ger.
155 E1 Regência Brazil
102 C2 Regensburg Ger.
114 C2 Reggane Alg.
109 C3 Reggio di Calabria Italy
108 B2 Reggio nell'Emilia Italy
110 B1 Reghin Romania
129 D2 Regina Can.
154 C2 Registro Brazil
122 A1 Rehoboth Namibia
80 B2 Rehovot Israel
101 F2 Reichenbach Ger.
141 E1 Reidsville U.S.A.
99 C4 Reigate U.K.
105 C2 Reims France
153 A5 Reina Adelaida, Archipiélago de la *is* Chile
101 E1 Reinbek Ger.
129 D2 Reindeer *r.* Can.
129 E2 Reindeer Island Can.
129 D2 Reindeer Lake Can.
92 F2 Reine Norway
106 C1 Reinosa Spain
100 C3 Reinsfeld Ger.
123 C2 Reitz S. Africa
122 B2 Reivilo S. Africa
129 D1 Reliance Can.
107 D2 Relizane Alg.
52 A2 Remarkable, Mount *h.* Austr.
79 C2 Remeshk Iran
105 C2 Remiremont France
100 C2 Remscheid Ger.
102 B1 Rendsburg Ger.
139 D1 Renfrew Can.
60 B2 Rengat Indon.
90 B2 Reni Ukr.
52 B2 Renmark Austr.
48 H5 Rennell *i.* Solomon Is
101 D2 Rennerod Ger.
104 B2 Rennes France
129 D1 Rennie Lake Can.
108 B2 Reno *r.* Italy
135 C3 Reno U.S.A.
70 A3 Renshou China
138 B2 Rensselaer U.S.A.
75 C2 Renukut India
54 B2 Renwick N.Z.
61 D2 Reo Indon.
136 D3 Republican *r.* U.S.A.
127 G2 Repulse Bay Can.
150 B3 Requena Peru
107 C2 Requena Spain
60 B2 Resag, Gunung *mt.* Indon.
111 B2 Resen Macedonia
154 B2 Reserva Brazil
91 C2 Reshetylivka Ukr.
152 C2 Resistencia Arg.
110 B1 Reşiţa Romania
126 F2 Resolute Can.
127 H2 Resolution Island Can.
155 D1 Resplendor Brazil
145 C3 Retalhuleu Guat.
98 C3 Retford U.K.
105 C2 Rethel France
111 B3 Rethymno Greece
113 I8 Réunion *terr.* Indian Ocean
107 D1 Reus Spain
Reut *r.* Moldova *see* Răut
102 B2 Reutlingen Ger.
86 C2 Revda Rus. Fed.
128 C2 Revelstoke Can.
144 A3 Revillagigedo, Islas *is* Mex.
128 A2 Revillagigedo Island U.S.A.
75 C2 Rewa India
134 D2 Rexburg U.S.A.
135 B3 Reyes, Point U.S.A.
160 M4 Reykjanes Ridge N. Atlantic Ocean
92 □A3 Reykjanestá *pt* Iceland
92 □A3 Reykjavik Iceland
145 C2 Reynosa Mex.
Reza'īyeh Iran *see* Urmia
Reza'īyeh, Daryächeh-ye *salt l.* Iran *see* Urmia, Lake
88 C2 Rēzekne Latvia
Rheims France *see* Reims
Rhein *r.* Ger. *see* Rhine
100 C1 Rheine Ger.
101 F1 Rheinsberg Ger.
100 C2 Rhin *r.* France *see* Rhine
Rhine *r.* Europe
138 B1 Rhinelander U.S.A.
101 F1 Rhinluch *marsh* Ger.

101 F1 Rhinow Ger.
108 A1 Rho Italy
139 E2 Rhode Island *state* U.S.A.
111 C3 Rhodes Greece
111 C3 Rhodes *i.* Greece
Rhodesia *country* Africa *see* Zimbabwe
111 B2 Rhodope Mountains Bulg./Greece
105 C3 Rhône *r.* France/Switz.
Rhum U.K. *see* Rum
Rhuthun U.K. *see* Ruthin
98 B3 Rhyl U.K.
155 D1 Riacho dos Machados Brazil
154 C1 Rialma Brazil
154 C1 Rianópolis Brazil
60 B1 Riau, Kepulauan *is* Indon.
106 B1 Ribadeo Spain
106 B1 Ribadesella Spain
154 B2 Ribas do Rio Pardo Brazil
121 C2 Ribáuè Moz.
98 B3 Ribble *r.* U.K.
93 E4 Ribe Denmark
154 C2 Ribeira *r.* Brazil
154 C2 Ribeirão Preto Brazil
104 C2 Ribérac France
152 B1 Riberalta Bol.
90 B2 Ribniţa Moldova
102 C1 Ribnitz-Damgarten Ger.
138 A1 Rice Lake U.S.A.
123 D2 Richards Bay S. Africa
143 D2 Richardson U.S.A.
126 C2 Richardson Mountains Can.
135 D3 Richfield U.S.A.
134 C1 Richland U.S.A.
138 A2 Richland Center U.S.A.
53 D2 Richmond N.S.W. Austr.
51 D2 Richmond Qld Austr.
134 B1 Richmond Can.
54 B2 Richmond N.Z.
122 B3 Richmond S. Africa
98 C2 Richmond U.K.
138 C3 Richmond IN U.S.A.
138 C3 Richmond KY U.S.A.
139 D3 Richmond VA U.S.A.
141 D2 Richmond Hill U.S.A.
53 D1 Richmond Range *hills* Aus...
77 E1 Ridder Kazakh.
130 C3 Rideau Lakes Can.
135 C3 Ridgecrest U.S.A.
140 B2 Ridgeland U.S.A.
102 C1 Riesa Ger.
101 D1 Rieste Ger.
123 B2 Riet *r.* S. Africa
101 D2 Rietberg Ger.
122 B2 Rietfontein S. Africa
108 B2 Rieti Italy
136 B3 Rifle U.S.A.
88 B2 Riga Latvia
88 B2 Riga, Gulf of Estonia/Latvi...
79 C2 Rīgān Iran
134 D2 Rigby U.S.A.
131 E2 Rigolet Can.
93 H3 Riihimäki Fin.
55 D2 Riiser-Larsen Ice Shelf Antarctica
108 B1 Rijeka Croatia
134 C2 Riley U.S.A.
105 C2 Rillieux-la-Pape France
78 B2 Rimah, Wādī ar *watercours...* Saudi Arabia
103 E2 Rimavská Sobota Slovakia
128 C2 Rimbey Can.
108 B2 Rimini Italy
Rîmnicu Sărat Romania *se...* Râmnicu Sărat
Rîmnicu Vîlcea Romania *s...* Râmnicu Vâlcea
131 D3 Rimouski Can.
144 B2 Rincón de Romos Mex.
93 F3 Ringebu Norway
93 E4 Ringkøbing Denmark
92 G2 Ringvassøya *i.* Norway
99 C4 Ringwood U.K.
101 D1 Rinteln Ger.
154 B3 Rio Azul Brazil
150 B3 Riobamba Ecuador
155 D2 Rio Bonito Brazil
150 C4 Rio Branco Brazil
154 C3 Rio Branco do Sul Brazil
154 B2 Rio Brilhante Brazil
155 D2 Rio Casca Brazil
154 C2 Rio Claro Brazil
153 B3 Río Colorado Arg.
153 B3 Río Cuarto Arg.
155 D2 Rio de Janeiro Brazil
155 D2 Rio de Janeiro *state* Brazil
153 B5 Río Gallegos Arg.
153 B5 Río Grande Arg.

152 C3 Rio Grande Brazil
144 B2 Río Grande Mex.
143 D2 Rio Grande r. Mex./U.S.A.
143 D3 Rio Grande City U.S.A.
150 B1 Riohacha Col.
150 B3 Rioja Peru
145 D2 Rio Lagartos Mex.
151 F3 Rio Largo Brazil
105 C2 Riom France
152 B1 Río Mulatos Bol.
154 C3 Rio Negro Brazil
155 D1 Rio Pardo de Minas Brazil
154 C1 Rio Preto, Serra do hills Brazil
142 B1 Rio Rancho U.S.A.
150 B3 Rio Tigre Ecuador
64 A3 Rio Tuba Phil.
154 B1 Rio Verde Brazil
145 C2 Rio Verde Mex.
154 B1 Rio Verde de Mato Grosso Brazil
90 C1 Ripky Ukr.
99 C3 Ripley U.K.
107 D1 Ripoll Spain
98 C2 Ripon U.K.
66 D1 Rishiri-tō i. Japan
Rîşnov Romania see Râşnov
93 E4 Risør Norway
122 B2 Ritchie S. Africa
73 D3 Ritchie's Archipelago is India
134 C1 Ritzville U.S.A.
152 B2 Rivadavia Arg.
108 B1 Riva del Garda Italy
146 B3 Rivas Nic.
152 C3 Rivera Uru.
114 B4 River Cess Liberia
129 D2 Riverhurst Can.
52 C2 Riverina reg. Austr.
122 B3 Riversdale S. Africa
135 C4 Riverside U.S.A.
136 B2 Riverton U.S.A.
131 D3 Riverview Can.
105 C3 Rivesaltes France
131 D3 Rivière-au-Renard Can.
131 D3 Rivière-du-Loup Can.
139 F1 Rivière-Ouelle Can.
90 B1 Rivne Ukr.
108 A1 Rivoli Italy
120 B2 Rivungo Angola
54 B2 Riwaka N.Z.
78 B2 Riyadh Saudi Arabia
81 C1 Rize Turkey
70 B2 Rizhao China
105 C2 Roanne France
138 D3 Roanoke U.S.A.
141 E1 Roanoke r. U.S.A.
141 E1 Roanoke Rapids U.S.A.
135 E3 Roan Plateau U.S.A.
146 B3 Roatán Hond.
52 A3 Robe Austr.
52 B2 Robe, Mount h. Austr.
101 F1 Röbel Ger.
130 C2 Robert-Bourassa, Réservoir resr Can.
53 D1 Roberts, Mount Austr.
92 H3 Robertsfors Sweden
122 A3 Robertson S. Africa
114 A4 Robertsport Liberia
Robert Williams Angola see Caála
130 C3 Roberval Can.
50 A2 Robinson Ranges hills Austr.
52 B2 Robinvale Austr.
129 D2 Roblin Can.
128 C2 Robson, Mount Can.
143 D3 Robstown U.S.A.
153 C3 Rocha Uru.
98 B3 Rochdale U.K.
154 B1 Rochedo Brazil
100 B2 Rochefort Belgium
104 B2 Rochefort France
52 D3 Rochester Austr.
137 E2 Rochester MN U.S.A.
139 D2 Rochester NH U.S.A.
139 D2 Rochester NY U.S.A.
94 A2 Rockall i. N. Atlantic Ocean
160 L4 Rockall Bank N. Atlantic Ocean
138 B2 Rockford U.S.A.
51 E2 Rockhampton Austr.
141 D2 Rock Hill U.S.A.
50 A3 Rockingham Austr.
138 A2 Rock Island U.S.A.
136 B2 Rock Springs MT U.S.A.
143 C3 Rocksprings U.S.A.
136 B2 Rock Springs WY U.S.A.

141 D1 Rockwood U.S.A.
136 C3 Rocky Ford U.S.A.
141 E1 Rocky Mount U.S.A.
128 C2 Rocky Mountain House Can.
124 F4 Rocky Mountains Can./U.S.A.
100 B3 Rocroi France
102 C1 Rødbyhavn Denmark
131 E2 Roddickton Can.
104 C3 Rodez France
Rodi i. Greece see Rhodes
89 F2 Rodniki Rus. Fed.
Rodos Greece see Rhodes
Rodos i. Greece see Rhodes
104 B2 Rodosto Turkey see Tekirdağ
50 A2 Roebourne Austr.
50 B1 Roebuck Bay Austr.
123 C1 Roedtan S. Africa
100 B2 Roermond Neth.
100 A2 Roeselare Belgium
127 G2 Roes Welcome Sound sea chan. Can.
128 C2 Rogers U.S.A.
140 B1 Rogers U.S.A.
138 C3 Rogers, Mount U.S.A.
138 C1 Rogers City U.S.A.
157 G8 Roggeveen Basin S. Pacific Ocean
122 B3 Roggeveldberge esc. S. Africa
92 G2 Rognan Norway
134 B2 Rogue r. U.S.A.
74 B2 Rohtak India
49 M5 Roi Georges, Îles du is Fr. Polynesia
88 B2 Roja Latvia
60 B1 Rokan r. Indon.
88 B1 Rokiškis Lith.
90 B1 Rokytne Ukr.
154 B2 Rolândia Brazil
137 E3 Rolla U.S.A.
143 C2 Rolling Prairies reg. U.S.A.
51 D2 Roma Austr.
123 C2 Roma Lesotho
Roma Italy see Rome
141 E2 Romain, Cape U.S.A.
110 C1 Roman Romania
158 E5 Romanche Gap sea feature S. Atlantic Ocean
59 C3 Romang, Pulau i. Indon.
110 B1 Romania country Europe
91 C3 Romanivka Rus. Fed.
69 D1 Romanovka Rus. Fed.
105 D2 Romans-sur-Isère France
105 D2 Rombas France
64 B2 Romblon Phil.
108 B2 Rome Italy
141 C2 Rome GA U.S.A.
139 D2 Rome NY U.S.A.
99 D4 Romford U.K.
105 C2 Romilly-sur-Seine France
91 C1 Romny Ukr.
104 C2 Romorantin-Lanthenay France
99 C4 Romsey U.K.
134 D1 Ronan U.S.A.
96 □ Ronas Hill U.K.
151 D4 Roncador, Serra do hills Brazil
106 B2 Ronda Spain
154 B2 Rondon Brazil
154 B1 Rondonópolis Brazil
71 A3 Rong'an China
Rongcheng China see Rongxian
71 A3 Rongjiang China
62 A1 Rongklang Range mts Myanmar
Rongmei China see Hefeng
71 B3 Rongxian China
Rongxian China see Jianli
93 F4 Rønne Denmark
93 G4 Ronneby Sweden
55 A2 Ronne Ice Shelf Antarctica
101 D1 Ronnenberg Ger.
100 A2 Ronse Belgium
74 B2 Roorkee India
100 B2 Roosendaal Neth.
135 E2 Roosevelt U.S.A.
55 N2 Roosevelt Island Antarctica
104 B3 Roquefort France
150 B2 Roraima, Mount Guyana
93 F3 Røros Norway

90 C2 Ros' r. Ukr.
142 B3 Rosa, Punta pt Mex.
106 B2 Rosal de la Frontera Spain
153 B3 Rosario Arg.
144 A1 Rosario Mex.
144 B2 Rosario Mex.
144 B2 Rosario Mex.
147 C3 Rosario Venez.
152 B2 Rosario de la Frontera Arg.
151 D4 Rosário Oeste Brazil
144 A1 Rosarito Mex.
142 A3 Rosarito Mex.
144 A2 Rosarito Mex.
109 C3 Rosarno Italy
143 C2 Roscoe U.S.A.
104 B2 Roscoff France
97 B2 Roscommon Ireland
97 C2 Roscrea Ireland
147 D3 Roseau Dominica
137 D1 Roseau U.S.A.
134 B2 Roseburg U.S.A.
96 C2 Rosehearty U.K.
143 D3 Rosenberg U.S.A.
101 D1 Rosengarten Ger.
102 C2 Rosenheim Ger.
129 D2 Rosetown Can.
122 A2 Rosh Pinah Namibia
110 C2 Roşiori de Vede Romania
93 F4 Roskilde Denmark
89 D3 Roslavl' Rus. Fed.
134 B1 Roslyn U.S.A.
109 C3 Rossano Italy
97 B1 Rossan Point Ireland
51 E1 Rossel Island P.N.G.
55 N Ross Ice Shelf Antarctica
131 D3 Rossignol, Lake Can.
128 C3 Rossland Can.
97 C2 Rosslare Ireland
97 C2 Rosslare Harbour Ireland
101 F2 Roßlau Ger.
101 E2 Roßleben Ger.
114 A3 Rosso Maur.
105 D3 Rosso, Capo c. France
99 B4 Ross-on-Wye U.K.
91 D1 Rossosh' Rus. Fed.
128 A1 Ross River Can.
55 N2 Ross Sea Antarctica
92 F2 Røssvatnet l. Norway
81 D3 Rostāq Iran
129 D2 Rosthern Can.
102 C1 Rostock Ger.
89 E2 Rostov Rus. Fed.
91 D2 Rostov-na-Donu Rus. Fed.
Rostov-on-Don Rus. Fed. see Rostov-na-Donu
104 B2 Rostrenen France
92 G2 Røsvik Norway
92 H2 Rosvik Sweden
142 C2 Roswell U.S.A.
59 D2 Rota i. N. Mariana Is
59 C3 Rote i. Indon.
101 D1 Rotenburg (Wümme) Ger.
101 E3 Roth Ger.
98 C2 Rothbury U.K.
101 E3 Rothenburg ob der Tauber Ger.
98 C3 Rotherham U.K.
96 C2 Rothes U.K.
96 B3 Rothesay U.K.
53 C2 Roto Austr.
105 D3 Rotondo, Monte mt. France
54 C1 Rotorua N.Z.
54 C1 Rotorua, Lake N.Z.
101 E2 Rottenbach Ger.
102 C2 Rottenmann Austria
100 B2 Rotterdam Neth.
102 B2 Rottweil Ger.
49 I5 Rotuma i. Fiji
105 C2 Roubaix France
104 C2 Rouen France
Roulers Belgium see Roeselare
53 D2 Round Mountain Austr.
131 E3 Round Pond l. Can.
143 D2 Round Rock U.S.A.
134 D1 Roundup U.S.A.
96 C1 Rousay i. U.K.
130 C3 Rouyn-Noranda Can.
92 I2 Rovaniemi Fin.
91 D2 Roven'ki Rus. Fed.
91 D2 Roven'ky Ukr.
108 B1 Rovereto Italy
63 B2 Rôviĕng Tbong Cambodia
108 B1 Rovigo Italy
108 B1 Rovinj Croatia
53 C1 Rowena Austr.
Równe Ukr. see Rivne
64 B2 Roxas Phil.

64 A2 Roxas Phil.
64 B2 Roxas Phil.
52 A2 Roxby Downs Austr.
142 C1 Roy NM U.S.A.
134 D2 Roy UT U.S.A.
138 B1 Royale, Isle i. U.S.A.
104 B2 Royan France,
99 C3 Royston U.K.
90 C2 Rozdil'na Ukr.
91 C2 Rozdol'ne Ukr.
103 E2 Rožňava Slovakia
100 B3 Rozoy-sur-Serre France
87 D3 Rtishchevo Rus. Fed.
Ruanda country Africa see Rwanda
54 C1 Ruapehu, Mount vol. N.Z.
54 A3 Ruapuke Island N.Z.
89 D2 Ruba Belarus
79 B3 Rub' al Khālī des. Saudi Arabia
119 D3 Rubeho Mountains Tanz.
91 D2 Rubizhne Ukr.
77 E1 Rubtsovsk Rus. Fed.
126 B2 Ruby U.S.A.
135 C2 Ruby Mountains U.S.A.
76 C3 Rūdbār Afgh.
66 C2 Rudnaya Pristan' Rus. Fed.
89 D3 Rudnya Rus. Fed.
76 C1 Rudnyy Kazakh.
Rudol'f, Lake salt l. Eth./Kenya see Turkana, Lake
82 E1 Rudol'fa, Ostrov i. Rus. Fed.
Rudolph Island i. Rus. Fed. see Rudol'fa, Ostrov
101 E2 Rudolstadt Ger.
116 B3 Rufa'a Sudan
119 D3 Rufiji r. Tanz.
153 B3 Rufino Arg.
121 B2 Rufunsa Zambia
70 C2 Rugao China
99 C3 Rugby U.K.
136 C1 Rugby U.S.A.
102 C1 Rügen i. Ger.
119 C3 Ruhengeri Rwanda
101 E2 Ruhla Ger.
88 B2 Ruhnu i. Estonia
100 C2 Ruhr r. Ger.
71 C3 Rui'an China
142 B2 Ruidoso U.S.A.
144 B2 Ruiz Mex.
119 D3 Rukwa, Lake Tanz.
96 A2 Rum i. U.K.
109 C1 Ruma Serbia
78 B2 Rumāh Saudi Arabia
117 A4 Rumbek S. Sudan
147 C2 Rum Cay i. Bahamas
139 E2 Rumford U.S.A.
103 D1 Rumia Pol.
105 D2 Rumilly France
50 C1 Rum Jungle Austr.
121 C2 Rumphi Malawi
66 D2 Rumoi Japan
54 B2 Runanga N.Z.
98 B3 Runcorn U.K.
120 A2 Rundu Namibia
119 C2 Rungu Dem. Rep. Congo
119 D3 Rungwa Tanz.
68 B2 Ruoqiang China
130 C2 Rupert r. Can.
134 D2 Rupert U.S.A.
130 C2 Rupert Bay Can.
Rusaddir N. Africa see Melilla
121 C2 Rusape Zimbabwe
110 C2 Ruse Bulg.
137 E1 Rush City U.S.A.
121 C2 Rushinga Zimbabwe
77 D3 Rushon Tajik.
136 C2 Rushville U.S.A.
53 C3 Rushworth Austr.
129 D2 Russell Can.
54 B1 Russell N.Z.
137 D3 Russell U.S.A.
140 C2 Russellville AL U.S.A.
140 B1 Russellville AR U.S.A.
138 B3 Russellville KY U.S.A.
101 D2 Rüsselsheim Ger.
82 F2 Russian Federation country Asia/Europe
81 C1 Rustavi Georgia
123 C2 Rustenburg S. Africa
140 B2 Ruston U.S.A.
98 B3 Ruthin U.K.
139 E2 Rutland U.S.A.
Rutog China see Dêrub
119 D3 Rutshuru Dem. Rep. Congo
119 E4 Ruvuma r. Moz./Tanz.
79 C2 Ruweis U.A.E.

Ruza

89 E2 Ruza Rus. Fed.
77 C1 Ruzayevka Kazakh.
87 D3 Ruzayevka Rus. Fed.
119 C3 Rwanda country Africa
89 E3 Ryazan' Rus. Fed.
89 F3 Ryazhsk Rus. Fed.
86 C2 Rybachiy, Poluostrov pen. Rus. Fed.
Rybach'ye Kyrg. see Balykchy
89 E2 Rybinsk Rus. Fed.
89 E2 Rybinskoye Vodokhranilishche resr Rus. Fed.
103 D1 Rybnik Pol.
Rybnitsa Moldova see Ribniţa
89 E3 Rybnoye Rus. Fed.
99 D4 Rye U.K.
Rykovo Ukr. see Yenakiyeve
89 D3 Ryl'sk Rus. Fed.
Ryojun China see Lüshunkou
67 C3 Ryōtsu Japan
69 E3 Ryukyu Islands is Japan
89 D3 Ryzhikovo Rus. Fed.
103 E1 Rzeszów Pol.
91 E1 Rzhaksa Rus. Fed.
89 D2 Rzhev Rus. Fed.

S

79 C2 Sa'ādatābād Iran
101 E2 Saale r. Ger.
101 E2 Saalfeld Ger.
134 B1 Saanich Can.
100 C3 Saar r. Ger.
102 B2 Saarbrücken Ger.
88 B2 Sääre Estonia
88 B2 Saaremaa i. Estonia
92 I2 Saarenkylä Fin.
93 I3 Saarijärvi Fin.
100 C3 Saarlouis Ger.
80 B2 Sab' Ābār Syria
107 D1 Sabadell Spain
67 C3 Sabae Japan
61 C1 Sabah state Malaysia
61 C2 Sabalana i. Indon.
146 B2 Sabana, Archipiélago de is Cuba
150 B1 Sabanalarga Col.
60 A1 Sabang Indon.
155 D1 Sabará Brazil
108 B2 Sabaudia Italy
122 B3 Sabelo S. Africa
119 D2 Sabena Desert Kenya
115 D2 Sabhā Libya
123 D2 Sabie r. Moz./S. Africa
123 D2 Sabie S. Africa
145 B2 Sabinas Mex.
145 B2 Sabinas Hidalgo Mex.
143 E3 Sabine r. U.S.A.
131 D3 Sable, Cape Can.
141 D3 Sable, Cape U.S.A.
131 E3 Sable Island Can.
106 B1 Sabugal Port.
78 B3 Şabyā Saudi Arabia
76 B3 Sabzevār Iran
137 D2 Sac City U.S.A.
120 A2 Sachanga Angola
65 B3 Sacheon S. Korea
130 A2 Sachigo Lake Can.
126 D2 Sachs Harbour Can.
154 C1 Sacramento Brazil
135 B3 Sacramento U.S.A.
135 B3 Sacramento r. U.S.A.
142 B2 Sacramento Mountains U.S.A.
135 B2 Sacramento Valley U.S.A.
110 B1 Săcueni Romania
123 C3 Sada S. Africa
107 C1 Sádaba Spain
Sá da Bandeira Angola see Lubango
78 B3 Şa'dah Yemen
63 B3 Sadao Thai.
63 B2 Sa Đec Vietnam
74 B2 Sadiqabad Pak.
72 D2 Sadiya India
67 C3 Sadoga-shima i. Japan
107 D2 Sa Dragonera i. Spain
81 D2 Safāshahr Iran
93 F4 Säffle Sweden
142 B2 Safford U.S.A.
99 D3 Saffron Walden U.K.
114 B1 Safi Morocco
155 D1 Safiras, Serra das mts Brazil
86 D2 Safonovo Rus. Fed.

89 D2 Safonovo Rus. Fed.
75 C2 Saga China
67 B4 Saga Japan
62 A1 Sagaing Myanmar
67 C3 Sagamihara Japan
74 B2 Sagar India
Sagarmatha mt. China/Nepal see Everest, Mount
138 C2 Saginaw U.S.A.
138 C2 Saginaw Bay U.S.A.
Saglouc Can. see Salluit
106 B2 Sagres Port.
146 B2 Sagua la Grande Cuba
139 F1 Sagueney r. Can.
107 C2 Sagunto Spain
76 B2 Sagyndyk, Mys pt Kazakh.
106 B1 Sahagún Spain
114 C3 Sahara des. Africa
Şaḩarā el Gharbîya des. Egypt see Western Desert
Şaḩarā el Sharqîya des. Egypt see Eastern Desert
Saharan Atlas mts Alg. see Atlas Saharien
74 B2 Saharanpur India
75 C2 Saharsa India
114 B3 Sahel reg. Africa
74 B1 Sahiwal Pak.
144 B2 Sahuayo Mex.
78 B2 Şāḩūq reg. Saudi Arabia
114 C1 Saïda Alg.
Saïda Lebanon see Sidon
75 C2 Saidpur Bangl.
Saigon Vietnam see Ho Chi Minh City
75 D2 Saiha India
70 B1 Saihan Tal China
67 B4 Saiki Japan
93 I3 Saimaa l. Fin.
144 B2 Sain Alto Mex.
96 C3 St Abb's Head hd U.K.
131 E3 St Alban's Can.
99 C4 St Albans U.K.
138 C3 St Albans U.S.A.
St Alban's Head hd U.K. see St Aldhelm's Head
99 B4 St Aldhelm's Head hd U.K.
St-André, Cap c. Madag. see Vilanandro, Tanjona
96 C2 St Andrews U.K.
134 D2 St Anthony Can.
52 B3 St Anthony U.S.A.
131 E2 St-Augustin Can.
131 E2 St-Augustin r. Can.
141 D3 St Augustine U.S.A.
99 A4 St Austell U.K.
104 C2 St-Avertin France
147 D3 St-Barthélemy terr. West Indies
98 B2 St Bees Head hd U.K.
105 D3 St-Bonnet-en-Champsaur France
99 A4 St Bride's Bay U.K.
104 B2 St-Brieuc France
130 C3 St Catharines Can.
141 D2 St Catherines Island U.S.A.
99 C4 St Catherine's Point U.K.
137 E3 St Charles U.S.A.
138 C2 St Clair, Lake Can./U.S.A.
105 D2 St-Claude France
99 A4 St Clears U.K.
137 E1 St Cloud U.S.A.
138 A1 St Croix r. U.S.A.
147 D3 St Croix i. Virgin Is (U.S.A.)
99 A4 St David's U.K.
99 A4 St David's Head hd U.K.
113 I8 St-Denis Réunion
104 C2 St-Denis France
St-Denis-du-Sig Alg. see Sig
105 D2 St-Dié-des-Vosges France
105 D2 St-Dizier France
130 C3 Ste-Adèle Can.
131 D3 Ste-Anne-des-Monts Can.
139 E1 Ste-Foy Can.
105 D2 St-Égrève France
128 A1 St Elias Mountains Can.
131 D2 Ste-Marguerite r. Can.
139 E1 Ste-Marie Can.
Ste-Marie, Cap c. Madag. see Vohimena, Tanjona
Sainte-Marie, Île i. Madag. see Boraha, Nosy
Ste-Rose-du-Dégelé Can. see Dégelis
129 E2 Sainte Rose du Lac Can.
104 B2 Saintes France
105 C2 St-Étienne France

104 C2 St-Étienne-du-Rouvray France
130 C3 St-Félicien Can.
97 D1 Saintfield U.K.
105 D3 St-Florent France
105 C2 St-Flour France
136 C3 St Francis U.S.A.
104 C3 St-Gaudens France
53 C1 St George Austr.
135 D3 St George U.S.A.
141 D3 St George Island U.S.A.
131 C3 St-Georges Can.
147 D3 St George's Grenada
131 E3 St George's Bay Can.
97 C3 St George's Channel Ireland/U.K.
105 D2 St Gotthard Pass pass Switz.
113 C7 St Helena terr. S. Atlantic Ocean
122 A3 St Helena Bay S. Africa
122 A3 St Helena Bay b. S. Africa
98 B3 St Helens U.K.
134 B1 St Helens, Mount vol. U.S.A.
95 C4 St Helier Channel Is
100 B2 St-Hubert Belgium
139 E1 St-Hyacinthe Can.
138 C1 St Ignace U.S.A.
130 B3 St Ignace Island Can.
99 A4 St Ives U.K.
St Jacques, Cap Vietnam see Vung Tau
128 A2 St James, Cape Can.
130 C3 St-Jean, Lac l. Can.
104 B2 St-Jean-d'Angély France
104 B3 St-Jean-de-Luz France
130 C3 St-Jean-de-Monts France
139 E1 St-Jérôme Can.
134 C1 St Joe r. U.S.A.
131 D3 Saint John Can.
137 D3 St John U.S.A.
139 F1 St John r. U.S.A.
147 D3 St John's Antigua
131 E3 St John's Can.
142 B2 St Johns U.S.A.
141 D2 St Johns r. U.S.A.
139 E2 St Johnsbury U.S.A.
137 E3 St Joseph U.S.A.
130 A2 St Joseph, Lake Can.
St-Joseph-d'Alma Can. see Alma
130 B3 St Joseph Island Can.
139 E1 St-Jovité Can.
104 C2 St-Junien France
94 B2 St Kilda is U.K.
147 D3 St Kitts and Nevis country West Indies
151 D2 St-Laurent-du-Maroni Fr. Guiana
131 E3 St Lawrence Can.
131 D3 St Lawrence inlet Can.
131 D3 St Lawrence, Gulf of Can.
126 A2 St Lawrence Island U.S.A.
104 B2 St-Lô France
114 A3 St-Louis Senegal
137 E3 St Louis U.S.A.
137 E1 St Louis r. U.S.A.
147 D3 St Lucia country West Indies
147 D3 St Lucia Channel Martinique/St Lucia
123 D2 St Lucia Estuary S. Africa
96 □ St Magnus Bay U.K.
104 B2 St-Malo France
104 B2 St-Malo, Golfe de g. France
147 C3 St-Marc Haiti
St Mark's S. Africa see Cofimvaba
147 D3 St-Martin terr. West Indies
122 A3 St Martin, Cape S. Africa
129 E2 St Martin, Lake Can.
139 D2 St Marys U.S.A.
124 A3 St Matthew Island U.S.A.
59 D3 St Matthias Group is P.N.G.
130 C3 St-Maurice r. Can.
130 C3 St-Michel-des-Saints Can.
104 B2 St-Nazaire France
104 C1 St-Omer France
137 E2 St Paul U.S.A.
156 A8 St-Paul, Île i. Indian Ocean
137 E2 St Peter U.S.A.
95 C4 St Peter Port Channel Is
89 D2 St Petersburg Rus. Fed.
141 D3 St Petersburg U.S.A.
131 E3 St-Pierre St Pierre and Miquelon
139 E1 St-Pierre, Lac l. Can.
131 E3 St Pierre and Miquelon terr. N. America

104 B2 St-Pierre-d'Oléron Franc
105 C2 St-Pourçain-sur-Sioule France
131 D3 St Quentin Can.
105 C2 St-Quentin France
105 D3 St-Raphaël France
122 B3 St Sebastian Bay S. Africa
104 B2 St-Sébastien-sur-Loire France
131 D3 St-Siméon Can.
129 E2 St Theresa Point Can.
130 B3 St Thomas Can.
105 D3 St-Tropez France
105 D3 St-Tropez, Cap de c. France
St Vincent, Cape c. Port. see São Vicente, Cabo de
52 A3 St Vincent, Gulf Austr.
147 D3 St Vincent and the Grenadines country West Indies
147 D3 St Vincent Passage St Lucia/St Vincent
100 C2 St-Vith Belgium
129 D2 St Walburg Can.
104 C2 St-Yrieix-la-Perche France
59 D1 Saipan i. N. Mariana Is
152 B1 Sajama, Nevado mt. Bol.
122 B2 Sak watercourse S. Africa
67 C4 Sakai Japan
67 B4 Sakaide Japan
78 B2 Sakākah Saudi Arabia
136 C1 Sakakawea, Lake U.S.A.
Sakarya Turkey see Adapazarı
111 D2 Sakarya r. Turkey
66 C3 Sakata Japan
65 B1 Sakchu N. Korea
66 D1 Sakhalin i. Rus. Fed.
123 C2 Sakhile S. Africa
81 C1 Şäki Azer.
88 B3 Šakiai Lith.
69 E3 Sakishima-shotō is Japan
62 B2 Sakon Nakhon Thai.
74 A2 Sakrand Pak.
122 B3 Sakrivier S. Africa
67 D3 Sakura Japan
91 C2 Saky Ukr.
93 G4 Sala Sweden
130 C3 Salaberry-de-Valleyfield Can.
88 B2 Salacgrīva Latvia
109 C2 Sala Consilina Italy
135 C4 Salada, Laguna salt l. Mex.
152 C2 Saladas Arg.
152 B3 Salado r. Arg.
153 B3 Salado r. Arg.
145 C2 Salado r. Mex.
114 B4 Salaga Ghana
122 B1 Salajwe Botswana
79 C2 Salakh, Jabal mt. Oman
115 D3 Salal Chad
78 A2 Salāla Sudan
79 C3 Şalālah Oman
145 B2 Salamanca Mex.
106 B1 Salamanca Spain
139 D2 Salamanca U.S.A.
106 B1 Salas Spain
63 B2 Salavan Laos
59 C3 Salawati i. Indon.
157 G7 Sala y Gómez, Isla i. S. Pacific Ocean
Salazar Angola see N'dalatando
104 C2 Salbris France
88 C3 Šalčininkai Lith.
106 C1 Saldaña Spain
122 A3 Saldanha S. Africa
88 B2 Saldus Latvia
53 C3 Sale Austr.
86 F2 Salekhard Rus. Fed.
73 B3 Salem India
138 B3 Salem IL U.S.A.
137 E3 Salem MO U.S.A.
134 B2 Salem OR U.S.A.
137 D2 Salem SD U.S.A.
96 B2 Salen U.K.
109 B2 Salerno Italy
108 B2 Salerno, Golfo di g. Italy
98 B3 Salford U.K.
151 E2 Salgado r. Brazil
103 D2 Salgótarján Hungary
151 F3 Salgueiro Brazil
136 B3 Salida U.S.A.
111 C3 Salihli Turkey
88 C3 Salihorsk Belarus
121 C2 Salima Malawi
121 C2 Salimo Moz.
137 D3 Salina KS U.S.A.
135 D3 Salina UT U.S.A.
109 B3 Salina, Isola i. Italy

145 C3 Salina Cruz Mex.
155 D1 Salinas Brazil
144 B2 Salinas Mex.
135 B3 Salinas U.S.A.
135 B3 Salinas r. U.S.A.
107 D2 Salines, Cap de ses c. Spain
151 E3 Salinópolis Brazil
99 C4 Salisbury U.K.
139 D3 Salisbury MD U.S.A.
141 D1 Salisbury NC U.S.A.
Salisbury Zimbabwe see Harare
99 B4 Salisbury Plain U.K.
151 E3 Salitre r. Brazil
92 I2 Salla Fin.
143 E1 Sallisaw U.S.A.
127 G2 Salluit Can.
75 C2 Sallyana Nepal
81 C2 Salmās Iran
128 C3 Salmo Can.
134 D1 Salmon U.S.A.
134 C1 Salmon r. U.S.A.
128 C2 Salmon Arm Can.
134 C2 Salmon River Mountains U.S.A.
100 C3 Salmtal Ger.
118 B2 Salo C.A.R.
93 H3 Salo Fin.
105 D3 Salon-de-Provence France
Salonica Greece see Thessaloniki
110 B1 Salonta Romania
87 D4 Sal'sk Rus. Fed.
122 B3 Salt watercourse S. Africa
107 D1 Salt Spain
142 A2 Salt r. U.S.A.
152 B2 Salta Arg.
99 A4 Saltash U.K.
96 B3 Saltcoats U.K.
145 B2 Saltillo Mex.
134 D2 Salt Lake City U.S.A.
154 C2 Salto Brazil
152 C3 Salto Uru.
155 E1 Salto da Divisa Brazil
154 B2 Salto del Guairá Para.
135 C4 Salton Sea salt l. U.S.A.
154 B3 Salto Osório, Represa de resr Brazil
154 B3 Salto Santiago, Represa de resr Brazil
141 D2 Saluda U.S.A.
76 B3 Sälük, Küh-e mt. Iran
108 A2 Saluzzo Italy
151 F4 Salvador Brazil
79 C2 Salwah Saudi Arabia
62 A2 Salween r. China/Myanmar
62 A2 Salween r. China/Myanmar
81 C2 Salyan Azer.
138 C3 Sayersville U.S.A.
122 A1 Salzbrunn Namibia
102 C2 Salzburg Austria
101 E1 Salzgitter Ger.
101 D2 Salzkotten Ger.
101 E1 Salzwedel Ger.
144 B1 Samalayuca Mex.
66 D2 Samani Japan
64 B2 Samar i. Phil.
87 E3 Samara Rus. Fed.
Samarahan Malaysia see Sri Aman
59 E3 Samarai P.N.G.
61 C2 Samarinda Indon.
77 C2 Samarqand Uzbek.
81 C2 Sämarrä' Iraq
81 C1 Şamaxı Azer.
119 C3 Samba Dem. Rep. Congo
61 C1 Sambaliung mts Indon.
75 C2 Sambalpur India
60 C2 Sambar, Tanjung pt Indon.
60 B1 Sambas Indon.
121 DE2 Sambava Madag.
74 B2 Sambhar India
90 A2 Sambir Ukr.
61 C2 Samboja Indon.
153 C3 Samborombón, Bahía b. Arg.
65 B2 Samcheok S. Korea
Samch'ŏnp'o S. Korea see Sacheon
81 C2 Samdi Dag mt. Turkey
119 D3 Same Tanz.
121 B2 Samfya Zambia
78 B2 Samirah Saudi Arabia
65 B1 Samjiyŏn N. Korea
Sam Neua Laos see Xam Nua
48 J5 Samoa country S. Pacific Ocean
156 E6 Samoa Basin S. Pacific Ocean

Samoa i Sisifo country S. Pacific Ocean see Samoa
109 C1 Samobor Croatia
110 B2 Samokov Bulg.
111 C3 Samos i. Greece
Samothrace i. Greece see Samothraki
111 C3 Samothraki Greece
111 C2 Samothraki i. Greece
61 C2 Sampit Indon.
119 C3 Sampwe Dem. Rep. Congo
143 E2 Sam Rayburn Reservoir U.S.A.
62 B2 Sâm Sơn Vietnam
80 B1 Samsun Turkey
81 C1 Samt'redia Georgia
63 B3 Samui, Ko i. Thai.
63 B2 Samut Songkhram Thai.
114 B3 San Mali
78 B3 San'ā' Yemen
118 A2 Sanaga r. Cameroon
81 C2 Sanandaj Iran
146 B3 San Andrés, Isla de i. Caribbean Sea
106 B1 San Andrés del Rabanedo Spain
142 B2 San Andres Mountains U.S.A.
145 C3 San Andrés Tuxtla Mex.
143 C2 San Angelo U.S.A.
143 D3 San Antonio U.S.A.
135 C4 San Antonio, Mount U.S.A.
152 B2 San Antonio de los Cobres Arg.
153 B4 San Antonio Oeste Arg.
108 B2 San Benedetto del Tronto Italy
144 A3 San Benedicto, Isla i. Mex.
135 C4 San Bernardino U.S.A.
135 C4 San Bernardino Mountains U.S.A.
142 B3 San Blas Mex.
141 C3 San Blas, Cape U.S.A.
146 C4 San Blas, Punta pt Panama
152 B1 San Borja Bol.
144 B2 San Buenaventura Mex.
146 C4 San Carlos Phil.
147 D4 San Carlos Venez.
153 A4 San Carlos de Bariloche Arg.
147 C4 San Carlos del Zulia Venez.
104 C2 Sancerrois, Collines du hills France
135 C4 San Clemente U.S.A.
135 C4 San Clemente Island U.S.A.
105 C2 Sancoins France
48 H5 San Cristobal i. Solomon Is
150 B2 San Cristóbal Venez.
145 C3 San Cristóbal de las Casas Mex.
146 C2 Sancti Spíritus Cuba
61 C1 Sandakan Malaysia
93 E3 Sandane Norway
111 B2 Sandanski Bulg.
114 A3 Sandaré Mali
96 C1 Sanday i. U.K.
143 C2 Sanderson U.S.A.
150 C4 Sandia Peru
135 C4 San Diego U.S.A.
111 D3 Sandıklı Turkey
93 E4 Sandnes Norway
92 F2 Sandnessjøen Norway
118 C3 Sandoa Dem. Rep. Congo
103 E1 Sandomierz Pol.
89 E2 Sandovo Rus. Fed.
94 B1 Sandoy i. Faroe Is
134 C1 Sandpoint U.S.A.
94 B1 Sandur Faroe Is
138 C2 Sandusky U.S.A.
122 A3 Sandverhaar Namibia
93 F3 Sandvika Norway
93 G3 Sandviken Sweden
131 E2 Sandwich Bay Can.
135 D2 Sandy U.S.A.
140 C2 Sandy Bay Can.
130 A2 Sandy Cape Austr.
130 A2 Sandy Lake Can.
130 A2 Sandy Lake l. Can.
141 D2 Sandy Springs U.S.A.
144 A1 San Felipe Mex.
145 B2 San Felipe Mex.
150 C1 San Felipe Venez.
144 C2 San Fernando Mex.
64 B2 San Fernando Phil.
64 B2 San Fernando Phil.
106 B2 San Fernando Spain

147 D3 San Fernando Trin. and Tob.
150 C2 San Fernando de Apure Venez.
141 D3 Sanford FL U.S.A.
139 E2 Sanford ME U.S.A.
141 E1 Sanford NC U.S.A.
152 B3 San Francisco Arg.
135 B3 San Francisco U.S.A.
74 B3 Sangamner India
83 J2 Sangar Rus. Fed.
108 A3 San Gavino Monreale Italy
101 E2 Sangerhausen Ger.
61 C1 Sanggau Indon.
118 B3 Sangha r. Congo
109 C3 San Giovanni in Fiore Italy
59 C2 Sangir i. Indon.
59 C2 Sangir, Kepulauan is Indon.
65 B2 Sangju S. Korea
63 A2 Sangkhla Buri Thai.
61 C1 Sangkulirang Indon.
73 B3 Sangli India
118 B2 Sangmélima Cameroon
121 C3 Sango Zimbabwe
San Gottardo, Passo del pass Switz. see St Gotthard Pass
136 B3 Sangre de Cristo Range mts U.S.A.
75 C2 Sangsang China
144 A2 San Hipólito, Punta pt Mex.
145 D3 San Ignacio Belize
152 B1 San Ignacio Bol.
144 A2 San Ignacio Mex.
130 C2 Sanikiluaq Can.
71 A3 Sanjiang China
Sanjiang China see Jinping
67 C3 Sanjō Japan
135 B3 San Joaquin r. U.S.A.
153 B4 San Jorge, Golfo de g. Arg.
146 B4 San José Costa Rica
64 B2 San Jose Phil.
135 B3 San Jose Phil.
144 A2 San José, Isla i. Mex.
64 B2 San José de Bavicora Mex.
64 B2 San Jose de Buenavista Phil.
144 A2 San José de Comondú Mex.
144 B2 San José del Cabo Mex.
150 B2 San José del Guaviare Col.
152 B3 San Juan Arg.
146 B3 San Juan r. Costa Rica/Nic.
147 C3 San Juan Dom. Rep.
147 D3 San Juan Puerto Rico
135 D3 San Juan r. U.S.A.
152 C2 San Juan Bautista Para.
145 C3 San Juan Bautista Tuxtepec Mex.
147 D4 San Juan de los Morros Venez.
145 C2 San Juan del Río Mex.
134 B1 San Juan Islands U.S.A.
144 B2 San Juanito Mex.
136 B3 San Juan Mountains U.S.A.
153 B4 San Julián Arg.
75 C2 Sankh r. India
63 B2 San Khao Phang Hoei mts Thai.
100 C2 Sankt Augustin Ger.
105 D2 Sankt Gallen Switz.
105 D2 Sankt Moritz Switz.
Sankt-Peterburg Rus. Fed. see St Petersburg
102 C2 Sankt Veit an der Glan Austria
100 C3 Sankt Wendel Ger.
80 B2 Şanlıurfa Turkey
142 B3 San Lorenzo Mex.
106 B2 Sanlúcar de Barrameda Spain
153 B3 San Luis Arg.
145 B2 San Luis de la Paz Mex.
142 A2 San Luisito Mex.
135 B3 San Luis Obispo U.S.A.
135 B3 San Luis Obispo Bay U.S.A.
145 B2 San Luis Potosí Mex.
144 A1 San Luis Río Colorado Mex.
143 D3 San Marcos U.S.A.
108 B2 San Marino country Europe
108 B2 San Marino San Marino
144 B2 San Martín de Bolaños Mex.
153 A4 San Martín de los Andes Arg.
135 B3 San Mateo U.S.A.
153 B4 San Matías, Golfo g. Arg.
70 B2 Sanmenxia China

146 B3 San Miguel El Salvador
152 B2 San Miguel de Tucumán Arg.
135 B4 San Miguel Island U.S.A.
145 C3 San Miguel Sola de Vega Mex.
71 B3 Sanming China
153 B3 San Nicolás de los Arroyos Arg.
135 C4 San Nicolas Island U.S.A.
110 B1 Sânnicolau Mare Romania
123 C2 Sannieshof S. Africa
114 B4 Sanniquellie Liberia
103 E2 Sanok Pol.
64 B2 San Pablo Phil.
144 B2 San Pablo Balleza Mex.
152 B2 San Pedro Arg.
152 B1 San Pedro Bol.
114 B4 San-Pédro Côte d'Ivoire
144 A2 San Pedro Mex.
142 A2 San Pedro watercourse U.S.A.
106 B2 San Pedro, Sierra de mts Spain
144 B2 San Pedro de las Colonias Mex.
152 C2 San Pedro de Ycuamandyyú Para.
142 A3 San Pedro el Saucito Mex.
146 B3 San Pedro Sula Hond.
108 A3 San Pietro, Isola di i. Italy
96 C3 Sanquhar U.K.
144 A1 San Quintín, Cabo c. Mex.
153 B3 San Rafael Arg.
108 A2 San Remo Italy
143 D2 San Saba r. U.S.A.
147 C2 San Salvador i. Bahamas
146 B3 San Salvador El Salvador
152 B2 San Salvador de Jujuy Arg.
107 C1 San Sebastián Spain
108 B2 Sansepolcro Italy
109 C2 San Severo Italy
109 C2 Sanski Most Bos.-Herz.
152 B1 Santa Ana Bol.
146 B3 Santa Ana El Salvador
144 A1 Santa Ana Mex.
135 C4 Santa Ana U.S.A.
152 B1 Santa Ana de Yacuma Bol.
144 B2 Santa Bárbara Mex.
135 C4 Santa Barbara U.S.A.
154 B2 Santa Bárbara, Serra de hills Brazil
152 B2 Santa Catalina Chile
135 C4 Santa Catalina Island U.S.A.
154 B3 Santa Catarina state Brazil
150 C3 Santa Clara Col.
146 C2 Santa Clara Cuba
135 B3 Santa Clara U.S.A.
107 D1 Santa Coloma de Gramenet Spain
Santa Comba Angola see Waku-Kungo
109 C3 Santa Croce, Capo c. Italy
153 B5 Santa Cruz r. Arg.
152 B1 Santa Cruz Bol.
64 B2 Santa Cruz Phil.
135 B3 Santa Cruz U.S.A.
145 C3 Santa Cruz Barillas Guat.
155 E1 Santa Cruz Cabrália Brazil
107 C2 Santa Cruz de Moya Spain
114 A2 Santa Cruz de Tenerife Islas Canarias
152 C2 Santa Cruz do Sul Brazil
135 C4 Santa Cruz Island U.S.A.
48 H5 Santa Cruz Islands Solomon Is
107 D2 Santa Eulalia del Río Spain
152 B3 Santa Fe Arg.
142 B1 Santa Fe U.S.A.
154 B2 Santa Fé do Sul Brazil
154 B1 Santa Helena de Goiás Brazil
153 B3 Santa Isabel Arg.
Santa Isabel Equat. Guinea see Malabo
48 G4 Santa Isabel i. Solomon Is
154 B1 Santa Luisa, Serra de hills Brazil
151 E3 Santa Luzia Brazil
144 A2 Santa Margarita, Isla i. Mex.
152 C2 Santa Maria Brazil
144 B1 Santa María r. Mex.
135 B4 Santa Maria U.S.A.
123 D2 Santa Maria, Cabo de c. Moz.
106 B2 Santa Maria, Cabo de c. Port.

Santa Maria, Chapadão de

155 C1 Santa Maria, Chapadão de hills Brazil
151 E3 Santa Maria das Barreiras Brazil
109 C3 Santa Maria di Leuca, Capo c. Italy
155 D1 Santa Maria do Suaçuí Brazil
150 B2 Santa Marta Col.
135 C4 Santa Monica U.S.A.
151 E4 Santana Brazil
110 B1 Sântana Romania
106 C1 Santander Spain
108 A3 Sant'Antioco Italy
108 A3 Sant'Antioco, Isola di i. Italy
107 D2 Sant Antoni de Portmany Spain
151 D3 Santarém Brazil
106 B2 Santarém Port.
154 B1 Santa Rita do Araguaia Brazil
153 B3 Santa Rosa Arg.
152 C2 Santa Rosa Brazil
135 B3 Santa Rosa CA U.S.A.
142 C2 Santa Rosa NM U.S.A.
146 B3 Santa Rosa de Copán Hond.
135 B4 Santa Rosa Island CA U.S.A.
140 C2 Santa Rosa Island FL U.S.A.
144 A2 Santa Rosalía Mex.
134 C2 Santa Rosa Range mts U.S.A.
106 B1 Santa Uxía de Ribeira Spain
154 B1 Santa Vitória Brazil
107 D1 Sant Carles de la Ràpita Spain
135 C4 Santee U.S.A.
107 D2 Sant Francesc de Formentera Spain
152 C2 Santiago Brazil
153 A3 Santiago Chile
147 C3 Santiago Dom. Rep.
144 B2 Santiago Mex.
146 B4 Santiago Panama
64 B2 Santiago Phil.
106 B1 Santiago de Compostela Spain
146 C2 Santiago de Cuba Cuba
144 B2 Santiago Ixcuintla Mex.
144 B2 Santiago Papasquiaro Mex.
106 C1 Santillana Spain
107 D2 Sant Joan de Labritja Spain
107 D1 Sant Jordi, Golf de g. Spain
155 D2 Santo Amaro de Campos Brazil
154 B2 Santo Anastácio Brazil
155 C2 Santo André Brazil
152 C2 Santo Angelo Brazil
154 B2 Santo Antônio da Platina Brazil
151 F4 Santo Antônio de Jesus Brazil
150 D3 Santo Antônio do Içá Brazil
155 C2 Santo Antônio do Monte Brazil
147 D3 Santo Domingo Dom. Rep.
144 A2 Santo Domingo Mex.
142 B1 Santo Domingo Pueblo U.S.A.
111 C3 Santorini i. Greece
155 C2 Santos Brazil
155 D2 Santos Dumont Brazil
157 I7 Santos Plateau S. Atlantic Ocean
152 C2 Santo Tomé Arg.
153 A4 San Valentín, Cerro mt. Chile
146 B3 San Vicente El Salvador
144 A1 San Vicente Mex.
64 B2 San Vicente Phil.
150 B4 San Vicente de Cañete Peru
108 B2 San Vincenzo Italy
108 B3 San Vito, Capo c. Italy
71 A4 Sanya China
155 C2 São Bernardo do Campo Brazil
152 C2 São Borja Brazil
154 C2 São Carlos Brazil
155 D1 São Felipe, Serra de hills Brazil
151 D4 São Félix Brazil
151 D3 São Félix Brazil
155 D2 São Fidélis Brazil
155 D1 São Francisco Brazil
151 F4 São Francisco r. Brazil
154 C3 São Francisco, Ilha de i. Brazil
154 C3 São Francisco do Sul Brazil
152 C3 São Gabriel Brazil

155 D2 São Gonçalo Brazil
155 C1 São Gonçalo do Abaeté Brazil
155 C1 São Gotardo Brazil
154 B1 São Jerônimo, Serra de hills Brazil
155 D2 São João da Barra Brazil
155 C2 São João da Boa Vista Brazil
106 B1 São João da Madeira Port.
155 D1 São João da Ponte Brazil
155 D2 São João del Rei Brazil
155 D1 São João do Paraíso Brazil
155 D1 São João Evangelista Brazil
155 D2 São João Nepomuceno Brazil
154 C2 São Joaquim da Barra Brazil
152 D2 São José Brazil
154 C2 São José do Rio Preto Brazil
155 C2 São José dos Campos Brazil
154 C3 São José dos Pinhais Brazil
154 A1 São Lourenço Brazil
155 C2 São Lourenço Brazil
151 E3 São Luís Brazil
154 C1 São Manuel Brazil
151 E3 São Marcos r. Brazil
151 E3 São Marcos, Baía de b. Brazil
155 E1 São Mateus Brazil
154 B3 São Mateus do Sul Brazil
105 C2 Saône r. France
155 C2 São Paulo Brazil
154 C1 São Paulo state Brazil
155 D2 São Pedro da Aldeia Brazil
151 E3 São Raimundo Nonato Brazil
155 C1 São Romão Brazil
São Salvador Angola see M'banza Congo
São Salvador do Congo Angola see M'banza Congo
155 C2 São Sebastião, Ilha do i. Brazil
154 C2 São Sebastião do Paraíso Brazil
154 B1 São Simão Brazil
154 B1 São Simão, Barragem de resr Brazil
59 C2 Sao-Siu Indon.
113 D5 São Tomé
113 D5 São Tomé i. São Tomé and Príncipe
155 D2 São Tomé, Cabo de c. Brazil
113 D5 São Tomé and Príncipe country Africa
155 C2 São Vicente Brazil
106 B2 São Vicente, Cabo de c. Port.
59 C3 Saparua Indon.
107 D2 Sa Pobla Spain
89 F3 Sapozhok Rus. Fed.
66 D2 Sapporo Japan
109 C2 Sapri Italy
143 D1 Sapulpa U.S.A.
81 C2 Saqqez Iran
81 C2 Sarāb Iran
63 B2 Sara Buri Thai.
Saragossa Spain see Zaragoza
89 F3 Sarai Rus. Fed.
109 C2 Sarajevo Bos.-Herz.
87 E3 Saraktash Rus. Fed.
62 A1 Saramati mt. India/Myanmar
139 E2 Saranac Lake U.S.A.
109 D3 Sarandë Albania
64 B3 Sarangani Islands Phil.
87 D3 Saransk Rus. Fed.
87 E3 Sarapul Rus. Fed.
141 D3 Sarasota U.S.A.
90 B2 Saratov Ukr.
136 B2 Saratoga U.S.A.
139 E2 Saratoga Springs U.S.A.
61 C1 Saratok Malaysia
81 D2 Saratov Rus. Fed.
79 D2 Sarāvān Iran
111 C2 Sarawak state Malaysia
111 C2 Saray Turkey
111 C2 Sarayköy Turkey
79 D2 Sarbāz Iran
76 B3 Sarbīsheh Iran
74 B2 Sardarshahr India
Sardegna i. Italy see Sardinia
108 A2 Sardinia i. Italy
92 G2 Sarektjåkkå mt. Sweden
77 C3 Sar-e Pul Afgh.

158 C3 Sargasso Sea sea N. Atlantic Ocean
74 B1 Sargodha Pak.
115 D4 Sarh Chad
79 D2 Sarhad reg. Iran
81 D2 Sārī Iran
111 C3 Sarıgöl Turkey
81 C1 Sarıkamış Turkey
61 C1 Sarikei Malaysia
51 D2 Sarina Austr.
65 B2 Sariwŏn N. Korea
111 C2 Sarıyer Turkey
78 B2 Sark, Safrā' as esc. Saudi Arabia
77 D2 Sarkand Kazakh.
111 C2 Şarköy Turkey
104 C3 Sarlat-la-Canéda France
59 D3 Sarmi Indon.
153 B4 Sarmiento Arg.
138 C2 Sarnia Can.
90 B1 Sarny Ukr.
60 B2 Sarolangun Indon.
111 B3 Saronikos Kolpos g. Greece
111 C2 Saros Körfezi b. Turkey
103 E2 Sárospatak Hungary
87 D3 Sarov Rus. Fed.
Sarpan i. N. Mariana Is see Rota
105 D2 Sarrebourg France
106 B1 Sarria Spain
107 C1 Sarrión Spain
105 D3 Sartène France
Sartu China see Daqing
111 C3 Saruhanlı Turkey
103 D2 Sárvár Hungary
81 D3 Sarvestān Iran
77 D1 Saryarka plain Kazakh.
76 B2 Sarykamyshskoye Ozero salt l. Turkm./Uzbek.
77 D2 Saryozek Kazakh.
77 C2 Saryshagan Kazakh.
77 D2 Sarysu watercourse Kazakh.
77 D3 Sary-Tash Kyrg.
75 C2 Sasaram India
67 A4 Sasebo Japan
129 C2 Saskatchewan prov. Can.
129 D2 Saskatchewan r. Can.
129 D2 Saskatoon Can.
83 I2 Saskylakh Rus. Fed.
123 C2 Sasolburg S. Africa
87 D3 Sasovo Rus. Fed.
114 B4 Sassandra Côte d'Ivoire
108 A2 Sassari Italy
102 C1 Sassnitz Ger.
114 A3 Satadougou Mali
136 C3 Satanta U.S.A.
73 B3 Satara India
123 D1 Satara S. Africa
87 E3 Satka Rus. Fed.
74 B2 Satluj r. India/Pak.
75 C2 Satna India
77 C2 Satpayev Kazakh.
74 B2 Satpura Range mts India
67 B4 Satsuma-Sendai Japan
63 B2 Sattahip Thai.
110 B1 Satu Mare Romania
63 B3 Satun Thai.
144 B2 Saucillo Mex.
93 E4 Sauda Norway
92 □B2 Sauðárkrókur Iceland
78 B2 Saudi Arabia country Asia
105 C3 Saugues France
137 E1 Sauk Center U.S.A.
105 C2 Saulieu France
130 B3 Saulkrasti Latvia
138 C1 Sault Sainte Marie Can.
138 C1 Sault Sainte Marie U.S.A.
77 C1 Saumalkol' Kazakh.
59 C3 Saumlakki Indon.
104 B2 Saumur France
120 A1 Saurimo Angola
109 C1 Sava r. Europe
49 J5 Sava'i i. Samoa
91 E1 Savala r. Rus. Fed.
114 C4 Savalou Benin
141 D2 Savannah GA U.S.A.
140 C1 Savannah TN U.S.A.
141 D2 Savannah r. U.S.A.
63 B2 Savannakhét Laos
130 A2 Savant Lake Can.
111 C3 Savaştepe Turkey
114 C4 Savè Benin
105 D2 Saverne France
89 F2 Savino Rus. Fed.
86 D2 Savinskiy Rus. Fed.
108 A2 Savona Italy
93 I3 Savonlinna Fin.
105 D2 Savoie i. France see Savoy
105 D2 Savoy reg. France
93 F4 Sävsjö Sweden

59 C3 Savu i. Indon.
92 I2 Savukoski Fin.
Savu Sea sea Indon. see Laut Sawu
74 B2 Sawai Madhopur India
62 A1 Sawan Myanmar
62 A2 Sawankhalok Thai.
136 B3 Sawatch Range mts U.S.A.
Sawhāj Egypt see Sūhāj
121 B2 Sawmills Zimbabwe
79 C3 Şawqirah, Dawḥat b. Oma...
Şawqirah Bay b. Oman see Şawqirah, Dawḥat
53 D2 Sawtell Austr.
134 C2 Sawtooth Range mts U.S.A.
68 C1 Sayano-Shushenskoye Vodokhranilishche resr Rus. Fed.
76 C3 Saýat Turkm.
79 C3 Saybūt Yemen
93 I3 Säynätsalo Fin.
69 D2 Saynshand Mongolia
139 D2 Sayre U.S.A.
144 B3 Sayula Mex.
145 C3 Sayula Mex.
128 B2 Sayward Can.
89 E2 Sazonovo Rus. Fed.
114 B2 Sbaa Alg.
98 B2 Scafell Pike h. U.K.
109 C3 Scalea Italy
96 □ Scalloway U.K.
108 B2 Scandicci Italy
96 C1 Scapa Flow inlet U.K.
96 B2 Scarba i. U.K.
130 C3 Scarborough Can.
147 D3 Scarborough Trin. and To...
98 C2 Scarborough U.K.
64 A2 Scarborough Shoal sea feature S. China Sea
96 A2 Scarinish U.K.
Scarpanto i. Greece see Karpathos
100 B2 Schaerbeek Belgium
105 D2 Schaffhausen Switz.
100 B1 Schagen Neth.
102 C2 Schärding Austria
100 A2 Scharendijke Neth.
101 D1 Scharhörn i. Ger.
102 C1 Scheeßel Ger.
131 D2 Schefferville Can.
135 D3 Schell Creek Range mts U.S.A.
139 E2 Schenectady U.S.A.
143 D3 Schertz U.S.A.
101 E1 Scheßlitz Ger.
100 C1 Schiermonnikoog i. Neth.
101 E1 Schilde Belgium
108 B1 Schio Italy
101 F2 Schkeuditz Ger.
101 E1 Schladen Ger.
102 C2 Schladming Austria
102 C2 Schleiz Ger.
102 B1 Schleswig Ger.
101 D2 Schloß Holte-Stukenbro... Ger.
101 D2 Schlüchtern Ger.
101 E3 Schlüsselfeld Ger.
101 E2 Schmalkalden, Kurort Ger.
101 D2 Schmallenberg Ger.
Schmidt Island i. Rus. Fed. see Shmidta, Ostrov
101 F2 Schmölln Ger.
101 D1 Schneverdingen Ger.
101 E1 Schönebeck (Elbe) Ger.
101 E1 Schöningen Ger.
100 B2 Schoonhoven Neth.
59 D3 Schouten Islands P.N.G.
97 B3 Schull Ireland
101 E2 Schwabach Ger.
102 B2 Schwäbische Alb mts Ger.
101 F3 Schwandorf Ger.
61 C2 Schwaner, Pegunungan m... Indon.
101 E1 Schwarzenbek Ger.
101 F2 Schwarzenberg Ger.
122 A2 Schwarzrand mts Namibia
Schwarzwald mts Ger. see Black Forest
102 C2 Schwaz Austria
101 F1 Schwedt an der Oder Ger.
101 E2 Schweinfurt Ger.
101 E1 Schwerin Ger.
101 E1 Schweriner See l. Ger.
105 D2 Schwyz Switz.
108 B3 Sciacca Italy
95 B4 Scilly, Isles of U.K.
138 C3 Scioto r. U.S.A.
136 B1 Scobey U.S.A.
53 D2 Scone Austr.

110 B2 Scorniceşti Romania
55 C3 Scotia Ridge
S. Atlantic Ocean
149 F8 Scotia Sea S. Atlantic Ocean
96 C2 Scotland admin. div. U.K.
128 B2 Scott, Cape Can.
123 D3 Scottburgh S. Africa
136 C3 Scott City U.S.A.
136 C2 Scottsbluff U.S.A.
140 C2 Scottsboro U.S.A.
96 B1 Scourie U.K.
139 D2 Scranton U.S.A.
98 C3 Scunthorpe U.K.
105 E2 Scuol Switz.
Scutari Albania see Shkodër
99 D4 Seaford U.K.
98 C2 Seaham U.K.
129 E2 Seal r. Can.
122 B3 Seal, Cape S. Africa
52 B3 Sea Lake Austr.
143 D3 Sealy U.S.A.
140 B1 Searcy U.S.A.
98 B2 Seascale U.K.
134 B1 Seattle U.S.A.
139 E2 Sebago Lake U.S.A.
144 A2 Sebastián Vizcaíno, Bahía
b. Mex.
Sebastopol Ukr. see
Sevastopol'
Sebenico Croatia see
Šibenik
110 B1 Sebeş Romania
60 B2 Sebesi i. Indon.
88 C2 Sebezh Rus. Fed.
80 B1 Şebinkarahisar Turkey
141 D3 Sebring U.S.A.
61 C2 Sebuku i. Indon.
128 B3 Sechelt Can.
150 A3 Sechura Peru
73 B3 Secunderabad India
137 E3 Sedalia U.S.A.
105 C2 Sedan France
54 B2 Seddon N.Z.
114 A3 Sédhiou Senegal
142 A2 Sedona U.S.A.
101 E2 Seeburg Ger.
101 E1 Seehausen (Altmark) Ger.
122 A2 Seeheim Namibia
104 C2 Sées France
98 C2 Seesen Ger.
101 E1 Seevetal Ger.
123 C1 Sefare Botswana
76 C3 Sefid Küh, Selseleh-ye mts
Afgh.
93 F3 Segalstad Norway
60 B1 Segamat Malaysia
86 C2 Segezha Rus. Fed.
114 B3 Ségou Mali
106 C1 Segovia Spain
86 C2 Segozerskoye
Vodokhranilishche resr
Rus. Fed.
115 D2 Séguédine Niger
114 B4 Séguéla Côte d'Ivoire
143 D3 Seguin U.S.A.
107 C2 Segura r. Spain
106 C2 Segura, Sierra de mts Spain
120 B3 Sehithwa Botswana
93 H3 Seinäjoki Fin.
104 C2 Seine r. France
104 B2 Seine, Baie de b. France
105 C2 Seine, Val de val. France
103 E1 Sejny Pol.
60 B2 Sekayu Indon.
114 B4 Sekondi Ghana
59 C3 Selaru i. Indon.
61 C2 Selatan, Tanjung pt Indon.
126 B2 Selawik U.S.A.
61 D2 Selayar, Pulau i. Indon.
98 C3 Selby U.K.
136 B1 Selby U.S.A.
111 C3 Selçuk Turkey
120 B3 Selebi-Pikwe Botswana
Selebi-Phikwe Botswana see
Selebi-Pikwe
105 D2 Sélestat France
Seletyteniz, Oz.
salt l. Kazakh. see
Siletiteniz, Ozero
92 □A3 Selfoss Iceland
14 A3 Séliláeli Maur.
42 A1 Seligman U.S.A.
116 A2 Selima Oasis Sudan
111 C3 Selimiye Turkey
114 B3 Sélingué, Lac de l. Mali
89 D2 Selizharovo Rus. Fed.
93 E4 Seljord Norway
29 E2 Selkirk Can.
96 C3 Selkirk U.K.

128 C2 Selkirk Mountains Can.
142 A2 Sells U.S.A.
140 C2 Selma AL U.S.A.
135 C3 Selma CA U.S.A.
105 D2 Selongey France
99 C4 Selsey Bill hd U.K.
89 D3 Sel'tso Rus. Fed.
Selukwe Zimbabwe see
Shurugwi
150 B3 Selvas reg. Brazil
134 C1 Selway r. U.S.A.
129 D1 Selwyn Lake Can.
128 A1 Selwyn Mountains Can.
51 C2 Selwyn Range hills Austr.
60 B2 Semangka, Teluk b. Indon.
59 D3 Semarang Indon.
60 B1 Sematan Malaysia
118 B2 Sembé Congo
81 C2 Şemdinli Turkey
91 C1 Semenivka Ukr.
87 D3 Semenov Rus. Fed.
61 C2 Semeru, Gunung vol. Indon.
77 E1 Semey Kazakh.
91 E2 Semikarakorsk Rus. Fed.
89 E3 Semiluki Rus. Fed.
136 B2 Seminoe Reservoir U.S.A.
143 C2 Seminole U.S.A.
141 D2 Seminole, Lake U.S.A.
61 C1 Semitau Indon.
Sem Kolodezey Ukr. see
Lenine
81 D2 Semnän Iran
61 C1 Semporna Malaysia
105 C2 Semur-en-Auxois France
Semyonovskoye Rus. Fed.
see Bereznik
Semyonovskoye Rus. Fed.
see Ostrovskoye
150 C3 Sena Madureira Brazil
120 B2 Senanga Zambia
67 D3 Sendai Japan
141 D2 Seneca U.S.A.
114 A3 Senegal country Africa
114 A3 Sénégal r. Maur./Senegal
102 C1 Senftenberg Ger.
119 D3 Sengerema Tanz.
61 D2 Sengkang Indon.
151 E4 Senhor do Bonfim Brazil
103 D2 Senica Slovakia
108 B2 Senigallia Italy
109 B2 Senj Croatia
92 G2 Senja i. Norway
122 B2 Senlac S. Africa
105 C2 Senlis France
63 B2 Senmonorom Cambodia
116 B3 Sennar Sudan
130 C3 Senneterre Can.
123 C3 Senqu r. Lesotho
105 C2 Sens France
109 D1 Senta Serbia
128 B2 Sentinel Peak Can.
123 C1 Senwabarwana S. Africa
65 B3 Seocheon S. Korea
65 B2 Seongnam S. Korea
75 B2 Seoni India
65 B2 Seosan S. Korea
65 B2 Seoul S. Korea
155 D2 Sepetiba, Baía de b. Brazil
59 D3 Sepik r. P.N.G.
61 C1 Sepinang Indon.
131 D2 Sept-Îles Can.
87 D4 Serafimovich Rus. Fed.
100 B2 Seraing Belgium
59 C3 Seram i. Indon.
60 B2 Serang Indon.
60 B1 Serasan, Selat sea chan.
Indon.
109 D2 Serbia country Europe
76 B3 Serdar Turkm.
117 C3 Serdo Eth.
89 E3 Serebryanyye Prudy
Rus. Fed.
60 B1 Seremban Malaysia
119 D3 Serengeti Plain Tanz.
121 C2 Serenje Zambia
90 B2 Seret r. Ukr.
87 D3 Sergach Rus. Fed.
86 F2 Sergino Rus. Fed.
89 E2 Sergiyev Posad Rus. Fed.
Sergo Ukr. see Stakhanov
74 A1 Serhetabat Turkm.
61 C1 Seria Brunei
61 C1 Serian Malaysia
111 B3 Serifos i. Greece
80 B2 Serik Turkey
59 C3 Sermata, Kepulauan is
Indon.
Sernyy Zavod Turkm. see
Kükürtli
86 F3 Serov Rus. Fed.

120 B3 Serowe Botswana
106 B2 Serpa Port.
Serpa Pinto Angola see
Menongue
89 D3 Serpukhov Rus. Fed.
155 D2 Serra Brazil
155 C1 Serra das Araras Brazil
108 A3 Serramanna Italy
154 B1 Serranópolis Brazil
100 A3 Serre r. France
111 B2 Serres Greece
151 F4 Serrinha Brazil
155 D1 Sêrro Brazil
154 C2 Sertãozinho Brazil
59 D3 Serui Indon.
120 B3 Serule Botswana
61 C2 Seruyan r. Indon.
68 C2 Sêrxü China
120 A2 Sesfontein Namibia
108 B2 Sessa Aurunca Italy
108 A2 Sestri Levante Italy
105 C3 Sète France
155 D1 Sete Lagoas Brazil
92 G2 Setermoen Norway
93 E4 Setesdal val. Norway
115 C1 Sétif Alg.
67 B4 Seto-naikai sea Japan
114 B1 Settat Morocco
98 B2 Settle U.K.
106 B2 Setúbal Port.
106 B2 Setúbal, Baía de b. Port.
130 A2 Seul, Lac l. Can.
81 C1 Sevan Armenia
76 A2 Sevan, Lake Armenia
Sevana Lich l. Armenia see
Sevan, Lake
91 C3 Sevastopol' Ukr.
Seven Islands Can. see
Sept-Îles
131 D2 Seven Islands Bay Can.
99 D4 Sevenoaks U.K.
105 C3 Sévérac-le-Château France
130 B2 Severn r. Can.
122 B2 Severn S. Africa
99 B4 Severn r. U.K.
86 D2 Severnaya Dvina r. Rus. Fed.
83 H1 Severnaya Zemlya is
Rus. Fed.
86 D2 Severnyy Rus. Fed.
86 F2 Severnyy Rus. Fed.
82 E2 Severnyy, Ostrov i.
Rus. Fed.
83 I3 Severobaykal'sk Rus. Fed.
86 C2 Severodvinsk Rus. Fed.
83 L3 Severo-Kuril'sk Rus. Fed.
92 J2 Severomorsk Rus. Fed.
86 C2 Severoonezhsk Rus. Fed.
83 H2 Severo-Yeniseyskiy
Rus. Fed.
91 D3 Sevsk Rus. Fed.
135 D3 Sevier r. U.S.A.
135 D3 Sevier Lake U.S.A.
Sevilla Spain see Seville
106 B2 Seville Spain
Sevlush Ukr. see
Vynohradiv
89 D3 Sevsk Rus. Fed.
126 C2 Seward U.S.A.
126 B2 Seward Peninsula U.S.A.
128 A2 Sewell Inlet Can.
128 C2 Sexsmith Can.
144 B2 Sextín r. Mex.
86 G1 Seyakha Rus. Fed.
113 I6 Seychelles country
Indian Ocean
92 □C2 Seyðisfjörður Iceland
80 B2 Seyhan Turkey see Adana
91 C1 Seym r. Rus. Fed./Ukr.
83 L2 Seymchan Rus. Fed.
53 C3 Seymour Austr.
123 C3 Seymour S. Africa
138 B3 Seymour IN U.S.A.
143 D2 Seymour TX U.S.A.
105 C2 Sézanne France
108 B2 Sezze Italy
110 C1 Sfântu Gheorghe Romania
115 D1 Sfax Tunisia
Sfîntu Gheorghe Romania
see Sfântu Gheorghe
's-Gravenhage Neth. see
The Hague
96 A2 Sgurr Alasdair h. U.K.
70 A2 Shaanxi prov. China
Shabani Zimbabwe see
Zvishavane
91 D2 Shabel'skoye Rus. Fed.
77 D3 Shache China
55 C1 Shackleton Range mts
Antarctica

86 F3 Shadrinsk Rus. Fed.
99 B4 Shaftesbury U.K.
126 B2 Shageluk U.S.A.
Shāhābād Iran see
Eslāmābād-e Gharb
74 A2 Shahdad Kot Pak.
75 C2 Shahdol India
77 C3 Shāh Fōlādī mt. Afgh.
75 B2 Shahjahanpur India
76 B3 Shāh Kūh mt. Iran
81 D2 Shahr-e Bābak Iran
81 D2 Shahr-e Kord Iran
81 D2 Shāhrezā Iran
77 C3 Shahrisabz Uzbek.
Shāhrūd Iran see Emāmrūd
79 B2 Shaj'ah, Jabal h.
Saudi Arabia
89 E2 Shakhovskaya Rus. Fed.
Shakhterskoye Ukr. see
Pershotravens'k
Shakhty Rus. Fed. see
Gusinoozersk
91 E2 Shakhty Rus. Fed.
Shakhtyorskoye Ukr. see
Pershotravens'k
86 D3 Shakhun'ya Rus. Fed.
114 C4 Shaki Nigeria
66 C2 Shakotan-hantō pen. Japan
66 C2 Shakotan-misaki c. Japan
76 B2 Shalkar Kazakh.
76 C2 Shalkar Kazakh.
Shalkarteniz, Solonchak
salt marsh Kazakh.
68 C2 Shaluli Shan mts China
129 E2 Shamattawa Can.
143 C1 Shamrock U.S.A.
70 B2 Shancheng China
70 A2 Shandan China
70 B2 Shandong prov. China
70 C2 Shandong Bandao pen.
China
121 B3 Shangani Zimbabwe
121 B2 Shangani r. Zimbabwe
70 B1 Shangdu China
70 C2 Shanghai China
70 C2 Shanghai mun. China
71 B3 Shanghang China
70 A2 Shangluo China
70 B2 Shangnan China
71 B3 Shangrao China
70 B2 Shangshui China
77 E2 Shangyou Shuiku resr
China
70 C2 Shangyu China
69 E1 Shangzhi China
Shangzhou China see
Shangluo
134 B1 Shaniko U.S.A.
97 B2 Shannon r. Ireland
97 B2 Shannon, Mouth of the
Ireland
62 A1 Shan Plateau Myanmar
Shansi prov. China see
Shanxi
71 B3 Shantou China
Shantung prov. China see
Shandong
70 B2 Shanxi prov. China
71 B3 Shaoguan China
71 B3 Shaowu China
71 C3 Shaoxing China
71 B3 Shaoyang China
96 C1 Shapinsay i. U.K.
116 C2 Shaqrā' Saudi Arabia
90 B2 Sharhorod Ukr.
79 C2 Sharjah U.A.E.
88 C3 Sharkawshchyna Belarus
50 A2 Shark Bay Austr.
78 A2 Sharm ash Shaykh Egypt
138 C2 Sharon U.S.A.
86 D3 Shar'ya Rus. Fed.
121 B3 Shashe r. Botswana/
Zimbabwe
117 B4 Shashemenë Eth.
Shashi China see Jingzhou
134 B2 Shasta, Mount vol. U.S.A.
134 B2 Shasta Lake U.S.A.
115 D2 Shāţi', Wādī ash watercourse
Libya
Shatilki Belarus see
Svyetlahorsk
89 E2 Shatura Rus. Fed.
129 D3 Shaunavon Can.
138 B2 Shawano U.S.A.
130 C3 Shawinigan Can.
143 D1 Shawnee U.S.A.
83 M2 Shaybovreym r. Rus. Fed.
50 B2 Shay Gap (abandoned)
Austr.
89 E3 Shchekino Rus. Fed.
89 E2 Shchelkovo Rus. Fed.

223

Shcherbakov Rus. Fed. see Rybinsk
Shcherbinovka Ukr. see Dzerzhyns'k
89 E3 Shchigry Rus. Fed.
91 C1 Shchors Ukr.
88 B3 Shchuchyn Belarus
91 D1 Shebekino Rus. Fed.
117 C4 Shebelē Wenz, Wabē r. Ethiopia/Somalia
138 B2 Sheboygan U.S.A.
91 D3 Shebsh r. Rus. Fed.
97 C2 Sheelin, Lough l. Ireland
136 B3 Sheep Mountain U.S.A.
99 D4 Sheerness U.K.
98 C3 Sheffield U.K.
143 C2 Sheffield U.S.A.
Sheikh Othman Yemen see Ash Shaykh 'Uthman
Shekhem West Bank see Nāblus
89 E2 Sheksna Rus. Fed.
89 E2 Sheksninskoye Vodokhranilishche resr Rus. Fed.
83 M2 Shelagskiy, Mys pt Rus. Fed.
131 D3 Shelburne Can.
138 B2 Shelby MI U.S.A.
134 D1 Shelby MT U.S.A.
141 D1 Shelby NC U.S.A.
138 B3 Shelbyville IN U.S.A.
140 C1 Shelbyville TN U.S.A.
83 L2 Shelikhova, Zaliv g. Rus. Fed.
126 B3 Shelikof Strait U.S.A.
129 D2 Shellbrook Can.
Shelter Bay Can. see Port-Cartier
134 B1 Shelton U.S.A.
137 D2 Shenandoah U.S.A.
139 D3 Shenandoah r. U.S.A.
139 D3 Shenandoah Mountains U.S.A.
118 A2 Shendam Nigeria
Shengli Feng mt. China/Kyrg. see Pobeda Peak
86 D2 Shenkursk Rus. Fed.
70 B2 Shenmu China
Shensi prov. China see Shaanxi
70 C1 Shenyang China
71 B3 Shenzhen China
90 B1 Shepetivka Ukr.
53 C3 Shepparton Austr.
99 D4 Sheppey, Isle of i. U.K.
131 D3 Sherbrooke N.S. Can.
130 C3 Sherbrooke Que. Can.
97 C2 Shercock Ireland
116 B3 Shereiq Sudan
136 B2 Sheridan U.S.A.
52 A2 Sheringa Austr.
86 F2 Sherkaly Rus. Fed.
143 D2 Sherman U.S.A.
100 B2 's-Hertogenbosch Neth.
96 □ Shetland Islands U.K.
76 B2 Shetpe Kazakh.
Shevchenko Kazakh. see Aktau
91 D2 Shevchenkove Ukr.
137 D1 Sheyenne r. U.S.A.
79 B3 Shibām Yemen
67 C3 Shibata Japan
66 D2 Shibetsu Japan
77 C3 Shibirghān Afgh.
71 B3 Shicheng China
70 C2 Shidao China
96 B2 Shiel, Loch l. U.K.
Shigatse China see Xigazê
77 E2 Shiezi China
Shihkiachwang China see Shijiazhuang
Shijiao China see Fogang
70 B2 Shijiazhuang China
Shijiusuo China see Rizhao
74 A2 Shikarpur Pak.
67 B4 Shikoku i. Japan
66 D2 Shikotsu-ko l. Japan
86 D2 Shilega Rus. Fed.
75 C2 Shiliguri India
77 D2 Shilik Kazakh.
97 C2 Shillelagh Ireland
75 D2 Shillong India
89 F3 Shilovo Rus. Fed.
69 E1 Shimanovsk Rus. Fed.
117 C4 Shimbiris mt. Somalia
67 C3 Shimizu Japan
74 B1 Shimla India
67 C4 Shimoda Japan

73 B3 Shimoga India
66 D2 Shimokita-hantō pen. Japan
67 B4 Shimonoseki Japan
89 D2 Shimsk Rus. Fed.
96 B1 Shin, Loch l. U.K.
62 A1 Shingbwiyang Myanmar
67 C4 Shingū Japan
123 D1 Shingwedzi S. Africa
123 D1 Shingwedzi r. S. Africa
66 D3 Shinjō Japan
119 D3 Shinyanga Tanz.
67 D3 Shiogama Japan
67 C4 Shiono-misaki c. Japan
71 A3 Shiping China
98 C3 Shipley U.K.
142 B1 Shiprock U.S.A.
71 A3 Shiqian China
Shiqizhen China see Zhongshan
70 A2 Shiquan China
78 B2 Shi'r, Jabal h. Saudi Arabia
67 C3 Shirane-san mt. Japan
67 C3 Shirane-san vol. Japan
81 D3 Shīrāz Iran
66 D2 Shiretoko-misaki c. Japan
66 D2 Shiriya-zaki c. Japan
76 B3 Shīrvān Iran
74 B2 Shiv India
Shivamogga India see Shimoga
74 B2 Shivpuri India
70 B2 Shiyan China
77 C2 Shiyeli Kazakh.
70 B2 Shizhong China
70 A2 Shizuishan China
67 C4 Shizuoka Japan
89 D3 Shklow Belarus
109 C2 Shkodër Albania
83 H1 Shmidta, Ostrov i. Rus. Fed.
67 B4 Shōbara Japan
158 F8 Sholapur India see Solapur
Shona Ridge S. Atlantic Ocean
62 A1 Sho'rchi Uzbek.
74 B1 Shorkot Pak.
135 C3 Shoshone CA U.S.A.
134 D2 Shoshone ID U.S.A.
135 C3 Shoshone Mountains U.S.A.
123 C1 Shoshong Botswana
91 C1 Shostka Ukr.
70 B2 Shouxian China
78 A3 Showak Sudan
142 A2 Show Low U.S.A.
91 C2 Shpola Ukr.
140 B2 Shreveport U.S.A.
99 B3 Shrewsbury U.K.
77 D2 Shu Kazakh.
Shuangjiang China see Tongdao
62 A1 Shuangjiang China
Shuangxi China see Shunchang
66 B1 Shuangyashan China
87 E4 Shubarkuduk Kazakh.
116 B1 Shubrā al Khaymah Egypt
89 D2 Shugozero Rus. Fed.
Shuidong China see Dianbai
120 B2 Shumba Zimbabwe
110 C2 Shumen Bulg.
87 F3 Shumikha Rus. Fed.
88 C2 Shumilina Belarus
143 C3 Shumla U.S.A.
89 D3 Shumyachi Rus. Fed.
78 B3 Shunchang China
126 B2 Shungnak U.S.A.
78 B3 Shuqrah Yemen
121 C2 Shurugwi Zimbabwe
89 F2 Shushkodom Rus. Fed.
81 C2 Shushtar Iran
128 C2 Shuswap Lake Can.
89 F2 Shuya Rus. Fed.
89 F2 Shuyskoye Rus. Fed.
62 A1 Shwebo Myanmar
62 A1 Shwedwin Myanmar
62 A2 Shwegun Myanmar
62 A2 Shwegyin Myanmar
62 A1 Shweli r. Myanmar
77 D2 Shyganak Kazakh.
77 C2 Shymkent Kazakh.
74 B1 Shyok r. India/Pak.
74 B1 Shyok India/Pak.
90 C2 Shyroke Ukr.
59 C3 Sia Indon.
74 A2 Siahan Range mts Pak.
74 B1 Sialkot Pak.
Siam country Asia see Thailand

Sian China see Xi'an
64 B3 Siargao i. Phil.
64 B3 Siasi Phil.
88 B2 Šiauliai Lith.
123 D1 Sibasa S. Africa
109 C2 Šibenik Croatia
83 G2 Siberia reg. Rus. Fed.
60 A2 Siberut i. Indon.
74 A2 Sibi Pak.
Sibir' reg. Rus. Fed. see Siberia
118 B3 Sibiti Congo
110 B1 Sibiu Romania
60 A1 Sibolga Indon.
61 C1 Sibu Malaysia
118 B2 Sibut C.A.R.
64 A3 Sibutu i. Phil.
64 B3 Sibuyan i. Phil.
64 B2 Sibuyan Sea Phil.
128 C2 Sicamous Can.
63 A3 Sichon Thai.
70 A2 Sichuan prov. China
70 A3 Sichuan Pendi basin China
105 D3 Sicié, Cap c. France
Sicilia i. Italy see Sicily
108 B3 Sicilian Channel Italy/Tunisia
108 B3 Sicily i. Italy
150 B4 Sicuani Peru
111 C3 Sideros, Akrotirio pt Greece
122 B3 Sidesaviwa S. Africa
75 C2 Sidhi India
74 B2 Sidhpur India
107 D2 Sidi Aïssa Alg.
107 D2 Sidi Ali Alg.
114 B1 Sidi Bel Abbès Alg.
114 A2 Sidi Ifni Morocco
114 B1 Sidi Kacem Morocco
60 A1 Sidikalang Indon.
111 B2 Sidirokastro Greece
96 C2 Sidlaw Hills U.K.
99 B4 Sidmouth U.K.
134 B1 Sidney Can.
136 C1 Sidney MT U.S.A.
136 C2 Sidney NE U.S.A.
138 C2 Sidney OH U.S.A.
141 D2 Sidney Lanier, Lake U.S.A.
61 D1 Sidoan Indon.
80 B2 Sidon Lebanon
154 B2 Sidrolândia Brazil
103 E1 Siedlce Pol.
100 C2 Sieg r. Ger.
101 D2 Siegen Ger.
63 B2 Siĕmréab Cambodia
Siem Reap Cambodia see Siĕmréab
108 B2 Siena Italy
103 D1 Sieradz Pol.
142 B2 Sierra Blanca U.S.A.
153 B4 Sierra Grande Arg.
114 A4 Sierra Leone country Africa
158 E4 Sierra Leone Basin N. Atlantic Ocean
158 E4 Sierra Leone Rise N. Atlantic Ocean
144 B2 Sierra Mojada Mex.
142 A2 Sierra Vista U.S.A.
105 D2 Sierre Switz.
116 C3 Sīfenī Eth.
111 B3 Sifnos i. Greece
107 C2 Sig Alg.
127 I2 Sigguup Nunaa pen. Greenland
110 B1 Sighetu Marmaţiei Romania
110 B1 Sighişoara Romania
60 A1 Sigli Indon.
92 □B2 Siglufjörður Iceland
102 B2 Sigmaringen Ger.
100 B3 Signy-l'Abbaye France
106 C1 Sigüenza Spain
114 B3 Siguiri Guinea
88 B2 Sigulda Latvia
63 B2 Sihanoukville Cambodia
92 I3 Siilinjärvi Fin.
81 C2 Siirt Turkey
60 B2 Sijunjung Indon.
74 B2 Sikar India
74 A1 Sīkaram mt. Afgh.
114 B3 Sikasso Mali
137 F3 Sikeston U.S.A.
66 B2 Sikhote-Alin' mts Rus. Fed.
111 C3 Sikinos i. Greece
103 D2 Siklós Hungary
65 A2 Sikuaishi China
88 B2 Šilalė Lith.
144 B2 Silao Mex.
101 D1 Silberberg h. Ger.
75 D2 Silchar India

77 D1 Siletyteniz, Ozero salt l. Kazakh.
75 C2 Silgarhi Nepal
80 B2 Silifke Turkey
75 C1 Siling Co salt l. China
110 C2 Silistra Bulg.
111 C2 Silivri Turkey
93 F3 Siljan l. Sweden
93 E4 Silkeborg Denmark
88 C2 Sillamäe Estonia
98 B2 Silloth U.K.
140 B1 Siloam Springs U.S.A.
123 D2 Silobela S. Africa
60 B1 Siluas Indon.
88 B2 Šilutė Lith.
81 C2 Silvan Turkey
154 C1 Silvânia Brazil
74 B2 Silvassa India
137 E1 Silver Bay U.S.A.
142 B2 Silver City U.S.A.
136 B3 Silverton U.S.A.
62 B1 Simao China
130 C3 Simard, Lac l. Can.
111 C3 Simav Turkey
111 C3 Simav Dağları mts Turkey
118 C2 Simba Dem. Rep. Congo
138 C2 Simcoe Can.
139 D2 Simcoe, Lake Can.
78 A3 Simēn Eth.
60 A1 Simeulue i. Indon.
91 C3 Simferopol' Ukr.
75 C2 Simikot Nepal
135 C4 Simi Valley U.S.A.
Simla India see Shimla
110 B1 Şimleu Silvaniei Romania
100 C3 Simmern (Hunsrück) Ger.
92 I2 Simo Fin.
129 D2 Simonhouse Can.
60 B2 Simpang Indon.
51 C2 Simpson Desert Austr.
93 F4 Simrishamn Sweden
60 A1 Sinabang Indon.
116 B2 Sinai pen. Egypt
105 E3 Sinalunga Italy
71 A3 Sinan China
65 B2 Sinanju N. Korea
62 A1 Sinbo Myanmar
62 A1 Sinbyugyun Myanmar
150 B2 Sincelejo Col.
60 B2 Sindangbarang Indon.
111 C3 Sındırgı Turkey
86 E2 Sindor Rus. Fed.
111 C2 Sinekçi Turkey
106 B2 Sines Port.
106 B2 Sines, Cabo de c. Port.
116 B3 Singa Sudan
75 C2 Singahi India
60 B1 Singapore country Asia
61 C2 Singaraja Indon.
63 B2 Sing Buri Thai.
119 D3 Singida Tanz.
62 A1 Singkaling Hkamti Myanmar
60 B1 Singkawang Indon.
60 B2 Singkep i. Indon.
60 A1 Singkil Indon.
53 D2 Singleton Austr.
Sin'gosan N. Korea see Kosan
62 A1 Singu Myanmar
Sining China see Xining
108 A2 Siniscola Italy
109 C2 Sinj Croatia
61 D2 Sinjai Indon.
116 B3 Sinkat Sudan
151 D2 Sinnamary Fr. Guiana
Sinnicolau Mare Romania see Sânnicolau Mare
Sinoia Zimbabwe see Chinhoyi
80 B1 Sinop Turkey
65 B1 Sinp'o N. Korea
61 C1 Sintang Indon.
100 B2 Sint Anthonis Neth.
100 A2 Sint-Laureins Belgium
147 D3 Sint-Maarten terr. West Indies
100 B2 Sint-Niklaas Belgium
143 D3 Sinton U.S.A.
65 A1 Sinŭiju N. Korea
64 B3 Siocon Phil.
103 D2 Siófok Hungary
105 D2 Sion Switz.
137 D2 Sioux Center U.S.A.
137 D2 Sioux City U.S.A.
137 D2 Sioux Falls U.S.A.
130 A2 Sioux Lookout Can.
65 A1 Siping China
129 E2 Sipiwesk Lake Can.
55 P2 Siple, Mount Antarctica

55 P2	Siple Island Antarctica
	Sipolilo Zimbabwe see Guruve
60 A2	Sipura i. Indon.
64 B3	Siquijor Phil.
93 E4	Sira r. Norway
	Siracusa Italy see Syracuse
51 C1	Sir Edward Pellew Group is Austr.
10 C1	Siret Romania
10 C1	Siret r. Romania
78 A1	Sirhān, Wādī an watercourse Saudi Arabia
79 C2	Sīrīk Iran
61 C1	Sirik, Tanjung pt Malaysia
62 B2	Siri Kit, Khuan Thai.
28 B1	Sir James MacBrien, Mount Can.
79 C2	Sīrjān Iran
81 C2	Şırnak Turkey
74 B2	Sirohi India
60 A1	Sirombu Indon.
74 B2	Sirsa India
15 D1	Sirte Libya
15 D1	Sirte, Gulf of Libya
81 C2	Şirvan Azer.
88 B2	Širvintos Lith.
09 C1	Sisak Croatia
11 C2	Sisaket Thai.
45 C2	Sisal Mex.
22 B2	Sishen S. Africa
81 C2	Sisian Armenia
27 I2	Sisimiut Greenland
29 D2	Sisipuk Lake Can.
63 B2	Sisŏphŏn Cambodia
05 D3	Sisteron France
	Sitang China see Sinan
75 C2	Sitapur India
11 C3	Siteia Greece
23 D2	Siteki Swaziland
28 A2	Sitka U.S.A.
00 B2	Sittard Neth.
62 A1	Sittaung Myanmar
62 A2	Sittaung r. Myanmar
61 C2	Sittwe Myanmar
61 C2	Situbondo Indon.
80 B2	Sivas Turkey
62 A1	Sivasagar India
11 C3	Sivaslı Turkey
80 B2	Siverek Turkey
88 D2	Siverskiy Rus. Fed.
80 B2	Sivrihisar Turkey
16 A2	Sīwah Egypt
75 B1	Siwalik Range mts India/Nepal
	Siwa Oasis oasis Egypt see Wāḥāt Sīwah
05 D3	Six-Fours-les-Plages France
70 B2	Sixian China
23 C2	Siyabuswa S. Africa
	Sjælland i. Denmark see Zealand
09 D2	Sjenica Serbia
92 G2	Sjøvegan Norway
91 C2	Skadovs'k Ukr.
93 F4	Skagen Denmark
93 E4	Skagerrak str. Denmark/Norway
34 B1	Skagit r. U.S.A.
28 A2	Skagway U.S.A.
92 G2	Skaland Norway
93 F4	Skara Sweden
96 C1	Skara Brae tourist site U.K.
74 B1	Skardu Pak.
03 E1	Skarżysko-Kamienna Pol.
03 D2	Skawina Pol.
14 A2	Skaymat Western Sahara
28 B2	Skeena r. Can.
28 B2	Skeena Mountains Can.
98 D3	Skegness U.K.
92 H3	Skellefteå Sweden
92 H3	Skellefteälven r. Sweden
97 C2	Skerries Ireland
93 F4	Ski Norway
11 B3	Skiathos i. Greece
97 B3	Skibbereen Ireland
92 □B2	Skíðadals-jökull glacier Iceland
98 B2	Skiddaw h. U.K.
93 E4	Skien Norway
03 E1	Skierniewice Pol.
15 C1	Skikda Alg.
52 B3	Skipton Austr.
98 B2	Skipton U.K.
93 E4	Skive Denmark
92 □B2	Skjervøy Norway
	Skobelev Uzbek. see Farg'ona
11 B3	Skopelos i. Greece
89 E3	Skopin Rus. Fed.
111 B2	Skopje Macedonia
111 C3	Skoutaros Greece
93 F4	Skövde Sweden
83 J3	Skovorodino Rus. Fed.
139 F2	Skowhegan U.S.A.
92 H2	Skröven Sweden
88 B2	Skrunda Latvia
128 A1	Skukum, Mount Can.
123 D1	Skukuza S. Africa
88 B2	Skuodas Lith.
90 B2	Skvyra Ukr.
96 A2	Skye i. U.K.
111 B3	Skyros Greece
111 B3	Skyros i. Greece
93 F4	Slagelse Denmark
60 B2	Slamet, Gunung vol. Indon.
97 C2	Slaney r. Ireland
88 C2	Slantsy Rus. Fed.
109 C1	Slatina Croatia
110 B2	Slatina Romania
143 C2	Slaton U.S.A.
129 C1	Slave r. Can.
114 C4	Slave Coast Africa
128 C2	Slave Lake Can.
77 D1	Slavgorod Rus. Fed.
88 C2	Slavkovichi Rus. Fed.
	Slavonska Požega Croatia see Požega
109 C1	Slavonski Brod Croatia
90 B1	Slavuta Ukr.
90 C1	Slavutych Ukr.
66 B2	Slavyanka Rus. Fed.
	Slavyanskaya Rus. Fed. see Slavyansk-na-Kubani
91 D2	Slavyansk-na-Kubani Rus. Fed.
89 D3	Slawharad Belarus
103 D1	Sławno Pol.
99 C3	Sleaford U.K.
97 A2	Slea Head hd Ireland
130 C2	Sleeper Islands Can.
97 D1	Slieve Donard h. U.K.
	Slieve Gamph hills Ireland see Ox Mountains
96 A2	Sligachan U.K.
97 B1	Sligo Ireland
97 B1	Sligo Bay Ireland
93 G4	Slite Sweden
110 C2	Sliven Bulg.
	Sloboda Rus. Fed. see Ezhva
110 C2	Slobozia Romania
88 C3	Slonim Belarus
100 B1	Sloten Neth.
99 C4	Slough U.K.
103 D2	Slovakia country Europe
108 B1	Slovenia country Europe
91 D2	Slov"yans'k Ukr.
102 C1	Słubice Pol.
90 B1	Sluch r. Ukr.
100 A2	Sluis Neth.
103 D1	Słupsk Pol.
88 C3	Slutsk Belarus
97 A2	Slyne Head hd Ireland
68 C1	Slyudyanka Rus. Fed.
131 D2	Smallwood Reservoir Can.
88 C3	Smalyavichy Belarus
88 C3	Smarhon' Belarus
129 D2	Smeaton Can.
109 D2	Smederevo Serbia
109 D2	Smederevska Palanka Serbia
91 C2	Smila Ukr.
88 C3	Smilavichy Belarus
88 C2	Smiltene Latvia
137 D3	Smith Center U.S.A.
128 B2	Smithers Can.
141 E1	Smithfield NC U.S.A.
134 D2	Smithfield UT U.S.A.
139 D3	Smith Mountain Lake U.S.A.
130 C2	Smiths Falls Can.
53 D2	Smithton Austr.
53 D2	Smoky Cape Austr.
137 D3	Smoky Hills U.S.A.
92 E3	Smøla i. Norway
89 D3	Smolensk Rus. Fed.
89 D3	Smolensko-Moskovskaya Vozvyshennost' hills Belarus/Rus. Fed.
111 B2	Smolyan Bulg.
66 B2	Smolyaninovo Rus. Fed.
130 B3	Smooth Rock Falls Can.
	Smyrna Turkey see İzmir
91 D2	Smyrnove Ukr.
92 □B3	Snæfell mt. Iceland
98 A2	Snaefell h. Isle of Man
128 A1	Snag (abandoned) Can.
134 C1	Snake r. U.S.A.
134 D2	Snake River Plain U.S.A.
	Snare Lakes Can. see Wekweètì
92 F3	Snåsvatn l. Norway
100 B1	Sneek Neth.
97 B3	Sneem Ireland
122 B3	Sneeuberge mts S. Africa
	Snegurovka Ukr. see Tetiyiv
103 D1	Sněžka mt. Czech Rep.
108 B1	Snežnik mt. Slovenia
103 E1	Śniardwy, Jezioro l. Pol.
	Sniečkus Lith. see Visaginas
91 C2	Snihurivka Ukr.
93 E3	Snøhetta mt. Norway
	Snovsk Ukr. see Shchors
129 D1	Snowbird Lake Can.
99 A3	Snowdon mt. U.K.
	Snowdrift Can. see Łutselk'e
129 C1	Snowdrift r. Can.
142 A2	Snowflake U.S.A.
129 D2	Snow Lake Can.
134 C1	Snowshoe Peak U.S.A.
52 A2	Snowtown Austr.
53 C3	Snowy r. Austr.
53 C3	Snowy Mountains Austr.
143 C2	Snyder U.S.A.
121 □D2	Soalala Madag.
121 □D2	Soanierana-Ivongo Madag.
90 B2	Sob r. Ukr.
65 B2	Sobaek-sanmaek mts S. Korea
117 B4	Sobat r. S. Sudan
89 F2	Sobinka Rus. Fed.
151 E4	Sobradinho, Barragem de resr Brazil
151 E3	Sobral Brazil
91 D3	Sochi Rus. Fed.
49 L5	Society Islands Fr. Polynesia
150 B2	Socorro Col.
142 B2	Socorro NM U.S.A.
142 B2	Socorro TX U.S.A.
144 A3	Socorro, Isla i. Mex.
56 B4	Socotra i. Yemen
63 B3	Soc Trang Vietnam
106 C2	Socuéllamos Spain
92 I2	Sodankylä Fin.
134 D2	Soda Springs U.S.A.
93 G4	Söderhamn Sweden
93 G4	Södertälje Sweden
116 A3	Sodiri Sudan
117 B4	Sodo Eth.
93 G3	Södra Kvarken str. Fin./Sweden
	Soerabaia Indon. see Surabaya
	Soest Ger.
53 C2	Sofala Austr.
110 B2	Sofia Bulg.
121 □D2	Sofia r. Madag.
	Sofiya Bulg. see Sofia
	Sofiyevka Ukr. see Vil'nyans'k
75 D1	Sog China
93 E3	Sognefjorden inlet Norway
111 D2	Söğüt Turkey
	Sohâg Egypt see Sūhāj
	Sohar Oman see Şuḩār
100 B2	Soignies Belgium
105 C2	Soissons France
90 A1	Sokal' Ukr.
65 B2	Sokcho S. Korea
111 C3	Söke Turkey
81 C1	Sokhumi Georgia
114 C4	Sokodé Togo
89 F2	Sokol Rus. Fed.
101 F2	Sokolov Czech Rep.
115 C3	Sokoto Nigeria
115 C3	Sokoto r. Nigeria
90 B2	Sokyryany Ukr.
73 B3	Solapur India
135 B3	Soledad U.S.A.
89 F2	Soligalich Rus. Fed.
99 C3	Solihull U.K.
86 E3	Solikamsk Rus. Fed.
87 E3	Sol'-Iletsk Rus. Fed.
100 C2	Solingen Ger.
122 A1	Solitaire Namibia
92 G3	Sollefteå Sweden
93 G4	Sollentuna Sweden
107 D2	Sóller Spain
101 D2	Solling hills Ger.
89 E2	Solnechnogorsk Rus. Fed.
60 B2	Solok Indon.
48 H4	Solomon Islands country S. Pacific Ocean
48 G4	Solomon Sea S. Pacific Ocean
61 D2	Solor, Kepulauan is Indon.
105 D2	Solothurn Switz.
81 D2	Soltānābād Iran
101 D1	Soltau Ger.
89 D2	Sol'tsy Rus. Fed.
96 C3	Solway Firth est. U.K.
120 B2	Solwezi Zambia
111 C3	Soma Turkey
117 C4	Somalia country Africa
117 C4	Somaliland terr. Somalia
61 C2	Somba Indon.
120 B1	Sombo Angola
109 C1	Sombor Serbia
144 B2	Sombrerete Mex.
138 C3	Somerset U.S.A.
123 C3	Somerset East S. Africa
126 F2	Somerset Island Can.
122 A3	Somerset West S. Africa
110 B1	Someş r. Romania
101 E2	Sömmerda Ger.
146 B3	Somoto Nic.
75 C2	Son r. India
65 C1	Sŏnbong N. Korea
93 E5	Sønderborg Denmark
101 E2	Sondershausen Ger.
	Søndre Strømfjord inlet Greenland see Kangerlussuaq
108 A1	Sondrio Italy
63 B3	Sông Câu Vietnam
62 B1	Sông Đa, Hồ resr Vietnam
119 D4	Songea Tanz.
65 B1	Sŏnggan N. Korea
65 B1	Songhua Hu resr China
65 B1	Songjianghe China
	Sŏngjin N. Korea see Kimch'aek
63 B3	Songkhla Thai.
65 B2	Songnim N. Korea
120 A1	Songo Angola
121 C2	Songo Moz.
	Songololo Dem. Rep. Congo see Mbanza-Ngungu
69 E1	Songyuan China
	Sonid Youqi China see Saihan Tal
74 B2	Sonipat India
89 E2	Sonkovo Rus. Fed.
62 B1	Sơn La Vietnam
74 A2	Sonmiani Pak.
74 A2	Sonmiani Bay Pak.
101 E2	Sonneberg Ger.
142 A2	Sonoita Mex.
144 A2	Sonora r. Mex.
135 B3	Sonora CA U.S.A.
143 C2	Sonora TX U.S.A.
146 B3	Sonsonate El Salvador
	Soochow China see Suzhou
117 A4	Sopo watercourse S. Sudan
103 D2	Sopron Hungary
74 B1	Sopur India
108 B2	Sora Italy
130 C3	Sorel Can.
51 D4	Sorell Austr.
106 C1	Soria Spain
90 B2	Soroca Moldova
154 C2	Sorocaba Brazil
87 E3	Sorochinsk Rus. Fed.
	Soroki Moldova see Soroca
59 D2	Sorol atoll Micronesia
59 C3	Sorong Indon.
119 D2	Soroti Uganda
92 H1	Sørøya i. Norway
108 B2	Sorrento Italy
92 G2	Sorsele Sweden
64 B2	Sorsogon Phil.
86 C2	Sortavala Rus. Fed.
92 G2	Sortland Norway
123 C2	Soshanguve S. Africa
89 E3	Sosna r. Rus. Fed.
153 B3	Sosneado mt. Arg.
86 E2	Sosnogorsk Rus. Fed.
86 D2	Sosnovka Rus. Fed.
88 C2	Sosnovyy Bor Rus. Fed.
103 D1	Sosnowiec Pol.
91 C1	Sosnytsya Ukr.
86 F3	Sos'va Rus. Fed.
91 D2	Sosyka r. Rus. Fed.
145 C2	Soto la Marina Mex.
118 B2	Souanké Congo
111 B3	Souda Greece
104 C3	Souillac France
	Soûl S. Korea see Seoul
104 B2	Soulac-sur-Mer France
104 B3	Soulom France
	Soûr Lebanon see Tyre
107 D2	Sour el Ghozlane Alg.
129 D3	Souris Man. Can.
131 D3	Souris P.E.I. Can.
129 D3	Souris r. Can.
151 F3	Sousa Brazil

115 D1 Sousse Tunisia
104 B3 Soustons France
122 B3 South Africa, Republic of country Africa
99 C4 Southampton U.K.
129 F1 Southampton, Cape Can.
129 F1 Southampton Island Can.
73 D1 South Andaman i. India
52 A1 South Australia state Austr.
140 B2 Southaven U.S.A.
142 B2 South Baldy mt. U.S.A.
130 B3 South Baymouth Can.
138 B2 South Bend U.S.A.
141 D2 South Carolina state U.S.A.
58 B2 South China Sea N. Pacific Ocean
South Coast Town Austr. see Gold Coast
136 C2 South Dakota state U.S.A.
99 C4 South Downs hills U.K.
159 E6 Southeast Indian Ridge Indian Ocean
55 O2 Southeast Pacific Basin S. Pacific Ocean
129 D2 Southend Can.
99 D4 Southend-on-Sea U.K.
54 B2 Southern Alps mts N.Z.
50 A3 Southern Cross Austr.
129 E2 Southern Indian Lake Can.
159 D7 Southern Ocean
141 E1 Southern Pines U.S.A.
Southern Rhodesia country Africa see Zimbabwe
96 B3 Southern Uplands hills U.K.
55 J2 South Geomagnetic Pole Antarctica
149 G8 South Georgia terr. S. Atlantic Ocean
149 G8 South Georgia and the South Sandwich Islands terr. S. Atlantic Ocean
138 B2 South Haven U.S.A.
129 E1 South Henik Lake Can.
119 D2 South Horr Kenya
54 B2 South Island N.Z.
65 B2 South Korea country Asia
135 B3 South Lake Tahoe U.S.A.
55 L3 South Magnetic Pole Antarctica
149 F9 South Orkney Islands S. Atlantic Ocean
136 C2 South Platte r. U.S.A.
98 B3 Southport U.K.
141 E2 Southport U.S.A.
130 C3 South River Can.
96 C1 South Ronaldsay i. U.K.
123 D3 South Sand Bluff pt S. Africa
149 H8 South Sandwich Islands S. Atlantic Ocean
55 C4 South Sandwich Trench S. Atlantic Ocean
129 D2 South Saskatchewan r. Can.
129 E2 South Seal r. Can.
149 E9 South Shetland Islands Antarctica
98 C2 South Shields U.K.
117 A4 South Sudan country Africa
54 B1 South Taranaki Bight b. N.Z.
156 C8 South Tasman Rise Southern Ocean
130 C2 South Twin Island Can.
96 A2 South Uist i. U.K.
South-West Africa country Africa see Namibia
Southwest Peru Ridge S. Pacific Ocean see Nazca Ridge
53 D2 South West Rocks Austr.
99 D3 Southwold U.K.
109 C3 Soverato Italy
88 B2 Sovetsk Rus. Fed.
86 F2 Sovetskiy Rus. Fed.
91 C2 Sovyets'kyy Ukr.
123 C2 Soweto S. Africa
66 D1 Sōya-misaki c. Japan
65 B2 Soyang-ho l. S. Korea
104 C2 Soyaux France
90 C1 Sozh r. Europe
110 C2 Sozopol Bulg.
100 B2 Spa Belgium
106 C1 Spain country Europe
Spalato Croatia see Split
99 C3 Spalding U.K.
135 D2 Spanish Fork U.S.A.
Spanish Guinea country Africa see Equatorial Guinea
97 B2 Spanish Point Ireland

Spanish Sahara terr. Africa see Western Sahara
146 C3 Spanish Town Jamaica
108 B3 Sparagio, Monte mt. Italy
135 C3 Sparks U.S.A.
138 A2 Sparta U.S.A.
141 D2 Spartanburg U.S.A.
111 B3 Sparti Greece
109 C3 Spartivento, Capo c. Italy
89 D3 Spas-Demensk Rus. Fed.
89 F2 Spas-Klepiki Rus. Fed.
66 B2 Spassk-Dal'niy Rus. Fed.
89 F3 Spassk-Ryazanskiy Rus. Fed.
111 B3 Spatha, Akrotirio pt Greece
96 B2 Spean Bridge U.K.
136 C2 Spearfish U.S.A.
143 C1 Spearman U.S.A.
Spence Bay Can. see Taloyoak
137 D2 Spencer IA U.S.A.
134 D2 Spencer ID U.S.A.
52 A2 Spencer Gulf est. Austr.
98 C2 Spennymoor U.K.
54 B2 Spenser Mountains N.Z.
101 E2 Spessart reg. Ger.
96 C2 Spey r. U.K.
101 D3 Speyer Ger.
100 C1 Spiekeroog i. Ger.
100 B2 Spijkenisse Neth.
100 B3 Spincourt France
128 C2 Spirit River Can.
103 E2 Spišská Nová Ves Slovakia
82 C1 Spitsbergen i. Svalbard
102 C2 Spittal an der Drau Austria
93 E3 Spjelkavik Norway
109 C2 Split Croatia
129 E2 Split Lake Can.
129 E2 Split Lake l. Can.
134 C1 Spokane U.S.A.
138 A1 Spooner U.S.A.
102 C1 Spree r. Ger.
122 A2 Springbok S. Africa
131 E3 Springdale Can.
140 B1 Springdale U.S.A.
101 D2 Springe Ger.
142 C1 Springer U.S.A.
142 B2 Springerville U.S.A.
136 C3 Springfield CO U.S.A.
138 B3 Springfield IL U.S.A.
139 E2 Springfield MA U.S.A.
137 E3 Springfield MO U.S.A.
138 C3 Springfield OH U.S.A.
134 B2 Springfield OR U.S.A.
140 C1 Springfield TN U.S.A.
123 C3 Springfontein S. Africa
131 D3 Springhill Can.
141 D3 Spring Hill U.S.A.
54 B2 Springs Junction N.Z.
51 D2 Springsure Austr.
98 D3 Spurn Head hd U.K.
128 B3 Squamish Can.
109 C3 Squillace, Golfo di g. Italy
Srbija country Europe see Serbia
108 C2 Srebrenica Bos.-Herz.
110 C2 Sredets Bulg.
83 L3 Sredinnyy Khrebet mts Rus. Fed.
83 L2 Srednekolymsk Rus. Fed.
Sredne-Russkaya Vozvyshennost' hills Rus. Fed. see Central Russian Upland
Sredne-Sibirskoye Ploskogor'ye plat. Rus. Fed. see Central Siberian Plateau
110 B2 Srednogorie Bulg.
69 D1 Sretensk Rus. Fed.
61 C1 Sri Aman Malaysia
73 B4 Sri Jayewardenepura Kotte Sri Lanka
73 C3 Srikakulam India
73 C4 Sri Lanka country Asia
74 B1 Srinagar India
73 B3 Srivardhan India
101 D1 Stade Ger.
101 E1 Stadensen Ger.
100 C1 Stadskanaal Neth.
101 D2 Stadtallendorf Ger.
101 D1 Stadthagen Ger.
101 E2 Staffelstein Ger.
99 B3 Stafford U.K.
99 C4 Staines-upon-Thames U.K.
91 D2 Stakhanov Ukr.
Stakhanovo Rus. Fed. see Zhukovskiy
Stalin Bulg. see Varna

Stalinabad Tajik. see Dushanbe
Stalingrad Rus. Fed. see Volgograd
Staliniri Georgia see Tskhinvali
Stalino Ukr. see Donets'k
Stalinogorsk Rus. Fed. see Novomoskovsk
Stalinogród Pol. see Katowice
Stalinsk Rus. Fed. see Novokuznetsk
103 E1 Stalowa Wola Pol.
138 B1 Stambaugh U.S.A.
99 C3 Stamford U.K.
139 E2 Stamford CT U.S.A.
143 D2 Stamford TX U.S.A.
Stampalia i. Greece see Astypalaia
122 A1 Stampriet Namibia
92 F2 Stamsund Norway
123 C2 Standerton S. Africa
138 C2 Standish U.S.A.
Stanislav Ukr. see Ivano-Frankivs'k
Stanke Dimitrov Bulg. see Dupnitsa
153 C5 Stanley Falkland Is
136 C1 Stanley U.S.A.
Stanleyville Dem. Rep. Congo see Kisangani
Stann Creek Belize see Dangriga
111 B3 Stanos Greece
83 I3 Stanovoy Nagor'ye mts Rus. Fed.
83 J3 Stanovoy Khrebet mts Rus. Fed.
53 D1 Stanthorpe Austr.
137 D1 Staples U.S.A.
103 E1 Starachowice Pol.
Stara Planina mts Bulg./Serbia see Balkan Mountains
89 D2 Staraya Russa Rus. Fed.
89 D2 Staraya Toropa Rus. Fed.
110 C2 Stara Zagora Bulg.
49 L4 Starbuck Island Kiribati
103 D1 Stargard Szczeciński Pol.
89 D2 Staritsa Rus. Fed.
141 D3 Starke U.S.A.
140 C2 Starkville U.S.A.
102 C2 Starnberg Ger.
91 D2 Starobil's'k Ukr.
89 D3 Starodub Rus. Fed.
103 D1 Starogard Gdański Pol.
90 B2 Starokostyantyniv Ukr.
91 D2 Starominskaya Rus. Fed.
91 D2 Staroshcherbinovskaya Rus. Fed.
91 D2 Starotitarovskaya Rus. Fed.
89 F3 Staroyur'yevo Rus. Fed.
99 B4 Start Point U.K.
88 C3 Staryya Darohi Belarus
86 G2 Staryy Nadym Rus. Fed.
89 F2 Staryy Oskol Rus. Fed.
101 E2 Staßfurt Ger.
103 E1 Staszów Pol.
139 D2 State College U.S.A.
141 D2 Statesboro U.S.A.
141 D1 Statesville U.S.A.
160 L1 Station Nord Greenland
139 D3 Staunton U.S.A.
93 E4 Stavanger Norway
87 D4 Stavropol' Rus. Fed.
Stavropol'-na-Volge Rus. Fed. see Tol'yatti
87 D4 Stavropol'skaya Vozvyshennost' hills Rus. Fed.
52 B3 Stawell Austr.
123 C2 Steadville S. Africa
136 B2 Steamboat Springs U.S.A.
101 E2 Stedten Ger.
128 C2 Steen River Can.
134 C1 Steens Mountain U.S.A.
100 C1 Steenwijk Neth.
126 E2 Stefansson Island Can.
110 B1 Ştei Romania
101 E3 Şteigerwald mts Ger.
100 B2 Stein Neth.
129 E3 Steinbach Can.
100 C1 Steinfurt Ger.
120 A3 Steinhausen Namibia
92 F3 Steinkjer Norway
122 A2 Steinkopf S. Africa

122 B2 Stella S. Africa
122 A3 Stellenbosch S. Africa
105 D3 Stello, Monte mt. France
105 D2 Stenay France
101 E1 Stendal Ger.
Steornabhagh U.K. see Stornoway
Stepanakert Azer. see Xankändi
52 B2 Stephens Creek Austr.
129 E2 Stephens Lake Can.
131 E3 Stephenville Can.
143 D2 Stephenville U.S.A.
Stepnoy Rus. Fed. see Elista
122 B3 Sterling S. Africa
136 C2 Sterling CO U.S.A.
138 B2 Sterling IL U.S.A.
136 C1 Sterling ND U.S.A.
138 C2 Sterling Heights U.S.A.
87 E3 Sterlitamak Rus. Fed.
101 E1 Sternberg Ger.
128 C2 Stettler Can.
138 C2 Steubenville U.S.A.
99 C4 Stevenage U.K.
129 E2 Stevenson Lake Can.
138 B2 Stevens Point U.S.A.
126 C2 Stevens Village U.S.A.
134 D1 Stevensville U.S.A.
128 B2 Stewart r. Can.
128 A1 Stewart r. Can.
54 A3 Stewart Island N.Z.
127 G2 Stewart Lake Can.
123 C3 Steynsburg S. Africa
102 C2 Steyr Austria
122 B3 Steytlerville S. Africa
128 A2 Stikine r. Can.
128 A2 Stikine Plateau Can.
122 B3 Stilbaai S. Africa
137 E1 Stillwater MN U.S.A.
143 D1 Stillwater OK U.S.A.
135 C3 Stillwater Range mts U.S.A.
109 D2 Ştip Macedonia
96 C2 Stirling U.K.
52 A2 Stirling North Austr.
92 F3 Stjørdalshalsen Norway
103 D2 Stockerau Austria
93 G4 Stockholm Sweden
98 B3 Stockport U.K.
135 B3 Stockton U.S.A.
98 C2 Stockton-on-Tees U.K.
143 C2 Stockton Plateau U.S.A.
63 B2 Stœng Trêng Cambodia
96 B1 Stoer, Point of U.K.
99 B3 Stoke-on-Trent U.K.
98 C2 Stokesley U.K.
92 F2 Stokmarknes Norway
110 B2 Stol mt. Serbia
109 C2 Stolac Bos.-Herz.
100 C2 Stolberg (Rheinland) Ger.
82 E2 Stolbovoy Rus. Fed.
88 C3 Stolin Belarus
101 F2 Stollberg Ger.
101 D1 Stolzenau Ger.
96 C2 Stonehaven U.K.
99 C4 Stonehenge tourist site U.
129 E2 Stonewall Can.
129 D2 Stony Rapids Can.
92 G2 Storavan l. Sweden
Store Bælt sea chan. Denmark see Great Belt
92 F3 Støren Norway
92 I1 Storfjordbotn Norway
92 F2 Storforshei Norway
126 E2 Storkerson Peninsula Can.
137 D2 Storm Lake U.S.A.
93 E3 Stornosa mt. Norway
96 A1 Stornoway U.K.
86 E2 Storozhevsk Rus. Fed.
90 B2 Storozhynets' Ukr.
92 F3 Storsjön l. Sweden
92 G2 Storuman Sweden
92 G2 Storuman l. Sweden
99 C3 Stour r. England U.K.
99 C4 Stour r. England U.K.
99 D4 Stour r. England U.K.
130 A2 Stout Lake Can.
88 C3 Stowbtsy Belarus
99 D3 Stowmarket U.K.
97 C1 Strabane U.K.
102 C2 Strakonice Czech Rep.
102 C1 Stralsund Ger.
122 A3 Strand S. Africa
93 E3 Stranda Norway
97 D1 Strangford Lough inlet U.
96 B3 Stranraer U.K.
105 D2 Strasbourg France
130 B3 Stratford Can.
54 B1 Stratford N.Z.
143 C1 Stratford U.S.A.

99 C3 Stratford-upon-Avon U.K.
28 C2 Strathmore Can.
96 C2 Strathspey val. U.K.
02 C2 Straubing Ger.
34 C2 Strawberry Mountain U.S.A.
51 C3 Streaky Bay Austr.
38 B2 Streator U.S.A.
99 B4 Street U.K.
10 B2 Strehaia Romania
94 B1 Streymoy i. Faroe Is
82 G2 Strezhevoy Rus. Fed.
01 F3 Stříbro Czech Rep.
53 B4 Stroeder Arg.
01 D1 Ströhen Ger.
09 C3 Stromboli, Isola i. Italy
96 B2 Stromeferry U.K.
96 C1 Stromness U.K.
92 G3 Strömsund Sweden
96 C1 Stronsay i. U.K.
53 D2 Stroud Austr.
99 B4 Stroud U.K.
00 C1 Strücklingen (Saterland) Ger.
11 B2 Struga Macedonia
88 C2 Strugi-Krasnyye Rus. Fed.
22 B3 Struis Bay S. Africa
11 B2 Struma r. Bulg.
99 A3 Strumble Head hd U.K.
11 B2 Strumica Macedonia
22 B2 Strydenburg S. Africa
11 B2 Strymonas r. Greece
93 E3 Stryn Norway
90 A2 Stryy Ukr.
90 A2 Stryy r. Ukr.
28 B2 Stuart Lake Can.
53 C2 Stuart Town Austr.
Stuchka Latvia see Aizkraukle
Stučka Latvia see Aizkraukle
30 A2 Stull Lake Can.
89 E3 Stupino Rus. Fed.
55 M3 Sturge Island Antarctica
38 B2 Sturgeon Bay U.S.A.
30 C3 Sturgeon Falls Can.
30 A3 Sturgeon Lake Can.
38 B2 Sturgis MI U.S.A.
36 C2 Sturgis SD U.S.A.
52 B1 Sturt, Mount h. Austr.
50 B1 Sturt Creek watercourse Austr.
50 C1 Sturt Plain Austr.
52 B1 Sturt Stony Desert Austr.
23 C3 Stutterheim S. Africa
02 B2 Stuttgart Ger.
40 B2 Stuttgart U.S.A.
92 ☐A2 Stykkishólmur Iceland
90 B1 Styr r. Belarus/Ukr.
55 D1 Suaçuí Grande r. Brazil
16 B3 Suakin Sudan
71 C3 Su'ao Taiwan
78 A3 Suara Eritrea
61 B1 Subi Besar i. Indon.
09 C1 Subotica Serbia
10 C1 Suceava Romania
Suchan Rus. Fed. see Partizansk
97 B2 Suck r. Ireland
52 B1 Sucre Bol.
54 B2 Sucuriú r. Brazil
Suczawa Romania see Suceava
89 E2 Suda Rus. Fed.
91 C3 Sudak Ukr.
16 A3 Sudan country Africa
30 B3 Sudbury Can.
99 D3 Sudbury U.K.
17 A4 Sudd swamp S. Sudan
89 F2 Sudislavl' Rus. Fed.
89 F2 Sudogda Rus. Fed.
94 B1 Suðuroy i. Faroe Is
89 E3 Sudzha Rus. Fed.
07 C2 Sueca Spain
16 B2 Suez Egypt
16 B2 Suez, Gulf of Egypt
80 B2 Suez Canal canal Egypt
39 D3 Suffolk U.S.A.
40 A3 Sugar Land U.S.A.
39 E1 Sugarloaf Mountain U.S.A.
53 D2 Sugarloaf Point Austr.
70 A2 Suhait China
16 B2 Sūhāj Egypt
79 C2 Şuḩār Oman
68 D1 Sühbaatar Mongolia
01 E2 Suhl Ger.
09 C1 Suhopolje Croatia
70 B2 Suide China
66 B2 Suifenhe China
69 E1 Suihua China
70 A2 Suining China
70 B2 Suiping China

97 C2 Suir r. Ireland
Suixian China see Suizhou
70 B2 Suiyang China
70 B2 Suizhou China
74 B2 Sujangarh India
74 B1 Sujanpur India
74 A2 Sujawal Pak.
60 B2 Sukabumi Indon.
60 B2 Sukadana Indon.
67 D3 Sukagawa Japan
60 C2 Sukaraja Indon.
Sukarnapura Indon. see Jayapura
Sukarno, Puntjak mt. Indon. see Jaya, Puncak
89 E3 Sukhinichi Rus. Fed.
89 F2 Sukhona r. Rus. Fed.
62 A2 Sukhothai Thai.
74 A2 Sukkur Pak.
89 E2 Sukromny Rus. Fed.
59 C3 Sula, Kepulauan is Indon.
74 A1 Sulaiman Range mts Pak.
Sulawesi i. Indon. see Celebes
150 A3 Sullana Peru
138 B3 Sullivan IN U.S.A.
137 E3 Sullivan MO U.S.A.
140 B2 Sulphur U.S.A.
143 D2 Sulphur Springs U.S.A.
138 C1 Sultan Can.
Sultanabad Iran see Arāk
64 B3 Sulu Archipelago is Phil.
64 A3 Sulu Sea N. Pacific Ocean
101 E3 Sulzbach-Rosenberg Ger.
79 C2 Sumāil Oman
Sumatera i. Indon. see Sumatra
60 A1 Sumatra i. Indon.
61 D2 Sumba i. Indon.
61 C2 Sumba, Selat sea chan. Indon.
61 C2 Sumbawa i. Indon.
61 C2 Sumbawabesar Indon.
119 D3 Sumbawanga Tanz.
120 A2 Sumbe Angola
96 ☐ Sumburgh U.K.
96 ☐ Sumburgh Head hd U.K.
119 C2 Sumeih Sudan
61 C2 Sumenep Indon.
67 D4 Sumisu-jima i. Japan
131 D3 Summerside Can.
138 C3 Summersville U.S.A.
141 D2 Summerville U.S.A.
137 D1 Summit U.S.A.
128 B2 Summit Lake Can.
103 D2 Šumperk Czech Rep.
81 C1 Sumqayıt Azer.
141 D2 Sumter U.S.A.
91 C1 Sumy Ukr.
75 D2 Sunamganj Bangl.
65 B2 Sunan N. Korea
79 C2 Şunaynah Oman
52 B3 Sunbury Austr.
139 D2 Sunbury U.S.A.
65 B3 Suncheon S. Korea
65 B3 Sunch'ŏn N. Korea
123 C2 Sun City S. Africa
93 H3 Sund Fin.
60 B2 Sunda, Selat str. Indon.
136 C2 Sundance U.S.A.
75 C2 Sundarbans coastal area Bangl./India
74 B1 Sundarnagar India
60 B2 Sunda Strait str. Indon. see Sunda, Selat
Sunda Trench Indian Ocean see Java Trench
98 C2 Sunderland U.K.
128 C2 Sundre Can.
93 G3 Sundsvall Sweden
123 D2 Sundumbili S. Africa
60 B2 Sungailiat Indon.
60 B2 Sungaipenuh Indon.
60 B1 Sungai Petani Malaysia
80 B1 Sungurlu Turkey
75 C2 Sun Kosi r. Nepal
93 E3 Sunndalsøra Norway
134 C1 Sunnyside U.S.A.
135 B3 Sunnyvale U.S.A.
141 D3 Sunrise U.S.A.
83 I2 Suntar Rus. Fed.
74 A2 Suntsar Pak.
114 B4 Sunyani Ghana
92 I3 Suomussalmi Fin.
67 B4 Suō-nada b. Japan
86 C2 Suoyarvi Rus. Fed.
142 A2 Superior AZ U.S.A.
137 D2 Superior NE U.S.A.

138 A1 Superior WI U.S.A.
138 B1 Superior, Lake Can./U.S.A.
63 B2 Suphan Buri Thai.
81 C2 Süphan Dağı mt. Turkey
89 D3 Suponevo Rus. Fed.
81 C2 Süq ash Shuyūkh Iraq
70 B2 Suqian China
78 A2 Süq Suwayq Saudi Arabia
Suqutrā i. Yemen see Socotra
79 C2 Şür Oman
74 A2 Surab Pak.
61 C2 Surabaya Indon.
61 C2 Surakarta Indon.
74 B2 Surat India
74 B2 Suratgarh India
63 A3 Surat Thani Thai.
89 D3 Surazh Rus. Fed.
109 D2 Surdulica Serbia
100 C3 Sûre r. Lux.
74 B2 Surendranagar India
82 F2 Surgut Rus. Fed.
64 B3 Surigao Phil.
63 B2 Surin Thai.
151 D2 Suriname country S. America
75 C2 Surkhet Nepal
Surt Libya see Sirte
Surt, Khalīj g. Libya see Sirte, Gulf of
60 B2 Surulangun Indon.
81 C2 Süsangerd Iran
89 F2 Susanino Rus. Fed.
135 B2 Susanville U.S.A.
80 B1 Suşehri Turkey
139 D3 Susquehanna r. U.S.A.
131 D3 Sussex Can.
101 D1 Süstedt Ger.
100 C1 Sustrum Ger.
83 K2 Susuman Rus. Fed.
111 C3 Susurluk Turkey
74 B1 Sutak India
53 D2 Sutherland Austr.
122 B3 Sutherland S. Africa
136 C2 Sutherland U.S.A.
134 B2 Sutherlin U.S.A.
138 C3 Sutton U.S.A.
99 C3 Sutton Coldfield U.K.
98 C3 Sutton in Ashfield U.K.
66 D2 Suttsu Japan
49 I5 Suva Fiji
Suvalki Pol. see Suwałki
89 E3 Suvorov Rus. Fed.
90 B2 Suvorove Ukr.
103 E1 Suwałki Pol.
141 D3 Suwannee Sound b. U.S.A.
63 B2 Suwannaphum Thai.
141 D3 Suwannee r. U.S.A.
Suways, Qanāt as canal Egypt see Suez Canal
Suweis, Qanā el canal Egypt see Suez Canal
65 B2 Suwon S. Korea
79 C2 Sūzā Iran
89 F2 Suzdal' Rus. Fed.
89 D3 Suzemka Rus. Fed.
70 B2 Suzhou Anhui China
70 C2 Suzhou Jiangsu China
67 C3 Suzu Japan
67 C3 Suzu-misaki pt Japan
82 B1 Svalbard terr. Arctic Ocean
90 A2 Svalyava Ukr.
92 H2 Svappavaara Sweden
91 D2 Svatove Ukr.
63 B2 Svay Riĕng Cambodia
93 F3 Sveg Sweden
88 C2 Švenčionys Lith.
93 F4 Svendborg Denmark
Sverdlovsk Rus. Fed. see Yekaterinburg
111 B2 Sveti Nikole Macedonia
66 C1 Svetlaya Rus. Fed.
88 B3 Svetlogorsk Rus. Fed.
87 D4 Svetlograd Rus. Fed.
88 B3 Svetlyy Rus. Fed.
93 I3 Svetogorsk Rus. Fed.
103 D2 Svidník Slovakia
111 C2 Svilengrad Bulg.
110 B2 Svinecea Mare, Vârful mt. Romania
110 C2 Svishtov Bulg.
88 B3 Svislach Belarus
103 D2 Svitavy Czech Rep.
91 C2 Svitlovods'k Ukr.
69 E1 Svobodnyy Rus. Fed.
110 B2 Svoge Bulg.
92 F2 Svolvær Norway
88 C3 Svyetlahorsk Belarus
141 D2 Swainsboro U.S.A.
120 A3 Swakopmund Namibia

52 B3 Swan Hill Austr.
128 C2 Swan Hills Can.
129 D2 Swan Lake Can.
97 C1 Swanlinbar Ireland
129 D2 Swan River Can.
53 D2 Swansea Austr.
99 B4 Swansea U.K.
122 B3 Swartkolkvloer salt pan S. Africa
123 C2 Swartruggens S. Africa
Swatow China see Shantou
123 D2 Swaziland country Africa
93 G3 Sweden country Europe
143 C2 Sweetwater U.S.A.
136 B2 Sweetwater r. U.S.A.
122 B3 Swellendam S. Africa
103 D1 Świdnica Pol.
103 E1 Świdwin Pol.
103 D1 Świebodzin Pol.
103 D1 Świecie Pol.
129 D2 Swift Current Can.
97 C1 Swilly, Lough inlet Ireland
99 C4 Swindon U.K.
102 C1 Świnoujście Pol.
105 D2 Switzerland country Europe
97 C2 Swords Ireland
88 C3 Syanno Belarus
89 D1 Syas'stroy Rus. Fed.
89 D2 Sychevka Rus. Fed.
53 D2 Sydney Austr.
131 D3 Sydney Can.
131 D3 Sydney Mines Can.
91 D2 Syeverodonets'k Ukr.
111 B2 Sykia Greece
86 E2 Syktyvkar Rus. Fed.
140 C2 Sylacauga U.S.A.
75 D2 Sylhet Bangl.
102 B1 Sylt i. Ger.
138 C2 Sylvania U.S.A.
51 C1 Sylvester, Lake imp. l. Austr.
111 C3 Symi i. Greece
91 D2 Synel'nykove Ukr.
90 C2 Synyukha r. Ukr.
109 C3 Syracuse Sicilia Italy
136 C3 Syracuse KS U.S.A.
139 D2 Syracuse NY U.S.A.
77 C2 Syrdar'ya r. Asia
80 B2 Syria country Asia
80 B2 Syrian Desert Asia
111 B3 Syros i. Greece
91 D2 Syvash, Zatoka lag. Ukr.
91 C2 Syvas'ke Ukr.
87 D3 Syzran' Rus. Fed.
102 C1 Szczecin Pol.
103 D1 Szczecinek Pol.
103 E1 Szczytno Pol.
Szechwan prov. China see Sichuan
103 E2 Szeged Hungary
103 D2 Székesfehérvár Hungary
103 D2 Szekszárd Hungary
103 E2 Szentes Hungary
103 E2 Szentgotthárd Hungary
103 E2 Szerencs Hungary
103 D2 Szigetvár Hungary
103 E2 Szolnok Hungary
103 D2 Szombathely Hungary
103 D2 Sztálinváros Hungary see Dunaújváros

T

64 B2 Tabaco Phil.
78 B2 Tābah Saudi Arabia
108 A3 Tabarka Tunisia
76 B3 Tabas Iran
79 C1 Tabāsīn Iran
81 D3 Tābask, Kūh-e mt. Iran
150 D3 Tabatinga Brazil
154 C2 Tabatinga Brazil
114 B2 Tabelbala Alg.
128 C3 Taber Can.
64 B2 Tablas i. Phil.
102 C2 Tábor Czech Rep.
119 D3 Tabora Tanz.
114 B4 Tabou Côte d'Ivoire
81 C2 Tabrīz Iran
48 L3 Tabuaeran atoll Kiribati
78 A2 Tabūk Saudi Arabia
93 G4 Täby Sweden
77 D2 Tacheng China
103 D1 Tachov Czech Rep.
64 B2 Tacloban Phil.
150 B4 Tacna Peru
134 B1 Tacoma U.S.A.
152 C3 Tacuarembó Uru.
142 B3 Tacupeto Mex.
114 C2 Tademaït, Plateau du Alg.

Tadjikistan

		Tadjikistan country Asia see Tajikistan
117	C3	Tadjourah Djibouti
80	B2	Tadmur Syria
129	E2	Tadoule Lake Can.
		Tadzhikskaya S.S.R. country Asia see Tajikistan
65	B2	Taebaek S. Korea
65	B2	Taebaek-sanmaek mts N. Korea/S. Korea
		Taech'ŏn S. Korea see Boryeong
		Taegu S. Korea see Daegu
107	C1	Tafalla Spain
152	B2	Tafí Viejo Arg.
79	D2	Taftān, Kūh-e mt. Iran
91	D2	Taganrog Rus. Fed.
91	D2	Taganrog, Gulf of Rus. Fed./Ukr.
62	A1	Tagaung Myanmar
64	B2	Tagaytay City Phil.
64	B3	Tagbilaran Phil.
64	B2	Tagudin Phil.
51	E1	Tagula Island P.N.G.
64	B3	Tagum Phil.
106	B2	Tagus r. Port./Spain
60	B1	Tahan, Gunung mt. Malaysia
115	C2	Tahat, Mont mt. Alg.
69	E1	Tahe China
49	M5	Tahiti i. Fr. Polynesia
143	E1	Tahlequah U.S.A.
135	B3	Tahoe, Lake U.S.A.
135	B3	Tahoe City U.S.A.
126	E2	Tahoe Lake Can.
115	C3	Tahoua Niger
79	C2	Tahrūd Iran
128	B3	Tahsis Can.
116	B2	Ṭahṭā Egypt
64	B3	Tahuna Indon.
70	B2	Tai'an China
70	A2	Taibai Shan mt. China
71	C3	Taibei Taiwan see Taibus Qi China see Baochang
70	B2	Taihang Shan mts China
54	C1	Taihape N.Z.
71	B3	Taihe China
70	C2	Tai Hu l. China
52	A3	Tailem Bend Austr.
71	C3	Tainan Taiwan
111	B3	Tainaro, Akra c. Greece
155	D1	Taiobeiras Brazil
		Taiping China see Chongzuo
60	B1	Taiping Malaysia
		Tairbeart U.K. see Tarbert
71	B3	Taishan China
70	B2	Tai Shan hills China
119	D3	Taita Hills Kenya
153	A4	Taitao, Península de pen. Chile
71	C3	T'aitung Taiwan
92	I2	Taivalkoski Fin.
92	H2	Taivaskero h. Fin.
71	C3	Taiwan country Asia
		Taiwan Shan mts Taiwan see Zhongyang Shanmo
71	B3	Taiwan Strait China/Taiwan
77	C1	Taiynsha Kazakh.
70	B2	Taiyuan China
71	C3	Taizhong China
70	B2	Taizhou Jiangsu China
71	C3	Taizhou Zhejiang China
78	B3	Ta'izz Yemen
77	D3	Tajikistan country Asia
74	B2	Taj Mahal tourist site India
		Tajo r. Spain see Tagus
145	C3	Tajumulco, Volcán de vol. Guat.
63	A2	Tak Thai.
54	B2	Takaka N.Z.
115	C2	Takalous, Oued watercourse Alg.
67	B4	Takamatsu Japan
67	C3	Takaoka Japan
54	B1	Takapuna N.Z.
67	C3	Takasaki Japan
122	B1	Takatokwane Botswana
122	B1	Takatshwaane Botswana
67	C3	Takayama Japan
60	A1	Takengon Indon.
63	B2	Takêv Cambodia
		Takhiatash Uzbek. see Taxiatosh
63	B2	Ta Khmau Cambodia
74	B1	Takht-i-Sulaiman mt. Pak.
66	D2	Takikawa Japan
128	B2	Takla Lake Can.
128	B2	Takla Landing Can.

		Takla Makan des. China see Taklimakan Desert
77	E3	Taklimakan Desert
		Taklimakan Desert des. China
		Taklimakan Shamo des. China see Taklimakan Desert
128	A2	Taku r. Can./U.S.A.
63	A3	Takua Pa Thai.
115	C4	Takum Nigeria
88	C3	Talachyn Belarus
74	B1	Talagang Pak.
146	B4	Talamanca, Cordillera de mts Costa Rica
150	A3	Talara Peru
59	C2	Talaud, Kepulauan is Indon.
106	C2	Talavera de la Reina Spain
153	A3	Talca Chile
153	A3	Talcahuano Chile
89	E2	Taldom Rus. Fed.
77	D2	Taldykorgan Kazakh.
		Taldy-Kurgan Kazakh. see Taldykorgan
59	C3	Taliabu i. Indon.
64	B2	Talisay Phil.
61	C2	Taliwang Indon.
81	C2	Tall 'Afar Iraq
141	D2	Tallahassee U.S.A.
53	C3	Tallangatta Austr.
88	B2	Tallinn Estonia
140	B2	Tallulah U.S.A.
104	B2	Talmont-St-Hilaire France
90	C2	Tal'ne Ukr.
117	B3	Talodi Sudan
74	A1	Tāloqān Afgh.
89	F2	Talovaya Rus. Fed.
126	F2	Taloyoak Can.
88	B2	Talsi Latvia
152	A2	Taltal Chile
129	C1	Taltson r. Can.
60	A1	Talu Indon.
53	C1	Talwood Austr.
114	B3	Tamale Ghana
115	C2	Tamanrasset Alg.
99	A4	Tamar r. U.K.
		Tamatave Madag. see Toamasina
144	B2	Tamazula Mex.
145	C2	Tamazunchale Mex.
114	A3	Tambacounda Senegal
60	B1	Tambelan, Kepulauan is Indon.
86	G1	Tambey Rus. Fed.
61	C1	Tambisan Malaysia
61	C2	Tambora, Gunung vol. Indon.
91	E1	Tambov Rus. Fed.
119	C2	Tambura S. Sudan
62	A1	Tamenglong India
145	C2	Tamiahua, Laguna de lag. Mex.
		Tammerfors Fin. see Tampere
141	D3	Tampa U.S.A.
141	D3	Tampa Bay U.S.A.
93	H3	Tampere Fin.
145	C2	Tampico Mex.
69	D1	Tamsagbulag Mongolia
102	C2	Tamsweg Austria
53	D2	Tamworth Austr.
99	C3	Tamworth U.K.
119	E3	Tana r. Kenya
		Tana, Lake l. Eth. see Lake Tana
67	C4	Tanabe Japan
92	I1	Tana Bru Norway
60	A2	Tanahbala i. Indon.
61	C2	Tanahgrogot Indon.
61	D2	Tanahjampea i. Indon.
60	A2	Tanahmasa i. Indon.
50	C1	Tanami Desert Austr.
63	B2	Tân An Vietnam
126	B2	Tanana U.S.A.
		Tananarive Madag. see Antananarivo
108	A1	Tanaro r. Italy
65	B1	Tanch'ŏn N. Korea
64	B3	Tandag Phil.
110	C2	Ţăndărei Romania
153	C3	Tandil Arg.
74	A2	Tando Adam Pak.
74	A2	Tando Muhammad Khan Pak.
52	B2	Tandou Lake imp. l. Austr.
67	B4	Tanega-shima i. Japan
114	B2	Tanezrouft reg. Alg./Mali
119	D3	Tanga Tanz.
75	C2	Tangail Bangl.
		Tanganyika country Africa see Tanzania

119	C3	Tanganyika, Lake Africa
55	F3	Tange Promontory hd Antarctica
114	B1	Tanger Morocco
101	E1	Tangermünde Ger.
75	D1	Tanggulashan China
75	C1	Tanggula Shan mts China
		Tangier Morocco see Tanger
75	C1	Tangra Yumco salt l. China
70	B2	Tangshan China
68	C2	Taniantaweng Shan mts China
59	C3	Tanimbar, Kepulauan is Indon.
64	B3	Tanjay Phil.
61	C2	Tanjung Indon.
60	A1	Tanjungbalai Indon.
		Tanjungkarang-Telukbetung Indon. see Bandar Lampung
60	B2	Tanjungpandan Indon.
60	B1	Tanjungpinang Indon.
61	C1	Tanjungredeb Indon.
61	C1	Tanjungselor Indon.
74	B1	Tank Pak.
48	H5	Tanna i. Vanuatu
115	C3	Tanout Niger
75	C2	Tansen Nepal
116	B1	Ţanţā Egypt
114	A2	Tan-Tan Morocco
145	C2	Tantoyuca Mex.
119	D3	Tanzania country Africa
		Tao'an China see Taonan
		Taocheng China see Yongchun
		Taolanaro Madag. see Tôlañaro
69	E1	Taonan China
109	C3	Taormina Italy
142	B1	Taos U.S.A.
114	B2	Taoudenni Mali
114	B1	Taounate Morocco
114	B1	Taourirt Morocco
88	C2	Tapa Estonia
145	C3	Tapachula Mex.
151	D3	Tapajós r. Brazil
60	A1	Tapaktuan Indon.
145	C3	Tapanatepec Mex.
150	C3	Tapauá Brazil
152	C2	Tapera Brazil
114	B4	Tapeta Liberia
139	D3	Tappahannock U.S.A.
74	B2	Tapti r. India
54	B2	Tapuaenuku mt. N.Z.
150	C3	Tapurucuara Brazil
154	B1	Taquaral, Serra do hills Brazil
154	A1	Taquarí r. Brazil
154	B1	Taquari, Serra do hills Brazil
154	C2	Taquaritinga Brazil
53	D1	Tara Austr.
115	D4	Taraba r. Nigeria
		Ţarābulus Libya see Tripoli
54	C1	Taradale N.Z.
61	C1	Tarakan Indon.
111	D2	Taraklı Turkey
88	A3	Taran, Mys pt Rus. Fed.
54	B1	Taranaki, Mount vol. N.Z.
106	C1	Tarancón Spain
109	C2	Taranto Italy
109	C2	Taranto, Golfo di g. Italy
150	B3	Tarapoto Peru
90	C2	Tarashcha Ukr.
91	E2	Tarasovskiy Rus. Fed.
150	B3	Tarauacá Brazil
150	C3	Tarauacá r. Brazil
48	I3	Tarawa atoll Kiribati
54	C1	Tarawera N.Z.
77	D2	Taraz Kazakh.
107	C1	Tarazona Spain
77	E2	Tarbagatay, Khrebet mts Kazakh.
96	C2	Tarbat Ness pt U.K.
97	B2	Tarbert Ireland
96	A2	Tarbert Scotland U.K.
96	B3	Tarbert Scotland U.K.
104	C3	Tarbes France
96	B2	Tarbet U.K.
141	E1	Tarboro U.S.A.
51	C3	Tarcoola Austr.
53	D2	Taree Austr.
110	C2	Târgovişte Romania
110	C1	Târgu Frumos Romania
110	B1	Târgu Jiu Romania
110	B1	Târgu Lăpuş Romania
110	B1	Târgu Mureş Romania
110	C1	Târgu Neamţ Romania
110	C1	Târgu Ocna Romania

79	C2	Tarif U.A.E.
152	B2	Tarija Bol.
79	B3	Tarīm Yemen
77	E3	Tarim Basin basin China
119	D3	Tarime Tanz.
77	E2	Tarim He r. China
		Tarim Pendi basin China see Tarim Basin
77	C3	Tarīn Kōt Afgh.
59	D3	Taritatu r. Indon.
123	C3	Tarkastad S. Africa
82	G2	Tarko-Sale Rus. Fed.
114	B4	Tarkwa Ghana
64	B2	Tarlac Phil.
68	C2	Tarlag China
105	C3	Tarn r. France
92	G2	Tärnaby Sweden
77	C3	Tarnak Rōd r. Afgh.
110	B1	Târnăveni Romania
103	E1	Tarnobrzeg Pol.
		Tarnopol Ukr. see Ternopil
103	E1	Tarnów Pol.
75	C1	Taro Co salt l. China
51	D2	Taroom Austr.
114	B1	Taroudannt Morocco
141	D3	Tarpon Springs U.S.A.
69	E1	Tarqi China
108	B2	Tarquinia Italy
107	D1	Tarragona Spain
107	D1	Tàrrega Spain
80	B2	Tarsus Turkey
152	B2	Tartagal Arg.
104	B3	Tartas France
88	C2	Tartu Estonia
80	B2	Ţarţūs Syria
155	D1	Tarumirim Brazil
89	E3	Tarusa Rus. Fed.
90	B2	Tarutyne Ukr.
108	B1	Tarvisio Italy
		Tashauz Turkm. see Daşoguz
81	D3	Tashk, Daryācheh-ye l. Iran
		Tashkent Uzbek. see Toshkent
74	A1	Tāshqurghān Afgh.
130	C2	Tasialujjuaq, Lac l. Can.
130	C2	Tasiat, Lac l. Can.
131	D2	Tasiujaq Can.
76	B1	Taskala Kazakh.
77	E2	Taskesken Kazakh.
54	B2	Tasman Bay N.Z.
51	D4	Tasmania state Austr.
54	B2	Tasman Mountains N.Z.
156	D8	Tasman Sea S. Pacific Ocean
61	D2	Tataba Indon.
103	D2	Tatabánya Hungary
90	B2	Tatarbunary Ukr.
83	K3	Tatarskiy Proliv str. Rus. Fed.
		Tatar Strait str. Rus. Fed. see Tatarskiy Proliv
67	C4	Tateyama Japan
128	C1	Tathlina Lake Can.
78	B3	Tathlith Saudi Arabia
78	B2	Tathlīth, Wādī watercourse Saudi Arabia
53	C3	Tathra Austr.
62	A1	Tatkon Myanmar
128	B2	Tatla Lake Can.
		Tatra Mountains mts Pol./Slovakia see Tatry
103	D2	Tatry Pol./Slovakia
154	C2	Tatuí Brazil
143	C2	Tatum U.S.A.
81	C2	Tatvan Turkey
151	E3	Taua Brazil
155	C2	Taubaté Brazil
101	D3	Tauberbischofsheim Ger.
54	C1	Taumarunui N.Z.
122	B2	Taung S. Africa
62	A1	Taunggyi Myanmar
62	A2	Taung-ngu Myanmar
62	A2	Taungup Myanmar
74	B1	Taunsa Pak.
99	B4	Taunton U.K.
101	C2	Taunus hills Ger.
54	C1	Taupo N.Z.
54	C1	Taupo, Lake N.Z.
88	B2	Tauragė Lith.
54	C1	Tauranga N.Z.
139	F1	Taureau, Réservoir resr Can.
80	B2	Taurus Mountains mts Turkey
111	C3	Tavas Turkey
86	F3	Tavda Rus. Fed.
106	B2	Tavira Port.
99	A4	Tavistock U.K.
63	A2	Tavoy Myanmar
111	C3	Tavşanlı Turkey

99 A4 Taw r. U.K.
138 C2 Tawas City U.S.A.
61 C1 Tawau Malaysia
64 A3 Tawi-Tawi i. Phil.
145 C3 Taxco Mex.
76 B2 Taxiatosh Uzbek.
77 D3 Taxkorgan China
96 C2 Tay r. U.K.
96 C2 Tay, Firth of est. U.K.
96 B2 Tay, Loch l. U.K.
76 C3 Tāybād Iran
128 B2 Taylor Can.
138 C2 Taylor MI U.S.A.
143 D2 Taylor TX U.S.A.
138 B3 Taylorville U.S.A.
78 A2 Taymā' Saudi Arabia
83 H2 Taymura r. Rus. Fed.
83 H2 Taymyr, Ozero l. Rus. Fed.
Taymyr, Poluostrov pen. Rus. Fed. see Taymyr Peninsula
83 G2 Taymyr Peninsula pen. Rus. Fed.
63 B2 Tây Ninh Vietnam
96 C2 Tayport U.K.
64 A2 Taytay Phil.
82 G2 Taz r. Rus. Fed.
114 B1 Taza Morocco
129 D2 Tazin Lake Can.
86 G2 Tazovskaya Guba sea chan. Rus. Fed.
81 C1 Tbilisi Georgia
91 E2 Tbilisskaya Rus. Fed.
118 B3 Tchibanga Gabon
115 C3 Tchin-Tabaradene Niger
118 B2 Tcholliré Cameroon
103 D1 Tczew Pol.
44 B2 Teacapán Mex.
54 A3 Te Anau N.Z.
54 A3 Te Anau, Lake N.Z.
45 C3 Teapa Mex.
54 C1 Te Awamutu N.Z.
115 C1 Tébessa Alg.
60 B2 Tebingtinggi Indon.
60 A1 Tebingtinggi Indon.
144 A1 Tecate Mex.
114 B4 Techiman Ghana
144 B2 Tecomán Mex.
144 B3 Tecoripa Mex.
145 B3 Técpan Mex.
144 B2 Tecuala Mex.
110 C1 Tecuci Romania
68 C1 Teeli Rus. Fed.
98 C2 Tees r. U.K.
111 C3 Tefenni Turkey
60 B2 Tegal Indon.
146 B3 Tegucigalpa Hond.
115 C3 Teguidda-n-Tessoumt Niger
129 E1 Tehek Lake Can.
Teheran Iran see Tehrān
114 B4 Téhini Côte d'Ivoire
81 D2 Tehrān Iran
145 C3 Tehuacán Mex.
Tehuantepec, Golfo de g. Mex. see
145 C3 Tehuantepec, Gulf of Mex.
Tehuantepec, Gulf of g. Mex.
145 C3 Tehuantepec, Istmo de isth. Mex.
114 A2 Teide, Pico del vol. Islas Canarias
99 A3 Teifi r. U.K.
Teixeira de Sousa Angola see Luau
76 B2 Tejen Turkm.
76 B2 Tejen r. Turkm.
Tejo r. Port. see Tagus
145 B3 Tejupilco Mex.
54 B2 Tekapo, Lake N.Z.
145 B3 Tekax Mex.
116 B3 Tekezē Wenz r. Eritrea/Eth.
111 C2 Tekirdağ Turkey
54 C1 Te Kuiti N.Z.
75 C2 Tel r. India
81 C1 Telavi Georgia
80 B2 Tel Aviv-Yafo Israel
145 C2 Telchac Puerto Mex.
128 A2 Telegraph Creek Can.
154 B2 Telêmaco Borba Brazil
61 C1 Telen r. Indon.
50 B2 Telfer Mining Centre Austr.
99 B3 Telford U.K.
128 B2 Telkwa Can.
60 A2 Telo Indon.
86 E2 Telpoziz, Gora mt. Rus. Fed.
88 B2 Telšiai Lith.
60 B2 Telukbatang Indon.

60 A1 Telukdalam Indon.
60 B1 Teluk Intan Malaysia
114 C4 Tema Ghana
130 C3 Temagami Lake Can.
60 C2 Temanggung Indon.
123 C2 Temba S. Africa
83 H2 Tembenchi r. Rus. Fed.
60 B2 Tembilahan Indon.
123 C2 Tembisa S. Africa
120 A1 Tembo Aluma Angola
Tembué Moz. see Chifunde
99 B3 Teme r. U.K.
135 C4 Temecula U.S.A.
63 B3 Temengor, Tasik resr Malaysia
60 B1 Temerluh Malaysia
77 D1 Temirtau Kazakh.
139 D1 Témiscamingue, Lac l. Can.
53 C2 Temora Austr.
142 A2 Tempe U.S.A.
143 D2 Temple U.S.A.
97 C2 Templemore Ireland
102 C1 Templin Ger.
145 C2 Tempoal Mex.
120 A2 Tempué Angola
91 D2 Temryuk Rus. Fed.
91 D2 Temryukskiy Zaliv b. Rus. Fed.
153 A3 Temuco Chile
54 B3 Temuka N.Z.
145 C2 Tenabo Mex.
143 E2 Tenaha U.S.A.
73 C3 Tenali India
63 A2 Tenasserim Myanmar
99 A4 Tenby U.K.
117 C3 Tendaho Eth.
105 D3 Tende France
105 D3 Tende, Col de pass France/Italy
73 D4 Ten Degree Channel India
114 A3 Ten-n-Dghâmcha, Sebkhet salt marsh Maur.
67 D3 Tendō Japan
114 B3 Ténenkou Mali
115 C2 Ténéré reg. Niger
115 D3 Ténéré, Erg du des. Niger
115 D2 Ténéré du Tafassâsset des. Niger
114 A2 Tenerife i. Islas Canarias
107 D2 Ténès Alg.
61 C2 Tengah, Kepulauan is Indon.
Tengcheng China see Tengxian
62 A1 Tengchong China
61 C2 Tenggarong Indon.
70 A2 Tengger Shamo des. China
77 C1 Tengiz, Ozero salt l. Kazakh.
71 B3 Tengxian China
119 C4 Tenke Dem. Rep. Congo
114 B3 Tenkodogo Burkina Faso
51 C1 Tennant Creek Austr.
140 C1 Tennessee r. U.S.A.
140 C1 Tennessee state U.S.A.
61 C1 Tenom Malaysia
145 C3 Tenosique Mex.
61 D2 Tenteno Indon.
53 D1 Tenterfield Austr.
141 D3 Ten Thousand Islands U.S.A.
154 B3 Teodoro Sampaio Brazil
155 D1 Teófilo Otoni Brazil
145 C3 Teopisca Mex.
59 C3 Tepa Indon.
144 B2 Tepache Mex.
54 B1 Te Paki N.Z.
144 B3 Tepalcatepec Mex.
109 D2 Tepatitlán Mex.
144 B2 Tepehuanes Mex.
109 D2 Tepelenë Albania
102 C1 Teplice Czech Rep.
89 E3 Teploye Rus. Fed.
90 B2 Teplyk Ukr.
54 C1 Te Puke N.Z.
144 B2 Tequila Mex.
49 K3 Teraina i. Kiribati
108 B2 Teramo Italy
52 B3 Terang Austr.
89 E3 Terbuny Rus. Fed.
90 B2 Terebovlya Ukr.
87 D4 Terek r. Rus. Fed.
154 B2 Terenos Brazil
151 E3 Teresina Brazil
155 D2 Teresópolis Brazil
63 A3 Teressa Island India
100 A3 Tergnier France
80 B1 Terme Turkey
108 B3 Termini Imerese Italy

145 C3 Términos, Laguna de lag. Mex.
77 C3 Termiz Uzbek.
109 B2 Termoli Italy
59 C2 Ternate Indon.
100 A2 Terneuzen Neth.
66 C1 Terney Rus. Fed.
108 B2 Terni Italy
90 B2 Ternopil' Ukr.
52 A2 Terowie Austr.
69 F1 Terpeniya, Mys c. Rus. Fed.
69 F1 Terpeniya, Zaliv g. Rus. Fed.
128 B2 Terrace Can.
130 B3 Terrace Bay Can.
122 B2 Terra Firma S. Africa
140 B3 Terrebonne Bay U.S.A.
138 B3 Terre Haute U.S.A.
131 E3 Terrenceville Can.
100 B1 Terschelling i. Neth.
108 A3 Tertenia Italy
107 C1 Teruel Spain
92 H2 Tervola Fin.
109 C2 Tešanj Bos.-Herz.
116 B3 Teseney Eritrea
66 D2 Teshio Japan
66 D2 Teshio-gawa r. Japan
128 A1 Teslin Can.
128 A1 Teslin Lake Can.
154 B1 Tesouro Brazil
115 C3 Tessaoua Niger
99 C4 Test r. U.K.
121 C2 Tete Moz.
90 C1 Teteriv r. Ukr.
101 F1 Teterow Ger.
90 B2 Tetiyiv Ukr.
114 B1 Tétouan Morocco
110 B2 Tetovo Macedonia
Tetyukhe Rus. Fed. see Dal'negorsk
Teuchezhsk Rus. Fed. see Adygeysk
152 B2 Teuco r. Arg.
144 B2 Teul de González Ortega Mex.
101 D1 Teutoburger Wald hills Ger.
Tevere r. Italy see Tiber
54 A3 Teviot N.Z.
96 C3 Teviot r. U.K.
96 C3 Teviothead U.K.
61 C2 Tewah Indon.
51 E2 Tewantin Austr.
54 C2 Te Wharau N.Z.
99 B4 Tewkesbury U.K.
143 C2 Texarkana U.S.A.
53 D1 Texas Austr.
143 D2 Texas state U.S.A.
143 E3 Texas City U.S.A.
145 C3 Texcoco Mex.
100 B1 Texel i. Neth.
143 D2 Texoma, Lake U.S.A.
123 C2 Teyateyaneng Lesotho
89 F2 Teykovo Rus. Fed.
89 F2 Teza r. Rus. Fed.
75 D2 Tezpur India
72 D2 Tezu India
129 E1 Tha-anne r. Can.
123 C2 Thabana-Ntlenyana mt. Lesotho
123 C2 Thaba Nchu S. Africa
123 C2 Thaba Putsoa mt. Lesotho
123 C1 Thabazimbi S. Africa
123 C2 Thabong S. Africa
63 A2 Thagyettaw Myanmar
62 B1 Thai Binh Vietnam
63 B2 Thailand country Asia
63 B2 Thailand, Gulf of Asia
62 B1 Thai Nguyên Vietnam
62 B2 Thakhèk Laos
74 B1 Thal Pak.
63 A3 Thalang Thai.
74 B1 Thal Desert Pak.
101 E2 Thale (Harz) Ger.
62 B2 Tha Li Thai.
53 C1 Thallon Austr.
123 C1 Thamaga Botswana
78 B3 Thamar, Jabal mt. Yemen
79 C3 Thamarit Oman
130 B3 Thamen r. Can.
54 C1 Thames N.Z.
99 D4 Thames est. U.K.
99 D4 Thames r. U.K.
79 B3 Thamüd Yemen
Thana India see Thane
63 A2 Thanbyuzayat Myanmar
62 A2 Thandwe Myanmar
74 B3 Thane India
62 B2 Thanh Hoa Vietnam
73 B3 Thanjavur India
63 A2 Thanlyin Myanmar

74 A2 Thano Bula Khan Pak.
62 B1 Than Uyên Vietnam
74 A2 Thar Desert India/Pak.
52 B1 Thargomindah Austr.
81 C2 Tharthār, Buḩayrat ath l. Iraq
111 B2 Thasos Greece
111 B2 Thasos i. Greece
63 B2 Thât Khê Vietnam
62 A2 Thaton Myanmar
62 A2 Thatta Pak.
62 A1 Thaungdut Myanmar
62 A2 Thayawadi Myanmar
63 A2 Thayetchaung Myanmar
62 A2 Thayetmyo Myanmar
62 A1 Thazi Myanmar
146 C2 The Bahamas country West Indies
98 B2 The Cheviot h. U.K.
134 B1 The Dalles U.S.A.
136 C2 Thedford U.S.A.
53 D2 The Entrance Austr.
99 C3 The Fens reg. U.K.
114 A3 The Gambia country Africa
52 B3 The Grampians mts Austr.
The Great Oasis oasis Egypt see Wāḩāt al Khārijah
79 C2 The Gulf Asia
100 B1 The Hague Neth.
129 E1 Thelon r. Can.
101 E2 Themar Ger.
123 C2 Thembalihle S. Africa
96 A1 The Minch sea chan. U.K.
97 A1 The Mullet b. Ireland
99 C4 The Needles stack U.K.
150 C3 Theodore Roosevelt r. Brazil
129 D2 The Pas Can.
111 B2 Thermaïkos Kolpos g. Greece
136 B2 Thermopolis U.S.A.
53 C3 The Rock Austr.
The Skaw spit Denmark see Grenen
99 C4 The Solent str. U.K.
130 B3 Thessalon Can.
111 B2 Thessaloniki Greece
99 D3 Thetford U.K.
131 C3 Thetford Mines Can.
62 A1 The Triangle mts Myanmar
147 D3 The Valley Anguilla
131 D2 Thévenet, Lac l. Can.
99 D3 The Wash b. U.K.
99 D4 The Weald reg. U.K.
143 D2 The Woodlands U.S.A.
140 B3 Thibodaux U.S.A.
129 E2 Thicket Portage Can.
137 D1 Thief River Falls U.S.A.
Thiel Neth. see Tiel
105 C2 Thiers France
114 A3 Thiès Senegal
119 D3 Thika Kenya
73 B4 Thiladhunmathi Maldives
75 C2 Thimphu Bhutan
105 D2 Thionville France
Thira i. Greece see Santorini
98 C2 Thirsk U.K.
73 B4 Thiruvananthapuram India
93 E4 Thisted Denmark
129 E1 Thlewiaza r. Can.
63 B3 Thô Chu, Đao i. Vietnam
62 A2 Thoen Thai.
123 C3 Thohoyandou S. Africa
101 E1 Thomasburg Ger.
97 C2 Thomastown Ireland
140 C2 Thomasville AL U.S.A.
141 D2 Thomasville GA U.S.A.
100 C2 Thommen Belgium
129 E2 Thompson Can.
128 E3 Thompson r. U.S.A.
134 C1 Thompson Falls U.S.A.
128 B2 Thompson Sound Can.
51 D2 Thomson watercourse Austr.
141 D2 Thomson U.S.A.
63 B2 Thon Buri Thai.
63 A2 Thongwa Myanmar
Thoothukudi India see Tuticorin
142 B1 Thoreau U.S.A.
96 C3 Thornhill U.K.
136 C3 Thornton U.S.A.
55 F2 Thorshavnheiane reg. Antarctica
123 C2 Thota-ea-Moli Lesotho
104 B2 Thouars France
62 A1 Thoubal India
134 D1 Three Forks U.S.A.
128 C2 Three Hills Can.
63 A2 Three Pagodas Pass Myanmar/Thai.

114 B4 Three Points, Cape Ghana
138 B2 Three Rivers *MI* U.S.A.
143 D3 Three Rivers *TX* U.S.A.
73 B3 Thrissur India
63 B2 Thu Dâu Môt Vietnam
100 B2 Thuin Belgium
127 H1 Thule Greenland
121 B3 Thuli Zimbabwe
102 B2 Thun Switz.
130 B3 Thunder Bay Can.
130 B3 Thunder Bay *b.* Can.
63 A3 Thung Song Thai.
101 E2 Thüringer Becken *reg.* Ger.
101 E2 Thüringer Wald *mts* Ger.
Thuringian Forest *mts* Ger.
see **Thüringer Wald**
97 C2 Thurles Ireland
59 D3 Thursday Island Austr.
96 C1 Thurso U.K.
96 C1 Thurso *r.* U.K.
55 R2 Thurston Island Antarctica
101 D1 Thüster Berg *h.* Ger.
121 C2 Thyolo Malawi
Thysville Dem. Rep. Congo
see **Mbanza-Ngungu**
79 C2 Tiäb Iran
151 E3 Tianguá Brazil
70 B2 Tianjin China
70 B2 Tianjin *mun.* China
71 A3 Tianlin China
70 B2 Tianmen China
69 E2 Tianshan China
70 A2 Tianshui China
70 A2 Tianzhu China
107 D2 Tiaret Alg.
114 B4 Tiassalé Côte d'Ivoire
154 B2 Tibagi Brazil
154 B2 Tibagi *r.* Brazil
118 B2 Tibati Cameroon
108 B2 Tiber *r.* Italy
Tiberias, Lake *i.* Israel see
Galilee, Sea of
115 D2 Tibesti *mts* Chad
75 C1 Tibet *aut. reg.* China
68 B2 Tibet, Plateau of China
115 D2 Tibīstī, Sarīr *des.* Libya
52 B1 Tibooburra Austr.
144 A2 Tiburón, Isla *i.* Mex.
114 B3 Tichît Maur.
114 A2 Tichît, Dhar *hills* Maur.
114 A2 Tichla Western Sahara
105 D2 Ticino *r.* Italy/Switz.
139 E2 Ticonderoga U.S.A.
145 D2 Ticul Mex.
114 A3 Tidjikja Maur.
100 B2 Tiel Neth.
70 C1 Tieling China
75 B1 Tielongtan China
100 A2 Tielt Belgium
100 B2 Tienen Belgium
68 B2 Tien Shan *mts* China/Kyrg.
Tientsin China see **Tianjin**
Tientsin *mun.* China see
Tianjin
93 G3 Tierp Sweden
145 C2 Tierra Blanca Mex.
145 C3 Tierra Colorada Mex.
153 B5 Tierra del Fuego, Isla
Grande de *i.* Arg./Chile
106 B1 Tiétar, Valle del *val.* Spain
154 C2 Tietê Brazil
154 B2 Tietê *r.* Brazil
138 C2 Tiffin U.S.A.
Tiflis Georgia see **Tbilisi**
141 D2 Tifton U.S.A.
90 B2 Tighina Moldova
75 C2 Tigiria India
118 B2 Tignère Cameroon
131 D3 Tignish Can.
150 B3 Tigre *r.* Ecuador/Peru
81 C2 Tigris *r.* Asia
114 A3 Tiguent Maur.
114 A3 Tiguesmat *hills* Maur.
115 D3 Tigui Chad
145 C2 Tihuatlán Mex.
144 A1 Tijuana Mex.
91 E2 Tikhoretsk Rus. Fed.
89 D2 Tikhvin Rus. Fed.
89 D2 Tikhvinskaya Gryada *ridge*
Rus. Fed.
157 F7 Tiki Basin S. Pacific Ocean
54 C1 Tikokino N.Z.
81 C2 Tikrīt Iraq
83 J2 Tiksi Rus. Fed.
100 B2 Tilburg Neth.
152 B2 Tilcara Arg.
52 B1 Tilcha (abandoned) Austr.
114 C3 Tîlemsi, Vallée du
watercourse Mali
Tilimsen Alg. see **Tlemcen**

114 C3 Tillabéri Niger
134 B1 Tillamook U.S.A.
63 A3 Tillanchong Island India
111 C3 Tilos *i.* Greece
52 B2 Tilpa Austr.
86 F2 Til'tim Rus. Fed.
89 E3 Tim Rus. Fed.
86 D2 Timanskiy Kryazh *ridge*
Rus. Fed.
54 B2 Timaru N.Z.
91 D2 Timashevsk Rus. Fed.
Timashevskaya Rus. Fed. see
Timashevsk
114 B3 Timbedgha Maur.
50 C1 Timber Creek Austr.
114 B3 Timbuktu Mali
115 C3 Timia Niger
114 C2 Timimoun Alg.
111 B2 Timiou Prodromou,
Akrotirio *pt* Greece
110 B1 Timiş *r.* Romania
110 B1 Timişoara Romania
130 B3 Timmins Can.
89 E2 Timokhino Rus. Fed.
151 E3 Timon Brazil
59 C3 Timor *i.* East Timor/
Indonesia
Timor-Leste *country* Asia see
East Timor
59 C3 Timor Sea Austr./Indon.
Timor Timur *country* Asia
see **East Timor**
93 G3 Timrå Sweden
78 B2 Tin, Jabal at *mt.*
Saudi Arabia
114 B2 Tindouf Alg.
53 D1 Tingha Austr.
75 C2 Tingri China
93 F4 Tingsryd Sweden
92 E3 Tingvoll Norway
Tingzhou China see
Changting
59 D2 Tinian *i.* N. Mariana Is
Tinnelvelly India see
Tirunelveli
152 B2 Tinogasta Arg.
111 C3 Tinos Greece
111 C3 Tinos *i.* Greece
100 A3 Tinqueux France
115 C2 Tinrhert, Hamada de Alg.
62 A1 Tinsukia India
52 B3 Tintinara Austr.
136 C1 Tioga U.S.A.
107 D2 Tipasa Alg.
97 B2 Tipperary Ireland
151 E3 Tiracambu, Serra do *hills*
Brazil
109 C2 Tirana Albania
Tiranë Albania see **Tirana**
108 B1 Tirano Italy
52 A1 Tirari Desert Austr.
90 B2 Tiraspol Moldova
122 A2 Tiraz Mountains Namibia
111 C3 Tire Turkey
96 A2 Tiree *i.* U.K.
Tîrgovişte Romania see
Târgovişte
Tîrgu Frumos Romania see
Târgu Frumos
Tîrgu Jiu Romania see
Târgu Jiu
Tîrgu Lăpuş Romania see
Târgu Lăpuş
Tîrgu Mureş Romania see
Târgu Mureş
Tîrgu Neamţ Romania see
Târgu Neamţ
Tîrgu Ocna Romania see
Târgu Ocna
74 B1 Tirich Mir *mt.* Pak.
Tîrnăveni Romania see
Târnăveni
155 C1 Tiros Brazil
118 C2 Tiroungoulou C.A.R.
73 B3 Tiruchchirappalli India
73 B4 Tirunelveli India
73 B3 Tirupati India
73 B3 Tiruppattur India
73 B3 Tiruppur India
Tisa *r.* Hungary see **Tisza**
109 D1 Tisa *r.* Serbia
129 D2 Tisdale Can.
107 D2 Tissemsilt Alg.
75 C2 Tista *r.* India
103 E2 Tisza *r.* Hungary
55 L1 Titan Dome Antarctica
Titicaca, Lago *l.* Bol./Peru
see **Titicaca, Lake**
152 B1 Titicaca, Lake *l.* Bol./Peru
75 C2 Titlagarh India

110 C2 Titu Romania
141 D3 Titusville U.S.A.
99 B4 Tiverton U.K.
108 B2 Tivoli Italy
79 C2 Ţiwī Oman
63 A2 Ti-ywa Myanmar
145 D2 Tizimín Mex.
107 D2 Tizi Ouzou Alg.
114 B2 Tiznit Morocco
92 G2 Tjaktjajaure *l.* Sweden
Tjirebon Indon. see **Cirebon**
145 C3 Tlacotalpán Mex.
144 B2 Tlahualilo Mex.
145 C3 Tlalnepantla Mex.
145 C3 Tlapa Mex.
145 C3 Tlaxcala Mex.
145 C3 Tlaxiaco Mex.
114 B1 Tlemcen Alg.
123 C1 Tlokweng Botswana
128 B2 Toad River Can.
121 □D2 Toamasina Madag.
60 A1 Toba, Danau *l.* Indon.
Toba, Lake *l.* Indon. see
Toba, Danau
74 A1 Toba and Kakar Ranges *mts*
Pak.
147 D3 Tobago *i.* Trin. and Tob.
59 C2 Tobelo Indon.
130 B3 Tobermory Can.
96 A2 Tobermory U.K.
129 D2 Tobin Lake Can.
60 B2 Toboali Indon.
86 F3 Tobol'sk Rus. Fed.
Tobruk Libya see **Tubruq**
76 C1 Tobyl *r.* Kazakh./Rus. Fed.
151 E3 Tocantinópolis Brazil
151 E3 Tocantins *r.* Brazil
141 D2 Toccoa U.S.A.
108 A1 Toce *r.* Italy
152 A2 Tocopilla Chile
53 C3 Tocumwal Austr.
59 C3 Todeli Indon.
108 B2 Todi Italy
144 A2 Todos Santos Mex.
128 B3 Tofino Can.
96 □ Toft U.K.
61 D2 Togian *i.* Indon.
61 D2 Togian, Kepulauan *is*
Indon.
Togliatti Rus. Fed. see
Tol'yatti
114 C4 Togo *country* Africa
74 B2 Tohana India
141 D3 Tohopekaliga, Lake U.S.A.
126 C2 Tok U.S.A.
116 B3 Tokar Sudan
69 E3 Tokara-rettō *is* Japan
91 E1 Tokarevka Rus. Fed.
80 B1 Tokat Turkey
49 J4 Tokelau *terr.*
S. Pacific Ocean
91 D2 Tokmak Ukr.
77 D2 Tokmok Kyrg.
54 C1 Tokoroa N.Z.
68 B2 Toksun China
67 B4 Tokushima Japan
67 C3 Tōkyō Japan
121 □D3 Tôlañaro Madag.
Tolbukhin Bulg. see
Dobrich
154 B2 Toledo Brazil
106 C2 Toledo Spain
138 C2 Toledo U.S.A.
106 C2 Toledo, Montes de *mts*
Spain
140 B2 Toledo Bend Reservoir
U.S.A.
121 □D3 Toliara Madag.
Toling China see **Zanda**
61 D1 Tolitoli Indon.
108 B1 Tolmezzo Italy
108 B1 Tolmin Slovenia
103 D2 Tolna Hungary
104 B3 Tolosa Spain
145 C3 Toluca Mex.
87 D3 Tol'yatti Rus. Fed.
138 A2 Tomah U.S.A.
138 B1 Tomahawk U.S.A.

Titograd Montenegro see
Podgorica
Titova Mitrovica Kosovo see
Mitrovicë
Titovo Užice Serbia see
Užice
Titovo Velenje Slovenia see
Velenje
Titov Veles Macedonia see
Veles
Titov Vrbas Serbia see
Vrbas

66 D2 Tomakomai Japan
61 D2 Tomali Indon.
61 C1 Tomani Malaysia
106 B2 Tomar Port.
103 E1 Tomaszów Lubelski Pol.
103 E1 Tomaszów Mazowiecki P
144 B3 Tomatlán Mex.
154 C2 Tomazina Brazil
140 C2 Tombigbee *r.* U.S.A.
120 A1 Tomboco Angola
155 D2 Tombos Brazil
Tombouctou Mali see
Timbuktu
142 A2 Tombstone U.S.A.
120 A2 Tombua Angola
123 C1 Tom Burke S. Africa
106 C2 Tomelloso Spain
53 C2 Tomingley Austr.
61 D2 Tomini, Teluk *g.* Indon.
109 C2 Tomislavgrad Bos.-Herz.
103 D2 Tompa Hungary
50 A2 Tom Price Austr.
82 G3 Tomsk Rus. Fed.
93 F4 Tomtabacken *h.* Sweden
83 K2 Tomtor Rus. Fed.
145 C3 Tonalá Mex.
150 C3 Tonantins Brazil
99 D4 Tonbridge U.K.
59 C2 Tondano Indon.
49 J5 Tonga *country*
S. Pacific Ocean
49 J6 Tongatapu Group *is* Tong
71 B3 Tongcheng China
70 A2 Tongchuan China
71 A3 Tongdao China
100 B2 Tongeren Belgium
71 A3 Tonghai China
65 B1 Tonghua China
65 B2 Tongjosŏn-man *b.* N. Kor
62 B1 Tongking, Gulf of China/
Vietnam
69 E2 Tongliao China
70 B2 Tongling China
52 B2 Tongo Austr.
Tongquan China see
Malong
71 A3 Tongren China
Tongshan China see
Xuzhou
Tongshi China see
Wuzhishan
Tongtian He *r.* China see
Yangtze
96 B2 Tongue U.K.
71 A3 Tongxian China see
Tongzhou
65 B3 Tongyeong S. Korea
69 E2 Tongyu China
65 A1 Tongyuanpu China
70 B2 Tongzhou China
71 A3 Tongzi China
119 C2 Tonj S. Sudan
74 B2 Tonk India
81 D2 Tonkābon Iran
Tônlé Sab *l.* Cambodia see
Tonle Sap
63 B2 Tonle Sap *l.* Cambodia
135 C3 Tonopah U.S.A.
93 F4 Tønsberg Norway
135 D2 Tooele U.S.A.
52 B3 Tooleybuc Austr.
53 D1 Toowoomba Austr.
137 D3 Topeka U.S.A.
144 B2 Topia Mex.
144 B2 Topolobampo Mex.
86 C2 Topozero, Ozero *l.*
Rus. Fed.
134 B1 Toppenish U.S.A.
111 C3 Torbalı Turkey
76 B3 Torbat-e Ḥeydarīyeh Iran
76 C3 Torbat-e Jām Iran
131 E3 Torbay Can.
106 C1 Tordesillas Spain
107 C1 Tordesilos Spain
92 H2 Töre Sweden
107 D1 Torelló Spain
100 B1 Torenberg *h.* Neth.
Toretam Kazakh. see
Baykonyr
101 F2 Torgau Ger.
76 C2 Torgay Kazakh.
100 A2 Torhout Belgium
Torino Italy see **Turin**
67 D4 Tori-shima *i.* Japan
117 B4 Torit S. Sudan
154 B1 Torixoréu Brazil
89 D2 Torkovichi Rus. Fed.
106 B1 Tormes *r.* Spain
92 H2 Torneälven *r.* Sweden
131 D2 Torngat Mountains Can.

92 H2	Tornio Fin.
106 B1	Toro Spain
130 C3	Toronto Can.
89 D2	Toropets Rus. Fed.
119 D2	Tororo Uganda
	Toros Dağları mts Turkey see Taurus Mountains
52 B3	Torquay Austr.
99 B4	Torquay U.K.
135 C4	Torrance U.S.A.
106 B2	Torrão Port.
106 B1	Torre mt. Port.
107 D1	Torreblanca Spain
106 C1	Torrecerredo mt. Spain
106 B1	Torre de Moncorvo Port.
106 C1	Torrelavega Spain
106 C2	Torremolinos Spain
52 A2	Torrens, Lake imp. l. Austr.
107 C2	Torrent Spain
144 B2	Torreón Mex.
106 B2	Torres Novas Port.
156 C6	Torres Strait Austr.
106 B2	Torres Vedras Port.
107 C2	Torrevieja Spain
96 B2	Torridon U.K.
96 B2	Torridon, Loch b. U.K.
106 C2	Torrijos Spain
139 E2	Torrington CT U.S.A.
136 C2	Torrington WY U.S.A.
107 D1	Torroella de Montgrí Spain
94 B1	Tórshavn Faroe Is
76 C2	To'rtko'l Uzbek.
108 A3	Tortolì Italy
108 A2	Tortona Italy
107 D1	Tortosa Spain
81 D2	Torūd Iran
103 D1	Toruń Pol.
97 B1	Tory Island Ireland
97 B1	Tory Sound sea chan. Ireland
89 D2	Torzhok Rus. Fed.
67 B4	Tosashimizu Japan
122 B2	Tosca S. Africa
77 C2	Toscano, Arcipelago is Italy
77 C2	Toshkent Uzbek.
89 D2	Tosno Rus. Fed.
68 C1	Tosontsengel Mongolia
152 B2	Tostado Arg.
101 D1	Tostedt Ger.
80 B1	Tosya Turkey
86 D3	Tot'ma Rus. Fed.
151 D2	Totness Suriname
67 B3	Tottori Japan
114 B4	Touba Côte d'Ivoire
114 B3	Touboro Cameroon
114 B3	Tougan Burkina Faso
115 C1	Touggourt Alg.
105 D2	Toul France
105 D3	Toulon France
104 C3	Toulouse France
	Tourane Vietnam see Đa Nẵng
104 B2	Tourlaville France
100 A2	Tournai Belgium
115 D2	Tourndo, Oued watercourse Alg./Niger
105 C2	Tournus France
151 F3	Touros Brazil
104 C2	Tours France
115 D2	Tousside, Pic mt. Chad
122 B3	Touwsrivier S. Africa
150 B2	Tovar Venez.
66 D2	Towada Japan
134 D1	Townsend U.S.A.
51 D1	Townsville Austr.
61 D2	Towori, Teluk b. Indon.
77 E2	Toxkan He r. China
66 D2	Tōya-ko l. Japan
67 C3	Toyama Japan
67 C4	Toyohashi Japan
67 C3	Toyota Japan
115 C1	Tozeur Tunisia
81 C1	T'q'varcheli Georgia
	Trablous Lebanon see Tripoli
80 B1	Trabzon Turkey
106 B2	Trafalgar, Cabo de c. Spain
128 C3	Trail Can.
88 B3	Trakai Lith.
97 B2	Tralee Ireland
	Trá Lí Ireland see Tralee
97 C2	Tramore Ireland
93 F4	Tranås Sweden
63 A3	Trang Thai.
59 C3	Trangan i. Indon.
55 L2	Transantarctic Mountains Antarctica
90 B2	Transnistria terr. Moldova
110 B1	Transylvanian Alps mts Romania
108 B3	Trapani Italy
53 C3	Traralgon Austr.
75 D2	Trashigang Bhutan
63 B2	Trat Thai.
102 C2	Traunstein Ger.
54 B2	Travers, Mount N.Z.
138 B2	Traverse City U.S.A.
63 B3	Tra Vinh Vietnam
103 D2	Třebíč Czech Rep.
109 C2	Trebinje Bos.-Herz.
103 E2	Trebišov Slovakia
	Trebizond Turkey see Trabzon
109 C1	Trebnje Slovenia
	Trefynwy U.K. see Monmouth
153 C3	Treinta y Tres Uru.
153 B4	Trelew Arg.
93 F4	Trelleborg Sweden
130 C3	Tremblant, Mont h. Can.
109 C2	Tremiti, Isole is Italy
134 D2	Tremonton U.S.A.
107 D1	Tremp Spain
103 D2	Trenčín Slovakia
153 B3	Trenque Lauquén Arg.
	Trent Italy see Trento
98 C3	Trent r. U.K.
108 B1	Trento Italy
139 D2	Trenton Can.
137 E2	Trenton MO U.S.A.
139 E2	Trenton NJ U.S.A.
131 E3	Trepassey Can.
153 B3	Tres Arroyos Arg.
155 C2	Três Corações Brazil
154 B2	Três Irmãos, Represa resr Brazil
154 B2	Três Lagoas Brazil
153 A4	Tres Lagos Arg.
155 C1	Três Marias, Represa resr Brazil
155 C2	Três Pontas Brazil
153 B4	Tres Puntas, Cabo c. Arg.
155 D2	Três Rios Brazil
101 F1	Treuenbrietzen Ger.
	Treves Ger. see Trier
108 A1	Treviglio Italy
108 B1	Treviso Italy
99 A4	Trevose Head hd U.K.
109 C3	Tricase Italy
	Trichinopoly India see Tiruchchirappalli
	Trichur India see Thrissur
53 C2	Trida Austr.
100 C3	Trier Ger.
108 B1	Trieste Italy
108 B1	Triglav mt. Slovenia
111 B3	Trikala Greece
59 D3	Trikora, Puncak mt. Indon.
97 C2	Trim Ireland
73 C4	Trincomalee Sri Lanka
154 C1	Trindade Brazil
148 H5	Trindade, Ilha da i. S. Atlantic Ocean
147 D3	Trinidad Bol.
136 C3	Trinidad i. Trin. and Tob.
147 D3	Trinidad U.S.A.
	Trinidad and Tobago country West Indies
143 E3	Trinity r. U.S.A.
131 E3	Trinity Bay Can.
111 B3	Tripoli Greece
80 B2	Tripoli Lebanon
115 D1	Tripoli Libya
75 D2	Tripura state India
113 B9	Tristan da Cunha i. S. Atlantic Ocean
	Trivandrum India see Thiruvananthapuram
108 B2	Trivento Italy
103 D2	Trnava Slovakia
59 E3	Trobriand Islands P.N.G.
92 F2	Trofors Norway
109 C2	Trogir Croatia
109 C2	Troia Italy
100 C2	Troisdorf Ger.
106 C2	Trois Fourches, Cap des c. Morocco
130 C3	Trois-Rivières Can.
87 F3	Troitsk Rus. Fed.
86 E2	Troitsko-Pechorsk Rus. Fed.
151 D3	Trombetas r. Brazil
	Tromelin Island i. Micronesia see Fais
123 C3	Trompsburg S. Africa
92 G2	Tromsø Norway
92 F3	Trondheim Norway
75 D2	Trongsa Bhutan
96 B3	Troon U.K.
89 E3	Trosna Rus. Fed.
97 C1	Trostan h. U.K.
128 C2	Trout Lake Alta Can.
128 B1	Trout Lake l. N.W.T. Can.
130 A2	Trout Lake l. Ont. Can.
99 B4	Trowbridge U.K.
111 C3	Troy tourist site Turkey
140 C2	Troy AL U.S.A.
139 E2	Troy NY U.S.A.
105 C2	Troyes France
135 C3	Troy Peak U.S.A.
109 D2	Trstenik Serbia
89 D3	Trubchevsk Rus. Fed.
	Truc Giang Vietnam see Bến Tre
106 B1	Truchas Spain
	Trucial Coast country Asia see United Arab Emirates
79 C2	Trucial States U.A.E.
	Trucial Coast country Asia see United Arab Emirates
146 B3	Trujillo Hond.
150 B3	Trujillo Peru
106 B2	Trujillo Spain
147 C4	Trujillo Venez.
	Trujillo, Monte mt. Dom. Rep. see Duarte, Pico
140 B1	Trumann U.S.A.
131 D3	Truro Can.
99 A4	Truro U.K.
61 C1	Trus Madi, Gunung mt. Malaysia
142 B2	Truth or Consequences U.S.A.
103 D1	Trutnov Czech Rep.
93 F3	Trysil Norway
103 D1	Trzebiatów Pol.
103 D1	Trzebnica Pol.
68 B1	Tsagaannuur Mongolia
	Tsaidam Basin basin China see Qaidam Pendi
121 □D2	Tsaratanana, Massif du mts Madag.
110 C2	Tsarevo Bulg.
122 A2	Tsaukaib Namibia
	Tselinograd Kazakh. see Astana
	Tsementnyy Rus. Fed. see Fokino
122 A2	Tses Namibia
122 B1	Tsetseng Botswana
68 C1	Tsetserleg Mongolia
68 C1	Tsetserleg Mongolia
122 B2	Tshabong Botswana
122 B1	Tshane Botswana
118 B3	Tshela Dem. Rep. Congo
118 C3	Tshikapa Dem. Rep. Congo
118 C3	Tshikapa r. Dem. Rep. Congo
123 C2	Tshing S. Africa
123 D1	Tshipise S. Africa
118 C3	Tshitanzu Dem. Rep. Congo
120 B3	Tshootsha Botswana
118 C3	Tshuapa r. Dem. Rep. Congo
	Tshwane S. Africa see Pretoria
87 D4	Tsimlyanskoye Vodokhranilishche resr Rus. Fed.
	Tsinan China see Jinan
	Tsingtao China see Qingdao
	Tsining China see Ulan Qab
121 □D2	Tsiroanomandidy Madag.
	Tsitsihar China see Qiqihar
76 A2	Tskhinvali Georgia
123 C3	Tsomo S. Africa
67 C4	Tsu Japan
67 D3	Tsuchiura Japan
66 D2	Tsugarū-kaikyō str. Japan
	Tsugaru Strait str. Japan see Tsugarū-kaikyō
120 A2	Tsumeb Namibia
122 A1	Tsumis Park Namibia
120 B2	Tsumkwe Namibia
67 C3	Tsuruga Japan
66 C3	Tsuruoka Japan
67 A4	Tsushima is Japan
67 A4	Tsushima i. Japan
67 B3	Tsuyama Japan
123 C2	Tswelelang S. Africa
88 C3	Tsyelyakhany Belarus
91 C2	Tsyurupyns'k Ukr.
	Tthenaagoo Can. see Nahanni Butte
59 C3	Tual Indon.
97 B2	Tuam Ireland
54 C1	Tuapeka Mouth N.Z.
91 D3	Tuapse Rus. Fed.
54 A3	Tuatapere N.Z.
96 A1	Tuath, Loch a' U.K.
142 A1	Tuba City U.S.A.
61 C2	Tuban Indon.
152 D2	Tubarão Brazil
102 B2	Tübingen Ger.
115 E1	Tubruq Libya
49 L6	Tubuai Islands is Fr. Polynesia
144 A1	Tubutama Mex.
152 C1	Tucavaca Bol.
128 B1	Tuchitua Can.
142 A2	Tucson U.S.A.
143 C1	Tucumcari U.S.A.
150 C2	Tucupita Venez.
151 E3	Tucuruí Brazil
151 E3	Tucuruí, Represa de resr Brazil
107 C1	Tudela Spain
118 A1	Tudun Wada Nigeria
106 B1	Tuela r. Port.
157 F2	Tufts Abyssal Plain N. Pacific Ocean
64 B2	Tuguegarao Phil.
106 B1	Tui Spain
59 C3	Tukangbesi, Kepulauan is Indon.
126 D2	Tuktoyaktuk Can.
88 B2	Tukums Latvia
119 D3	Tukuyu Tanz.
145 C2	Tula Mex.
89 E3	Tula Rus. Fed.
	Tulach Mhór Ireland see Tullamore
145 C2	Tulancingo Mex.
135 C3	Tulare U.S.A.
142 B2	Tularosa U.S.A.
110 C1	Tulcea Romania
90 B2	Tul'chyn Ukr.
	Tuléar Madag. see Toliara
129 E1	Tulemalu Lake Can.
143 C2	Tulia Can.
128 B1	Tulita Can.
140 C1	Tullahoma U.S.A.
53 C2	Tullamore Austr.
97 C2	Tullamore Ireland
104 C2	Tulle France
97 C2	Tullow Ireland
51 D1	Tully Austr.
143 D1	Tulsa U.S.A.
126 B2	Tuluksak U.S.A.
83 H3	Tulun Rus. Fed.
150 B2	Tumaco Col.
123 C2	Tumahole S. Africa
	Tumakuru India see Tumkur
93 G4	Tumba Sweden
118 B3	Tumba, Lac l. Dem. Rep. Congo
61 C2	Tumbangtiti Indon.
53 C3	Tumbarumba Austr.
150 A3	Tumbes Peru
128 B2	Tumbler Ridge Can.
52 A2	Tumby Bay Austr.
65 B1	Tumen China
150 C2	Tumereng Guyana
64 A3	Tumindao i. Phil.
73 B3	Tumkur India
74 A2	Tump Pak.
151 D2	Tumucumaque, Serra hills Brazil
53 C3	Tumut Austr.
99 D4	Tunbridge Wells, Royal U.K.
80 B2	Tunceli Turkey
53 D2	Tuncurry Austr.
119 D4	Tunduru Tanz.
110 C2	Tundzha r. Bulg.
128 B1	Tungsten (abandoned) Can.
115 D1	Tunis Tunisia
108 B3	Tunis, Golfe de g. Tunisia
115 C1	Tunisia country Africa
150 B2	Tunja Col.
92 F3	Tunnsjøen l. Norway
	Tunxi China see Huangshan
154 B2	Tupã Brazil
154 C1	Tupaciguara Brazil
140 C2	Tupelo U.S.A.
152 B2	Tupiza Bol.
76 B2	Tupkaragan, Mys pt Kazakh.
83 H2	Tura Rus. Fed.
86 F3	Tura r. Rus. Fed.
78 B2	Turabah Saudi Arabia
83 J3	Turana, Khrebet mts Rus. Fed.
54 C1	Turangi N.Z.
74 B2	Turan Lowland Asia
77 D2	Turar Ryskulov Kazakh.
78 A1	Turayf Saudi Arabia
88 B2	Turba Estonia

74 A2 **Turbat** Pak.
150 B2 **Turbo** Col.
110 B1 **Turda** Romania
Turfan China see Turpan
76 C1 **Turgayskaya Stolovaya Strana** reg. Kazakh.
110 C2 **Türgovishte** Bulg.
111 C3 **Turgutlu** Turkey
80 B1 **Turhal** Turkey
107 C2 **Turia** r. Spain
108 A1 **Turin** Italy
86 F3 **Turinsk** Rus. Fed.
90 A1 **Turiya** r. Ukr.
90 A1 **Turiys'k** Ukr.
90 A2 **Turka** Ukr.
119 D2 **Turkana, Lake** salt l. Eth./Kenya
103 E2 **Túrkeve** Hungary
80 B2 **Turkey** country Asia/Europe
50 B1 **Turkey Creek** Austr.
77 C2 **Turkistan** Kazakh.
76 C3 **Türkmenabat** Turkm.
76 B2 **Türkmenbaşy** Turkm.
76 B2 **Turkmenistan** country Asia
Turkmeniya country Asia see Turkmenistan
Turkmenskaya S.S.R. country Asia see Turkmenistan
147 C2 **Turks and Caicos Islands** terr. West Indies
147 C2 **Turks Islands** Turks and Caicos Is
93 H3 **Turku** Fin.
119 D2 **Turkwel** watercourse Kenya
135 B3 **Turlock** U.S.A.
155 D1 **Turmalina** Brazil
54 C2 **Turnagain, Cape** N.Z.
146 B3 **Turneffe Islands** Belize
100 B2 **Turnhout** Belgium
129 D2 **Turnor Lake** Can.
Türnovo Bulg. see Veliko Türnovo
110 B2 **Turnu Măgurele** Romania
68 B2 **Turpan** China
96 C2 **Turriff** U.K.
64 A3 **Turtle Islands** Malaysia/Phil.
77 D2 **Turugart Pass** China/Kyrg.
82 G2 **Turukhansk** Rus. Fed.
140 C2 **Tuscaloosa** U.S.A.
140 C2 **Tuskegee** U.S.A.
81 C2 **Tutak** Turkey
89 E2 **Tutayev** Rus. Fed.
73 B4 **Tuticorin** India
121 C1 **Tutubu** Tanz.
49 J5 **Tutuila** i. American Samoa
120 B3 **Tutume** Botswana
93 H3 **Tuusula** Fin.
49 I4 **Tuvalu** country S. Pacific Ocean
78 B2 **Tuwayq, Jabal** hills Saudi Arabia
78 B2 **Tuwayq, Jabal** mts Saudi Arabia
78 A2 **Tuwwal** Saudi Arabia
144 B2 **Tuxpan** Mex.
145 C2 **Tuxpan** Mex.
145 C3 **Tuxtla Gutiérrez** Mex.
62 B1 **Tuyên Quang** Vietnam
63 B2 **Tuy Hoa** Vietnam
80 B2 **Tuz, Lake** salt l. Turkey
Tuz Gölü salt l. Turkey see Tuz, Lake
81 C2 **Tuz Khurmātū** Iraq
109 C2 **Tuzla** Bos.-Herz.
91 E2 **Tuzlov** r. Rus. Fed.
89 E2 **Tver'** Rus. Fed.
98 B2 **Tweed** r. U.K.
53 D1 **Tweed Heads** Austr.
122 A2 **Twee Rivier** Namibia
135 C4 **Twentynine Palms** U.S.A.
131 E3 **Twillingate** Can.
134 D2 **Twin Falls** U.S.A.
54 B2 **Twizel** N.Z.
137 E1 **Two Harbors** U.S.A.
128 C2 **Two Hills** Can.
Tyddewi U.K. see St David's
143 D2 **Tyler** U.S.A.
83 J3 **Tynda** Rus. Fed.
Tyndinskiy Rus. Fed. see Tynda
96 B2 **Tyndrum** U.K.
98 C2 **Tyne** r. England U.K.
95 C2 **Tyne** r. Scotland U.K.
93 F3 **Tynset** Norway
80 B2 **Tyre** Lebanon
69 E1 **Tyrma** Rus. Fed.
111 B3 **Tyrnavos** Greece
52 B3 **Tyrrell, Lake** dry lake Austr.

108 B2 **Tyrrhenian Sea** France/Italy
87 E3 **Tyul'gan** Rus. Fed.
86 F3 **Tyumen'** Rus. Fed.
83 J2 **Tyung** r. Rus. Fed.
Tyuratam Kazakh. see Baykonyr
99 A4 **Tywi** r. U.K.
123 D1 **Tzaneen** S. Africa

U

Uaco Congo Angola see Waku-Kungo
120 B2 **Uamanda** Angola
150 C3 **Uarini** Brazil
150 C3 **Uaupés** Brazil
155 D2 **Ubá** Brazil
155 D1 **Ubaí** Brazil
151 F4 **Ubaitaba** Brazil
118 B3 **Ubangi** r. C.A.R./Dem. Rep. Congo
Ubangi-Shari country Africa see Central African Republic
67 B4 **Ube** Japan
106 C2 **Úbeda** Spain
154 C1 **Uberaba** Brazil
154 C1 **Uberlândia** Brazil
106 B1 **Ubiña, Peña** mt. Spain
123 D2 **Ubombo** S. Africa
63 B2 **Ubon Ratchathani** Thai.
119 C3 **Ubundu** Dem. Rep. Congo
150 B3 **Ucayali** r. Peru
100 B2 **Uccle** Belgium
74 B2 **Uch** Pak.
66 D2 **Uchiura-wan** b. Japan
76 C2 **Uchquduq** Uzbek.
83 J3 **Uchur** r. Rus. Fed.
99 D4 **Uckfield** U.K.
128 B3 **Ucluelet** Can.
83 I2 **Udachnyy** Rus. Fed.
74 B2 **Udaipur** India
91 C1 **Uday** r. Ukr.
93 F4 **Uddevalla** Sweden
92 G2 **Uddjaure** l. Sweden
100 B2 **Uden** Neth.
74 B1 **Udhampur** India
108 B1 **Udine** Italy
89 E2 **Udomlya** Rus. Fed.
62 B2 **Udon Thani** Thai.
73 B3 **Udupi** India
83 K3 **Udyl', Ozero** l. Rus. Fed.
67 C3 **Ueda** Japan
61 D2 **Uekuli** Indon.
118 C2 **Uele** r. Dem. Rep. Congo
83 N2 **Uelen** Rus. Fed.
101 E1 **Uelzen** Ger.
119 C2 **Uere** r. Dem. Rep. Congo
87 E3 **Ufa** Rus. Fed.
119 D3 **Ugalla** r. Tanz.
119 D2 **Uganda** country Africa
69 F1 **Uglegorsk** Rus. Fed.
89 E2 **Uglich** Rus. Fed.
89 D2 **Uglovka** Rus. Fed.
66 B2 **Uglovoye** Rus. Fed.
89 D3 **Ugra** Rus. Fed.
103 D2 **Uherské Hradiště** Czech Rep.
Uibhist a' Deas i. U.K. see South Uist
Uibhist a' Tuath i. U.K. see North Uist
101 E2 **Uichteritz** Ger.
96 A2 **Uig** U.K.
120 A1 **Uíge** Angola
65 B2 **Uijeongbu** S. Korea
65 A1 **Uiju** N. Korea
135 D2 **Uinta Mountains** U.S.A.
65 B2 **Uiseong** S. Korea
120 A3 **Uis Mine** Namibia
123 C3 **Uitenhage** S. Africa
100 C1 **Uithuizen** Neth.
131 D2 **Uivak, Cape** Can.
Ujiyamada Japan see Ise
74 B2 **Ujjain** India
Ujung Pandang Indon. see Makassar
89 F3 **Ukholovo** Rus. Fed.
86 E2 **Ukhta** Rus. Fed.
135 B3 **Ukiah** U.S.A.
127 I2 **Ukkusissat** Greenland
88 B2 **Ukmergė** Lith.
90 C2 **Ukraine** country Europe
Ukrayinska S.S.R. country Europe see Ukraine
Ulaanbaatar Mongolia see Ulan Bator
68 C1 **Ulaangom** Mongolia

59 E3 **Ulamona** P.N.G.
70 A2 **Ulan** China
69 D1 **Ulan Bator** Mongolia
Ulanhad China see Chifeng
69 E1 **Ulanhot** China
87 D4 **Ulan-Khol** Rus. Fed.
70 B1 **Ulan Qab** China
69 D1 **Ulan-Ude** Rus. Fed.
75 D1 **Ulan Ul Hu** l. China
Uleåborg Fin. see Oulu
88 C2 **Ülenurme** Estonia
73 B3 **Ulhasnagar** India
69 D1 **Uliastai** China
68 C1 **Uliastay** Mongolia
59 D2 **Ulithi** atoll Micronesia
65 B2 **Uljin** S. Korea
53 D3 **Ulladulla** Austr.
96 B2 **Ullapool** U.K.
65 C2 **Ulleung-do** i. S. Korea
98 B2 **Ullswater** l. U.K.
102 B2 **Ulm** Ger.
65 B2 **Ulsan** S. Korea
96 □ **Ulsta** U.K.
97 C1 **Ulster** reg. Ireland/U.K.
52 B3 **Ultima** Austr.
145 D3 **Ulúa** r. Hond.
111 C3 **Ulubey** Turkey
111 D3 **Uluborlu** Turkey
111 C2 **Uludağ** mt. Turkey
126 E2 **Ulukhaktok** Can.
123 D2 **Ulundi** S. Africa
77 E2 **Ulungur Hu** l. China
50 C2 **Uluru** h. Austr.
98 B2 **Ulverston** U.K.
90 C2 **Ul'yanovka** Ukr.
87 D3 **Ul'yanovsk** Rus. Fed.
136 C3 **Ulysses** U.S.A.
90 C2 **Uman'** Ukr.
86 C2 **Umba** Rus. Fed.
59 D3 **Umboi** i. P.N.G.
59 D3 **Umbukul** P.N.G.
92 H3 **Umeå** Sweden
92 H3 **Umeälven** r. Sweden
123 D2 **uMhlanga** S. Africa
127 J2 **Umiiviip Kangertiva** inlet Greenland
126 E2 **Umingmaktok (abandoned)** Can.
123 D2 **Umlazi** S. Africa
78 A2 **Umm al Birak** Saudi Arabia
79 C2 **Umm as Samīm** salt flat Oman
116 A3 **Umm Keddada** Sudan
78 A2 **Umm Lajj** Saudi Arabia
78 A2 **Umm Mukhbār, Jabal** mt. Saudi Arabia
116 B3 **Umm Ruwaba** Sudan
115 E1 **Umm Sa'ad** Libya
134 B2 **Umpqua** r. U.S.A.
120 A2 **Umpulo** Angola
Umtali Zimbabwe see Mutare
123 D3 **Umtentweni** S. Africa
154 B2 **Umuarama** Brazil
123 C3 **Umzimkulu** S. Africa
109 C1 **Una** r. Bos.-Herz./Croatia
155 E1 **Una** Brazil
154 C1 **Unaí** Brazil
126 B2 **Unalakleet** U.S.A.
78 B2 **'Unayzah** Saudi Arabia
136 B3 **Uncompahgre Peak** U.S.A.
52 B3 **Underbool** Austr.
136 C1 **Underwood** U.S.A.
89 D3 **Unecha** Rus. Fed.
53 C2 **Ungarie** Austr.
52 A2 **Ungarra** Austr.
127 H2 **Ungava, Péninsule d'** pen. Can.
131 D2 **Ungava Bay** Can.
Ungeny Moldova see Ungheni
90 B2 **Ungheni** Moldova
Unguja i. Tanz. see Zanzibar Island
119 E3 **Ungwana Bay** Kenya
154 B3 **União da Vitória** Brazil
150 C3 **Unini** r. Brazil
134 C1 **Union** U.S.A.
140 C1 **Union City** U.S.A.
122 B3 **Uniondale** S. Africa
139 D3 **Uniontown** U.S.A.
79 C2 **United Arab Emirates** country Asia
United Arab Republic country Africa see Egypt
95 C3 **United Kingdom** country Europe
United Provinces state India see Uttar Pradesh

133 B3 **United States of America** country N. America
129 D2 **Unity** Can.
100 C2 **Unna** Ger.
96 □ **Unst** i. U.K.
101 E2 **Unstrut** r. Ger.
89 E3 **Upa** r. Rus. Fed.
119 C3 **Upemba, Lac** l. Dem. Rep. Congo
122 B2 **Upington** S. Africa
74 B2 **Upleta** India
49 J5 **'Upolu** i. Samoa
134 B2 **Upper Alkali Lake** U.S.A.
128 C2 **Upper Arrow Lake** Can.
54 C2 **Upper Hutt** N.Z.
134 B2 **Upper Klamath Lake** U.S.A.
128 B1 **Upper Liard** Can.
97 C1 **Upper Lough Erne** l. U.K.
137 E1 **Upper Red Lake** U.S.A.
Upper Tunguska r. Rus. Fed. see Angara
Upper Volta country Africa see Burkina Faso
93 G4 **Uppsala** Sweden
78 B2 **'Uqlat aş Şuqūr** Saudi Arabia
Urad Qianqi China see Xishanzui
76 B2 **Ural** r. Kazakh./Rus. Fed.
53 D2 **Uralla** Austr.
87 E3 **Ural Mountains** Rus. Fed.
76 B1 **Ural'sk** Kazakh.
Ural'skiy Khrebet mts Rus. Fed. see Ural Mountains
119 D3 **Urambo** Tanz.
53 C3 **Urana** Austr.
129 D2 **Uranium City** Can.
86 F2 **Uray** Rus. Fed.
98 C2 **Ure** r. U.K.
86 D3 **Uren'** Rus. Fed.
82 G2 **Urengoy** Rus. Fed.
144 A2 **Ures** Mex.
Urfa Turkey see Şanlıurfa
76 C2 **Urganch** Uzbek.
74 A1 **Urgün-e Kalān** Afgh.
100 B1 **Urk** Neth.
111 C3 **Urla** Turkey
110 C2 **Urlaţi** Romania
81 C2 **Urmia** Iran
81 C2 **Urmia, Lake** salt l. Iran
Uroševac Kosovo see Ferizaj
144 B2 **Uruáchic** Mex.
151 E4 **Uruaçu** Brazil
144 B3 **Uruapan** Mex.
150 B4 **Urubamba** r. Peru
151 D3 **Urucara** Brazil
151 E3 **Uruçuí** Brazil
151 E3 **Uruçuí, Serra do** hills Brazil
151 D3 **Urucurituba** Brazil
152 C2 **Uruguaiana** Brazil
153 C3 **Uruguay** country S. America
Urumchi China see Ürümqi
68 B2 **Ürümqi** China
Urundi country Africa see Burundi
53 D2 **Urunga** Austr.
119 D3 **Uruwira** Tanz.
110 C2 **Urziceni** Romania
67 B4 **Usa** Japan
86 E2 **Usa** r. Rus. Fed.
111 C3 **Uşak** Turkey
120 A3 **Usakos** Namibia
88 C2 **Ushachy** Belarus
82 G1 **Ushakova, Ostrov** i. Rus. Fed.
77 E2 **Usharal** Kazakh.
77 D2 **Ushtobe** Kazakh.
Ush-Tyube Kazakh. see Ushtobe
153 B5 **Ushuaia** Arg.
86 E2 **Usinsk** Rus. Fed.
99 B4 **Usk** r. U.K.
88 C3 **Uskhodni** Belarus
89 E3 **Usman'** Rus. Fed.
86 D2 **Usogorsk** Rus. Fed.
104 C2 **Ussel** France
66 C1 **Ussuri** r. China/Rus. Fed.
66 B2 **Ussuriysk** Rus. Fed.
Ust'-Abakanskoye Rus. Fed. see Abakan
Ust'-Balyk Rus. Fed. see Nefteyugansk
91 E2 **Ust'-Donetskiy** Rus. Fed.
108 B3 **Ustica, Isola di** i. Italy
83 H3 **Ust'-Ilimsk** Rus. Fed.
86 E2 **Ust'-Ilych** Rus. Fed.
102 C1 **Ústí nad Labem** Czech Rep.
Ustinov Rus. Fed. see Izhevsk

103 D1 Ustka Pol.
83 L3 Ust'-Kamchatsk Rus. Fed.
77 E2 Ust'-Kamenogorsk Kazakh.
86 F2 Ust'-Kara Rus. Fed.
86 E2 Ust'-Kulom Rus. Fed.
83 I3 Ust'-Kut Rus. Fed.
91 D2 Ust'-Labinsk Rus. Fed.
   Ust'-Labinskaya Rus. Fed. see Ust'-Labinsk
88 C2 Ust'-Luga Rus. Fed.
83 K2 Ust'-Nem Rus. Fed.
83 K2 Ust'-Nera Rus. Fed.
83 I2 Ust'-Olenek Rus. Fed.
83 K2 Ust'-Omchug Rus. Fed.
83 H3 Ust'-Ordynskiy Rus. Fed.
103 E2 Ustrzyki Dolne Pol.
86 E2 Ust'-Tsil'ma Rus. Fed.
86 D2 Ust'-Ura Rus. Fed.
76 B2 Ustyurt Plateau Kazakh./Uzbek.
89 E2 Usyuzhna Rus. Fed.
146 B3 Usulután El Salvador
   Usumbura Burundi see Bujumbura
89 D2 Usvyaty Rus. Fed.
135 D3 Utah state U.S.A.
135 D2 Utah Lake U.S.A.
88 C2 Utena Lith.
119 D3 Utete Tanz.
63 B2 Uthai Thani Thai.
74 A2 Uthal Pak.
123 D2 uThukela r. S. Africa
139 D2 Utica U.S.A.
107 C2 Utiel Spain
128 C2 Utikuma Lake Can.
93 G4 Utlängan i. Sweden
100 B1 Utrecht Neth.
123 D2 Utrecht S. Africa
106 B2 Utrera Spain
92 I2 Utsjoki Fin.
67 C3 Utsunomiya Japan
87 D4 Utta Rus. Fed.
62 B2 Uttaradit Thai.
75 B1 Uttarakhand state India
75 B2 Uttar Pradesh state India
   Uummannaq Greenland see Dundas
127 I2 Uummannaq Greenland
127 I2 Uummannaq Fjord inlet Greenland
93 H3 Uusikaupunki Fin.
120 A2 Uutapi Namibia
143 D3 Uvalde U.S.A.
119 D3 Uvinza Tanz.
123 D3 Uvongo S. Africa
68 C1 Uvs Nuur salt l. Mongolia
67 B4 Uwajima Japan
78 A2 'Uwayriḍ, Ḥarrat al lava field Saudi Arabia
116 A2 Uweinat, Jebel mt. Sudan
83 H3 Uyar Rus. Fed.
115 C4 Uyo Nigeria
79 B2 Uyun Saudi Arabia
152 B2 Uyuni Bol.
152 B2 Uyuni, Salar de salt flat Bol.
76 C2 Uzbekistan country Asia
   Uzbekskaya S.S.R. country Asia see Uzbekistan
   Uzbek S.S.R. country Asia see Uzbekistan
104 C2 Uzerche France
105 C3 Uzès France
90 C1 Uzh r. Ukr.
90 A2 Uzhhorod Ukr.
   Uzhorod Ukr. see Uzhhorod
89 E3 Užice Serbia
89 E3 Uzlovaya Rus. Fed.
111 C3 Üzümlü Turkey
111 C2 Uzunköprü Turkey

V

123 B2 Vaal r. S. Africa
92 I3 Vaala Fin.
123 C2 Vaal Dam S. Africa
123 C1 Vaalwater S. Africa
92 H3 Vaasa Fin.
103 D2 Vác Hungary
152 C2 Vacaria Brazil
154 B2 Vacaria, Serra hills Brazil
135 B3 Vacaville U.S.A.
74 B2 Vadodara India
92 I1 Vadsø Norway
105 D2 Vaduz Liechtenstein
94 B1 Vágar i. Faroe Is
94 B1 Vágar Faroe Is
103 D2 Váh r. Slovakia
49 I4 Vaiaku Tuvalu

88 B2 Vaida Estonia
136 B3 Vail U.S.A.
77 C3 Vakhsh Tajik.
   Vakhstroy Tajik. see Vakhsh
79 C2 Vakīlābād Iran
108 B1 Valdagno Italy
   Valdai Hills hills Rus. Fed. see Valdayskaya Vozvyshennost'
89 D2 Valday Rus. Fed.
89 D2 Valdayskaya Vozvyshennost' hills Rus. Fed.
106 B2 Valdecañas, Embalse de resr Spain
93 G4 Valdemarsvik Sweden
106 C2 Valdepeñas Spain
153 B4 Valdés, Península pen. Arg.
126 C2 Valdez U.S.A.
153 A3 Valdivia Chile
130 C3 Val-d'Or Can.
141 D2 Valdosta U.S.A.
128 C2 Valemount Can.
152 E1 Valença Brazil
105 C3 Valence France
107 C2 Valencia Spain
107 C2 Valencia reg. Spain
150 C1 Valencia Venez.
107 D2 Valencia, Golfo de g. Spain
106 B1 Valencia de Don Juan Spain
97 A3 Valencia Island Ireland
105 C1 Valenciennes France
136 C2 Valentine U.S.A.
64 B2 Valenzuela Phil.
150 B2 Valera Venez.
88 C2 Valga Estonia
109 C2 Valjevo Serbia
88 C2 Valka Latvia
93 H3 Valkeakoski Fin.
100 B2 Valkenswaard Neth.
91 D2 Valky Ukr.
55 G2 Valkyrie Dome Antarctica
145 D2 Valladolid Mex.
106 C1 Valladolid Spain
93 E4 Valle Norway
145 C2 Vallecillos Mex.
145 C2 Valle de la Pascua Venez.
150 B1 Valledupar Col.
145 C2 Valle Hermoso Mex.
135 B3 Vallejo U.S.A.
152 A2 Vallenar Chile
84 F5 Valletta Malta
137 D1 Valley City U.S.A.
134 B2 Valley Falls U.S.A.
128 C2 Valleyview Can.
107 D1 Valls Spain
129 D3 Val Marie Can.
88 C2 Valmiera Latvia
104 B2 Valognes France
88 C3 Valozhyn Belarus
154 B2 Valparaíso Brazil
153 A3 Valparaíso Chile
105 C3 Valréas France
59 D3 Vals, Tanjung c. Indon.
74 B2 Valsad India
122 B2 Valspan S. Africa
91 D1 Valuyki Rus. Fed.
106 B2 Valverde del Camino Spain
81 C2 Van Turkey
81 C2 Van, Lake salt l. Turkey
81 C1 Vanadzor Armenia
83 H2 Vanavara Rus. Fed.
140 B1 Van Buren AR U.S.A.
139 F1 Van Buren ME U.S.A.
   Van Buren U.S.A. see Kettering
128 B3 Vancouver Can.
134 B1 Vancouver U.S.A.
128 B3 Vancouver Island Can.
138 B3 Vandalia IL U.S.A.
138 C3 Vandalia OH U.S.A.
123 C2 Vanderbijlpark S. Africa
128 B2 Vanderhoof Can.
122 B3 Vanderkloof Dam resr S. Africa
50 C1 Van Diemen Gulf Austr.
88 C2 Vändra Estonia
   Väner, Lake l. Sweden see Vänern
93 F4 Vänern l. Sweden
93 F4 Vänersborg Sweden
121 □D3 Vangaindrano Madag.
   Van Gölü salt l. Turkey see Van, Lake
142 C2 Van Horn U.S.A.
59 D3 Vanimo P.N.G.
83 K3 Vanino Rus. Fed.
104 B2 Vannes France
   Vannovka Kazakh. see Turar Ryskulov

59 D3 Van Rees, Pegunungan mts Indon.
122 A3 Vanrhynsdorp S. Africa
93 H3 Vantaa Fin.
49 I5 Vanua Levu i. Fiji
48 H5 Vanuatu country S. Pacific Ocean
138 C2 Van Wert U.S.A.
122 B3 Van Wyksvlei S. Africa
122 B2 Van Zylsrus S. Africa
75 C2 Varanasi India
92 I1 Varangerfjorden sea chan. Norway
92 I1 Varangerhalvøya pen. Norway
88 C2 Varapayeva Belarus
109 C1 Varaždin Croatia
93 F4 Varberg Sweden
111 B3 Varda Greece
111 B2 Vardar r. Macedonia
93 E4 Varde Denmark
92 J1 Vardø Norway
101 D1 Varel Ger.
88 B3 Varéna Lith.
139 E1 Varennes Can.
108 A1 Varese Italy
155 C2 Varginha Brazil
93 I3 Varkaus Fin.
110 C2 Varna Bulg.
93 F4 Värnamo Sweden
155 D1 Várzea da Palma Brazil
86 C2 Varzino Rus. Fed.
   Vasa Fin. see Vaasa
88 C3 Vasilyevichy Belarus
88 C2 Vasknarva Estonia
110 C1 Vaslui Romania
93 G4 Västerås Sweden
93 G3 Västerdalälven r. Sweden
88 A2 Västerhaninge Sweden
93 G4 Västervik Sweden
108 B2 Vasto Italy
91 D2 Vasylivka Ukr.
90 C1 Vasyl'kiv Ukr.
91 D2 Vasyl'kivka Ukr.
104 C2 Vatan France
111 B3 Vatheia Greece
108 B2 Vatican City Europe
92 □B3 Vatnajökull Iceland
110 C1 Vatra Dornei Romania
   Vätter, Lake l. Sweden see Vättern
93 F4 Vättern l. Sweden
142 B2 Vaughn U.S.A.
105 C3 Vauvert France
121 □D2 Vavatenina Madag.
49 I5 Vava'u Group is Tonga
88 B3 Vawkavysk Belarus
93 F4 Växjö Sweden
   Vayenga Rus. Fed. see Severomorsk
86 E1 Vaygach, Ostrov i. Rus. Fed.
154 C1 Vazante Brazil
101 D1 Vechta Ger.
110 C2 Vedea r. Romania
100 C1 Veendam Neth.
100 B1 Veenendaal Neth.
92 F2 Vega i. Norway
129 C2 Vegreville Can.
106 B2 Vejer de la Frontera Spain
93 E4 Vejle Denmark
110 B2 Velbǔzhdki Prokhod pass Bulg./Macedonia
122 A3 Velddrif S. Africa
100 B2 Veldhoven Neth.
109 C2 Velebit mts Croatia
100 C2 Velen Ger.
109 C1 Velenje Slovenia
109 D2 Velika Plana Serbia
89 D2 Velikaya r. Rus. Fed.
89 D2 Velikiye Luki Rus. Fed.
89 D2 Velikiy Novgorod Rus. Fed.
86 D2 Velikiy Ustyug Rus. Fed.
110 C2 Veliko Tŭrnovo Bulg.
108 B2 Veli Lošinj Croatia
89 D2 Velizh Rus. Fed.
108 B2 Velletri Italy
73 B3 Vellore India
86 D2 Vel'sk Rus. Fed.
91 D1 Velykyy Burluk Ukr.
   Velykyy Tokmak Ukr. see Tokmak
159 E4 Vema Trench Indian Ocean
108 B2 Venafro Italy
105 C2 Venceslau Bráz Brazil
104 C2 Vendôme France
108 B1 Veneta, Laguna lag. Italy

89 E3 Venev Rus. Fed.
   Venezia Italy see Venice
150 C2 Venezuela country S. America
150 B1 Venezuela, Golfo de g. Venez.
108 B1 Venice Italy
141 D3 Venice U.S.A.
108 B1 Venice, Gulf of Europe
100 C2 Venlo Neth.
93 E4 Vennesla Norway
100 B2 Venray Neth.
88 B2 Venta r. Latvia/Lith.
88 B2 Venta Lith.
123 C2 Ventersburg S. Africa
123 C3 Venterstad S. Africa
108 A2 Ventimiglia Italy
99 C4 Ventnor U.K.
88 B2 Ventspils Latvia
135 C4 Ventura U.S.A.
143 C3 Venustiano Carranza, Presa resr Mex.
107 C2 Vera Spain
154 C2 Vera Cruz Brazil
145 C3 Veracruz Mex.
74 B2 Veraval India
108 A1 Verbania Italy
108 A1 Vercelli Italy
105 D3 Vercors reg. France
92 F3 Verdalsøra Norway
154 B1 Verde r. Brazil
154 B2 Verde r. Brazil
144 B2 Verde r. Mex.
142 A2 Verde r. U.S.A.
155 D1 Verde Grande r. Brazil
101 D1 Verden (Aller) Ger.
154 B1 Verdinho, Serra do mts Brazil
105 D3 Verdon r. France
105 D2 Verdun France
123 C2 Vereeniging S. Africa
106 B1 Verín Spain
91 D3 Verkhnebakanskiy Rus. Fed.
89 D3 Verkhnedneprovskiy Rus. Fed.
92 J2 Verkhnetulomskiy Rus. Fed.
92 J2 Verkhnetulomskoye Vodokhranilishche resr Rus. Fed.
87 D4 Verkhniy Baskunchak Rus. Fed.
91 E1 Verkhniy Mamon Rus. Fed.
86 D2 Verkhnyaya Toyma Rus. Fed.
89 E3 Verkhov'ye Rus. Fed.
90 A2 Verkhovyna Ukr.
83 J2 Verkhoyanskiy Khrebet mts Rus. Fed.
129 C2 Vermilion Can.
137 D2 Vermillion U.S.A.
130 A3 Vermillion Bay Can.
139 E2 Vermont state U.S.A.
135 E2 Vernal U.S.A.
122 B2 Verneuk Pan salt pan S. Africa
143 D2 Vernon Can.
141 D3 Vernon U.S.A.
108 B1 Vero Beach U.S.A.
138 B2 Veroia Greece
108 B1 Verona Italy
104 C2 Versailles France
100 B2 Vertou France
105 C2 Verulam S. Africa
105 C2 Verviers Belgium
105 C2 Vervins France
105 D3 Vescovato France
91 C2 Veselaya, Gora mt. Rus. Fed.
91 E2 Vesele Ukr.
91 E2 Veselyy Rus. Fed.
105 C2 Vesoul France
140 C2 Vestavia Hills U.S.A.
92 F2 Vesterålen is Norway
92 F2 Vestfjorden sea chan. Norway
92 B1 Vestmanna Faroe Is
92 □A3 Vestmannaeyjar Iceland
92 □A3 Vestmannaeyjar is Iceland
93 E3 Vestnes Norway
   Vesuvio vol. Italy see Vesuvius
108 B2 Vesuvius vol. Italy
89 E2 Ves'yegonsk Rus. Fed.
103 D2 Veszprém Hungary
93 G4 Vetlanda Sweden
86 D3 Vetluga Rus. Fed.
86 D3 Vetluzhskiy Rus. Fed.
100 A2 Veurne Belgium
119 D2 Veveno r. S. Sudan
105 D2 Vevey Switz.

Veydelevka

91 D1 Veydelevka Rus. Fed.
80 B1 Vezirköprü Turkey
　 Vialar Alg. see Tissemsilt
152 C3 Viamao Brazil
151 E3 Viana Brazil
106 B1 Viana do Castelo Port.
　 Viangchan Laos see Vientiane
62 B1 Viangphoukha Laos
111 C3 Viannos Greece
154 C1 Vianópolis Brazil
108 B2 Viareggio Italy
93 E4 Viborg Denmark
　 Viborg Rus. Fed. see Vyborg
109 C3 Vibo Valentia Italy
107 D1 Vic Spain
144 A1 Vicente Guerrero Mex.
108 B1 Vicenza Italy
89 F2 Vichuga Rus. Fed.
105 C2 Vichy France
140 B2 Vicksburg U.S.A.
155 D2 Viçosa Brazil
52 A3 Victor Harbor Austr.
50 C1 Victoria r. Austr.
52 B3 Victoria state Austr.
　 Victoria Cameroon see Limbe
128 B3 Victoria Can.
153 A3 Victoria Chile
　 Victoria Malaysia see Labuan
113 I6 Victoria Seychelles
143 D3 Victoria U.S.A.
119 D3 Victoria, Lake Africa
52 B2 Victoria, Lake Austr.
62 A1 Victoria, Mount Myanmar
59 D3 Victoria, Mount P.N.G.
154 B3 Victoria, Sierra de la hills Arg.
120 B2 Victoria Falls waterfall Zambia/Zimbabwe
120 B2 Victoria Falls Zimbabwe
126 E2 Victoria Island Can.
55 M2 Victoria Land coastal area Antarctica
50 C1 Victoria River Downs Austr.
139 E1 Victoriaville Can.
122 B3 Victoria West S. Africa
135 C4 Victorville U.S.A.
141 D2 Vidalia U.S.A.
110 C2 Videle Romania
92 □A2 Víðidalsá Iceland
110 B2 Vidin Bulg.
74 B2 Vidisha India
140 B2 Vidor U.S.A.
153 B4 Viedma Arg.
153 A4 Viedma, Lago l. Arg.
102 C2 Viehberg mt. Austria
107 D1 Vielha Spain
100 B2 Vielsalm Belgium
101 E2 Vienenburg Ger.
103 D2 Vienna Austria
138 C3 Vienna U.S.A.
105 C2 Vienne France
104 C2 Vienne r. France
62 B2 Vientiane Laos
100 C2 Viersen Ger.
104 C2 Vierzon France
144 B2 Viesca Mex.
109 C2 Vieste Italy
62 B2 Vietnam country Asia
62 B1 Việt Tri Vietnam
64 B2 Vigan Phil.
108 A1 Vigevano Italy
106 B1 Vigo Spain
　 Viipuri Rus. Fed. see Vyborg
73 C3 Vijayawada India
92 □B3 Vík Iceland
111 B2 Vikhren mt. Bulg.
128 C2 Viking Can.
92 F3 Vikna i. Norway
　 Vila Alferes Chamusca Moz. see Guija
　 Vila Arriaga Angola see Bibala
　 Vila Bugaço Angola see Camanongue
　 Vila Cabral Moz. see Lichinga
　 Vila da Ponte Angola see Kuvango
　 Vila de Aljustrel Angola see Cangamba
　 Vila de Almoster Angola see Chiange
　 Vila de João Belo Moz. see Xai-Xai
　 Vila de Trego Morais Moz. see Chókwè
　 Vila Fontes Moz. see Caia

106 B2 Vila Franca de Xira Port.
106 B1 Vilagarcía de Arousa Spain
123 D1 Vila Gomes da Costa Moz.
106 B1 Vilalba Spain
　 Vila Luísa Moz. see Marracuene
　 Vila Marechal Carmona Angola see Uíge
　 Vila Miranda Moz. see Macaloge
121 □D2 Vilanandro, Tanjona c. Madag.
88 C2 Viļāni Latvia
106 B1 Vila Nova de Gaia Port.
107 D1 Vilanova i la Geltrú Spain
　 Vila Paiva de Andrada Moz. see Gorongosa
　 Vila Pery Moz. see Chimoio
106 B1 Vila Real Port.
106 B1 Vilar Formoso Port.
　 Vila Salazar Angola see N'dalatando
　 Vila Salazar Zimbabwe see Sango
　 Vila Teixeira de Sousa Angola see Luau
155 D2 Vila Velha Brazil
150 B4 Vilcabamba, Cordillera mts Peru
82 F1 Vil'cheka, Zemlya i. Rus. Fed.
92 G3 Vilhelmina Sweden
150 C4 Vilhena Brazil
88 C2 Viljandi Estonia
123 C2 Viljoenskroon S. Africa
88 B3 Vilkaviškis Lith.
83 H1 Vil'kitskogo, Proliv str. Rus. Fed.
144 B1 Villa Ahumada Mex.
106 B1 Villablino Spain
102 C2 Villach Austria
　 Villa Cisneros Western Sahara see Dakhla
144 B2 Villa de Cos Mex.
152 B3 Villa Dolores Arg.
145 C3 Villa Flores Mex.
153 C3 Villa Gesell Arg.
145 C2 Villagrán Mex.
145 C3 Villahermosa Mex.
144 A2 Villa Insurgentes Mex.
107 C2 Villajoyosa-La Vila Joiosa Spain
152 B3 Villa María Arg.
153 B3 Villa Mercedes Arg.
152 B2 Villa Montes Bol.
144 B2 Villanueva Mex.
152 B3 Villanueva de la Serena Spain
106 C2 Villanueva de los Infantes Spain
152 C2 Villa Ocampo Arg.
142 B3 Villa Ocampo Mex.
108 A3 Villaputzu Italy
152 C2 Villarrica Para.
106 C2 Villarrobledo Spain
　 Villasalazar Zimbabwe see Sango
152 B2 Villa Unión Arg.
144 B2 Villa Unión Mex.
144 B2 Villa Unión Mex.
150 B2 Villavicencio Col.
152 B2 Villazon Bol.
104 C3 Villefranche-de-Rouergue France
105 C2 Villefranche-sur-Saône France
107 C2 Villena Spain
100 A2 Villeneuve-d'Ascq France
104 C3 Villeneuve-sur-Lot France
140 B2 Ville Platte U.S.A.
100 A3 Villers-Cotterêts France
105 C2 Villeurbanne France
123 C2 Villiers S. Africa
102 B2 Villingen Ger.
137 E2 Villisca U.S.A.
88 C3 Vilnius Lith.
91 C2 Vil'nohirs'k Ukr.
91 D2 Vil'nyans'k Ukr.
100 B2 Vilvoorde Belgium
88 C3 Vilyeyka Belarus
83 J2 Vilyuy r. Rus. Fed.
93 G4 Vimmerby Sweden
153 A3 Viña del Mar Chile
107 D1 Vinarós Spain
138 B3 Vincennes U.S.A.
55 J3 Vincennes Bay Antarctica
139 D3 Vineland U.S.A.
62 B2 Vinh Vietnam
63 B2 Vinh Long Vietnam
143 D1 Vinita U.S.A.

90 B2 Vinnytsya Ukr.
55 R2 Vinson Massif mt. Antarctica
93 E3 Vinstra Norway
105 E2 Vipiteno Italy
64 B2 Virac Phil.
74 B2 Viramgam India
80 B2 Viranşehir Turkey
129 D3 Virden Can.
104 B2 Vire France
120 A2 Virei Angola
155 D1 Virgem da Lapa Brazil
142 A1 Virgin r. U.S.A.
123 C2 Virginia S. Africa
137 E1 Virginia U.S.A.
139 D3 Virginia state U.S.A.
139 D3 Virginia Beach U.S.A.
135 C3 Virginia City U.S.A.
147 D3 Virgin Islands (U.K.) terr. West Indies
147 D3 Virgin Islands (U.S.A.) terr. West Indies
63 B2 Viróchey Cambodia
109 C1 Virovitica Croatia
100 B3 Virton Belgium
88 B2 Virtsu Estonia
73 B4 Virudhunagar India
109 C2 Vis i. Croatia
88 C2 Visaginas Lith.
135 C3 Visalia U.S.A.
74 B2 Visavadar India
64 B2 Visayan Sea Phil.
93 G4 Visby Sweden
126 E2 Viscount Melville Sound sea chan. Can.
151 E3 Viseu Brazil
106 B1 Viseu Port.
110 B1 Vişeu de Sus Romania
73 C3 Vishakhapatnam India
88 C2 Viški Latvia
109 C2 Visoko Bos.-Herz.
103 D1 Vistula r. Pol.
　 Vitebsk Belarus see Vitsyebsk
108 B2 Viterbo Italy
49 I5 Viti Levu i. Fiji
83 I3 Vitim r. Rus. Fed.
155 D2 Vitória Brazil
151 E4 Vitória da Conquista Brazil
106 C1 Vitoria-Gasteiz Spain
104 B2 Vitré France
105 C2 Vitry-le-François France
89 D2 Vitsyebsk Belarus
105 D2 Vittel France
108 B3 Vittoria Italy
108 B1 Vittorio Veneto Italy
106 B1 Viveiro Spain
136 C2 Vivian U.S.A.
　 Vizagapatam India see Vishakhapatnam
142 A3 Vizcaíno, Desierto de des. Mex.
144 A2 Vizcaíno, Sierra mts Mex.
111 C2 Vize Turkey
73 C3 Vizianagaram India
100 B2 Vlaardingen Neth.
87 D4 Vladikavkaz Rus. Fed.
89 F2 Vladimir Rus. Fed.
66 B2 Vladivostok Rus. Fed.
109 D2 Vlasotince Serbia
100 B1 Vlieland i. Neth.
100 A2 Vlissingen Neth.
109 C2 Vlorë Albania
102 C1 Vltava r. Czech Rep.
102 C2 Vöcklabruck Austria
109 C2 Vodice Croatia
　 Vogelkop Peninsula pen. Indon. see Doberai, Jazirah
101 D2 Vogelsberg hills Ger.
　 Vohémar Madag. see Iharaña
　 Vohibinany Madag. see Ampasimanolotra
　 Vohimarina Madag. see Iharaña
121 □D3 Vohimena, Tanjona c. Madag.
121 □D3 Vohipeno Madag.
119 D3 Voi Kenya
105 D2 Voiron France
131 D2 Voisey's Bay Can.
109 C1 Vojvodina prov. Serbia
92 J3 Volcano Bay b. Japan see Uchiura-wan
69 F3 Volcano Islands is Japan
90 B2 Volchansk Ukr. see Vovchans'k
89 E2 Volga Rus. Fed.
89 F2 Volga r. Rus. Fed.

87 D4 Volgodonsk Rus. Fed.
87 D4 Volgograd Rus. Fed.
87 D4 Volgogradskoye Vodokhranilishche resr Rus. Fed.
89 D2 Volkhov Rus. Fed.
89 D1 Volkhov r. Rus. Fed.
101 E2 Volkstedt Ger.
91 D2 Volnovakha Ukr.
90 B2 Volochys'k Ukr.
91 D2 Volodars'ke Ukr.
　 Volodarskoye Kazakh. see Saumalkol'
90 B1 Volodars'k-Volyns'kyy Ukr.
90 B1 Volodymyrets' Ukr.
90 A1 Volodymyr-Volyns'kyy Ukr.
89 E2 Vologda Rus. Fed.
89 E2 Volokolamsk Rus. Fed.
91 D1 Volokonovka Rus. Fed.
111 B3 Volos Greece
88 C2 Volosovo Rus. Fed.
89 D2 Volot Rus. Fed.
89 E3 Volovo Rus. Fed.
87 D3 Vol'sk Rus. Fed.
114 C4 Volta r. Ghana
114 C4 Volta, Lake resr Ghana
155 D2 Volta Redonda Brazil
110 C2 Voluntari Romania
87 D4 Volzhskiy Rus. Fed.
92 □C2 Vopnafjörður Iceland
88 C3 Voranava Belarus
160 K3 Voring Plateau N. Atlantic Ocean
86 F2 Vorkuta Rus. Fed.
88 B2 Vormsi i. Estonia
83 G2 Vorogovo Rus. Fed.
89 E3 Voronezh Rus. Fed.
89 E3 Voronezh r. Rus. Fed.
91 E1 Vorontsovka Rus. Fed.
　 Voroshilov Rus. Fed. see Ussuriysk
　 Voroshilovgrad Ukr. see Luhans'k
　 Voroshilovsk Rus. Fed. see Stavropol'
　 Voroshilovsk Ukr. see Alchevs'k
91 C2 Vorskla r. Rus. Fed.
88 C2 Võrtsjärv l. Estonia
88 C2 Võru Estonia
122 B3 Vosburg S. Africa
105 D2 Vosges mts France
89 E2 Voskresensk Rus. Fed.
93 E3 Voss Norway
86 C2 Vostochnaya Litsa Rus. Fed.
　 Vostochno-Sibirskoye More sea Rus. Fed. see East Siberian Sea
83 H3 Vostochnyy Sayan mts Rus. Fed.
66 C1 Vostok Rus. Fed.
49 L5 Vostok Island Kiribati
86 E3 Votkinsk Rus. Fed.
86 E3 Votkinskoye Vodokhranilishche resr Rus. Fed.
154 C2 Votuporanga Brazil
105 C2 Vouziers France
91 D1 Vovchans'k Ukr.
91 C2 Voznesens'k Ukr.
76 B2 Vozrozhdenya Island pen. Kazakh./Uzbek.
93 E4 Vrådal Norway
90 C2 Vradiyivka Ukr.
66 B2 Wrangel' Rus. Fed.
　 Vrangelya, Ostrov i. Rus. Fed. see Wrangel Island
109 D2 Vranje Serbia
110 C2 Vratnik pass Bulg.
110 B2 Vratsa Bulg.
109 C1 Vrbas r. Bos.-Herz.
109 C1 Vrbas Serbia
122 A3 Vredenburg S. Africa
122 A3 Vredendal S. Africa
100 B3 Vresse Belgium
100 C1 Vriezenveen Neth.
109 D1 Vršac Serbia
122 B2 Vryburg S. Africa
123 D2 Vryheid S. Africa
89 D1 Vsevolozhsk Rus. Fed.
109 C1 Vukovar Croatia
86 E2 Vuktyl Rus. Fed.
123 C2 Vukuzakhe S. Africa
90 B2 Vulcăneşti Moldova
109 B3 Vulcano, Isola i. Italy

Vulkaneshty Moldova *see*
Vulcăneşti
63 B2 Vung Tau Vietnam
92 H2 Vuollerim Sweden
92 I2 Vuotso Fin.
109 D2 Vushtrri Kosovo
74 B2 Vyara India
Vyarkhowye Belarus *see*
Ruba
Vyatka Rus. Fed. *see* Kirov
89 D2 Vyaz'ma Rus. Fed.
88 C1 Vyborg Rus. Fed.
88 C1 Vyborgskiy Zaliv *b.*
Rus. Fed.
86 D2 Vychegda *r.* Rus. Fed.
88 C2 Vyerkhnyadzvinsk Belarus
89 D3 Vyetka Belarus
89 D3 Vygonichi Rus. Fed.
86 C2 Vygozero, Ozero *l.* Rus. Fed.
87 D3 Vyksa Rus. Fed.
90 B2 Vylkove Ukr.
90 A2 Vynohradiv Ukr.
89 D2 Vypolzovo Rus. Fed.
89 D2 Vyritsa Rus. Fed.
91 D2 Vyselki Rus. Fed.
90 C1 Vyshhorod Ukr.
89 D2 Vyshnevolotskaya Gryada
ridge Rus. Fed.
89 D2 Vyshniy-Volochek Rus. Fed.
103 D2 Vyškov Czech Rep.
89 E2 Vysokovsk Rus. Fed.
86 C2 Vytegra Rus. Fed.

W

114 B3 Wa Ghana
100 B2 Waal *r.* Neth.
100 B2 Waalwijk Neth.
119 D2 Waat S. Sudan
128 C2 Wabasca *r.* Can.
128 C2 Wabasca-Desmarais Can.
138 B3 Wabash *r.* U.S.A.
129 E2 Wabowden Can.
103 D1 Wąbrzeźno Pol.
141 D3 Waccasassa Bay U.S.A.
101 D2 Wächtersbach Ger.
143 D2 Waco U.S.A.
115 D2 Waddān Libya
Waddeneilanden *is* Neth.
see West Frisian Islands
Wadden Islands *is* Neth. *see*
West Frisian Islands
100 B1 Waddenzee *sea chan.* Neth.
128 B2 Waddington, Mount Can.
100 B1 Waddinxveen Neth.
129 D2 Wadena Can.
137 D1 Wadena U.S.A.
50 B1 Wadeye Austr.
74 A2 Wadh Pak.
Wadhwan India *see*
Surendranagar
116 B2 Wadi Halfa Sudan
116 B3 Wad Medani Sudan
70 C2 Wafangdian China
100 B2 Wageningen Neth.
127 G2 Wager Bay Can.
53 C3 Wagga Wagga Austr.
137 D2 Wagner U.S.A.
74 B1 Wah Pak.
116 A2 Wāḩāt ad Dākhilah Egypt
116 A2 Wāḩāt al Baḩrīyah Egypt
116 A2 Wāḩāt al Farāfirah Egypt
116 B2 Wāḩāt al Khārijah Egypt
116 A2 Wāḩāt Sīwah Egypt
137 D2 Wahoo U.S.A.
137 D1 Wahpeton U.S.A.
54 B2 Waiau *r.* N.Z.
59 C3 Waigeo *i.* Indon.
61 C2 Waikabubak Indon.
54 C1 Waikaremoana, Lake N.Z.
52 A2 Waikerie Austr.
54 B2 Waimate N.Z.
75 B3 Wainganga *r.* India
61 D2 Waingapu Indon.
129 C2 Wainwright Can.
126 B2 Wainwright U.S.A.
54 C1 Waiouru N.Z.
54 B2 Waipara N.Z.
54 C1 Waipawa N.Z.
54 B2 Wairau *r.* N.Z.
54 C1 Wairoa N.Z.
54 B1 Waitaki *r.* N.Z.
54 B1 Waitara N.Z.
54 B1 Waiuku N.Z.
67 C3 Wajima Japan
119 E2 Wajir Kenya
67 C3 Wakasa-wan *b.* Japan
54 A3 Wakatipu, Lake N.Z.

129 D2 Wakaw Can.
67 C4 Wakayama Japan
136 D3 WaKeeney U.S.A.
54 B2 Wakefield N.Z.
98 C3 Wakefield U.K.
Wakeham Can. *see*
Kangiqsujuaq
48 H2 Wake Island *terr.*
N. Pacific Ocean
66 D1 Wakkanai Japan
123 D2 Wakkerstroom S. Africa
120 A2 Waku-Kungo Angola
103 D1 Wałbrzych Pol.
53 D2 Walcha Austr.
100 C1 Walchum Ger.
103 D1 Wałcz Pol.
99 B3 Wales *admin. div.* U.K.
53 C2 Walgett Austr.
119 C3 Walikale Dem. Rep. Congo
135 C3 Walker Lake U.S.A.
134 C1 Wallace *ID* U.S.A.
141 E2 Wallace *NC* U.S.A.
52 A2 Wallaroo Austr.
98 B3 Wallasey U.K.
134 C1 Walla Walla U.S.A.
101 D3 Walldürn Ger.
122 A3 Wallekraal S. Africa
53 C2 Wallendbeen Austr.
49 J5 Wallis, Îles *is*
Wallis and Futuna Is
49 J5 Wallis and Futuna Islands
terr. S. Pacific Ocean
96 □ Walls U.K.
98 B2 Walney, Isle of *i.* U.K.
99 C3 Walsall U.K.
136 C3 Walsenburg U.S.A.
101 D1 Walsrode Ger.
141 D2 Walterboro U.S.A.
120 A3 Walvis Bay Namibia
158 F6 Walvis Ridge
S. Atlantic Ocean
119 C2 Wamba Dem. Rep. Congo
52 B1 Wanaaring Austr.
54 A2 Wanaka N.Z.
54 A2 Wanaka, Lake N.Z.
71 B3 Wan'an China
130 B3 Wanapitei Lake Can.
154 B3 Wanda Arg.
66 B1 Wanda Shan *mts* China
62 A1 Wanding China
Wandingzhen China *see*
Wanding
54 C1 Wanganui N.Z.
54 B1 Wanganui *r.* N.Z.
53 C3 Wangaratta Austr.
65 B1 Wangqing China
62 A1 Wan Hsa-la Myanmar
Wankie Zimbabwe *see*
Hwange
71 B4 Wanning China
100 B2 Wanroij Neth.
99 C4 Wantage U.K.
70 A2 Wanyuan China
70 A2 Wanzhou China
73 B3 Warangal India
101 D2 Warburg Ger.
50 B2 Warburton Austr.
52 A1 Warburton *watercourse*
Austr.
74 B2 Wardha India
96 C1 Ward Hill U.K.
128 B2 Ware Can.
101 F1 Waren Ger.
101 C2 Warendorf Ger.
53 D1 Warialda Austr.
122 A2 Warmbad Namibia
135 C3 Warm Springs U.S.A.
134 C2 Warner Lakes U.S.A.
134 B2 Warner Mountains U.S.A.
141 D2 Warner Robins U.S.A.
152 B1 Warnes Bol.
52 B3 Warracknabeal Austr.
53 C3 Warrandyte Austr.
117 A4 Warrap S. Sudan
53 C2 Warrego *r.* Austr.
53 C2 Warren Austr.
140 B2 Warren *AR* U.S.A.
138 C2 Warren *OH* U.S.A.
139 D2 Warren *PA* U.S.A.
97 C1 Warrenpoint U.K.
137 E3 Warrensburg U.S.A.
122 B2 Warrenton S. Africa
115 C4 Warri Nigeria
98 B3 Warrington U.K.
52 B3 Warrnambool Austr.
103 E1 Warsaw Pol.
138 B3 Warsaw U.S.A.
Warszawa Pol. *see* Warsaw
103 C1 Warta *r.* Pol.
53 D1 Warwick Austr.

99 C3 Warwick U.K.
139 E2 Warwick U.S.A.
134 D3 Wasatch Range *mts* U.S.A.
135 C3 Wasco U.S.A.
136 C1 Washburn U.S.A.
139 D3 Washington *DC* U.S.A.
137 E2 Washington *IA* U.S.A.
138 B2 Washington *IL* U.S.A.
138 B3 Washington *IN* U.S.A.
137 E3 Washington *MO* U.S.A.
141 E1 Washington *NC* U.S.A.
138 C2 Washington *PA* U.S.A.
135 D3 Washington *UT* U.S.A.
134 B1 Washington *state* U.S.A.
139 E2 Washington, Mount U.S.A.
138 C3 Washington Court House
U.S.A.
74 A2 Washuk Pak.
130 C2 Waskaganish Can.
129 E2 Waskaiowaka Lake Can.
122 A2 Wasser Namibia
101 D2 Wasserkuppe *h.* Ger.
130 C3 Waswanipi, Lac *l.* Can.
61 D2 Watampone Indon.
Watenstadt-Salzgitter Ger.
see Salzgitter
139 E2 Waterbury U.S.A.
129 D2 Waterbury Lake Can.
97 C2 Waterford Ireland
97 C2 Waterford Harbour Ireland
100 B2 Waterloo Belgium
137 E2 Waterloo U.S.A.
99 C4 Waterlooville U.K.
123 C1 Waterpoort S. Africa
139 D2 Watertown *NY* U.S.A.
137 D2 Watertown *SD* U.S.A.
138 B2 Watertown *WI* U.S.A.
97 A3 Waterville Ireland
139 F2 Waterville U.S.A.
99 C4 Watford U.K.
136 C1 Watford City U.S.A.
129 D2 Wathaman *r.* Can.
Watling Island *i.* Bahamas
see San Salvador
143 D1 Watonga U.S.A.
129 D2 Watrous Can.
119 C2 Watsa Dem. Rep. Congo
138 B2 Watseka U.S.A.
118 C3 Watsi Kengo
Dem. Rep. Congo
128 B1 Watson Lake Can.
135 B3 Watsonville U.S.A.
59 C3 Watubela, Kepulauan *is*
Indon.
59 D3 Wau P.N.G.
117 A4 Wau S. Sudan
53 D2 Wauchope Austr.
138 B2 Waukegan U.S.A.
138 B2 Waukesha U.S.A.
143 D2 Waurika U.S.A.
138 B2 Wausau U.S.A.
99 D3 Waveney *r.* U.K.
137 E2 Waverly U.S.A.
130 B3 Wawa Can.
114 B4 Wawa Can.
141 D2 Waycross U.S.A.
137 D2 Wayne U.S.A.
141 D2 Waynesboro *GA* U.S.A.
139 D3 Waynesboro *VA* U.S.A.
137 E3 Waynesville *MO* U.S.A.
141 D1 Waynesville *NC* U.S.A.
74 B1 Wazirabad Pak.
60 A1 We, Pulau *i.* Indon.
98 C2 Wear *r.* U.K.
143 D1 Weatherford *OK* U.S.A.
143 D2 Weatherford *TX* U.S.A.
134 B2 Weaverville U.S.A.
143 D3 Webb U.S.A.
130 B2 Webequie Can.
137 D1 Webster U.S.A.
137 E2 Webster City U.S.A.
55 C3 Weddell Abyssal Plain
Southern Ocean
55 B3 Weddell Sea Antarctica
55 A3 Weenen S. Africa
100 B2 Weert Neth.
53 C2 Weethalle Austr.
53 C2 Wee Waa Austr.
100 C2 Wegberg Ger.
103 E1 Węgorzewo Pol.
103 E1 Węgrów Pol.
70 B1 Weichang China
101 F3 Weiden in der Oberpfalz
Ger.
Weidongmen China *see*
Qianjin
70 B2 Weifang China
70 C2 Weihai China
70 B2 Wei He *r.* China
53 C1 Weilmoringle Austr.
101 E2 Weimar Ger.

143 D3 Weimar U.S.A.
70 A2 Weinan China
71 A3 Weining China
51 D1 Weipa Austr.
53 C1 Weir *r.* Austr.
138 C2 Weirton U.S.A.
62 B1 Weishan China
101 E2 Weiße Elster *r.* Ger.
101 E2 Weißenfels Ger.
102 C2 Weißkugel *mt.* Austria/Italy
71 A3 Weixin China
Weizhou China *see*
Weichuan
103 D1 Wejherowo Pol.
128 C1 Wekweètì Can.
138 C3 Welch U.S.A.
117 B3 Weldiya Eth.
123 C2 Welkom S. Africa
99 C3 Welland *r.* U.K.
51 C1 Wellesley Islands Austr.
99 C3 Wellingborough U.K.
53 C2 Wellington Austr.
54 B2 Wellington N.Z.
122 A3 Wellington S. Africa
136 B2 Wellington CO U.S.A.
137 D3 Wellington KS U.S.A.
135 D3 Wellington UT U.S.A.
153 A4 Wellington, Isla *i.* Chile
53 C3 Wellington, Lake Austr.
128 B2 Wells Can.
99 B4 Wells U.K.
134 D2 Wells U.S.A.
50 B2 Wells, Lake *imp. l.* Austr.
54 B1 Wellsford N.Z.
99 D3 Wells-next-the-Sea U.K.
142 A2 Wellton U.S.A.
102 C2 Wels Austria
99 B3 Welshpool U.K.
Welwitschia Namibia *see*
Khorixas
123 C2 Wembesi S. Africa
130 C2 Wemindji Can.
134 B1 Wenatchee U.S.A.
71 B4 Wencheng China
114 B4 Wenchi China
Wenchow China *see*
Wenzhou
70 A2 Wenchuan China
70 C2 Wendeng China
101 E1 Wendisch Evern Ger.
117 B4 Wendo Eth.
135 D2 Wendover U.S.A.
71 B3 Wengyuan China
Wenhua China *see* Weishan
Wenlan China *see* Mengzi
Wenlin China *see* Renshou
71 C3 Wenling China
Wenquan China *see*
Yingshan
71 A3 Wenshan China
52 B2 Wentworth Austr.
71 C3 Wenxing China
123 C2 Wepener S. Africa
122 B2 Werda Botswana
101 F2 Werdau Ger.
101 F1 Werder Ger.
101 F3 Wernberg-Köblitz Ger.
101 E2 Wernigerode Ger.
101 D2 Werra *r.* Ger.
52 B2 Werrimull Austr.
53 D2 Werris Creek Austr.
101 D3 Wertheim Ger.
100 C2 Wesel Ger.
101 E1 Wesendorf Ger.
101 D1 Weser *r.* Ger.
101 D1 Weser *sea chan.* Ger.
51 C1 Wessel, Cape Austr.
51 C1 Wessel Islands Austr.
123 C2 Wesselton S. Africa
55 P2 West Antarctica *reg.*
Antarctica
156 B7 West Australian Basin
Indian Ocean
80 B2 West Bank *terr.* Asia
138 B2 West Bend U.S.A.
75 C2 West Bengal *state* India
99 C3 West Bromwich U.K.
139 E2 Westbrook U.S.A.
137 E2 West Des Moines U.S.A.
100 C2 Westerburg Ger.
100 C1 Westerholt Ger.
139 E2 Westerly U.S.A.
50 B2 Western Australia *state*
Austr.
122 B3 Western Cape *prov.*
S. Africa
116 A2 Western Desert Egypt
Western Dvina *r.* Europe *see*
Zapadnaya Dvina
73 B3 Western Ghats *mts* India

114 A2 Western Sahara *terr.* Africa
Western Samoa *country*
S. Pacific Ocean *see* Samoa
Western Sayan Mountains
reg. Rus. Fed. *see*
Zapadnyy Sayan
100 A2 Westerschelde *est.* Neth.
100 C1 Westerstede Ger.
101 C2 Westerwald *hills* Ger.
153 B5 West Falkland *i.* Falkland Is
138 B3 West Frankfort U.S.A.
100 B1 West Frisian Islands Neth.
96 C2 Westhill U.K.
55 I3 West Ice Shelf Antarctica
147 D2 West Indies *is*
Caribbean Sea
100 A2 Westkapelle Neth.
96 A1 West Loch Roag *b.* U.K.
128 C2 Westlock Can.
100 B2 Westmalle Belgium
Westman Islands *is* Iceland
see Vestmannaeyjar
53 C1 Westmar Austr.
156 C4 West Mariana Basin
N. Pacific Ocean
140 B1 West Memphis U.S.A.
138 C3 Weston U.S.A.
99 B4 Weston-super-Mare U.K.
141 D3 West Palm Beach U.S.A.
137 E3 West Plains U.S.A.
137 D2 West Point U.S.A.
54 B2 Westport N.Z.
97 B2 Westport Ireland
129 D2 Westray Can.
96 C1 Westray *i.* U.K.
82 G2 West Siberian Plain *plain*
Rus. Fed.
100 B1 West-Terschelling Neth.
136 A2 West Thumb U.S.A.
West Town Ireland *see*
An Baile Thiar
135 D2 West Valley City U.S.A.
138 C3 West Virginia *state* U.S.A.
53 C2 West Wyalong Austr.
134 D2 West Yellowstone U.S.A.
59 C3 Wetar *i.* Indon.
128 C2 Wetaskiwin Can.
119 D3 Wete Tanz.
101 D2 Wetzlar Ger.
59 D3 Wewak P.N.G.
97 C2 Wexford Ireland
97 C2 Wexford Harbour *b.* Ireland
129 D2 Weyakwin Can.
129 D3 Weyburn Can.
101 D1 Weyhe Ger.
99 B4 Weymouth U.K.
55 C1 Whakaari *i.* N.Z.
54 C1 Whakatane N.Z.
129 E1 Whale Cove Can.
96 □ Whalsay *i.* U.K.
54 B1 Whangamomona N.Z.
54 B1 Whangaparaoa N.Z.
54 B1 Whangarei N.Z.
98 C3 Wharfe *r.* U.K.
143 D3 Wharton U.S.A.
128 C1 Whati Can.
136 B2 Wheatland U.S.A.
138 B2 Wheaton U.S.A.
140 C2 Wheeler Lake *resr* U.S.A.
142 B1 Wheeler Peak *NM* U.S.A.
135 D3 Wheeler Peak *NV* U.S.A.
138 C2 Wheeling U.S.A.
98 B2 Whernside *h.* U.K.
128 B2 Whistler Can.
98 C2 Whitby U.K.
128 A1 White *r.* Can./U.S.A.
140 B2 White *r. AR* U.S.A.
138 B3 White *r. IN* U.S.A.
50 B2 White, Lake *imp. l.* Austr.
131 E3 White Bay Can.
136 C1 White Butte *mt.* U.S.A.
52 B2 White Cliffs Austr.
128 C2 Whitecourt Can.
134 D1 Whitefish U.S.A.
98 B2 Whitehaven U.K.
97 D1 Whitehead U.K.
128 A1 Whitehorse Can.
140 B3 White Lake U.S.A.
51 D4 Whitemark Austr.
135 C3 White Mountain Peak
U.S.A.
116 B3 White Nile *r.* Africa
White Russia *country*
Europe *see* Belarus
86 C2 White Sea Rus. Fed.
134 D1 White Sulphur Springs
U.S.A.
141 E2 Whiteville U.S.A.
114 B3 White Volta *r.*
Burkina Faso/Ghana

136 B3 Whitewater U.S.A.
142 B2 Whitewater Baldy *mt.* U.S.A.
130 B2 Whitewater Lake Can.
129 D2 Whitewood Can.
96 B3 Whithorn U.K.
54 C1 Whitianga N.Z.
135 C3 Whitney, Mount U.S.A.
99 D4 Whitstable U.K.
51 D2 Whitsunday Island Austr.
53 C3 Whittlesea Austr.
52 A2 Whyalla Austr.
62 A2 Wiang Pa Pao Thai.
100 A2 Wichelen Belgium
137 D3 Wichita U.S.A.
143 D2 Wichita Falls U.S.A.
143 D2 Wichita Mountains U.S.A.
96 C1 Wick U.K.
142 A2 Wickenburg U.S.A.
97 C2 Wicklow Ireland
97 D2 Wicklow Head *hd* Ireland
97 C2 Wicklow Mountains
Ireland
98 B3 Widnes U.K.
101 D1 Wiehengebirge *hills* Ger.
100 C2 Wiehl Ger.
103 D1 Wieluń Pol.
Wien Austria *see* Vienna
103 D2 Wiener Neustadt Austria
100 B1 Wieringerwerf Neth.
101 D2 Wiesbaden Ger.
100 C1 Wiesmoor Ger.
103 D1 Wieżyca *h.* Pol.
98 B3 Wigan U.K.
99 C3 Wight, Isle of *i.* U.K.
96 B3 Wigtown U.K.
100 B2 Wijchen Neth.
130 B3 Wikwemikong Can.
52 B2 Wilcannia Austr.
Wilczek Land *i.* Rus. Fed.
see Vil'cheka, Zemlya
123 C3 Wild Coast S. Africa
101 D1 Wildeshausen Ger.
136 C2 Wild Horse Hill *mt.* U.S.A.
123 C2 Wilge *r.* S. Africa
48 F4 Wilhelm, Mount P.N.G.
101 D1 Wilhelmshaven Ger.
139 D2 Wilkes-Barre U.S.A.
55 K3 Wilkes Land *reg.* Antarctica
129 D2 Wilkie Can.
134 B1 Willamette *r.* U.S.A.
134 B1 Willapa Bay U.S.A.
142 B2 Willcox U.S.A.
100 B2 Willebroek Belgium
147 D3 Willemstad Curaçao
52 B3 William, Mount Austr.
52 A1 William Creek Austr.
142 A1 Williams U.S.A.
138 C3 Williamsburg *KY* U.S.A.
139 D3 Williamsburg *VA* U.S.A.
128 B2 Williams Lake Can.
138 C3 Williamson U.S.A.
139 D2 Williamsport U.S.A.
141 E1 Williamston U.S.A.
122 B3 Williston S. Africa
136 C1 Williston U.S.A.
128 B2 Williston Lake Can.
135 B3 Willits U.S.A.
137 D1 Willmar U.S.A.
122 B3 Willowmore S. Africa
135 B3 Willows U.S.A.
123 C3 Willowvale S. Africa
50 B2 Wills, Lake *imp. l.* Austr.
52 A3 Willunga Austr.
52 A2 Wilmington Austr.
139 D3 Wilmington *DE* U.S.A.
141 E2 Wilmington *NC* U.S.A.
138 C3 Wilmington *OH* U.S.A.
141 D2 Wilmington Island U.S.A.
Wilno Lith. *see* Vilnius
101 D2 Wilnsdorf Ger.
101 D1 Wilseder Berg *h.* Ger.
141 E1 Wilson U.S.A.
53 C3 Wilson's Promontory *pen.*
Austr.
100 B3 Wiltz Lux.
50 B2 Wiluna Austr.
99 D4 Wimereux France
123 C2 Winburg S. Africa
99 B4 Wincanton U.K.
99 C4 Winchester U.K.
138 C3 Winchester *KY* U.S.A.
139 D3 Winchester *VA* U.S.A.
98 B2 Windermere *l.* U.K.
122 A1 Windhoek Namibia
137 D2 Windom U.S.A.
51 D2 Windorah Austr.
136 B2 Wind River Range *mts*
U.S.A.
53 D2 Windsor Austr.
130 B3 Windsor Can.

147 D3 Windward Islands
Caribbean Sea
147 C3 Windward Passage Cuba/
Haiti
137 D3 Winfield U.S.A.
100 A2 Wingene Belgium
53 D2 Wingham Austr.
130 B2 Winisk *r.* Can.
130 B2 Winisk (abandoned) Can.
130 B2 Winisk Lake Can.
63 A2 Winkana Myanmar
129 E3 Winkler Can.
114 B4 Winneba Ghana
138 B2 Winnebago, Lake U.S.A.
134 C2 Winnemucca U.S.A.
136 C2 Winner U.S.A.
140 B2 Winnfield U.S.A.
137 E1 Winnibigoshish, Lake
U.S.A.
129 E3 Winnipeg Can.
129 E2 Winnipeg *r.* Can.
129 E2 Winnipeg, Lake Can.
129 D2 Winnipegosis Can.
139 C2 Winnipesaukee, Lake U.S.A.
140 B2 Winnsboro U.S.A.
137 E2 Winona *MN* U.S.A.
140 C2 Winona *MS* U.S.A.
100 C1 Winschoten Neth.
101 D1 Winsen (Aller) Ger.
101 E1 Winsen (Luhe) Ger.
142 A1 Winslow U.S.A.
141 D1 Winston-Salem U.S.A.
101 D2 Winterberg Ger.
141 D3 Winter Haven U.S.A.
100 C2 Winterswijk Neth.
105 D2 Winterthur Switz.
51 D2 Winton Austr.
54 A3 Winton N.Z.
52 A2 Wirrabara Austr.
52 A2 Wirraminna Austr.
99 D3 Wisbech U.K.
138 A2 Wisconsin *r.* U.S.A.
138 B2 Wisconsin *state* U.S.A.
138 B2 Wisconsin Rapids U.S.A.
Wisła *r.* Pol. *see* Vistula
101 E1 Wismar Ger.
122 A2 Witbooisvlei Namibia
98 D3 Witham *r.* U.K.
98 D3 Withernsea U.K.
100 B1 Witmarsum Neth.
99 C4 Witney U.K.
123 D2 Witrivier S. Africa
123 C3 Witteberg *mts* S. Africa
101 F2 Wittenberg, Lutherstadt
Ger.
101 E1 Wittenberge Ger.
101 E1 Wittenburg Ger.
50 A2 Wittenoom Austr.
101 E1 Wittingen Ger.
100 C3 Wittlich Ger.
100 C1 Wittmund Ger.
101 F1 Wittstock Ger.
122 A1 Witvlei Namibia
101 D2 Witzenhausen Ger.
103 D1 Władysławowo Pol.
103 D1 Włocławek Pol.
53 C3 Wodonga Austr.
59 C3 Wokam *i.* Indon.
62 A1 Wokha India
99 C4 Woking U.K.
101 F2 Wolfen Ger.
101 E1 Wolfenbüttel Ger.
101 D2 Wolfhagen Ger.
136 B1 Wolf Point U.S.A.
101 E1 Wolfsburg Ger.
100 C3 Wolfstein Ger.
131 D3 Wolfville Can.
102 C1 Wolgast Ger.
102 C1 Wolin Pol.
129 D2 Wollaston Lake Can.
129 D2 Wollaston Lake *l.* Can.
126 E2 Wollaston Peninsula
Can.
53 D2 Wollongong Austr.
101 E2 Wolmirsleben Ger.
101 E1 Wolmirstedt Ger.
100 C1 Wolvega Neth.
99 B3 Wolverhampton U.K.
65 B2 Wonju S. Korea
128 B2 Wonowon Can.
65 B2 Wŏnsan N. Korea
53 C3 Wonthaggi Austr.
52 A2 Woocalla Austr.
51 C1 Woodah, Isle *i.* Austr.
99 D3 Woodbridge U.K.
134 B1 Woodburn U.S.A.
136 B3 Woodland Park U.S.A.
138 A3 Wood River U.S.A.
50 A2 Woodroffe, Mount Austr.
51 C1 Woods, Lake *imp. l.* Austr.

129 E3 Woods, Lake of the Can./
U.S.A.
53 C3 Woods Point Austr.
131 B2 Woodstock N.B. Can.
138 C2 Woodstock Ont. Can.
54 C2 Woodville N.Z.
143 D1 Woodward U.S.A.
98 B2 Wooler U.K.
53 D2 Woolgoolga Austr.
52 A2 Woomera Austr.
138 C2 Wooster U.S.A.
99 C4 Wootton Bassett, Royal U.K.
122 A3 Worcester S. Africa
99 B3 Worcester U.K.
139 E2 Worcester U.S.A.
102 C2 Wörgl Austria
98 B2 Workington U.K.
98 C3 Worksop U.K.
136 B2 Worland U.S.A.
101 D3 Worms Ger.
99 A4 Worms Head *hd* U.K.
122 A1 Wortel Namibia
99 C4 Worthing U.K.
137 D2 Worthington U.S.A.
61 D2 Wotu Indon.
61 D2 Wowoni *i.* Indon.
83 N2 Wrangel Island Rus. Fed.
128 A2 Wrangell U.S.A.
96 B1 Wrath, Cape U.K.
136 C2 Wray U.S.A.
122 A2 Wreck Point S. Africa
Wrecsam U.K. *see* Wrexham
99 B3 Wrexham U.K.
136 B2 Wright U.S.A.
63 A2 Wrightmyo India
142 A2 Wrightson, Mount U.S.A.
128 B1 Wrigley Can.
103 D1 Wrocław Pol.
103 D1 Września Pol.
70 B2 Wu'an China
Wuchow China *see* Wuzhou
70 A2 Wuhai China
70 B2 Wuhan China
70 B2 Wuhu China
71 A3 Wu Jiang *r.* China
Wujin China *see*
Changzhou
115 C4 Wukari Nigeria
75 D1 Wuli China
62 B1 Wuliang Shan *mts* China
59 C3 Wuliaru *i.* Indon.
118 B2 Wum Cameroon
71 A3 Wumeng Shan *mts* China
101 D1 Wümme *r.* Ger.
130 B2 Wunnummin Lake Can.
101 F2 Wunsiedel Ger.
101 D1 Wunstorf Ger.
62 A1 Wuntho Myanmar
100 C2 Wuppertal Ger.
122 A3 Wuppertal S. Africa
101 E2 Wurzbach Ger.
101 E3 Würzburg Ger.
101 F2 Wurzen Ger.
101 D2 Wüstegarten *h.* Ger.
59 D3 Wuvulu Island P.N.G.
70 A2 Wuwei China
70 A2 Wuxi *Chongqing* China
70 C2 Wuxi *Jiangsu* China
Wuxing China *see* Huzhou
71 A3 Wuxuan China
Wuyang China *see*
Zhenyuan
69 E1 Wuyiling China
71 B3 Wuyishan China
71 B3 Wuyi Shan *mts* China
70 A1 Wuyuan China
71 A4 Wuzhishan China
70 A2 Wuzhong China
71 B3 Wuzhou China
51 D2 Wyandra Austr.
53 C2 Wyangala Reservoir Austr.
52 B3 Wycheproof Austr.
99 B4 Wye *r.* U.K.
50 B1 Wyndham Austr.
52 B3 Wyndham-Werribee Austr.
140 B1 Wynne U.S.A.
129 D2 Wynyard Can.
138 B2 Wyoming U.S.A.
136 B2 Wyoming *state* U.S.A.
53 D2 Wyong Austr.
103 E1 Wyszków Pol.
138 C3 Wytheville U.S.A.

X

117 D3 Xaafuun Somalia
62 B2 Xaignabouli Laos
121 C3 Xai-Xai Moz.

145 C3 Xalapa Mex.
70 A1 Xamba China
62 A1 Xamgyi'nyilha China
62 B1 Xam Nua Laos
120 A1 Xá-Muteba Angola
120 A2 Xangongo Angola
81 C2 Xankändi Azer.
111 B2 Xanthi Greece
154 B3 Xanxerê Brazil
150 C4 Xapuri Brazil
107 C2 Xàtiva Spain
120 B3 Xhumo Botswana
  Xiaguan China see Dali
71 B3 Xiamen China
70 A2 Xi'an China
70 A3 Xianfeng China
62 A1 Xiangcheng China
  Xiangfan China see Xiangyang
  Xianghuang Qi China see Xin Bulag
  Xiangjiang China see Huichang
71 B3 Xiang Jiang r. China
71 B3 Xiangtan China
70 B2 Xiangyang China
71 B3 Xiangyin China
70 B3 Xianning China
70 B2 Xiantao China
70 A2 Xianyang China
70 B2 Xiaogan China
69 E1 Xiao Hinggan Ling mts China
70 C2 Xiaoshan China
70 B2 Xiaowutai Shan mt. China
  Xiayingpan China see Liuzhi
  Xibu China see Dongshan
71 A3 Xichang China
145 C2 Xicohténcatl Mex.
71 A3 Xifeng China
  Xifengzhen China see Qingyang
75 C2 Xigazê China
71 A3 Xilin China
69 D2 Xilinhot China
68 C2 Ximiao China
70 B1 Xin Bulag China
70 B2 Xincai China
  Xincun China see Dongchuan
  Xindi China see Honghu
71 B3 Xing'an China
  Xingba China see Lhünzê
68 C2 Xinghai China
70 B2 Xinghua China
71 B3 Xinging China
70 A2 Xingping China
70 B2 Xingtai China
151 B3 Xingu r. Brazil
151 D3 Xinguara Brazil
71 A3 Xingyi China
71 B3 Xinhua China
70 A2 Xining China
75 C1 Xinjiang aut. reg. China
  Xinjing China see Jingxi
69 D2 Xinkou China
65 A1 Xinmin China
71 B3 Xinning China
71 A3 Xinping China
  Xinshiba China see Ganluo
70 B2 Xintai China
  Xinxian China see Xinzhou
70 B2 Xinxiang China
70 B2 Xinyang China
71 A4 Xinyi China
71 C3 Xinying Taiwan
71 B3 Xinyu China
77 E2 Xinyuan China
70 B2 Xinzhou China
71 C3 Xinzhu Taiwan
106 B1 Xinzo de Limia Spain
  Xiongshan China see Zhenghe
  Xiongzhou China see Nanxiong
70 A2 Xiqing Shan mts China
151 E4 Xique Xique Brazil
70 A1 Xishanzui China
71 A3 Xiushan China
  Xiushan China see Tonghai
71 B3 Xiushui China
71 B3 Xiuying China
70 B2 Xixia China
  Xixón Spain see Gijón
76 B2 Xo'jayli Uzbek.
70 B2 Xuanchang China
70 B1 Xuanhua China
71 A3 Xuanwei China
  Xuanzhou China see Xuancheng
70 B2 Xuchang China
  Xucheng China see Xuwen
117 C4 Xuddur Somalia
  Xuefeng China see Mingxi
  Xujiang China see Guangchang
71 B3 Xun Jiang r. China
71 B3 Xunwu China
107 C2 Xúquer, Riu r. Spain
71 B3 Xuwen China
71 A3 Xuyong China
70 B2 Xuzhou China
111 B3 Xylokastro Greece

Y

70 A2 Ya'an China
117 B4 Yabēlo Eth.
69 D1 Yablonovyy Khrebet mts Rus. Fed.
141 D1 Yadkin r. U.S.A.
75 C2 Yadong China
70 A1 Yagan China
55 A4 Yaghan Basin S. Atlantic Ocean
89 E2 Yagnitsa Rus. Fed.
83 K2 Yagodnoye Rus. Fed.
118 B1 Yagoua Cameroon
128 C3 Yahk Can.
91 C1 Yahotyn Ukr.
144 B2 Yahualica Mex.
80 B2 Yahyalı Turkey
67 C4 Yaizu Japan
134 B1 Yakima U.S.A.
134 C1 Yakima r. U.S.A.
74 A2 Yakmach Pak.
114 B3 Yako Burkina Faso
66 D2 Yakumo Japan
67 B4 Yaku-shima i. Japan
126 C3 Yakutat U.S.A.
128 A2 Yakutat Bay U.S.A.
83 J2 Yakutsk Rus. Fed.
91 D2 Yakymivka Ukr.
63 B3 Yala Thai.
118 C2 Yalinga C.A.R.
53 C3 Yallourn Austr.
111 C2 Yalova Turkey
90 B2 Yalpuh, Ozero l. Ukr.
91 C3 Yalta Ukr.
65 A1 Yalu Jiang r. China/N. Korea
86 F3 Yalutorovsk Rus. Fed.
67 D3 Yamagata Japan
67 B4 Yamaguchi Japan
  Yamal, Poluostrov pen. Rus. Fed. see Yamal Peninsula
86 F1 Yamal Peninsula pen. Rus. Fed.
  Yamankhalinka Kazakh. see Makhambet
53 D1 Yamba Austr.
150 B2 Yambi, Mesa de hills Col.
117 A4 Yambio S. Sudan
110 C2 Yambol Bulg.
59 C3 Yamdena i. Indon.
62 A1 Yamethin Myanmar
88 C2 Yamm Rus. Fed.
51 D2 Yamma Yamma, Lake imp. l. Austr.
114 B4 Yamoussoukro Côte d'Ivoire
91 C1 Yampil' Ukr.
90 B2 Yampil' Ukr.
75 C2 Yamuna r. India
62 A1 Yamzho Yumco l. China
83 K2 Yana r. Rus. Fed.
70 A2 Yan'an China
150 B4 Yanaoca Peru
78 A2 Yanbu' al Bahr Saudi Arabia
70 A2 Yancheng China
50 A3 Yanchep Austr.
114 B3 Yanfolila Mali
118 C2 Yangambi Dem. Rep. Congo
71 B3 Yangchun China
65 B2 Yangdok N. Korea
71 B3 Yangjiang China
  Yangön Myanmar see Rangoon
70 B2 Yangquan China
71 B3 Yangshuo China
63 B2 Yang Sin, Chu' mt. Vietnam
62 B1 Yangtouyan China
70 C23 Yangtze r. China
70 C2 Yangtze, Mouth of the China
  Yangtze Kiang r. China see Yangtze
70 A2 Yangxian China
70 B2 Yangzhou China
65 B1 Yanji China
137 D2 Yankton U.S.A.
83 K2 Yano-Indigirskaya Nizmennost' lowland Rus. Fed.
70 B1 Yanqing China
71 A3 Yanshan China
83 K2 Yanskiy Zaliv g. Rus. Fed.
53 C1 Yantabulla Austr.
70 C2 Yantai China
118 B2 Yaoundé Cameroon
59 D2 Yap i. Micronesia
59 D3 Yapen i. Indon.
59 D3 Yapen, Selat sea chan. Indon.
144 A2 Yaqui r. Mex.
51 D2 Yaraka Austr.
86 D3 Yaransk Rus. Fed.
48 H4 Yaren Nauru
78 B3 Yarim Yemen
  Yarkand China see Shache
  Yarkant China see Shache
77 D3 Yarkant He r. China
  Yarlung Zangbo r. China see Brahmaputra
131 D3 Yarmouth Can.
142 A2 Yarnell U.S.A.
86 F2 Yarono Rus. Fed.
89 E2 Yaroslavl' Rus. Fed.
66 B2 Yaroslavskiy Rus. Fed.
53 C3 Yarra Junction Austr.
53 C3 Yarram Austr.
89 D2 Yartsevo Rus. Fed.
89 E3 Yasnogorsk Rus. Fed.
63 B2 Yasothon Thai.
53 C2 Yass Austr.
81 D2 Yāsūj Iran
111 C3 Yatağan Turkey
119 D3 Yata Plateau Kenya
129 E1 Yathkyed Lake Can.
67 B4 Yatsushiro Japan
150 C3 Yavari r. Brazil/Peru
73 B2 Yavatmal India
90 A2 Yavoriv Ukr.
67 B4 Yawatahama Japan
62 A1 Yawng-hwe Myanmar
  Yaxian China see Sanya
81 D2 Yazd Iran
140 B2 Yazoo r. U.S.A.
140 B2 Yazoo City U.S.A.
111 B3 Ydra Greece
111 B3 Ydra i. Greece
63 A2 Ye Myanmar
77 D3 Yecheng China
107 C2 Yecla Spain
144 B2 Yécora Mex.
  Yedintsy Moldova see Edineţ
89 E3 Yefremov Rus. Fed.
91 E2 Yegorlykskaya Rus. Fed.
89 E2 Yegor'yevsk Rus. Fed.
117 B4 Yei S. Sudan
86 F3 Yekaterinburg Rus. Fed.
  Yekaterinodar Rus. Fed. see Krasnodar
  Yekaterinoslav Ukr. see Dnipropetrovs'k
  Yekaterinovskaya Rus. Fed. see Krylovskaya
77 D1 Yekibastuz Kazakh.
  Yelenovskiye Kar'yery Ukr. see Dokuchayevs'k
89 E3 Yelets Rus. Fed.
89 D2 Yeligovo Rus. Fed.
114 A3 Yélimané Mali
  Yelizavetgrad Ukr. see Kirovohrad
96 ☐ Yell i. U.K.
128 C1 Yellowknife Can.
53 C2 Yellow Mountain h. Austr.
70 B2 Yellow River r. China
69 E2 Yellow Sea N. Pacific Ocean
136 C1 Yellowstone r. U.S.A.
136 A2 Yellowstone Lake U.S.A.
88 C3 Yel'sk Belarus
78 B3 Yemen country Asia
90 B1 Yemil'chyne Ukr.
86 E2 Yemva Rus. Fed.
91 D2 Yenakiyeve Ukr.
62 A1 Yenangyaung Myanmar
62 B1 Yên Bái Vietnam
114 B4 Yendi Ghana
111 C3 Yenice Turkey
111 C3 Yenifoça Turkey
68 C1 Yenisey r. Rus. Fed.
65 B2 Yeongdeok S. Korea
65 B2 Yeongju S. Korea
  Yeotmal India see Yavatmal
53 C2 Yeoval Austr.
99 B4 Yeovil U.K.
51 E2 Yeppoon Austr.
  Yeraliyev Kazakh. see Kuryk
83 I2 Yerbogachen Rus. Fed.
81 C1 Yerevan Armenia
77 D1 Yereymentau Kazakh.
  Yereymentau Kazakh. see Yereymentau
143 C3 Yermo Mex.
135 C4 Yermo U.S.A.
89 D3 Yershichi Rus. Fed.
87 D3 Yershov Rus. Fed.
  Yertis r. Kazakh./Rus. Fed. see Irtysh
150 B4 Yerupaja mt. Peru
  Yerushalayim Israel/West Bank see Jerusalem
65 B2 Yesan S. Korea
77 C1 Yesil' Kazakh.
77 D1 Yesil' r. Kazakh./Rus. Fed.
111 C3 Yeşilova Turkey
83 H2 Yessey Rus. Fed.
99 A4 Yes Tor h. U.K.
53 D1 Yetman Austr.
62 A1 Ye-U Myanmar
104 B2 Yeu, Île d' i. France
87 D4 Yevlax Azer.
91 C2 Yevpatoriya Ukr.
  Yexian China see Laizhou
91 D2 Yeya r. Rus. Fed.
91 D2 Yeysk Rus. Fed.
88 C2 Yezyaryshcha Belarus
  Y Fenni U.K. see Abergavenny
154 A2 Ygatimí Para.
71 A3 Yibin China
70 B2 Yichang China
69 E1 Yichun Heilong. China
71 B3 Yichun Jiangxi China
  Yidu China see Qingzhou
66 A1 Yilan China
110 C2 Yıldız Dağları mts Turkey
80 B2 Yıldızeli Turkey
  Yilong China see Shiping
70 A2 Yinchuan China
65 B1 Yingchengzi China
71 B3 Yingde China
72 D2 Yingkiong India
70 C1 Yingkou China
70 B2 Yingshan Hubei China
70 A2 Yingshan Sichuan China
71 B3 Yingtan China
  Yining China see Xiushui
77 E2 Yining China
62 A1 Yinmabin Myanmar
70 A1 Yin Shan mts China
117 B4 Yirga Alem Eth.
119 D2 Yirga Ch'efē Eth.
119 D2 Yirol S. Sudan
  Yishan China see Yizhou
70 B2 Yishui China
62 A1 Yi Tu, Nam r. Myanmar
68 C2 Yiwu China
70 C1 Yixian China
70 B2 Yixing China
71 B3 Yiyang China
71 B3 Yizhang China
71 A3 Yizhou China
  Yizhou China see Yixian
92 I2 Yli-Kitka l. Fin.
92 H2 Ylitornio Fin.
92 H3 Ylivieska Fin.
93 H3 Ylöjärvi Fin.
83 K2 Ynykchanskiy Rus. Fed.
  Ynys Môn i. U.K. see Anglesey
61 C2 Yogyakarta Indon.
118 B2 Yokadouma Cameroon
118 B2 Yoko Cameroon
67 C3 Yokohama Japan
66 D3 Yokote Japan
115 D4 Yola Nigeria
67 D3 Yonezawa Japan
71 B3 Yong'an China
  Yongbei China see Yongsheng
71 C3 Yongchang China
70 A2 Yongdeng China
65 B2 Yŏnghŭng N. Korea
  Yongjing China see Xifeng
71 C3 Yongkang China
  Yongle China see Zhen'an
62 B1 Yongren China
62 B1 Yongsheng China
71 B3 Yongzhou China

139 E2 Yonkers U.S.A.
105 C2 Yonne r. France
150 B2 Yopal Col.
50 A3 York Austr.
98 C3 York U.K.
140 C2 York AL U.S.A.
137 D2 York NE U.S.A.
139 D3 York PA U.S.A.
51 D1 York, Cape Austr.
52 A3 Yorke Peninsula Austr.
52 A3 Yorketown Austr.
98 C3 Yorkshire Wolds hills U.K.
129 D2 Yorkton Can.
87 D3 Yoshkar-Ola Rus. Fed.
97 C3 Youghal Ireland
53 C2 Young Austr.
52 A3 Younghusband Peninsula Austr.
138 C2 Youngstown U.S.A.
114 B3 Youvarou Mali
71 A3 Youyang China
77 E2 Youyi Feng mt. China/Rus. Fed.
80 B2 Yozgat Turkey
154 A2 Ypé-Jhú Para.
134 B2 Yreka U.S.A.
76 C2 Yrgyz Kazakh.
Yr Wyddfa mt. U.K. see Snowdon
59 D3 Ysabel Channel P.N.G.
105 C2 Yssingeaux France
93 F4 Ystad Sweden
Ysyk-Köl Kyrg. see Balykchy
77 D2 Ysyk-Köl salt l. Kyrg.
Y Trallwng U.K. see Welshpool
92 □A3 Ytri-Rangá r. Iceland
83 J2 Ytyk-Kyuyel' Rus. Fed.
71 A3 Yuanbao Shan mt. China
71 A3 Yuanjiang China
62 B1 Yuan Jiang r. China
71 B3 Yuanling China
71 A3 Yuanmou China
70 B2 Yuanping China
135 B3 Yuba City U.S.A.
66 D2 Yūbari Japan
145 C3 Yucatán pen. Mex.
146 B2 Yucatan Channel Cuba/Mex.
Yuci China see Jinzhong
50 C2 Yuendumu Austr.
71 C3 Yueqing China
71 B3 Yueyang China
86 F2 Yugorsk Rus. Fed.
71 B3 Yujiang China
83 L2 Yukagirskoye Ploskogor'ye plat. Rus. Fed.
89 E3 Yukhnov Rus. Fed.
128 A1 Yukon admin. div. Can.
126 B2 Yukon r. Can./U.S.A.
143 D1 Yukon U.S.A.
50 C2 Yulara Austr.
71 B3 Yulin Guangxi China
70 A2 Yulin Shaanxi China
62 B1 Yulong Xueshan mt. China
142 A2 Yuma AZ U.S.A.
136 C2 Yuma CO U.S.A.
135 D4 Yuma Desert U.S.A.
Yumen China see Laojunmiao
80 B2 Yunak Turkey
70 B2 Yuncheng China
71 B3 Yunfu China
71 A3 Yungui Gaoyuan plat. China
Yunjinghong China see Jinghong
Yunling China see Yunxiao
71 A3 Yunnan prov. China
52 A2 Yunta Austr.
71 B3 Yunxiao China
70 B2 Yunyang China
Yuping China see Libo
71 A3 Yuping China
82 G3 Yurga Rus. Fed.
66 D3 Yurihonjō Japan
150 B3 Yurimaguas Peru
75 C1 Yurungkax He r. China
Yuryev Estonia see Tartu
71 C3 Yu Shan mt. Taiwan
70 B2 Yushe China
68 C2 Yushu China
Yushuwan China see Huaihua
81 C1 Yusufeli Turkey
75 C1 Yutian China
71 A3 Yuxi China
89 F2 Yuzha Rus. Fed.
83 K3 Yuzhno-Sakhalinsk Rus. Fed.
91 C2 Yuzhnoukrayins'k Ukr.

82 E1 Yuzhnyy, Ostrov i. Rus. Fed.
70 B2 Yuzhou China
Yuzovka Ukr. see Donets'k
105 D2 Yverdon Switz.
104 C2 Yvetot France

Z

100 B1 Zaandam Neth.
69 D1 Zabaykal'sk Rus. Fed.
119 C2 Zabia Dem. Rep. Congo
78 B3 Zabid Yemen
79 D1 Zābol Iran
79 D2 Zāboli Iran
145 D3 Zacapa Guat.
144 B3 Zacapu Mex.
144 B2 Zacatecas Mex.
145 C3 Zacatepec Mex.
145 C3 Zacatlán Mex.
111 B3 Zacharo Greece
91 D2 Zachepylivka Ukr.
144 B2 Zacoalco Mex.
145 C2 Zacualtipán Mex.
109 C2 Zadar Croatia
63 A3 Zadetkyi Kyun i. Myanmar
89 E3 Zadonsk Rus. Fed.
80 B3 Za'farānah Egypt
106 B2 Zafra Spain
Zagazig Egypt see Az Zaqāzīq
114 B1 Zagora Morocco
Zagorsk Rus. Fed. see Sergiyev Posad
109 C1 Zagreb Croatia
Zagros, Kūhhā-ye mts Iran see Zagros Mountains
81 C2 Zagros Mountains mts Iran
79 D2 Zāhedān Iran
80 B2 Zahlé Lebanon
78 B3 Zahrān Saudi Arabia
Zaire country Africa see Congo, Democratic Republic of the
109 D2 Zaječar Serbia
121 C3 Zaka Zimbabwe
89 E3 Zakharovo Rus. Fed.
81 C2 Zākhō Iraq
86 C2 Zakhrebetnoye Rus. Fed.
111 B3 Zakynthos Greece
111 B3 Zakynthos i. Greece
103 D2 Zalaegerszeg Hungary
110 B1 Zalău Romania
78 B2 Zalim Saudi Arabia
116 A3 Zalingei Sudan
90 B2 Zalishchyky Ukr.
78 A2 Zalmā, Jabal az mt. Saudi Arabia
128 C2 Zama City Can.
Zambeze r. Moz. see Zambezi
120 C2 Zambezi r. Africa
120 B2 Zambezi Zambia
120 B2 Zambezi Escarpment Zambia/Zimbabwe
120 B2 Zambia country Africa
64 B3 Zamboanga Phil.
64 B3 Zamboanga Peninsula Phil.
103 E1 Zambrów Pol.
106 B1 Zamora Spain
144 B3 Zamora de Hidalgo Mex.
103 E1 Zamość Pol.
Zamost'ye Pol. see Zamość
75 B1 Zanda China
100 B2 Zandvliet Belgium
100 B1 Zandvoort Neth.
138 C2 Zanesville U.S.A.
77 D3 Zangguy China
75 C1 Zangsêr Kangri mt. China
81 C2 Zanjān Iran
74 B1 Zanskar Mountains India
Zante i. Greece see Zakynthos
119 D3 Zanzibar Tanz.
119 D3 Zanzibar Island Tanz.
89 E3 Zaokskiy Rus. Fed.
115 C2 Zaouatallaz Alg.
Zaouet el Kahla Alg. see Bordj Omer Driss
70 B2 Zaoyang China
83 H3 Zaozernyy Rus. Fed.
89 D2 Zapadnaya Dvina r. Europe
89 D2 Zapadnaya Dvina Rus. Fed.
Zapadno-Sibirskaya Ravnina plain Rus. Fed. see West Siberian Plain
68 B1 Zapadnyy Sayan reg. Rus. Fed.
143 D3 Zapata U.S.A.

92 J2 Zapolyarnyy Rus. Fed.
91 D2 Zaporizhzhya Ukr.
101 E2 Zappendorf Ger.
81 C1 Zaqatala Azer.
Zara Croatia see Zadar
80 B2 Zara Turkey
145 B2 Zaragoza Mex.
107 C1 Zaragoza Spain
74 A1 Zarah Sharan Afgh.
79 C1 Zarand Iran
76 C3 Zaranj Afgh.
88 C2 Zarasai Lith.
89 E3 Zaraysk Rus. Fed.
150 C2 Zaraza Venez.
115 C3 Zaria Nigeria
90 B1 Zarichne Ukr.
81 D3 Zarqān Iran
66 B2 Zarubino Rus. Fed.
103 D1 Żary Pol.
115 D1 Zarzis Tunisia
88 C3 Zaslawye Belarus
123 C3 Zastron S. Africa
121 C3 Zavala Moz.
Zavitaya Rus. Fed. see Zavitinsk
69 E1 Zavitinsk Rus. Fed.
89 F2 Zavolzhsk Rus. Fed.
Zavolzh'ye Rus. Fed. see Zavolzhsk
103 D1 Zawiercie Pol.
115 E1 Zāwiyat Masūs Libya
77 E2 Zaysan Kazakh.
77 E2 Zaysan, Lake l. Kazakh.
Zaysan, Ozero l. Kazakh. see Zaysan, Lake
90 B2 Zbarazh Ukr.
90 B1 Zdolbuniv Ukr.
93 F4 Zealand i. Denmark
100 A2 Zedelgem Belgium
100 A2 Zeebrugge Belgium
123 C2 Zeerust S. Africa
101 F1 Zehdenick Ger.
50 C2 Zeil, Mount Austr.
101 F2 Zeitz Ger.
109 C2 Zelena Gora mt. Bos.-Herz.
87 D3 Zelenodol'sk Rus. Fed.
88 C1 Zelenogorsk Rus. Fed.
89 E2 Zelenograd Rus. Fed.
88 B3 Zelenogradsk Rus. Fed.
88 B3 Zel'va Belarus
119 C2 Zémio C.A.R.
107 D2 Zemmora Alg.
145 C3 Zempoaltépetl, Nudo de mt. Mex.
109 D2 Zemun Serbia
65 B1 Zengfeng Shan mt. China
109 C2 Zenica Bos.-Herz.
107 D2 Zenzach Alg.
101 F2 Zerbst Ger.
105 D2 Zermatt Switz.
91 E2 Zernograd Rus. Fed.
Zernovoy Rus. Fed. see Zernograd
101 E2 Zeulenroda Ger.
101 D1 Zeven Ger.
100 C2 Zevenaar Neth.
100 B2 Zevenbergen Neth.
83 J3 Zeya Rus. Fed.
79 C2 Zeydābād Iran
79 C2 Zeynalābād Iran
83 J3 Zeyskoye Vodokhranilishche resr Rus. Fed.
103 D1 Zgierz Pol.
88 B3 Zhabinka Belarus
Zhabye Ukr. see Verkhovyna
Zhaksy Sarysu watercourse Kazakh. see Sarysu
76 A2 Zhalpaktal Kazakh.
77 C1 Zhaltyr Kazakh.
Zhambyl Kazakh. see Taraz
76 B2 Zhanakala Kazakh.
76 B2 Zhanaozen Kazakh.
Zhanatas Kazakh. see Ayteke Bi
66 A1 Zhangguangcai Ling mts China
71 C3 Zhanghua Taiwan
71 B3 Zhangjiajie China
70 B1 Zhangjiakou China
71 B3 Zhangping China
71 B3 Zhangshu China
65 A1 Zhangwu China
70 A2 Zhangxian China
68 C2 Zhangye China
71 B3 Zhangzhou China
76 A2 Zhanibek Kazakh.
71 B3 Zhanjiang China

71 B3 Zhao'an China
69 E1 Zhaodong China
Zhaoge China see Qixian
71 B3 Zhaoqing China
71 A3 Zhaotong China
75 C1 Zhari Namco salt l. China
77 E2 Zharkent Kazakh.
89 D2 Zharkovskiy Rus. Fed.
77 E2 Zharma Kazakh.
90 C2 Zhashkiv Ukr.
Zhaxi China see Weixin
77 D2 Zhayrem Kazakh.
Zhayyk r. Kazakh./Rus. Fed. see Ural
Zhdanov Ukr. see Mariupol
71 C3 Zhejiang prov. China
82 F1 Zhelaniya, Mys c. Rus. Fed.
Zheleznodorozhnyy Rus. Fed. see Yemva
Zheleznodorozhnyy Uzbek. see Qo'ng'irot
89 E3 Zheleznogorsk Rus. Fed.
Zheltyye Vody Ukr. see Zhovti Vody
76 B2 Zhem r. Kazakh.
70 A2 Zhen'an China
70 A2 Zhenba China
71 A3 Zheng'an China
71 B3 Zhenghe China
70 B2 Zhengzhou China
70 B2 Zhengzhou China
71 A3 Zhenyuan China
91 E1 Zherdevka Rus. Fed.
86 D2 Zheshart Rus. Fed.
77 D2 Zhetysuskiy Alatau mts China/Kazakh.
77 C2 Zhezkazgan Kazakh.
77 C2 Zhezkazgan Kazakh.
62 A1 Zhigang China
83 J2 Zhigansk Rus. Fed.
70 B2 Zhijiang China
76 C1 Zhitikara Kazakh.
89 D3 Zhizdra Rus. Fed.
88 D3 Zhlobin Belarus
90 B2 Zhmerynka Ukr.
74 A1 Zhob Pak.
88 C3 Zhodzina Belarus
83 L1 Zhokhova, Ostrov i. Rus. Fed.
Zholkva Ukr. see Zhovkva
Zhongba China see Jiangyou
75 C2 Zhongba China
Zhongduo China see Youyang
Zhonghe China see Xiushan
Zhongning China
Zhongping China see Huize
70 A2 Zhongning China
71 B3 Zhongshan China
Zhongshan China see Liupanshui
70 A2 Zhongwei China
Zhongxin China see Xamgyi'nyilha
71 C3 Zhongyang Shanmo mts Taiwan
76 C2 Zhosaly Kazakh.
70 B2 Zhoukou China
70 C2 Zhoushan China
90 A1 Zhovkva Ukr.
91 C2 Zhovti Vody Ukr.
65 A2 Zhuanghe China
70 B2 Zhucheng China
89 D3 Zhukovka Rus. Fed.
89 E2 Zhukovskiy Rus. Fed.
70 B2 Zhumadian China
Zhuoyang China see Suiping
71 B3 Zhuzhou Hunan China
71 B3 Zhuzhou Hunan China
90 A2 Zhydachiv Ukr.
88 C3 Zhytkavichy Belarus
90 B1 Zhytomyr Ukr.
103 D2 Žiar nad Hronom Slovakia
70 B2 Zibo China
103 D1 Zielona Góra Pol.
100 A2 Zierikzee Neth.
62 A1 Zigaing Myanmar
115 C2 Zighan Libya
71 A3 Zigong China
Zigui China see Guojiaba
114 A3 Ziguinchor Senegal
144 B3 Zihuatanejo Mex.
103 D2 Žilina Slovakia
115 D2 Zillah Libya
83 H3 Zima Rus. Fed.
145 C2 Zimapán Mex.
121 B2 Zimbabwe country Africa
114 A4 Zimmi Sierra Leone
110 C2 Zimnicea Romania

Zyryanovsk

86 C2	**Zimniy Bereg** *coastal area* Rus. Fed.	
115 C3	**Zinder** Niger	
78 B3	**Zinjibār** Yemen	
91 C1	**Zin'kiv** Ukr.	
	Zinoyevsk Ukr. *see* Kirovohrad	
150 B2	**Zipaquirá** Col.	
76 C4	**Zirah, Gōd-e** *depr.* Afgh.	
103 D2	**Zirc** Hungary	
75 D2	**Ziro** India	
79 C2	**Zīr Rūd** Iran	
103 D2	**Zistersdorf** Austria	
145 B3	**Zitácuaro** Mex.	
103 C1	**Zittau** Ger.	
87 E3	**Zlatoust** Rus. Fed.	
103 D2	**Zlín** Czech Rep.	
115 D1	**Zlīṭan** Libya	
103 D1	**Złotów** Pol.	
89 D3	**Zlynka** Rus. Fed.	
89 E3	**Zmiyevka** Rus. Fed.	
91 D2	**Zmiyiv** Ukr.	
89 E3	**Znamenka** Rus. Fed.	
91 E1	**Znamenka** Rus. Fed.	
91 C2	**Znam"yanka** Ukr.	
103 D2	**Znojmo** Czech Rep.	
122 B3	**Zoar** S. Africa	
70 A2	**Zoigê** China	
91 D1	**Zolochiv** Ukr.	
90 A2	**Zolochiv** Ukr.	
91 C2	**Zolotonosha** Ukr.	
89 E3	**Zolotukhino** Rus. Fed.	
121 C2	**Zomba** Malawi	
118 B2	**Zongo** Dem. Rep. Congo	
80 B1	**Zonguldak** Turkey	
105 D3	**Zonza** France	
114 B3	**Zorgho** Burkina Faso	
114 B4	**Zorzor** Liberia	
115 D2	**Zouar** Chad	
114 A2	**Zouérat** Maur.	
109 D1	**Zrenjanin** Serbia	
89 D2	**Zubtsov** Rus. Fed.	
105 D2	**Zug** Switz.	
81 C1	**Zugdidi** Georgia	
	Zuider Zee *l.* Neth. *see* IJsselmeer	
106 B2	**Zújar** *r.* Spain	
100 C2	**Zülpich** Ger.	
100 A2	**Zulte** Belgium	
121 C2	**Zumbo** Moz.	
145 C3	**Zumpango** Mex.	
142 B1	**Zuni Mountains** U.S.A.	
71 A3	**Zunyi** China	
109 C1	**Županja** Croatia	
105 D2	**Zürich** Switz.	
105 D2	**Zürichsee** *l.* Switz.	
100 C1	**Zutphen** Neth.	
115 D1	**Zuwārah** Libya	
90 C2	**Zvenyhorodka** Ukr.	
121 C3	**Zvishavane** Zimbabwe	
103 D2	**Zvolen** Slovakia	
109 C2	**Zvornik** Bos.-Herz.	
114 B4	**Zwedru** Liberia	
123 C3	**Zwelitsha** S. Africa	
103 D2	**Zwettl** Austria	
101 F2	**Zwickau** Ger.	
100 C1	**Zwolle** Neth.	
83 L2	**Zyryanka** Rus. Fed.	
77 E2	**Zyryanovsk** Kazakh.	

Acknowledgements

pages 34–5

Köppen classification map: Kottek, M., J. Grieser, C. Beck, B. Rudolf, and F. Rubel, 2006: World Map of the Köppen-Geiger climate classification updated. Meteorol. Z., **15**, 259–263.
http://koeppen-geiger.vu-wien.ac.at

pages 36–37

Land cover map: © ESA 2010 and UCLouvain
Arino, O., Ramos, J., Kalogirou, V., Defourny, P., Achard, F., 2010. GlobCover 2009. ESA Living Planet Symposium 2010, 28th June - 2nd July, Bergen, Norway, SP-686, ESA.
www.esa.int/due/globcover
http://due.esrin.esa.int/prjs/Results/20110202183257.pdf

pages 38–39

Population map data:
Gridded Population of the World (GPW), Version 3.
Palisades, NY: CIESN, Columbia University. Available at
http://sedac.ciesin.columbia.edu/plue/gpw

Cover image

Great Sand Dunes National Park and Preserve, Colorado, USA.
Jim Parkin/Shutterstock

Greenland

Ice extent data: © Geological Survey of Denmark and Greenland (GEUS).
Ice margins produced by PROMICE - Programme for Monitoring of the Greenland Ice Sheet, updated to 2011 based on MODIS imagery from NASA

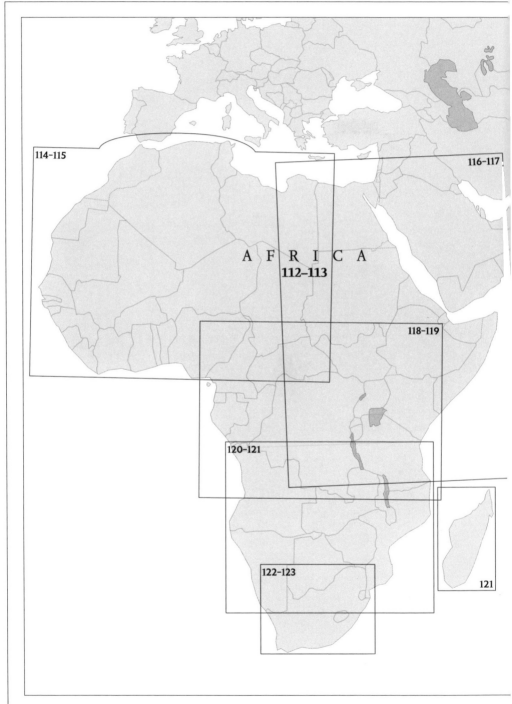